AF248724

FILGRASTIM
(r-metHuG-CSF)
IN CLINICAL PRACTICE

FILGRASTIM
(r-metHuG-CSF)
IN CLINICAL PRACTICE
SECOND EDITION

edited by

GEORGE MORSTYN
*Amgen Inc.
Thousand Oaks, and
UCLA School of Medicine
Los Angeles, California*

T. MICHAEL DEXTER
*Paterson Institute for Cancer Research
Christie Hospital
Manchester, England*

MARYANN FOOTE
*Amgen Inc.
Thousand Oaks, California*

MARCEL DEKKER, INC.   NEW YORK · BASEL

Library of Congress Cataloging-in-Publication Data

Filgrastim (r-metHuG-CSF) in clinical practice / edited by George Morstyn, T. Michael
  Dexter, MaryAnn Foote.—2nd ed.
        p. cm.
      Includes bibliographical references and index.
      ISBN 0-8247-0057-0 (alk. paper)
      1. Figrastim—Therapeutic use. I. Morstyn, George. II. Dexter, T. Michael. III.
  Foote, MaryAnn.
      [DNLM: 1. Filgrastim—therapeutic use. 2. Filgrastim—pharmacology. QV 180 F481
  1998]
      RM666.F49F55 1998
      616.99'4061—dc21
      DNLM/DLC
      for Library of Congress   98-3974
                                                                                              CIP

This book is printed on acid-free paper.

Headquarters
Marcel Dekker, Inc.
270 Madison Avenue, New York, NY 10016
tel: 212-696-9000; fax: 212-685-4540

Eastern Hemisphere Distribution
Marcel Dekker AG
Hutgasse 4, Postfach 812, CH-4001 Basel, Switzerland
tel: 44-61-261-8482; fax: 44-61-261-8896

World Wide Web
http://www.dekker.com

The publisher offers discounts on this book when ordered in bulk quantities. For more information, write to Special Sales/Professional Marketing at the headquarters address above.

Current printing (last digit):
10 9 8 7 6 5 4 3

PRINTED IN THE UNITED STATES OF AMERICA

Preface to the Second Edition

The development of hematopoietic growth factors, especially Filgrastim, has altered the treatment of many diseases. Since the publication of the first edition of *Filgrastim (r-metHuG-CSF) in Clinical Practice,* our basic knowledge of granulocyte colony-stimulating factor (G-CSF) and its mechanisms of action has advanced.

Each chapter has been revised and the book has been expanded. The book starts with an overview of the biology, structure, and pharmacology of G-CSF and Filgrastim. Extensive new information on the use of Filgrastim in oncology and other settings is provided.

There are chapters on the use of Filgrastim in malignancies commonly associated with neutropenic infections during therapy, such as breast cancer, lymphoma, and leukemia. This book also includes expanded chapters on neutrophil transfusions and transplantation.

New data are included on the potential use of Filgrastim in AIDS, in non-neutropenic infections, and in treating the neutropenia associated with nonmalignant conditions, such as autoimmune disease. There is provocative information on the potential of peripheral blood progenitor cell (PBPC) transplantation in such diseases as rheumatoid arthritis. The book concludes with chapters on pharmacoeconomics and a safety update.

In the clinical setting, Filgrastim-mobilized PBPC have produced rapid engraftment in patients receiving myeloablative chemotherapy. The use of Filgrastim and Filgrastim-mobilized PBPC has also allowed clinicians to study the benefits of chemotherapy dose intensity. In areas other than oncology, Filgrastim has been shown to reduce morbidity and to reduce hospital days and their inherent costs. In patients with severe chronic neutropenia (SCN), Filgrastim decreases the morbidity due to infections; in patients with AIDS, Filgrastim decreases the morbidity associated with the disease as well as the morbidity associated with antiretroviral drugs. Filgrastim has been shown to be safe in the treatment of non-neutropenic patients with infections.

Filgrastim remains a promising agent whose potential has not yet been fully realized. Exciting results of preclinical studies have opened new research pathways for the use of Filgrastim and it will be interesting to see if any of these pathways lead to new therapies.

Filgrastim is becoming an important part of treatment for both patients with cancer and those with nonmalignant disease. This book should be of interest to physicians and medical students who prescribe the agent. Its development serves as a bellwether of how a basic scientific discovery can be developed into a very useful therapeutic agent and the importance of collaboration between scientists and clinicians in this development.

George Morstyn
T. Michael Dexter
MaryAnn Foote

Preface to the First Edition

It has been known for at least 25 years that hemopoiesis is under the control of specific factors that act on early cells in the hemopoietic system to produce mature cells. The isolation, purification, and cloning of these factors have led to a new class of therapeutic agents. These factors include colony-stimulating factors and interleukins and they have rapidly become a new class of pharmaceuticals. This book is devoted to granulocyte colony-stimulating factor (G-CSF), a protein that acts on the neutrophil lineage. Neutrophils are the body's major defense against infections. Filgrastim is the bacterially synthesized recombinant protein form of G-CSF. The name "Filgrastim" is the generic name given by the U.S. Adopted Name Council and refers to the function of the product. The "grastim" suffix means that the product works on granulocytes; "fil" refers (phonetically) to neutrophils.

Filgrastim was approved in February 1991 for ameliorating one of the major side effects of cancer chemotherapy. Its use has led to reduced infections and hospital admissions for patients with cancer. In the period 1991 to 1993, Filgrastim has treated more than 200,000 patients with cancer. Filgrastim has had a major impact in improving cancer therapy by reducing side effects and the requirement for hospitalization. Filgrastim has also transformed the lives of children with severe chronic neutropenia.

This book describes the biochemical nature of Filgrastim, its biological activity, and its impact both on cancer therapy and the larger problem of severe infections in immunocompromised patients.

Several chapters also discuss peripheral blood progenitor cell support and the use of Filgrastim to release these cells.

There are extensive discussions on new areas made possible by the use of Filgrastim, such as delivery of full-dose chemotherapy and high-dose chemotherapy. There is also an extensive discussion of the potential anti-cancer effects of Filgrastim.

The greatest potential for the use of Filgrastim may be in non-neutropenic patients whose neutrophils function abnormally, perhaps because of the patient's underlying illness, age, or because the patient is immunocompromised. Extensive discussion of the potential of Filgrastim in immunocompromised patients is presented.

The authors include some of the early clinical investigators of Filgrastim and staff from Amgen Inc., which manufactures Filgrastim. We have tried to achieve balance in the presentation of clinical information by melding information from the early clinical investigators with the vast knowledge about Filgrastim that resides within Amgen, Kirin, and Roche. Some overlap among chapters has been retained so that several perspectives on some topics are presented. It is our hope that the reader will find the presentation varied yet balanced.

George Morstyn
T. Michael Dexter

Contents

Contributors

Douglas R. Adkins, M.D.   Assistant Professor of Medicine, Department of Internal Medicine, Division of Bone Marrow Transplantation, Washington University School of Medicine, St. Louis, Missouri

Magdy Ahmad, M.D.   Fellow, Department of Pediatrics, Division of Neonatology, University Medical Center, SUNY at Stony Brook, Stony Brook, New York

Jeff Andresen, Ph.D.   Research Scientist, Department of Pharmacology, Amgen Inc., Thousand Oaks, California

Andrea Bacigalupo, M.D.   Head, Division of Immunology, Azienda Ospedale San Martino, Genoa, Italy

Alan J. Barge, M.D., M.R.C.P.   Medical Director, Hematology/Oncology, Department of Clinical Development, Amgen Ltd., Cambridge, England

William I. Bensinger, M.D.   Professor, Department of Medicine, University of Washington, and Member, Division of Oncology, Fred Hutchinson Cancer Research Center, Seattle, Washington

Thomas C. Boone, Ph.D.   Director, Department of Process Science, Amgen Inc., Thousand Oaks, California

Candace Breen, M.A.   Clinical Instructor/Coordinator, Department of Clinical Laboratory Sciences, Division of Health Technology and Management, SUNY at Stony Brook, Stony Brook, New York

Miguel H. Bronchud, M.D., Ph.D.   Head, Department of Internal Medicine, Division of Medical Oncology, General Hospital of Granollers, Barcelona, Spain

Randy A. Brown, M.D.   Assistant Professor, Department of Internal Medicine, Division of Bone Marrow Transplantation, Washington University School of Medicine, St. Louis, Missouri

Sherri L. Brown, M.D. Director and Senior Safety Officer, Department of International Clinical Safety, Amgen Inc., Thousand Oaks, California

C. Dean Buckner, M.D. Scientific Director, Clinical Trials Division, Response Oncology, Inc., Seattle, Washington

Jane Campbell, R.N., B.N. Director of Education, Haematology Oncology Clinics of Australasia, Brisbane, Australia

Ellen N. Cheung, Ph.D. Research Scientist 3, Department of Pharmacokinetics and Drug Metabolism, Amgen Inc., Thousand Oaks, California

Jeffrey Crawford, M.D. Associate Professor of Medicine and Director of Clinical Research, Duke Comprehensive Cancer Center and Duke University Medical Center, Durham, North Carolina

David C. Dale, M.D. Professor, Department of Medicine, University of Washington, Seattle, Washington

T. Michael Dexter, Ph.D., D.Sc., M.R.C.Path., F.R.S., Hon. M.R.C.P. Professor and Director, Paterson Institute for Cancer Research, Christie Hospital, Manchester, England

John F. DiPersio, M.D., Ph.D. Chief, Department of Medicine, Division of Bone Marrow Transplantation and Stem Cell Biology, Washington University School of Medicine, St. Louis, Missouri

Peter Dreger, M.D. Second Department of Internal Medicine, University Hospital, Kiel, Germany

Michael Edmonds, M.D., F.R.C.P. Consultant Diabetologist, Department of Diabetes, Kings College Hospital, London, England

M. Haim Erder, Ph.D. Associate Director and Head, Department of Health Economics, Amgen Inc., Thousand Oaks, California

Carol Fier, R.N., M.S.N. Manager, SCN International Registry, Department of International Clinical Safety, Amgen Inc., Thousand Oaks, California

Howard B. Fleit, Ph.D. Associate Professor, Department of Pathology, University Medical Center, SUNY at Stony Brook, Stony Brook, New York

MaryAnn Foote, Ph.D. Associate Director and Head, Department of Medical Writing, Amgen Inc., Thousand Oaks, California

Moshe Fridman, Ph.D. Assistant Professor, Department of Information and Operations Management, Marshall School of Business, University of Southern California, Los Angeles, California

Janice Gabrilove, M.D.   Associate Member, Department of Medicine, Memorial Sloan-Kettering Cancer Center, New York, New York

Marc G. Golightly, Ph.D.   Associate Professor, Department of Pathology, University Medical Center, SUNY at Stony Brook, Stony Brook, New York

Lawrence T. Goodnough, M.D.   Professor, Departments of Medicine and Pathology, Washington University School of Medicine, St. Louis, Missouri

Andrew Gough, M.B., Ch.B., M.R.C.P.   Clinical Research Fellow, Department of Diabetes, Kings College Hospital, London, England

Peter L. Greenberg, M.D.   Professor of Medicine, Hematology Division, Stanford Medical Center, Stanford, and VA Palo Alto Health Care System, Palo Alto, California

Thomas Hartung, M.D., Ph.D.   Assistant Professor, Department of Biochemical Pharmacology, University of Konstanz, Konstanz, Germany

Gerhard Heil, M.D., Ph.D.   Deputy Head, Department of Hematology/Oncology, Hanover Medical School, Hanover, Germany

Gisela Helm, M.D.   Attending Physician, Department of Medicine III, University of Erlangen–Nuernberg, Erlangen, Germany

Prabhakar Kocherlakota, M.D.   Fellow in Neonatal-Perionatal Medicine, Department of Pediatrics, University Medical Center, SUNY at Stony Brook, Stony Brook, New York

Irene Kuter, M.D., D.Phil.   Associate Physician, Hematology–Oncology Unit, Massachusetts General Hospital Cancer Center, Massachusetts General Hospital, Boston, Massachusetts

Edmund F. LaGamma, M.D.   Professor of Pediatrics and Neurobiology, Department of Pediatrics, University Medical Center, SUNY at Stony Brook, Stony Brook, New York

Fa-Chyi Lee, M.D.   Fellow, Hematology/Oncology, Department of Medicine, UCLA, Los Angeles, California

W. Conrad Liles, M.D., Ph.D.   Assistant Professor, Department of Medicine, University of Washington, Seattle, Washington

Shirley Y. Masuda, B.S.N.   Project Manager, Department of International Clinical Safety, Amgen Inc., Thousand Oaks, California

Ronald T. Mitsuyasu, M.D., F.A.C.P.   Associate Professor, Department of Medicine, UCLA, Los Angeles, California

Graham Molineux, Ph.D.   Research Scientist, Department of Pharmacology, Amgen Inc., Thousand Oaks, California

George Morstyn, M.B., Ph.D., F.R.A.C.P.   Associate Clinical Professor, UCLA School of Medicine, Los Angeles, and Vice President, Clinical Development, and Chief Medical Officer, Amgen Inc., Thousand Oaks, California

Hassan Movahhed, M.S.   Associate Director, Department of Clinical Affairs, Amgen Inc., Thousand Oaks, California

Robert S. Negrin, M.D.   Associate Professor, Department of Medicine, Stanford University, Stanford, California

Steve Nelson, M.D., F.C.C.P.   John H. Seabury Professor of Medicine, Department of Pulmonary/Critical Care Medicine, Louisiana State University Medical Center, New Orleans, Louisiana

Craig Nichols, M.D.   Professor, Department of Medicine, Division of Hematology/Oncology, Indiana University School of Medicine, Indianapolis, Indiana

Timothy Osslund, Ph.D.   Research Scientist, Department of Molecular Structure, Amgen Inc., Thousand Oaks, California

Karen L. Patterson, B.S.N.   Manager, Department of International Clinical Safety, Amgen Inc., Thousand Oaks, California

David L. Pitrak, M.D.   Associate Professor, Department of Medicine, University of Ilinois College of Medicine at Chicago, Chicago, Illinois

Aruna Raghavachar, M.D.   Vice Chairman, Department of Medicine III, University of Ulm, Ulm, Germany

Roland P. Repp, M.D.   Department of Medicine III, University of Erlangen–Nuernberg, Erlangen, Germany

Lorin K. Roskos, Ph.D.   Research Scientist, Department of Pharmacokinetics and Drug Metabolism, Amgen Inc., Thousand Oaks, California

Eric Keith Rowinsky, M.D., F.A.C.P.   Clinical Professor of Medicine, University of Texas Health Sciences Center at San Antonio, and Director, Clinical Research, Institute for Drug Development, Cancer Therapy and Research Center, San Antonio, Texas

Scott Saxman, M.D.   Assistant Professor of Medicine, Indiana University School of Medicine, Indianapolis, Indiana

J. Howard Scarffe, M.D., F.R.C.P. Professor, Department of Medical Oncology, University of Manchester, Christie Hospital NHS Trust, Manchester, England

Norbert Schmitz, M.D., Ph.D. Second Department of Internal Medicine, University Hospital, Kiel, Germany

Hubert Schrezenmeier, M.D. Department of Medicine III, University of Ulm, Ulm, Germany

William P. Sheridan, M.B., F.R.A.C.P. Associate Professor, Department of Medicine, UCLA School of Medicine, Los Angeles, and Director, Clinical Hematology, Amgen Inc., Thousand Oaks, California

Jeffrey H. Silber, M.D., Ph.D. Associate Professor of Pediatrics, Anesthesiology, and Health Care Systems, University of Pennsylvania School of Medicine, and Director, Center for Outcomes Research, Children's Hospital of Philadelphia, Philadelphia, Pennsylvania

Bernhard Stockmeyer, M.D. Department of Medicine III, University of Erlangen–Nuernberg, Erlangen, Germany

Thomas Valerius, M.D. Department of Medicine III, Division of Hematology/Oncology, University of Erlangen–Nuernberg, Erlangen, Germany

Juan W. Valle, M.B., Ch.B., M.R.C.P. Research Fellow, Department of Medical Oncology, University of Manchester, Christie Hospital NHS Trust, Manchester, England

J. G. J. van de Winkel, Ph.D. Professor, Department of Immunology, University Hospital Utrecht and Medarex Europe, Utrecht, The Netherlands

Martha Vincent, Ph.D. Vice President, Department of Clinical Research, Amgen Inc., Thousand Oaks, California

Karl Welte, M.D. Professor of Pediatrics and Head, Department of Pediatric Hematology and Oncology, Medizinische Kinderklinik, Hanover Medical School, Hanover, Germany

Albrecht Wendel, Ph.D. Professor, Department of Biology, University of Konstanz, Konstanz, Germany

1
Biology of G-CSF

Graham Molineux
Amgen Inc., Thousand Oaks, California

T. Michael Dexter
*Paterson Institute for Cancer Research, Christie Hospital,
Manchester, England*

I. THE HEMATOPOIETIC SYSTEM

All circulating blood cells are the progeny of populations of primitive
hematopietic precursor cells present in the bone marrow, and each of the
different types of mature cells have different specialized functions and
different life spans. To maintain adequate numbers of circulating blood
cells, they must be replaced at a rate equal to their rate of loss and some
5×10^{11} myeloid cells need to be produced per day in the human to account
only for loss through normal aging processes. This output is sustained by
the pluripotent hematopoietic stem cells present in the bone marrow and the
considerable amplification that occurs among their more developmentally
restricted progeny. Clearly, the regulation of the balance between maintenance of stem cell numbers via the process of self-renewal, and the loss of
stem cells through differentiation leading to the production of various
kinds of mature cells, is an issue central to the understanding of blood
cell production.

Some of the regulatory pathways that control hematopoiesis have
become better understood in the last 30 years because of the development
of in vitro assays for clonogenic precursor cells present largely in bone
marrow and other hematopoietic tissues. Initially, these assays detected only
progenitor cells for neutrophils and macrophages (1,2), but the methods
subsequently were extended to facilitate detection of precursors for all the
various blood cell types. In the early studies, however, it was realized

that the formation of colonies containing these mature blood cells absolutely required the presence of medium, previously conditioned by non-hematopoietic tissues, that contained "colony-stimulating factors."

As knowledge of these colony-stimulating factors (CSF) advanced, it became apparent that there existed a large number of glycoproteins, hormone-like growth factors, that were responsible for, and are required for, growth and development of hematopoietic cells in vitro. Some of these have a broad spectrum of target cells (eg, interleukin [IL]-3), others mainly influence the survival and/or development of primitive non–lineage-restricted multipotent progenitor cells (eg, stem cell factor [SCF], leukemia-inhibitory factor [LIF]), while others (eg, granulocyte colony-stimulating factor [G-CSF] and macrophage colony-stimulating factor [M-CSF, CSF-I]) only or preferentially promote the growth of cells that have more restricted development potential. The network of these growth-stimulatory molecules, and other cytokines that can inhibit the proliferation of hematopoietic stems cells and their progeny (eg, transforming growth factor-beta [TGF-β], macrophage inflammatory protein 1-alpha [MIP-1α]), together with the bone-marrow stromal complex (the cellular and extracellular components of the hematopoietic microenvironment which support and direct hematopoiesis), form the effectors of the homeostatic regulation of blood cell production (**Figure 1**).

II. THE ISOLATION AND CHARACTERIZATION OF G-CSF

A. Murine G-CSF

One distinctive property of G-CSF (other than its ability to selectively stimulate the production of neutrophils in vitro) distinguished it from the other CSF and facilitated its purification, subsequent molecular cloning, and large-scale production in prokaryotic cells: this was its ability to induce terminal differentiation of the murine leukemic cell line WEHI-3BD[+]. After observing that serum from endotoxin-treated mice could cause the differentiation of WEHI-3B myelomonocytic leukemic cells (3), the molecule responsible was named GM-DF. When subsequently analyzed, it was found that the serum from these mice contained, in addition to granulocyte-macrophage colony-stimulating factor (GM-CSF), another activity (G-CSF) that copurified with the differentiation-inducing activity. Granulocyte colony-stimulating factor was then further purified from medium conditioned by the lung tissue of endotoxin-treated mice (4) and it was shown not only to possess an ability to stimulate differentiation of WEHI-3BD[+] cells, but also to be active upon normal hematopoietic progenitor cells,

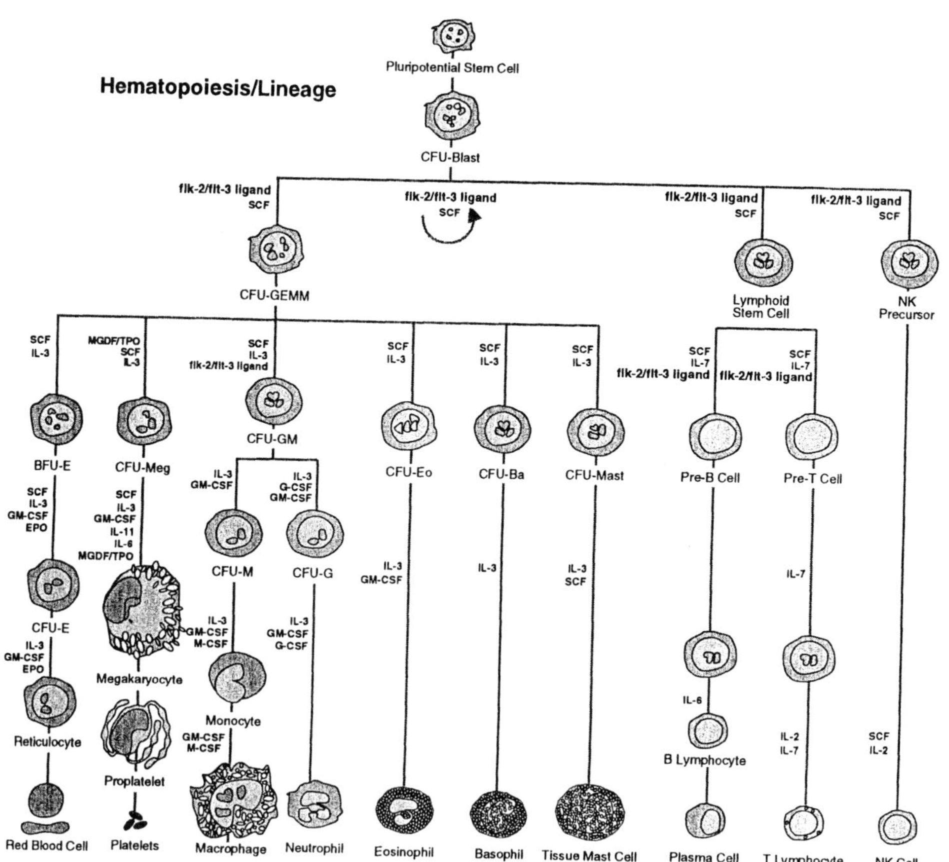

Figure 1. Scheme of hematopoiesis including some of the growth factors that influence the production of blood cells. CFU-GEMM, multilineage colony-forming cell; BFU-E, erythroid burst-forming cell; CFU-E, erythroid colony-forming cell; CFU-GM, granulocyte-macrophage colony-forming cell; CFU-M, macrophage colony-forming cell; CFU-E, eosinophil colony-forming cell; CFU-Ba, basophil colony-forming cell; IL, interleukin; SCF, stem cell factor, G-CSF, granulocyte colony-stimulating factor; GM-CSF, granulocyte-macrophage colony-stimulating factor; M-CSF, macrophage colony-stimulating factor; TPO, thrombopoietin; MGDF, megakaryocyte growth and development factor; EPO, erythropoietin. (Figure courtesy Amgen Inc., Thousand Oaks, CA.)

supporting the formation in vitro of numerous small colonies containing neutrophils, at similar concentrations to those required for WEHI-3B differentiation (4).

B. Human G-CSF

Using assays similar to those developed for the characterization of murine G-CSF (induction of differentiation of WEHI-3B cells and stimulation of the growth of normal progenitor cells in vitro), an activity was isolated from medium conditioned by the growth of the 5637 bladder carcinoma cell line and named "pluripoietin" by Welte et al (5) because of its ability to stimulate the proliferation and development of multipotent myeloid progenitor cells in vitro in addition to its effects on differentiation of WEHI-3BD$^+$ cells. When this activity was molecularly cloned and expressed as a recombinant protein it was then shown (using enriched bone marrow progenitor cells) that the "pluripoietin" ability was lost and only (or mainly) neutrophil colonies developed. This resulted in the molecule being renamed G-CSF (eg, Ref. 6).

III. THE ROLE OF G-CSF IN NORMAL HEMATOPOIESIS

Analysis of the physiological role of the CSF has not been a straightforward exercise. In the case of G-CSF, however, several different approaches have yielded a considerable body of information relating to this question.

A. Circulating Levels

Unlike many other putative regulatory molecules, G-CSF is detectable in the blood, although normally the level is less than 10 pg/mL of plasma (7). Higher levels have been documented in various disease states. Neutropenic patients tend towards higher circulating levels, while bacteremic and febrile neutropenic patients may have levels as high as 1000 to 100,000 pg/mL in the blood (8). Similarly, patients with aplastic anemia, but with no obvious signs of infection, have been shown to have approaching 100 pg/mL, while those showing indications of infection may have up to 1000 pg/mL or higher (7). It is therefore likely that G-CSF, along with other cytokines, is involved in the response to neutropenia and/or bacterial infection. Moreover, a role for G-CSF in baseline neutrophil production is indicated by studies in dogs (9). These animals generate neutralizing antibodies directed against injected human G-CSF and then acquire a significant neutropenia, presumably because these antibodies cross-react with endogenous canine G-CSF. If plasma from these dogs is then transferred to untreated animals, they too develop neutropenia, implying that endogenous G-CSF influences baseline neutrophil production.

G-CSF is also present at very high concentrations in various fluids linked to pregnancy and the early stages of infancy. Amniotic fluid contains high levels of G-CSF (10) and the concentration is further elevated in infectious episodes, hypertension, and in premature delivery (11,12). Maternal plasma also contains G-CSF, especially during the third trimester of pregnancy (13), though it remains a point of debate whether the physiological neutrophilia of pregnancy is linked to G-CSF levels (14). A high concentration of G-CSF also has been noted in normal umbilical cord blood (15,16). In addition, high concentrations of G-CSF have been seen in women with chorioamnionitis, and umbilical cord blood from these deliveries contains very high G-CSF concentrations (17). After childbirth, both maternal and infant plasma G-CSF concentrations are high one day after birth (18–20), though lower than the >50,000 pg/mL level noted in normal full-term cord blood (21). While the significance of this finding is unclear, one study shows strong correlations between elevated neonatal G-CSF concentrations and various complications during pregnancy: infections ($p < 0.01$), fetal distress ($p < 0.01$), premature rupture of membranes ($p < 0.05$), neonatal asphyxia ($p < 0.01$), and meconium staining of the amniotic fluid ($p < 0.01$) (22). Taken together, these data represent a substantial body of information implying a role for G-CSF during pregnancy and in early infancy. However, whether or not the elevations in G-CSF concentrations associated with pregnancy and early infancy have any functional significance other than the prophylactic benefit of increased neutrophil numbers associated with impending tissue injury remains to be investigated.

Further evidence for a role of G-CSF in hematopoiesis has arisen from studies of the rare disease cyclic neutropenia, where there is an inverse correlation between the number of neutrophils and the circulating levels of G-CSF at various points in the cycle, ie, at times of low neutrophil numbers G-CSF concentrations are high, and conversely when neutrophil numbers attain high levels, G-CSF levels are reduced (23).

B. Gene-Deletion Studies

Using retrovirus-mediated targeted disruption of the G-CSF gene in embryonal stem (ES) cells, Lieschke et al (24) generated "knockout" mice lacking expression of G-CSF in their tissues. Despite the absence of G-CSF, these animals had approximately 20% to 30% of normal neutrophil numbers: clearly other factors in addition to G-CSF can influence production of neutrophils. However, the important observation made from these mice was that the absence of G-CSF resulted in a 70% to 80% reduction in circulating neutrophil numbers and an impaired ability to counter bacterial infection, which was regained upon injection of recombinant G-CSF. In other words, a pivotal role for G-CSF in steady-state neutropoiesis is suggested by these studies, in addition to a function in infection. Interestingly, G-CSF $^{-/-}$ mice also showed reduced macrophage/monocyte progenitor cells and reduced circulating monocyte numbers in elderly mice, implying, as do the kinetic studies discussed below, that the activities of G-CSF, at least in vivo, are not restricted to neutrophil production.

C. Overexpression of G-CSF

Two significant studies have examined the role of G-CSF in chronic overexpression systems. The first, by Chang et al (25), uses retroviral vector-mediated transfer of G-CSF cDNA to bone marrow cells which were then used to rescue lethally irradiated mice. In the second study, Yamada et al (26) produced mice transgenic for human G-CSF. Obviously, in both studies the range of cell types in which G-CSF overexpression occurs is a significant concern, bearing in mind that over production in cell types not normally associated with high levels of G-CSF expression may lead to inappropriate responses and erroneous conclusions. Therefore, while valuable insights into the actions of excess G-CSF can be studied in these systems, a level of caution is required in interpreting the results.

Bearing in mind the reservations expressed above, however, the data from both systems are very similar. Excess G-CSF (and both systems documented high circulating levels of G-CSF) caused a significant neutrophilia, neutrophilic infiltration of various tissues, increases in the progenitor content of bone marrow (only in transgenics), spleen and peripheral blood, and monocytosis and lymphocytosis (only in transgenics). There were, however, no deleterious effects reported despite the high numbers of neutrophils in inappropriate sites. These data are in contrast to the serious side effects documented in mice chronically overexpressing erythropoietin (EPO) (27,28), GM-CSF (29), IL-3 (30,31), and LIF (32,33).

IV. THE EFFECTS OF G-CSF IN VITRO

The initial assertion that G-CSF represented a pluripoietin (5) has been revised in the light of more recent data using highly enriched target-cell populations and serum-depleted growth conditions (34). Indeed, this work became a paradigm in that it is now clear that to confidently ascribe a direct activity to any cytokine, including G-CSF, requires the elimination of the products of accessory cells and serum factors that may influence the response obtained. Under serum-deprived growth conditions and using target cells purified to apparent homogeneity, G-CSF stimulates predominantly the development of neutrophil colonies; however, the initial proliferation of granulocyte, macrophage, eosinophil, and erythroid colony-formers may also be promoted by G-CSF (35), but this growth is not sustained and does not result in the production of mixed-lineage colonies.

In addition to its colony-stimulating activity on progenitors of the neutrophil/monocyte lineages, G-CSF also can act as a proliferative stimulus for relatively mature granulocytic cells. For instance, neutrophilic promyelocytes, when transiently exposed to high doses of G-CSF, can undergo one or two cell divisions (36–39). It has been reported that the survival of human peripheral blood neutrophils (post-mitotic) also is enhanced by G-CSF in vitro; however, eosinophils are not similarly affected (40). Mature neutrophils also are functionally activated by G-CSF, for example, by increased antibody-dependent cell-mediated cytotoxicity (ADCC) against antibody-coated thymoma cells and also are primed to respond to the chemotactic peptide N-formyl-methionyl-leucyl-phenylalanine (FMLP) by prior exposure to G-CSF, which accelerates the generation of superoxide anions (41). Although it is apparent that G-CSF has direct effects upon neutrophils and their precursor cells when used on its own, several studies have shown that in combination with other growth factors, G-CSF can exert powerful synergistic effects. Primitive multipotent stem cells, for example, show a very limited response to either G-CSF or GM-CSF, but when these CSF are combined, multipotent stem cells are stimulated to proliferate and develop into colonies containing mature neutrophils and macrophages (42). G-CSF also can synergize with M-CSF to recruit proliferation and development of multipotent cells leading to production of cells predominantly of the monocyte-macrophage lineage (43). Studies have shown that the combination of G-CSF and SCF results in stimulation of monocyte progenitors from whole bone marrow (44) whereas the combination of *flt-3* ligand and G-CSF promotes the growth of primitive mouse $(Thy^{lo}Sca-1^+)$(45) and human (Lin^-CD34^+)(46) progenitor cells. Human pluripotent progenitor cells purified from the blood of post-chemotherapy patients also were more responsive to the combination of IL-3 and G-CSF than to either factor alone (47).

What these data suggest is that receptors for G-CSF (as well as a variety of other hematopoietic growth factors) are expressed before the multipotent cells undergo lineage restriction, but that G-CSF on its own is not sufficient to influence proliferation and/or differentiation of such cells. In combination with the other growth factors, however, G-CSF can provide a powerful growth/developmental stimulus for such cells. This has clear implications for therapeutic applications involving combinations of hemato-poietic cell growth factors.

In summary, G-CSF has been shown to have at least three separate roles in vitro: 1) a direct influence upon proliferation and development of neutrophil progenitor cells; 2) stimulation of proliferative and functional activation of more mature cells of the neutrophil lineage; and 3) ability to synergize with other hematopoietic growth factors.

V. EFFECTS IN VIVO: PRECLINICAL STUDIES

A. Elevation of Neutrophil Counts

When injected in vivo, recombinant G-CSF increases the number of mature neutrophils in the circulation. This has been shown in rats (48), mice (49 [human G-CSF]; and 50 [using murine G-CSF]), hamsters (51,52), dogs (53), non-human primates (54,55), pigs (56), guinea pigs (57), rabbits (58,59), and humans (60). The increase in neutrophil counts succeeds a transitory neutropenia, which occurs within 1 hour of injecting Filgrastim (r-metHuG-CSF) into patients (61); thereafter numbers increase rapidly (51,62). The rapid 2.5-fold increase in neutrophil numbers within a few hours of injection suggests that the early response may be due to migration of mature neutro-phils from sites such as the bone marrow (63).

It has been shown that immunosuppressed mice can be used as a quantitative model for G-CSF activity based upon the accumulation of neutrophils in the blood in response to infused rHuG-CSF. The correlation coefficient of this relationship exceeds 0.9 (64). These same authors showed a lack of linearity when untreated mice were treated with similar doses of rHuG-CSF, and argued that 100 mg/kg of cyclophosphamide was an essen-tial pretreatment to obtain a linear-dose response in the range of 0.1 to 10 μg rHuG-CSF/mouse/day (about 4 to 400 μg/kg body weight/day). In a separate study, however, we have shown that the response of normal mice to Filgrastim is linear if the injections are done twice daily instead of once (65). The number of circulating mature neutrophils is dependent on the dose of rHuG-CSF administered, and does not depend on the route of administration (even oral) (66).

Cynomolgus monkeys also showed dramatic (up to 10-fold) increases in blood neutrophils when injected with between 1 and 100 μg rHuG-CSF/kg body weight/day for up to 27 days (67). The neutrophils of these non-human primates also showed normal or elevated functional activities, and there was evidence of enhanced myelopoiesis in bone marrow, lymph nodes, and spleen, but not in other organs.

With these effects upon neutrophil production the most obvious and widely exploited use of rHuG-CSF has been in the treatment of neutropenia associated with iatrogenic therapy for malignancy: however, uses in non-hematological settings also are being developed, for example, to alleviate the neutropenia associated with the toxicity of clozapine, a psychoactive drug used for serious mental illness (68).

B. Accelerated Recovery of Neutrophils and Morbidity

In view of the ability of rHuG-CSF to elevate blood neutrophil numbers, Tamura et al (63) administered rHuG-CSF to mice that had been previously myelosuppressed by a single injection of cyclophosphamide. A dose of 2.5 μg (100 μg/kg) rHuG-CSF/day in these animals not only prevented the usual depression in blood neutrophil numbers, but increased counts (up to 10-fold) above their starting values. These responses to rHuG-CSF after cyclophosphamide were later shown to have a profound influence upon the ability of the mice to resolve infections with a range of different bacteria (69). In nearly all cases tested 1 μg (40 μg/kg) rHuG-CSF per day completely restored the resistance of cyclophosphamide-treated mice to bacterial infections to the levels seen in mice untreated with cytotoxic drugs.

Enhanced recovery from cyclophosphamide treatment also has been noted in monkeys treated with Filgrastim (67), where the period of neutropenia was significantly reduced, and marrow cellularity recovered more quickly.

In a canine model of cyclic neutropenia, administration of rHuG-CSF did not eliminate cycling in neutrophil counts, but the neutrophil counts were maintained at a higher value, ie, without the neutropenia (53; see also Chapter 5). In this context, it is worth noting that GM-CSF does not alter the cycling levels (53), but may increase the peaks without affecting the depths of the nadirs. In gray collie dogs with cyclic neutropenia (70) the number and activity of G-CSF receptors are normal. Thus, the inference from these data is that an event downstream of G-CSF receptor/ligand binding is responsible for the unusual response to G-CSF and that production of the regulator is not the fundamental problem.

C. Lack of Pathological Changes

Exposure of animals to high doses of G-CSF for protracted periods does not appear to have deleterious effects. For example, high constitutive expression of G-CSF, attained after bone-marrow transplantation with hematopoietic cells that had been infected with a retroviral expression vector containing the gene encoding G-CSF, resulted in a sustained neutrophilia and infiltration of tissues, especially lung and liver, with neutrophils (25). Similar infiltration was noted in mice that had been repeatedly injected with Filgrastim (65). In both studies, however, this did not result in morbidity and no obvious pathological changes were seen, with the possible exception of thrombocytopenia (see discussion below). This is in contrast to the situation in mice producing a sustained excess of IL-3 (29) or GM-CSF (30) as a result of gene transfection, where potentially life-threatening consequences occur.

Furthermore, the generation of transgenic mice which presumably overexpress G-CSF from early fetal life confirm the effects seen with repeated-injection regimens. These changes include persistent granulocytosis, infiltration of various tissues with neutrophils, mobilization of hematopoietic progenitor cells to the blood, and an increase in splenic hematopoiesis. However, these studies also indicate the lack of any significant morbidity in association with these conditions. Yamada et al (26) documented a lymphocytosis of both B and T lineages in their G-CSF transgenic mice, though the significance of this finding is not apparent.

D. Treatment of Animals Can Lead to a Redistribution of Hematopoiesis

During treatment with rHuG-CSF, mice show substantially reduced CFC-S (multipotent spleen colony-forming cells), CFC-GM, and CFC-Mix (multilineage in vitro colony-forming cells) numbers in the bone marrow, and concurrently increased numbers in the spleen (up to 100-fold) of CFC-GM, CFC-MK (megakaryocyte colony-forming cells), CFC-E, BFC-E (erythroid burst-forming cells), and CFC-S (63,71). Hamsters, on the other hand, develop a marrow hypercellularity (51) with increased colony-forming cells in bone marrow and also increased S-phase cells, though the doses used were not directly comparable. Clearly the effects of rHuG-CSF, in terms of marrow cellularity and progenitor cell distribution, show species variation.

E. Influence on the Production of Lymphocytes and/or Monocytes

The production of lymphocytes in vivo is largely unaffected by the administration of rHuG-CSF in humans and mice (60,71); however, lymphocytosis has been reported in cats treated with rHuG-CSF (72) and increased lymphocyte counts also have been noted in irradiated dogs treated with rHuG-CSF (73) compared with those which received irradiation only. It is possible that these effects are due to a direct stimulatory effect of G-CSF on lymphoid precursor cells because G-CSF receptors have been documented on CD10$^+$ B-cell precursors (74).

Transgenic mice which chronically overexpress G-CSF show a persistent lymphocytosis involving both B220$^+$ B-lineage and CD3$^+$ T-lineage cells (26). Despite this, these transgenic mice have reduced B-cell lymphoid populations in the bone marrow. This confirms data that indicate inhibition of B-lymphopoiesis in vitro in the presence of stroma and reduced marrow lymphopoiesis in mice either injected with rHuG-CSF or bearing tumors which produce excess G-CSF (75). These data appear to contrast with in vitro data (76,77) that document the support offered to B-cell precursors by combinations of growth factors including G-CSF. The systems used by these different authors vary significantly: the stromal-dependent cultures used by Lee et al differ from the methylcellulose-based anchorage-independent system used by Hirayama et al, and may offer some explanation for these differences based on the documented influence of stroma on lymphoid-cell development. Thus, it would appear that G-CSF, in concert with other factors, can positively influence B-cell development when used in the absence of stroma. In vivo, or in the presence of stroma, there would appear to be a suppression of bone-marrow lymphopoiesis, but stimulation of splenic and perhaps thymic lymphocyte production. This may lead to lymphocytosis, but a further complication is introduced by the antigenicity of G-CSF in some species which may itself be the cause of excessive lymphocyte counts (see Ref. 78).

In clinical studies a matter of particular interest is the manipulation of lymphoid numbers in acquired immunodeficiency syndrome (AIDS), and a preliminary report by Stricker et al (79) indicates that rHuG-CSF, administered to treat successfully the neutropenia associated with this disease, also caused significant increases in natural killer (NK) cells and CD8 and CD4 T cells in the absence of any change in viral load. This report follows anecdotal evidence of improvement in bacterial infections in patients with AIDS treated with Filgrastim (80). Furthermore, in patients with ovarian cancer, rHuG-CSF also has some beneficial effects on lymphoid reconstitution after chemotherapy, improving overall lymphoid recovery without affecting the proportion of the various lymphocyte subsets (81). Taken together, these data suggest a role for G-CSF in lymphopoiesis. While some of these effects may well be indirect, data from in vitro studies indicate the possibility of a direct effect of G-CSF on lymphoid progenitor growth.

Lieschke et al (24) discuss the reduced monocyte counts in G-CSF knockout mice (see above) and Lord et al (82) show the stimulatory effect of injected rHuG-CSF on monocytopoiesis. It is perhaps surprising therefore that there are not more references to effects on the monocyte lineage from published clinical work. A monocytosis has been documented in 8 of 33 patients receiving Filgrastim (83), a similar fraction (4/13) to those receiving only rHuGM-CSF in the same study. A small but significant increase in monocyte counts also has been documented in normal healthy human volunteers injected with rHuG-CSF (84). Some insight into the mechanism involved has been gained by the work of Gilmore et al (85) who injected rHuG-CSF–treated mice with a neutralizing antibody directed against M-CSF. In this setting the G-CSF–related monocytosis was eliminated, implying either an indirect action of G-CSF via M-CSF or, perhaps, a synergistic action between the two regulators. The generation of M-CSF/GM-CSF/G-CSF triple knockout mice (discussed in Ref. 24) should be particularly informative in further addressing this issue.

F. Response of Non-Hematopoietic Cells

1. Non-Hematopoietic Tissues

It has been claimed that G-CSF, in common with GM-CSF, promotes the growth and migration of vascular endothelium in vitro (86,87), findings that may have some bearing on, for example, angiogenesis. Other authors, however, claim that G-CSF shows no effect on primary or passaged umbilical vein endothelial cells (88). Receptors have been found for G-CSF on human placenta and trophoblastic cells (89), though the significance of these receptors is yet to be defined.

Studies involving a relatively small number of patients with breast cancer have shown that the incidence of oral mucositis is markedly reduced in patients given rHuG-CSF in addition to Adriamycin® chemotherapy (90). It is as yet unclear whether the documented benefit is obtained via an indirect mechanism involving the elevated neutrophil counts in patients receiving G-CSF, or a direct effect of G-CSF on the oral epithelium via unknown mechanisms. Some insight into a possible mechanism is gained from the studies of Denzlinger et al (91), who note that, in contrast to IL-3 or GM-CSF, rHuG-CSF reduces the production of leukotrienes in treated cancer patients. In addition, TNF-α is known to stimulate endogenous leukotriene production in humans (92) and previous work has shown that GM-CSF and IL-3 increase TNF-α production (93) while G-CSF reduces TNF-α production (91,94).

2. Malignant Cells

A widely expressed concern is that neoplastic cells may be stimulated to grow by rHuG-CSF in a manner similar to normal hematopoietic cells. If so, this could preclude the use of rHuG-CSF, at least in some circumstances.

It has been reported that blast cells from patients with acute myeloid leukemia (AML) have surface receptors for G-CSF (95–97) and that, in at least some cases, leukemic blast cells proliferate in response to applied G-CSF (98–100). It should be noted that this finding does not necessarily obviate the use of rHuG-CSF in patients with AML since it may invoke cycling of the leukemic stem cells and therefore increase their sensitivity to cycle-specific cytotoxic agents (101), a mechanism exploited by Mori et al (102) to induce a temporary remission in several cases of AML treated with G-CSF and Ara-C. In the context of treatment of patients with AML, it has also been reported that rHuG-CSF can induce the differentiation of leukemic cells when used alone (6) or in combination with retinoic acid (103), although larger studies have shown that rHuG-CSF is not a strong differentiation stimulus, at least in vitro, for a majority of AML cells (104,105). Nevertheless, it is of obvious therapeutic advantage to induce differentiation (and therefore death) of leukemic cells without proliferation, and Irvine et al (106) have retained the differentiation-inducing ability of G-CSF without the concurrent induction of proliferation by combining interferon (IFN)-γ with G-CSF. In this respect, it is interesting that in a murine radiation-induced leukemia model, rHuG-CSF promoted the growth of leukemic cells in vitro. Yet when mice were injected with the leukemic cells and then in the succeeding period with rHuG-CSF, survival was improved over non-rHuG-CSF–treated tumor-bearing animals (107). In a related study, Hayashi et al (108) used a WEHI-3BD^{+} subclone which was leukemogenic when injected into syngeneic mice, but differentiated terminally when the mice were also given rHuG-CSF.

The majority of transformed cell lines which express receptors for G-CSF are of hematopoietic (myeloid) origin (109,110) though it has been shown that the growth of several small-cell lung cancer (SCLC) cell lines also is promoted by G-CSF (111) and two colon adenocarcinoma lines show more in vitro colony growth in agar in response to G-CSF (96). It is possible that these responses are peculiar to cell lines. However, there is an accumulating literature indicating direct promotion of certain tumors by administered rHuG-CSF. Kawachi et al (112) report a case of Richter syndrome which, when treated with G-CSF for neutropenia, developed generalized lymph node enlargement which regressed on cessation of rHuG-CSF therapy. When rHuG-CSF was restarted the lymphadenopathy reappeared, and post-mortem examination (after death from an unrelated cause) revealed the presence of small lymphocytic leukemia cells that expressed surface receptors for G-CSF. Izumi et al (113) discuss a patient with aplastic anemia who received rHuG-CSF for neutropenia and immediately developed signs of AML which also responded to rHuG-CSF in vitro. In addition, Tachibana et al (114) elucidate a mechanism of autocrine/paracrine growth stimulation involving rHuG-CSF in a case of transitional cell carcinoma of the bladder. These data need, however, to be viewed in the context of rHuG-CSF–induced remission in occasional unexpected diseases such as acute lymphoblastic leukemia (ALL) reported by Hatta et al (115) and the significant antitumor effect of rHuG-CSF when used in combination with TNF in a nude mouse model of human medulloblastoma xenografting (116). Furthermore, from the large number of patients (in excess of 1.3 million) who have been treated with Filgrastim, it is clear that the risk of enhanced tumor growth is very small indeed.

G. Effects of G-CSF on Erythropoiesis

Erythropoiesis is suppressed in mice after prolonged treatment with rHuG-CSF (71,117,118). In the first few days of treatment (4 to 5 days) the deficiency is seen only in marrow differential counts and radioactive iron incorporation into the bone marrow, but this shortfall of red cell production is offset (to a large degree) by the increased erythropoietic activity noted in the spleen (65,71). In splenectomized mice, however, protracted administration of rHuG-CSF resulted in anemia. This anemia was present in splenectomized animals not only after doses of 250 μg/kg/day, but also in mice treated with 10 μg/kg/day, a dose in routine clinical use. This observation raises the question, is this suppression likely to be significant in patients treated with rHuG-CSF?

The adult mouse differs in several important areas from the adult human; first, in mice it is a common reaction to hematopoietic stress for the spleen to assume a hematopoietic role (119–121), unlike the human, where extramedullary hematopoiesis is rare (although splenomegaly has been reported in patients with congenital agranulocytosis who received Filgrastim for up to 13 months [122]). Second, most of the adult rodent skeleton contains erythropoietically active red marrow, unlike in the human where there are more restricted skeletal sites of red cell production and even these areas decrease with time and are replaced with hematopoietically quiescent yellow marrow. This species difference may allow expansion of erythropoiesis within the human skeleton without the need to establish red cell production at extraskeletal sites such as the spleen. Certainly, it should be stressed that prolonged (>4 years) treatment of patients with Filgrastim has, at least so far, not resulted in anemia (see Chapter 5). A single preliminary report by Pojda et al (123) indicated that a 5-month-old child treated with rHuGM-CSF or Filgrastim showed some signs of impaired erythropoiesis during treatment, but the causes and significance of this finding have yet to be established.

A report by Bungart et al (124) provides some information on the mechanisms that may be involved in the suppression of erythropoiesis in rodents. Using C57Bl/6 mice which had been treated with 150 μg rHuG-CSF/kg body weight/day in a twice-daily regimen, Bungart and colleagues confirmed the suppression of marrow erythropoiesis by rHuG-CSF, but they then extended their studies by injecting concurrently isologous packed red blood cells to artificially elevate the hematocrit to approximately 70% (compared with the low 40% level in control animals). Mice rendered polycythemic in this manner have been shown to stop production of EPO which, in turn, leads to suppressed production of red blood cells. If there were competition between the granulopoietic and erythropoietic lineages, then the neutrophil response to G-CSF might be expected to be enhanced in such circumstances; however, they found no increase in granulopoietic response to administered rHuG-CSF, and contend that this argues against the idea that any suppression of erythropoiesis (under rHuG-CSF treatment) results from interlineage competition. Further data from Nijhof and colleagues (125,126) extend these observations. This group administered simultaneously rHuG-CSF and rHuEPO to splenectomized mice and analyzed the mutual inhibition by the two regulators. Thus, EPO suppressed, to a large degree, the neutrophil response to rHuG-CSF and conversely rHuG-CSF reduced the erythroid response to rHuEPO. In contrast to these data are initial studies from Locatelli et al (127), Pierelli et al (128), and Negrin et al (129), who document a significant benefit from administering combined rHuG-CSF and rHuEPO to children recovering from bone marrow transplantation, in adults recovering from chemotherapy for ovarian carcinoma, or in patients with myelodysplastic syndromes. Despite this considerable body of literature, little further insight has been gained into the mechanism involved in the interlineage effect.

Thus the mechanism for the erythroid suppression that occurs in rodents in response to G-CSF remains unclear, but the observation that erythropoiesis in the spleen is not suppressed (indeed, it is enhanced) indicates that only bone marrow red cell production is affected. Also a preliminary report by Udupa et al (130) shows that under similar conditions rHuGM-CSF may also be capable of suppressing erythropoiesis, perhaps indicating that the effect upon red cell production is a more general feature of growth factor–stimulated granulopoiesis.

H.   The Effects of G-CSF on Thrombopoiesis

Data collected from clinical studies of patients treated with rHuG-CSF have not indicated a significant impact on platelet counts (reviewed in Ref. 131). However, recent data from Baer et al (132), who administered rHuG-

CSF to otherwise untreated patients with AML, indicate a reduction in platelet counts in response to rHuG-CSF. In addition to this, treatment of neutropenic neonatal children with rHuG-CSF has been reported to exacerbate the concurrent thrombocytopenia (133), data confirmed by studies in neonatal rats (134). Recombinant GM-CSF also has been implicated in the prolongation of thrombocytopenia (reviewed in Ref. 131).

Studies in mice using thrombopoietin (TPO), a recently identified regulator of thrombopoiesis, have extended these data. Murine recipients of either peripheral blood progenitor cells (PBPC) or marrow grafts that receive TPO show accelerated platelet recovery and a significant reduction in the period of thrombocytopenia (135,136). The inclusion of rHuG-CSF in the post-transplant support regimen offered to marrow recipients predictably accelerates neutrophil recovery, but prolongs the duration of thrombocytopenia. Simultaneous administration of both growth factors led to a platelet response which, though slower than noted in TPO-supported recipients, was significantly faster than seen in transplanted mice receiving no growth factor support. In other words, the inclusion of rHuG-CSF in the post-transplant support of marrow recipients delayed platelet recovery and reduced the acceleration of platelet recovery obtained with TPO. This dampening of platelet recovery was dependent on the dose of rHuG-CSF, and even the highest doses of rHuG-CSF given along with TPO still resulted in enhanced platelet recovery. Recombinant HuGM-CSF, on the other hand, suppressed platelet recovery markedly, and the inclusion of rHuGM-CSF in the TPO-supported groups reduced platelet recovery to slower than that seen in unsupported marrow recipients.

Interestingly, irradiated mice that received a graft of PBPC mobilized with rHuG-CSF benefited from the rHuG-CSF/TPO combination in terms of both enhanced neutrophil and enhanced platelet recovery. The inclusion of rHuG-CSF in the post-transplant support of PBPC recipients did not compromise platelet recovery as it had done in marrow recipients. Recombinant HuGM-CSF–treated PBPC recipients showed suppressed platelet recovery, just as the marrow recipients had (137). Thus, the situation appears to reflect to some extent a cross-lineage effect of both G-CSF and GM-CSF manifest on the platelet lineage in addition to the effects on erythropoiesis discussed above. G-CSF, with perhaps greater lineage fidelity than GM-CSF, has less impact on platelets, and PBPC recipients are more refractory to the interlineage effects than marrow recipients. We are left with the question of a mechanism for this effect which may be difficult to elucidate in the absence of any in vitro correlate of the data. It is interesting, however, that PBPC recipients who receive greater numbers of progenitor cells are more resistant to the interlineage effects than marrow recipients.

I. Hematopoietic Cell Kinetics

Autoradiographic studies have shown a greater than normal proportion of bone-marrow myeloblasts and promyelocytes to be in the S-phase of the cell cycle of mice (82,138) and humans (139) treated with rHuG-CSF. Myelocytes also show an increased labeling index during rHuG-CSF treatment and, in common with myeloblasts and promyelocytes, a decreased duration of the S-phase and significantly accelerated cell cycle. In combination with accelerated release from the marrow and an unchanged peripheral half-life of mature neutrophils under rHuG-CSF treatment, the reported increases in peripheral blood neutrophil count can be accounted for by the inclusion of relatively few extra amplification divisions in the post–CFC-GM compartments. Thus, little "stress" is placed on the more primitive cells of the marrow during treatment with rHuG-CSF, since the majority of the extra cells recruited to the blood under the influence of G-CSF are accounted for by the increased proliferative activity within the maturing compartments of the marrow (**Table 1**).

Interestingly, the kinetics of monocyte production also are affected profoundly during G-CSF treatment in mice (82). Despite relatively little change in their numbers in peripheral blood, the drastically reduced half-life (from 60 hours in control animals to 6.5 hours in rHuG-CSF–treated mice) conceals a greatly enhanced rate of output merely to maintain num-

Table 1. The Kinetics of Neutrophil Production Under Normal Conditions and G-CSF Stimulation.

	Human		Mouse	
	Control	G-CSF	Control	G-CSF
Fold increase in neutrophil count	1	10.25	1	14.3
Time of peak appearance in peripheral blood (days)	4*	1–2	3–4	1
Peripheral half-life: $t^{1/2}$ (hr.)	7–8†	7.6	8	9.8
Amplification factor‡	1	9.4	1	14.6
Extra maturation divisions§	0	3.2	0	3.8

* From Refs. 197–199.

† From Refs. 200 and 201.

‡ The amplification factor is the factor by which neutrophil production is increased over baseline to account for the elevated neutrophil count and taking account of the neutrophil loss over the period.

§ The number of extra maturational divisions is the number of supplementary divisions required to account for the amplification of neutrophil production.

Table 2. The Kinetics of Monocyte Production Under Normal Conditions and G-CSF Stimulation.

	Control	G-CSF	IL-3	GM-CSF
Fold increase in monocyte count	1	7.8	1.1	1.5
Time of peak appearance in peripheral blood (days)	24	12–15	15–20	15–20
Peripheral half-life: $t^{1/2}$ (hr.)	60	6.5	15.1	23.3
Amplification factor	1	30	3.1	3.4
Extra maturational divisions	0	4.9	1.7	1.8

bers (**Table 2**). It is noteworthy that Metcalf (44) has shown G-CSF to be a growth stimulus for macrophage progenitors indicating that this effect may well be a direct stimulation of monocyte growth by G-CSF.

Greater numbers of the more primitive hematopoietic progenitor populations of the bone marrow are recruited to the S-phase of the cell cycle during rHuG-CSF treatment. It was shown by Cohen et al (51) that increased cycling activity within the CFC-GM compartment was a feature of the G-CSF response in hamsters. However, Bronchud et al (60) showed no such increase in S-phase cells in the equivalent populations in patients treated with Filgrastim. Roberts and Metcalf (140) document marginally increased cycling activity in the bone marrow of mice injected with rHuG-CSF (from 39 ± 6.5% to 47 ± 5%). They also show that PBPC are universally quiescent, irrespective of the species they were obtained from or the growth factor treatment they had received. It would appear therefore that the increased neutrophil output sustained under the influence of Filgrastim results from only minimally increased cycling within the bone-marrow progenitor pool, which makes a relatively minor contribution to overall neutrophil production, the majority being the result of increased cycling activity and reduced transit time in the post-progenitor populations.

J. Mobilization of Peripheral Blood Progenitor Cells

The relocation of transplantable hematopoietic progenitor cells, from the bone marrow to the blood stream, under the influence of rHuG-CSF is an established procedure in clinical and experimental systems. These circulating cells have the advantages that they are more easily accessible and show superior performance to conventionally harvested bone marrow cells in terms of short-term recovery. Detailed discussion of the clinical data obtained with Filgrastim-mobilized PBPC in various disease states appears elsewhere in this volume (Chapters 13 and 14). This unusual effect of both rHuG-CSF and rHuGM-CSF (141) was not anticipated from in vitro studies. It should also be stressed that not only does the peripheral blood become a more effective source of transplantable progenitors under the influence of rHuG-CSF, but bone marrow also may benefit from rHuG-CSF pretreatment. Slowman et al (142) showed that the number of nucleated cells and the number of CD34$^+$/CD33$^-$ cells harvested from the bone marrow of patients with cancer is increased after rHuG-CSF or rHuGM-CSF. In this study, the increased progenitor numbers did not translate to any clinical advantage, but recipients also received rHuG-CSF in the post-transplant phase that may have masked any benefit from the pre-harvest conditioning by maximizing the post-transplant recovery rate. By measuring competitive repopulating ability of mouse bone marrow and blood at various times after rHuG-CSF treatment, Bodine et al (143) showed that at the time rHuG-CSF was stopped the number of repopulating cells in the blood was high while that in the bone marrow was low. However, 14 days later, the competitive repopulating ability of the bone marrow had increased to 10-fold greater than normal bone marrow. This overshoot phenomenon, if confirmed by studies in humans, may be exploited in patients to increase the number of cells available for transplant.

1. Long-Term Reconstituting Cells

Although cells capable of reconstituting the hematopoietic system of myeloablated recipients are normally present in the peripheral blood, their concentration is low. The numbers of these transplantable cells are known to be increased after chemotherapy or after rHuG-CSF or rHuGM-CSF therapy and these cells have been used successfully as a source of hematopoietic tissue in transplantation to support chemotherapy or irradiation for various disease conditions. An obvious question that arises from these studies, however, relates to the "quality" of the primitive cells that can be harvested during these procedures. Are pluripotent stem cells present in the peripheral blood and are they able to produce sustained engraftment? Our own studies, using peripheral blood cells harvested from mice treated with Filgrastim, indicate that primitive hematopoietic cells are indeed present in the blood of these animals and that these cells can reestablish lymphomyelopoiesis and lead to long-term engraftment when transferred to syngeneic irradiated recipients (120). In these studies, sufficient PBPC were contained in 10 μL blood from mice treated with Filgrastim to rescue 100% of the recipients and establish full chimerism in all measured hematopoietic cell populations. Detailed studies of these recipients showed that long-term regeneration occurred as a result of proliferation and differentiation of the donor stem cells rather than as a result of endogenous recovery of host stem cells. On the other hand, when peripheral blood stem cells were harvested from mice that had not been pretreated with Filgrastim, we found that 3000 μL were required to obtain sufficient PBPC to rescue irradiated recipient animals and provide full long-term engraftment. Thus, our data clearly show that while a few stem cells capable of regenerating hematopoiesis are present in the peripheral blood of normal individuals, the number of such cells can be greatly amplified following treatment with Filgrastim.

Filgrastim also can work in concert with several other materials to synergistically mobilize a quantitatively superior PBPC product. The augmentation of chemotherapy-mediated PBPC mobilization by Filgrastim is routinely used in the clinic, but interactions with different hematopoietic regulators also have been tested in both clinical and experimental systems. Stem cell factor has been shown to be a useful cofactor in mobilization protocols in both patients and experimental animals (144–147) and more limited studies with *flt-3* ligand (148), MIP-1α (149), and IL-6 (150) also have been reported to increase the number of progenitors, or the timeliness of their appearance in the blood, or both. It is as yet unclear whether the cells obtained are qualitatively different in addition to the demonstrated quantitative or temporal advantages gained by these multifactor combinations.

2.   Stem-Cell Assays

By analogy to the situation in rodents it is reasonable to suggest that, in patients, rHuG-CSF may well elicit similar effects and recruit more stem cells into the circulation. It must be stressed, however, that the actual nature of the cells which effect long-term engraftment in humans has not yet been adequately defined, although the important cells reside in a minority population, comprising <1% of bone-marrow cells (151–154). In this respect, it is of interest that the number of CD34$^+$ cells infused has been shown to correlate well with early recovery of both neutrophils and platelets, and the number of transplanted myeloid progenitors (CFC-GM) with the rate of neutrophil recovery (155,156). In more recent studies long-term culture initiating cells (LT-CIC), candidates for the most primitive of hematopoietic cells, have been shown to correlate with CD34$^+$ numbers in PBPC harvests (157). However, a more detailed phenotypic analysis shows that cobblestone area–forming cells (CAFC), a similar, perhaps overlapping population to LT-CIC, may mobilize with different kinetics than the overall CD34$^+$ populations (158). Also, CD34$^+$/CD33$^-$ cells may also mobilize on a different timescale than CD34$^+$/CD33$^+$ cells that include progenitor cells identified in many of the more familiar semisolid assay systems (159). The question still remains over whether "stem cells" with significant long-term reconstituting potential are present in the mobilized PBPC product. Despite the development of numerous surrogate assays for stem cell performance (eg, detailed phenotyping [160], high proliferation potential-CFC [HPP-CFC] assay [161], CAFC assay [162], LT-CIC assay [163]) in humans and circumstantial evidence derived from long-term survivors of autologous PBPC transplants, there is as yet no definitive proof that long-term reconstituting cells are contained in a PBPC product.

3. CD34 Analysis

Expression of the membrane antigen CD34 is the most widely used marker for hematopoietic progenitor cells. $CD34^+$ cells have been shown to correlate with the incidence of CFC-GM in PBPC populations (164), and CD34 alone has also been shown to predict engraftment in allogeneic or autologous transplants of marrow or blood populations (165,166). The usefulness of CD34 as a sole marker for monitoring hematopoietic graft quality is supported by these studies and since CD34 sorting can be performed in real time, it is possible to monitor product quality and quantity during an ongoing leukapheresis.

4. Short-Term Reconstituting/Progenitor Cells

The peripheral blood compartment, under the influence of injected rHuG-CSF, contains, in addition to long-term reconstituting cells, progenitors committed to all hematopoietic lineages (167). Under certain circumstances, these less primitive progenitor populations may be more effective than their more primitive ancestors, for instance, in support of increasing chemotherapy dose intensity (168) or for repeated transplants (169). At present it is unclear which of the candidate assays may be best used to assess the quality of a PBPC product. It has been shown, for instance, that CFC-GM numbers and CFC-MK correlate equally well with platelet recovery after transplant (170), despite the absence of platelets among the progeny of the cells identified in the CFC-GM assay.

5. Lymphoid Cells

Leukapheresis products from rHuG-CSF–treated donors invariably contain many more lymphoid cells than comparable bone marrow harvests, and early concerns were raised over the possibility of increasing the incidence of graft-versus-host disease (GVHD) should these cells be used in the allogeneic setting. However, data indicate that though the harvests may contain several logs more lymphoid cells when apheresed from rHuG-

CSF–treated donors, these cells are significantly less able to launch a GVHD response, and the majority of them are passively removed by procedures such as CD34 enrichment (171–174). Graft-versus-tumor effects have also been documented, particularly in chronic myeloid leukemia (CML) (eg, Ref. 175), but it is yet to be determined which, if any, of the peripheral blood lymphoid populations are responsible for this phenomenon.

It is also worth noting that the pattern of immune reconstitution after transplantation of blood-derived cells is different from that after bone-marrow transplantation. Immune recovery as measured by lymphocyte subsets is more rapid after PBPC transplantation (176–180). Some studies indicate more rapid recovery of T cells, NK cells, and helper/suppressor T-cell ratios after blood cell–derived transplantation (176,178,181). Theoretically, this recovery should reduce the risk of opportunistic infection. Whether or not it also contributes to the anti-tumor effects (seen with bone-marrow transplantation) is unclear.

6. Use of rHuG-CSF in Allogeneic Transplantation

The acceptable toxicity profile and the documented benefits to the recipient of receiving a PBPC graft has prompted the suggestion that PBPC may be useful as an alternative to bone marrow for allogeneic transplantation (eg, Ref. 182). Several studies using disease-free donors, including children and the elderly, have shown that this approach is feasible (183–186). One advantage, of course, is that collection of PBPC does not require a general anesthetic, while the use of allografts gives the recipient the advantages of a disease-free graft (see Chapter 14).

7. Mobilization of Tumor Cells

Despite the increasing acceptance of rHuG-CSF-mobilized PBPC transfer in nonmalignant disease, the major application continues to be in oncology and the support of chemotherapy-dose escalation or intensification. Because of this, much emphasis has been placed on determining the extent of contamination of the PBPC product with circulating (or mobilized) tumor cells. From these studies it is clear that some tumor cell populations may be found in the blood after mobilization protocols. However, it is not yet clear if rHuG-CSF can specifically mobilize malignant cells. The sensitivity with which tumor cells can be detected in PBPC products varies widely from Southern blot analysis (at approximately 2% to 5%) to polymerase chain reaction (PCR) techniques at perhaps 1:10,000 to 1:1,000,000 tumor cells (187) and monoclonal antibody-staining methods at approximately 1:400,000 cells (188) to clonal assays of tumor growth (eg, Ref. 189) of mostly unknown sensitivity. Since tumor populations are heterogeneous and molecular methods that differentiate between clonogenic and non-clonogenic cells have yet to be developed, it is not known if the tumor cells detected by these techniques are equivalent in terms of their ability to reestablish tumor growth. In vitro clonogenic assays may reflect a more significant biological end point, but technical considerations such as the inability to grow certain primary tumor cell types and unknown cloning efficiency of those that can be grown make interpretation of these data difficult. For a comprehensive review of available methodologies refer to Campana and Pui (190).

Martiat et al (191) have reported that mobilized PBPC have higher levels of tumor cell contamination than either bone marrow or blood. However, not all tumors react in the same way to mobilization protocols (187) and it is very difficult to draw general conclusions. Significantly, it has been shown that recipients of mobilized (with a variety of growth factors) PBPC are no more likely to show progressive disease than are recipients of non-mobilized PBPC (192).

The data presented by Brugger et al (188) illustrate some of the frustrations of detecting tumor-cell contamination in peripheral blood samples. These investigators used immunostaining techniques which detected tumor cells at concentrations as low as 1 cell per 400,000 mononuclear cells (MNC). They obtained baseline data from 46 patients with various tumors and concluded that 4 of the patients had a detectable tumor-cell load in the blood that was above the minimum level of detection. Obviously tumor cells may have been present in the blood of the other patients as well, but it would have been <1 cell in 400,000 that the assay was capable of detecting. After chemotherapy, the number of patients with detectable tumor cells in the blood increased to 13 of the 46. The immediate conclusion is of course that more patients now have tumor cells in the blood. However, it is possible that the 7 patients who now had detectable tumor cells in the blood did not in fact have more tumor cells, but merely had fewer MNC. That is, instead of increasing the numerator of the 1/400,000 fraction, what has been detected is a decrease in the denominator. Chemotherapy with VIP-16 is extremely myelosuppressive, and blood populations can be reduced to <0.1% of their starting levels (193). Data indicating absolute increases in circulating tumor cell numbers have yet to be published. The studies published to date also fail to differentiate between tumor cells mobilized by chemotherapy and those mobilized by rHuG-CSF.

8. Mechanism of Stem-Cell Mobilization

The mechanism of the enhanced release of stem cells from the bone marrow into the peripheral blood under the influence of rHuG-CSF is not known.

Data have suggested a role for extracellular adhesion molecules and/or their receptors in the adhesion of CD34$^+$ cells to marrow stromal elements (194). VLA-4 and LFA-1 (CD11a) are receptors for vascular CAM-1 and intercellular CAM-1 (ICAM-1, CD54), respectively, and PECAM-1 (CD31) is an adhesion molecule in its own right. It was found (195) that therapeutically mobilized CD34$^+$ cells (using rHuG-CSF or chemotherapy) show reduced expression of VLA-4 and LFA-1 but unchanged levels of PECAM-1. There are preliminary data to suggest that changes in the adhesive properties of CD34$^+$ cells in response to rHuG-CSF or other mobilizing agents may mediate their relocation to the circulating compartment. As yet it remains unclear whether these changes are common to all CD34$^+$ cells mobilized by various means, whether they occur equally on all the various mobilized progenitor populations, or whether they are the cause or the result of the relocation process.

Interestingly, preliminary data have been presented (196) that indicate that administration of anti-VLA-4 antibody in addition to rHuG-CSF and rHuSCF augments the mobilization of PBPC. These data imply that interference with the VLA-4–mediated adhesion process is not the sole mechanism by which G-CSF $\pm$ SCF causes the emigration of PBPC from the marrow to the blood. Clearly, further studies are indicated.

REFERENCES

1. Bradley, T. R., and Metcalf, D. (1966). The growth of mouse bone marrow cells in vitro. *Aust J Exp Biol Med Sci 44*:287–299.
2. Pluznick, D., and Sachs, L. (1965). The cloning of normal "mast" cells in tissue culture. *J Cell Physiol 66*:319–324.
3. Metcalf, D. (1980). Clonal extinction of myelomonocytic leukemic cells by serum from mice injected with endotoxin. *Int J Cancer 25*:225–233.
4. Nicola, N. A., Metcalf, D., Matsumoto, M., and Johnson, G. R. (1983). Purification of a factor inducing differentiation in murine myelomonocytic leukemia cells. Identification as granulocyte colony-stimulating factor (G-CSF). *J Biol Chem 258*:9017–9023.
5. Welte, K., Platzer, E., Lu, L., et al (1985). Purification and biochemical characterization of human pluripotent hematopoietic colony-stimulating factor. *Proc Natl Acad Sci USA 82*:1526–1530.
6. Souza, L. M., Boone, T. C., Gabrilove, J., et al (1986). Recombinant human granulocyte colony stimulating factor: effects on normal and leukemic myeloid cells. *Science 232*:61–65.
7. Kojima, S., Matsuyama, T., Kodera, Y., et al (1996). Measurement of endogenous plasma granulocyte-colony-stimulating factor in patients with acquired aplastic anemia by a sensitive chemiluminescent immunoassay. *Blood 87*:1303–1308.

8. Cebon, J., Layton, J. E., Maher, D., and Morstyn, G. (1994). Endogenous haemopoietic growth factors in neutropenia and infection. *Br J Haematol 86:*265–274.

9. Torres-Gomez, A., Jimenez, M. A., Alvarez, M. A., et al (1995). Optimal timing of granulocyte colony-stimulating factor (G-CSF) administration after bone marrow transplantation. A prospective randomized study. *Ann Hematol 71:*65–70.

10. Weimann, E., Reisbach, G., Reinsberg, J., and Lentze, M. J. (1995). IL-6 and G-CSF levels in amniotic fluid during the second trimester in normal and abnormal pregnancies. *Arch Gynecol Obstet 256:*125–130.

11. Saito, S., Kasahara, T., Kato, Y., Ishihara, Y., and Ichijo, M. (1993). Elevation of amniotic fluid interleukin 6 (IL-6), IL-8 and granulocyte colony stimulating factor (G-CSF) in term and preterm parturition. *Cytokine 5:*81–88.

12. Russell, A. R., Davies, E. G., McGuigan, S., Scopes, G. J., Daly, S., and Gordon-Smith, E. C. (1994). Plasma granulocyte-colony stimulating factor concentrations (G-CSF) in the early neonatal period. *Br J Haematol 86:*642–644.

13. Umesaki, N., Fukumasu, H., Miyama, M., Kawabata, M., Ogita, S. (1995). Plasma granulocyte colony stimulating factor concentrations in pregnant women. *Gynecol Obstet Invest 40:*5–7.

14. Cincotta, R., Balloch, A., Metz, J., Layton, J. E., Lieschke, G. J. (1995). Physiological neutrophilia of pregnancy is not associated with a rise in plasma granulocyte colony-stimulating factor (G-CSF). *Am J Hematol 48:*288.

15. Shimoda, K., Okamura, S., Harada, N., Omori, F., and Niho, Y. (1992). Serum granulocyte colony-stimulating factor levels in umbilical cord blood of normal full-term neonates. *Biomed Pharmacother 46:*337–341.

16. Laver, J., Duncan, E., Abboud, M., et al (1990). High levels of granulocyte and granulocyte-macrophage colony-stimulating factors in cord blood of normal full-term neonates. *J Pediatr 116:*627–632.

17. Li, Y., Ohls, R. K., Rosa, C., Shah, M., Richards, D. S., and Christensen, R. D. (1995). Maternal and umbilical serum concentrations of granulocyte colony-stimulating factor and its messenger RNA during clinical chorioamnionitis. *Obstet Gynecol 86:*428–432.

18. Ishii, E., Masuyama, T., Yamaguchi, H., et al (1995). Production and expression of granulocyte- and macrophage-colony-stimulating factors in newborns: their roles in leukocytosis at birth. *Acta Haematol 94:*23–31.

19. Nakayama, H., Kukita, J., Ohga, S., and Ueda, K. (1995). Granulocyte-colony stimulating factor levels in cord blood and neonatal peripheral blood. *Acta Paediatr Jpn 37:*237–239.

20. Gessler, P., Kirchmann, N., Kientsch-Engel, R., Haas, N., Lasch, P., and Kachel, W. (1993). Serum concentrations of granulocyte colony-stimulating factor in healthy term and preterm neonates and in those with various diseases including bacterial infections. *Blood 82:*3177–3182.

21. Wilimas, J. A., Wall, J. E., Fairclough, D. L., et al (1995). A longitudinal study of granulocyte colony-stimulating factor levels and neutrophil counts in newborn infants. *J Pediatr Hematol Oncol 17:*176–179.

22. Ikeno, K. (1994). Increased granulocyte-colony stimulating factor (G-CSF) levels in neonates with perinatal complications. *Acta Paediatr Jpn 36:* 366–370.

23. Watari, K., Asano, S., Shirafuji, N., et al (1989). Serum granulocyte colony-stimulating factor levels in healthy volunteers and patients with various disorders as estimated by enzyme immunoassay. *Blood 73:*117–122.

24. Lieschke, G. J., Grail, D., Hodgson, G., et al (1994). Mice lacking granulocyte colony-stimulating factor have chronic neutropenia, granulocyte and macrophage progenitor cell deficiency, and impaired neutrophil mobilization. *Blood 84:*1737–1746.

25. Chang, J. M., Metcalf, D., Gonda, T. J., and Johnson, G. R. (1989). Long-term exposure to retrovirally expressed granulocyte-colony-stimulating factor induces a nonneoplastic granulocytic and progenitor cell hyperplasia without tissue damage in mice. *J Clin Invest 84:*1488–1496.

26. Yamada, T., Kaneko, H., Iizuka, K., Matsubayashi, Y., Kokai, Y., Fujimoto, J. (1996). Elevation of lymphocyte and hematopoietic stem-cell numbers in mice transgenic for human granulocyte CSF. *Lab Invest 74:*384–394.

27. Semenza, G. L., Traystman, M. D., Gearhart, J. D., and Antonarakis, S. E. (1989). Polycythemia in transgenic mice expressing the human erythropoietin gene. *Proc Natl Acad Sci USA 86:*2301–2305.

28. Metcalf, D. (1990). The consequences of excess levels of haemopoietic growth factors. *Br J Haematol 75:*1–3.

29. Lang, R., Metcalf, D., Cuthbertson, R. A., et al (1987). Transgenic mice expressing a hematopoietic growth factor gene (GM-CSF) develop accumulations of macrophages, blindness, and a fatal syndrome of tissue damage. *Cell 51:*675–686.

30. Wong, P., Chung, S. W., Dunbar, C., Bodine, D., Ruscetti, S., and Nienhuis, A. (1989). Retrovirus-mediated transfer and expression of the interleukin-3 gene in mouse hematopoietic results in a myeloproliferative disorder. *Mol Cell Biol 9:*798–808.

31. Chang, J. M., Metcalf, D., Lang, R. A., Gonda, T. J., and Johnson, G. R. (1989). Nonneoplastic hematopoietic myeloproliferative syndrome induced by dysregulated multi-CSF (IL-3) expression. *Blood 73:*1487–1497.

32. Metcalf, D., and Gearing, D. P. (1989). Fatal syndrome in mice engrafted with cells producing high levels of the leukemia inhibitory factor. *Proc Natl Acad Sci USA 86:*5948–5952.

33. Metcalf, D., and Gearing, D. P. (1989). A myelosclerotic syndrome in mice engrafted with cells producing high levels of leukemia inhibitory factor (LIF). *Leukemia 3:*847–852.

34. Eliason, J. F. (1986). Granulocyte-macrophage colony formation in serum-free culture: effects of purified colony-stimulating factors and modulation by hydrocortisone. *J Cell Physiol 128:*231–238.

35. Metcalf, D., and Nicola, N. A. (1983). Proliferative effects of purified granulocyte colony-stimulating factor (G-CSF) on normal mouse hematopoietic cells. *J Cell Physiol 116:*198–206.

36. Morstyn, G., Nicola, N. A., and Metcalf, D. (1981). Separate actions of different colony stimulating factors from human placental conditioned medium on human hemopoietic progenitor cell survival and proliferation. *J Cell Physiol 109*:133–142.
37. Begley, C. G., Metcalf, D., Lopez, A. F., and Nicola, N. A. (1985). Fractionated populations of normal human marrow cells respond to both human colony-stimulating factors with granulocyte-macrophage activity. *Exp Hematol 13*:956–962.
38. Metcalf, D., Begley, C. G., and Nicola, N. A. (1985). The proliferative effects of human GM-CSF alpha and beta and murine G-CSF in microwell cultures of fractionated human marrow cells. *Leuk Res 9*:521–527.
39. Begley, C. G., Nicola, N. A., and Metcalf, D. (1988). Proliferation of normal human promyelocytes and myelocytes after a single pulse stimulation by purified GM-CSF or G-CSF. *Blood 71*:640–645.
40. Begley, C. G., Lopez, A. F., Nicola, N. A., Warren, D. J., Vadas, M. A., and Sanderson, C. J. (1986). Purified colony-stimulating factors enhance the survival of human neutrophils and eosinophils in vitro: a rapid and sensitive microassay for colony-stimulating factors. *Blood 68*:162–166.
41. Vadas, M. A., and Lopez, A. F. (1985). Regulation of granulocyte function by colony-stimulating factors and monoclonal antibodies. *Lymphokines 12*:179–200.
42. Heyworth, C. M., Ponting, I. L., and Dexter, T. M. (1988). The response of haemopoietic cells to growth factors: developmental implications of synergistic interactions. *J Cell Science 91*:239–247.
43. Ponting, I. L. O., Heyworth, C. M., Cormier, F., and Dexter, T. M. (1991). Serum-free culture of enriched murine haemopoietic stem cells II: effects of growth factors and haemin on development. *Growth Factors 4*:165–173.
44. Metcalf, D. (1991) Lineage commitment of hematopoietic progenitor cells in developing blast cell colonies: influence of colony-stimulating factors. *Proc Natl Acad Sci USA 88*:11310–11314.
45. Hudak, S., Hunte, B., Culpepper, J., et al (1995). FLT3/FLK2 ligand promotes the growth of murine stem cells and the expansion of colony-forming cells and spleen colony-forming units. *Blood 85*:2747–2755.
46. Piacibello, W., Fubini, L., Sanavio, F., et al (1995). Effects of human FLT3 ligand on myeloid leukemia cell growth: heterogeneity in response and synergy with other hematopoietic growth factors. *Blood 86*:4105–4114.
47. Takaue, Y., Kawano, Y., Reading, C. L., et al (1990). Effects of recombinant GCSF, GM-CSF, IL-3 and IL-Iα on the growth of purified human peripheral blood progenitors. *Blood 76*:330–335.
48. Ulich, T. R., del Castillo, J., and Souza, L. (1988). Kinetics and mechanisms of recombinant human granulocyte-colony stimulating factor-induced neutrophilia. *Am J Pathol 133*:630–638.
49. Broxmeyer, H. E., Williams, D. E., Cooper, S., Hangoc, G., and Ralph, P. (1988). Recombinant human granulocyte-colony stimulating factor and recombinant human macrophage-colony stimulating factor synergize in vivo

to enhance proliferation of granulocyte-macrophage, erythroid, and multipotential progenitor cells in mice. *J Cell Biochem 38*:127–136.

50. Suda, T., Suda, J., Kajigaya, S., et al (1987). Effects of recombinant murine granulocyte colony-stimulating factor on granulocyte-macrophage and blast colony formation. *Exp Hematol 15*:958–965.

51. Cohen, A. M., Zsebo, K. M., Inoue, H., et al (1987). In vivo stimulation of granulopoiesis by recombinant human granulocyte colony-stimulating factor. *Proc Natl Acad Sci USA 84*:2484–2488.

52. Citarrella, P., Gebbia, V., Miceli, S., Palmeri, S., and Rausa, L. (1990). Effect of human granulocyte colony-stimulating factor (G-CSF) and reduced glutathione (GSH) on drug-induced leukopenia in golden Syrian hamsters. *Pharmacol Res 22*:99–100.

53. Lothrop, C. J., Warren, D. J., Souza, L. M., Jones, J. B., and Moore, M. A. (1988). Correction of canine cyclic hematopoiesis with recombinant human granulocyte colony-stimulating factor. *Blood 72*:1324–1328.

54. Welte, K., Bonilla, M. A., Gillio, A. P., Gabrilove, J. L., O'Reilly, R. J., and Souza, L. M. (1988). Recombinant human granulocyte-colony stimulating factor: in vivo effects on myelopoiesis in primates. *Behring Inst Mitt 83*:102–106.

55. Gillio, A. P., Bonilla, M. A., Potter, G. K., et al (1987). Effect of recombinant human granulocyte colony-stimulating factor on hematopoietic reconstitution after autologous bone marrow transplantation in primates. *Transplant Proc 19*:153–156.

56. Fink, M. P., O'Sullivan, B. P., Menconi, M. J., et al (1993). Effect of granulocyte colony-stimulating factor on systemic and pulmonary responses to endotoxin in pigs. *J Trauma 34*:571–577.

57. Kanazawa, M., Ishizaka, A., Hasegawa, N., Suzuki, Y., and Yokoyama, T. (1992). Granulocyte colony-stimulating factor does not enhance endotoxin-induced acute lung injury in guinea pigs. *Am Rev Respir Dis 145*:1030–1035.

58. Smith, W. S., Sumnicht, G. E., Sharpe, R. W., Samuelson, D., and Millard, F. E. (1995). Granulocyte colony-stimulating factor versus placebo in addition to penicillin G in a randomized blinded study of gram-negative pneumonia sepsis: analysis of survival and multisystem organ failure. *Blood 86*:1301–1309.

59. Gratwohl, A., Baldomero, H., John, L., et al (1995). Transplantation of G-CSF mobilized allogeneic peripheral blood stem cells in rabbits. *Bone Marrow Transplant 16*:63–68.

60. Bronchud, M. H., Scarffe, J. H., Thatcher, N., et al (1987). Phase I/II study of recombinant human granulocyte colony-stimulating factors in patients receiving intensive chemotherapy for small cell lung cancer. *Br J Cancer 56*:809–813.

61. Morstyn, G., Campbell, L., Souza, L. M., et al (1988). Effect of granulocyte colony stimulating factor on neutropenia induced by cytotoxic chemotherapy. *Lancet 1*:667–672.

62. Tanaka, H., Satake-Ishikawa, R., Matsuki, S., and Asano, K. (1991). Pharmacokinetics of recombinant human granulocyte colony-stimulating factor conjugated to polyethylene glycol in rats. *Cancer Res 51*:3710–3714.

63. Tamura, M., Hattori, K., Nomura, H., et al (1987). Induction of neutrophilic granulocytosis in mice by administration of purified human native granulocyte colony-stimulating factor (G-CSF). *Biochem Biophys Res Commun 142*:454–460.

64. Hattori, K., Shimizu, K., Takahashi, M., et al (1990). Quantitative in vivo assay of human granulocyte colony-stimulating factor using cyclophosphamide-induced neutropenia in mice. *Blood 75*:1228–1233.

65. Pojda, Z., Molineux, G., and Dexter, T. M. (1990). Hematopoietic effects of short-term in vivo treatment of mice with various doses of rhG-CSF. *Exp Hematol 18*:27–31.

66. Takada, K., Tohyama, Y., Oohashi, M., et al (1989). Is recombinant human granulocyte colony stimulating factor (G-CSF) orally available in rats? *Chem Pharm Bull Tokyo 37*:838–839.

67. Welte, K., Bonilla, M. A., Gillio, A. P., et al (1987). Recombinant human granulocyte colony stimulating factor. Effects on hematopoiesis in normal and cyclophosphamide treated primates. *J Exp Med 165*:941–948.

68. Gerson, S. L. (1994). G-CSF and the management of clozapine-induced agranulocytosis. *J Clin Psychiatry 55, Suppl B*:139–142.

69. Matsumoto, M., Matsubara, S., Matsuno, T., et al (1987). Protective effect of human granulocyte colony-stimulating factor on microbial infection in neutropenic mice. *Infect Immunol 55*:2715–2720.

70. Avalos, B. R., Broudy, V. C., Ceselski, S. K., Druker, B. J., Griffin, J. D., and Hammond, W. P. (1994). Abnormal response to granulocyte colony-stimulating factor (G-CSF) in canine cyclic hematopoiesis is not caused by altered G-CSF receptor expression. *Blood 84*:789–794.

71. Molineux, G., Pojda, Z., and Dexter, T. M. (1990). A comparison of hematopoiesis in normal and splenectomized mice treated with granulocyte colony-stimulating factor. *Blood 75*:563–569.

72. Fulton, R., Gasper, P. W., Ogilvie, G. K., Boone, T. C., and Dornsife, R. E. (1991). Effect of recombinant human granulocyte colony-stimulating factor on hematopoiesis in normal cats. *Exp Hematol 19*:759–767.

73. Schuening, F. G., Storb, R., Goehle, S., et al (1990). Recombinant human granulocyte colony-stimulating factor accelerates haematopoietic recovery after DLA-identical littermate marrow transplants in dogs. *Blood 76*:636–640.

74. Inukai, T., Sugita, K., Iijima, K., et al (1995). Expression of granulocyte colony-stimulating factor receptor on CD10-positive human B-cell precursors. *Br J Haematol 89*:623–626.

75. Lee, M. Y., Fevold, K. L., Dorshkind, K., Fukunaga, R., Nagata, S., and Rosse, C. (1993). In vivo and in vitro suppression of primary B lymphocytopoiesis by tumor-derived and recombinant granulocyte colony-stimulating factor. *Blood 82*:2062–2068.

76. Woodward, T. A., McNiece, I. K., Witte, P. L., et al (1990). Further studies on growth factor production by the TC-I stromal cell line: pre-B stimulating activity. *Blood 75*:2130–2136.

77. Hirayama, F., and Ogawa, M. (1994). Cytokine regulation of early B-lymphopoiesis assessed in culture. *Blood Cells 20*:341–346.

78. Kato, Y., Yamamoto, M., Ikenaga, T., et al (1991). In vivo effect of human granulocyte colony-stimulating factor derivatives on hematopoiesis in primates. *Acta Haematol 86*:70–78.

79. Stricker, R. B., and Goldberg, B. (1995). Increase in lymphocyte subsets following treatment of HIV-associated neutropenia with granulocyte colony-stimulating factor. *Clin Immunol Immunopathol 79*:194–196.

80. Hengge, U. R., Brockmeyer, N. H., and Goos, M. (1992). Granulocyte colony-stimulating factor treatment in AIDS patients. *Clin Invest 70*:922–926.

81. Bohme, M., Schmidt, D., Radke, J., Morenz, J., and Weise, W. (1994). Effect of granulocyte colony stimulating factor (G-CSF) on peripheral blood leukocytes and lymphocytes in patients with chemotherapy-induced leukopenia. *Geburtshilfe Frauenheilkd 54*:670–674.

82. Lord, B. I., Molineux, G., Pojda, Z., Souza, L. M., Mermod, J. J., and Dexter, T. M. (1991). Myeloid cell kinetics in mice treated with recombinant interleukin-3, granulocyte colony-stimulating (CSF), or granulocyte-macrophage CSF in vivo. *Blood 77*:2154–2159.

83. Schmitz, L. L., McClure, J. S., Litz, C. E., et al (1994). Morphologic and quantitative changes in blood and marrow cells following growth factor therapy. *Am J Clin Pathol 101*:67–75.

84. Pollmacher, T., Korth, C., Mullington, J., et al (1996). Effects of granulocyte-colony-stimulating factor on plasma cytokine and cytokine receptor levels and on the in-vivo host response to endotoxin in healthy men. *Blood 87*:900–905.

85. Gilmore, G. L., DePasquale, D. K., Fischer, B. C., and Shadduck, R. K. (1995). Enhancement of monocytopoiesis by granulocyte colony-stimulating factor: evidence for secondary cytokine effects in vivo. *Exp Hematol 23*:1319–1323.

86. Bussolino, F., Wang, J. M., Defilippi, P., et al (1989). Granulocyte- and granulocyte-macrophage–colony stimulating factors induce human endothelial cells to migrate and proliferate. *Nature 337*:471–473.

87. Bussolino, F., Ziche, M., Wang, J. M., et al (1991). In vitro and in vivo activation of endothelial cells by colony-stimulating factors. *J Clin Invest 87*:986–995.

88. Yong, K., Cohen, H., Khwaja, A., Jones, H. M., and Linch, D. C. (1991). Lack of effect of granulocyte-macrophage and granulocyte colony-stimulating factors on cultured human endothelial cells. *Blood 77*:1675–1680.

89. Uzumaki, H., Okabe, T., Sasaki, N., et al (1989). Identification and characterization of receptors for granulocyte colony-stimulating factor on human placenta and trophoblastic cells. *Proc Natl Acad Sci USA 86*:9323–9326.

90. Katano, M., Nakamura, M., Matsuo, T., Iyama, A., and Hisatsugu, T. (1995). Effect of granulocyte colony-stimulating factor (G-CSF) on chemotherapy-induced oral mucositis. *Surg Today 25*:202–206.

91. Denzlinger, C., Holler, E., Reisbach, G., Hiller, E., and Wilmanns, W. (1994). Granulocyte colony-stimulating factor inhibits the endogenous leukotriene production in tumour patients. *Br J Haematol 86*:881–882.

92. Moore, K. P., Sheron, N., Ward, P., Taylor, G. W., Alexander, G. J., and Williams, R. (1991). Leukotriene and prostaglandin production after infusion of tumour necrosis factor in man. *Eicosanoids 4*:115–118.
93. Cannistra, S. A., Vellenga, E., Groshek, P., Rambaldi, A., and Griffin, J. D. (1988). Human granulocyte-monocyte colony-stimulating factor and interleukin 3 stimulate monocyte cytotoxicity through a tumor necrosis factor-dependent mechanism. *Blood 71*:672–676.
94. Goergen, I., Hartung, T., Leist, M., et al (1992). Granulocyte colony-stimulating factor treatment protects rodents against lipopolysaccharide-induced toxicity via suppression of systemic tumor necrosis factor-alpha. *J Immunol 149*:918–924.
95. Begley, C. G., Metcalf, D., and Nicola, N. A. (1987). Primary human myeloid leukemia cells: comparative responsiveness to proliferative stimulation by GM-CSF or G-CSF and membrane expression of CSF receptors. *Leukemia 1*:1–8.
96. Berdel, W. E., Danhauser, Riedel, S., Steinhauser, G., and Winton, E. F. (1989). Various human hematopoietic growth factors (interleukin-3, GM-CSF, G-CSF) stimulate clonal growth of nonhematopoietic tumor cells. *Blood 73*:80–83.
97. Kondo, S., Okamura, S., Asano, Y., Harada, M., Niho, Y. (1991). Human granulocyte colony-stimulating factor receptors in acute myelogenous leukaemia. *Eur J Haematol 46*:223–230.
98. Delwel, R., Salem, M., Pellens, C., et al (1988). Growth regulation of human acute myeloid leukemia: effects of five recombinant hematopoietic factors in a serum- free culture system. *Blood 72*:1944–1949.
99. Itoh, K., Bessho, M., and Hirashima, K. (1989). Effects of recombinant human G-CSF and GM-CSF on primary human leukemic cells. *Nippon Ketsueki Gakkai Zasshi 52*:988–995.
100. Carlo-Stella, C., Mangoni, L., Almici, C., Frassoni, F., Fiers, W., and Rizzoli, V. (1990). Growth of CD34$^+$ acute myeloblastic leukemia colony-forming cells in response to recombinant hematopoietic growth factors. *Leukemia 4*:561–566.
101. Santini, V., Nooter, K., Delwel, R., and Lowenberg, B. (1990). Susceptibility of acute myeloid leukemia (AML) cells from clinically resistant and sensitive patients to daunomycin (DNR): assessment in vitro after stimulation with colony stimulating factors (CSFs). *Leuk Res 14*:377–380.
102. Mori, H., Kuriyama, K., Tawara, M., et al (1995). Trial of combined cytosine arabinoside with granulocyte colony-stimulating factor therapy or refractory acute myeloid leukemia. *Rinsho Ketsueki 36*:648–656.
103. Santini, V., Colombat, P., Delwel, R., van Gurp, R., Touw, I., and Lowenberg, B. (1991). Induction of granulocytic maturation in acute myeloid leukemia by G-CSF and retinoic acid. *Leuk Res 15*:341–350.
104. Vellenga, E., Ostapovicz, D., O'Rourke, B., and Griffin, J. D. (1987). Effects of recombinant IL-3, GM-CSF, and G-CSF on proliferation of leukemic clonogenic cells in short-term and long-term cultures. *Leukemia 1*:584–589.

105. Vellenga, E., Young, D. C., Wagner, K., Wiper, D., Ostapovicz, D., and Griffin, J. D. (1987). The effects of GM-CSF and G-CSF in promoting growth of clonogenic cells in acute myeloblastic leukemia. *Blood 69*:1771–1776.

106. Irvine, A. E., Berney, J. J., and Francis, G. E. (1990). Dissociation of the proliferation and differentiation stimuli of granulocyte colony-stimulating factor (GCSF). *Leukemia 4*:203–209.

107. Tamura, M., Hattori, K., Ono, M., et al (1989). Effects of recombinant human granulocyte colony-stimulating factor (rG-CSF) on murine myeloid leukemia: stimulation of proliferation of leukemic cells in vitro and inhibition of development of leukemia in vivo. *Leukemia 3*:853–858.

108. Hayashi, M., Okabe-Kado, J., and Hozumi, M. (1994). Flow-cytometric analysis of in vivo induction of differentiation of WEHI-3B myelomonocytic leukemia cells by recombinant granulocyte colony-stimulating factor. *Exp Hematol 22*:393–398.

109. Park, L. S., Waldron, P. E., Friend, D., et al (1989). Interleukin-3, GM-CSF, and G-CSF receptor expression on cell lines and primary leukemia cells: receptor heterogeneity and relationship to growth factor responsiveness. *Blood 74*:56–65.

110. Kuwaki, T., Hosoi, T., Hanazono, Y., et al (1990). Distribution of human granulocyte-colony stimulating factor receptors on hematopoietic and nonhematopoietic tumor cell lines. *Jpn J Cancer Res 81*:560–563.

111. Avalos, B. R., Gasson, J. C., Hedvat, C., et al (1990). Human granulocyte colony-stimulating factor: biologic activities and receptor characterization on hematopoietic cells and small cell lung cancer cell lines. *Blood 75*:851–857.

112. Kawachi, Y., Ozaki, S., Sakamoto, Y., et al (1994). Richter's syndrome showing pronounced lymphadenopathy in response to administration of granulocyte colony-stimulating factor. *Leuk Lymph 13*:509–514.

113. Izumi, T., Muroi, K., Takatoku, M., Imagawa, S., Hatake, K., and Miura, Y. (1994). Development of acute myeloblastic leukaemia in a case of aplastic anaemia treated with granulocyte colony-stimulating factor. *Br J Haematol 87*:666–668.

114. Tachibana, M., Miyakawa, A., Tazaki, H., Nakamura, K., Kubo, A., Hata, J., et al (1995). Autocrine growth of transitional cell carcinoma of the bladder induced by granulocyte-colony stimulating factor. *Cancer Res 55*:3438–3443.

115. Hatta, Y., Iwata, T., Takeuchi, J., Ohshima, T., and Horie, T. (1995). Complete remission in a patient with hypoplastic acute lymphoblastic leukemia induced by granulocyte-colony-stimulating factor. *Acta Haematol 94*:39–43.

116. Maeda, H., Uozumi, T., Kurisu, K., Matsuoka, T., Kawamoto, K., Kiya, K., et al (1994). Combined antitumour effects of TNF and G-CSF on a human medulloblastoma xenograft line. *J Neurooncol 21*:203–213.

117. Cronkite, E. P., Burlington, H., Shimosaka, A., Bullis, J. E., and Pappas, N. (1993). Anemia induced in splenectomized mice by administration of rhG-CSF. *Exp Hematol 21*:319–325.

118. Moore, M. A., Muench, M. O., Warren, D. J., and Laver, J. (1990). Cytokine networks involved in the regulation of haemopoietic stem cell proliferation and differentiation. *Ciba Found Symp 148*:43–58.

119. Lord, B. I., Molineux, G., Testa, N. G., Kelly, M., Spooncer, E., and Dexter, T. M. (1986). The kinetic response of haemopoietic cells, in vivo, to highly purified recombinant interleukin-3. *Lymphokine Res 5*:97–104.

120. Molineux, G., Pojda, Z., Hampson, I. N., Lord, B. I., and Dexter, T. M. (1990). Transplantation potential of peripheral blood stem cells induced by granulocyte colony-stimulating factor. *Blood 76*:2153–2158.

121. Pojda, Z., Molineux, G., and Dexter, T. M. (1989). Effects of long-term in vivo treatment of mice with purified murine recombinant GM-CSF. *Exp Hematol 17*:1100–1104.

122. Bonilla, M. A., Gillio, A. P., Ruggiero, M., et al (1989). Effects of recombinant human granulocyte colony-stimulating factor on neutropenia in patients with congenital agranulocytosis. *N Engl J Med 320*:1574–1580.

123. Pojda, Z., Machaj, E., Gesicka-Wozniakowska, T., et al (1992). Effects of granulocyte macrophage colony-stimulating factor and granulocyte colony-stimulating factor in patient with severe congenital neutropenia. *Exp Hematol 20*:777 (abstr 278).

124. Bungart, B., Loeffler, M., Goris, H., Dontje, B., Diehl, V., and Nijhof, W. (1990). Differential effects of recombinant human colony stimulating factor (rhG-CSF) on stem cells in marrow, spleen, and peripheral blood in mice. *Br J Haematol 76*:174–179.

125. Nijhof, W., De Haan, G., Dontje, B., and Loeffler, M. (1994). Effects of G-CSF on erythropoiesis. *Ann NY Acad Sci 718*:312–324.

126. de Haan, G., Engel, C., Dontje, B., Nijhof, W., and Loeffler, M. (1994). Mutual inhibition of murine erythropoiesis and granulopoiesis during combined erythropoietin, granulocyte colony-stimulating factor, and stem cell factor administration: in vivo interactions and dose-response surfaces. *Blood 84*:4157–4163.

127. Locatelli, F., Zecca, M., Ponchio, L., et al (1994). Pilot trial of combined administration of erythropoietin and granulocyte colony-stimulating factor to children undergoing allogeneic bone marrow transplantation. *Bone Marrow Transplant 14*:929–935.

128. Pierelli, L., Menichella, G., Scambia, G., et al (1994). In vitro and in vivo effects of recombinant human erythropoietin plus recombinant human G-CSF on human haemopoietic progenitor cells. *Bone Marrow Transplant 14*:23–30.

129. Negrin, R. S., Stein, R., Doherty, K., et al (1996). Maintenance treatment of the anemia of myelodysplastic syndromes with recombinant human granulocyte colony-stimulating factor and erythropoietin: evidence for in vivo synergy. *Blood 87*:4076–4081.

130. Udupa, K. B., and Sharma, B. G. (1992). Response of erythropoietin and granulocyte/macrophage colony-stimulating factor in exhypoxic polycythaemic mice. *Exp Hematol 20*:793 (abstr 344).

131.   Bishop, J. F. (1994). Platelet support and the use of cytokines. *Stem Cells 12*:370–377.

132.   Baer, M. R., Bernstein, S. H., Brunetto, V. L., et al (1996). Biological effects of recombinant human granulocyte colony-stimulating factor in patients with untreated acute myeloid leukemia. *Blood 87*:1484–1494.

133.   Russell, A. R., Davies, E. G., Ball, S. E., and Gordon-Smith, E. (1995). Granulocyte colony stimulating factor treatment for neonatal neutropenia. *Arch Dis Child Fetal Neonatal Ed 72*:53–54.

134.   Vandeven, C., Fernandez, G. W., Herbst, T., Knoppel, A., Vanwinkle, P., Qian, J., and Cairo, M. S. (1996). Thrombopoietin (tpo) significantly increase neonatal (nb) rat myelopoiesis and thrombopoiesis—the addition of rhG-CSF is additive relative to myelopoiesis but abrogates the effect of tpo on platelet production. *Pediat Res 39*:1734–1734.

135.   Molineux, G., Hartley, C., McElroy, P., McCrea, C., and McNiece, I. K. (1996). Megakaryocyte growth and development factor accelerates platelet recovery in peripheral blood progenitor cell transplant recipients. *Blood 88*:366–376.

136.   Molineux, G., Hartley, C. A., McElroy, P., McCrea, C., and McNiece, I. K. (1996). Megakaryocyte growth and development factor stimulates enhanced platelet recovery in mice after bone marrow transplantation. *Blood 88*:1509–1514.

137.   Molineux, G., Hartley, C., McElroy, P., McCrea, C., Kerzic, P., and McNiece, I. (1997). An analysis of the effects of combined treatment with rmGM-CSF and PEG-rHuMGDF in murine bone marrow transplant recipients. *Stem Cells 15*:43–49.

138.   Uchida, T., and Yamagiwa, A. (1992). Kinetics of rG-CSF-induced neutrophilia in mice. *Exp Hematol 20*:152–155.

139.   Lord, B. I., Bronchud, M. H., Owens, S., et al (1989). The kinetics of human granulopoiesis following treatment with granulocyte colony-stimulating factor in vivo. *Proc Natl Acad Sci USA 86*:9499–9503.

140.   Roberts, A. W., and Metcalf, D. (1994) Granulocyte colony-stimulating factor induces selective elevations of progenitor cells in the peripheral blood of mice. *Exp Hematol 22*:1156–1163.

141.   Bolwell, B. J., Goormastic, M., Yanssens, T., et al (1994). Comparison of G-CSF with GM-CSF for mobilizing peripheral blood progenitor cells and for enhancing marrow recovery after autologous bone marrow transplant. *Bone Marrow Transplant 14*:913–918.

142.   Slowman, S., Danielson, C., Graves, V., et al (1994). Administration of GM-/G-CSF prior to bone marrow harvest increases collection of CD34$^+$ cells. *Prog Clin Biol Res 389*:363–369.

143.   Bodine, D. M., Seidel, N. E., and Orlic, D. (1996). Bone marrow collected 14 days after in vivo administration of graulocyte colony-stimulating factor and stem cell factor to mice has 10-fold more repopulating ability than untreated bone marrow. *Blood 88*:89–97.

144. McNiece, I. K., Briddell, R. A., Hartley, C. A., and Andrews, R. G. (1993). The role of stem cell factor in mobilization of peripheral blood progenitor cells: synergy with G-CSF. *Stem Cells 11* (suppl 3):83–88.

145. Drize, N., Chertkov, J., Samoilina, N., and Zander, A. (1996). Effect of cytokine treatment (granulocyte colony-stimulating factor and stem cell factor) on hematopoiesis and the circulating pool of hematopoietic stem cells in mice. *Exp Hematol 24*:816–822.

146. Donahue, R. E., Kirby, M. R., Metzger, M. E., Agricola, B. A., Sellers, S. E., and Cullis, H. M. (1996). Peripheral-blood $CD34^+$ cells differ from bone-marrow $CD34^+$ cells in Thy-1 expression and cell cycle status in nonhuman primates mobilized or not mobilized with granulocyte colony-stimulating factor and/or stem cell factor. *Blood 87*:1644–1653.

147. Glaspy, J., LeMaistre, C. F., Lill, M., et al (1995). Dose-response of 7 day administration of recombinant methionyl human stem cell factor (SCF) in combination with filgrastim (G-CSF) for progenitor cell mobilization in patients with stage II-IV breast cancer. *Blood 86*:463a (abstr 1837).

148. Brasel, K., McKenna, H. J., Charrier, K., Morrissey, P., Williams, D. E., and Lyman, S. D. (1995). Synergistic effects in vivo of Flt3 ligand with GM-CSF or G-CSF in mobilization of colony forming cells in mice. *Blood 86*:499a.

149. Parker, A. N., Simms, A. R., Tsang, M. L.-S., Clark, S. C., Graham, G. J., and Pragnell, I. B. (1994). MIP-1α: mobilization of peripheral blood cells. *Blood 84*:738a (abstr 2937).

150. Pettengell, R., Luft, T., deWynter, E., Coutinho, L., Young, R., Fitzsimmons, L., Scarffe, J. H., and Testa, N. G. (1995). Effects of interleukin-6 on mobilization of primitive hematopoietic cells into the circulation. *Br J Haematol 89*:237–242.

151. Andrews, R. G., Singer, J. W., and Bernstein, I. D. (1989). Precursors of colony-stimulating cells in humans can be distinguished from colony forming cells by expression of the CD33 and CD34 antigens and light scatter properties. *J Exp Med 169*:1721–1731.

152. Ema, H., Suda, T., Miura, Y., and Nakauchi, H. (1990). Colony formation of clone-sorted human hematopoietic progenitors. *Blood 75*:1941–1946.

153. Siena, S., Bregni, M., Ravagnani, F., et al (1990). Heterogeneity of circulating hematopoietic progenitors in cancer patients treated with high-dose cyclophosphamide and recombinant human granulocyte macrophage colony-stimulating factor. *Haematologica 75*:6–10.

154. Terstappen, L. W., Huang, S., Safford, M., Lansdorp, P. M., and Loken, M. R. (1991). Sequential generations of hematopoietic colonies derived from single nonlineage-committed $CD34^+$ CD38-progenitor cells. *Blood 77*:1218–1227.

155. Ravagnani, F., Bregni, M., Siena, S., Sciorelli, G., Gianni, A. M., and Pellegris, G. (1990). Role of recombinant human granulocyte-macrophage colony stimulating factor for large scale collection of peripheral blood stem cells for autologous transplantation. *Haematologica 75*:22–25.

156. Siena, S., Bregni, M., Brando, B., et al (1991). Clinical criteria for harvesting peripheral blood hematopoietic progenitors (PBHP) for autologous transplantation in cancer patients. *Proc Am Soc Clin Oncol 10:*(abstr 19).

157. Bender, J. G., Lum, L., Unverzagt, K. L., et al (1994). Correlation of colony-forming cells, long-term culture initiating cells and CD34$^+$ cells in apheresis products from patients mobilized for pheripheral blood progenitors with different regimens. *Bone Marrow Transplant 13:*479–485.

158. Murray, L., Chen, B., Galy, A., et al (1995). Enrichment of human hematopoietic stem cell activity in the CD34$^+$ Thy-1$^+$Lin-subpopulation from mobilized peripheral blood. *Blood 85:*368–378.

159. Croockewit, S., Raymakers, R., Trilsbeek, C., Dolstra, H., Pennings, A., and de Witte, T. (1994). Primitive multilineage progenitor cells predominate in peripheral blood early after mobilization with high-dose cyclophosphamide and GM-CSF or G-CSF. *Leukemia 8:*2194–2199.

160. Baum, C. M., Weissman, I. L., Tsukamoto, A. S., Buckle, A. M., and Peault, B. (1992). Isolation of a candidate human hematopoietic stem-cell population. *Proc Natl Acad Sci USA 89:*2804–2808.

161. McNiece, I. K., Stewart, F. M., Deacon, D. M., et al (1989). Detection of a human CFC with a high proliferative potential. *Blood 74:*609–612.

162. Breems, D. A., Blokland, E. A., Neben, S., and Ploemacher, R. E. (1994). Frequency analysis of human primitive hematopoietic stem cell subsets using a cobblestone area forming cell assay. *Leukemia 8:*1095–1104.

163. Tong, J., Gianni, A. M., Siena, S., Srour, E. F., Bregni, M., and Hoffman, R. (1994). Primitive hematopoietic progenitor cells are present in peripheral blood autografts. *Blood Cells 20:*351–362.

164. Min, Y. H., Lee, S. T., Hahn, J. S., and Ko, Y. W. (1994). Effects of granulocyte-macrophage colony-stimulating factor (GM-CSF) on the kinetics of circulating CD34$^+$ cells, CD34$^+$ subsets, and myeloid progenitor cells during chemotherapy-induced mobilization in acute leukemia. *Blood 84:*584a (abstr 2322).

165. Rybka, W. B., Kiss, J. E., Lister, J., Winkelstein, A., Donnenberg, A. D., and Ball, E. D. (1994). Graft CD34$^+$ cell content predicts engraftment following allogeneic and autologous hematopoietic stem cell transplantation. *Blood 84:*87a (abstr 338).

166. Pecora, A. L., Brochstein, J. A., Jennis, A., Michelis, M., Grazioso, L., Zahos, K., et al (1994). Total CD34$^+$ cell content and CD34$^+$CD33- subset predict time to neutrophil and platelet engraftment and hospital discharge in patients following high-dose chemotherapy (HDC) with primed peripheral blood progenitor cells (PBPC). *Blood 84:*88a (abstr 342).

167. Roberts, A. W., and Metcalf, D. (1995). Noncycling state of peripheral blood progenitor cells mobilized by granulocyte colony-stimulating factor and other cytokines. *Blood 86:*1600–1605.

168. Pettengell, R., Woll, P. J., Thatcher, N., Dexter, T. M., and Testa, N. G. (1995). Multicyclic, dose-intensive chemotherapy supported by sequential reinfusion of hematopoietic progenitors in whole blood. *J Clin Oncol 13:*148–156.

169. Ghalie, R., Richman, C. M., Bender, J. G., et al (1994). Sequential transplants using mobilized peripheral blood progenitor cells. *J Clin Apheresis 9*:176–182.

170. Takamatsu, Y., Harada, M., Teshima, T., et al (1995). Relationship of infused CFU-GM and CFU-Mk mobilized by chemotherapy with or without G-CSF to platelet recovery after autologous blood stem cell transplantation. *Exp Hematol 23*:8–13.

171. Pan, L., Delmonte, Jr., J., Jalonen, C. K., and Ferrara, J. L. (1995). Pretreatment of donor mice with granulocyte colony-stimulating factor polarizes donor T lymphocytes toward type 2 cytokine production and reduces severity of experimental graft versus host disease. *Blood 86*:4422–4429.

172. Dreger, P., Steinmann, J., Drost, R., et al (1994). Enrichment of G-CSF-mobilized peripheral blood progenitor cells for allogeneic transplantation by negative selection. *Prog Clin Biol Res. 389*:407–413.

173. Dreger, P., Haferlach, T., Eckstein, V., et al (1994). G-CSF-mobilized peripheral blood progenitor cells for allogeneic transplantation: safety, kinetics of mobilization, and composition of the graft. *Br J Haematol 87*:609–613.

174. Dreger, P., Viehmann, K., Steinmann, J., et al (1995). G-CSF-mobilized peripheral blood progenitor cells for allogeneic transplantation: comparison of T cell depletion strategies using different CD34$^+$ selection systems or CAMPATH-1. *Exp Hematol 23*:147–54.

175. Helg, C., Roux, E., Beris, P., et al (1993). Adoptive immunotherapy for recurrent CML after BMT. *Bone Marrow Transplant 12*:125–129.

176. Shiobara, S., Harada, M., Mori, T., et al (1982). Difference in post-transplant recovery of immune reactivity between allogeneic and autologous bone marrow transplantation. *Transplant Proc 14*:429–433.

177. Kiesel, S., Pezzutto, A., Korbling, M., et al (1989). Autologous peripheral blood stem cell transplantation: analysis of autografted cells and lymphocyte recovery. *Transplant Proc 21*:3084–3088.

178. Henon, P., Debecker, A., Lepers, M., Kandel, G., and Eisenmann, J. C. (1988). Hematopoietic and immune reconstitution following peripheral blood stem cell autografting in acute leukemia. *Bone Marrow Transplant 3*:171–172.

179. To, L. B., Russell, J., Moore, S., and Juttner, C. A. (1987). Residual leukaemia cannot be detected in very early remission peripheral blood stem cell collections in acute nonlymphoblastic leukemia. *Leuk Res 11*:327–329.

180. To, L. B., and Juttner, C. A. (1987). Peripheral blood stem cell autografting: a new therapeutic option for AML. *Br J Haematol 66*:285–288.

181. Gale, R. P., Opelz, G., Mickey, M. R., Graze, P. R., and Saxon, A. (1978). Immunodeficiency following allogenic bone marrow transplantation. *Transplant Proc 10*:223–227.

182. Baumann, I., Testa, N. G., Lange, C., et al (1993). Haemopoietic cells mobilised into the circulation by lenograstim as alternative to bone marrow for allogeneic transplants. *Lancet 341*:369.

183. Chatta, G. S., Price, T. H., Allen, R. C., and Dale, D. C. (1994). Effects of in vivo recombinant methionyl human granulocyte colony-stimulating factor on the neutrophil response and peripheral blood colony-forming cells in healthy young and elderly adult volunteers. *Blood 84*:2923–2929.

184. de Haas, M., Kerst, J. M., van der Schoot, C. E., et al (1994). Granulocyte colony-stimulating factor administration to healthy volunteers: analysis of the immediate activating effects on circulating neutrophils. *Blood 84:*3885–3894.

185. Kanold, J., Rapatel, C., Berger, M., et al (1994). Use of G-CSF alone to mobilize peripheral blood stem cells for collection from children. *Br J Haematol 88:*633–635.

186. Sato, N., Sawada, K., Takahashi, T. A., et al (1994). A time course study for optimal harvest of peripheral blood progenitor cells by granulocyte colony-stimulating factor in healthy volunteers. *Exp Hematol 22:*973–978.

187. Craig, JI, Langlands, K., Parker, AC, Anthony, RS. (1994). Molecular detection of tumor contamination in peripheral blood stem cell harvests. *Exp Hematol 22:*898–902.

188. Brugger, W., Bross, K. J., Glatt, M., Weber, F., Mertelsmann, R., and Kanz, L. (1994). Mobilization of tumor cells and hematopoietic progenitor cells into peripheral blood of patients with solid tumors. *Blood 83:*636–640.

189. Moss, TJ. (1994). Detection of minimal residual disease in autologous grafts. *Immunomethods 5:*226–231.

190. Campana, D., and Pui, C. H. (1995). Detection of minimal residual disease in acute leukemia: methodologic advances and clinical significance. *Blood 85:*1416–1434.

191. Martiat, P., Van Daele, S., Ferrant, A., Straetmans, N., Van Den Neste, E., Marevoet, M., et al (1995). Frequent contamination of PBSC mobilized using chemotherapy and G-CSF in intermediate and high-grade lymphomas, even without bone marrow involvement. *Blood 84:*645a (abstr 2564).

192. Kessinger, A., Anderson, J., Bierman, P., Vose, J., Bishop, M., and Armitage, J. (1994). Mobilized versus non-mobilized peripheral stem-cell transplantation after high-dose therapy for low grade non-Hodgkin-lymphoma: effect on progression-free survival. *Blood 84:*396a (abstr 1568).

193. Brugger, W., Frisch, J., Schulz, G., Pressler, K., Mertelsmann, R., and Kanz, L. (1992). Sequential administration of interleukin-3 and granulocyte-macrophage colony-stimulating factor following standard-dose combination chemotherapy with etoposide, ifosfamide, and cisplatin. *J Clin Oncol 10:*1452–1459.

194. Moehle, R., Haas, R., and Hunstein, W. (1993). Expression of adhesion molecules and c-kit on CD34+ hematopoietic progenitor cells: comparison of cytokine-mobilized blood stem cells with normal bone marrow and peripheral blood. *J Hematother 2:*483–489.

195. Leavesley, D. I., Oliver, J. M., Swart, B. W., Berndt, M. C., Haylock, D. N., and Simmons, P. J. (1994). Signals from platelet/endothelial cell adhesion molecule enhance the adhesive activity of the very late antigen-4 integrin of human CD34+ hematopoietic progenitor cells. *J Immunol 153:*4673–4683.

196. Craddock, C., Nakamoto, B., and Papayannopoulou, T. (1995). The VLA$_4$/CS-1 pathway does not actively participate in VLA$_4$ mediated hematopoietic progenitor trafficking in vivo. *Blood 86:*975a (abstr 3889).

197. Cronkite, E. P., Fliedner, T. M., Bond, V. P., and Rubini, J. R. (1959). Dynamics of hematopoietic differentiation in man and mice studied by ^{3}H-thymidine incorporation into DNA. *Ann NY Acad Sci 77:*803–820.

198. Fliedner, T. M., Cronkite, E. P., and Robertson, J. S. (1964). Granulocytopoiesis I. Senescence and random loss of neutrophilic granulocytes in human beings. *Blood 24*:402–414.

199. Fliedner, T. M., Cronkite, E. P., Killman, S. A., and Bond, V. P. (1964). Granulocytopoiesis II. Emergence and pattern of labeling of neutrophilic granulocytes in humans. *Blood 24*:683–700.

200. Vincent, P. C., Cronkite, M. L., Greenberg, M. L., Kirsten, C., Schiffer, L. M., and Stryckmans, P. A. (1977). Leukocyte kinetics in chronic leukaemia: I. DNA synthesis time in blood and marrow myelocytes. *Clin Haematol 6*:695–716.

201. Athens, J. W., Haab, O. P., Raab, S. O., et al (1961). Leukokinetic studies IV. The total blood circulating and marginal granulocyte pools and the granulocyte turnover rate in normal subjects. *J Clin Invest 40*:989–996.

2

Biochemistry and Structure of Filgrastim (r-metHuG-CSF)

Timothy Osslund and Thomas C. Boone
Amgen Inc., Thousand Oaks, California

I. INTRODUCTION

Cytokines are a family of glycoproteins that have biological activities essential for the proliferation and differentiation of hematopoietic stem and progenitor cells. They work independently and synergistically to focus the production of myeloid and leukoid cells according to the biological challenges of the organism (1,2). Granulocyte colony-stimulating factor (G-CSF) is one type of cytokine. A recombinant G-CSF, Filgrastim (r-metHuG-CSF), is 175 amino acids long (included the initiating methionine) and is the subject of this study.

Other members of the cytokine family include erythropoietin (EPO), granulocyte-macrophage CSF (GM-CSF), interleukin (IL)-2, IL-3, and IL-6 (3). The carbohydrate component of members of the cytokine family is varied in the extent and the type of glycosylation. While many cytokines are highly glycosylated (4), Filgrastim has only one O-linked glycosylation site. This post-translational modification is suspected to function to inhibit proteolytic degradation and increase half-life in vivo (3,5).

Two different genes encoding the amino acid sequence of G-CSF were isolated independently from different tissue sources. Souza et al (6) isolated a G-CSF cDNA from a bladder carcinoma cell line that coded for a predicted amino-acid sequence of 174 amino acids. Nagata et al (7) reported a G-CSF cDNA from a squamous cell line that encoded for a polypeptide of 177 amino acids. The difference between these two isolates is caused by an alternative mRNA splicing site. Three additional amino acids are inserted between residues 32 and 33 of the 174-amino-acid polypeptide which greatly diminishes the biological activity of the longer squamous cell line-derived G-CSF. Large quantities of the bacterially expressed recombinant G-CSF (ie, Filgrastim) have been produced and are highly active both in vitro and in vivo (8).

The amino-acid identities between pairs of aligned sequences of human, murine, bovine, and canine G-CSF are between 70% and 80%. One notable difference between murine recombinant G-CSF and Filgrastim is a fifth cysteine residue found in the human but not the murine molecule (9). The murine G-CSF has a serine at this position which, when simulated in the human G-CSF, causes no decrease in biological activity. This cysteine at position 17 does not form a disulfide linkage and is inaccessible to modification with iodoacetic acid (10). Biological activity with a mutation at position 17 with alanine is at least as good as the unaltered recombinant protein.

II. THE G-CSF RECEPTOR

Receptors have been identified for G-CSF (11,12) (**Figure 1**). The G-CSF receptor is expressed on cells of the neutrophil lineage from myeloblasts to the mature neutrophil as well as on subsets of cells of the monocyte lineage. Receptors for G-CSF also have been found on non-hematopoietic cell lines (13). The receptors are single, high-affinity receptors, specific to G-CSF (12). Receptor density varies from 50 to 500 per cell, and the receptor affinity equals approximately 100 pM. Studies in mice have shown that the number of G-CSF receptors increases as cells mature (11,12).

With regard to the biochemical structure, the receptor is a 130- to 150-kD glycoprotein that can be downregulated by lipopolysaccharide (LPS), N-formyl-methionyl-leucyl-phenylalanine (FMLP), or GM-CSF (12). The human G-CSF receptor is encoded by a single gene or chromosome 1p35-p34.3 (14) and contains 17 exons that span approximately 16.5 kilobase pairs (kbp) of DNA (15). The G-CSF receptor gene does not contain a classical TATA box, but the major transcription initiation site is 20 bp upstream of the cap site with a minor initiation site at 10 bp upstream of the cap site (15).

The murine and human G-CSF receptors have a single transmembrane domain of 813 and 812 amino acids, respectively. Within the extracellular domain of the G-CSF receptors, there are at least three regions that have statistically significant identity to other proteins. These are the Ig-like do-

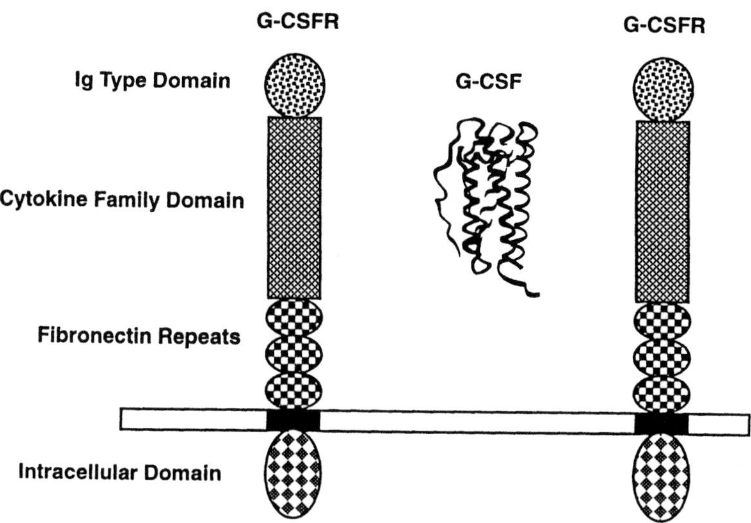

Figure 1. Schematic representation of the structure of Filgrastim receptor. A ribbon diagram of Filgrastim has been inserted between two molecules of Filgrastim receptor suggesting that signal transduction is related to binding of the ligand to two molecules of the receptor.

main, three fibronectin type-3 repeat units, and the WSXWS motif common to the cytokine-receptor family.

III. STRUCTURE OF FILGRASTIM

Secondary structure prediction analyses predict the cytokine family members to have substantial helical infrastructures. The x-ray crystallographic structures of related cytokines IL-2 and human growth hormone (hGH) exhibit a four-alpha-helical bundle (**Figure 2**). Circular dichromic analysis which predicts 69% alpha helix for Filgrastim and secondary structure analysis has suggested to Bazan (16) and to others (17) that Filgrastim has a four-helical-bundle motif.

The x-ray crystallographic atomic structure of Filgrastim was determined at high resolution (18). The crystal structure confirmed the predicted topology and showed that Filgrastim has four antiparallel helices with crossing angles of about 18° which is very close to that expected for an ideal left-handed helical bundle. In addition to the core bundle, there is a

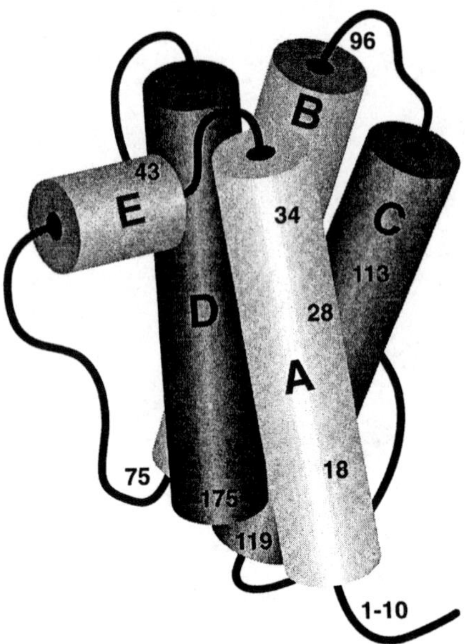

Figure 2.   Schematic representation of the structure of Filgrastim. Filgrastim has four anti-parallel alpha helices arranged in a manner to form a helical bundle, the constituents of which are labeled A, B, C, and D. There is a fifth helix, E, which is a short 3 to 10 type helix. The helices are connected with two long loops and one short loop. The relative positions of the amino acids are numbered and indicated on the figure.

short 3 to 10 helix immediately after the first disulfide bond. There are two long loops and one short loop connecting the helical core bundle. The topology has been described as being "up-up-down-down" and is consistent with the structures of GM-CSF and hGH.

IV.   CORRELATION OF THE BIOCHEMISTRY OF FILGRASTIM WITH ITS STRUCTURE

The deletion of the first 10 residues of Filgrastim or addition of tyrosines to the amino terminus have little effect on biological activity (19,20). The crystal structure of Filgrastim indicates that this region is extended from the core helical bundle and mobile, since the first 10 N-terminal residues are not visible in the electron density map. The mobility of this region would be consistent with suggestions that the first 10 residues do not participate in the overall function or stability of the protein. Reported increases in proteolytic resistance of Filgrastim truncated at the amino terminus may be explained by the availability of the first 10 residues to enzymatic degradation.

Filgrastim has two short disulfide loops, and the kinetics of the formation of the two disulfide bonds is dramatically different. The first disulfide bond of Filgrastim between residues of 37 and 43 is between the first helix and the helical bundle and the short "fifth" helix E. In contrast, the second disulfide bond is positioned at the carboxy terminal end of the flexible E to B loop (between residues 65 and 75). In the process of folding and oxidation of Filgrastim, the protein rapidly undergoes a formation of an intermediate that is the partially oxidized form of Filgrastim containing a single disulfide bond between Cys^{37} and Cys^{43}. The rate-limiting step in the formation of an active form of Filgrastim is the formation of the second disulfide bond (21). The difference in rate may be caused by the greater mobility of the second disulfide loop.

Both Filgrastim and hGH have long (20 to 30 residues) antiparallel helices, and they each have flexible interconnecting loops between helix C and helix D. It is tempting to speculate that the flexibility of these loops is a requirement for the proper folding of these proteins. Unlike GM-CSF, which has shorter helices and larger crossing angles, the greater length and more parallel nature of the Filgrastim bundle may demand a more pliable extended segment. The flexibility required by the folding might cause the C-D loop to be more vulnerable to proteolytic attack. To retain the extended loop, but to prevent proteolytic cleavage, native human G-CSF has positioned its only O-linked glycosylation in the middle of the C-D loop, presumably to mechanically impede proteolysis.

The nonbonded fifth cysteine of Filgrastim has been the subject of mutational and biochemical modification. Mutation of this cysteine (Cys^{18}) to either alanine or serine has little effect on the biological activity and presumably none on the structure of Filgrastim. Replacement of Cys^{18} with bulkier or charged residues lowers the activity of Filgrastim. The recent observation (21) that this cysteine can be iodinated with DTNB (5,5′ dithiobis [2-nitrobenzoic acid]), but not iodoacetic acid, is consistent with the positioning of the side chain of cysteine in the hydrophobic core of the helical bundle.

The two versions of human G-CSF created by the alternative mRNA splicing sites have dramatically different biological activity. Analysis of the structure with respect to the different gene products suggests a region of the molecule that directly affects receptor binding. The product of the alternatively spliced human G-CSF mRNA (ie, the mRNA having a 177-amino-acid translation product) adds three additional residues at positions 33 to 35. On the basis of the molecular structure, these residues would extend helix A one additional turn. The molecule would be extended 4.5 Å, and the long loop between helices E and B would be forced into a more extended conformation. The decrease in biological activity of this version is related to the position of the insertion which probably disrupts a potential receptor-binding site.

The packing of the long helical bundle of Filgrastim also induces an unusual structure in the short loop connecting helices B and C. The B-C loop which begins at residue Ala[92] and ends at the amino terminus of helix C at Pro[102] has a short stretch of left-handed helix. Three residues (94, 95, and 96) are required to have this conformation because the amino terminus of helix D has taken the space required at positions 150 and 151 to prevent collision with the leucine at position 93. Human growth hormone also uses an omega turn at the B-C loop, suggesting this loop may have significant conformational requirements that cannot be accommodated by the more usual secondary structure. Correlation of the Met[138]>>>Met[119] (data on file) with the Filgrastim structure suggests that these residues are exposed to very different environments. Presumably, the difference in oxidation is caused by shielding of the methionines by other residues. Methionine[127] is located on the amino terminus of loop C-D and is totally exposed to solvent. Methionine[138] is partially buried and fits in a hydrophobic pocket bounded by the carboxy terminal end of the B helix and the carboxy terminal end of the C-D loop. Methionine[119] is located in helix C and its side chain is positioned within the helical core bundle.

V. POTENTIAL RECEPTOR-BINDING SITES

Studies using epitope mapping data have suggested that the receptor-binding region of Filgrastim is found between residues 20 and 56 (22). This region is comprised of the carboxy terminus of helix A, the first disulfide loop, and a portion of helix E. One neutralizing antibody was mapped to a smaller region that included residues 36 to 46. This smaller segment includes only the disulfide loop residues (36 to 43) and a portion of the first turn of helix E (residues 44 to 46). In light of the structure of Filgrastim, there are four charged residues which might be contributing to receptor binding. These residues are Lys[41], His[44], Glu[46], and Glu[47], which have $C\alpha$s within 10 Å of each other. Mutational analyses show that a replacement of Glu[47] with Ala reduces 90% of the biological activity of Filgrastim.

Several lines of evidence also implicate the extended portion of the D helix (residues 165 to 175) as part of a receptor-binding domain. This evidence includes direct mutational analysis and indirect comparative protein structure analysis. Mutation of the carboxy terminus of Filgrastim abolishes activity in a similar manner to the loss of activity in hGH (23; data on file). Related cytokines, such as IL-6 and GM-CSF, predicted to have similar topology to Filgrastim also center their biological activity at the carboxy terminus of the D helix (16). In the hGH and receptor complex, Vos et al (24) suggest that there are two independent binding sites that function to dimerize the hGH receptor. The first receptor binding site of hGH corresponds to the region of Filgrastim bounded by the carboxy terminus of helix A, the first disulfide loop, the amino terminus of helix E, and the carboxy terminus of helix D. The second binding site of hGH corresponds to the area formed at the interface between helix A and helix C. The biochemical, mutational, and structural comparative analysis of Filgrastim confirms that this is a receptor-binding face that functions in a manner similar to hGH in its receptor complex.

Confirmation of a second binding site on Filgrastim analogous to that proposed for hGH is more difficult. In contrast to Filgrastim, hGH has amino acids in the amino terminus of the molecule that participate in the proposed second-receptor binding interface; however, the first 10 residues of Filgrastim can be deleted without affecting biological activity. Similar to hGH, Filgrastim does have a hydrophilic interface between helices A (residues 20 to 28) and C (residues 110 to 119) that could participate in a receptor-ligand complex. Glu^{20} and Lys^{24} are located on the A helix of Filgrastim and their side chains are pointed towards the C helix. Independent replacement of either of these two residues with alanine dramatically lowers the biological activity of Filgrastim (data on file). The position of the side chains of these residues in Filgrastim and their position relative to hGH is suggestive that these residues participate in a second receptor binding interface. Further analysis of this region of Filgrastim, coupled with cocrystallization of the receptor, should confirm whether this second proposed binding site is general for this class of cytokines or specific for hGH.

REFERENCES

1. Nicola, N. A. (1990). Granulocyte colony-stimulating factor. *Immunol Ser* 49:77–109.
2. Golde, D. W., Gasson, J. C. (1988). Hormones that stimulate the growth of blood cells. *Sci Am* 259:62–71.
3. Metcalf, D. (1989). The molecular control of cell division, differentiation, commitment, and maturation in hemopoietic cells. *Nature* 339:27–30.
4. Browne, J. K., Cohen, A. M., Egrie, J. C., et al (1986). Erythropoietin: gene cloning, protein structure, and biological properties. *Cold Spring Harbor Symp Quant Biol* 51:693.

5. Metcalf, D. (1989). Haemopoietic growth factors. *Lancet 1*:825–827.
6. Souza, L. M., Boone, T. C., Gabrilove, J., et al (1986). Recombinant human granulocyte colony-stimulating factor: effects on normal and leukemic myeloid cells. *Science 232*:61–65.
7. Nagata, S., Tsuchiya, M., Asano, S., et al (1986). Molecular cloning and expression of cDNA for human granulocyte colony-stimulating factor. *Nature 319*:415–418.
8. Scarffe, J. H., Kamthan, A. (1990). Clinical studies of granulocyte colony stimulating factor (G-CSF). *Cancer Surveys 9*:115–130.
9. Kuga, T., Komatsu, Y., Motoo, Y., et al (1989). Mutagenesis of human granulocyte colony stimulating factor. *Biochem Biophys Res Comm 159*:103–111.
10. Lu, S. H., Boone, T. C., Souza, L. M., Lai, P. (1989). Disulfide and secondary structure of recombinant human granulocyte colony stimulating factor. *Arch Biochem Biophys 268*:81–92.
11. Demetri, G. D., Griffin, J. D. (1991). Granulocyte colony-stimulating factor and its receptor. *Blood 78*:2791–2802.
12. Nicola, N. A. (1989). Hemopoietic cell growth factors and their receptors. *Annu Rev Biochem. 58*:45–77.
13. Mazanet, R., Griffin, J. D. (1992). Hematopoietic growth factors. In *High-Dose Cancer Chemotherapy.* J. O. Armitage and K. H. Antman, eds. Williams and Wilkins, Baltimore, pp. 289–313.
14. Inazawa, J., Fukunaga, R., Seto, Y., et al (1991). Assignment of the human granulocyte colony-stimulating factor receptor gene (CSF3R) to chromosome 1 at region 35-p34.3. *Genomics 10*:1075–1078.
15. Seto, Y., Fukunaga, R., Nagata, S., et al (1992). Chromosomal gene organization of the human granulocyte colony-stimulating factor receptor. *J Immunol 148*:259–266.
16. Bazan, J. F. (1990). Haemopoietic receptors and helical cytokines. *Immunol Today 11*:350–354.
17. Parry, D. A. (1988). Conformational homologies among cytokines: interleukins and colony stimulating factors. *J Mol Recog 1*:107–110.
18. Hill, C. P., Osslund, T. D., Eisenberg, D. S. (1993). The structure of granulocyte-colony stimulating factor (r-hu-G-CSF) and its relationship to other growth factors. *Proc Natl Acad Sci USA 90*:5167–5171.
19. Okabe, M., Asano, M., Kuga, T., et al (1990). In vitro and in vivo hematopoietic effect of mutant human granulocyte colony-stimulating factor. *Blood 75*:1788–1793.
20. Hanazono, Y., Hosoi, T., Kuwaki, T., et al (1990). Structural analysis of the receptors for granulocyte colony-stimulating factor on neutrophils. *Exp Hematol 18*:1097–1103.
21. Lu, H. S., Clogston, C. L., Narhi, L. O., Merewether, L. A., Pearl, W. R., Boone, T. C. (1992). Folding and oxidation of recombinant human granulocyte colony-stimulating factor produced in Escherichia coli. Characterization of the disulfide-reduced intermediates and cysteine-serine analogs. *J Biol Chem 267*:8770–8777.

22. Layton, J. E., Morstyn, G., Fabri, L. J., et al (1991). Identification of a functional domain of human granulocyte colony-stimulating factor using neutralizing monoclonal antibodies. *J Biol Chem 266*:23815–23823.
23. Cunningham, B. C., Wells, J. A. (1989). High-resolution epitope mapping of hGH-receptor interactions by alanine-scanning mutagenesis. *Science 244*: 1081–1084.
24. Vos, A. M., Ultsch, M., Kossiakoff, A. A. (1992). Human growth hormone and extracellular domain of its receptor: crystal structure of the complex. *Science 255*:306–312.

3
Pharmacology of Filgrastim (r-metHuG-CSF)

Lorin K. Roskos, Ellen N. Cheung, Martha Vincent, and MaryAnn Foote
Amgen Inc., Thousand Oaks, California

George Morstyn
UCLA School of Medicine, Los Angeles, and Amgen Inc., Thousand Oaks, California

I. INTRODUCTION

The pharmacokinetics and pharmacodynamics of Filgrastim have been studied in animals, normal volunteers, and patients with acute and chronic neutropenia. The results of these studies have shown that Filgrastim has a consistent and predictable pharmacological profile when given intravenously (IV) or subcutaneously (SC) over a wide dosing range. The clinical efficacy of Filgrastim is related to the concentration-dependent effects of the cytokine on mobilization and proliferation of neutrophils and neutrophil precursors.

Filgrastim is a hematopoietic growth factor that selectively stimulates granulopoietic cells of the neutrophilic lineage (1). A schematic illustration of hematopoiesis is presented in Chapter 1 of this volume. Filgrastim regulates the number of circulating neutrophils by binding to a specific cell-surface receptor that results in the proliferation, differentiation, and accelerated maturation of neutrophilic precursor cells, as well as enhancement of mature cell function. The receptors for Filgrastim are specific; they have been found on all cells of the neutrophil granulocyte series as well as on monocytes and monocyte precursors. Receptors for Filgrastim have not been found on lymphocytes, eosinophils, basophils, megakaryocytes, or their progeny (2). The number of human G-CSF receptors on neutrophilic precursors increases slightly as the cells mature, and binding of Filgrastim to the receptors decreases the number of available receptors as the Filgrastim-receptor complex is internalized and degraded. Receptor binding and internalization is required for Filgrastim to exert its effects.

Filgrastim exerts two major effects on neutrophil cytokinetics that increase neutrophil mass within the circulating granulocyte pool (**Figure 1**). The increase in absolute neutrophil count (ANC) that begins within 2 hours of the first dose of Filgrastim is due to an effect on cytokinesis, the movement of cells between body compartments. Neutrophil influx into the circulating granulocyte pool is accelerated by mobilization of band cells and polymorphonuclear neutrophils (PMN) from a marrow storage compartment, and possibly by demargination of cells from a marginal pool. Therefore, Filgrastim rapidly increases the total number of circulating blood neutrophils. Filgrastim also mobilizes hematopoietic progenitor cells into the blood, which increases the yield of CD34[+] cells collected by leukapher-

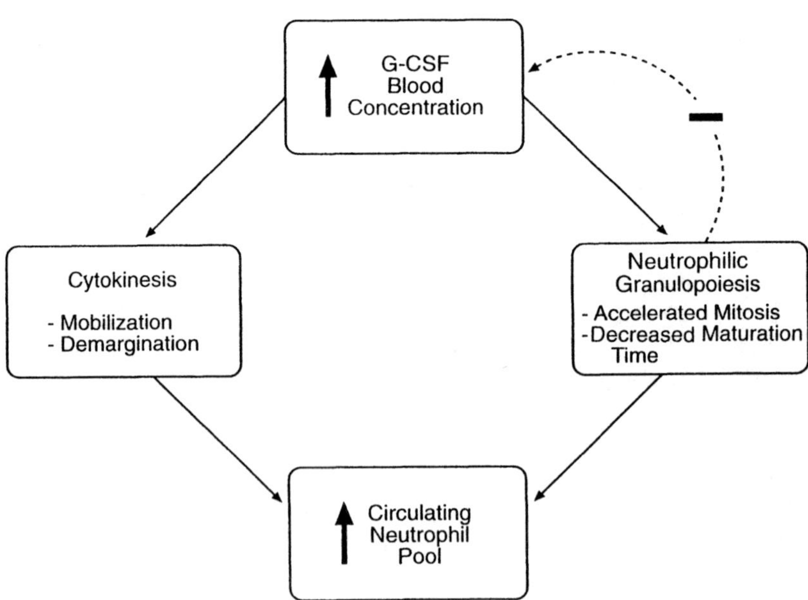

Figure 1. Filgrastim increases absolute neutrophil count by two mechanisms: mobilization of mature neutrophils into the circulating neutrophil pool and acceleration of granulopoiesis.

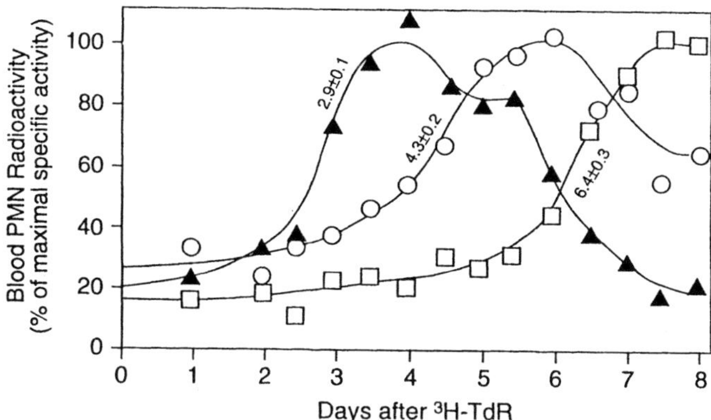

Figure 2. Filgrastim causes a dose-dependent acceleration of precursor maturation time. Solid triangles, 300 μg; open circles, 30 μg; open squares, no drug. (Adapted from Ref. 6.)

esis for use in peripheral blood progenitor cell (PBPC) transplantation (3,4) (see also Chapters 13 and 14).

The second major pharmacological effect of Filgrastim is stimulation of neutrophilic granulopoiesis. Pulse labeling studies have indicated that in healthy subjects, the mean transit time of precursor cells from the mitotic compartment to emergence of labeled PMN in blood is approximately 7 days (5). **Figure 2** shows that Filgrastim causes a dose-dependent acceleration of precursor maturation time (6). Mean emergence time of labeled PMN into blood is reduced to 2.9 days after daily SC dosing with 300 μg Filgrastim. Filgrastim also stimulates proliferation of neutrophilic precursors within the mitotic compartment. The marrow myeloid pool undergoes mitotic expansion primarily at the promyelocyte and myelocyte stages (7–9). Mitotic expansion of the marrow pool produces an increase in the total number of G-CSF receptor-bearing cells. Consequently, marrow expansion acts as a negative feedback mechanism for granulopoiesis by increasing the receptor-mediated clearance of Filgrastim. Ultimately, a dose-dependent net increase in granulopoiesis and steady-state neutrophil count is achieved.

II. PRECLINICAL PHARMACOLOGY

A. Pharmacokinetics

The pharmacokinetics of Filgrastim have been studied in mice, hamsters, rats, and monkeys. After IV dosing, the initial distribution volume approximates plasma volume. The volume of distribution is linear with dose. The elimination half-life ranges from 1 to 2 hours in all species. In rats and monkeys, clearance is concentration-dependent: clearance decreases and half-life increases with increasing dose. The dose-dependent elimination is due to saturation of G-CSF receptor-mediated clearance at high concentrations of Filgrastim. Clearance in rats and monkeys progressively increases with repeated dosing due to increasing neutrophil counts, neutrophil precursor mass, and subsequent acceleration of receptor-mediated clearance.

Pharmacokinetics in mice have been studied after single IV doses of 100 μg/kg (10). Serum concentrations of Filgrastim declined in a biexponential manner. The α-phase half-life was 5.2 minutes and the β-phase half-life was 122 minutes.

The dose-ranging pharmacokinetics of Filgrastim have been evaluated in rats after single IV and SC doses (11). After IV dosing, clearance decreased with increasing dose. After doses of 1, 15, 10, and 100 μg/kg, clearance was 79.8, 70.3, 55.6, and 53.1 mL/hr/kg, respectively. Elimination half-life increased with increasing dose, ranging from 1.08 hour at 5 μg/kg to 1.70 hour at 100 μg/kg. Initial dilution volumes (47.8 to 53.9 mL/kg) were independent of dose. After SC dosing, Filgrastim was rapidly absorbed. Peak concentrations were attained in approximately 2 hours. Relative SC exposures (AUC_{SC}/AUC_{IV}) were 0.64, 0.64, 0.51, and 0.70, indicating that Filgrastim was well absorbed. (The AUC_{SC}/AUC_{IV} ratio underestimates absolute bioavailability when clearance is dose dependent since the IV C_{max} is higher than the SC C_{max}, resulting in greater saturation of clearance after IV dosing). A nonlinear increase in AUC also was observed after SC administration.

Pharmacokinetics of Filgrastim in rats after 7 days of repeated IV and SC dosing (5 μg/kg/day) have been compared with the first-dose pharmacokinetics (12). After the first IV dose, C_{max}, V_C, AUC, clearance, and elimination half-life were 142 ng/mL, 35.2 mL/kg, 118 ng*hr/mL, 42.4 mL/hr/kg, and 1.05 hour, respectively. After the first SC dose, C_{max}, AUC, and apparent clearance (CL/F), and relative SC exposure (AUC_{SC}/AUC_{IV}) were 17.2 ng/mL, 79.2 ng*hr/mL, 63.1 mL/hr/kg, and 0.67, respectively. After the seventh dose, systemic clearance increased by 17.4%, and CL/F increased by 25.6%. Initial dilution volume after the seventh IV dose (V_C) did not change significantly (35.2 mL/kg). Relative SC exposure was 60%. The increased clearance after multiple dosing was attributed to increased receptor-mediated endocytosis and degradation of Filgrastim.

The influence of renal and hepatic failure on the pharmacokinetics of Filgrastim was evaluated in rats after IV doses of 10 μg/kg (13). In the renal failure models, clearance was 44.5 mL/hr/kg in sham-operated control animals, 24.1 mL/hr/kg (54% of control) after unilateral nephrectory, and

9.43 mL/hr/kg (21% of control) after bilateral nephrectomy. In the hepatic failure models, clearance was 42.1 mL/hr/kg in sham-operated control animals and 31.9 mL/hr/kg (76% of control) in a 70% partial hepatectomy model. At the dose of 10 μg/kg IV used in this study, receptor-mediated clearance was probably largely saturated. The results of this study suggest that in the absence of receptor-mediated clearance, renal clearance is the predominant pathway of Filgrastim elimination.

Repeated dosing of Filgrastim (10 μg/kg/day SC) for 7 days in cynomolgus monkeys resulted in 112% increase in apparent clearance of the seventh dose relative to the first dose (E. Cheung, unpublished data). Apparent clearance and elimination half-life were 59.6 mL/hr/kg and 1.63 hour after the first dose, and 127 mL/hr/kg and 1.07 hour after the seventh dose. Compartmental pharmacokinetic modeling projects that clearance would have been induced by 182% at steady-state.

Tissue distribution and excretion studies have been conducted in hamsters and rats after IV and SC dosing with ^{35}S- or ^{125}I-labeled Filgrastim at doses of 5.75 to 57.5 μg/kg. The distribution of Filgrastim was primarily to the bone marrow, adrenal glands, kidneys, and liver. Regardless of the route of administration, more than 50% of the administered radioactivity was found in the urine within 12 hours. Accumulation was not observed in any of the major tissues.

B. Pharmacodynamics

Pharmacodynamic studies in rabbits have been reported by Stevens et al (14). The pharmacodynamics of a single SC dose of Filgrastim 10 μg/kg was monitored and correlated with serum concentrations, specific oxygenation capacity of circulating PMN as measured by luminol-dependent chemiluminescence using C3b/C3bi-opsonized zymosan as a stimulus, ANC, and estimation of PMN-myeloperoxidase content by peroxidase staining. After administration of Filgrastim, the concentration peaked at 2 to 4 hours with an accompanying 5-fold decrease in ANC and a 2.5-fold increase in chemiluminescence, but no change in PMN-myeloperoxidase content. The ANC returned to normal by 6 hours, at which time the chemiluminescence was within normal range. Between 24 and 48 hours with Filgrastim concentrations at baseline values, the ANC increased 6-fold, chemiluminescence increased a second time to 3-fold greater than normal, and the PMN-myeloperoxidase content concomitantly increased 6-fold. The results of this preclinical study showed a biphasic increase in PMN oxygenation capacity that consisted of an early phase which likely occurs because of increased CR1 and CR3 opsonin-receptor expression due to the direct exposure of the peripheral blood PMN to Filgrastim and a second phase which correlates with increased PMN-myeloperoxidase. In addition to the increase in the

ANC, the effects of Filgrastim to increase the PMN oxygenation capacity may provide a proportional therapeutic advantage with regard to PMN microbicidal capacity.

III. CLINICAL PHARMACOLOGY

A. Overview

Filgrastim produces dose-dependent increases in the total circulating neutrophil pool by two major mechanisms: cytokinesis and accelerated granulopoiesis. The initial increase in ANC that occurs on the first day of dosing is predominantly due to mobilization of mature neutrophils from a marrow storage pool. Sustained increase of neutrophil count after repeated dosing is due to acceleration of neutrophil granulopoiesis by mitotic expansion and decreased maturation time of neutrophil precursors. Mitotic expansion occurs at the myeloblast, promyelocyte, and myelocyte stages (5,6,9). Neutrophil survival time after Filgrastim administration is not altered (5).

Numerous studies have confirmed that during the first 60 minutes after administration of Filgrastim there is a transient decrease in the number of peripheral neutrophils. This decrease is followed consistently by a rapid and significant increase in the number of neutrophils to greater than normal counts within 4 to 5 hours (15–17).

The initial transient reduction in the number of measurable neutrophils in the peripheral circulation may be due to neutrophil margination to endothelial cells that is followed by neutrophil demargination and, finally, by an increase in neutrophil proliferation and accelerated release of mature neutrophils from the bone marrow (15–16). When Filgrastim is discontinued, there is a progressive decrease in the number of neutrophils in the peripheral blood over the next 24 to 48 hours (7,16). **Figure 3** shows the effects of Filgrastim on neutrophil counts of 15 patients who received Filgrastim IV for 5 days before receiving chemotherapy. Absolute neutrophil counts were measured 24 hours after administration of Filgrastim. Increases to greater than 10-fold were achieved and were sustained during the Filgrastim administration period. **Figure 4** illustrates the dose-dependent increase in white blood cell counts in a second patient population treated with the same doses of Filgrastim given as a 30-minute IV infusion once daily from day 1 to day 14. Similar increases in the ANC were seen in the first 24-hour period after dosing and increases continued (after doses of 30 and 60 μg/kg) thereafter. The rapid initial increases in ANC within 24 hours of Filgrastim administration may be due predominantly to the release of neutrophils from the bone marrow, but the sustained increase of neutrophils is due to Filgrastim's ability to stimulate bone marrow-cell

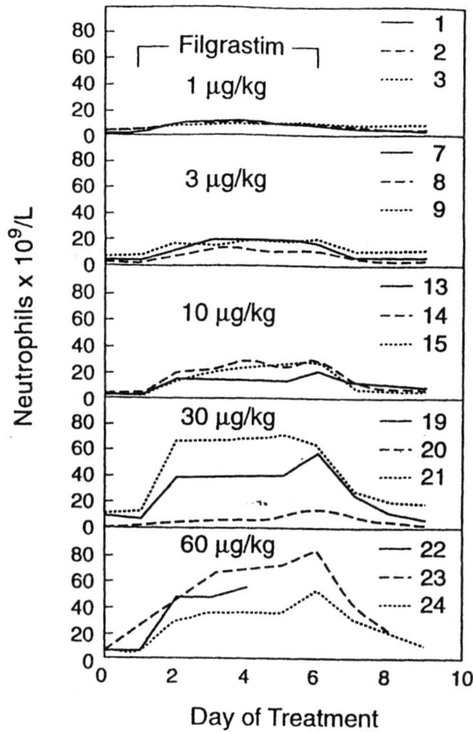

Figure 3. The effects of Filgrastim in individual patients given doses of 1, 3, 10, 30, or 60 µg/kg intravenously on day 1 to day 5. Neutrophil counts were done daily. (Adapted from Ref. 16.)

proliferation. This is evidenced by the increased myeloid:erythroid ratio, the higher proportion of early myeloid cells, and the increase in the proportion of cells in the S-phase (7).

B. Pharmacokinetics

The clinical pharmacokinetics of Filgrastim have been studied in normal volunteers, cancer patients prechemotherapy and postchemotherapy, and patients with severe chronic neutropenia. The pharmacokinetics have been described after IV infusion, SC bolus, and SC infusion. After SC administration, Filgrastim is rapidly absorbed, and peak serum concentrations are attained 2 to 8 hours after dosing. Elimination half-life after IV and SC dosing is usually between 2 and 4 hours. Clearance and half-life are dependent on dose and neutrophil count: G-CSF receptor-mediated clearance is

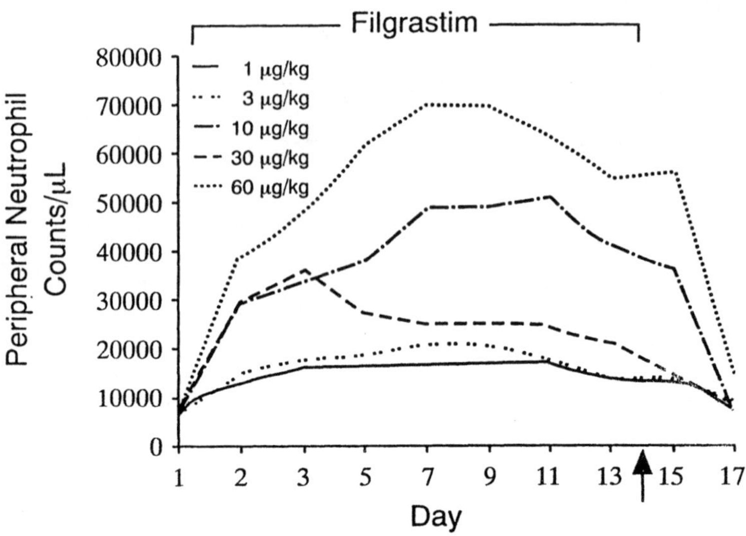

Figure 4.   White bood cell counts in patients treated with 1, 3, 10, 30, and 60 μg/ kg Filgrastim administered as a 30-minute intravenous infusion once daily from day 1 to day 14. The arrow indicates the last dosage of Filgrastim. The mean of three patients was used for each dose level. (Adapted from Ref. 21).

saturated by high concentrations of Filgrastim ($>$20 ng/mL) and is diminished by neutropenia. After discontinuation of dosing, Filgrastim concentrations decrease to endogenous levels within 24 hours.

Pharmacokinetics in healthy volunteers have been studied after IV and SC dosing. Filgrastim pharmacokinetics in normal recipients are summarized in **Table 1**. In a dose-ranging study in healthy male volunteers (Amgen, data on file), Filgrastim was administered SC at 75 to 600 μg/day for 10 days. Serum concentration profiles were collected after the first and tenth dose. Dose-dependent, autoinduction of Filgrastim clearance was evident: the ratios of the first dose AUC to the tenth dose AUC were 2.5, 3.3, 5.9, and 7.2 after 75, 150, 300, and 600 μg/day, respectively. The ratios of predose ANC on day 1 to predose ANC on day 10 were 4.0, 5.5, 7.8, and 13.7, respectively. The magnitude of autoinduction appeared closely related to the degree of neutrophilia in the recipients, which is consistent with increased receptor-mediated clearance by the expanded neutrophil pool.

The pharmacokinetics of Filgrastim in cancer patients are summarized in **Table 2**. Dose-linear clearance has been reported after high doses of Filgrastim (7,8) and during severe neutropenia (3). However, nonlinear pharmacokinetics have been reported after low IV doses administered

Table 1. Summary of Clinical Pharmacokinetic Studies in Normal Volunteers.

Study	Dose and route	Results
8711 (N = 5)	3.45 μg/kg IV, single 30-minute infusion	• Elimination by first-order kinetics $t_{1/2}\beta$ = 2.7 hr V_d = 162 mL/kg clearance = 0.6 mL/min/kg
91170 (N = 30)	75, 150, 300, or 600 μg per individual, SC, multiple dosing for 10 days	• Decrease in serum levels upon multiple dosing. • C_{max} (ng/mL)

C_{max} (ng/mL)

	Day 1	Day 10
75 μg	1.65 ± 0.80	0.99 ± 0.28
150 μg	4.60 ± 3.78	1.84 ± 1.29
300 μg	14.79 ± 6.55	3.64 ± 1.63
600 μg	16.28 ± 7.14	3.46 ± 0.52

t_{max} (hr)

	Day 1	Day 10
75 μg	5.5 ± 1.8	3.5 ± 0.9
150 μg	5.0 ± 2.0	4.5 ± 1.0
300 μg	4.0 ± 0.0	3.5 ± 1.0
600 μg	5.8 ± 1.3	3.7 ± 0.8

AUC_{0-24} (ng·hr/mL)

	Day 1	Day 10
75 μg	14.3 ± 4.3	5.7 ± 1.6
150 μg	33.2 ± 16.9	10.2 ± 4.5
300 μg	119.0 ± 41.7	20.3 ± 8.8
600 μg	209.4 ± 107.3	29.1 ± 5.3

before chemotherapy (15,18). Clearance of IV Filgrastim prechemotherapy was more rapid after an IV infusion of 1.73 μg/kg than after 5.75 to 34.5 μg/kg (15). Filgrastim clearance appeared linear at higher doses, presumably due to saturation of receptor-mediated clearance. When receptor-mediated clearance is saturated by high Filgrastim concentrations or is diminished by neutropenia, the linear clearance pathway predominates and the pharmacokinetics appear linear.

Homeostatic regulation of Filgrastim clearance was first described in cancer patients receiving Filgrastim before and after chemotherapy (18). Patients receiving 10 μg/kg/day Filgrastim by SC infusion for 5 days before chemotherapy experienced declines in serum concentrations before discon-

Table 2. Summary of Clinical Pharmacokinetic Studies in Cancer Patients.

Population	Dose and route	Results
Advanced breast cancer (N = 5) (7,8)	11.5, 34.5, or 69 μg/kg IV over 30 minutes, single dose	Positive linear correlation between Filgrastim dose and both peak serum concentration (R^2 = .98) and total systemic amount of Filgrastim (as measured by AUC) (R^2 = .95) Clearance 0.62, 0.14, and 0.17 mL/min/kg after 11.5, 34.5, and 69 μg/kg, respectively $t_{1/2}\beta$ ranged from 1.8 to 5.9 hr V_d ranged from 47 to 98 mL/kg
Various cancer diagnoses (N = 21) (15)	1.73 to 34.5 μg/kg IV over 30 minutes	Filgrastim clearance in the 1.73 μg/kg group was more rapid than in the higher dose groups For the 1.73 μg/kg dose, $t_{1/2}\beta$ was 1.43 $\pm$ 0.60 hr; for doses >1.73 μg/kg, $t_{1/2}\beta$ ranged from 3.5 to 4.5 hr V_d ranged from 76 to 215 mL/kg and was independent of dose
	3.45 or 11.5 μg/kg SC bolus	For 3.45 μg/kg: C_{max} = 4 ($\pm$2.6) ng/mL at 4.4 ($\pm$1.3) hours For 11.5 μg/kg: C_{max} = 49 ($\pm$40) ng/mL at 5.1 ($\pm$2.3) hours V_d/F ranged from 46 to 385 mL/kg
	3.45 or 11.5 μg/kg SC infusion over 24 hours	$t_{1/2}\beta$ = ~3.5 hours
Poor prognosis acute lymphoblastic leukemia or non-Hodgkin's lymphoma (N = 11) (3)	23 μg/kg/day continuous SC infusion beginning 1 day after bone marrow transplant for 11 to 20 days; reduced stepwise to 5.75, then to 1.15 μg/kg/day when ANC increased >1.0 $\times$ 10^9/L	Within 24 hours of starting the infusion, serum concentrations of Filgrastim were $\geq$20 ng/mL in all patients. At the 23-μg/kg dose level, daily concentrations of Filgrastim remained in the range of 19 to 96 ng/mL at the end of each 24 hours. There was no evidence that repeated daily dosing in neutropenic patients resulted in an increased clearance rate of Filgrastim. Reduction of the Filgrastim dose to 5.75, then to 1.15 μg/kg resulted in immediate and rapid decreases in serum concentrations of Filgrastim.

Metastatic malignancy after chemotherapy and bone marrow transplant (N = 5) (25)	4.6 to 73.6 μg/kg/day continuous IV infusion; dose reduced by 25% when ANC increased above 2.5 × 10^9/L	Concentrations of Filgrastim were sustained during the period of infusion and were highly correlated with the daily dose (R^2 = .98). No evidence that repeated dosing with Filgrastim over this dose range resulted in altered clearance rates of Filgrastim. Twenty-four hours after the end of the infusion, Filgrastim was either undetectable or only detectable in low concentrations. Median $t_{1/2}\beta$ = 3.4 hours (range: 1.9 to 7.8 hours)
Various cancer diagnoses (18)	1.0 to 60 μg/kg IV over 20 to 30 minutes prechemotherapy	Progressive increase in elimination half-life with increasing dose. 1 μg/kg: $t_{1/2}\beta$ = 1.4 ± 0.2 hr 60 μg/kg: $t_{1/2}\beta$ = 4.2 ± 0.2 hr
	10 μg/kg/day SC infusion for 5 days prechemotherapy	Peak Filgrastim concentrations attained 2 to 3 days after start of infusion. Filgrastim levels decreased markedly over last 2 days of infusion.
	20 μg/kg/day SC infusion for at least 10 days postchemotherapy	Plateau concentrations of Filgrastim maintained until onset of neutrophil recovery. A rapid decline in Filgrastim concentrations during infusion was associated with increased ANC.
Children (age 1.2 to 9.4 years) with advanced neuroblastoma (N = 15) (26)	5, 10, or 15 μg/kg SC bolus for 10 days after chemotherapy	First dose clearance (0.31 ± 0.13 mL/min/kg) was lower than tenth dose clearance (0.71 ± 0.66 mL/min/kg). First dose $t_{1/2}\beta$ (5.8 ± 2.1 hr) was longer than tenth dose $t_{1/2}\beta$(4.5 ± 2.1 hr). AUC after first and tenth doses were significantly related to dose and ANC.

tinuation of dosing; however, patients receiving 20 μg/kg/day Filgrastim by SC infusion after high-dose chemotherapy maintained steady-state concentrations for 10 days without autoinduction of clearance (**Figure 5**). In patients receiving Filgrastim at 10 μg/kg/day by SC infusion after an injection of melphalan, plateau serum concentrations were maintained until onset of hematopoietic recovery. Subsequent declines in serum concentrations corresponded closely with neutrophil recovery (**Figure 6**).

Filgrastim pharmacokinetics also have been studied in children with severe chronic neutropenia (19). Eleven children were given 6 to 48 μg/kg twice daily by SC injection. Peak serum concentrations occurred 2 to 8 hours after dosing. A strong relationship between ANC and Filgrastim clearance was described using a sigmoid model. At low neutrophil counts ($<0.1 \times 10^9$/L), clearance approached a minimum value of 0.29 mL/min/kg. Maximum clearance approached 2 mL/min/kg at high values of ANC ($>17.0 \times 10^9$/L). Mean half-life was 4.7 hours at low ANC and less than 2 hours at ANC $>17.0 \times 10^9$/L.

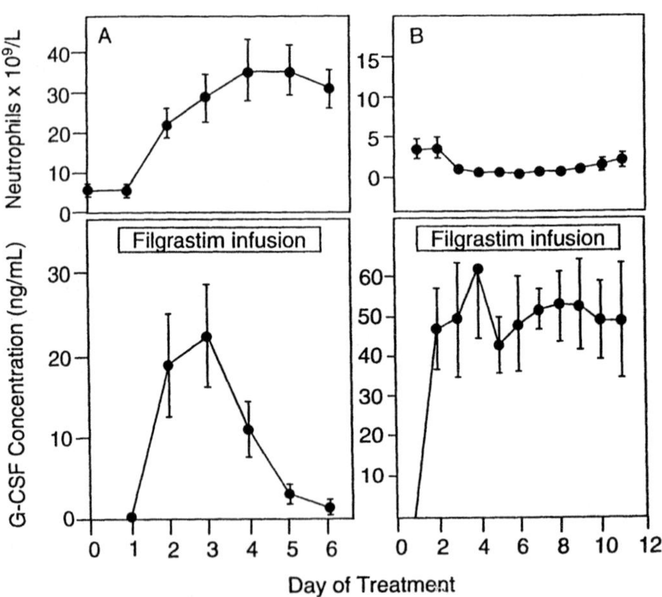

Figure 5. Levels of G-CSF and neutrophils in patients receiving a continuous infusion of Filgrastim. Panel A, six patients who received Filgrastim 10 μg/kg/day for 5 days before chemotherapy. Panel B, three patients who received 20 μg/kg/day for at least 10 days after high-dose chemotherapy. (Adapted from Ref. 18.)

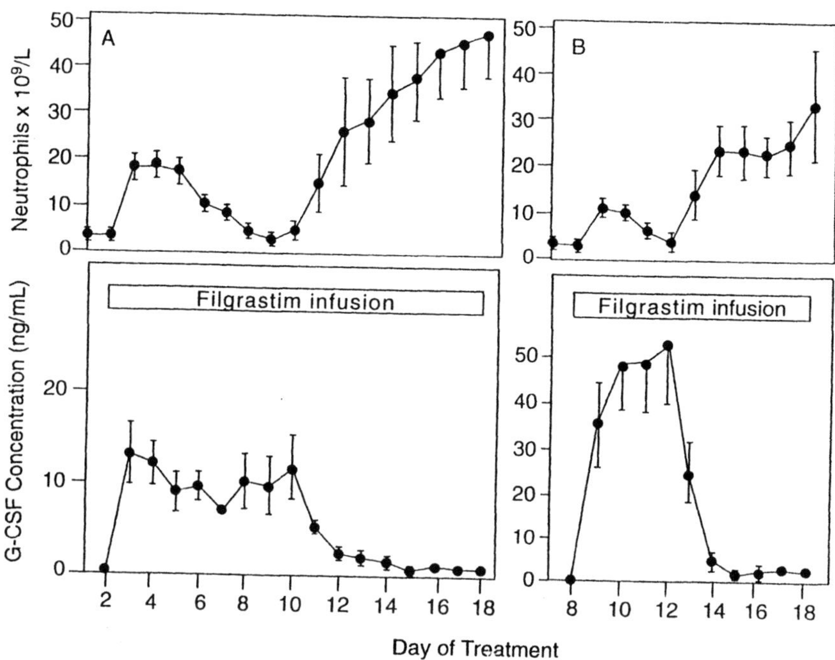

Figure 6. Levels of G-CSF and neutrophils in patients receiving a continuous infusion of 10 µg/kg/day after chemotherapy. Panel A, three patients who received Filgrastim from day 2 to day 18, and three patients who received Filgrastim from day 8 to day 16. Panel B, data from all patients combined. (Adapted from Ref. 18.)

C. Pharmacodynamics

The efficacy of Filgrastim is related directly to its pharmacodynamic effects; Filgrastim acts selectively on cells of the neutrophil lineage.

A summary of the reported findings in peripheral blood smears from patients receiving Filgrastim is listed in **Table 3**. Consistent reports include a dose-dependent increase in the number of banded neutrophils and late metamyelocytes. Occasionally, morphologically normal myelocytes and promyelocytes and, rarely, morphologically normal myeloblasts are seen. Other dose-dependent effects include the appearance in the myeloid series of densely staining secondary granulation, cytoplasmic vacuolations, and cytoplasmic Döhle bodies. No morphological changes in platelets or erythrocytes are seen. Doses greater than 10 µg/kg/day have yielded dose-dependent increases in monocytes and small increases in lymphocytes, which remain within the normal range. Eosinophilia and basophilia have not been seen.

Table 3. Summary of Reported Findings in Peripheral Blood Smears in Patients Treated with Filgrastim.

Neutrophilia*
Marked left shift*, characterized by:
- Increase in late metamyelocyte and banded neutrophils
- Occasional appearance of morphologically normal myelocytes and promyelocytes and, in rare instances, morphologically normal myeloblasts

Mophological changes in myeloid series*, characterized by:
- Densely staining secondary granulation
- Cytoplasmic vacuolation
- Cytoplasmic Döhle bodies

Appearance of macropolycytes (<5% of circulating myeloid cells)
Occasional appearance of normoblasts*
Mild monocytosis*
Mild lymphocytosis
No morphological changes in platelets or erythrocytes

---

* Dose-dependent effect

The predominant changes in the bone marrow of patients enrolled in phase 1 studies were dose-dependent increases in the proportion of early myeloid cells (myeloblasts and/or promyelocytes) which yielded an increase in the early-to-late myeloid cell ratio.

In other studies, there were small and inconsistent decreases in other cell types, specifically in the number of red blood cells and platelets in some patients enrolled in these studies. No consistent effect of Filgrastim on hemoglobin or hematocrit has been noted (7); there was a small decrease in treatment values, but all measurements remained within the normal range. In addition, Filgrastim led to a 2-fold dose-independent increase in the number of circulating lymphocytes and a slight increase in the number of monocytes and the myeloid-to-erythroid ratio in the bone marrow. The latter effect may be due to a change in the proportion of early myeloid cells in the marrow (7). In one study, there was a slight decrease in platelet number in patients receiving 14 days of doses >11.5 μg/kg; however, the platelet number reductions were transient, with recovery during continued Filgrastim administration.

In normal subjects, elevations of the ANC >10 × 10^9/L were observed after a single IV injection of Filgrastim at doses of 1.15 and 3.45 μg/kg. In patients with cancer, elevations of ANC above this value were consistent after repeated daily doses of ≥3.45 μg/kg Filgrastim, regardless of the route of administration. Four of 17 patients (24%) treated with doses of 1.15 μg/kg/day or less did not achieve an ANC >10 × 10^9/L. In all studies, the increase in ANC was characterized by a dose-dependent increase in banded neutrophils in the peripheral blood. There were also dose-dependent appearances of secondary cytoplasmic granules and Döhle bodies in the neutrophils, the same bodies observed during infections and which are indicative of functionally primed neutrophils.

The increase in ANC was observed within 5 hours of Filgrastim administration in normal subjects. Studies in cancer patients have shown a decrease in circulating neutrophil counts during the first 30 to 60 minutes, followed by a steep increase, with increases above normal levels within 4 hours (15,20–22). The precise mechanism of this early depression of ANC, which is selective for neutrophils, is not known, although it may be caused by margination to endothelial cells. Since the increase in neutrophil counts occurs within 24 hours, it is possible that neutrophil demargination and mobilization of neutrophils from marrow may contribute to the first observed ANC increase after Filgrastim administration.

With repeated IV doses of Filgrastim, the ANC increased each day of administration until a peak response ANC plateau was achieved. This plateau was reported after 2 to 4 days at doses of <11.5 μg/kg/day, and after 6 to 8 days at higher doses. The response was dose-dependent for any given route and schedule of administration. The ANC increased from baseline by approximately 3-, 4-, and 6-fold at IV doses of 1.15, 3.45, and 11.5 μg/kg/day, respectively. The dose-response relationship was less clear at doses of >11.5 μg/kg/day, although an increase in response was still observed in all studies up to 69.0 μg/kg/day. Data reported from patients given Filgrastim SC also included a positive dose-response relationship, with 3-fold, 4-to-5-fold, and 12-fold increases at doses of 1.15, 3.45, and 11.50 μg/kg/day, respectively. There did not appear to be any difference in ANC response between those patients who had previously received chemotherapy or radiotherapy and those patients who had not previously received any myelosuppressive therapy.

In all studies, the ANC was reported to decrease within 24 hours of the end of Filgrastim treatment. Neutrophil counts returned to normal within 1 to 7 days after the end of administration, depending on the peak ANC achieved. There was no evidence of withdrawal effects after Filgrastim discontinuation at doses up to 69.0 μg/kg/day; in particular, there was no evidence of rebound neutropenia.

Neutrophils produced during a proliferative response to Filgrastim have been reported to possess normal or enhanced functional properties as shown in assays of phagocytosis and chemotaxis (23,24). Clinical observations are consistent with these results since patients treated with Filgrastim resolved infection with no evidence of impaired neutrophil response.

D. Pharmacokinetic-Pharmacodynamic Modeling

The effects of Filgrastim on neutrophilic granulopoiesis and cytokinesis are directly related to the concentrations of Filgrastim achieved in blood. Therefore, a biomathematical model can be constructed to describe concentration-response relationships. The mathematical relationship is complex since a number of pharmacological responses to Filgrastim must be considered. The complexity of the concentration-effect relationship, as well as heterogeneity in patient populations, has often led to a misconception that the pharmacodynamics of response are not predicted by the pharmacokinetics of Filgrastim. However, a predictive pharmacokinetic-pharmacodynamic model may be constructed by building a mathematical model based on the pharmacology of Filgrastim. The pharmacokinetic-pharmacodynamic relationship is best illustrated in healthy subjects receiving Filgrastim, where there is less variability in response due to marrow status. The data used in the following example were collected in a phase 1 study of Filgrastim in healthy male subjects (Amgen, data on file).

A model for response to Filgrastim must consider effects on cytokinesis, marrow precursor expansion, and increased receptor-mediated clearance of Filgrastim. **Figure 7** illustrates the mean pharmacokinetic and ANC profiles after the first and tenth daily SC dose of Filgrastim. The transient decrease in the first hour after dosing is due to rapid redistribution of neutrophils into the marginal blood pool. The margination process is extremely rapid; hence, margination may be modeled as a Filgrastim concentration-dependent shift in neutrophil dilution volume. The increase in neutrophil count above the predose ANC is modeled as a Filgrastim concentration-dependent flux into the circulating neutrophil pool. The differential equation describing neutrophil influx into the circulating neutrophil pool and neutrophil survival may be defined as follows:

$$\dot{N} = I_0 + k_{eff}(t) * G - \frac{N}{MST}$$

I_0 is the baseline influx rate of neutrophils, MST is the mean survival time of neutrophils, and $k_{eff}(t)$ is a time-dependent function that relates Filgrastim concentration (G) to neutrophil influx rate. Neutrophil influx rate will depend on the number of neutrophils in the marrow storage compartment as well as the concentration of Filgrastim. Since the number of neutrophils available for influx increases due to mitotic expansion of the precursor pool, the proportionality term (k_{eff}) increases with time. A parallel increase in Filgrastim clearance of identical magnitude is assumed to occur. Therefore, the time-dependent functions for cell influx and Filgrastim

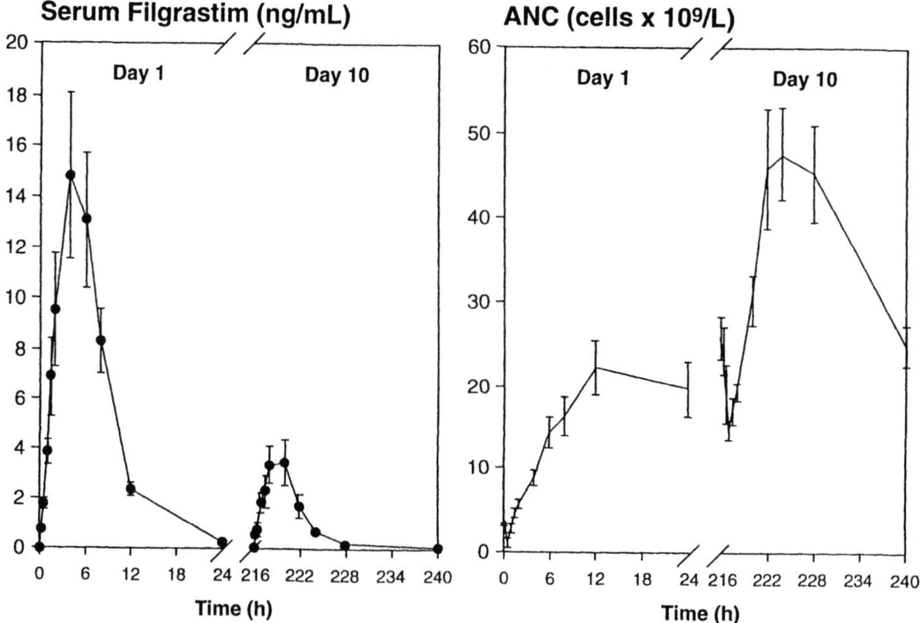

Figure 7. Serum Filgrastim levels and absolute neutrophil counts (ANC) in normal volunteers after the first and the tenth daily subcutaneous dose (300 μg/day).

clearance can be modeled as asymptotic functions that reflect expansion of the storage pool to a new steady-state multiple (m) of baseline cell mass:

$$t \leq \tau_{mit}:$$

$$k_{eff}(t) = k_{eff}^0$$

$$CL(t) = CL^0$$

$$t > \tau_{mit}:$$

$$k_{eff}(t) = m * k_{eff}^0 + (k_{eff}^0 - m * k_{eff}^0)e^{-k_{mit}(t-\tau_{mit})}$$

$$CL(t) = m * CL^0 + (CL^0 - m * CL^0)e^{-k_{mit}(t-\tau_{mit})}$$

Onset of mitotic expansion is assumed to begin at time τ_{mit} following the first dose of Filgrastim. A compartmental representation of the linked pharmacokinetic-pharmacodynamic model is presented in **Figure 8**. The biomathematical model can be fit to the pharmacokinetic and ANC data by nonlinear regression. **Figure 9** illustrates simultaneous modeling of serum Filgrastim concentrations and absolute neutrophil counts. The model accurately describes induction of receptor-mediated Filgrastim clearance and accession of ANC to steady-state levels.

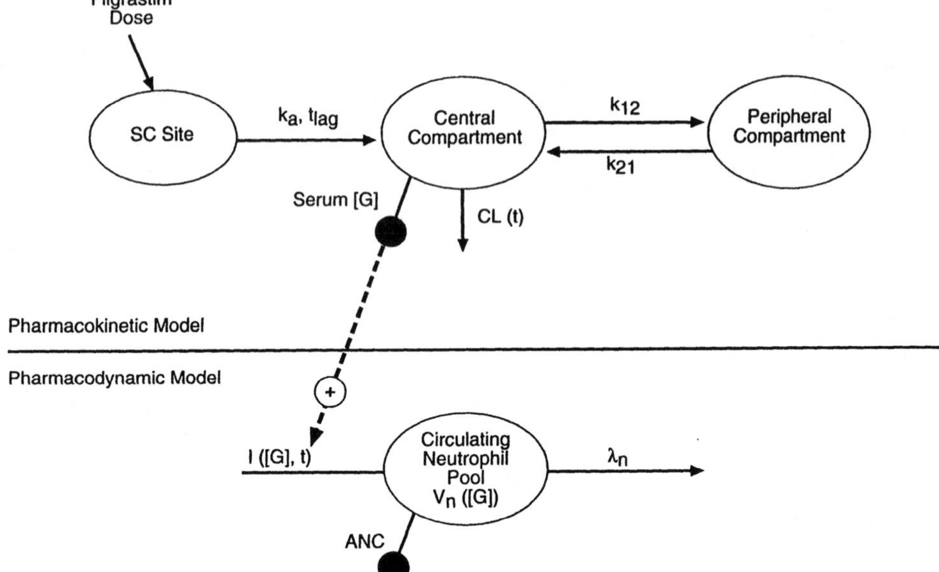

Figure 8.   The pharmacokinetic and pharmacodynamic models of Filgrastim are linked through serum levels of G-CSF.

IV.   CONCLUSIONS

A consistent and predictable pharmacological response to Filgrastim has been shown in preclinical and clinical studies. Filgrastim produces rapid, specific, and concentration-dependent increases in circulating neutrophils regardless of the route of administration. The mechanisms of action are mobilization of neutrophils from marrow, decreased maturation time of post-mitotic precursor cells, and mitotic expansion of the myeloblast, pro-myelocyte, and myelocyte pools in bone marrow. Neutrophils formed after administration of Filgrastim have normal or enhanced function. Filgrastim also mobilizes PBPC into peripheral circulation, which facilitates peripheral blood progenitor cell collection by apheresis.

Because of the unique pharmacology and low adverse event profile, Filgrastim has been proven clinically to promote hematopoietic recovery in conditions of acute and chronic neutropenia. In the United States, at the time of this writing, Filgrastim is licensed for amelioration of chemotherapy-induced neutropenia, severe chronic neutropenia, bone marrow transplan-tation, and mobilization of PBPC. In other parts of the world, Filgrastim is licensed for treatment of aplastic anemia, myelodysplastic syndromes,

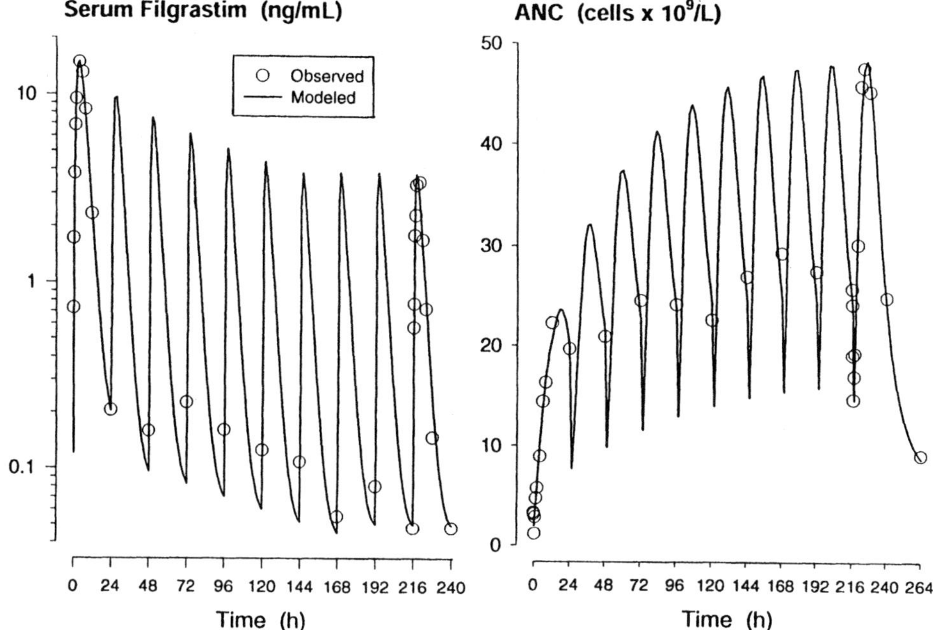

Figure 9. Simultaneous pharmacokinetic-pharmacodynamic modeling of mean absolute neutrophil count (ANC) response in normal volunteers receiving subcutaneous (SC) Filgrastim (300 μg/day) for 10 days.

acute leukemia, and AIDS. Some of these areas are discussed in other chapters in this volume.

ACKNOWLEDGMENTS

The authors gratefully acknowledge the contributions of Jan C. C. Borleffs, Keith M. Borkett, and Maria K. Small in the conduction of the normal volunteer study.

REFERENCES

1. Metcalf, D., and Nicola, N. A. (1983). Proliferation effects of purified granulocyte colony-stimulating factor (G-CSF) on normal mouse hematopoietic cells. *J Cell Physiol 116:*198–206.
2. Nicola, N. A. (1987). Why do hemopoietic growth factor receptors interact with each other? *Immunol Today 8:*134–140.

3. Sheridan, W. P., Wolf, M., Lusk, J., et al (1989). Granulocyte colony-stimulating factor hastens granulocyte recovery after high-dose chemotherapy and autologous bone marrow transplantations in Hodgkin's disease. *Lancet 2*:891–894.

4. Chao, N. J., Schriber, J. B., Grimes, J., et al (1993). Granulocyte colony-stimulating factor "mobilized" peripheral blood progenitor cells accelerate granulocyte and platelet recovery after high-dose chemotherapy. *Blood 81*:2031–2035.

5. Lord, B. I., Bronchud, M. H., Owens, S., et al (1989). The kinetics of human granulopoiesis following treatment with granulocyte colony-stimulating factor in vivo. *Proc Natl Acad Sci USA 86*:9499–9503.

6. Price, T. H., Chatta, G. S., and Dale, D. C. (1996). Effect of recombinant granulocyte colony-stimulating factor on neutrophil kinetics in normal young and elderly humans. *Blood 88*:335–340.

7. Gabrilove, J. L., Jakubowski, A., Fain, K., et al (1988). Phase I study of granulocyte colony-stimulating factor in patients with transitional cell carcinoma of the urothelium. *J Clin Invest 82*:1454–1461.

8. Gabrilove, J. L., Jakubowski, A., Scher, H., et al (1988). Effect of granulocyte colony-stimulating factor on neutropenia and associated morbidity due to chemotherapy for transitional-cell carcinoma of the urothelium. *N Engl J Med 318*:1414–1422.

9. Chatta, G. S., Price, T. H., Allen, R. C., and Dale, D. C. (1994). Effects of in vivo recombinant methionyl human granulocyte colony-stimulating factor on the neutrophil response and peripheral blood colony-forming cells in healthy young and elderly volunteers. *Blood 84*:2923–2929.

10. Tanaka, H., and Kaneko, T. (1992). Pharmacokinetic and pharmacodynamic comparisons between human granulocyte colony-stimulating factor purified from human bladder carcinoma cell line 5637 culture medium and recombinant human granulocyte colony-stimulating factor produced in *Escherichia coli. J Pharmacol Exp Ther 262*:439–444.

11. Tanaka, H., and Tokiwa, T. (1990). Pharmacokinetics of recombinant human granulocyte colony-stimulating factor studied in the rat by a sandwich enzyme-linked immunosorbent assay. *J Pharmacol Exp Ther 255*:724–729.

12. Tanaka, H., and Kaneko, T. (1991). Pharmacokinetics of recombinant human granulocyte colony-stimulating factor in the rat. Single and multiple dosing studies. *Drug Metab Dispos 19*:200–204.

13. Tanaka, H., and Tokiwa, T. (1990). Influence of renal and hepatic failure on the pharmacokinetics of recombinant human granulocyte colony-stimulating factor (KRN8601) in the rat. *Cancer Res 50*:6615–6619.

14. Stevens, P., Shatzer, E. M., Hanson, E. S., and Allen, R. C. (1991). Pharmacodynamics of recombinant human G-CSF with respect to an increase of neutrophil oxidative metabolism. *J Leucocyte Biol 2*:40.

15. Morstyn, G., Campbell, L., Souza, L. M., et al (1988). Effect of granulocyte colony stimulating factor on neutropenia induced by cytotoxic chemotherapy. *Lancet 1*:667–672.

16. Morstyn, G., Campbell, L., Dührsen, U., et al (1988). Clinical studies with granulocyte colony stimulating factor (G-CSF) in patients receiving cytotoxic chemotherapy. *Behring Inst Mitt 83*:234–239.

17. Bronchud, M. H., Potter, M. R., Morgenstern, G., et al (1988). In vitro and in vivo analysis of the effects of recombinant human granulocyte colony-stimulating factor in patients. *Br J Cancer 58:*64–69.
18. Layton, J. E., Hockman, H., Sheridan, W. P., and Morstyn, G. (1989). Evidence for a novel in vivo control mechanism of granulopoiesis: mature cell-related control of a regulatory growth factor. *Blood 74:*1303–1307.
19. Kearns, C. M., Wang, W. C., Stute, N., Ihle, J. N., and Evans, W. E. (1993). Disposition of recombinant human granulocyte colony-stimulating factor in children with severe chronic neutropenia. *Ped Pharmacol Thera 123:*471–479.
20. Morstyn, G., Campbell, L., Lieschke, G., et al (1989). Treatment of chemotherapy-induced neutropenia by subcutaneously administered granulocyte colony-stimulating factor with optimization of dose and duration of therapy. *J Clin Oncol 7:*1554–1562.
21. Lindeman, A., Herrmann, F., Oster, W., et al (1989). Hematologic effects of recombinant human granulocyte colony-stimulating factor in patients with malignancy. *Blood 74:*2644–2651.
22. Bronchud, M. H., Howell, A., Crowther, D., et al (1989). Phase I/II study of recombinant human granulocyte colony-stimulating factor to increase the intensity of treatment with doxorubicin in patients with advanced breast and ovarian cancer. *Br J Cancer 60:*121–128.
23. Cohen, A. M., Zsebo, K. M., Inoue, H., et al (1987). In vivo stimulation of granulopoiesis by recombinant human granulocyte colony-stimulating factor. *Proc Natl Acad Sci USA 84:*2484–2488.
24. Columbo, M. P., Ferrari, G., Stoppacciaro, A., et al (1991). Granulocyte colony-stimulating factor gene transfer suppresses tumorigenicity of a murine adenocarcinoma in vivo. *J Exp Med 173:*889–897.
25. Peters, W. P., Kutzberg, J., Atwater, S., et al. (1989). Comparative effects of rHuG-CSF and rHuGM-CSF on hematopoietic reconstitution and granulocyte function following high dose chemotherapy and autologous bone marrow transplantation (ABMT). *Proc Am Soc Clin Oncol 8:*181a.
26. Stute, N., Santana, V. M., Rodman, J. H., Schell, M. J., Ihle, J. N., and Evans, W. E. (1992). Pharmacokinetics of subcutaneous recombinant human granulocyte colony-stimulating factor in children. *Blood 79:*2849–2854.

4

Use of Filgrastim (r-metHuG-CSF) to Enhance Fc Receptor Function

Thomas Valerius and Bernhard Stockmeyer
University of Erlangen–Nuernberg, Erlangen, Germany

J. G. J. van de Winkel
*University Hospital Utrecht and Medarex Europe, Utrecht,
The Netherlands*

I. INTRODUCTION

The spectrum of indications for the clinical use of Filgrastim has become broader over the last years and is reviewed in various chapters of this book. Receptors for the constant part of immunoglobulins (Fc receptors) link the antigen-specific humoral immune system, constituted by antibodies, with the cellular part of the defense system. After a short overview on the biology of these receptors, we will discuss the clinical potential of using Filgrastim in combination with therapeutic antibodies. The rationale for this combination is that Filgrastim dramatically increases the numbers of Fc receptor–bearing effector cells, and, at the same time, enhances the Fc receptor expression and function of these cells. Clinical efforts currently are concentrated on several infectious diseases and on antibody-based therapies in oncology.

II. STRUCTURE AND FUNCTION OF Fc RECEPTORS

On phagocytic cells, Fc receptors (1–3) mediate a plethora of important functions such as phagocytosis, endocytosis, induction of the respiratory burst, and antibody-dependent cellular cytotoxicity (ADCC), all of which are relevant for directly destroying and clearing pathogens. In addition, Fc receptors serve relevant modulatory functions in the immune response such as regulating antibody production, release of inflammatory mediators, and cytokines. Most Fc receptors belong to the family of multichain immune recognition receptors (MIRR). These receptors consist of hetero-oligomeric complexes of distinct ligand-binding α-chains, which associate with shared signaling components, and in the case of Fc receptors, named γ-chain, β-chain, or ζ-chain (**Figure 1**). The γ- and β-chains were originally identified as part of the high-affinity IgE receptor (FcεRI) complex, whereas the ζ-chains were first described as one of the signaling components of the T-cell receptor (TCR). Depending on their specificity for the heavy chains

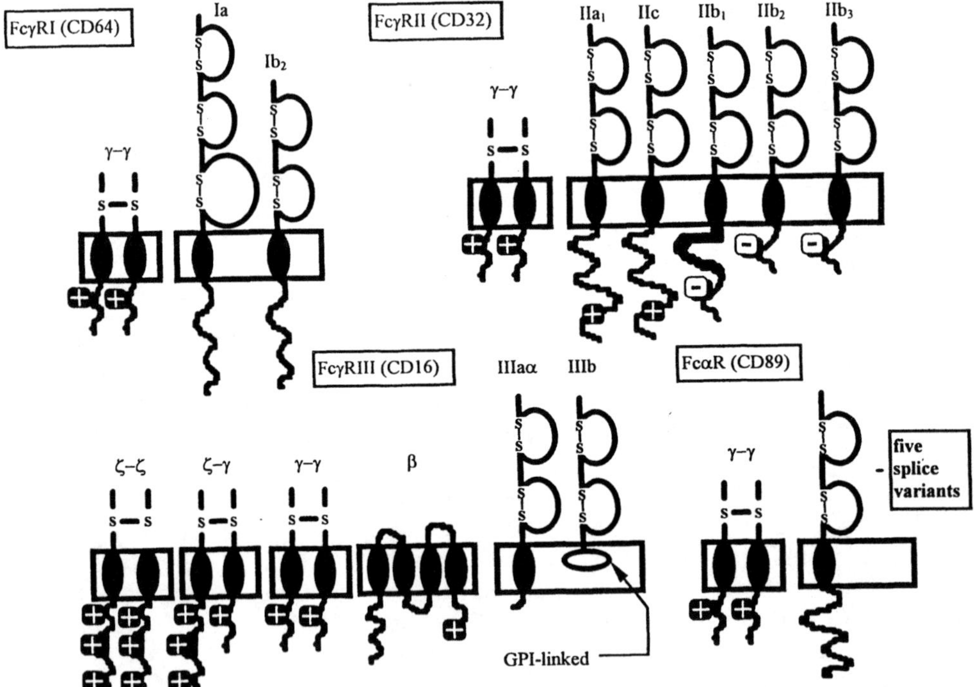

Figure 1. Schematic representation of the membrane-bound leukocyte IgG and IgA Fc receptor α-chains, and their associated signaling chains—designated β-, γ- or ζ-chain. All receptors belong to the immunoglobulin superfamily, with their extracellular regions composed of disulfide-bonded domains. Except for FcγRIIIb, which is a glycosyl-phosphatidyl-inositol (GPI)–linked protein, all other receptors are transmembrane molecules. A + or − denotes a stimulatory ITAM− or an inhibitory ITIM− motif within in the intracytoplasmic domains, respectively.

of IgA, IgE, or IgG, Fc receptors are grouped as Fcα, Fcε, or Fcγ receptors, respectively (**Table 1**). The existence of specific receptors for IgM or IgD has been postulated from functional assays, but they have not been defined in molecular terms. Another type of receptor, termed FcRB (in tribute to Professor F.W.R. Brambell, who first suggested the existence of this type of receptor) differs structurally from other Fc receptors, because it is composed of an HLA class I–related α-chain, which associates with β_2-microglobulin (4). This receptor is involved in both maternal-fetal transfer of IgG and protection of IgG from catabolism and, thereby, regulates the serum half-life of therapeutic IgG antibodies. Today, a total of 12 different Fc receptor α-chain genes have been identified (FcαRI,

Table 1. General Characteristics of Human Leukocyte Fcα and Fcγ Receptors.

		FcαRI (CD89)	FcγRI (CD64)	FcγRII (CD32)	FcγRIII (CD16)
Molecular weight		55–100 kD	72 kD	40 kD	50–80 kD
Chromosomal location		19q13.4	1q21.1	1q23–24	1q23–24
Genes		FcαRI	FcγRIA, -IB, -IC	FcγRIIA, -IIB, -IIC	FcγRIIIA, -IIIB
Isoforms		5 splice variants	Ia, sIb1, 1b2, sIc	IIa1, sIIa2, IIb1, IIb2, IIb3, IIc	IIIa, IIIb
Polymorphisms		unknown	unknown	FcγRIIa-H131	FcγRIIIa[high]
				RcγRIIa-R131	FcγRIIIa[low]
					FcγRIIIa-R48
					FcγRIIIa-L48
					FcγRIIIa-H48
					FcγRIIIa-F158
					FcγRIIIa-V158
					FcγRIIIb-NA1
					FcγRIIIb-NA2
Affinity		$5 \times 10^7 \ \text{M}^{-1}$	$10^8 - 10^9 \ \text{M}^{-1}$	$< 10^7 \ \text{M}^{-1}$	IIIa: $3-5 \times 10^7 \ \text{M}^{-1}$
					IIIb: $< 10^7 \ \text{M}^{-1}$
Isotype specificity	human Ig	IgA1 = IgA2	3 > 1 > 4 ≫ 2	IIa-R131: 3 > 1 ≫ 2 = 4	1 = 3 ≫ 2 = 4
				IIa-H131: 3 > 1 = 2 ≫ 4	
	mouse Ig	no binding	2a = 3 ≫ 1 = 2b	IIa-R131: 2a = 2b = 1	3 > 2a > 2b ≫ 1
				IIa-H131: 2a = 2b ≫ 1	

FcεRI, FcεRII, FcγRI [A, B, and C], FcγRII [A, B, and C], FcγRIII [A and B], and FcRB). With exception of the low-affinity IgE receptor (FcεRII, CD23), a member of the lectin family, all other Fc receptors belong to the immunoglobulin superfamily of proteins. The extracellular parts of FcγRIa and FcRB bear three Ig-like domains, whereas all other Ig receptors have two. Based on their affinity for ligand, classes of high-, intermediate-, and low-affinity receptors are distinguishd ($K_a \sim 10^8 - 10^9$, 3 to 5 $\times$ 10^7, and $<10^7$ M^{-1}, respectively). Low-affinity receptors can only interact with complexed or aggregated immunoglobulins, whereas high-, and to some extent intermediate-, affinity receptors can induce passive sensitization of effector cells. As effects of granulocyte colony-stimulating factor (G-CSF) have only been documented for select Fcα and Fcγ receptors, we will focus on these two receptor classes.

A. Fcγ Receptors

For IgG, three classes of leukocyte receptors (FcγRI, FcγRII, and FcγRIII, clustered as CD64, CD32, and CD16, respectively) are distinguished based on genetic, biochemical, and functional characteristics. These three classes are composed of 12 receptor isoforms, several of which result from alternative splice events. Soluble receptors have been described for all three classes and are generated either by alternate mRNA splicing (sFcγRIIa2), by proteolytic cleavage of membrane-bound forms (sFcγRIIIa and IIIb), or by the translation of stop codons within the corresponding genes (sFcγRIb1 and Ic). In comparison with the membrane-bound isoforms, the biological role of soluble receptors is less well established (5).

FcγRI, the human high-affinity IgG receptor, has a molecular weight of 72 kD; is constitutively expressed on monocytes/macrophages, myeloid progenitors, and dendritic cells; and expression can be induced on neutrophils by interferon (IFN)-γ or G-CSF. From cDNA analysis, three highly homologous genes (IA, IB, and IC) are predicted to encode two transmembrane (FcγRIa and Ib2) and two soluble receptor isoforms (FcγRIb1 and Ic). Only FcγRIa contains the third Ig-like extracellular domain, considered crucial for high-affinity ligand binding. FcγRI binds monomeric human IgG1, IgG3, and IgG4 isotypes (in decreasing order) and cross-reacts with murine IgG2a and IgG3 and with rat IgG2b antibodies.

The family of 40 kD FcγRII (CD32) molecules has the broadest cell distribution, found on most leukocytes, but also on Langerhans cells, tissue macrophages, and platelets, and on non-hematopoietic cells such as different populations of endothelial cells. Again, three genes (IIA, IIB, and IIC) were identified in humans, with IIC resulting from an unequal cross-over between IIA and IIB. The FcγRII genes encode six receptor isoforms (FcγRIIa1, a2, b1, b2, b3, and c). All FcγRII isoforms have low affinity for IgG and contain homologous extracellular regions. Intracellular tails of these molecules, however, vary resulting in fundamentally different signaling roles. Myeloid cells predominantly express FcγRIIa, but mRNA for FcγRIIb isoforms also has been found in these cells. On mature B cells, on the other hand, FcγRIIb is the predominant FcγR, but immature or malignant B cells also express message for the myeloid FcγRIIa isoform. All FcγRII isoforms preferentially bind human IgG1 and IgG3 and the murine IgG1 and IgG2b isotypes. FcγRIIa represents the sole Fc receptor that can bind human IgG2. However, this interaction, as well as binding of murine IgG1 or rat IgG2b, is profoundly affected by a genetic polymorphism of this receptor at amino acid position 131 (either arginine or histidine). Notably, this polymorphism seems of clinical relevance in situations where IgG2 antibodies are generated (eg, against encapsulated bacteria) (6) or when mIgG1 antibodies are used therapeutically. Expression levels of FcγRII remain largely unaffected by cytokines. However, granulocyte-macrophage colony-stimulating factor (GM-CSF) has been demonstrated to induce an activated state of FcγRII affecting ligand-binding characteristics.

The molecular weight of FcγRIII molecules ranges from 50 to 80 kD. Two genes, IIIA and IIIB, specify two types of receptors, FcγRIIIa and IIIb, respectively. FcγRIIIa is expressed by natural killer (NK) cells, macrophages, subpopulations of T cells and freshly isolated blood monocytes, immature thymocytes, and placental trophoblasts. Three distinct polymorphisms of FcγRIIIa were described, one based on expression levels on NK cells, which were either high (approximately 14,000/cell) or low (approximately 5500/cell). A second one is a triallelic polymorphism determined by a single nucleotide difference with either leucine, histidine, or arginine at position 48 of FcγRIIIa. Either phenylalanine or valine at amino acid position 158 in the membrane-proximal, IgG-binding domain determines a third biallelic polymorphism of this receptor shown to affect binding characteristics for IgG. Clinical consequences of FcγRIIIa polymorphisms are suggested by case reports, but need to be confirmed in larger patient populations. FcγRIIIb, exclusively found on neutrophils, differs from all other Fc receptors by its glycosyl-phosphatidyl-inositol (GPI) linkage to the outer leaflet of the cell membrane, and, therefore, shows decreased expression on polymorphonuclear leukocytes (PMN) from patients with paroxysmal nocturnal hemoglobinuria. FcγRIIIb bears a polymorphism, called neutrophil antigen (NA) 1 and NA2, with neutrophils from homozygous NA2 donors displaying lower phagocytic activity in different in vitro assays. The isotype specificities of both FcγRIII are the same, with preferential binding of human IgG1 and IgG3, as well as murine IgG3 and IgG2a isotypes. However, the affinity of FcγRIIIb for IgG is lower than that of FcγRIIIa ($K_a < 10^7$ versus approximately 3×10^7 M^{-1}). FcγRIII expression on monocytes/macrophages is up-regulated by transforming growth factor (TGF)-β, and down-regulated on PMN by a variety of stimuli, some of which also lead to apoptosis of PMN.

B. Fcα Receptor

The myeloid receptor for IgA (FcαRI; CD89), expressed on monocytes/ macrophages, PMN, and eosinophils, is a transmembrane glycoprotein of 55 to 100 kD, which binds both IgA1 and IgA2 isotypes with similar affinity (approximately 5×10^7 M^{-1}) (7,8). To date, one human FcαR gene has been identified, mapped to chromosome 19, in contrast to the FcγR genes that are located on chromosome 1. Five alternatively spliced isoforms of this gene were characterized from human PMN, all of which may represent transmembrane proteins. Expression levels of FcαRI increase upon exposure of cells to chemokines or pro-inflammatory agents such as N-formyl-methionyl-leucyl-phenylalanine (FMLP), C5a, TNF-α, or interleukin (IL)-8. Interestingly, GM-CSF up-regulates FcαRI expression on monocytes, but not on PMN, suggesting fundamental differences in the regulation of FcαRI expression between these cell types. Both, GM-CSF and G-CSF were shown to induce increased IgA-binding to PMN (9), which is not understood in molecular terms.

C. Signaling by Fc Receptors

During the past few years, rapid progress has been made in elucidating signaling mechanisms of Fc receptors (10,11). Except for FcγRII, other Fc receptor-family members do not contain recognized signaling motifs in their cytoplasmic tails and require association with signaling components to mediate their functions. Fc receptor γ-, ζ-, and β-chains contain well-characterized immunoreceptor tyrosine-based activation motifs (ITAM) (12) in their cytoplasmic regions. Similar ITAM have been identified in the δ- and ε-chains of the TCR, in the cytoplasmic tails of Ig-α and Ig-β from the B-cell receptor complex, and in selected activator NK cell receptors. In Fc receptor-expressing cells, the most widely distributed of these molecules represents the γ-chain that can physically associate with FcγRIa, FcγRIIIa, and FcαRI. Interestingly, FcγRIIa already bears a similar, non-canonical ITAM in its tail that can trigger some, but not all, FcγRIIa-mediated functions. Accordingly, FcγRIIa also requires association with the γ-chain for full functional activity. Cross-linking of FcR results in tyrosine-phosphorylation of ITAM by Src-family protein tyrosine kinases, followed by association of ITAM with a Syk-family protein tyrosine kinase. This leads to activation of various downstream enzymes such as protein kinase C, phospholipase Cγ1, and MAP kinases, which finally results in complex cellular functions like phagocytosis or ADCC. In contrast to FcγRIIa, FcγRIIb molecules contain an inhibitory motif named ITIM (immunoreceptor tyrosine-based inhibition motif), which is furthermore found in the KIR (killer-cell inhibitory receptor) types of NK cell receptors, in the CD22 molecule of B cells, and in CTLA-4 in activated T cells. Phosphorylation of ITIM leads to recruitment and activation of tyrosine phosphatases SHP-1 and SHP-2, and inositol phosphatase SHIP, thereby attenuating activator signals via ITAM (13). FcγRIIIa can interact with γ- or ζ-chain homodimers and heterodimers in NK cells, and with the β-chain in mast cells. Expression of these signaling components of FcR has not been analyzed in PMN after exposure to Filgrastim, but functional data suggest a simultaneous up-regulation of signaling chain and FcR expression, as has been shown for FcγRIa and γ-chain after IFN-γ exposure. In contrast to other FcR, the GPI-linked FcγRIIIb on PMN does not associate with the "classical" signaling chains, but may require association with other transmembrane proteins, such as CR3 (CD11b/CD18), for signaling and function.

III. INFLUENCE OF FILGRASTIM ON Fc RECEPTOR EXPRESSION AND FUNCTION

In addition to increasing neutrophil production from bone marrow precursor cells, in vitro incubation of PMN with Filgrastim was shown to stimulate several functions of mature neutrophils, most of which also were enhanced when PMN collected from patients during Filgrastim treatment were analyzed ex vivo (14,15) (**Table 2**). As Fc receptors were known to be involved in some of these functions, expression of these receptors was investigated. Healthy donor neutrophils and neutrophils incubated with Filgrastim in vitro for up to 24 hours expressed both the low-affinity FcγRIIa and FcγRIIIb receptors, and only very low levels of the high-affinity FcγRIa (16). Remarkably, however, PMN collected during Filgrastim therapy also stained positive with antibodies against FcγRI (17–19). Northern blot analysis showed mRNA for FcγRIa in Filgrastim-primed PMN, and by reverse transcription polymerase chain reaction (RT-PCR) mRNA for both membrane-bound isoforms, FcγRIa and FcγRIb2, was demonstrated (20). Kinetic studies after a single injection of 300 μg Filgrastim to healthy

Table 2. Polymorphonuclear Leukocyte Functions Triggered by Fcα and Fcγ Receptors.

	FcαRI (CD89)	FcγRI (CD64)	FcγRII (CD32)	FcγRIII (CD16)
Ca^{2+} influx	+	+	+	+
Phagocytosis	+ +	+ +*	+ +	+
Superoxide production	+	+ +	+ +	+
Degranulation	+ +	+	+ +	+
ADCC	+ +	+ +*	+ +	−

* Indicates documented enhancement of this function by G-CSF
ADCC = antibody-dependent cellular cytotoxicity
+ = present
− = absent

volunteers revealed that neutrophil counts increased significantly after 4 hours, peaked at 12 hours, and returned to baseline after 3 days. This increase in neutrophil numbers was paralleled by expression of FcγRI on PMN beginning at 8 hours, and peaking between 24 and 48 hours after Filgrastim administration (21). Studies of the mechanism of FcγRI induction by G-CSF showed that Filgrastim required prolonged in vitro incubation to induce FcγRI expression on mature PMN (22) or on initially CD64^{-}/ CD34^{+} bone marrow cells (21). On the other hand, there is a pool of FcγRI-positive neutrophil precursors in the bone marrow (23), and G-CSF is known to reduce the maturation time of bone marrow cells (24). Thus, the presence of FcγRI-positive PMN soon after administration of Filgrastim is likely caused by early release of FcγRI-positive PMN from this maturation pool within the bone marrow.

Expression of FcγRI on PMN also has been described in patients with pharyngitis (25), and during the recovery phase after intensive chemotherapy (17), situations in which endogenous G-CSF levels are increased (26–28). Interestingly, after removal of a G-CSF–producing tumor initially increased FcγRI expression on PMN decreased to baseline levels (29). The molecular mechanism of FcγRI induction on myeloid cells by G-CSF is largely unknown, and elucidation is complicated by the fact that no myeloid cell lines have been observed to up-regulate FcγRI expression after incubation with Filgrastim. However in human neutrophils, Filgrastim induces binding of STAT1 and STAT3 to the IFN-γ response region (GRR) within the promoter of the FcγRI gene (30), suggesting G-CSF and IFN-γ may use similar mechanisms for regulating FcγRI expression.

In addition to its effects on FcγRI, administration of Filgrastim in vivo is associated with reduced expression of FcγRIII on PMN (17,21). In contrast to monocytes/macrophages and NK cells, which express a transmembrane form of FcγRIII (FcγRIIIa), FcγRIIIb on PMN is GPI-linked and is shed upon PMN activation. However, lower levels of FcγRIII also are found on more immature PMN in the bone marrow (23). The kinetics of decreased cellular expression of FcγRIII paralleled those for up-regulation of FcγRI, but peak levels of soluble FcγRIII were observed as late as 6 days after a single injection of Filgrastim (21) and probably reflect the expanded pool of PMN in the bone marrow rather than active shedding. Filgrastim application also leads to small increases in monocyte/macrophage numbers in the peripheral blood that show enhanced expression of FcγRI and FcγRIII (19). Concerning FcγRII expression on PMN, small effects by exogenous G-CSF were described in one report (19), but not in another (21).

A functional role for in vivo–induced FcγRI on PMN during Filgrastim therapy was suggested by demonstrating cytophilic binding of human IgG on these cells (20). In vitro–induced FcγRI on Filgrastim-cultured CD34$^+$ bone marrow cells was shown to induce rosetting of these cells with mIgG2a-coated or anti-rhesus D-coated erythrocytes that could be blocked by antibodies to FcγRI (21). Not only binding of antibodies and opsonized particles, but also more complex IgG-mediated functions of PMN such as ADCC were enhanced by Filgrastim therapy. Thus, PMN were shown to mediate tumor-cell killing with target antibodies of mIgG1, mIgG2a, mIgG2b, mIgG3, or hIgG1 isotypes. Involvement of individual FcγR in ADCC, as indicated by blocking analyses with antibodies to select FcγR, was found to strongly depend on the isotype of targeting antibodies. Thus, the contribution of FcγRI in cytotoxicity was highest with sensitizing antibodies of mIgG3, mIgG2a, or hIgG1 isotypes. Correspondingly, comparison of PMN stimulated by Filgrastim and those from healthy donors showed the most dramatic enhancement of tumor cell killing in assays in which these three antibody isotypes were used. In contrast, when FcγRII was the predominant receptor for ADCC, as in assays using mIgG1 or mIgG2b isotypes, differences between FcγRI-expressing PMN and healthy donor PMN were less obvious or non-existing. These data suggest that the selection of appropriate antibody isotypes is critical for recruitment of FcγRI as additional cytotoxic trigger molecule. Irrespective of the target antibody isotype, no inhibition of ADCC by blocking FcγRIII was obtained, suggesting this FcγR on PMN does not mediate killing of tumor targets. Interestingly, F(ab′) fragments of FcγRIII-directed monoclonal antibody 3G8 stimulated cytotoxicity by antibody isotypes which mediated killing via FcγRII.

Reverse ADCC assays, which selectively probe the cytolytic capacity of individual Fc receptors (31), confirmed the functional role of FcγRI as cytotoxic trigger molecule on G-CSF–primed PMN (32). When hybridoma cells selected for high membrane expression of antibodies against FcγRI, FcγRII, FcγRIII, or CR3 were used as target cells in reverse ADCC (33), only rHuG-CSF–primed PMN, but not healthy donor PMN, were able to lyse the anti-FcγRI hybridoma 32.2. Both PMN from healthy donors and from Filgrastim-treated patients showed similar levels of killing against the IV.3 hybridoma (anti-FcγRII), whereas the anti-CR3 (OKM-1) or the anti-FcγRIII (3G8) hybridomas were not killed by either PMN population. In reverse cytotoxicity assays with bispecific antibodies—directed against FcγRI and different tumor target antigens—again only Filgrastim-primed PMN but not healthy donor PMN, were effective in inducing target cell lysis (32,34,35). Whereas G-CSF has not been reported to affect the number of IgA receptors on cells, it has been suggested to induce a higher affinity for binding of IgA (9). Similar to FcγR, FcαRI can be targeted by bispecific antibodies, and FcαR-dependent killing of breast cancer cells was enhanced in blood from Filgrastim-treated patients, probably due to increased numbers of FcαR-expressing PMN. These reverse ADCC data may have therapeutic relevance, because bispecific antibodies offer a promising approach to improve effector cell recruitment in vivo (31).

In contrast to ADCC, the role of G-CSF and Fc receptors in enhancing phagocytosis and killing of pathogens is less well established. Depending on important details of the assays, such as the investigated pathogens, different results were reported. For example, Filgrastim-primed PMN did not demonstrate increased phagocytosis or fungicidal activity against *Candida blastoconidia.* However, in the same study enhanced phagocytosis and bactericidal activity against *Staphylococcus aureus* was observed (36). In most of these studies the role of Fc receptors in comparison to other phagocytic receptors, such as complement, mannose and scavenger receptors or lectin-binding proteins, were not analyzed. In addition, many studies used polyclonal immune serum, but not monoclonal antibodies for opsonization of pathogens.

IV. ANIMAL MODELS TO STUDY THE INFLUENCE OF G-CSF ON Fc RECEPTORS

Three classes of FcγR are recognized in mice; however, molecules of each class are encoded by single genes in contrast to the more complex system in humans. Analogous to human FcγR, the murine receptors mFcγRI, mFcγRII, and mFcγRIII genetically and functionally resemble the human FcγRIa, FcγRIIb, and FcγRIIIa isoforms, respectively. Comparison between murine and human FcγR is further complicated by the lack of specific antibodies for individual murine FcγR. Thus, only one antibody (2.4G2) recognizing both the low-affinity receptors mFcγRII and mFcγRIII has been described, and no antibody against the high-affinity mFcγRI is available. Therefore, information on FcγR expression in mice is less detailed than in humans, and it is unknown whether mice respond to G-CSF by upregulating FcγRI on neutrophils. Recently, we observed dramatically increased ADCC activity in blood from mice treated with Filgrastim. However, whether this increase is caused solely by elevated neutrophil numbers, or whether mFcγRI is involved as additional FcγR, as demonstrated in humans, needs to be determined. In cooperation with Dr. M. Glennie (Tenovus Research Laboratory, Southampton, UK), we currently are investigating whether Filgrastim can enhance efficacy of therapeutic antibodies in syngenic B-cell lymphoma models.

One possibility to circumvent some of these technical limitations are transgenic animals. Human FcγRI transgenic mice were generated (37) that showed myeloid-specific expression of the transgene under its endogenous promoter. As in humans, FcγRIa expression in mice was properly regulated by exposure to the cytokines IFN-γ, G-CSF, IL-4, and IL-10, and neutrophil FcγRI was increased during inflammation. Importantly, human FcγRI was shown to be functional in murine effector cells as evidenced by phagocytosis and ADCC experiments with FcγRI-directed bispecific antibodies. Moreover, number and functional activity of FcγRIa-expressing effector cells were increased by treating mice with Filgrastim. These animals are valuable for further analyzing the in vivo potential of Filgrastim in combination with FcγRI-directed bispecific antibodies in infectious disease and tumor models.

V. THERAPEUTIC APPLICATION OF FILGRASTIM TO ENHANCE RECEPTOR FUNCTION

By increasing neutrophil numbers, Filgrastim may potentially enhance immune functions via FcαR and all three classes of FcγR expressed on these effector cells. As discussed above, however, the most dramatic effect of Filgrastim on Fc receptors is upregulation of FcγRI expression on PMN, and also to some extent on monocytes. As FcγRIa binds monomeric IgG with high affinity, Filgrastim-primed PMN are loaded with cytophilic serum IgG (20). The specificity of these immunoglobulins has not been analyzed in detail, but probably reflects the spectrum of serum IgG antibodies. Therefore, approaches that combine Filgrastim with conventional monoclonal antibodies will face the problem that "therapeutic" antibodies compete with serum immunoglobulins for binding to FcγRI. To enhance the therapeutic efficacy of FcγRI, it seems important to selectively sensitize PMN with antibodies directed to clinically relevant antigens. The most practical solution to this problem appears to be bispecific antibodies (38), which combine target antigen reactivity with an FcγRI-directed specificity. These constructs bind to epitopes on FcγRI that are distinct from the ligand-binding region and, therefore, do not compete with serum IgG. Nevertheless, they can trigger all FcγRI-mediated functions, sometimes even more effectively than conventional antibodies. The use of Fc receptor–directed bispecific antibodies has recently been reviewed (39). Here, we will concentrate on the role of Filgrastim to enhance Fc receptor function in infectious diseases and on its potential to synergize with tumor-directed antibodies in oncology.

A. Infectious Diseases

Neutrophils are the predominant effector cells against most extracellular bacteria, pathogenic fungi such as *Candida* and *Aspergillus* species, as well as selected parasites, and Filgrastim was shown to be effective in the treatment and prevention of these infections in neutropenic patients (15). As animal data suggested that Filgrastim also may be protective in infectious diseases in non-neutropenic hosts, several clinical trials currently evaluate the potential of Filgrastim in these conditions (see other chapters in this book). The contribution of enhanced Fc receptor function to the clinical efficacy of Filgrastim, in comparison to other variables such as increased neutrophil numbers, and general stimulation of their functional state, is unknown, but is addressed in ongoing clinical trials. For example, patients undergoing surgery of the esophagus who received prophylactic Filgrastim showed a lower incidence of infections than did patients who did not receive Filgrastim. This reduced incidence of infection was accompanied by a documented enhancement of granulocyte function, including IgG-dependent phagocytosis (40). Several studies have shown reduced granulocyte function in diabetic patients (41), who are at increased risk for certain bacterial infections. However, these patients responded to Filgrastim treatment with upregulation of FcγRI expression and enhanced FcR-mediated phagocytosis (unpublished observation). Whether these in vitro findings translate into improved healing of diabetic foot ulcers is currently being tested in a phase 2 trial, in which patients are randomized to receive either standard care alone or standard care plus Filgrastim (see Chapter 27).

Neutrophil-mediated phagocytosis and killing of pathogens can be enhanced in the presence of pathogen-directed antibodies. Combining Filgrastim with antibodies could be a logical next step to improve its efficacy in the treatment of infectious diseases. Today, fungi cause significant morbidity and mortality in oncology and other immunosuppressed patients. The only established therapy for severe fungal infections in these patients is amphotericin B, which, however, bears the risk of considerable side effects. In vitro, Filgrastim was shown to stimulate PMN-mediated killing of different *Candida* and *Aspergillus* species, which was further enhanced in the presence of immune serum (42). A combination of Filgrastim and fungi-directed antibodies is being tested in vitro and in animal models, and preliminary results look encouraging.

Cooperative effects between Filgrastim and Fc receptor function were suggested in a rodent malaria model. Mice treated either with Filgrastim or with immune-serum alone before challenge with the parasites had either no or only a minor delay in parasitemia compared with nontreated control animals. However, when both treatment modalities were combined, the period until fatal parasitemia was significantly delayed (43).

Although neutrophils are nonprofessional effector cells in virus infections, it could be clinically interesting to engage them in the host response against particular viruses. For example, Filgrastim-activated PMN were shown to mediate cytotoxicity toward HIV-infected cells (44), and targeting HIV to FcγR on phagocytes via bispecific antibodies reduced infectivity of HIV to T cells (45). [FcγRI $\times$ gp41] bispecific antibodies (MDX-240) have been tested in patients with HIV infection and were well tolerated without signs of increased virus production. Filgrastim has been safely given to many patients with HIV infection. A combination of both could be promising, especially because PMN, in contrast to other potential effector cells, are not infected by HIV and, therefore, could be safely stimulated and expanded by Filgrastim.

B. Oncology

Promising results from recent clinical trials (46–49) have led to renewed interest in monoclonal antibodies as therapeutic reagents in oncology. Tumor-directed antibodies provide a range of different mechanisms for controlling malignancies, including induction of programmed cell death, blockade of tumor growth factors, activating complement, or recruiting cytolytic effector cells for ADCC. The contribution of these individual mechanisms to the therapeutic efficacy of antibodies in vivo has not been determined; however, ADCC is considered among the most potent of them. ADCC can be mediated by all Fc receptor–expressing cells with the cytolytic potential to kill tumor cells, a capacity that is well documented for monocytes/macrophages, NK cells, and neutrophilic and eosinophilic granulocytes. The in vivo role of different cell populations for antibody efficacy is still unclear. In vitro, however, neutrophils, which can kill a broad spectrum of tumor-derived cell lines (50), were shown to be an important effector cell population in HER-2/neu antibody-mediated ADCC against breast cancer cells, especially when blood from Filgrastim-treated patients was analyzed (35). A function for PMN in the immune surveillance against tumor growth in vivo is suggested by several animal studies. Administration of the cytokine to mice during challenge with hematogenous or non-hematogenous tumor cells significantly prolonged survival of growth factor–treated animals (51). Similarly, mice rejected IL-2– or G-CSF–transfected, but not control-transfected tumor cells (52,53). Granulocyte-depletion experiments and histological examinations demonstrated PMN to be involved in this antitumor response (54). In humans, ^{111}In-labeled PMN actively invade a significant percentage of malignancies (55), a process that may be enhanced after induction of tumor necrosis by chemo- or radiotherapy. Administration of Filgrastim to patients, however, has not been associated with direct antitumor activity. This may be explained by the fact that tumor cell lysis by neutrophils usually requires tumor-directed antibodies which are not present in sufficient amounts in most tumor patients and which would compete with all the serum antibodies for binding to FcR.

In contrast to NK cells and lymphocytes, neutrophils do not express granzymes and perforins, important enzymes involved in T-cell– and NK-cell–mediated tumor cell lysis. Studies on the mechanism of neutrophil-mediated tumor cell killing showed, instead, defensins and cathepsins to be the most relevant molecules (56). Interestingly, a detailed comparison of antibodies to different target antigens on malignant B cells revealed that the selection of target antigens is a critical issue for PMN-mediated tumor cell lysis (32). Thus, only antibodies to HLA class II, but not antibodies to other more B-cell–specific antigens such as CD19, CD20, CD21, CD22, CD37, or CD38 induced target cell lysis with PMN effector cells. This target antigen restriction was not observed with mononuclear effector cells and could not be overcome by using bispecific or human/mouse chimeric antibodies instead of conventional murine antibodies. The mechanism underlying PMN's target antigen restriction is not clearly understood, but may involve upregulation of adhesion molecules on B cells after cross-linking of HLA class II molecules (57). This may render target cells more susceptible to PMN-mediated lysis. In the presence of HLA class antibodies, however, Filgrastim-primed neutrophils constitute an important effector cell population for the treatment of malignant B cells (58,59) that can be significantly expanded by Filgrastim (32). The clinical potential of this approach has recently been reviewed (60), and clinical trials with [FcγRI × HLA class II] bispecific antibodies in combination with Filgrastim are expected.

Against solid tumor cell lines, effective target cell killing by PMN has been demonstrated not only with polyclonal immune sera, but also in

ADCC assays with therapeutically relevant monoclonal antibodies (50). Thus, neutrophils killed target cells with antibodies directed against GD2 (34,56), GD3 (61), 17-1A (61), EGF-R (20), or HER-2/neu (35). HER-2/neu and EGF-R are members of the EGF (epidermal growth factor)–receptor family of transmembrane protein kinases, which are over-expressed in a significant percentage of epithelial tumors such as breast, ovarian, prostate, head and neck, or renal cancer. Because of their tumor specificity, they constitute interesting target molecules for antibody-based immunotherapy. In the presence of HER-2/neu antibodies, neutrophils were the predominant effector cells of peripheral blood for lysis of breast cancer cells. As expected, use of Filgrastim induced significant increases in the cytotoxic activity that was correlated with an increase in neutrophil numbers. Comparing bispecific antibodies to HER-2/neu and either FcγRI, FcγRII, or FcγRIII, FcγRI became the most effective cytotoxic FcγR during Filgrastim therapy, whereas in blood from non-Filgrastim–treated control donors, FcγRIII was found more effective. These promising preclinical data formed the basis for trials testing an [HER-2/neu × FcγRI] bispecific antibody (MDX-210) in combination with Filgrastim (35).

In a phase 1 trial at our institutions (62), patients with metastatic breast cancer receive subcutaneous injections of Filgrastim for 8 consecutive days. On day 4 of Filgrastim treatment, cohorts of three patients were treated with increasing single doses of MDX-210 infused intravenously over 2 hours. Filgrastim pretreatment increased the number of FcγRI-expressing effector cells from approximately 5 × 10^9/L to more than 30 × 10^9/L, meaning that an additional 25 × 10^9/L effector cells (of patient's blood) are available at the time of bispecific antibody infusion, and even several-fold more during the following days. This therapy is well tolerated, and, interestingly, the combination of MDX-210 and Filgrastim seemed not to induce increased toxicity compared with MDX-210 alone (63). Side effects observed in most patients consisted of nausea, vomiting, fever, and chills, all of which could be controlled by symptomatic treatment. Shortly after the infusion, high levels of pro-inflammatory cytokines such as IL-6 and TNF-α were measured in the serum. Several patients had pain in tumor areas and inflammatory reactions in skin metastases. Most interestingly, soluble HER-2/neu levels on day 30 decreased to below pretreatment levels in the majority of patients, suggesting antitumor activity of this approach. Results from another phase 1 trial, testing multiple injections of MDX-210 in combination with Filgrastim, suggest that repetitive doses are even better tolerated than the first injection (64). Statements concerning clinical efficacy of this approach must await results of phase 2/3 trials with MDX-210 in combination with Filgrastim, which are expected at the end of 1997.

In summary, enhancement of Fc receptor function may become an important indication for Filgrastim, especially in clinical situations where both antibodies and neutrophils are relevant components of the host's immune response.

REFERENCES

1. Ravetch, J. V., and Kinet, J. P. (1991). Fc receptors. *Annu Rev Immunol* 9:457–492.

2. Van de Winkel, J. G. (1993). Human IgG Fc receptor heterogeneity: molecular aspects and clinical implications. *Immunol Today 14:*215–221.

3. Kimberly, R. P., Salmon, J. E., and Edberg, J. C. (1995). Receptors for immunoglobulin G. Molecular diversity and implications for disease. *Arthritis Rheum 38:*306–314.

4. Junghans, R. P. (1997). The Brambell receptor (FcRB): mediator of transmission of immunity and protection from catabolism for IgG. *Immunol Res 16(1):*29–57.

5. Fridman W. H., Teillaud J. L., Bouchard C., et al (1993). Soluble Fcγ receptors. *J Leukocyte Biol 54:*504–512.

6. Rascu, A., Repp, R., Westerdaal, N. A. C., Kalden, J. R., and van de Winkel, J. G. J. Clinical relevance of Fcγ receptor polmorphism. *Ann NY Acad Sci* New York: New York Academy of Sciences; 1997, pp 282–295.

7. Shen, L. (1992). Receptors for IgA on phagocytic cells. *Immunol Res 11:*273–282.

8. Morton, H. C., van Egmond, M., and Van de Winkel, J. G. (1996). Structure and function of human IgA Fc receptors (FcαR). *Crit Rev Immunol 16:* 423.

9. Weisbart, R. H., Kacena, A., Schuh, A., and Golde, D. W. (1988). GM-CSF induces human neutrophil IgA-mediated phagocytosis by an IgA Fc receptor activation mechanism. *Nature 332:*647–648.

10. Cambier, J. C. (1995). New nomenclature for the Reth motif (or ARH1/TAM/ARAM/YXXL). *Immunol Today 16:*110.

11. Indik, Z. K., Park, J. G., Hunter, S., Mantaring, M., and Schreiber, A. D. (1995). Molecular dissection of Fcγ receptor-mediated phagocytosis. *Immunol Lett 44:*133–138.

12. Cambier, J. C. (1995). Antigen and Fc receptor signalling. *J Immunol 155:*3281–3285.

13. Scharenberg, A. M., and Kinet, J. P. (1996). The emerging field of receptor-mediated inhibitory signaling: SHP or SHIP? *Cell 87:*961.

14. Platzer, E. (1989). Human hematopoietic growth factors. *Eur J Haematol 42:*1–5.

15. Lieschke, G. J., and Burgess, A. W. (1992). Granulocyte colony-stimulating factor and granulocyte-macrophage colony-stimulating factor (2). *N Engl J Med 327:*99–106.

16. Erbe, D. V., Collins, J. E., Shen, L., Graziano, R. F., and Fanger, M. W. (1990). The effect of cytokines on the expression and function of Fc receptors for IgG on human myeloid cells. *Mol Immunol 27*:57–67.

17. Repp, R., Valerius, T., Sendler, A., et al (1991). Neutrophils express the high affinity receptor for IgG (Fc-γ RI, CD64) after in vivo application of recombinant human granulocyte colony-stimulating factor. *Blood 78*:885–889.

18. Elsner, J., Roesler, J., Emmendörfer, A., Zeidler, C., Lohmann-Matthes, M. L., and Welte, K. (1992). Altered function and surface marker expression of neutrophils induced by rhG-CSF treatment of in severe congenital neutropenia. *Eur J Haematol 48*:10–19.

19. Ohsaka, A., Saionji, K., Kuwaki, T., Takeshima, T., and Igari, J. (1995). Granulocyte colony-stimulating factor administration modulates the surface expression of effector cell molecules on human monocytes. *Br J Haematol 89*:465–472.

20. Valerius, T., Repp, R., de Wit, T. P., et al (1993). Involvement of the high affinity receptor for IgG (FcγRI; CD64) in enhanced tumor cell cytotoxicity of neutrophils during granulocyte colony-stimulating factor therapy. *Blood 82*:931–939.

21. Kerst, J. M., Van de Winkel, J. G., Evans, A. H., et al (1993). Granulocyte colony-stimulating factor induces hFcγRI (CD64 antigen)-positive neutrophils via an effect on myeloid precursor cells. *Blood 81*:1457–1464.

22. Gericke, G. H., Ericson, S. G., Mills, L. E., Guyre, P. M., Pan, L., and Ely, P. (1995). Mature polymorphonuclear leukocytes express high-affinity receptors for IgG (Fc γ RI) after stimulation with granulocyte colony-stimulating factor (G-CSF). *J Leukocyte Biol 57*:455–461.

23. Ball, E. D., Mc Dermott, J., Griffin, J. D., Davey, F. R., Davis, R., and Bloomfield, C. D. (1989). Expression of the three myeloid cell-associated immunoglobulin G Fc receptors defined by murine monoclonal antibodies on normal bone marrow and acute leukemia cells. *Blood 73*:1951.

24. Lord, I. L., Bronchud, M. H., Owens, S., et al (1989). The kinetics of human granulopoiesis following treatment with granulocyte colony-stimulating factor in vivo. *Proc Natl Acad Sci USA 86*:9499.

25. Guyre, P. M., Campbell, A. S., Kniffin, W. D., and Fanger, M. W. (1990). Monocytes and polymorphonuclear neutrophils of patients with streptococcal pharyngitis express increased numbers of type I IgG Fc receptors. *J Clin Invest 86*:1892–1896.

26. Pauksen, K., Elfman, L., Ulfgren, A. K., and Venge, P. (1994). Serum levels of granulocyte-colony stimulating factor (G-CSF) in bacterial and viral infections, and in atypical pneumonia. *Br J Haematol 88*:256–260.

27. Kawakami, M., Tsutsumi, H., Kumakawa, T., et al (1990). Levels of serum granulocyte colony-stimulating factor in patients with infections. *Blood 76*:1962–1964.

28. Cairo, M. S., Suen, Y., Sender, L., et al (1992). Circulating granulocyte colony-stimulating factor (G-CSF) levels after allogeneic and autologous bone marrow transplantation: endogenous G-CSF production correlates with myeloid engraftment. *Blood 79*:1869–1872.

                                                              Valerius et al.

29. Ohsaka, A., Saionji, K., Endo, K., et al (1995). Alterations of effector cell molecule expression on neutrophils in granulocyte colony-stimulating factor-producing tumour. *Br J Haematol 91*:571–574.

30. Bovolenta, C., Gasperini, S., and Cassatella, M. A. (1996). Granulocyte colony-stimulating factor induces the binding of STAT1 and STAT3 to the IFNγ response region within the promoter of the FcγRI/CD64 gene in human neutrophils. *FEBS Lett 386*:239–242.

31. Fanger, M. W., Shen, L., Graziano, R. F., and Guyre, P. M. (1989). Cytotoxicity mediated by human Fc receptors for IgG. *Immunol Today 10*:92–99.

32. Elsässer, D., Valerius, T., Repp, R., et al (1996). HLA class II as potential target antigen on malignant B cells for therapy with bispecific antibodies in combination with granulocyte colony-stimulating factor. *Blood 87*:3803–3812.

33. Graziano, R. F., and Fanger, M. W. (1987). Human monocyte-mediated cytotoxicity: the use of Ig-bearing hybridomas as target cells to detect trigger molecules on the monocyte cell surface. *J Immunol 138*:945–950.

34. Michon, J., Moutel, S., Barbet, J., et al (1995). In vitro killing of neuroblastoma cells by neurophils derived from granulocyte colony-stimulating factor-treated cancer patients using an anti-disialoganglioside/anti-FcγRI bispecific antibody. *Blood 86*:1124–1130.

35. Stockmeyer, B., Valerius, T., Repp, R., et al (1997). Preclinical studies with FcγR bispecific antibodies and G-CSF primed neutrophils as effector cells against HER-2/neu overexpressing breast cancer. *Cancer Res 57*:696–701.

36. Roilides, E., Walsh, T. J., Pizzo, P. A., and Rubin, M. (1991). Granulocyte colony-stimulating factor enhances the phagocytic and bactericidal activity of normal and defective human neutrophils. *J Infect Dis 163*:579–583.

37. Heijnen, I. A., van Vugt, M. J., Fanger, N. A., et al (1996). Antigen targeting to myeloid-specific human FcγRI/CD64 triggers enhanced antibody responses in transgenic mice. *J Clin Invest 97*:331.

38. Fanger, M. W., Morganelli, P. M., and Guyre, P. M. (1992). Bispecific antibodies. *Crit Rev Immunol 12*:101–124.

39. Deo, Y. M., Graziano, R. F., Repp, R., and Van de Winkel, J. G. (1997) Clinical significance of IgG receptors and FcγR-directed immunotherapies. *Immunol Today 18*:127–135.

40. Hübel, K., Schäfer, H., Bohlen, H., et al (1996). Stimulation of neutrophil function in patients with esophageal cancer using recombinant granulocyte colony stimulating factor (rhG-CSF) reduces infectious complications after surgery. *Ann Hematol A53* (abstr).

41. Bagdade, J. D., Root, R. K., and Bulger, R. J. (1974). Impaired leukocyte function in patients with poorly controlled diabetes. *Diabetes 23*:9–15.

42. Roilides, E., Holmes, A., Blake, C., Pizzo, P. A., and Walsh, T. J. (1995). Effects of granulocyte colony-stimulating factor and interferon-γ on antifungal activity of human polymorphonuclear neutrophils against pseudohyphae of different medically important *Candida* species. *J Leukocyte Biol 57*:651–656.

43. Waki, S. (1994). Antibody-dependent neutrophil-mediated parasite killing in non-lethal rodent malaria. *Parasite Immunol 16*:587–591.

44. Baldwin, G. C., Fuller, N. D., Roberts, R. L., Ho, D. D., and Golde, D. W. (1989). Granulocyte- and granulocyte-macrophage colony-stimulating factors enhance neutrophil cytotoxicity toward HIV-infected cells. *Blood 74*:1673–1677.

45. Howell, A. L., Guyre, P. M., You, K., and Fanger, M. W. (1994). Targeting HIV-1 to FcγR on human phagocytes via bispecific antibodies reduces infectivity of HIV-1 to T cells. *J Leukocyte Biol 55*:385–391.

46. Baselga, J., Tripathy, D., Mendelsohn, J., et al (1996). Phase II study of weekly intravenous recombinant humanized anti-p185^{HER2} monoclonal antibody in patients with HER2/neu overexpressing metastatic advanced breast cancer. *J Clin Oncol 14*:737–744.

47. Davis, T. A., Maloney, D. G., Liles, T. M., Czerwinski, D., and Levy, R. (1996). Long-term remissions in patients treated with anti-idiotype monoclonal antibodies for non-Hodgkin's lymphoma. *Proc Am Soc Clin Oncol 15*:444 (abstr 1383).

48. McLaughlin, P., Grillo-López, A. J., Czuczman, M., et al (1996). Preliminary report on a phase III pivotal trial of the anti-CD20 antibody (MAB) IDEC-C2B8 in patients (PTS) with relapsed low-grade or follicular lymphoma (LG/F NHL). *Proc Am Soc Clin Oncol 15*:417 (abstr 1281).

49. Riethmüller, G., Holz, E., Schlimock, G., et al (1996). Monoclonal antibody (MAB) adjuvant therapy of Dukes C colorectal carcinoma: 7-year update of a prospective randomized trial. *Proc Am Soc Clin Oncol 15*:444 (abstr 1385).

50. Lichtenstein, A. (1990). Granulocytes as possible effectors of tumor immunity. In *Human Cancer Immunology.* H. F. Oettgen, ed. W. B. Saunders, Philadelphia, p. 731.

51. Matsumoto, Y., Saiki, I., Murata, J., Okuyama, H., Tamura, M., and Azuma, I. (1991). Recombinant human granulocyte colony-stimulating factor inhibits the metastasis of hematogenous and non-hematogenous tumors in mice. *Int J Cancer 49*:444–449.

52. Cavallo, F., Giovarelli, M., Gulino, A., et al (1992). Role of neutrophils and CD4$^+$ T lymphocytes in the primary and memory response to nonimmunogenic murine mammary adenocarcinoma made immunogenic by IL-2 gene. *J Immunol 149*:3627–3635.

53. Colombo, M. P., Ferrari, G., Stoppacciaro, A., et al (1991). Granulocyte colony-stimulating factor gene transfer suppresses tumorigenicity of a murine adenocarcinoma in vivo. *J Exp Med 173*:889–897.

54. Stoppacciaro, A., Melani, C., Parenza, M., et al (1993). Regression of an established tumor genetically modified to release granulocyte colony-stimulated factor requires granulocyte - T cell cooperation and T cell-produced interferon γ. *J Exp Med 178*:151–161.

55. Schmidt, K. G., Rasmussen, J. W., Wedebye, I. M., Frederiksen, P. B., and Pedersen, N. T. (1988). Accumulation of Indium-111-labeled granulocytes in malignant tumors. *J Nucl Med 29*:479–484.

56. Barker, E., and Reisfeld, R. A. (1993). A mechanism for neutrophil-mediated lysis of human neuroblastoma cells. *Cancer Res 53*:362–367.

57. Mourad, W., Geha, R. S., and Chatila, T. (1990). Engagement of major histocompatibility complex class II molecules induces sustained, lymphocyte

function-associated molecule 1-dependent cell adhesion. *J Exp Med 172:* 1513–1516.

58. Cemerlic, D., Dadey, B., Han, T., and Vaickus, L. (1991). Cytokine influence on killing of fresh chronic lymphocytic leukemia cells by human leukocytes. *Blood 77:*2707–2715.

59. Ottonello, L., Morone, P., Dapino, P., and Dallegri, F. (1996). Monoclonal Lym-1 antibody-dependent lysis of B-lymphoblastoid tumor targets by human complement and cytokinine-exposed mononuclear and neutrophilic polymorphonuclear leukocytes. *Blood 87:*5171–5178.

60. Valerius, T., Elsässer, D., Repp, R., Van de Winkel, J. G. J., Gramatzki, M., and Glennie, M. (1997). HLA class II antibodies recruit G-CSF activated neutrophils for treatment of B cell malignancies. *Leuk Lymph 26:*261–269.

61. Ragnhammar, P., Frödin, J. E., Trotta, P. P., and Mellstedt, H. (1994). Cytotoxicity of white blood cells activated by granulocyte-colony-stimulating factor, granulocyte/macrophage-colony-stimulating factor and macrophage-colony-stimulating factor against tumor cells in the presence of various monoclonal antibodies. *Cancer Immunol Immunother 39:*254–262.

62. Repp, R., Valerius, T., Wieland, G., et al (1995). G-CSF stimulated PMN in immunotherapy of breast cancer with a bispecific antibody to FcγRI and to HER-2/neu (MDX-210). *J Hematother 4:*415–421.

63. Valone, F. H., Kaufman, P. A., Guyre, P. M., et al (1995). Phase Ia/Ib trial of bispecific antibody MDX-210 (anti-HER-2/neu x anti-FcγRI) in patients with advanced breast or ovarian cancer that overexpresses the proto-oncogene HER-2/neu. *J Clin Oncol 13:*2281–2292.

64. Weber, J. S., Spears, L., Deo, Y., and Link, J. L. (1996). Phase I study of a HER-2/neu bispecific antibody with G-CSF in patients with metastatic breast cancer. *Proc Am Soc Clin Oncol 15:*354 (abstr 1041).

5
Causes, Clinical Consequences, and Treatment of Neutropenia

W. Conrad Liles and David C. Dale
University of Washington, Seattle, Washington

I. INTRODUCTION

Neutrophils play a critical role in protecting the body from infection by the microorganisms that colonize the oropharynx, gastrointestinal tract, and the skin and in the acute response to most infectious and inflammatory diseases. Whenever there are breaks in the surface barriers, it is the capacity to recruit neutrophils to the site rapidly that determines if an overt infection will occur. At the inflammatory focus, neutrophils ingest and kill bacteria, limiting their production of endotoxins and allowing tissue repair to begin.

Neutropenia is defined as a blood neutrophil count $<1.8 \times 10^9$/L, or two standard deviations below the normal mean for adults (1). For children, a level of 1.5×10^9/L is often used (2). Granulocytopenia (ie, reduced numbers of blood granulocytes [neutrophils, eosinophils, and basophils]) is a term often used synonymously. Agranulocytosis, a term which implies the complete absence of granulocytes, is sometimes used for very severe neutropenia.

Neutropenia is a well-recognized factor predisposing to infections. In general, susceptibility to infection increases as the neutrophil count decreases. Individuals with counts of 1.0 to 1.8×10^9/L have "mild" neutropenia and have relatively little susceptibility to infections. With "moderate" neutropenia (ie, counts of 0.5 to 1.0×10^9/L), risk is somewhat greater. With "severe" neutropenia (ie, counts $<0.5 \times 10^9$/L), susceptibility to infections increases dramatically (1, 3–5). The frequency and severity of infections in patients with severe neutropenia depend on several factors. These include the integrity of the skin and mucous membranes; level and function of monocytes, lymphocytes, immunoglobulins and complement components; and function of the spleen and reticuloendothelial system. A patient's general well-being, including nutritional status, physical strength, capacity to cough and clear secretions, level of consciousness, previous antibiotic exposures, and many other factors, is also important in determining the occurrence and severity of infections. In general, a weak and debilitated person with a poor nutritional status, whose body surfaces are penetrated by catheters and colonized by resistant microorganisms after days of antibiotic treatment, is very prone to infection with even a transient period of neutropenia. On the other hand, a relatively healthy individual often can tolerate severe neutropenia for several days (6,7).

II. PATHOPHYSIOLOGICAL MECHANISMS OF NEUTROPENIA

Neutrophils are produced from hematopoietic stem cells in the bone marrow, ordinarily over a period of 10 to 14 days (8). For the first half of this period, the cells are in a mitotic pool (committed stem cells, myeloblasts, promyelocytes, and myelocytes). Subsequently they enter the post-mitotic or maturational pool (metamyelocytes, bands, and mature neutrophils). Once fully mature, neutrophils enter the blood. Approximately half the cells are freely flowing in the circulating pool; the other half are in the marginal pool, adhering loosely to the vascular endothelium or transiently trapped in the microcirculation. Neutrophils leave the circulation by migrating between vascular endothelial cells to tissue sites of inflammation. Neutropenia can develop because of abnormalities at any stage in this pathway, and many diseases affect neutrophil production and deployment at several stages (1).

The causes of neutropenia, divided as disorders of production, distribution, utilization, or turnover, are shown in **Table 1**. Disorders causing neutropenia also can be categorized in several other ways: congenital or acquired; acute or chronic; mild, moderate or severe; or occurring with or without other host defense abnormalities such as lymphocytopenia or moncytopenia. Another system categorizes neutropenia as primary, due to a hematological disease, or secondary (ie, due to a malignant, infectious, or immune disease). In general, disease processes affecting production, particularly the early phases of neutrophil development, lead to the most profound neutropenia and the greatest susceptibility to infection.

Table 1. Causes of Neutropenia.

Decreased production
Primary hematologic disorders
  Aplastic anemia
  Myelodysplasia and leukemia
  Congenital neutropenia: Kostmann's syndrome and related disorders
  Cyclic neutropenia
  Idiopathic neutropenia
  Myelokathexis
Secondary disorders
  Drugs: cytotoxic and idiosyncratic
  Radiation: to central, axial skeleton
  Chemicals: benzene, phenols, arsenic
  Infections: viral (hepatitis, parvovirus, HIV), bacterial (mycobacterial)
  Replacement: metastatic malignancies, myelofibrosis
Altered distribution in circulation
Increased margination: severe infections, endotoxemia, exposure of blood to foreign surfaces (eg, dialysis, blood oxygenators), acute antigen-antibody reactions, serum sickness, acute administration of colony-stimulating factors, leukoagglutinin transfusion reactions, pseudoneutropenia
Neutropenia associated with splenomegaly: infections (eg, malaria, tuberculosis, kala-azar), inflammation (eg, sarcoidosis), immunologic (eg, Felty's syndrome [rheumatoid arthritis], Sjögren's syndrome), congestive (eg, cirrhosis), hematologic malignancies
Increased utilization or turnover
Acute and overwhelming infections, usually with an abnormal marrow, eg, after cytotoxic chemotheraphy, chronic ethanol intoxication, or drug-related or immunologically mediated agranulocytosis
Disorders with decreased production and increased turnover
Autoimmune neutropenia
Isoimmune neutropenia
Pure leukocyte aplasia
Drug-induced neutropenia

III. DISEASES CAUSING NEUTROPENIA AND PATTERNS OF INFECTIOUS COMPLICATIONS

The pattern of neutropenia and susceptibility to infection for diseases causing neutropenia are briefly highlighted in this section, with references to more detailed reports.

A. Disorders of Production

1. Aplastic Anemia

Neutropenia may be a presenting feature or develop late in the course. It is often severe and accompanied by monocytopenia, due to loss of functional hematopoietic stem cells. Serious and fatal infections are common late in the disease (9,10).

2. Myelodysplasia and Leukemia

Varible degrees of neutropenia are common with the myelodysplastic syndromes. With severe neutropenia, fever and infectious complications are common, including perianal infection, pneumonia, and bacteremia. With acute myeloid leukemia (AML), severe neutropenia is common at diagnosis and worsens with initial treatment. With induction therapy, severe neutropenia often lasts for 2 to 4 weeks and is associated with severe susceptibility to infections (11–15).

3. Congenital Neutropenia

There are multiple causes for severe neutropenia at birth. The most severe disorder is Kostmann's syndrome, characterized by a severe deficiency in marrow myeloid cells beyond the promyelocyte stage of development, but with normal or near-normal numbers of monocytes, lymphocytes, and platelets. The susceptibility to infection is severe, and pneumonia, bacteria, and deep tissue abscesses are common (16).

4. Cyclic Neutropenia

This disorder causes oscillations of numbers of blood neutrophils, with periods of severe neutropenia at approximately 21-day intervals, accompanied by fever, oral ulcers, lymphadenopathy, and skin infections. Intestinal ulcerations with fatal clostridial bacteremia are an important complication (17).

5. Idiopathic and Autoimmune Neutropenia

This category includes patients with selective neutropenia of unknown cause in children and adults; the distinction between idiopathic neutropenia and autoimmune neutropenia can be difficult because of the unavailability and non-specificity of anti-neutrophil antibody tests. The susceptibility to infection is inversely related to the blood neutrophil counts. The most frequent events are oropharyngeal and skin infections (18–21).

6. Myelokathexis

In this rare disorder, there are abundant developing neutrophils in the marrow, but neutropenia is severe. Infection susceptibility varies; some patients have relatively few infections, while others have had skin, respiratory, and deep tissue infections (22).

7. Drugs

Neutropenia occurs as an idiosyncratic reaction to many drugs, presumably on an immune or toxic basis. The patterns vary greatly. Toxic responses, such as those occurring with the phenothiazine and anti-thyroid medications may occur gradually; others, such as those after administration of sulfonamides and penicillin, may occur precipitously. Neutropenia is severe, but recovery is the rule if the offending medication is stopped and supportive care, including antibiotics, is given. Patients usually present with fever and pharyngitis or other oropharyngeal infections (23,24).

Neutropenia is also a common consequence of many cancer chemotherapeutic agents and immunosuppressive drugs. These agents generally cause a dose-dependent neutropenia, with cumulative effects with repeated administration.

8. Radiation

Treatment causes a dose-dependent decrease in neutrophils, particularly with treatment to multiple areas of the central axial skeleton. Usually severe neutropenia is avoided by titration of the dose and schedule of treatments. The infection pattern is similar to that with chemotherapy, except that the risk of bacteremia from a gastrointestinal source is greater if the bowel has been in the field of radiation.

9. Chemicals

Many organic chemicals can induce aplastic anemia with neutropenia as a severe consequence. Severe neutropenia may occur with acute and intensive exposure, but chronic neutropenia is not a well-defined complication of chemical exposures (1).

10. Infectious Diseases

Viral infections (eg, infectious mononucleosis and human immunodeficiency virus [HIV] infection) can cause severe neutropenia accompanied by a marked increased susceptibility to bacterial infections due to suppressed production. The neutropenia arising during the course of self-limited viral syndromes usually does not place the affected patient at risk for opportunistic infection. With HIV infection, bacterial and fungal infections occur in an inverse relation to the neutrophil count with risk increasing with counts $<1.0 \times 10^9$/L (25–28). When mycobacterial infections and other granulomatous infections develop in this setting, the neutropenia is severe and prognosis very poor (29).

11. Myelofibrosis and Metastatic Malignancies

In these conditions, the marrow is replaced, and this causes variable patterns of anemia, thrombocytopenia, and neutropenia. Neutropenia often limits chemotherapy late in the course and may predispose to serious infections.

B. Neutropenia Due to Altered Distribution in the Circulation

Table 1 lists numerous causes for increased margination of neutrophils with acute neutropenia due to a redistribution of the cells from the circulating to the marginal pool. Nearly all of these conditions are transient, usually lasting for minutes to hours. Exposure of blood to dialysis membranes and other foreign surfaces is a well-studied example of the phenomenon (30).

Neutropenia due to redistribution of cells in the circulation also occurs with conditions causing splenomegaly, probably due to sequestration of neutrophils in the spleen. Infectious complications are generally far less than with comparable degrees of neutropenia because of disorders of neutrophil production. Some patients, however, may have both impaired production and distribution (eg, patients with Felty's syndrome, with greatly enhanced susceptibility to infections) (1).

C. Neutropenia Due to Increased Utilization

With acute or overwhelming infections, severe neutropenia can occur (31,32). With a normal and competent marrow, this complication is unusual. With previous injury to the marrow from radiation or cytotoxic drugs or with acute or chronic ethanol exposure, the supply of neutrophils in the post-mitotic pool may be deficient and the proliferative response of the mitotic pool blunted. In this situation, if bacteremia occurs, neutropenia lasting for hours to days often results, often with a fatal outcome (33,34).

IV. INFECTIOUS COMPLICATIONS OF NEUTROPENIA

A. Pathogens

1. Aerobic Bacteria

The majority of infections in patients with severe neutropenia are caused by aerobic gram-negative bacilli (especially *Escherichia coli, Klebsiella pneumoniae,* and *Pseudomonas aeruginosa*) and gram-positive cocci (particularly coagulase-negative staphylococci, α-hemolytic streptococci, and *Staphylococcus aureus*) (6,35). Because of the decline in frequency of infections caused by gram-negative bacilli over the past two decades, coagulase-negative staphylococci and *S. aureus* have emerged as the predominant isolates from neutropenic patients with cancer (6,35).

2. Anaerobic Bacteria

While anaerobic bacteria are relatively rare causes of infection in neutropenic patients (33), they can contribute to polymicrobial infections, such as necrotizing gingivitis, typhlitis (neutropenic colitis), and perianal cellulitis (6,35,36). Bacteremia with *Clostridium* spp, including *C. perfringens, C. septicum,* and *C. tertium,* is a recognized complication of leukemia and neutropenic colitis (37–40). Clinical suspicion of such infections warrants empiric administration of an anti-anerobic antibiotic, such as clindamycin, metronidazole, imipenem, or ampicillin-sulbactam. Suppression of the normal bacterial flora of the gut by broad-spectrum antimicrobial flora can result in overgrowth of *Clostridium difficile,* leading to the toxin-mediated syndrome of pseudomembranous colitis, characterized by diarrhea, abdominal pain, fever, and possible sepsis syndrome (41).

3. Mycobacteria

Mycobacteria have not been implicated as major causes of infection in patients with neutropenia. However, rapidly growing mycobacteria, such as *Mycobacterium chelonei* and *M. fortuitum,* have been associated with exit-site or tunnel infections around indwelling central venous catheters (42).

4. Mycoses

Invasive mycotic infections have emerged as one of the leading causes of morbidity and mortality in patients with neutropenia, especially when neutropenia is severe and prolonged (6,35,43–45). Secondary fungal infections are frequently encountered in neutropenic patients who have received extended courses of broad-spectrum antibiotics (35). In febrile neutropenic patients who fail to respond to a 7-day course of empiric broad-spectrum antibiotic therapy, approximately one-third will ultimately be found to have a fungal infection (46). Although the majority of these fungal infections continue to be caused by *Candida* spp and *Aspergillus* spp, prolonged neutropenia increases the risk of infection with unusual fungal species of low pathogenic potential in most circumstances. *Fusarium* spp are an example of such fungi increasingly encountered in high-risk patient groups (eg, bone marrow transplantation) (47–49). Distinct subpopulations of neutropenic patients appear to be at greater risk for the development of specific fungal infections. For example, invasive aspergillosis has emerged as a major cause of death in patients with aplastic anemia (9).

The clinical presentation of infection due to typical fungal pathogens may be altered in neutropenic patients. A notable example is hepatosplenic candidiasis, in which persistent fever and abdominal pain develop after recovery from chemotherapy-induced severe neutropenia. This diagnosis should be suspected when characteristic "bull's eye" lesions are detected in the liver and/or spleen by computed tomography (CT), magnetic resonance imaging (MRI), or ultrasonography (50,51). The spectrum of fungal infection also may be influenced by antifungal prophylaxis regimens (see below) currently used to lower the risk of fungal infection during prolonged neutropenia. Ketoconazole prophylaxis has been associated with increased risk of colonization and possible infection with either *Torulopsis glabrata* or *Candida tropicalis* (6). In one study of patients receiving bone marrow transplantation, *C. krusei* was isolated frequently from neutropenic patients receiving fluconazole prophylaxis (52).

5.   Viruses and *Pneumocystis carinii*

Although neutropenia per se is not a major risk factor for infections with viruses or *Pneumocystis carinii,* patients receiving bone marrow transplantation or intensive chemotherapy may develop significant defects in cell-mediated immunity. Infections with herpesviruses (eg, cytomegalovirus [CMV], herpes simplex virus, and varicella-zoster virus) and *P. carinii* can cause serious morbidity and mortality in these patient groups (6).

B.   Clinical Evaluation of the Neutropenic Patient with Fever

Patients with neutropenia or at risk for developing treatment-related neutropenia should be instructed to seek medical evaluation promptly for any febrile episode. To reduce the risk of inadvertent transmission of pathogens, healthcare providers should wash their hands carefully before contact with neutropenic patients (6,35,53). Because signs of inflammation at sites of infection may be markedly diminshed or absent during neutropenia, thorough clinical evaluation is mandatory. Primary sites of infection include mucosal surfaces, such as the alimentary tract (which may be damaged as a result of chemotherapy or radiation) and the skin, where vascular catheters may serve as portals for systemic infection (6,35,54). Special attention during physical examination should be directed to the most common sites implicated in infections of neutropenic hosts. These sites include the skin, gums, pharynx, lungs, perineum (including the anus), vascular catheter sites, bone marrow aspiration sites, venipuncture sites, and the nail beds (35). When documenting fever in neutropenic patients, rectal thermometers should be avoided because of the inherent risk of mucosal disruption during insertion and removal.

In general, two sets of blood cultures should be obtained from all neutropenic patients (35). When an indwelling intravenous (IV) catheter is present, cultures should be obtained from each lumen as well as a peripheral vein, if possible (6,35). Any drainage from a catheter exit site should be examined by gram stain and submitted for bacterial and fungal culture (35). Persistent or chronic drainage should be evaluated for the presence of mycobacteria (42). Suspicious skin lesions should be aspirated or biopsied for microscopic examination and culture (55). Urine cultures are indicated if signs or symptoms of urinary tract infection are present, bearing in mind that pyuria is an unreliable indicator of urinary tract inflammation in the neutropenic patient (35). If diarrhea is present, stool should be submitted for assay of *Clostridium difficile* toxin and culture of enteric bacterial pathogens (35). If the patient complains of any pulmonary symptom, including cough, hemoptysis, dyspnea, or pain, a chest radiograph should be obtained and sputum, if available, evaluated for potential bacterial and fungal pathogens by stain and culture. The clinician should be aware that the formation of radiographically demonstrable infiltrates suggestive of pneumonitis may be impaired in the neutropenic host (35). Lumbar puncture for examination of cerebrospinal fluid is not recommended as a routine procedure in the diagnostic evaluation of febrile neutropenia. However, it should be performed, after CT evaluation of the brain, if signs or symptoms of central nervous system dysfunction evolve (35). Similarly, CT analysis of the paranasal sinuses is indicated if signs or symptoms of sinusitis (eg, headache, facial swelling, or nasal discharge) are present (35). Depending on the clinical presentation of a particular patient, more extensive, and potentially invasive, diagnostic measures may be warranted, including bronchoscopy, endoscopy, ultrasonography, CT, and MRI.

To guide supportive care and monitor for potential toxicity from antimicrobial drug therapy (see below), complete blood counts and levels of electrolytes, creatinine, urea nitrogen, and serum aminotransferases should be performed at least every third day (35). Because of the risk of antibiotic-induced suppression of normal intestinal bacterial flora resulting in vitamin K deficiency and associated coagulopathy (56,57), periodic evaluation of the prothrombin time is advised.

C. Antimicrobial Treatment of Fever Associated with Neutropenia

Empiric treatment with broad-spectrum antibiotics is the cornerstone of initial management of neutropenic patients with fever (6,35,58–60). Serious bacterial infection should be suspected in all febrile neutropenic patients and therapy initiated with broad-spectrum antibiotics as soon as possible. In most cases, it is advised that patients be hospitalized to receive IV antibiotics administered at maximal dosages while diagnostic evaluation is in progress (6,35,58–60). Out-patient therapy may be considered in selected low-risk patients, such as patients with solid tumors with transient chemotherapy-related neutropenia, younger patients without comorbid diseases, and compliant patients with ready access to hospital facilities (61–66).

Although indwelling vascular devices (eg, Hickman and Groshong catheters) may remain in place during early management of most patients even if bacteremia is suspected or documented (54,68,69), central venous catheters should be removed from critically ill patients if at all possible (35,54). Evidence of septic thrombophlebitis, a subcutaneous (SC) tunnel infection, or a nonpatent catheter mandates immediate catheter removal (35,54,68). The majority of catheter-related infections are caused by coagulase-negative staphylococci and *S. aureus* (54,67,68). Infections due to coagulase-negative staphylococci generally respond well to antimicrobial treatment alone. In contrast, catheter removal in addition to antibiotic treatment is often required for resolution of catheter-related infections caused by *S. aureus, Bacillus* spp, *Corynebacterium jekeium,* aerobic gram-negative bacilli, or *Candida* spp (35,54,67,68). If an indwelling catheter is retained during treatment of a catheter-related infection, then administration of antimicrobial therapy should be rotated to each lumen of a multiple-lumen catheter (6).

An array of broad-spectrum antimicrobial regimens have been advocated for empiric treatment of the febrile neutropenic patient. When choosing a regimen for an individual patient, it is important to consider the antimicrobial susceptibilities of the predominant pathogens at that particular institution responsible for infections in neutropenic patients

(9,10,69). Potential toxicity of a given antimicrobial regimen in a specific patient should also be considered (6,35,59). A concensus panel assembled by the Infectious Diseases Society of America (IDSA) in 1990 advocated four different regimens for initial empiric treatment of febrile neutropenic patients (**Figure 1**): an aminoglycoside plus an anti-pseudomonal β-lactam; a combination of two β-lactam drugs; single-drug therapy (monotherapy); and vancomycin plus an aminoglycoside and an anti-pseudomonal penicillin (or cephalosporin).

The regimen of an aminoglycoside plus an anti-pseudomonal β-lactam has received extensive clinical use, provides antimicrobial activity against

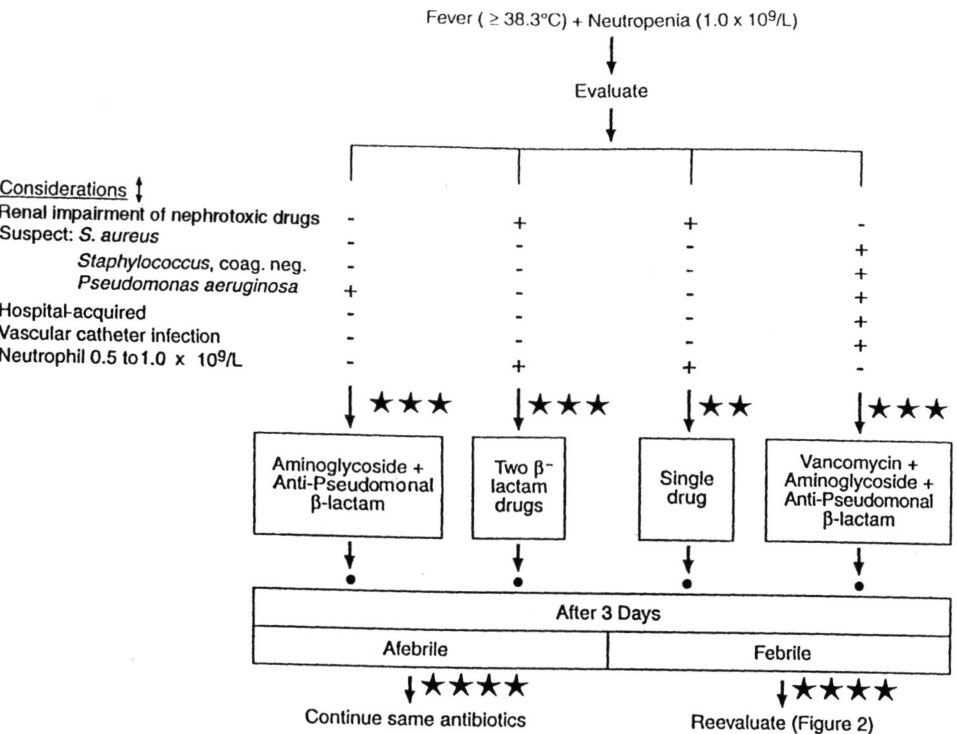

Figure 1.   Guide to the initial management of febrile neutropenic patients. Stars indicate rating of recommendations by the Working Committee, Infectious Diseases Society of America: ****, definite choice, strong support; ***, choice adequately supported with data; **, promising but not proven. These recommendations represent guides towards selection of one of four general antibiotic regimens and do not necessarily preclude use of any regimen. (Figure reprinted from Ref. 35 with permission from The University of Chicago Press.)

anaerobic bacteria, and demonstrates synergistic bactericidal activity against some aerobic gram-negative bacilli (70,71). Drawbacks to this regimen include lack of activity against some gram-positive bacteria, such as *S. aureus,* and potential aminoglycoside-related toxicity, including nephrotoxicity and ototoxicity. If this regimen is used, aminoglycoside levels should be monitored and dosages adjusted properly. This regimen is especially recommended for patients at high risk with *Pseudomonas aeruginosa,* which includes patients known to be previously colonized with this organism and patients with severe chemotherapy-related mucositis (72).

The combination of a third-generation cephalosporin, such as ceftazidime, and an extended spectrum penicillin, such as piperacillin or mezlocillin, has been used successfully at several centers with extensive experience in the management of neutropenic patients (73–77). Low potential for treatment-related toxicity is a major advantage of this regimen. Disadvantages include the relatively high cost, the potential for selection of resistant organisms, and possible antimicrobial antagonism with certain bacterial infections (79).

Several studies have documented the efficacy of empiric monotherapy with either ceftazidime or imipenem for initial management of febrile neutropenia (80–84). Although there is concern that empiric monotherapy may be inadequate for patients with prolonged courses of neutropenia (81,84), it appears safe and efficacious for a large proportion of patients who develop fever during transient neutropenia after chemotherapy. Ceftazidime is usually considered the drug of choice for monotherapy, with imipenem reserved for patients with either suspected polymicrobial infection involving anaerobic bacteria or suspected infection due to methicillin-sensitive *S. aureus.*

The regimen consisting of vancomycin plus an aminoglycoside and an antipseudomonal penicillin (or a third-generation cephalosporin) should be selected as initial empiric therapy for patients in whom coagulase-negative staphylococci, methicillin-resistant *S. aureus* (MRSA), *Corynebacterium* spp., and α-hemolytic streptococci are relatively common. Such patients include neutropenic patients with obvious Hickman line infections. Mortality rates have been relatively low, however, in infections caused by all of these organisms except MRSA when vancomycin was not included in the initial treatment regimen (85). In general, ultimate mortality and morbidity do not appear to be adversely affected by delaying administration of vancomycin until its use is supported by microbiologic culture data (79,82). On the other hand, several studies suggest that inclusion of vancomycin in the initial empiric treatment regimen for febrile neutropenia may shorten duration of fever, reduce days of documented bacteremia, and decrease rate of treatment failure (86,88). Overall, given the associated risk of promoting colonization and possible subsequent infection with resistant organisms, such as vancomycin-resistant enterococci (89–93), it appears unwise to routinely include vancomycin in antimicrobial regimens used for initial empiric treatment of febrile neutropenia.

If fever resolves within 72 hours of the start of therapy, it is advised that antibiotics be continued for a minimum of 7 days and until significant signs and symptoms of infection resolve (35). While a therapeutic regimen may be modified based on identification and susceptibility testing of a pathogen, broad-spectrum antibiotic therapy should be maintained for the duration of neutropenia (6,35). Although antibiotic therapy is recommended until the neutrophil count increases above 0.5×10^9/L, discontinuation of antimicrobial treatment before recovery from neutropenia can be considered in selected patients, especially those in whom the course of neutropenia is prolonged and signs and symptoms of infection have resolved (**Figure 2**) (35,95–97).

Fever persisting for greater than 72 hours despite administration of empiric broad-spectrum antibiotics suggests a nonbacterial etiology, an

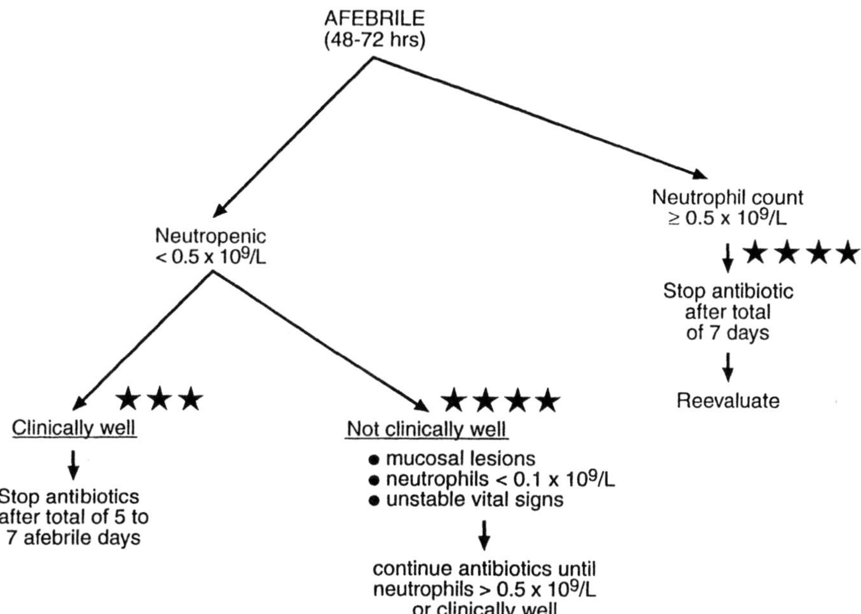

Figure 2. Management of neutropenic patient with persistent fever after 3 days of treatment. Refer to Figure 1 for explanation of the star rating system formulated by the Working Committee, Infectious Diseases Society of America. (Figure reprinted from Ref. 35 with permission from The University of Chicago Press.)

infection with a bacterial species resistant to the chosen antibiotic regimen, a superinfection with a second, resistant bacterial species, inadequate serum and/or tissue levels of the antibiotic(s), or localized infection at avascular sites (ie, abscess) (35). Neutropenic patients with persistent fevers should be rigorously reevaluated on day 4 or 5. If doing well clinically, the patient can be maintained on the initial antibiotic regimen (**Figure 3**). If clinical deterioration is noted, the antibiotic regimen should be altered (**Figure 3**). Empiric addition of vancomycin to the regimen may be justified and should be considered (**Figure 3**). Because occult fungal infection represents a

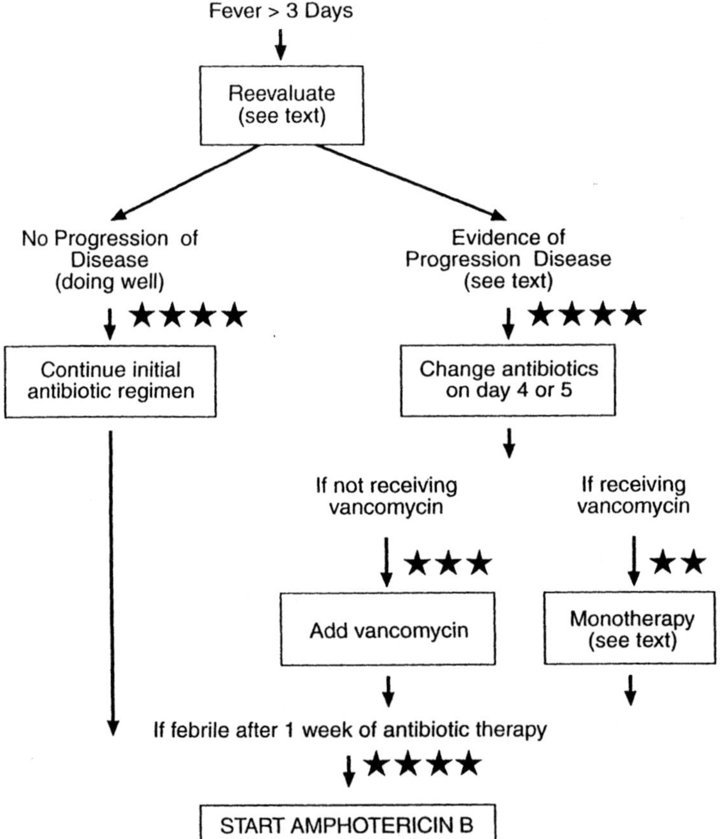

Figure 3.   Discontinuation of antibiotics if fever resolves on initial empiric antibiotic(s). Refer to Figure 1 for explanation of the star rating system formulated by the Working Committee, Infectious Diseases Society of America. (Figure reprinted from Ref. 35 with permission from The University of Chicago Press.)

significant cause of persistent fever during neutropenia, empiric treatment with amphotericin B should be strongly considered in neutropenic patients who remain febrile despite broad-spectrum antibiotic therapy for 7 days (**Figure 3**) (6,35,59,97). If fever fails to resolve after 3 weeks of empiric antimicrobial therapy (including 2 weeks of amphotericin B), discontinuation of all antimicrobial agents should be considered, and the patient should be carefully reevaluated for the cause of fever (**Figure 4**) (6,35).

Antiviral agents are not routinely recommended for empirical management of fever in the neutropenic host (35). However, systemic therapy with either acyclovir or ganciclovir (as appropriate) is indicated for neutropenic patients who develop active infection with herpes simplex virus, varicella-zoster virus, or cytomegalovirus (35).

D. Role of Antimicrobial Prophylaxis During Neutropenia

The role of antibiotic prophylaxis in nonfebrile neutropenic patients remains controversial but should be considered for patients expected to be profoundly neutropenic ($<0.1 \times 10^9$/L neutrophils) for longer than 7 days. When deciding whether or not antibiotic prophylaxis should be started, potential toxicity must be weighed against perceived benefit for each patient. Both oral non-absorbable (ie, selective decontamination) and absorb-

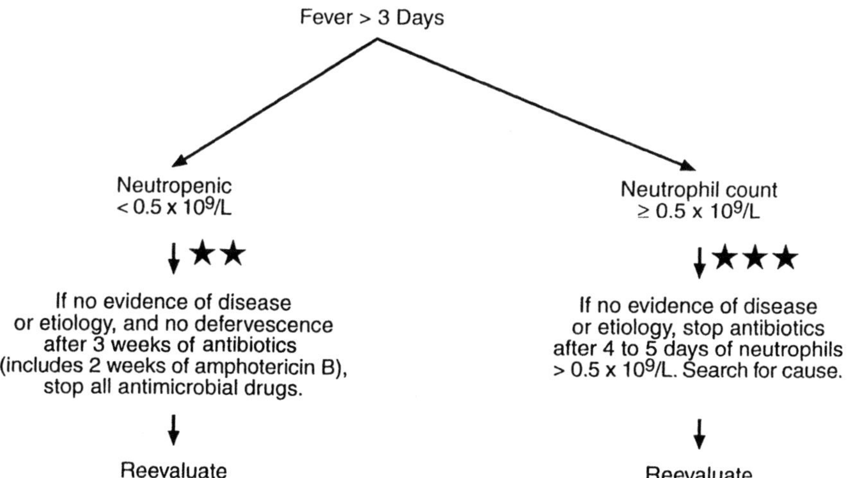

Figure 4. Discontinuation of antibiotics if fever persists for longer than 3 days. Refer to Figure 1 for explanation of the star rating system formulated by the Working Committee, Infectious Diseases Society of America. (Figure reprinted from Ref. 35 with permission from The University of Chicago Press.)

able antibiotic regimens have been employed for this purpose. Two types of oral absorbable antibiotics have been studied: trimethoprim-sulfamethoxasole (TMP-SMX) and fluoroquinolones (35,98–105).

Infection by opportunistic fungi continues to be a major cause of morbidity and mortality in patients with severe, prolonged neutropenia (53). Consequently, significant effort has been focused recently on the role of antifungal prophylaxis in selected neutropenic patient populations. Compulsory hand washing by hospital personnel caring for neutropenic patients remains a critical component in any infection-control program designed to reduce systemic candidiasis (53,106). High-efficiency particulate air (HEPA) filtration is the only measure with proven efficacy for the prevention of invasive aspergillosis in severely neutropenic patients (53,107,108), although it appears reasonable in the care of neutropenic patients to recommend simple environmental measures, such as removing plants from patients' rooms and sealing of construction areas, to minimize the risk of exposure to *Aspergillus* spores (53,109).

Antifungal prophylaxis with either amphotericin B or systemic azoles also has been evaluated in selected neutropenic patient populations at especially high risk for developing fungal infections, such as after bone marrow transplantation (110–120). Routine use of amphotericin B, the only reliably fungicidal systemic agent currently available for clinical use, has been compromised by infusion-related toxicity, nephrotoxicity, anemia, and electrolyte abnormalities. In an effort to reduce toxicity related to conventional doses of daily amphotericin B, investigators have studied prophylactic administration of systemic low-dose amphotericin B, but this approach has been associated with failure to prevent invasive aspergillosis (113,116,121). Aerosolized and intranasal amphotericin B regimens are attractive but, as yet, unproven modes of prophylaxis against invasive aspergillosis in high-risk patients (117,122–128). Newer liposomal preparations of amphotericin B, while less toxic than conventional amphotericin B, have yet to be widely adopted for antifungal prophylaxis because of their high cost (118,129).

Fluconazole, while much less toxic than amphotericin B, lacks activity against opportunistic molds, such as *Aspergillus* and *Rhizopus* spp, and has been associated with the emergence of resistant *Candida* infections when used for antifungal prophylaxis (52). Widespread prophylactic use of itraconazole, which has in vitro activity against both *Candida* and *Aspergillus* spp, has been limited by erratic bioavailability after oral administration and the current lack of a parenteral formulation (115,117,119).

At present, systemic antifungal prophylaxis is not recommended for routine use in the management of neutropenia. Ongoing clinical studies, however, may define specific indications for antifungal prophylaxis in selected neutropenic patient populations in the future.

E. Role of Granulocyte Transfusion Therapy

Despite improvements in therapeutic antimicrobial regimens over the last several decades (6,35,60), systemic infections with bacteria or opportunistic fungi remain major causes of morbidity and mortality among patients with prolonged neutropenia (6,60,130). Theoretically, restoration of sufficient levels of biologically active polymorphonuclear leukocytes (PMN) represents another therapeutic modality with potential benefit for the management of neutropenic patients with life-threatening infections (60,130–132).

The discovery and subsequent commercial development of myeloid growth factors (133), particularly recombinant human granulocyte colony-stimulating factor (rHuG-CSF) and granulocyte-macrophage colony-stimulating factor (rHuGM-CSF), have changed the approach to prevention and treatment of neutropenia (1,60,134–139). Both rHuG-CSF (139–145) and rHuGM-CSF (146–148) have been shown to accelerate hematopoietic recovery after myelosuppressive chemotherapy (149). Certain patients, however, continue to experience prolonged neutropenia secondary to slow hematopoietic recovery despite the clinical use of growth factor therapy. Granulocyte transfusion therapy represents a potentially useful adjunctive modality for either prophylaxis or treatment of infection in this group of patients (130–132, 150, 151).

At present, the clinical role for granulocyte transfusion therapy in the management of neutropenic patients with infection remains controversial. The clinical utility of neutrophil transfusion therapy has been compromised by the inability to obtain sufficient quantities of functional neutrophils from donors (130–132). Glucocorticoids have generally been used to stimulate mobilization of PMN in the peripheral blood of granulocyte donors (152,153). In an effort to optimize the collection of sufficient quantities of neutrophils, Filgrastim has been administered to granulocyte donors in an attempt to augment the yield of neutrophils obtained by centrifugation leukapheresis (152–156). A recent study demonstrated that dexamethasone significantly increases the level of neutrophilia induced in normal subjects by Filgrastim, thereby affording enhanced yields of neutrophils by leukapheresis (156). Subsequent analysis revealed that these cells retained biological activity when examined in vitro and migrated to tissue sites in vivo when infused (152). These encouraging results prompted initiation of an ongoing clinical trial, using both dexamethasone and Filgrastim for neutrophil mobilization in community donors, to determine the clinical benefit of granulocyte transfusion therapy in neutropenic patients with life-threatening bacterial or fungal infections after bone marrow transplantation. Similar trials in other neutropenic patient populations may establish specific indications for granulocyte transfusion therapy in the future.

V. SPECIFIC TREATMENT OF NEUTROPENIA

No specific treatment is required for brief, self-limited neutropenia (149). Patients with sustained or prolonged severe neutropenia are at significant risk for infection by bacteria and opportunistic fungi. Before the development of the hematopoietic growth factors, rHuG-CSF and rHuGM-CSF, lithium (157) and androgens (1) were administered to patients to stimulate neutrophil production for the treatment of a variety of diseases associated with neutropenia, including cyclic neutropenia (158–161), and after cancer chemotherapy (162,163). Corticosteroids have also been reported to reduce the oscillations in some cases of cyclic neutropenia (164,165). The use of lithium, androgens, and corticosteroids for the specific treatment of severe neutropenia has largely been supplanted by rHuG-CSF and rHuGM-CSF. Therapy with either of these two myeloid growth factors has emerged as the treatment of choice of prolonged, severe neutropenia. The use of Filgrastim for the treatment of specific disorders of neutropenia is addressed in other chapters contained in this volume.

VI. CONCLUSIONS AND SUMMARY

Opportunistic infections due to bacteria and opportunistic fungi continue to pose a major risk for patients with neutropenia. The risk of serious infection for a particular patient varies according to the etiology of neutropenia but, in general, increases with the duration and severity of neutropenia. Proper clinical management of the neutropenic management involves prompt treatment of fever with appropriate empiric antibiotics. The roles of antimicrobial prophylaxis and granulocyte transfusion therapy are evolving, and specific indications for their use will depend upon the results of ongoing and future clinical trials. The discovery and development of the hematopoietic growth factors, rHuG-CSF and rHuGM-CSF, have revolutionized our approach to clinically significant neutropenia. Growth factor therapy has emerged as the treatment of choice for patients with sustained or prolonged severe neutropenia.

REFERENCES

1. Dale, D. C. Neutropenia. In: Williams, W. J., Beutler, E., Erslev, A. J., and Lichtman, M. A., eds. *Hematology,* 5th ed. New York: McGraw-Hill, 1995.

2. Coates, T. D., and Baehner, R. Leukocytosis and leukopenia. In: R. Hoffman, E. J. Benz Jr., S. J. Shattil, B. Furie, H. J. Cohen, and L. E. Silberstein, eds, *Hematology,* 2nd ed. New York: Churchill Livingstone, 1995. pp. 769–784.

3. Bodey, G. P., Buckley, M., Sathe, Y. S., and Freirich, E. J. (1966). Quantitative relationships between circulating leukocytes and infection in patients with acute leukemia. *Ann Intern Med 64:*328–340.

4. Dale, D. C., Guerry, D., Wewerka, J. R., et al (1979). Chronic neutropenia. *Medicine 58:*128–144.

5. Weetman, R. M., and Boxer, L. A. (1980). Childhood neutropenia. *Pediatr Clin North Am 27:*361–375.

6. Pizzo, P. A. (1993). Management of fever in patients with cancer and treatment-induced neutropenia. *N Engl J Med 328:*1323–1332.

7. Freifeld, A. G., and Pizzo, P. A. (1996). The outpatient management of febrile neutropenia in cancer patients. *Oncology-Huntington 10:*599–606.

8. Babior, B. M., and Golde, D. W. Production, distribution and fate of neutrophils. In: E. Beutler, M. A. Lichtman, B. S. Coller, and T. J. Kipps, eds. *Williams Hematology,* 5th ed. New York: McGraw-Hill, 1995. pp. 773–779.

9. Weinberger, M., Elattar, I., Marshall, D., et al (1992). Patterns of infection with aplastic anemia and the emergence of *Aspergillus* as a major cause of death. *Medicine 71:*24–43.

10. Weinberger, M. (1993). Approach to management of fever and infection in patients with primary bone marrow failure and hemoglobinopathies. *Hematol Oncol Clin North Am 7:*865–885.

11. Wade, J. C. (1993). Management of infection in patients with acute leukemia. *Hematol Oncol Clin North Am 7:*293–315.

12. Schimpff, S. C., Scott, D. A., and Wade, J. C. (1994). Infections in cancer patients: some controversial issues. *Supportive Care Cancer 2:*94–104.

13. Feusner, J. H., and Hastings, C. A. (1995). Infections in children with acute myelogenous leukemia. *J Pediatr Hematol/Oncol 17:*234–247.

14. Micozzi, A., Cartoni, C., Monaco, M., et al (1996). High incidence of infectious gastrointestinal complications observed in patients with acute myeloid leukemia receiving intensive chemotherapy for first induction of remission. *Supportive Care Cancer 4:*294–297.

15. Barton, T. D., and Schuster, M. G. (1996). The cause of fever following resolution of neutropenia in patients with acute leukemia. *Clin Infect Dis 22:*1064–1068.

16. Welte, K., and Dale, D. C. (1996). Pathophysiology and treatment of severe chronic neutropenia. *Ann Hematol 72:*158–165.

17. Dale, D. C., and Hammond, W. P. (1988). Cyclic neutropenia: a clinical review. *Blood Rev 2:*178–185.

18. Kyle, R. A. (1980). Natural history of chronic idiopathic neutropenia. *N Engl J Med 302:*908–909.

19. Conway, L. T., Clay, M. E., Kline, W. E., et al (1987). Natural history of primary autoimmune neutropenia in infancy. *Pediatrics 79:*728–733.

20. Logue, G. L., Shastri, K. A., Laughlin, M., et al (1991). Idiopathic neutropenia: antineutrophil antibodies and clinical correlations. *Am J Med 90:*211.

21. Bux, J., Kissel, K., Nowak, K., et al (1991). Clinical and laboratory studies in 143 patients. *Ann Hematol 63*:249–252.
22. Bassan, R., Viero, P., Minetti, B., et al (1984). Myelokathexis: a rare form of chronic benign granulocytopenia. *Br J Haematol 58*:115–117.
23. Strom, B. L., Carson, J. L., Schinnar, R., Snyder, E. S., and Shaw, M. (1992). Descriptive epidemiology of agranulocytosis. *Arch Int Med 152*:1475–1480.
24. Julia, A., Olona, M., Bueno, J., et al (1991). Drug-induced agranulocytosis: prognostic factors in a series of 168 episodes. *Br J Hematol 79*:366–371.
25. Hambleton, J., Aragon, Modin, G., Northfelt, D. W., and Sande, M. A. (1995). Outcome for hospitalized patients with fever and neutropenia who are infected with the human immunodeficiency virus. *Clin Infect Dis 20*:363–375.
26. Moore, R. D., Keruly, J. C., and Chaisson, R. E. (1995). Neutropenia and bacterial infection in acquired immunodeficiency syndrome. *Arch Intern Med 155*:1965–1970.
27. Frumkin, L., and Dale, D. C. (1996). The role of colony-stimulating factors in HIV disease. *Aids Reader Nov/Dec*:185–193.
28. Keiser, P., Higgs, E., and Smith, J. (1996). Neutropenia is associated with bacteremia in patients infected with the human immunodeficiency virus. *Am J Med Sci 312*:118–122.
29. Nichols, L., Florentine, B., Lewis, W., et al (1991). Bone marrow examination for the diagnosis of mycobacterial and fungal infections in the acquired immunodeficiency syndrome. *Arch Pathol Lab Med 115*:1125–1132.
30. Craddock, P. R., Hammerschmidt, D. E., Moldow, C. F., et al (1979). Granulocyte aggregation as a manifestation of membrane interactions with complement: possible role in leukocyte margination, microvascular occlusion, and endothelial damage. *Semin Hematol 16*:140–147.
31. Christianson, R. D., and Rothstein, G. (1980). Exhaustion of mature marrow neutrophils in neonates with sepsis. *J Pediatr 96*:316–320.
32. Gomez, J., Banos, V., Ruiz-G'omez, J., et al (1995). Clinical significance of pneumococcal bacteraemias in a general hospital: a prospective study. *J Antimicrob Chemother 36*:1021–1030.
33. Pizzo, P. A. (1987). After empiric therapy: What to do until the granulocyte comes back. *Rev Infect Dis 9*:214–219.
34. Klastersky, J. (1993). Febrile neutropenia. *Curr Opin Oncol 5*:625–632.
35. Hughes, W. T., Armstrong, D., Bodey, G. P., et al (1990). Guidelines for the use of antimicrobial agents in neutropenic patients with unexplained fever. *J Infect Dis 161*:381–396.
36. Glenn, J., Cotton, D., Wesley, R., and Pizzo, P. A. (1988). Anorectal infections in patients with malignancies. *Rev Infect Dis 10*:42–52.
37. Bodey, G. P., Rodriguez, S., Fainstein, V., et al (1991). Clostridial bacteremia in cancer patients. A 12-year experience. *Cancer 67*:1928–1942.
38. Rifkin, G. D. (1980). Neutropenic enterocolitis and *Clostridium septicum* infection in patients with agranulocytosis. *Arch Intern Med 140*:834–835.
39. Hopkins, D. G., and Kushner, J. P. (1983). Clostridial species in pathogenesis of necrotizing enterocolitis in patients with neutropenia. *Am J Hematol 14*:289–295.

40. Speirs, G., Warren, R. E., and Rampling, A. (1988). *Clostridium tertium* septicemia in patients with neutropenia. *J Infect Dis 158:*1336–1340.

41. McFarland, L. V., Mulligan, M. E., Kwok, R. Y., and Stamm, W. E. (1989). Nosocomial acquisition of *Clostridium difficile* infection. *N Engl J Med 320:*204–210.

42. Flynn, P. M., Van Hooser, B., and Gigliotti, F. (1988). Atypical mycobacterial infections of Hickman catheter exit sites. *Pediatr Infect Dis J 7:*510–513.

43. Pizzo, P. A., Robichaud, K. J., Gill, F. A., and Witebsky, F. G. (1982). Empiric antibiotic and antifungal therapy for cancer patients with prolonged fever and granulocytopenia. *Am J Med 72:*101–111.

44. Meunier-Carpentier, F., Kiehn, T. E., and Armstrong, D. (1981). Fungemia in the immunocompromised host. *Am J Med 71:*363–370.

45. Bodey, G. P. (1966). Fungal infections complicating acute leukemia. *J Chronic Dis 19:*667–687.

46. Walsh, T. J., and Pizzo, P. A. (1988). Treatment of systemic fungal infections: recent progress and current problems. *Eur J Clin Microbiol Infect Dis 7:*460–475.

47. Anaissie, E., Kantarjian, H., Ro, J., et al (1988). The emerging role of fusarium infections in patients with cancer. *Medicine 67:*77–83.

48. Rinaldi, M. G. (1989). Emerging opportunists. *Infect Dis Clin North Am 7:*460–475.

49. Anaissie, E. (1982). Opportunistic mycoses in the immunocompromised host: experience at a cancer center and review. *Clin Infect Dis 14:*S43–S53.

50. Haron, E., Feld, R., Tuffnell, X., et al (1987). Hepatic candidiasis: an increasing problem in immunocompromised patients. *Am J Med 83:*17–26.

51. Thaler, M., Pastakia, B., Shawker, T. H., et al (1988). Hepatic candidiasis in cancer patients: the evolving picture of the syndrome. *Ann Intern Med 108:*88–100.

52. Wingard, J. R., Merz, W. G., Rinaldi, M. G., et al (1991). Increase in *C. krusei* infection among patients with bone marrow transplantation and neutropenia treated prophylactically with fluconazole. *N Engl J Med 325:*1274–1277.

53. Uzun, O., and Anaissie, E. J. (1995). Antifungal prophylaxis in patients with hematologic malignancies: a reappraisal. *Blood 86:*2063–2072.

54. Press, O. W., Ramsey, P. G., Larson, E. B., et al (1984). Hickman catheter infections in patients with malignancies. *Medicine 63:*189–200.

55. Allen, R. C., Stevens, P. R., Price, T. H., Chatta, G. S., and Dale, D. C. (1997). In vivo effects of recombinant human granulocyte colony-stimulating factor on neutrophil oxidative functions in normal human volunteers. *J Infect Dis 175:*1189–1192.

56. Fainstein, V., Bodey, G. P., McCredie, K. B., et al (1983). Coagulation abnormalities induced by β-lactam antibiotics in cancer patients. *J Infect Dis 148:*745–750.

57. Conly, J. M., Chubb, H., Bow, E. J., and Louie, T. J. (1984). Hypothrombinemia in febrile neutropenic patients with cancer: association with antimicrobial suppression of intestinal microflora. *J Infect Dis 150:*202–212.

58. Armstrong, D. (1991). Empiric therapy for the immunocompromised host. *Rev Infect Dis 13*:S763–S769.
59. Bodey, G. P. (1993). Empirical antibiotic therapy for fever in neutropenic patients. *Clin Infect Dis 17(suppl 2)*:S378–S394.
60. Liles, W. C., and Dale, D. C. (1995). Current approach to the management of neutropenia. *J Intensive Care Med 10*:283–293.
61. Rubenstein, E. B., Rolston, K., Benjamin, R. S., et al (1993). Outpatient treatment of febrile episodes in low-risk neutropenic patients with cancer. *Cancer 71*:3640–3646.
62. Talcott, J. A., Siegel, R. D., Finberg, R., and Goldman, L. (1992). Risk assessment in cancer patients with fever and neutropenia: a prospective, two-center validation of a prediction rule. *J Clin Oncol 10*:316–322.
63. Talcott, J. A., Whalen, A., Clark, J., et al (1994). Home antibiotic therapy for low-risk cancer patients with fever and neutropenia—a pilot study of 30 patients based on a validated prediction rule. *J Clin Oncol 12*:107–114.
64. Malik, I., Khan, W. A., Aziz, A., and Karim, M. (1994). Self-administered antibiotic therapy for chemotherapy-induced low-risk febrile neutropenia in patients with non-hematologic neoplasms. *Clin Infect Dis 19*:522–527.
65. Malik, I. A., Khan, W. A., Karim, M., et al (1995). Feasibility of outpatient management of fever in cancer patients with low-risk neutropenia: results of a prospective randomized trial. *Am J Med 98*:224–231.
66. Lucas, K. G., Brown, A. E., Armstrong, D., et al (1996). The identification of febrile, neutropenic children with neoplastic disease at low risk for bacteremia and complications of sepsis. *Cancer 77*:791–798.
67. Benerza, D., Kiehn, T. E., Gold, J. W., et al (1988). Prospective study of infection in indwelling central venous catheters using quantitative blood cultures. *Am J Med 85*:495–498.
68. Dugdale, D. C., and Ramsey, P. G. (1990). *Staphylococcus aureus* bacteremia in patients with Hickman catheters. *Am J Med 89*: 137–141.
69. Schimpff, S. C. (1986). Empiric antibiotic therapy for granulocytopenic cancer patients. *Am J Med 80(Suppl 5C)*:13–20.
70. Klastersky, J., Vamecq, G., Cappel, R., et al (1972). Effects of the combination of gentamicin and carbenicillin on the bactericidal activity of serum. *J Infect Dis 125*:183–186.
71. Klastersky, J. (1993). Fever and neutropenia during intensive chemotherapy. *Ann Oncol 4*:603–616.
72. Bodey, G. P., Jadeja, L., and Elting, L. (1985). Pseudomonas bacteremia: retrospective analysis of 410 episodes. *Arch Intern Med 145*:1621–1629.
73. DeJongh, C. A., Joshi, J. H., Thompson, B. W., et al (1986). A double β-lactam combination versus an aminoglycoside-containing regimen as empiric antibiotic therapy for febrile granulocytopenic cancer patients. *Am J Med 80*:101–111.
74. Winston, D. J., Winston, G. H., Bruckner, D. A., et al (1988). Controlled trial of β-lactam therapy with cefoperazone plus piperacillin in febrile granulocytopenic patients. *Am J Med 85(suppl 1A)*:21–30.

75. Mortimer, J., Miller, S., Black, D., et al (1988). Comparison of cefoperazone and mezlocillin with imipenem as empiric therapy in febrile granulocytopenic patients. *Am J Med 85(suppl 1A)*:17–20.

76. Anaissie, E. J., Fainstein, V., Bodey, G. P., et al (1988). Randomized trial of β-lactam regimens in febrile neutropenic cancer patients. *Am J Med 84:*581–589.

77. Rotstein, C., Cimino, M., Winkey, K., et al (1988). Cefoperazone plus piperacillin versus mezlocillin plus tobramycin as empiric therapy for febrile episodes in neutropenic patients. *Am J Med 85(suppl 1A)*:36–39.

78. Gutmann, L., Williamson, R., Kitzic, M. D., and Acar, J. F. (1986). Synergism and antagonism in double β-lactam antibiotic combinations. *Am J Med 80:*21–29.

79. Pizzo, P. A., Hathorn, J. W., Himenez, J., et al (1986). A randomized trial comparing ceftazidime along with combination antibiotic therapy in cancer patients with fever and neutropenia. *N Engl J Med 315:*552–558.

80. Verhagen, C. S., de Pauw, B., de Witte, T., et al (1987). Randomized prospective study of ceftazidime plus cephalothin in empiric treatment of febrile episodes in severely neutropenic patients. *Antimicrob Agents Chemother 31:*191–196.

81. Wade, J. C., Johnson, D. E., and Bustamante, C. I. (1986). Monotherapy for empiric treatment of fever in granulocytopenic cancer patients. *Am J Med 80(suppl 5C)*:89–95.

82. Ramphal, R., Bolger, M., Oblon, D. J., et al (1992). Vancomycin is not an essential component of the initial empiric treatment regimen for febrile neutropenic patients receiving ceftazidime: a randomized prospective study. *Antimicrob Agents Chemother 36:*1062–1067.

83. Freifeld, A. G., Walsh, T., Marshall, D., et al (1995). Monotherapy for fever and neutropenia in cancer patients: a randomized comparison of ceftazidime versus imipenem. *J Clin Oncol 13:*165–176.

84. Granowetten, L., Wells, H., and Lange, B. J. (1988). Ceftazidime with or without vancomycin vs. cephalothin, carbenicillin and gentamicin as initial therapy of the febrile neutropenic pediatric cancer patient. *Pediatr Infect Dis J 7:*165–170.

85. Rubin, M., Hathorn, J. W., Marshall, D., et al (1988). Gram-positive infections and the use of vancomycin in 550 episodes of fever and neutropenia. *Ann Intern Med 108:*30–35.

86. Shenep, J., Hughes, W. T., Roberson, P. K., et al (1988). Vancomycin, ticarcillin, and amikacin compared with ticarcillin-clavulanate and amikacin in the empirical treatment of febrile, neutropenic children with cancer. *N Engl J Med 319:*1053–1058.

87. Karp, J. E., Hick, J. D., Angelopulos, C., et al (1986). Empiric use of vancomycin during prolonged treatment-induced granulocytopenia. *Am J Med 81:*237–242.

88. Kramer, B. S., Ramphal, R., and Rand, K. H. (1986). Randomized comparison between two ceftazidime-containing regimens and cephalothin-gentamicin-

carbenicillin in febrile granulocytopenic cancer patients. *Antimicrob Agents Chemother 30:*64–68.

89. Murray, B. E. (1991). Antibiotic resistance among enterococci: current problems and management strategies. *Curr Clin Top Infect Dis 11:*94–117.

90. Moellering, R. C. (1992). Emergence of enterococcus as a significant pathogen. *Clin Infect Dis 14:*1173–1178.

91. Leclercq, R., Dutka-Malen, S., Brisson-Noel, A., et al (1992). Resistance of enterococci to aminoglycosides and glycopeptides. *Clin Infect Dis 15:*495–501.

92. Shales, D. M., Binczewski, B., and Rice, L. B. (1993). Emerging antimicrobial resistance and the immunocompromised host. *Clin Infect Dis 17(suppl 2):*S527–S536.

93. Shay, D. K., Maloney, S. A., Montecalvo, M., et al (1995). Epidemiology and mortality risk of vancomycin-resistant enterococcal bloodstream infections. *J Infect Dis 172:*993–1000.

94. DiNubile, M. J. (1988). Stopping antibiotic therapy in neutropenic patients. *Ann Intern Med 108:*289–292.

95. Joshi, J. H., Schimpff, S. C., Tenney, J. H., et al (1979). Can antibacterial therapy be discontinued in persistently febrile granulocytopenic cancer patients? *Am J Med 76:*450–457.

96. Pizzo, P. A., Robichaud, K. J., Gill, F. A., et al (1979). Duration of empiric antibiotic therapy in granulocytopenic patients with cancer. *Am J Med 67:*194–200.

97. Sugar, A. M. (1990). Empiric treatment of fungal infections in the neutropenic host—review of the literature and guidelines for use. *Arch Intern Med 150:*2258–2264.

98. Karp, J. E., Mer, W. G., Hendricksen, C., et al (1987). Oral norfloxacin for prevention of gram-negative bacterial infections in patients with acute leukemia and granulocytopenia. A randomized, double-blind, placebo-controlled trial. *Ann Intern Med 106:*1–7.

99. Winston, D. J., Ho, W. G., Champlin, R. E., et al (1987). Norfloxacin for prevention of bacterial infections in granulocytopenic patients. *Am J Med 82:*40–46.

100. Dekker, A. W., Rozenberg-Arska, M., and Verhoef, J. (1987). Infection prophylaxis in acute leukemia: a comparison of ciprofloxacin with trimethoprim-sulfamethoxazole and colistin. *Ann Intern Med 106:*7–11.

101. Bow, E. J., Loewen, R., and Vaughan, D. (1996). Reduced requirement for gram-negative antibiotic therapy in febrile, neutropenic patients with cancer who are receiving antibacterial chemoprophylaxis with oral quinolones. *Clin Infect Dis 20:*907–912.

102. Bow, E. J., Mandell, L. A., Louie, T. J., et al (1996). Quinolone-based antibacterial chemoprophylaxis in neutropenic patients: effect of augmented gram-positive activity on infectious morbidity. *Ann Intern Med 125:*183–190.

103. Bow, E. J., Rayner, E., and Louie, T. J. (1988). Comparison of norfloxacin with cotrimoxazole for infection prophylaxis in acute leukemia. The trade-off for reduced gram-negative sepsis. *Am J Med 84:*847–854.

104. Winston, D. J., Ho, W. G., Bruckner, D. A., et al (1990). Ofloxacin versus vancomycin/polymyxin for prevention of infections in granulocytopenic patients. *Am J Med 88:*36–42.

105. Donnelly, J. P., Maschmeyer, G., Daenen, S. (1992). Selective oral antimicrobial prophylaxis for the prevention of infection in acute leukaemia-ciproflaxacin versus co-trimoxazole plus colistin. The EORTC-Gnotobiotic Project Group. *Eur J Cancer 28A:*873–878.

106. Burnie, J. P., Odds, F. C., Lee, W., Webster, C., and Williams, J. D. (1985). Outbreak of systemic *Candida albicans* in intensive care unit caused by cross infection. *Br Med J 290:*746–748.

107. Milliken, S. T., and Powles, R. L. (1990). Antifungal prophylaxis in bone marrow transplantation. *Rev Infect Dis 12(suppl 3)*:S374–S379.

108. Denning, D. W., Donnelly, J. P., Hellreigel, K. P., et al (1991). Antifungal prophylaxis during neutropenia or allogeneic bone marrow transplantation: What is state of the art? *Chemotherapy 38:*43–49.

109. Beyer, J., Schwartz, S., Heinemann, V., and Siegert, W. (1994). Strategies in prevention of invasive aspergillosis in immunosuppressed or neutropenic patients. *Antimicrob Agents Chemother 38:*911–917.

110. Walsh, T. J., and Lee, J. W. (1993). Prevention of invasive fungal infections in patients with neoplastic diseases. *Clin Infect Dis 17(suppl 2)*:S468–S480.

111. Menichetti, F., Del Favero, A., Martino, P., et al (1994). Preventing fungal infection in patients with acute leukemia: fluconazole compared with oral amphotericin B. *Ann Intern Med 120:*913–918.

112. Goodman, J. L., Winston, D. J., Greenfield, R. A., et al (1992). A controlled trial of fluconazole to prevent fungal infections in patients undergoing bone marrow transplantation. *N Engl J Med 326:*845–851.

113. Rousey, S. R., Russler, S., Gottlieb, M., and Ash, R. C. (1991). Low-dose amphotericin B prophylaxis against invasive *Aspergillus* infections in allogeneic transplantation. *Am J Med 91:*484–492.

114. Perfect, J. R., Klotman, M. E., Gilbert, C. C., et al (1992). Prophylactic intravenous amphotericin in B neutropenic autologous bone marrow transplant recipients. *J Infect Dis 165:*891–897.

115. Muhldorfer, S. M., and Konig, H. J. (1990). Prophylaxis of fungal infections with itraconazole in immunocompromised cancer patients. *Mycoses 33:* 291–295.

116. Singh, N., Mieles, L., Yu, V. L., and Gayowski, T. (1993). Invasive aspergillosis in liver transplant recipients: association with candidemia and consumption coagulopathy and failure of prophylaxis with low-dose amphotericin B. *Clin Infect Dis 17:*906–908.

117. Todeschini, G., Murari, C., Bonesi, R., et al (1993). Oral itraconazole plus nasal amphotericin B for prophylaxis of invasive aspergillosis in patients with hematological malignancies. *Eur J Clin Microbiol Infect Dis 12:*614–618.

118. Tollemar, J., Hockerstedt, K., Ericzon, B. G., et al (1995). Liposomal amphotericin B prevents invasive fungal infections in liver transplant recipients. A randomized, placebo-controlled study. *Transplantation 59:*45–50.

119. Van Cutsem, J. (1994). Prophylaxis of *Candida* and *Aspergillus* infections with oral administration of itraconazole. *Mycoses 37*:243–248.

120. O'Donnell, M. R., Schmidt, G. M., Tegtmeier, B. R., et al (1994). Prediction of systemic fungal infection in allogeneic marrow recipients: Impact of amphotericin prophylaxis in high-risk patients. *J Clin Oncol 12*:827–834.

121. Riley, D. K., Pavia, A. T., Beatty, P. G., et al (1994). The prophylactic use of low-dose amphotericin B in bone marrow transplant patients. *Am J Med 97*:509–514.

122. Conneally, E., Cafferkey, M. T., Daly, P. A., et al (1990). Nebulized amphotericin B as prophylaxis against invasive aspergillosis in granulocytopenic patients. *Bone Marrow Transplant 5*:403–406.

123. Jeffrey, G. M., Beard, M. E., Ikram, R. B., et al (1991). Intranasal amphotericin B reduces the frequency of invasive aspergillosis in neutropenic patients. *Am J Med 90*:685–692.

124. Myers, S. E., Devine, S. M., Topper, R. L., et al (1992). A pilot study of prophylactic aerosolized amphotericin B in patients at risk for prolonged neutropenia. *Leuk Lymph 8*:229–232.

125. Beyer, J., Baarzen, G., Risse, G., et al (1993). Aerosol amphotericin B for prevention of invasive pulmonary aspergillosis. *Antimicrob Agents Chemother 37*:1367–1369.

126. Beyer, J., Schwartz, S., Barzen, G., et al (1994). Use of amphotericin B aerosols for the prevention of pulmonary aspergillosis. *Infection 22*:143–148.

127. Hertenstein, B., Kern, W. W., Schmeiser, T., et al (1994). Low incidence of invasive fungal infections after bone marrow transplantation in patients receiving amphotericin B inhalations during neutropenia. *Ann Hematol 68*:21–26.

128. Behre, G. F., Schwartz, S., Lenz, K., et al (1995). Aerosol amphotericin B inhalations for prevention of invasive pulmonary aspergillosis in neutropenic cancer patients. *Ann Hematol 71*:287–291.

129. Tollemar, J., Ringden, O., Andersson, S., et al (1993). Prophylactic use of liposomal amphotericin B (AmBisome) against fungal infections: A randomized trial in bone marrow transplant patients. *Transplant Proc 25*:1495–1497.

130. Strauss, R. G. (1993). Therapeutic granulocyte transfusions in 1993. *Blood 81*:1675–1678.

131. Strauss, R. G. (1994). Granulocyte transfusion therapy. *Hematol Oncol Clin North Am 8*:1159–1166.

132. Strauss, R. G. (1995). Clinical perspectives of granulocyte transfusions: efficacy to date. *J Clin Apheresis 10*:114–118.

133. Metcalf, D. (1990). The colony stimulating factors: discovery, development, and clinical applications. *Cancer 65*:2185–2195.

134. Groopman, J. E., Molina, J. M., and Scadden, D. T. (1989). Hematopoietic growth factors: biology and clinical applications. *N Engl J Med 321*:1449–1459.

135. Weisbart, R. H., and Golde, D. W. (1989). Physiology of granulocyte and macrophage colony-stimulating factors in host disease. *Hematol Oncol Clin North Am 3*:401–409.

136. Demetri, G. D., and Griffin, J. D. (1991). Granulocyte colony-stimulating factor and its receptor. *Blood 78:*791–808.

137. Lieschke, G. J., and Burgess, A. (1992). Granulocyte colony-stimulating factor and granulocyte-macrophage colony-stimulating factor, Parts I and II. *N. Engl J Med 327:*28–35;99–106.

138. Gasson, J. C. (1991). Molecular physiology of granulocyte-macrophage colony-stimulating factor. *Blood 77:*1131–1145.

139. Morstyn, G., Campbell, L., Souza, L. M., et al (1988). Effect of granulocyte colony stimulating factor on neutropenia induced by cytotoxic chemotherapy. *Lancet i:*667–672.

140. Crawford, J., Ozer, H., Stoller, R., et al (1991). Reduction by granulocyte colony-stimulating factor of fever and neutropenia induced by chemotherapy in patients with small-cell lung cancer. *N Engl J Med 325:*164–170.

141. Gisselbrecht, C., Prentice, H. G., Bacigalupo, A., et al (1994). Placebo-controlled phase III trial of lenograstim in bone marrow transplantation. *Lancet 343:*696–700.

142. Gabrilove, J., Jakubowski, A., Scher, H., et al (1988). Granulocyte colony stimulating factor reduces neutropenia and associated morbidity of chemotherapy for transitional cell carcinoma of the urothelium. *N Engl J Med 318:*1414–1422.

143. Taylor, K., Jagannath, S., Spitzer, G., et al (1989). Recombinant human granulocyte colony-stimulating factor hastens granulocyte recovery after high-dose chemotherapy and autologous bone marrow transplantation in Hodgkin's disease. *J Clin Oncol 7:*1791–1799.

144. Dale, D. C., Bonilla, M. A., Davis, M. W., et al (1993). A randomized controlled phase III trial of recombinant human G-CSF for treatment of severe chronic neutropenia. *Blood 81:*2496–2502.

145. Maher, D. W., Lieschke, G. J., Green, M., et al (1994). Filgrastim in patients with chemotherapy-induced febrile neutropenia. A double-blind, placebo-controlled trial. *Ann Intern Med 121:*492–501.

146. Nemunaitis, J., Rabinowe, S., Singer, J., et al (1991). Recombinant granulo-cyte-macrophage colony-stimulating factor after autologous bone marrow transplantation for lymphoid cancer. *N Engl J Med 324:*1773–1778.

147. Gerhartz, H. H., Engelhard, M., Meusers, M., et al (1993). Randomized, double-blind, placebo-controlled, phase III study of recombinant human granulocyte-macrophage colony-stimulating factor as adjunct to induction treatment of high-grade malignant non-Hodgkin's lymphomas. *Blood 82:*2329–2339.

148. deVries, E. G., Biesma, B., Willemse, P. H., et al (1991). A double-blind placebo-controlled study with granulocyte-macrophage colony-stimulating factor during chemotherapy for ovarian carcinoma. *Cancer Res 51:*116–122.

149. Ozer, H. (for Am Soc Clin Oncol ad hoc CSF Guidelines Expert Panel) (1994). American Society of Clinical Oncology recommendations for the use of hematopoietic colony-stimulating factors: evidence-based clinical practice guidelines. *J Clin Oncol 12:*2471–2508.

150. Schiffer, C. A. (1990). Granulocyte transfusions: an overlooked therapeutic modality. *Transf Med Rev 4*:2–7.

151. Freireich, E. J. (1994). White cell transfusions born-again. *Leuk Lymph 11*:161–165.

152. Dale, D. C., Liles, W. C., Llewellyn, C., Rodger, E., Bowden, R., and Price, T. H. (1996). Neutrophil transfusion therapy: characteristics and kinetics of cells from donors treated with a combination of G-CSF and dexamethasone. *Blood 88*:627a (abstr 2495).

153. Price, T. H., Lee, M. Y., Dale, D. C., and Finch, C. A. (1979). Neutrophil kinetics in chronic neutropenia. *Blood 54*:581–594.

154. Bensinger, W. I., Price, T. H., Dale, D. C., et al (1993). The effects of daily recombinant human granulocyte colony-stimulating factor administration on normal granulocyte donors undergoing leukapheresis. *Blood 81*:1883–1888.

155. Caspar, C. B., Seger, R. A., Burger, J., et al (1993). Effective stimulation of donors for granulocyte transfusions with recombinant methionyl granulocyte colony-stimulating factor. *Blood 81*:2866–2871.

156. Liles, W. C., Huang, J. E., Llewellyn, C., SenGupta, D., Price, T. H., and Dale, D. C. (1997). A comparative trial of granulocyte colony-stimulating factor (G-CSF) and dexamethasone alone and in combination for the mobilization of neutrophils in the peripheral blood of normal volunteers. *Transfusion 37*:182–187.

157. Boggs, D. R., and Joyce, R. A. (1983). The hematopoietic effects of lithium. *Sem Hematol 20*:129–138.

158. Roozedaal, K. J., Dicke, K. A., and Boonzajer Flaes, M. L. (1981). Effect of oxymetholone on human cyclic hematopoiesis. *Br J Haematol 47*:185–193.

159. von Schulthess, G. K., Fehr, J., and Dahinden, C. (1983). Cyclic neutropenia: amplification of granulocyte oscillations by lithium and long-term suppression of cycling by plasmapheresis. *Blood 62*:320–326.

160. Williams, D. J., and Jones, J. V. (1984). Lithium carbonate treatment in familial cyclic neutropenia. *Am J Clin Pathol 81*:120–122.

161. Klutman, N. E. (1989). Lithium carbonate therapy for zidovudine-associated neutropenia in patients with acquired immunodeficiency syndrome. *Am J Med 87*:362–363.

162. Stein, R. S., Beaman, C., Ali, M. Y., et al (1977). Lithium carbonate attenuation of chemotherapy-induced neutropenia. *N Engl J Med 297*:430–431.

163. Lyman, G. H., Williams, C. C., Preston, P., et al (1983). Lithium carbonate in patients with small cell lung cancer receiving combination chemotherapy (and radiotherapy). *Am J Med 70*:1222–1229.

164. Wright, D. G., Fauci, A. S., Dale, D. C., and Wolff, S. M. (1978). Correction of human cyclic neutropenia with prednisolone. *N Engl J Med 298*:295–300.

165. Rodgers, G. M., and Shuman, M. A. (1982). Acquired cyclic neutropenia: successful treatment with prednisone. *Am J Hematol 13*:83–89.

6
Use of Filgrastim (r-metHuG-CSF) in Severe Chronic Neutropenia

David C. Dale
University of Washington, Seattle, Washington

Carol Fier
Amgen Inc., Thousand Oaks, California

Karl Welte
Medizinische Kinderklinik, Hanover Medical School, Hanover, Germany

I. INTRODUCTION

Chronic neutropenia usually is defined as a reduction in blood absolute neutrophil count (ANC) below the lower limits for normal individuals (ie, 1.8×10^9/L that lasts for more than a few weeks) (1). When the neutropenia is mild (ie, 1.0 to 1.8×10^9/L), there is little or no increased risk of infections (2,3). In some populations, neutrophil counts in this range should be regarded as normal (4).

With more severe neutropenia (ie, ANC $<1.0 \times 10^9$/L), problems with chronic inflammation and recurrent infections occur. These problems are increasingly frequent and severe, with progressively lower neutrophil counts because maintenance of the integrity of the mucosal surfaces of the body and handling minor breaks in the skin requires a steady supply of neutrophils. In neutropenic patients there is a direct relationship between the circulating neutrophil count and the number of these cells accumulating in oral secretions or at a site of acute cutaneous injury (5,6). It is, therefore, not surprising that deficiencies of neutrophils, either acute or chronic, result in oral ulcerations, gingivitis, and superficial skin infections caused by invasion of surface organisms.

Although mild-to-moderate neutropenia is a relatively common finding, severe chronic neutropenia (SCN) (ie, blood neutrophils $<0.5 \times 10^9$/L for more than 1 month) is much less common. Causes of chronic neutropenia include chronic viral infections (eg, infectious mononucleosis, viral hepatitis, and human immunodeficiency virus [HIV] infection), the autoimmune diseases (eg, systemic lupus erythematosus, rheumatoid arthritis, and Sjögren's syndrome), and several diseases associated with splenomegaly (eg, cirrhosis, sarcoidosis, and chronic parasitic infections). Tumors infiltrating the bone marrow, bone marrow ablation by radiation or chemotherapy, and progressive hematologic diseases (eg, leukemia, myelodysplasia, and aplastic anemia) also can cause chronic neutropenia (1).

In contrast to these conditions in which neutropenia, anemia, and thrombocytopenia often occur concomitantly, severe chronic neutropenia also occurs as a selective hematologic abnormality in a somewhat heterogeneous group of rare conditions. This group of disorders, which may be collectively referred to as severe chronic neutropenia, includes congenital neutropenia, cyclic neutropenia, and idiopathic neutropenia (2,3,7–11). Because of low ANC, these patients have repeated episodes of fever, oropharyngeal inflammation, gastrointestinal symptoms, perirectal inflammation, and cutaneous infections. Deep tissue infections (eg, pneumonia and liver abscess) and bacteremias occur less often. Serious or life-threatening infections occur in the most severely affected patients, ie, individuals with ANC $< 0.2 \times 10^9$/L. The pattern of these events, particularly the frequency of chronic inflammation and the infrequency of bacteremias, is probably attributable to the intactness of other host defense mechanisms, including normal or elevated amounts of blood monocytes and eosinophils, normal functions of the reticuloendothelial system, and normal amounts of immunoglobulins and complement (2). These patients rarely develop chronic viral infections or infections by intracellular pathogens (eg, tuberculosis, salmonellosis, and fungal infections) unless they are treated with corticosteroids, immunosuppressive drugs or prolonged antibiotics. In general, infections are more severe for congenital neutropenia than for cyclic or idiopathic neutropenia (3,8,11).

II. CONGENITAL NEUTROPENIA

Congenital neutropenia, or congenital agranulocytosis, is usually a severe disorder recognized at or soon after birth because of fever, infections, and severely diminished blood neutrophils (9). This condition was originally described in several Swedish families as an autosomal recessive disorder, but the family history is negative for the majority of reported cases (1).

The bone marrow shows normal or reduced cellularity with few neutrophil precursors beyond the promyelocyte stage. Increased eosinophils and monocytes in the bone marrow and blood may be present. Severe infections, ie, pneumonia, abscesses of the liver and other tissues, diarrhea, and bacteremia, occur frequently. With good medical care and frequent antibiotic treatments, survival beyond the first few months of life is now common, with many patients reaching adulthood. Throughout life, however, recurrent fevers and infections remain a major problem (3).

There are probably several forms of severe congenital neutropenia, but definitive description of distinctive subtypes is not yet possible because the cellular and molecular basis for this condition is not yet known (12). Congenital neutropenia may be confused with other disorders causing chronic infections, anemia, and neutropenia. For example, patients with combined immunodeficiency syndromes, ie, deficits of both T- and B-cell lineages (13,14) or congenital aplastic anemia (Fanconi's anemia and other types), may be severely neutropenic and anemic (15). Severe chronic neutropenia may be the predominant hematologic abnormality in some cases of the Shwachman-Diamond syndrome (congenital exocrine pancreatic insufficiency with hypoplastic hematologic abnormalities), whereas in other cases anemia and thrombocytopenia may be equally severe (16). The Chédiak-Higashi syndrome also causes severe anemia, thrombocytopenia, and neutropenia, but is distinguishable by the abnormal primary granules in leukocytes, both in the blood and bone marrow (17).

There are also more benign forms of congenital or childhood neutropenia, distinguished on clinical grounds by somewhat less severe infections, neutrophil counts which are slightly higher, and a less severe marrow defect in neutrophil precursor numbers. In some cases, usually labeled as myelokathexis, the neutrophils show cytoplasmic vacuoles and abnormal nuclear morphology (18). In other cases, sometimes called the lazy-leukocyte syndrome, the neutrophils appear morphologically normal (19). In both of these conditions there appears to be a substantially greater supply of bone-marrow neutrophils than blood neutrophils. Children with glycogen storage disease (Type 1b) also have severe neutropenia with a relative abundance of normal-appearing bone marrow neutrophils; the neutropenia in this disorder may become progressively more severe over the first few years of life (20). In neonatal isoimmune neutropenia, a condition attributable to transplacental transfer of maternally produced antineutrophil antibodies, the neutropenia generally resolves within the first few months of life (21). In other cases without these markers, neutropenia may be attributable to autoimmune neutrophil destruction (22), but accurate tests for antineutrophil antibodies are not widely available.

The clinical course and occurrence of splenomegaly in the congenital neutropenias is quite variable. Children with chronic inflammation and recurrent infections have intense immune stimulation which may result in lymphadenopathy, hypergammaglobulinemia, and splenomegaly. In general, the clinical severity of these disorders decreases with age, but the neutropenia persists. Remissions without recurrence are expected in isoimmune neutropenia; spontaneous remissions in other forms of congenital neutropenia have been reported, but are probably quite rare. Leukemic transformations have been reported in only a few cases of congenital neutropenia (23–25); leukemic transformations may be more frequent in Fanconi's anemia and the Shwachman-Diamond syndrome (26).

III. CYCLIC NEUTROPENIA

Cyclic neutropenia is a rare hematological disorder first described in 1910 (for a review see Ref. 8). It is characterized by regular oscillations of blood neutrophils and other leukocytes, as well as reticulocytes and platelets, with a usual periodicity of 21 days. It is also called cyclic hematopoiesis because of the cycling of all blood cell types and evidence that it can be transferred by bone marrow transplantation (27). Approximately two-thirds of cases have a family history suggesting inheritance as an autosomal dominant disorder. Cycling of neutrophils also can occur as an acquired disorder, sometimes associated with features suggesting an autoimmune disorder, eg, large granular lymphocytes or symptoms of arthritis (28). The clinical features of cyclic neutropenia are almost exclusively attributable to the oscillations of blood neutrophils and the recurrent severe neutropenia (usually ANC $< 0.1 \times 10^9$/L) which last for 3 to 6 days during every 21-day cycle. During these neutropenic periods the patients have fever, oral ulcerations, and cervical lymphadenopathy; these findings are particularly severe in children. Episodes of abdominal pain may occur due to ulcerations in the colon and can result in peritonitis and clostridial bacteremia (29–31).

Long-term follow-up of patients with cyclic neutropenia suggest that the characteristic pattern of cyclical symptoms lasts indefinitely but that symptoms tend to become milder in early adulthood. Splenomegaly, chronic anemia, or hypergammaglobulinemia are uncommon. Family studies suggest that the degree of oscillation of the counts and the severity of neutropenia may be less in adults than in children (32,33).

Because mouth ulcers and repeated bouts of inflammation occur in all forms of SCN, it is difficult to diagnose cyclic neutropenia on the basis of symptoms and only a few neutrophil counts; serial counts for several weeks are often required. Regular oscillations of blood cell counts can occur in other disorders, eg, chronic myelogenous leukemia (34) and the early phase of myeloid malignancies (35).

IV. CHRONIC IDIOPATHIC NEUTROPENIA

In both children and adults, severe neutropenia occurs in patients lacking a lifelong history of frequent infections and with previous normal blood cell counts. If there is no evidence of a hematologic malignancy (morphologic abnormalities of marrow cells, chromosomal abnormalities, etc), other blood cells are normal or near normal, and the neutropenia cannot be attributed to another disease, the condition is called chronic idiopathic neutropenia (11) or chronic neutropenia of childhood (36). This is a heterogeneous group of disorders; immunological testing (ie, antineutrophil antibody tests) has not yet provided an unequivocal way to distinguish which of these patients have neutropenia on an immune basis and which are due to another mechanism (37,38). Often the patients have leukopenia as well as neutropenia, with reduced blood lymphocytes, monocytes, and eosinophils, and varying degrees of splenomegaly. Long-term observations suggest that most of these patients will maintain a reasonably stable leukocyte and neutrophil count for long periods (2). They do not predictably evolve to leukemia, aplastic anemia, or the myelodysplastic syndromes. Unless the neutrophil count is $<0.5 \times 10^9$/L, infections are quite infrequent. Below this value, however, patients have recurrent fevers and chronic inflammation in the oropharynx and other areas.

V. THERAPY BEFORE THE AVAILABILITY OF HEMATOPOIETIC GROWTH FACTORS

Before the availability of the hematopoietic growth factors, there was no predictably effective therapy available for SCN. The mainstays of care were careful observation and prompt antibiotic treatment. Historically, a variety of agents were given short term without apparent benefit. Splenectomy for neutropenia was introduced in the 1930s, but generally proved ineffective except in the syndrome of rheumatoid arthritis, splenomegaly, and neutropenia, ie, Felty's syndrome (39–41). Glucocorticosteroids, usually prednisone up to 1 mg/kg/day, have been widely used. They are often either ineffective or unsuitable for long-term therapy because of their side effects (2). Lithium salts, which may increase endogenous levels of colony-stimulating factors (42), have proved to be ineffective as a long-term treatment of chronic neutropenia except in rare instances. A variety of immunosuppressive drugs have been tried in the idiopathic neutropenias, presuming an autoimmune mechanism (37,43). The benefits were very limited, with adverse sequelae, even deaths, reported. More recently, various gamma-globulin preparations have been used to treat SCN, particularly in children, with transient benefit in many cases (22). The long-term benefits from this treatment are uncertain. Bone marrow transplantation (BMT) has been used successfully for only a few patients with congenital neutropenia (44).

Many other therapies have been attempted (39,45,46). In one interesting case of cyclic neutropenia, a child was treated with his mother's plasma collected after endotoxin administration (47). This is the only known case in which any treatment altered neutrophil cycling before the availability of the hematopoietic growth factors. In dogs with cyclic neutropenia, lithium carbonate and endotoxin, presumably by elevating colony-stimulating factor amounts, are effective treatments (48,49).

VI. HEMATOPOIETIC GROWTH FACTOR THERAPY

The principal reasons for initiating clinical trials of Filgrastim (r-metHuG-CSF) for treatment of patients with SCN were the severity of this condition, the lack of any predictably efficacious therapy, and the observation that Filgrastim selectively stimulated neutrophil production (50,51).

Early in development of Filgrastim treatment for SCN, four phase 1 and 2 trials and one phase 3 trial were conducted (51–57) (**Table 1**) and numerous other cases reported (58–64).

Initial trials of Filgrastim for congenital neutropenia also showed its potential to increase blood neutrophil counts and to reduce chronic infection and inflammation (53,56) (**Figure 1**). The doses of Filgrastim required to increase blood neutrophils to the lower limits of normal were quite variable, but almost all patients responded. Even with only modest neutrophil increases, clinical improvement occurred. Given the nature of the bone-marrow defect in these patients, ie, often very few neutrophils beyond the promyelocyte stage, and the time normally required for neutrophils to transit the postmitotic compartment, ie, about 6 days, it was perhaps not surprising that it took several days for increases in blood neutrophil counts to occur. In some patients, particularly those with eosinophilia before therapy, the eosinophil counts also increased with treatment. Splenic enlargement was noteworthy both before and with treatment. The splenomegaly was generally not a cause for symptoms, but in some instances, there was an associated thrombocytopenia.

Table 1. Overview of the Filgrastim Severe Chronic Neutropenia Clinical Trials.

Reference no.	Country	Study design	Filgrastim treatment	Diagnosis	Patients evaluated for efficacy
Phase I/II Studies					
56	USA	Open-label, single center	0.58 to 6.90 μg/kg SC qd	Idiopathic neutropenia	3
54	USA	Open-label, single center	2.07 to 11.50 μg/kg iv 30 minutes qd, 0.58 to 5.75 μg/kg SC qd	Cyclic neutropenia	7
53	USA	Open-label, single center	3.45 to 69.00 μg/kg iv 30 minutes qd	Congenital neutropenia	13
56,76	Germany/France	Open-label two centers	72.0 to 145.00 μg/kg iv infusion, 0.25 to 145.00 μg/kg SC qd	Idiopathic cyclic, and congenital neutropenia	51
Phase III Study					
94	USA	Open-label, randomized controlled, comparative multicenter	0.14 to 100.00 μg/kg SC qd	Idiopathic, cyclic, and congenital neutropenia	123

iv = intravenous
SC = subcutaneous
qd = everyday

A

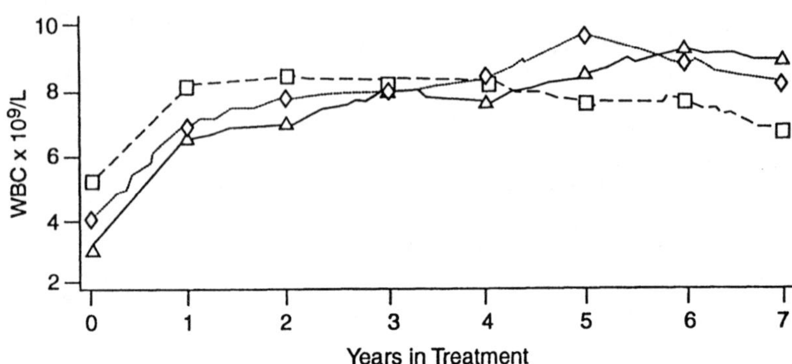

B

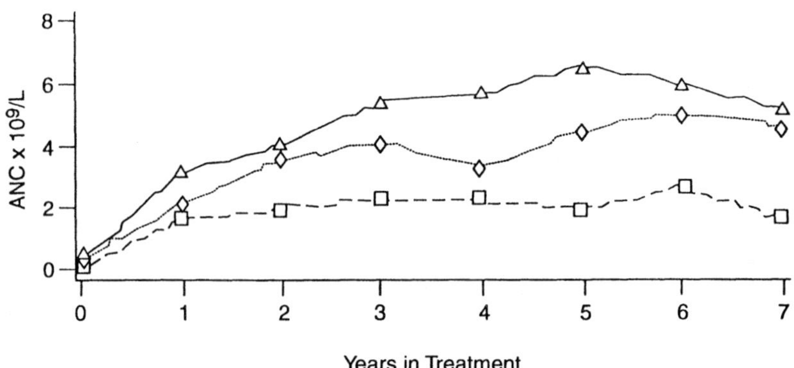

Figure 1. Hematologic response for Filgrastim-treated patients in the Severe Chronic Neutropenia International Registry. (A) median white blood cell count by year of Filgrastim treatment; (B) median neutrophil count by year of Filgrastim treatment; (C) median hemoglobin count by year of Filgrastim treatment; and (D) median platelet count by year of Filgrastim treatment. Open boxes, congenital neutropenia; open triangles, idiopathic neutropenia; open diamonds, cyclic neutropenia.

Seven patients with cyclic neutropenia were enrolled in another trial (54). The results of this trial showed that Filgrastim in doses of 3 to 5 μg/ kg/day administered subcutaneously could increase peak neutrophil counts to supranormal values, but that cyclical oscillations of neutrophil and other blood cell counts persisted. Major clinical benefit occurred, however, be-

C

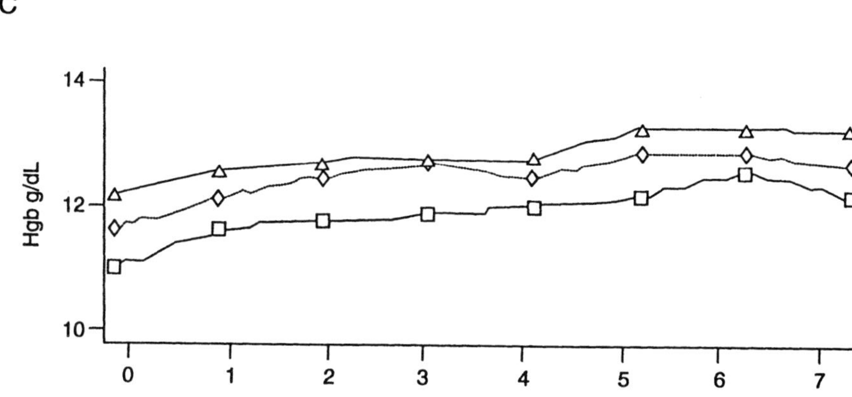

D

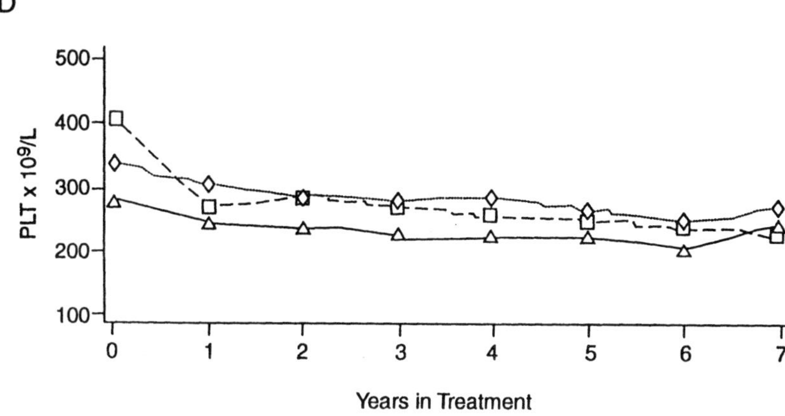

cause the duration of neutropenic periods was greatly reduced, ie, shortened from 4 to 6 days per cycle to usually about 1 day. Treatment also shortened the time between neutropenic periods from about 21 days to approximately 14 days. Other noteworthy observations with treatment were increased cellularity and percentage of postmitotic neutrophils in bone marrow samples, the normal half-life of blood neutrophils, and normal neutrophil inflammatory response, which were measured using a skin chamber technique. It also was recognized in this study that spleen size increased during treatment, but other adverse events were noted very infrequently. Treatment responses have been similar in other case reports of cyclic neutropenia (58,62–65).

Initially, a single case of chronic idiopathic neutropenia was reported to respond promptly to relatively low doses of daily Filgrastim administration (55). Subsequently, two other patients treated had similar responses, although one eventually underwent splenectomy because of massive splenic enlargement (unpublished data). In other patients diagnosed with idiopathic neutropenia, prompt response also occurred, although it was difficult in some of these cases to be sure the neutropenia was not drug induced (61,66).

VII. PHASE 3 TRIAL OF FILGRASTIM FOR SEVERE CHRONIC NEUTROPENIA

On the basis of these preliminary trials, a multicenter, randomized, phase 3 controlled trial of Filgrastim for the treatment of SCN was organized (57). The study design required patients to be at least 6 months old, a provision which excluded neonates with presumed isoimmune neutropenia. Patients enrolled were required to have a history of significant infections and either documented severe neutropenia with at least three neutrophil counts of $<0.5 \times 10^9$/L in the previous 6-month period, or regular neutrophil cycles with five consecutive days of neutrophils $<0.5 \times 10^9$/L with each cycle. To exclude hematologic malignancies and myelodysplasia, a pretreatment bone-marrow examination and cytogenetic analysis were required. Patients were excluded if they had a history or physical finding suggestive of Felty's syndrome or other autoimmune diseases, were receiving corticosteroids or gamma-globulin injections, or had neutropenia associated with circulating large granular lymphocytes $>5.0 \times 10^9$/L. Patients with pancytopenia, ie, a hematocrit consistently $<30\%$ or a platelet count consistently $<100 \times 10^9$/L, were excluded. Patients with other, potentially confounding medical conditions, eg, heart disease, pregnancy, or psychiatric disorders, also were excluded.

Patients who met the entry criteria were grouped by diagnosis and randomized from a central coordinating center to either (a) begin treatment with Filgrastim or (b) delay treatment until completion of a 4-month observation period, after which treatment would be initiated. A standard preparation of Filgrastim (300 μg/mL) was used. The first month of treatment was a dose-adjustment period, with stepwise increases in Filgrastim. The initial Filgrastim doses used were 3.6 μg/kg/day for idiopathic neutropenia, 6.0μg/kg/day for cyclic neutropenia, and 6.0 μg/kg/twice a day (BID) for congenital neutropenia. Patients not responding to a given dose within 2 weeks had their daily dose increased to the next dose level until they reached 12 μg/kg/BID; higher doses were used in a small number of patients. All doses were administered subcutaneously (SC) in the abdominal wall by the

Table 2. Severe Chronic Neutropenia Phase 3 Trial; Overall Significant Changes in Clinical Endpoints.*

	Control patients	Filgrastim-treated patients	p-Value
Incidence of infection	0.50	0.20	<0.001
Incidence of fever	0.25	0.20	<0.001
Incidence of oropharyngeal ulcers	0.26	0.00	0.001
Incidence of antibiotic use	0.49	0.20	<0.001

* Median incidence per 28-day period

patient or other care giver after careful instruction. Observations included physical and bone-marrow examinations, complete blood counts, radiographs, and other clinical assessments as indicated.

The results of this trial confirm the findings of the phase 1/2 trials (52,67,68). For the 123 patients evaluated, the median age was 12 years (range, 1 to 76 years). One hundred eight of the 120 patients who received Filgrastim showed a complete response, defined as an increase of the median ANC to $>1.5 \times 10^9$/L. Four additional patients showed partial responses, ie, median ANC $>0.5 \times 10^9$/L but $<1.5 \times 10^9$/L. In all responding patients, as well as others with lesser ANC increases, there was clinical improvement similar to that noted in the phase 1/2 trials. Mouth ulcers, presumed infections (defined as inflammatory symptoms treated with antibiotics), and total antibiotic use all decreased significantly (**Table 2**).

In this trial, bone pain early in therapy was the most noteworthy symptom, as in other Filgrastim trials (69,70). Splenic enlargement occurred and the degree of enlargement was quite variable. Splenomegaly occurred with greater frequency and severity in patients with congenital neutropenia. Other adverse effects included rashes (usually mild and remote from the injection site), hair loss (transient), vasculitis and glomerulonephritis (infrequent and not progressive), and platelet count reductions. Doses of Filgrastim were reduced in a few patients because of thrombocytopenia, but this alteration infrequently required discontinuation of Filgrastim therapy.

VIII. LONG-TERM FILGRASTIM THERAPY

Governmental agencies throughout the world reviewed the results of these trials and approved marketing of Filgrastim for the treatment of SCN. A steadily increasing number of patients with SCN were begun on treatment (71). In March 1994, the Severe Chronic Neutropenia International Registry

(SCNIR, or the Registry) opened to monitor the clinical course, treatment, and complications of patients with SCN, including patients receiving no treatment, Filgrastim, or other therapies (7,72,73). The Registry provides data on the safety and efficacy of Filgrastim treatment to regulatory agencies, to physicians caring for patients with SCN and, through liaison with patient support organizations, to patients themselves. The Registry maintains offices for centralized collection and analysis of data in Seattle, Washington and Hannover, Germany, and its activities are overseen by an Advisory Board of expert physicians from Australia, Europe, and North America. The Registry has become a unique resource to facilitate research on the genetic, molecular, and cellular mechanisms for SCN.

The Registry began its activities with patients previously enrolled in clinical trials who were continuing long-term treatment with Filgrastim. Over time, patients receiving no treatment or treatments other than Filgrastim as well as new patients receiving Filgrastim have been added. Enrollment requires documented SCN and other clinical findings consistent with the diagnosis of congenital, cyclic, or idiopathic neutropenia.

Through 1996, the Registry has collected long-term follow-up data on 420 patients in 12 countries. There are some variations by diagnosis in the age, sex, and ethnicity of the enrolled population (**Table 3**). The patients with congenital neutropenia are the youngest (mean age, 11.3 years), but the range of age in this group extends well into adulthood. The patients with cyclic and idiopathic neutropenia are somewhat older (mean age, 23.7 and 30.2 years, respectively). There is roughly an equal proportion of male and female registrants with congenital and cyclic neutropenia. For patients with idiopathic neutropenia, there are slightly more female registrants, as has been noted in other studies. Most enrolled subjects are white; the distribution by ethnicity for patients with congenital neutropenia is somewhat more diverse than for the other groups. Little data are available to indicate whether these diagnoses are equally prevalent in various racial or ethnic groups.

Before the initiation of clinical trials of Filgrastim therapy for SCN, there were no systematically collected data available on the natural history or the efficacy of treatments other than anecdotal reports. Clinical data for patients enrolled in the phase 1–2 and phase 3 trials indicated that some patients had previously received corticosteroids and recombinant human granulocyte-macrophage colony-stimulating factor (rHuGM-CSF), but these treatments were often discontinued because of a lack of efficacy or side effects (56,74). Currently because most patients with SCN receiving no treatment have experienced recurring symptoms and because of the efficacy of Filgrastim, most data that can now be collected relate to long-term treatment with this growth factor.

Table 3. Severe Chronic Neutropenia International Registry Demographics (as of 12/95).

	Congenital n = 219	Cyclic n = 78	Idiopathic n = 123	Total n = 420
Age				
mean (±SD)	11.3(15.4)	23.7(15.4)	30.2(23.1)	19.1(17.5)
median	10.2	21.0	27.8	13.7
range	0.2–40.6	1.4–72.8	0.6–77.5	0.2–77.5
Pediatric patients (age < 18 yrs) n(%)	180(82.2)	35(44.9)	48(39)	263(62.6)
Adult patients (age ≥ 18 years) n(%)	39(17.8)	43(55.1)	75(61)	157(37.4)
Sex n(%)				
Male	113(51.6)	34(43.6)	42(34.1)	189(45)
Female	106(48.4)	44(56.4)	79(64.2)	229(54.5)
Unknown	0	0	2(1.7)	2(0.5)
Race/Ethnicity n(%)				
White	181(82.6)	71(91)	103(83.7)	355(84.5)
Black	11(5)	1(1.3)	0	12(2.9)
Asian	6(2.7)	0	0	6(1.4)
Hispanic	8(3.7)	0	1(0.8)	9(2.1)
Unknown	13(6)	6(7.7)	19(15.5)	38(9)

The goal of treatment of SCN with Filgrastim is to reduce or prevent the occurrence of infections and symptoms of inflammation. Predictably this can be done by increasing the blood neutrophil count to a normal level. The dose and schedule of Filgrastim to achieve this response is quite variable. In general, patients well documented to have idiopathic or cyclic neutropenia require substantially lower doses than patients with congenital neutropenia (**Table 4**). Based on these data, a good plan for initiating

Table 4. Severe Chronic Neutropenia International Registry Filgrastim Dosing by Diagnosis (as of 12/95).

	Filgrastim dosing (μg/kg/day)			
	Congenital	Cyclic	Idiopathic	Total
Mean (±SD)	15.7(39.2)	2.0(1.5)	2.1(3.1)	9.5(29.6)
Median	5.6	1.7	1.0	3.1
Range	.05–242.4	.12–5.5	.1–15.8	.05–242.4

therapy for a patient not treated previously with Filgrastim is to begin with the median dose shown in Table 4 and then to titrate the dose up or down at 1- to 2-week intervals based on the individual patient's response. Alternate day or three-times-a-week treatment is often effective for patients responding to low doses of daily therapy (75).

A.   Effects of Filgrastim Treatment on Blood Counts

In the randomized clinical trial, more than 90% of patients responded to Filgrastim treatment, with normalization of the blood neutrophil count (7). Data collected through the Registry suggest even a higher proportion of patients respond, but there are still exceptions, particularly among patients with congenital neutropenia (72,76–78). Almost all patients achieving a good initial response will have a good long term-response as well. Loss of response with time through "marrow exhaustion" or other changes has not been an observed effect of long-term Filgrastim therapy. The effect of Filgrastim on median leukocytes, neutrophils, hemoglobin, and platelet values over a 7-year treatment period is shown in **Figure 1.** These data show a trend for the hemoglobin to increase with time, probably due to increased endogenous erythropoietin with reduced chronic inflammation. Platelet counts may decline initially, especially in patients with congenital neutropenia with previously elevated counts. Otherwise they usually remain within the normal range, as observed in the randomized trial. In general, as long as the patient's health status is otherwise unchanged, blood counts are relatively stable on a stable dose of Filgrastim. In cyclic neutropenia, oscillations of counts persist with only brief periods of neutropenia (**Figure 2**). Based on available data, monitoring blood cell counts at approximately monthly or bimonthly intervals is satisfactory for patients on long-term therapy.

B.   Effects on Infection and Inflammation

In the randomized controlled trial, Filgrastim treatment clearly reduced the occurrence of fever, mouth ulcers, and infections requiring antibiotic treatment. Over time, these same patients continued to show this benefit of therapy including decreased hospitalizations (76) (**Figure 3**). Although there is no contemporary control group for these observations, it appears that the reduction in inflammatory events can be sustained indefinitely. Development of antibodies to Filgrastim has not been observed. In many patients the reduction in chronic gingivitis and improvement in oral hygiene has been particularly noteworthy. The improvement in the quality of life of patients with SCN receiving long-term Filgrastim treatment probably

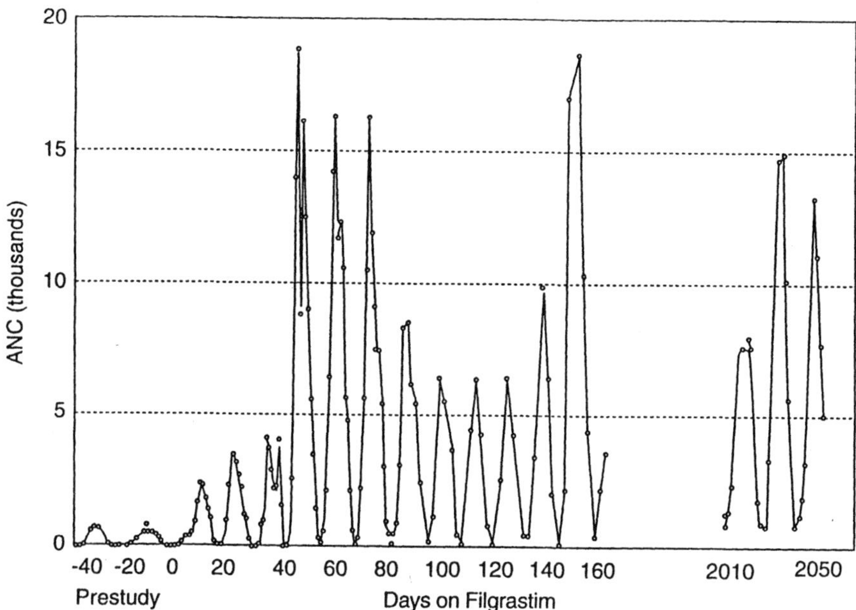

Figure 2. Serial neutrophil counts (ANC) for a patient with cyclic neutropenia treated with Filgrastim 3 μg/kg/day. Counts are shown for the pretreatment period (day-40 to day 0), beginning of treatment (day 1 to day 160), and a second period after 66 months of treatment (day 2010 to day 2050). (Adapted from Ref. 76.)

reflects the decreased occurrence of inflammatory events and reduction in the anxiety often associated with chronic neutropenia (79,80) (**Table 5**).

C. Adverse Events

Patients beginning Filgrastim treatment for SCN often experience bone pain, headache, and musculoskeletal symptoms early in treatment; these symptoms are attributable to expansion of the hematopoietic tissues that occurs as a direct effect of Filgrastim therapy. Predictably, they decline with continued therapy (**Figure 4**). When these symptoms occur, they are usually managed with acetaminophen or other mild analgesics.

D. Bone Remodeling and Osteoporosis

Because Filgrastim expands hematopoietic tissues and causes bone remodeling, patients receiving long-term therapy have been observed for the

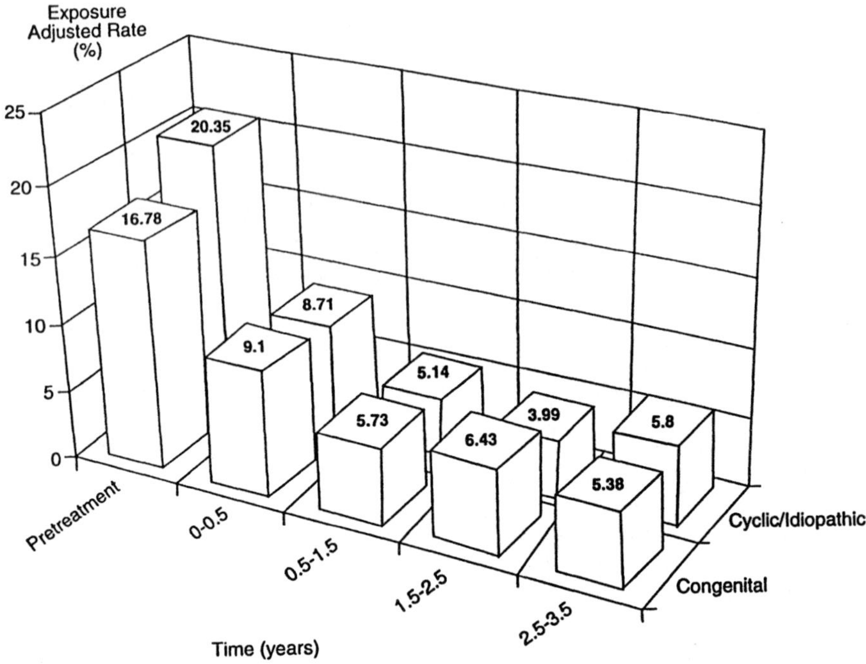

Figure 3. Phase 3 trial: incidence of hospitalization. Numbers represent days of hospitalization during a 100-month period.

development of osteoporosis or other bone changes. In some centers, repeated bone density measurements have been made, as well as observations for the occurrence of fractures. In a few patients with congenital neutropenia, there are reports of development of osteoporosis, some of these patients being osteoporotic before the initiation of Filgrastim (81–83). Interestingly, to date most of these reports have come from Europe. The occurrence of fractures or other sequelae of these changes are very infrequent. The height and weight of children have been monitored in all clinical trials and through the Registry. Analysis of height and weight data on a population basis for subjects in North America indicates that there is no apparent effect of long-term Filgrastim treatment on linear growth. Comparing children started on Filgrastim before or after 3 years of age, in fact, suggests that early treatment may enhance linear growth, presumably through reduction in the growth-retarding effects of frequent inflammation and infection in these years of rapid skeletal development.

Table 5. Nottingham Health Profile.

	Pretreatment	Posttreatment	p-Value*
Energy	44.0	6.4	.001
Pain	13.8	10.5	.138
Emotional reactions	25.2	7.3	.002
Sleep	29.8	17.7	.085
Spiritual isolation	20.3	4.1	.008
Mobility	8.9	7.4	.170

	Pretreatment	Posttreatment	p-Value◆
Job or work	86%	29%	.02–.05
Home	38%	10%	>.10
Social life	67%	10%	<.01
Personal Relationships	24%	10%	>.10
Sex life	5%	10%	>.10
Hobbies	62%	10%	.01–.02
Education	76%	14%	<.01

All values in this table represent arithmetic means for 21 respondents; a high number reflects a problem or concern for the patient.
* Wilcoxon signed ranks tests between pre- and posttreatment questionnaires.
◆ McNeman test between pre- and posttreatment questionnaires.

E. Splenomegaly

Preclinical trials and early studies of Filgrastim treatment for SCN established that enlargement of the spleen occurs as a direct effect of Filgrastim therapy, presumably because the spleen serves as a site for expansion of the hematopoietic tissues in response to this stimulus. The magnitude of the splenic enlargement is quite variable; often it is not detectable by physical examination and can only be recognized through imaging techniques (ultrasound, computed tomography [CT], or magnetic resonance imaging [MRI]). Except in rare cases, progressive splenomegaly with long-term treatment of SCN has not occurred. Data from the Registry now include 11 cases of splenectomy or progressive splenomegaly. Six of these cases have occurred in patients with some form of congenital neutropenia.

F. Hematological Sequelae—Abnormal Cytogenetics, Myelodysplasia, and Leukemic Transformations

Before the availability of Filgrastim and other hematopoietic growth factors, it was recognized that leukemic transformations can occur in patients with

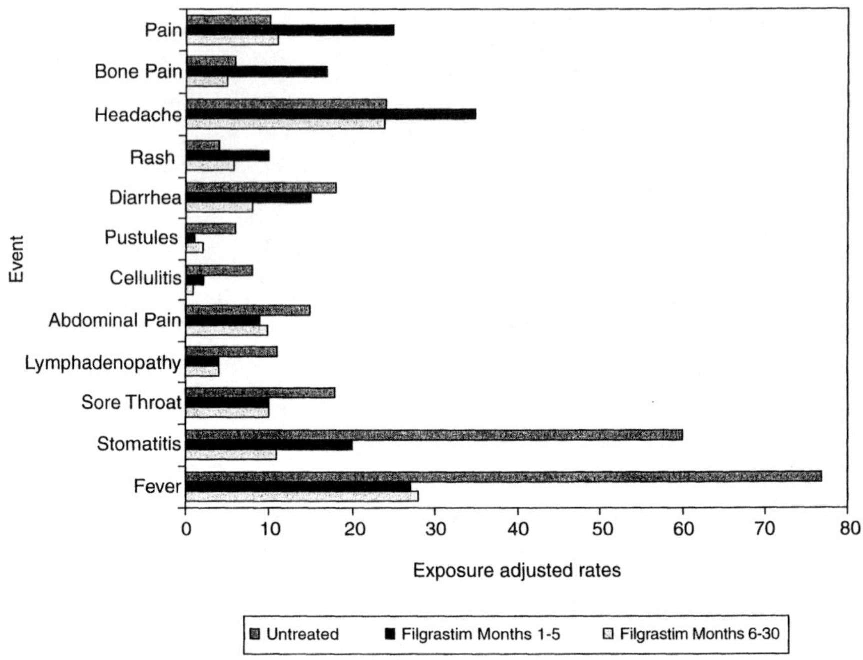

Figure 4.   Phase 3 trial: adverse event rates by exposure time.

congenital neutropenia. In addition to three published cases of Kostmann's syndrome developing acute myeloid leukemia (AML), three patients with SCN who had marrow aspiration and biopsy before or shortly after entering the Filgrastim clinical trials were found to have myelodysplasia or AML. One subgroup, patients with Shwachman-Diamond syndrome, appear to have a high frequency of evolution to AML (26), as to patients with congenital aplastic anemia.

During the era of treatment of SCN with Filgrastim, a clear pattern has emerged—patients with congenital neutropenia may develop cytogenetic changes, myelodysplasia, and AML (76,84,85). To date, such events have not appeared in other forms of neutropenia treated long-term with Filgrastim over this same period. Through 1996 there were 23 known cases of myelodysplasia or acute leukemia that occurred in a total population of 249 congenital neutropenia patients for whom the SCNIR has data from long-term Filgrastim treatment (**Table 6**). It has proven difficult to determine if there is a direct relationship between these events and therapy. It is difficult because the frequency of this evolution is low (ie, <2% per year of exposure), the underlying hematological disease of congenital neutro-

Table 6. Cases of Myelodysplasia/Acute Myeloid Leukemia (MDS/ AML) in Severe Chronic Neutropenia International Registry Data (Includes Patients from Clinical Trials).

	n	MDS/AML	Incidence
Congenital	249	23	9.2%
Cyclic	97	0	0
Idiopathic	160	0	0
Total	506	23	4.5%

n = number of patients in Registry and clinical trials.
MDS/AML = number of patients diagnosed with myelodysplasia or acute myeloid leukemia.

penia is rare (1 to 5 cases per million births), and the recognition of dysplasia in marrow samples from patients with congenital neutropenia can be difficult. It is not known to what degree the improved health and survival of patients with congenital neutropenia has allowed these individuals to survive to develop these changes.

Several important observations have been made for the cohort of congenital neutropenia patients on long-term treatment with Filgrastim. First, cytogenetic abnormalities have not been detected in patients with this diagnosis before Filgrastim treatment and have rarely been observed in less than 1 to 2 years of therapy. Second, patients evolving to AML have often followed a now predictable pathway: emergence of chromosomal abnormalities (most frequently monosomy 7), progressive increases in the proportion of cells showing single or multiple cytogenetic abnormalities, and finally increasing numbers of abnormal myeloblasts in the blood and marrow. Third, for most patients this evolution has been accompanied by development of pancytopenia and refractoriness of the neutrophil response to continuation of Filgrastim therapy. Finally, when evolution has occurred it has developed slowly over several months, not days or weeks, as documented by serial observations of patients' blood counts and periodic marrow examinations. Studies are now in progress to try and determine if there are definable sub-groups of patients with congenital neutropenia who are at greatest risk for these changes.

At present BMT or long-term antibiotics are the only alternatives to long-term Filgrastim treatment for patients with SCN. The risk of transplantation depends on the pretransplant health of the patient, the closeness of match of the potential marrow donor, and the chance of infection in the post-transplant period. At present most experts agree that these risks are greater than the risk of leukemic transformation with long-term Filgrastim therapy. However, patients observed on therapy to have chromosomal abnormalities such as monosomy 7 associated with development of myelodysplasia or acute leukemia are now recommended to undergo BMT if there is a suitable donor.

It is important to emphasize that patients with either cyclic neutropenia or idiopathic neutropenia do not appear to be at risk for development of myelodysplasia or AML. The biological basis for these differences is not yet known. Some studies now suggest that patients with Kostmann's type of congenital neutropenia may have underlying abnormalities of the receptor for G-CSF (86–90) and perhaps this predisposes to the development of AML (91). No such receptor abnormalities have been detected to date in patients with cyclic or idiopathic neutropenia. Mutations of the *Ras* oncogene also have been observed in patients with congenital neutropenia evolving to AML (92). It is not yet known if abnormalities of this or other genes involved in the proliferative response to Filgrastim are abnormal in this population before development of myelodysplasia.

G. Pregnancy and Fertility

Before the availability of Filgrastim, it was often difficult for women with SCN to carry a pregnancy to term because of the risk of uterine infection. For patients with cyclic neutropenia, it is estimated that less than 50% of pregnancies were carried to term. Data from the SCNIR indicate that at least 11 pregnancies have occurred in 9 women receiving long-term Filgrastim treatment. Five resulted in normal births and, one resulted in an infant with cyclic neutropenia, a disease known to be inherited as an autosomal dominant; four of these six women were treated with Filgrastim throughout pregnancy. In three other cases, common congenital abnormalities were detected either during gestation or at birth. In the other two cases, elective abortions were performed. There are no clear patterns in these observations and at present no recommendations can be made regarding the safety of the use of Filgrastim during pregnancy; however, Calhoun et al (93) suggest that Filgrastim can pass through the placenta to the fetus.

IX. SUMMARY

Filgrastim is a very effective treatment for long-term treatment of SCN. Treatment usually requires daily or alternate-day SC injections. The doses of Filgrastim can be titrated to increase the ANC to normal levels in more than 90% of patients. Based on clinical experience with hundreds of patients who have now received this therapy, the suggested initial treatment is

6 μg/kg/day for congenital neutropenia, 2 μg/kg/day for cyclic neutropenia, and 1 μg/kg/day for idiopathic neutropenia with dose titration to achieve a mean ANC within the normal range. The expected benefit is reduction in the occurrence of fever, mouth ulcers, gingivitis, and infections requiring antibiotic therapy.

Long-term treatment with Filgrastim is associated with relatively few adverse events. Bone pain noted early in treatment usually resolves spontaneously. Other blood cell counts usually are not affected. Children receiving long-term treatment with Filgrastim have normal growth and development. Mild splenic enlargement occurs frequently but is generally asymptomatic and of little clinical consequence. Long-term stimulation of hematopoiesis with Filgrastim is not associated with loss of responsiveness, either due to the development of antibodies to this drug or tolerance to the stimulatory effect of Filgrastim. Some patients with congenital neutropenia, but not patients with cyclic or idiopathic neutropenia, have evolved to have chromosomal abnormalities, myelodysplasia, or leukemia. The overall risk is currently estimated to be less than 2% per year of observation on treatment for this population; it is not known if this rate is greater or less than would be observed in a similar population not treated with Filgrastim. Recognizing the benefits of treatment, as well as this potential risk, it is recommended that patients with congenital neutropenia treated with Filgrastim have careful, long-term serial measurements of blood cell counts, marrow histology, and cytogenetics. In patients showing evolution to myelodysplasia, bone marrow transplantation is now recommended. In other patients a careful pretreatment evaluation including a bone marrow examination and regular blood counts on therapy is recommended.

The SCNIR maintains records on more than 500 patients with SCN who are either untreated, treated with Filgrastim, or receiving other therapies. The data base from this Registry provides a quantitative basis for current and future recommendations regarding treatment of this rare blood disorder and supports numerous investigators throughout the world interested in understanding the basic mechanisms of congenital, cyclic, and idiopathic neutropenia.

ACKNOWLEDGMENTS

The authors gratefully acknowledge the contributions of the SCNIR Advisory Board: Drs. Mary Ann Bonilla, Laurence Boxer, Patricia Catalano, Bonnie Cham, Melvin Freedman, George Kannourakis, Sally Kinsey, and Pier Giorgio Mori, and staff members Audrey Anna Bolyard, Tammy Cottle, Kristine Crusius, Dr. Thomas Weiberlenn, and Dr. Cornelia Zeidler.

The authors also acknowledge the many local treating physicians for their cooperation with the collection of Registry data.

REFERENCES

1. Dale, D. C. Neutropenia. In: Williams, W. J., Beutler, E., Erslev, A. J., and Lichtman, M. A., eds. *Hematology,* 5th ed. New York: McGraw-Hill, 1995, pp. 815–824.
2. Dale, D. C., Guerry, D., Werwerka, J. R., Bull, J. M., and Chusid, M. H. (1979). Chronic neutropenia. *Medicine 58:*128–144.
3. Weetman, R. M., and Boxer, L. A. (1980). Childhood neutropenia. *Pediatr Clin North Am 27:*361–375.
4. Berrebi, A., Melamed, Y., and Van Dam, U. (1987). Leukopenia in Ethiopian Jews. *N Engl J Med 316:*549.
5. Dale, D. C., and Wolff, S. M. (1971). Skin window studies of the acute inflammatory responses of neutropenia patients. *Blood 36:*138–142.
6. Wright, D. G., Meierovics, A. I., and Foxley, J. M. (1986). Assessing the delivery of neutrophils to tissues in neutropenia. *Blood 67:*1023–1030.
7. Welte, K., and Dale, D. (1996). Pathophysiology and treatment of severe chronic neutropenia. *Ann Hemotol 72:*158–165.
8. Dale, D. C., and Hammond, W. P. (1988). Cyclic neutropenia: a clinical review. *Blood Rev 2:*178–185.
9. Kostmann, R. (1956). Infantile genetic agranulocytosis. *Acta Paediatr Scand 45:*1.
10. Kostmann, R. (1975). Infantile genetic agranulocytosis. A review with presentation of ten new cases. *Acta Paediatr Scand 64:*362.
11. Kyle, R. A., and Linman, J. W. (1968). Chronic idiopathic neutropenia, a newly recognized entity? *N Engl J Med 279:*1015–1019.
12. Hestdal, K., Welte, K., Lie, S. O., Keller, J. R., Ruscetti, F. W., and Abrahamsen, T. G. (1993). Severe congenital neutropenia: abnormal growth and differentiation of myeloid progenitors to granulocyte colony-stimulating factor (G-CSF) but normal response to G-CSF plus stem cell factor. *Blood 82:*2991–2997.
13. Rosen, F. S., Cooper, M. D., and Wedgwood, R. J. P. (1984). The primary immunodeficiencies. *N Engl J Med 311:*235–238, 300–310.
14. Gasparetto, C., Smith, C., Firpo, M., et al (1994). Dyshematopoiesis in combined immune deficiency with congenital neutropenia. *Am J Hematol 45:*63–72.
15. Young, N. S. (1995). Pathogenesis and pathophysiology of aplastic anemia. In: R. Hoffman et al, eds. *Hematology Basic Principles and Practices.* New York, NY: Churchill Livingstone, 1995. pp. 299–336.
16. Aggett, P. J., Cavanagh, N. P., Matthew, D. J., et al (1980). Schwachman's syndrome. A review of 21 cases. *Arch Dis Child 55:*331–347.
17. Blume, R. S., and Wolff, S. M. (1972). The Chediak-Higashi syndrome: studies in four patients and a review of the literature. *Medicine 51:*247–280.

18. Bassan, R., Viero, P., Minetti, B., Comotti, B., and Barbui, T. (1984). Myelokathaxis: a rare form of chronic benign granulocytopenia. *Br J Haematol* 58:115–117.

19. Miller, M. E., Oski, F. A., and Harris, M. B. (1971). Lazy-leukocyte syndrome: a new disorder of neutrophil function. *Lancet 1*:665–669.

20. Kilpatrick, L., Garty, B. Z., Lundquist, K. F., et al (1990). Impaired metabolic function and signaling defects in phagocytic cells in glycogen storage disease type lb. *J Clin Invest 86*:196–202.

21. Lalezari, P., Murphy, G. B., and Allen, F. H., Jr. (1971). NBI, a new neutrophil specific antigen involved in the pathogenesis of neonatal neutropenia. *J Clin Invest 50*:1108–1115.

22. Bussel, J. B., and Abboud, M. R. (1987). Autoimmune neutropenia of childhood. *Crit Rev Oncol Hematol 7*:37–51.

23. DeVries, A., Peketh, L., and Joshua, H. (1958). Leukaemia and agranulocytosis in a member of a family with hereditary leukopenia. *Acta Med Orient 17*:26–32.

24. Gilman, P. A., Jackson, D. P., and Guild, H. G. (1970). Congenital agranulocytosis: prolonged survival and terminal acute leukemia. *Blood 36*:576–585.

25. Rosen, R. B., and Kung, S. J. (1979). Congenital agranulocytosis terminating in acute myelomonocytic leukemia. *J Pediatr 94*:406–408.

26. Woods, W. G., Roloff, J. S., Lukens, J. N., and Krivit, W. (1981). The occurrence of leukemia in patients with the Shwachman syndrome. *J Pediatr 99*:425–428.

27. Krance, R. A., Spruce, W. E., Forman, S. J., et al (1982). Human cyclic neutropenia transferred by allogeneic bone marrow grafting. *Blood 60*:1263–1266.

28. Loughran, T. P., and Hammond, W. P. (1986). Adult onset-cycle neutropenia is associated with increased large granular lymphocytes. *Blood 68*:1082–1087.

29. Hopkins, D. G., and Kushner, J. P. (1983). Clostridial species in the pathogenesis of necrotizing enterocolitis in patients with neutropenia. *Am J Hematol 14*:289–295.

30. Langer, J. C., Papa, M. Z., Hoffman, M. A., Loeff, D. S., Pearl, R. H., and Filler, R. M. (1990). Cyclic neutropenia with colonic perforation and nonhealing colocutaneous fistula. *J Pediatr Surg 25*:346–348.

31. O'Hanrahan, T., Dark, P., and Irving, M. H. (1991). Cyclic neutropenia—unusual cause of acute abdomen. Report of a case. *Dis Colon Rectum 34*:1125–1127.

32. Morley, A. A., Carew, J. P., and Baikie, A. G. (1967). Familial cyclical neutropenia. *Br J Haematol 13*:719–738.

33. Palmer, S. E., Stevens, K., and Dale, D. C. (1996). Genetics phenotype and natural history of autosomal dominant cyclic hematopoiesis. *Am J Med Genet 66*:413–422.

34. Chikkappa, G., Borner, G., Burlington, H., et al (1976). Periodic oscillation of blood leukocytes, platelets, and reticulocytes in a patient with chronic myelocytic leukemia. *Blood 47*:1023–1030.

35. Lensink, D. B., Barton, A., Appelbaum, F. R., and Hammond, W. P. (1986). Cyclic neutropenia as a premalignant manifestation of acute lymphoblastic leukemia. *Am J Hematol 22*:9–16.

36. Jonsson, O. G., and Buchanan, G. R. (1991). Chronic neutropenia during childhood. A 13-year experience in a single institution. *Am J Dis Child 145*:232–235.

37. Logue, G. L., Shastri, K. A., Laughlin, M., et al (1991). Idiopathic neutropenia: antineutrophil antibodies and clinical correlations. *Am J Med 90*:211–216.

38. Shastri, K. A., and Logue, G. L. (1993). Autoimmune neutropenia. *Blood 81*:1984–1995.

39. Monto, R. W., Shafer, H. C., Brennan, M. J., and Rebuck, J. W. (1952). Periodic neutropenia treated by adrenocorticotrophic hormone and splenectomy. *N Engl J Med 246*:893–896.

40. Moore, R. A., Brunner, C. M., Sandusky, W. R., and Leavell, B. S. (1971). Felty's syndrome: long-term follow-up after splenectomy. *Ann Intern Med 75*:381–385.

41. Campion, G., Maddison, P. J., Goulding, N., et al (1990). The Felty syndrome: a case-matched study of clinical manifestations and outcome, serologic features, and immunological associations. *Medicine 69*:69–80.

42. Harker, W. G., Rothstein, G., Clarkson, D., Athens, J. W., and Macfarlene, J. L. (1977). Enhancement of colony-stimulating activity production by lithium. *Blood 49*:263–267.

43. Fiechtner, J. J., Miller, D. R., and Starkebaum, G. (1989). Reversal of neutropenia with methotrexate treatment in patients with Felty's syndrome. *Arthritis Rheum 32*:194–201.

44. Rappeport, J. M., Parkman, R., Newburger, P., Camitta, B. M., and Chusid, M. J. (1980). Correction of infantile agranulocytosis (Kostmann's syndrome) by allogeneic bone marrow transplantation. *Am J Med 68*:605–609.

45. Proctor, S. J., Reid, M. M., and Low, W. T. (1979). Levamisole in the treatment of cyclical neutropenia. *Br Med J 55*:279–281.

46. Roosendaal, K. J., Dicke, K. A., and Boonzajer Flaes, M. L. (1981). Effect of oxymetholone on human cyclic haematopoiesis. *Br J Haematol 47*:185–193.

47. Pachman, L. M., Schwartz, A. D., and Barron, R. (1974). The effect of plasma with colony stimulating activity (SCA) in cyclic neutropenia (CN). *Pediatr Res 8*:407.

48. Hammond, W. P., Engelking, E. R., and Dale, D. C. (1979). Cyclic hematopoiesis: the effect of endotoxin on colony forming cells and colony stimulating activity in grey collie dogs. *J Clin Invest 63*:785–792.

49. Hammond, W. P., and Dale, D. C. (1980). Lithium therapy of canine cyclic hematopoiesis. *Blood 55*:26–28.

50. Souza, L. M., Boone, T. C., Gabrilove, J., et al (1986). Recombinant human granulocyte colony stimulating factor: effects on normal and leukemic myeloid cells. *Science 232*:61–64.

51. Welte, K., Bonilla, M. A., Gillio, A. P., et al (1987). Recombinant human granulocyte colony stimulating factor: effects on hematopoiesis in normal and cyclophosphamide-treated primates. *J Exp Med 16S*:941–948.

52. Bonilla, M. A. (1990). Clinical efficacy of recombinant human granulocyte colony stimulating factor (r-metHuG-CSF) in patients with severe chronic neutropenia. *Blood 76*:133a (abstr 523).

53. Bonilla, M. A., Gillio, A. P., Ruggeiro, M., et al (1989). Effects of recombinant human granulocyte colony-stimulating factor on neutropenia in patients with congenital agranulocytosis. *N Engl J Med 320*:1574–1580.

54. Hammond, W. P., Price, T. H., Souza, L. M., and Dale, D. C. (1989). Treatment of cyclic neutropenia with granulocyte colony-stimulating factor. *N Engl J Med 320*:1306–1311.

55. Jakubowski, A. A., Souza, L., Kelly, F., et al (1989). Effects of human granulocyte colony-stimulating factor in a patient with idiopathic neutropenia. *N Engl J Med 320*:38–42.

56. Welte, K., Zeidler, C., Reiter, A., et al (1990). Differential effects of granulocyte-macrophage colony-stimulating factor and granulocyte colony-stimulating factor in children with severe congenital neutropenia. *Blood 75*:1056–1063.

57. Dale, D. C., Bonilla, M. A., Davis, M. W., et al (1993). A randomized controlled phase III trial of recombinant human G-CSF for treatment of severe chronic neutropenia. *Blood 81*:2496–2502.

58. Hanada, T., Ono, I., and Nagasawa, T. (1990). Childhood cyclic neutropenia treated with recombinant human granulocyte colony stimulating factor. *Br J Haematol 75*:135–137.

59. Imashuku, S., Tsuchida, M., Sasaki, M., et al (1992). Recombinant human granulocyte-colony-stimulating factor in the treatment of patients with chronic benign granulocytopenia and congenital agranulocytosis (Kostmann's syndrome). *Acta Paediatr 81*:133–136.

60. Kobayashi, M., Yumiba, C., and Ueda, K. (1990). Effects of recombinant human colony-stimulating factors on granulopoiesis in patients with congenital neutropenia. *Med Pediatr Oncol 18*:519.

61. Sonoda, Y., Yashege, H., Fujii, H., Maekawa, T., and Abe, T. (1991). Treatment of idiopathic neutropenia in the elderly with recombinant human granulocyte colony-stimulating factor. *Acta Haematol 8S*:146–152.

62. Sugimoto, K., Togawal, A., Miyazono, K., et al (1990). Treatment of childhood onset cyclic neutropenia with recombinant human granulocyte colony stimulating factor. *Eur J Haematol 4S*:110–111.

63. Tsunogake, S., Nagashima, S., Maekawa, R., et al (1991). Myeloid progenitor cell growth characteristics and effect of G-CSF in a patient with congenital cyclic neutropenia. *Int J Hematol S4*:251–256.

64. Ueda, K., Hanawa, Y., Takaku, F., et al (1991). The effect of recombinant human granulocyte colony-stimulating factor (rG-CSF) on childhood neutropenias. *Rinsho Ketsueki 32*:212–220.

65. Heussner, P., Haase, D., Kanz, L., Fonatsch, C., Welte, K., and Freund, M. (1995). G-CSF in the long-term treatment of cyclic neutropenia and chronic idiopathic neutropenia. *Int J Hematol 62*:225–234.

66. Marlton, P. V., Wright, S. J., and Taylor, K. M. (1992). Granulocyte colony-stimulating factor in management of chronic neutropenia. *Med J Aust 156*:729–731.

67. Boxer, L. A., Hutchinson, R., and Emerson, E. (1992). Recombinant human granulocyte-colony-stimulating factor in the treatment of patients with neutropenia. *Clin Immunol Immunopathol 62*:S39–S46.

68. Weston, B., Todd, R. F., III, Axtell, R., et al (1991). Severe congenital neutropenia: clinical effects and neutrophil function during treatment with granulocyte colony-stimulating factor. *J Lab Clin Med 117*:282–290.

69. Crawford, J., Oger, H., Stoller, R., et al (1991). Reduction by granulocyte colony stimulating factor of fever and neutropenia in patients with small cell lung cancer. *N Engl J Med 315*:164–170.

70. Welte, K., Gabrilove, J., Bronchud, M. H., Platzer, E., and Morstyn, G. (1996). Filgrastim (r-metHuG-CSF): the first 10 years. *Blood 88*:1907–1929.

71. Boogaerts, M., Cavalli, F., Cort'es-Funes, H., et al (1995). Granulocyte growth factors: achieving a consensus. *Ann Oncol 6*:237–244.

72. Dale, D. C. (1995). Hematopoietic growth factors for the treatment of severe chronic neutropenia. Concise Review. *Stem Cells 13*:94–100.

73. Donadieu, J. (1996). Congenital and acquired neutropenia in children. *Presse Med 25*:293–298.

74. Wright, D. G., Kenney, R. F., Oette, D. H., LaRussa, V. F., Boxer, L. A., and Malech, H. L. (1994). Contrasting effects of recombinant human granulocyte-macrophage colony stimulating factor (CSF) and granulocyte CSF treatment on the cycling of blood elements in childhood-onset cyclic neutropenia. *Blood 84*:1257–1267.

75. Jayabose, S., Tugal, O., Sandoval, C., and Li, K. (1994). Recombinant human granulocyte colony-stimulating factor in cyclic neutropenia: use of a new 3-day-a-week regimen. *Am J Pediatr Hematol Oncol 16*:338–340.

76. Bonilla, M. A., Dale, D., Zeidler, C., et al (1994). Long-term safety of treatment with recombinant human granulocyte colony-stimulating factor (r-metHuG-CSF) in patients with severe congenital neutropenias. *Br J Haematol 88*:723–730.

77. Ryan, M., Will, A. M., Testa, N., Hayworth, C., and Darbyshire, P. J. (1995). Severe congenital neutropenia unresponsive to G-CSF. *Br J Haematol 91*:43–45.

78. Sandoval, C., Parganas, E., Wang, W., Ihle, J. N., and Adam-Graves, P. (1995). Lack of alterations in the cytoplasmic domains of the granulocyte colony-stimulating factor receptors in eight cases of severe congenital neutropenia. *Blood 85*:852–853.

79. Fazio, M. T., and Glaspy, J. A. (1991). The impact of granulocyte colony-stimulating factor on quality of life in patients with severe chronic neutropenia. *Oncol Nurs Forum 18*:1411–1414.

80. Jones, E. A., Bolyard, A. A., and Dale, D. C. (1993). Quality of life patients with severe chronic neutropenia receiving long-term treatment with granulocyte colony-stimulating factor. *JAMA 270*: 1132–1133.

81. Bishop, N. J., Williams, D. M., Compston, J. C., Stirling, D. M., and Prentice, A. (1995). Osteoporosis in severe congenital neutropenia with granulocyte colony-stimulating factor. *Br J Haemetol 89*:927–928.

82. Simon, M., Lengfelder, E., Reiter, S., and Hehlmann, R. (1996). Osteoporosis in severe congenital neutropenia; inherent to the disease or a sequela of G-CSF treatment? *Am J Hematol 52*:127.

83. Yakisan, E., Zeidler, C., Schirg, E., et al (1997). High incidence of significant bone loss in patients with severe congenital neutropenia (Kostmann's syndrome). *J Pediatr* (in press).
84. Naparstek, E. (1995). Granulocyte colony-stimulating factor, congenital neutropenia, and acute myeloid leukemia. *N Engl J Med 335*:516–518.
85. Weinblatt, M. E., Scimeca, P., James-Herry, A., Sahdev, I., and Kochen, J. (1995). Transformation of congenital neutropenia into monosomy 7 and acute nonlymphoblastic leukemia in a child treated with granulocyte colony-stimulating factor. *J Pediatr 126*:263–265.
86. Kyas, U., Pietsch, T., and Welte, K. (1992). Expression of receptors for granulocyte colony-stimulating factor on neutrophils from patients with severe congenital neutropenia and cyclic neutropenia. *Blood 79*:1144–1147.
87. Guba, S. C., Sartor, C. A., Hutchison, R., Boxer, L. A., and Emerson, S. G. (1994). Granulocyte colony-stimulating factor (G-CSF) production and G-CSF receptor structure in patients with congenital neutropenia. *Blood 83*:1486–1492.
88. Dong, F., Brynes, R. K., Tidow, N., Welte, K., Lowenberg, B., and Touw, I. P. (1995). Mutations in the gene for the granulocyte colony-stimulating factor receptor in patients with acute myeloid leukemia preceded by severe congenital neutropenia. *N Engl J Med 333*:487–493.
89. Dong, F., Dale, D. C., Bonilla, M. A., et al (1997). Mutations in the granulocyte colony-stimulating factor receptor gene in patients with severe congenital neutropenia. *Leukemia 11*:120–125.
90. Tidow, N., Pilz, C., Teichmann, B., et al (1997). Clinical relevance of point mutations in the cytoplasmic domain of the granulocyte colony-stimulating factor (G-CSF) receptor gene in patients with severe congenital neutropenia. *Blood 86*:2369–2375.
91. Hammond, W. P, Chatta, G. S., Andrewsa, R. G., and Dale, D. C. (1992). Abnormal responsiveness of granulocyte-committed progenitor cells in cyclic neutropenia. *Blood 79*:2536–2539.
92. Kalra, R., Dale, D., Freedman, M., et al (1995). Monosomy 7 activating RAS mutations accompany malignant transformations in patients with congenital neutropenia. *Blood 86*:4579–4586.
93. Calhoun, D. A., Rosa, C., and Christensen, R. D. (1996). Transplancental passage of recombinant human granulocyte colony-stimulating factor in women with an imminent extremely preterm delivery. *Am J Obstet Gynecol 174(4)*:1306–1311.

7
Use of Filgrastim (r-metHuG-CSF) in Neutrophil Transfusions

Douglas R. Adkins, Randy A. Brown, Lawrence T. Goodnough, and John F. DiPersio
Washington University School of Medicine, St. Louis, Missouri

I. INTRODUCTION

Infections are a frequent cause of morbidity and mortality after dose-intense myelosuppressive chemotherapy. Although many factors predispose patients receiving chemotherapy to infections, neutropenia is the principal risk factor. The frequency of infection correlates with the severity and duration of neutropenia, with the highest risk for infection occurring when the absolute neutrophil count (ANC) is $< 0.1 \times 10^9/L$ (1). Strategies to abbreviate the duration of neutropenia and decrease infectious complications after dose-intensive therapy include administration of myeloid growth factors such as recombinant human granulocyte colony-stimulating factor (rHuG-CSF) or rHu granulocyte-macrophage colony-stimulating factor (GM-CSF), and infusion of bone marrow or mobilized peripheral blood progenitor cells (PBPC); however, all patients so treated still develop severe neutropenia and most require antibiotics for fever or documented infections. Infusion of ex vivo–expanded, late progenitor cells and neutrophils during the expected period of treatment-related neutropenia, while attractive, requires technology not widely available. Improved methods to shorten the duration of neutropenia are needed.

Transfusion of donor granulocytes is a complementary investigational approach to the prevention and treatment of infections in neutropenic patients. Although prior controlled trials of granulocyte transfusions demonstrated mixed clinical results, the interpretation of these studies was hindered by several confounding variables. Perhaps the most important limitation in previous trials of granulocyte transfusion therapy was the inability to reliably collect and transfuse products containing adequate numbers of neutrophils. The recent availability of myeloid growth factors, such as rHuG-CSF, provided a new class of neutrophil-mobilizing agents that are more effective than previous drugs used for this purpose. This discussion will focus on the potential role of Filgrastim as an approach to further improve granulocyte collection yields from normal donors and neutrophil increments in recipients of such products.

II. HISTORY OF GRANULOCYTE TRANSFUSION THERAPY

In controlled trials of neutropenic dogs with experimental gram-negative sepsis (2–6), pneumonia (7,8), or candidemia (9–11), survival was increased with administration of granulocyte transfusions alone or in combination with antimicrobial therapy. These preclinical studies established the critical importance of the neutrophil cell dose in the transfused product on both survival from gram-negative sepsis and on post-transfusion granulocyte increments. Appelbaum et al (2), in a neutropenic dog model of *Pseudomonas* bacteremia, demonstrated that a threshold dose of 2×10^8 neutrophils/kg (collected by continuous flow centrifugation) assured protection from otherwise lethal infection with *Pseudomonas*. The cell dose-versus-survival curve was found to be steep, with mortality occurring uniformly in those animals given $<1.5 \times 10^8$ neutrophils/kg. The 1-hour post-transfusion neutrophil increments also were dependent on the cell dose of granulocytes infused. One hour post-infusion mean ANC increments of $\geq 0.2 \times 10^9$/L and $\geq 0.5 \times 10^9$/L required transfusion of at least 2 and 3×10^8 neutrophils/kg, respectively. These data presented strong evidence that control of infections and post-infusion ANC increments in neutropenic patients receiving granulocyte transfusions would depend also on the number of neutrophils administered.

Over the last two decades, a large number of published human trials examined the effects of granulocyte transfusions given to neutropenic patients on post-infusion increments in the ANC and on infection control or prevention (12–30); however, these studies failed to consistently demonstrate a threshold neutrophil cell dose that reliably resulted in significant and sustained increments in the recipient ANC and in control of infection. In addition, randomized and non-randomized trials of either prophylactic or therapeutic granulocyte transfusions showed mixed clinical results, with some studies suggesting either improved survival or resolution of infection or fevers compared with those patients receiving antibiotics alone (12–28), whereas other studies showed no benefit (15,16,19,22).

Granulocyte transfusions for the prevention or treatment of infections in neutropenic patients have not gained widespread acceptance, in part due to conflicting published data on potential benefits, toxicity, and the impracticality of the method. Limitations to this approach have included low neutrophil cell dose with previous techniques for mobilization and collection (19,22) the short half-life of resting neutrophils (31), poor and non-sustained post-transfusion increments in the ANC (12–15,17,18, 20,21,23–28), and the presence of factors known to result in increased clearance of neutrophils such as HLA and ABO incompatibility and established infections (10,17,32,33). In addition, the frequent occurrence of infusion-related toxicity, particularly febrile reactions, confounded the analysis of clinical benefit in neutropenic patients (19,31–34). Much of the difficulty in interpreting these studies also may be explained by the extreme heterogeneity in trial design observed both within and among studies (19,22,35,36), including variable selection of donor/recipient pairs, differing methods used for donor granulocyte mobilization and collection, and the wide range of numbers of granulocytes transfused per product. Study endpoints such as resolution or prevention of fever and/or infection were difficult to assess due to the presence of other concurrent confounding causes, besides neutropenia, and due to rapidly evolving improvements in supportive care, especially antibiotic therapy. Collectively, these limitations likely contributed to the poor and non-sustained post-transfusion increments in the ANC observed among most studies and the conflicting outcomes regarding control and prevention of infection and fever. Using effective approaches to adequately address each of these limitations in prior trials of granulocyte transfusion therapy may result in improved and sustained increments in the ANC of recipients, which may be a key precondition to demonstrating reproducible improvements in clinical outcomes with this strategy.

III. METHODS OF COLLECTING NEUTROPHIL PRODUCTS

Neutrophils may be collected by gravity leukapheresis, filtration leukapheresis, or continuous-flow centrifugation techniques. Gravity leukapheresis results in small neutrophil yields, is labor intensive, and is rarely used clinically today. Although filtration leukapheresis results in superior cell yields compared with continuous-flow centrifugation (37), neutrophil function after filtration leukapheresis is severely impaired and increments in the recipient ANC with infusion of granulocyte products collected by filtration leukapheresis is less than that observed with continuous-flow centrifugation (38). Thus, most centers today use continuous-flow centrifugation apheresis techniques to collect granulocytes from human donors.

Hetastarch (HES) is an effective red blood cell sedimenting agent and when administered to granulocyte donors improves cell yields by a factor of 2 (39). Although initially thought to be less toxic than HES (40), pentastarch was shown to result in inferior neutrophil cell yields and to have a similar (low) frequency of adverse side effects as HES (41). Hetastarch is thus the preferred agent for red cell sedimentation to improve granulocyte cell yields using continuous-flow centrifugation apheresis collection techniques.

IV. AGENTS OTHER THAN rHuG-CSF TO MOBILIZE NEUTROPHILS FOR COLLECTION

An important factor limiting the effectiveness of granulocyte-transfusion therapy has been the inability to reliably collect granulocyte products with adequate cell doses. At steady state and from normal donors, approximately 0.2 to 0.5×10^9 neutrophils are collected per liter of blood processed with the continuous-flow centrifugation technique (37); however, infusion of products containing $<1 \times 10^{10}$ granulocytes rarely results in a measurable increment in the recipient ANC at 1 hour postinfusion (30). Corticosteroid or etiocholanalone premedication increases the donor's prepheresis ANC and results in an improved cell yield (by a factor of 2) (37). Administration of corticosteroids and HES to normal donors, followed by continuous-flow centrifugation apheresis of 7 liters blood volume yields an average of 2.3×10^{10} granulocytes (41); however, using these standard and "best case" conditions for granulocyte collection infrequently results in cell yields in excess of 3.0×10^{10}. For the most part, infusion of granulocyte products with $<3 \times 10^{10}$ cells results in subtle or immeasurable increments in the recipient ANC at 1 hour post-infusion, which are rarely sustained.

V. DEFINITION OF OPTIMAL CELL DOSE AND THE ROLE FOR rHuG-CSF IN GRANULOCYTE COLLECTIONS

What is the optimal cell dose in humans desired for granulocyte-product transfusion therapy? Studies in humans failed to consistently demonstrate a threshold cell dose that reliably resulted in significant and sustained increments in the recipient ANC and in control of infection; however, data from animal models (2) as well as human recipients of chronic myelogenous leukemia (CML) granulocytes (42) demonstrated that the curve of neutrophil cell dose versus post-infusion ANC increment was steep, the relationship direct, and that a threshold cell dose between 1 and $10 \times 10^{10}/m^2$ neutrophils (in humans) was required to observe a significant increment in the 1 hour post-infusion ANC. Assuming, in human recipients of granulocyte transfusions, that a relationship exists between post-infusion increments in the ANC and protection from infection, administration of apheresis products with larger numbers of neutrophils would more reliably result in improvements in measured clinical outcomes.

Based on studies using donors with CML, $\geq 1 \times 10^{11}$ cells/m^2 likely represents the optimal cell dose for human trials investigating granulocyte transfusions. What strategy can be used in normal donors to achieve this goal? Prior studies demonstrated that increasing the donor ANC resulted in improved neutrophil yields with apheresis-collection techniques. Previously, the only agents available to increase the donor ANC included corticosteroids and etiocholanolone; however, the donor ANC rarely exceeded $15 \times 10^9/L$ with these agents and the neutrophil yield consistently was $<3.0 \times 10^{10}$ per apheresis procedure. Schiffer et al (43) demonstrated that an average of 1.3×10^{11} white blood cells/product can be collected from donors with CML with average peripheral white cell count of $62 \times 10^9/L$. Thus, measures to further increase the peripheral white blood cells of normal donors at the time of apheresis may result in larger neutrophil yields.

The recent availability of myeloid growth factors such as rHuG-CSF (Filgrastim) has provided another class of agents to improve granulocyte collection yields from normal donors. Filgrastim is a safe agent to administer to normal donors, with side effects limited to mild reversible bone pain (44,45). Filgrastim also inhibits apoptosis and prolongs neutrophil survival in vitro (46–49) and thus may further sustain increments in the ANC of recipients of granulocyte transfusions. In 1993, two published articles demonstrated that Filgrastim effectively mobilized neutrophils in normal donors undergoing granulocyte collection (44,45). Collectively, these reports described 30 normal donors who received Filgrastim (either single dose of 300 μg in 22 donors or 5 μg/kg/day for 9 to 14 days in 8 donors) and underwent continuous-flow centrifugation apheresis (5 to 12 liters blood volume processed) with either pentastarch or HES as red blood cell–sedimenting agents. After a single 300-μg dose of Filgrastim, the donor's peripheral ANC increased from a mean of 4.2 to $20.2 \times 10^9/L$, and approximately 4.4×10^{10} neutrophils were collected (45). With each normal donor receiving a constant daily dose of Filgrastim, Bensinger et al (44) observed that the granulocyte yield per product obtained during the initial 3 days of Filgrastim administration (when the peripheral white cell count was approximately 20×10^9/L) increased greater than 4-fold by day 9 of Filgrastim administration (when the peripheral white cell count was approximately 40×10^9/L). Thus, in these preliminary investigations, Filgrastim mobilized large numbers of neutrophils into the peripheral blood of normal donors and resulted in granulocyte yields consistently in excess of other mobilizing agents.

To date, we have collected Filgrastim-mobilized granulocyte products from 32 normal donors. The dose of Filgrastim administered varied between cohorts of donors and included 5, 10, and 15 μg/kg/day for 1 to 17 consecutive days. All donors underwent one or more 7-liter continuous-flow centrifugation apheresis procedures using HES. The mean (range) number of neutrophils collected from donors receiving 5 μg/kg/day Filgrastim was 5.1 (1.25 to 9.67) $\times 10^{10}$ (50). The cell yield improved in each donor with subsequent apheresis procedures and continued Filgrastim administration at the same daily dose. Increased doses (10 or 15 versus 5 μg/kg/day) of Filgrastim also resulted in larger granulocyte yields, frequently in excess of 1×10^{11} cells/product. Toxicity has been limited to mild reversible bone pain in approximately 25% of the normal donors.

Interestingly, we observed an inverse relationship between donor age and blood neutrophil response to Filgrastim and number of neutrophils collected per leukapheresis procedure (51). Ten healthy adult donors (mean age, 45 years; range, 25 to 63 years) received Filgrastim 5 μg/kg subcutaneously (SC) at noon daily for five consecutive days, beginning at day 0. Donor ANC were determined at baseline (day 0, before Filgrastim), then at 9 AM on days 1, 3, and 5 after starting Filgrastim. On day 1 (before the second dose of Filgrastim), the first leukapheresis was performed and repeated on days 3 and 5 (maximum of three procedures/donor). Each granulocyte collection was performed on a Baxter CS-3000$^+$ blood cell separator (standard procedure 2; 7 liters blood volume) using 6% HES (500 mL) to facilitate red blood cell sedimentation. When analyzed by simple regression analysis, older donors had significantly lower ANC compared with younger donors on days 1 (r=.65, p=.04) and 3 (r=.66, p=.04), but not at baseline (day 0, before first dose of Filgrastim) nor on day 5. Twenty-nine of the scheduled 30 leukapheresis procedures were completed (one not because the donor was anemic). An inverse relationship between donor age and the number of neutrophils collected per leukapheresis procedure was observed (r=.43, p=.02). Only 2 of 11 (18.1%) leukapheresis products collected from donors $\geq$50 years of age contained $\geq 5.0 \times 10^{10}$ neutrophils/product, as opposed to 11 of 17 (64.7%) products collected from donors <50 years of age. Thus, larger doses of Filgrastim may be required in older donors to consistently obtain granulocyte products with adequate cell doses.

VI. SUSTAINED NEUTROPHIL INCREMENTS ARE OBSERVED AFTER INFUSION OF FILGRASTIM-MOBILIZED HLA-MATCHED GRANULOCYTE APHERESIS PRODUCTS

Only two studies with small numbers of patients have examined post-infusion increments in the ANC of recipients of Filgrastim-mobilized donor granulocyte transfusions (44,45). In total, these two reports include a heterogeneous group of 10 neutropenic adult and pediatric patients who underwent syngeneic (n=5) or allogeneic (n=3) bone marrow transplantation or who received antimicrobial therapy for unresponsive infections in the setting of severe aplastic anemia (n=2). Irradiated (n=5) or nonirradiated (n=5) Filgrastim-stimulated donor granulocyte transfusions were given either as prophylaxis against (n=5) or treatment of (n=5) infections. Both reports found significant increments in the recipient ANC after infusion of Filgrastim-stimulated donor granulocyte products; however, recipient ANC between baseline and 24 hours post-infusion were evaluated in only two of these patients, and ANC values beyond 24 hours post-infusion were not determined. Eight of these 10 patients were reported by Bensinger et al (44), who observed a median recipient ANC of 0.57×10^9/L 24 hours post-infusion. Two components of this report by Bensinger et al cloud the interpretation of the data presented and confound the ability of this information to address the effect of infusion of larger numbers of mature neutrophils/apheresed product made possible by Filgrastim mobilization. First, post-infusion ANC increments could not be determined because preinfusion ANC values were not reported. In addition, 5 of the 8 recipients received non-irradiated Filgrastim-mobilized donor granulocyte products. The latter contain $CD34^+$ stem cells (52; D. Adkins et al, unpublished observations) and presumably late progenitors which, driven by Filgrastim administered to the recipients post-transplant, may have rapidly proliferated and differentiated to mature neutrophils and contributed to the observed post-infusion ANC values.

No published study has carefully assessed the kinetics of post-infusion neutrophil counts in recipients of Filgrastim-mobilized donor granulocyte transfusions. In our pilot study of 10 donor/recipient pairs uniformly managed, a significant increment in the mean recipient ANC occurred at 1 hour and at peak post-leukapheresis infusion compared with the baseline value on each leukapheresis infusion day (50). The primary objective of this study was to assess the kinetics of the recipient ANC with each of three granulocyte transfusions collected from sibling bone-marrow donors receiving Filgrastim and infused into recipients on alternate days beginning day +1 post-allogeneic bone marrow transplantation. To reduce the number of confounding variables that may adversely influence the kinetics of the recipient ANC with each infusion, all granulocyte donors were HLA matched and ABO compatible with the recipient. Filgrastim was selected as the neutrophil-mobilizing agent administered to normal donors as this is a safe and effective agent for this purpose, able to consistently result in neutrophil yields of 4 to 10×10^{10} cells/apheresis (44,45) as opposed to $<3 \times 10^{10}$ cells/apheresis with other mobilizing agents (37,53). Exposure of donor neutrophils to Filgrastim pre-collection and post-infusion was intended to take advantage of any potential in vivo benefit of the ability of this agent to inhibit neutrophil apoptosis and prolong neutrophil survival in vitro (46–49).

Recipients between the ages of 15 and 55 years undergoing related donor-allogeneic marrow transplantation for malignant disease were eligible. Bone marrow aspiration from the donor and infusion into the recipient occurred on day 0. On the day of (and subsequent to) the bone marrow aspiration, the donor received Filgrastim 5 μg/kg SC and continued daily for 5 consecutive days, at which time the Filgrastim was stopped. Leukapheresis of the donor began 1 day after the bone marrow harvest (day +1), and was repeated on day +3 and day +5 (maximum of three procedures per donor). Granulocyte collections were performed with an automated continuous-flow blood cell separator by means of standard procedure 2 (7 liter blood volume processed) on a CS-3000$^+$ Blood Cell Separator (Baxter, Deerfield, IL). Each leukapheresis product was irradiated (2500 cGy, ^{137}Cs source) before infusion into the recipient, to prevent transfusion-associated graft-versus-host disease (GVHD). Beginning the day after infusion of allogeneic bone marrow, each recipient received Filgrastim 7.5 μg/kg intravenously (IV) every 12 hours until the ANC recovered to $\geq 1.5 \times 10^9$/L for two consecutive days. Promptly after completion of each leukapheresis collection, the irradiated donor leukapheresis product was transfused IV over 1 hour into the recipient. The recipient's ANC and platelet counts were determined immediately before each leukapheresis infusion, then at 1, 4, 8, 12, 24, 30, 36, 42, and 48 hours after completion of each leukapheresis infusion.

This study demonstrated that Filgrastim-mobilized HLA-matched granulocyte transfusions result in significant and sustained increments in the ANC of neutropenic recipients. The mean 1 hour and peak post-infusion increments in the recipient ANC were 0.26 to 0.73×10^9/L and 0.63 to 1.2×10^9/L, respectively, depending on the marrow day of administration. In addition, the increments in the ANC post-infusion were sustained: The duration of mean post-infusion recipient ANC greater or equal to that of the baseline (pre-infusion) value, $\geq 0.1 \times 10^9$/L, and $\geq 0.5 \times 10^9$/L was 25 to 37, 41 to 46, and 16 to 38 hours, respectively, depending on the marrow day of administration. These impressive and sustained increments in the ANC of recipients of Filgrastim-mobilized donor granulocyte transfusions compare with the poor and non-sustained increments in the ANC of recipients of granulocytes collected from donors at steady-state or receiving corticosteroids or etiocholanolone for neutrophil mobilization (12,18,20,24,27–29). The latter reports usually observed no increment in the ANC at 1 or 24 hours post-infusion. If an increment in the recipient ANC was observed at 1 hour post-infusion in these studies, it was modest, typically 0.1 to 0.2 $\times$ 10^9/L, and not sustained beyond this time point.

The significant and sustained increment in the recipient ANC with Filgrastim-mobilized donor granulocyte transfusions observed in our study has previously been reported only in recipients of non-irradiated granulocytes obtained from donors with CML (42,54) or in small children given granulocyte products collected from adult donors (24); however, infusion of non-irradiated granulocyte products obtained from donors with CML may not be entirely analogous to infusion of irradiated granulocyte products obtained from normal donors receiving Filgrastim, as the former contain, in addition to mature neutrophils, viable committed myeloid progenitors that may proliferate and differentiate and add to the effects observed (43,55). Although 1 hour post-infusion ANC increments in excess of 0.5 $\times$ 10^9/L were reported in children given non-mobilized granulocyte products (mean 5.45 $\times$ 10^9 cells/collection) collected by apheresis from adult donors, significant increments did not persist at subsequent determinations $\geq$2 hours post-infusion (24).

VII. INFUSION OF FILGRASTIM-MOBILIZED HLA-MATCHED GRANULOCYTE PRODUCTS REDUCES THE DURATION OF NEUTROPENIA FOLLOWING ALLOGENEIC PBPC TRANSPLANT

Although transplantation of Filgrastim-mobilized, allogeneic PBPC results in rapid hematologic recovery, nearly all patients develop severe neutropenia (ANC < 0.1 $\times$ 10^9/L). To reduce the duration of profound neutropenia after allogeneic PBPC transplant, thus far 12 patients with ABO-compatible donors received infusions of irradiated granulocytes on days +3 and +6 post-transplant (56). Granulocytes were collected from PBPC donors on the same day of infusion and after 5 to 10 μg/kg Filgrastim (given on days +2 and +5). The mean daily ANC for patients who received granulocyte transfusions were compared with the daily mean ANC for 23 concurrent controls who underwent PBPC transplant without granulocytes. For patients who received granulocyte transfusions, the mean duration of severe neutropenia was 3 days (range, 1 to 5; ±SD, 1.5), significantly less than the mean duration of severe neutropenia for controls (mean, 6 days; range, 3 to 11; ±SD, 2.2) (p<.001). Thus, routine administration of Filgrastim-mobilized granulocyte transfusions collected from the patient's stem cell donor can significantly reduce the duration of profound neutropenia after allogeneic PBPC transplant.

VIII. FILGRASTIM-MOBILIZED GRANULOCYTE PRODUCTS LOCALIZE TO SITES OF INFLAMMATION WHEN INFUSED INTO NEUTROPENIC RECIPIENTS

Neutrophils migrate to sites of inflammation or infection to effect their critical functions. Preclinical and human studies of non-Filgrastim–mobilized granulocyte transfusions demonstrated that transfused neutrophils circulate and migrate to sites of inflammation or infection (9,57–60). Certain myeloid growth factors, including G-CSF, inhibit random migration of neutrophils in vitro and theoretically may alter neutrophil migration to these sites in vivo (61). To assess the distribution of Filgrastim-mobilized neutrophils in vivo, leukapheresis products collected from five normal donors receiving Filgrastim were labeled with indium[111] and infused into five neutropenic allogeneic marrow recipients. Both anterior and posterior whole body scintigraphic images of the recipient were obtained at 1 to 4, 24, and 48 hours after injection of indium-labeled donor white blood cells. Two recipients were found to have localization of activity in the mouth and nasal region on the scintigraphic images obtained at 24 and 48 hours post-injection, which correlated with the presence of grade II-III mucositis on physical examination. One recipient with concurrent grade II diarrhea was found to have intense localization of activity in the terminal ileum and cecum on the delayed images.

There are no reports documenting that transfused granulocytes obtained from donors receiving Filgrastim localize to sites of inflammation when infused into neutropenic human recipients. In the current study, localization of activity to sites of tissue damage such as the terminal ileum and the oral/nasopharynx on the delayed (24 and 48 hours post-injection) nuclear scan images was observed and supports the concept that Filgrastim-mobilized irradiated donor granulocytes are able to migrate to sites of inflammation in vivo. Currently available data do not support that altered random neutrophil migration by exposure to myeloid growth factors observed in vitro is clinically meaningful in vivo.

IX. ADDITIONAL BENEFITS OF GRANULOCYTE TRANSFUSIONS

Although platelets were known to be present in granulocyte products obtained by apheresis, the potential clinical relevance of this contaminating component had not been formally assessed or reported. In our pilot study of Filgrastim-mobilized granulocyte transfusions administered to allogeneic marrow recipients (50), the number of platelets present in each donor leukapheresis product (mean, 2.55 $\times$ 10^{11}; range, 1.17 to 4.52 $\times$ 10^{11}) was comparable to the number present within a unit of single-donor apheresed platelets. Significant increments from baseline in the mean recipient platelet count and corrected count increment were observed at 1 hour post-leukapheresis infusions given on marrow transplant days +1, +3, and +5. The magnitude of this increment in platelet count was comparable to that reported for recipients of unselected pooled or HLA-matched single donor platelet products (62–64). We also observed reduced requirements for platelet product transfusions in the study patients, compared with a similar historical control group who did not receive granulocyte transfusions.

X. TOXICITY OF FILGRASTIM-MOBILIZED GRANULOCYTE TRANSFUSIONS

To date, we have given more than 80 Filgrastim-mobilized HLA-matched granulocyte apheresis product transfusions to neutropenic transplant recipients. Acute toxicity in recipients due to the infusion of HLA-matched donor neutrophils was surprisingly minimal. One recipient developed transient cough and shortness of breath at the end of the second granulocyte product infusion, which did not recur with a third infusion. No other recipient developed respiratory symptoms with infusion of the granulocyte products. Respiratory distress with granulocyte transfusions has occasionally, but infrequently, been reported (21,30). Febrile reactions due to infusion of donor granulocytes were not observed in our study, although frequently reported in other studies, which usually did not use HLA matching in donor selection criteria (13,14,20,21,30). This observation may be due to the selection of only HLA-matched siblings as donors of granulocyte products and the use of continuous flow centrifugation apheresis, as opposed to filtration leukapheresis, in our study. No other acute toxicity due to the granulocyte product infusions was observed. Thus, infusions of Filgrastim–mobilized HLA matched granulocyte products into neutropenic subjects will not confound clinical outcome analysis of future trials by virtue of febrile transfusion reactions.

XI. CONCLUSION

We, and others, have demonstrated that Filgrastim is an effective and safe agent to mobilize neutrophils in normal donors, resulting in granulocyte products containing neutrophil-cell doses superior to that observed with other agents. Whether there are any long-term concerns with the use of Filgrastim in normal donors has not been defined, although it seems unlikely and the use of Filgrastim in normal donors is not a labeled indication. Larger doses of Filgrastim resulted in greater cell yields per collection without additional risks to donors; however, older donors receiving Filgrastim failed to mobilize neutrophils as well as younger individuals, and consequently had poorer granulocyte yields.

We observed significant and sustained increments in the ANC of neutropenic recipients after transfusion of Filgrastim–mobilized HLA-matched granulocyte products. Studies are in progress in both allogeneic and autologous transplant patients to address the clinical relevance of this finding. Further, transfusion of Filgrastim–mobilized HLA-matched granulocyte products do not cause febrile reactions, are functional in vivo, and provide supplemental platelet support by virtue of large numbers of contaminating passenger platelets present in the apheresis product.

Dose-intense chemotherapy, whether or not supported with myeloid growth factors and stem cell transplant, results in an obligatory neutropenic period of 7 to 10 days, which presumably reflects the time to maturation of progenitor cells. Thus, short of infusing mature neutrophils, no currently available strategy offers any likelihood of further abbreviating this obligatory period. The recently available data regarding Filgrastim–mobilized granulocyte apheresis products presented in this section provide strong evidence to investigate the potential supportive role of transfusions of these products into patients during the expected neutropenic period after dose-intense therapy.

REFERENCES

1. Bodey, G. P., Buckley, M., Sathe, Y. S., and Freireich, E. J. (1966). Quantitative relationships between circulating leukocytes and infection in patients with acute leukemia. *Ann Intern Med* 64:328–340.

2. Appelbaum, F. R., Bowles, C. A., Makuch, R. W., and Deisseroth, A. B. (1978). Granulocyte transfusion therapy of experimental *Pseudomonas* septicemia: study of cell dose and collection technique. *Blood 52*:323–331.
3. Debelak, K. M., Epstein, R. B., and Andersen, B. R. (1974). Granulocyte transfusions in leukopenic dogs: in vivo and in vitro function of granulocytes obtained by continuous-flow filtration leukopheresis. *Blood 43*:757–766.
4. Epstein, R. B., Clift, R. A., and Thomas, E. D. (1969). The effect of leukocyte transfusions on experimental bacteremia in the dog. *Blood 34*:782–790.
5. Epstein, R. B., Waxman, F. J., Bennet, B. T., and Andersen, B. R. (1974). *Pseudomonas* septicemia in neutropenic dogs. I. Treatment with granulocyte transfusions. *Transfusion 14*:51–57.
6. Westrick, M. A., Debelak-Fehir, K. M., and Epstein, R. B. (1977). The effect of prior whole blood transfusion on subsequent granulocyte support in leukopenic dogs. *Transfusion 17*:611–614.
7. Dale, D. C., Reynolds, H. Y., Pennington, J. E., Elin, R. J., Pitts, T. W., and Graw, R. G. Jr. (1974). Granulocyte transfusion therapy of experimental *Pseudomonas* pneumonia. *J Clin Invest 54*:664–671.
8. Dale, D. C., Reynolds, H. Y., Pennington, J. E., Elin, R. J., and Herzig, G. P. (1976). Experimental *Pseudomonas* pneumonia in leukopenic dogs: comparison of therapy with antibiotics and granulocyte transfusions. *Blood 47*:869–876.
9. Chow, H. S., Sarpel, S. C., and Epstein, R. B. (1980). Pathophysiology of *Candida albicans* meningitis in normal, neutropenic, and granulocyte transfused dogs. *Blood 55*:546–551.
10. Chow, H. S., Sarpel, S. C., and Epstein, R. B. (1982). Experimental candidiasis in neutropenic dogs: tissue burden of infection and granulocyte transfusion effects. *Blood 59*:328–333.
11. Epstein, R. B., and Chow, H. S. (1981). An analysis of quantitative relationships of granulocyte transfusion therapy in canines. *Transfusion 21*:360–362.
12. Alavi, J. B., Root, R. K., Djerassi, I., et al (1977). A randomized clinical trial of granulocyte transfusions for infection in acute leukemia. *N Engl J Med 296*:706–711.
13. Ambinder, E. P., Button, G. R., Cheung, T., Goldberg, J. D., and Holland, J. F. (1981). Filtration versus gravity leukapheresis in febrile granulocytopenic patients: a randomized prospective trial. *Blood 57*:836–841.
14. Clift, R. A., Sanders, J. E., Thomas, E. D., Williams, B., and Buckner, C. D. (1978). Granulocyte transfusions for the prevention of infection in patients receiving bone-marrow transplants. *N Engl J Med 298*:1052–1057.
15. Dahlke, M. B., Keashen, M., Alavi, J. B., Koch, P. A., and Eisenstaedt, R. (1982). Granulocyte transfusions and outcome of alloimmunized patients with gram-negative sepsis. *Transfusion 22*:374–378.
16. Fortuny, I. E., Bloomfield, C. D., Hadlock, D. C., Goldman, A., Kennedy, B. J., and McCullough, J. J. (1975). Granulocyte transfusion: a controlled study in patients with acute nonlymphocytic leukemia. *Transfusion 75*:548–558.
17. Graw, R. G. Jr., Herzig, G., Perry, S., and Henderson, E. S. (1972). Normal granulocyte transfusion therapy: treatment of septicemia due to gram-negative bacteria. *N Engl J Med 287*:367–371.

18. Herzig, R. H., Herzig, G. P., Graw, R. G. Jr., Bull, M. I., and Ray, K. K. (1977). Successful granulocyte transfusion therapy for gram-negative septicemia. A prospectively randomized controlled study. *N Engl J Med* 296:701–705.

19. Higby, D. J., and Burnette, D. (1980). Granulocyte transfusions: current status. *Blood* 55:2–8.

20. Higby, D. J., Yates, J. W., Henderson, E. S., and Holland, J. F. (1975). Filtration leukapheresis for granulocyte therapy. Clinical and laboratory studies. *N Engl J Med* 292:761–766.

21. Higby, D. J., Freeman, A., Henderson, E. S., Sinks, L., and Cohen, E. (1976). Granulocyte transfusions in children using filter-collected cells. *Cancer* 38:1407–1413.

22. Higby, D. J. (1981). Granulocyte transfusions: where now? *N Engl J Med* 305:636–638.

23. Lowenthal, R. M., Grossman, L., Goldman, J. M., et al (1975). Granulocyte transfusions in treatment of infections in patients with acute leukaemia and aplastic anaemia. *Lancet* 1:353–358.

24. Maybee, D. A., Millan, A. P., and Ruymann, F. B. (1977). Granulocyte transfusion therapy in children. *Southern Med J* 70:320–324.

25. Schwarzenberg, L., Mathe, G., DeGrouchy, J., et al (1965). White blood cell transfusions. *Isr J Med Sci* 1:925.

26. Strauss, R. G., Goedken, M. M., Maguire, L. C., Koepke, J. A., and Thompson, J. S. (1980). Gram-negative sepsis treated with neutrophils collected exclusively by intermittent flow centrifugation leukapheresis. *Transfusion* 20: 79–81.

27. Vallejos, C., McCredie, K. B., Bodey, G. P., Hester, J. P., and Freireich, E. J. (1975). White blood cell transfusions for control of infections in neutropenic patients. *Transfusion* 15:28–33.

28. Vogler, W. R., and Winton, E. F. (1977). A controlled study of the efficacy of granulocyte transfusions in patients with neutropenia. *Am J Med* 63:548–555.

29. Winston, D. J., Ho, W. G., Young, L. S., and Gale, R. P. (1980). Prophylactic granulocyte transfusions during human bone marrow transplantation. *Am J Med* 68:893–897.

30. Winston, D. J., Ho, W. G., and Gale, R. P. (1982). Therapeutic granulocyte transfusions for documented infections. A controlled trial in ninety-five infectious granulocytopenic episodes. *Ann Intern Med* 97:509–515.

31. Dancey, J. T., Deubelbeiss, K. A., Harker, L. A., and Finch, C. A. (1976). Neutrophil kinetics in man. *J Clin Invest* 58:705–715.

32. Graw, R. G., Jr., Goldstein, I. M., Eyre, H. J., and Terasaki, P. I. (1970). Histocompatibility testing for leukocyte transfusion. *Lancet* 1:77–78.

33. Higby, D. J., Mishler, J. M., Cohen, E., Rhomberg, W., Nicora, R. W., and Holland, J. F. (1974). Increased elevation of peripheral leukocyte counts by infusion of histocompatible granulocytes. *Vox Sang* 27:186–189.

34. French, J. E., Solomon, J. M., and Fratantoni, J. C. (1982). Survey on the current use of leukapheresis and the collection of granulocyte concentrates. *Transfusion* 22:220–225.

35. Boggs, D. R. (1977). Neutrophils in the blood bank. *N Engl J Med* 296:748–750.

36. Strauss, R. G. (1993). Therapeutic granulocyte transfusions in 1993. *Blood 81*:1675–1678.

37. MacPherson, J. L., Nusbacher, J., and Bennett, J. M. (1976). The acquisition of granulocytes by leukapheresis: a comparison of continuous flow centrifugation and filtration leukapheresis in normal and corticosteroid-stimulated donors. *Transfusion 16*:221–228.

38. Herzig, G. P., Root, R. K., and Graw, R. G. Jr. (1972). Granulocyte collection by continuous-flow filtration leukapheresis. *Blood 39*:554–567.

39. Mishler, J. M., Hadlock, D. C., Fortuny, I. E., Nicora, R. W., and McCullough, J. J. (1974). Increased efficiency of leukocyte collection by the addition of hydroxyethyl starch to the continuous flow centrifuge. *Blood 44*:571–581.

40. Strauss, R. G., Hester, J. P., Vogler, W. R., et al (1986). A multicenter trial to document the efficacy and safety of a rapidly excreted analog of hydroxyethyl starch for leukapheresis with a note on steroid stimulation of granulocyte donors. *Transfusion 26*:258–264.

41. Lee, J. H., Leitman, S. F., and Klein, H. G. (1995). A controlled comparison of the efficacy of hetastarch and pentastarch in granulocyte collections by centrifugal leukapheresis. *Blood 86*:4662–4666.

42. Freireich, E. J., Levin, R. H., Whang, J., Carbone, P. P., Bronson, W., and Morse, E. E. (1964). The function and fate of transfused leukocytes from donors with chronic myelocytic leukemia in leukopenic recipients. *Ann NY Acad Sci 113*:1081.

43. Schiffer, C. A., Aisner, J., Dutcher, J. P., and Wiernik, P. H. (1983). Sustained post-transfusion granulocyte count increments following transfusion of leukocytes obtained from donors with chronic myelogenous leukemia. *Am J Hematol 15*:65–74.

44. Bensinger, W. I., Price, T. H., Dale, D. C., et al (1993). The effects of daily recombinant human granulocyte colony-stimulating factor administration on normal granulocyte donors undergoing leukapheresis. *Blood 81*:1883–1888.

45. Caspar, C. B., Seger, R. A., Burger, J., and Gmur, J. (1993). Effective stimulation of donors for granulocyte transfusions with recombinant methionyl granulocyte colony-stimulating factor. *Blood 81*:2866–2871.

46. Begley, C. G., Lopez, A. F., Nicola, N. A., et al (1986). Purified colony-stimulating factors enhance the survival of human neutrophils and eosinophils in vitro: a rapid and sensitive microassay for colony stimulating factors. *Blood 68*:162–166.

47. Colotta, F., Re, F., Polentarutti, N., Sozzani, S., and Mantovani, A. (1992). Modulation of granulocyte survival and programmed cell death by cytokines and bacterial products. *Blood 80*:2012–2020.

48. Colotta, F., Re, F., and Mantovani, A. (1993). Granulocyte transfusions from granulocyte colony-stimulating factor-treated donors: also a question of cell survival? *Blood 82*:2258.

49. Liles, W. C., Rodger, E. R., and Dale, D. C. (1994). In vivo administration of granulocyte colony-stimulating factor (G-CSF) to normal human subjects: priming of the respiratory burst and inhibition of apoptosis in neutrophils. *Blood 84*:24a (abstr 85).

50. Adkins, D., Johnston, M., Cravens, D., et al (1995). Significant prolonged increments in the absolute neutrophil count (ANC) of allogeneic bone marrow transplant (Allo-BMT) recipients (R) occurs with prophylactic neutrophil infusions (INF) obtained from HLA matched donors (D) receiving granulocyte colony stimulating factor (G-CSF). *Blood 86:*582a (abstr 2317).

51. Adkins, D., Goodnough, T., Brown, R., and DiPersio, J. (1996). Inverse relationship between age of normal donors and blood neutrophil response to G-CSF and number of neutrophils collected per leukapheresis (LA) procedure. *Blood 88:*331a, (abstr 1314).

52. Adkins, D., Johnston, M., Cravens, D., et al (1994). Significant prolonged increments in the absolute neutrophil count (ANC) of allogenic bone marrow transplant (ABMT) recipients (R) occurs with prophylactic neutrophil infusions (INF) obtained from HLA matched donors (D) receiving granulocyte colony stimulating factor (G-CSF). *Proc ASCO 13:*437 (abstr 1504).

53. Shoji, M., and Vogler, W. R. (1974). Effects of hydrocortisone on the yield and bactericidal function of granulocytes collected by continuous-flow centrifugation. *Blood 44:*435–443.

54. Morse, E. E., Freireich, E. J., Carbone, P. P., Bronson, W., and Frei, E. (1966). The transfusion of leukocytes from donors with chronic myelocytic leukemia to patients with leukopenia. *Transfusion 6:*183.

55. Levin, R. H., Whang, J., Tjio, J. H., Carbone, P., Frei, E., and Freireich, E. J. (1963). Persistent mitosis of transfused homologous leukocytes in children receiving antileukemic therapy. *Science 142:*1305–1311.

56. Brown, R. A., Adkins, D. R., Goodnough, L., Haug, J., Todd, G., DiPersio, J. (1996). Infusion of granulocytes collected from HLA-identical sibling donors reduces the duration of neutropenia following allogeneic peripheral blood stem cell (PBSC) transplant. *Blood 88:*261a (abstr 3772).

57. Appelbaum, F. R., Norton, L., and Graw, R. G. Jr. (1977). Migration of transfused granulocytes in leukopenic dogs. *Blood 49:*483–488.

58. Dutcher, J. P., Schiffer, C. A., and Johnston, G. S. (1981). Rapid migration of Indium-labeled granulocytes to sites of infection. *N Engl J Med 304:*586–589.

59. Kauder, E., Boggs, D. R., Athens, J. W., et al (1965). Leukokinetic studies. XII. Kinetic studies of normal isologous neutrophilic granulocytes transfused into normal subjects. *Proc Soc Exp Biol Med 120:*595–599.

60. Price, T. H. Kinetics of transfused neutrophils. In: Vogler, W. R., ed. *Cytapheresis and Plasma Exchange: Clinical Indications.* New York: Alan R. Liss, Inc., 1982, p. 19.

61. Smith, W. B., Gamble, X. Jr., and Vadas, M. A. (1994). The role of granulocyte-macrophage and granulocyte colony-stimulating factors in neutrophil transendothelial migration: comparison with interleukin-8. *Exp Hematol 22:*329–334.

62. Bishop, J. F., McGrath, K., Wolf, M. M., et al (1988). Clinical factors influencing the efficacy of pooled platelet transfusions. *Blood 71:*383–387.

63. Moroff, G., Garratty, G., Heal, J. M., et al (1992). Selection of platelets for refractory patients by HLA matching and prospective crossmatching. *Transfusion 32:*633–640.

64. Murphy, S. Preservation and Clinical Use of Platelets. In: Williams, W. J., Beutler, E., Erstev, A. J., and Lichtman, M. A., eds. *Hematology.* 4th ed. New York: McGraw-Hill, 1996, p. 1654.

8

The Importance of Dose in Cancer Chemotherapy and the Role of Hematopoietic Growth Factors

Miguel H. Bronchud
General Hospital of Granollers, Barcelona, Spain

I. INTRODUCTION

Disseminated cancer can be cured only if all the malignant cells at the primary and metastatic sites are eradicated. To achieve this goal, active cytotoxic agents are required and it is a bonus if the host–defense mechanisms assist in the process. The development of effective chemotherapy agents and successful combination regimens was a milestone in oncology and resulted in previously fatal conditions becoming potentially curable. The majority of chemotherapy cures, however, occur in a small proportion of all patients with cancer, a fact that clearly indicates that there is still considerable scope and need for improvement for current treatment strategies.

The failure of chemotherapy treatment is blamed ultimately on the development of resistant clones in the malignant population. The resistance mechanism may be present immediately at the time of diagnosis or it may develop following exposure to chemotherapy. In the former case, the prognosis is poor without the development of new cytotoxic agents or novel therapeutic strategies. In the latter case, the aim of treatment should be (where possible) to eradicate all tumor cells before the emergence of resistance defeats chemotherapy. Successful chemotherapy regimens, with few exceptions, consist of a combination of active, non-cross–resistant drugs administered at an effective dose. Chemotherapy dose intensity is probably important to the success of treatment. Cure rate is sacrificed if the dose intensity is decreased. In clinical practice, however, doses of chemotherapy often are reduced or cycles delayed in deference to the toxic sequelae of treatment. Reductions in dose rates often result in minimal decreases in toxicity, while drastically altering the therapeutic response and hence the success of treatment. DeVita (1) has stated: "Both the physician and the patient must consider the risk of dying from cancer along with the transient benefits of reducing the side effects of treatment."

This chapter describes the experimental and clinical evidence supporting the importance of dose intensity in cancer chemotherapy and shows how outcome may be jeopardized through inadequate dosing even in curable malignancies. There is reasonable evidence to suggest that the success of chemotherapy in sensitive malignancies will be improved if chemotherapy regimens are delivered at full dose intensity and not compromised by modifications for toxicity. A major cause of dose modification with many chemotherapy regimens is myelosuppression, principally neutropenia; however, the use of hematopoietic growth factors that stimulate neutrophil recovery after chemotherapy alleviates drug-induced neutropenia and may allow the delivery of chemotherapy as scheduled. Furthermore, by overcoming the principal dose-limiting toxicity of many drugs, the hematopoietic growth factors may be used to test the hypothesis that dose intensification will produce a higher response rate.

II. PRINCIPLES OF DOSE AND RESPONSE

A. Cell Kinetics

The response to chemotherapy is affected by the biology of tumor growth. Pioneering experiments by Skipper (2) showed that in model tumors in animals, a given dose of a chemotherapeutic agent killed a constant fraction of cells regardless of the population size. The killing effects of cancer drugs thus follow log-kill kinetics. For example, if a given dose of a particular drug kills three log units of cells, it will reduce the tumor burden from 10^{10} to 10^7 cells, or from 10^9 to 10^6, or 10^8 to 10^5, and so on. From this observation comes the cardinal rule of chemotherapy: the invariable relationship between cell number and curability (1). In other words, the smaller the tumor burden, the higher the likelihood of cure at a given dose.

The logarithmic relationship between dose and cell kill is dependent on the tumor growing exponentially (first-order kinetics). Although model tumors such as L1210 murine leukemia grow exponentially, ie, with a constant doubling time, most human tumors do not appear to share the same kinetics. The growth curves of most human solid tumors are better explained by Gompertzian kinetics, where exponential growth is matched by exponential retardation of growth (3). This means that when tumors are small, they grow slowly but with a high growth fraction; when they are of intermediate size, they grow faster; and when they are large they reach a plateau state with a small growth rate and a small growth fraction. The rate of regression of a tumor after treatment should be closely related to the growth characteristics at the time of treatment. This suggests that tumors following a Gompertzian growth pattern regress in a Gompertzian fashion, with a slower rate of regression at small and larger tumor sizes than at intermediate sizes (3). Furthermore, such experimental observations imply that there are kinetic reasons for the failure of chemotherapy to cure larger tumors. On the other hand, in patients with a smaller tumor volume it should be possible to overcome kinetic reasons for chemotherapy failure.

Biochemical resistance to chemotherapy is the major impediment to successful treatment. In 1979 Goldie and Coldman proposed that the non-random cytogenetic changes associated with most human cancers were closely associated with the development of resistance to anticancer drugs. They developed a mathematical model that predicted that the rate of emergence of drug-resistant cells was intrinsic to the genetic instability of a particular tumor and that resistant cells would be present in population sizes as low as 10^3 to 10^6, well below clinically detectable levels. Thus, the probability that a tumor will contain resistant clones is a function of the mutation rate and the size of the tumor. If the mutation rate is only 10^{-6} (1 in 1,000,000), a tumor composed of 10^9 cells (a mass only 1 cm in size) would be virtually certain to contain at least one drug-resistant clone, although the absolute number of resistant cells would be relatively small. Applying this example to the clinical situation, such a tumor would appear to respond to treatment initially, but then reappear as the resistant clone expanded. The Coldman-Goldie hypothesis has provided an important argument in favor of using early dose-intensive chemotherapy (4).

B. General Principles

1. Dose Response

The relationship between the dose of chemotherapy and antitumor response was first demonstrated by Skipper (2) in experiments in animals using tumor models such as L1210 leukemia. In these experimental models, the dose-response curve for the majority of antitumor agents is usually steep in the linear phase (**Figure 1**) (5). Reduction of doses in the linear phase of the dose-response curve results first in a loss of capacity to cure the tumor before there is a diminution of response rates. In other words, complete remissions will continue to be observed in animals bearing palpable tumors,

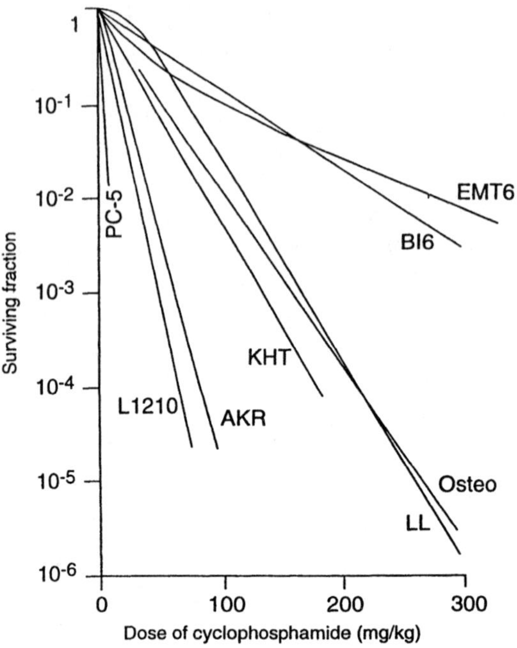

Figure 1.   Single-dose cyclophosphamide curves in various in vivo experimental tumors. (Adapted from Ref. 5.)

but the last few residual cells will not be ablated and relapse becomes inevitable. For example, in the Ridgway osteosarcoma tumor model, reduction in the average dose intensity of the combination of L-phenylalanine mustard (L-PAM) and cyclophosphamide was associated with a marked decrease in the cure rate before a significant reduction in the complete remission rate was noted (**Table 1**) (6). It is apparent from these results that a dose reduction of only approximately 20% led to a loss in cure rate in excess of 50%. Furthermore, in high growth-fraction tumors, a small increase in dose can result in a disproportionately large increase in cure rate. For example, a twofold increase in dose often leads to a tenfold (1-log) increase in tumor cell kill (1).

While it is not possible to extrapolate these results directly to human tumors, the general principle is transferable to the clinic and is ignored at great peril. To quote DeVita (1), "*Ad hoc* adjustment of dosing is probably the main reason for treatment failure in patients with drug sensitive human tumors undergoing their first treatment."

Table 1. Ridgway Osteogenic Sarcoma: Response to Different Dose Intensity of Two-Drug Combination of Cyclophosphamide and L-PAM.

Relative dose intensity				
CTX	L-PAM	Average	% CR	% Cures
0.38	0.82	0.60	100	60
0.75	0.18	0.47	100	44
0.25	0.55	0.44	100	10
0.50	0.12	0.31	10	0
0.17	0.36	0.27	0	0

CTX = Cyclophosphamide; L-PAM = L-phenylalanine mustard; CR = complete response. Tumors weighed 2 to 3 g.
Source: Adapted from Ref. 2.

An analogy with radiotherapy strongly supports this view. The realization that the radiation dose-response curve for Hodgkin's disease (HD) was steep led to the development of curative radiotherapy in appropriate patients, despite initial skepticism. In HD, and essentially all other cancers curable by radiotherapy, there is a critical dose that must be attained to achieve cure; any significant reduction from that dose compromises the cure rate substantially (5). The threshold dose below which it is not possible to achieve cure also applies to treatment with chemotherapy.

2. Dose, Dose Intensity, and Total Dose

There are several measurements used to express the amount of drug delivered. The way dose is expressed has important implications for assessing the quality and quantity of the response to anticancer treatment. The dose intensity is the amount of drug delivered per unit time, expressed as, for example, $mg/m^2/week$, regardless of the schedule or route of administration. The relative dose intensity is the amount of drug delivered per unit time relative to an arbitrarily chosen standard single drug, or for a combination regimen the decimal fraction of the ratio of the test regimen to the standard regimen. The total dose is the amount of drug given per dose (mg/m^2) multiplied by the total number of doses. The cumulative dose graphically represents both dose intensity and total drug delivery and is useful when analyzing trials designed to isolate treatment variables and improve the therapeutic index of available drugs (7). Hryniuk (8) described in detail the calculation of dose-intensity variables.

In animal models, an almost invariable direct relation exists between the individual dose level, dose intensity, and total dose of a given drug, and both the degree and duration of therapeutic response of a given tumor, as long as the staging and treatment schedule are constant and lethal toxicity or limiting drug-resistant tumor cells do not intervene (9). By retrospectively analyzing the results of thousands of experiments in animal models, Skipper was able to conclude that the dose intensity is the dominant treatment-design variable with respect to the degree of therapeutic response achievable (cure or nearness to cure at the nadir); however, total dose usually correlates best with the duration of response in treatment failures (ie, the duration of temporary partial and complete responses and the survival time of treatment failures) (**Figure 2**). These retrospective results from animal experiments indicate that high dose intensity for a relatively short duration is the best choice when cure is the goal and feasible, but lower dose intensity for longer periods may be a better choice when palliation is the goal (9).

These data have important implications in the design of clinical trials. Since dose intensity in rodents appears to be more important than total dose in the ability to cure, shorter intense regimens need to be studied in humans. In addition, when toxicity forces dose modification, it may be more appropriate to maintain dose and widen the intervals between chemotherapy, especially if tumor regrowth rates are altered by therapy (3).

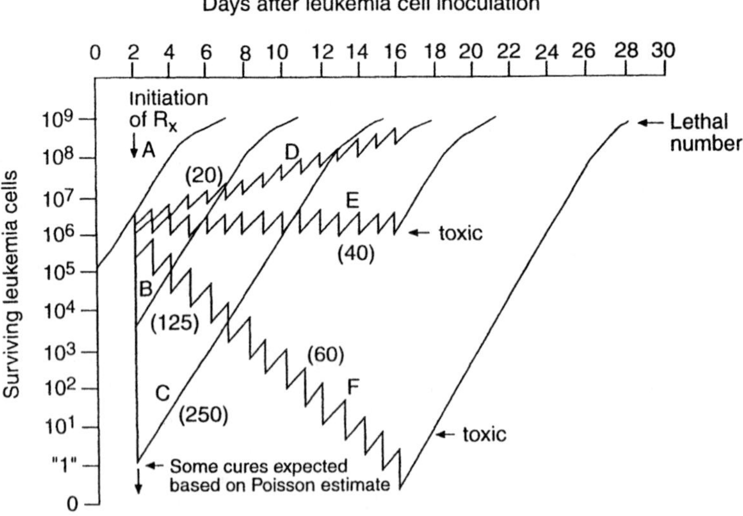

Figure 2. Experimental model showing relationship between dose and outcome. A = Control group; B and C = single dose; D, E, and F = daily programs giving lower dose intensity, but higher total dose. (Adapted from Ref. 3.)

C. Development of Combination Chemotherapy

1. Historical Perspective

In 1961, it was shown that combination chemotherapy with methotrexate and mercaptopurine improved the complete remission rate in children with acute leukemia (10). Subsequently, with vincristine and prednisone, a complete remission rate of nearly 90% was achieved (11). A number of combination regimens then were designed with curative intent for acute leukemias, HD, and non-Hodgkin's lymphoma (NHL) (12). These studies established that combination chemotherapy was superior to single-agent therapy in these diseases. The development of combination chemotherapy regimens also led to improved response rates in pediatric tumors, such as Ewing's sarcoma, embryonal rhabdomyosarcoma, Wilm's tumor, and osteogenic sarcoma, and in adult solid tumors including testicular cancer, ovarian cancer, breast cancer, and small-cell lung cancer (SCLC) (12).

2. The Importance of Dose Intensity

In chemosensitive malignancies where combination chemotherapy has improved response rates and overall survival, dose intensity is of fundamental importance. The impact of individual drugs or combinations of two or three drugs in a multidrug combination can be assessed separately, and this can help identify the most important drug in a combination (13). This provides important information that can help to avoid dose adjustments that could radically alter the effectiveness of a combination regimen. Alteration in dose intensity of the most effective drug in a combination has the greatest impact on outcome. For example, in the Ridgway osteogenic sarcoma model, decreases in the dose of L-PAM reduce the effectiveness of the L-PAM/6-mercaptopurine (6-MP) combination even when the dose of 6-MP is increased to compensate for the reduction in the dose of the more effective drug (**Table 2**). A decrease in the dose intensity of L-PAM to <55% of the optimal single dose schedule results in loss of capacity to cure animals regardless of the dose of 6-MP.

Actual calculations of dose intensity indicate that, for many cases, combination chemotherapy has resulted in significant compromises in the dose intensity of the most valuable agents used in these programs (13,14). This observation is well illustrated in the treatment of breast cancer where a comparison of several commonly used combination regimens shows a wide variation in relative dose intensities (**Table 3**). Furthermore, data from a number of studies have shown that in the adjuvant setting and in patients with advanced disease, treatment outcome is influenced by dose intensity (8,13,15).

Table 2.   Ridgway Osteogenic Sarcoma: Effect of Varying Dose Intensity of a More Effective Drug, L-PAM.

	Relative dose intensity			Observed	
L-PAM	6-MP	Ratio L-PAM to 6-MP	Average	% CR	% Cures
0.82	0.49	1.7	0.66	100	60
0.73	1.3	0.56	1.0	90	50
0.55	1.0	0.55	0.78	90	20
0.55	0.33	1.7	0.44	80	20
0.36	0.67	0.54	0.52	56	0
0.36	0.21	1.7	0.29	30	0
0.27	1.5	0.18	0.89	70	0
0.24	0.44	0.57	0.35	0	0
0.24	0.15	1.6	0.20	0	0
0.18	1.0	0.18	0.59	0	0
0.12	0.67	0.18	0.50	0	0
0.08	0.44	0.18	0.26	0	0

L-PAM = L-Phenylalanine mustard; 6-MP = 6-mercaptopurine; CR = complete response. Tumors weighed 2 to 3 g.
Source: Adapted from Ref. 2.

Table 3.   Breast Cancer: Relative Dose Intensity.

Regimen	Institution/ Investigators	Drug				RDI (3)
		C	Dox	5-FU	M	
CMFAV	Johns Hopkins	350	20	600	50	323
CMFVP	Cooper et al	560	—	294	17	284
CMF	Bonadonna et al	350	—	300	20	216
CAF	NCI	350	15	250	—	205
CAF	M.D. Anderson	167	16	167	—	116
CAMF	NCI	167	10	167	13	115
Super CMF	Dana-Farber	200	12	17	857	76

C = Cyclophosphamide; Dox = doxorubicin; M = methotrexate; 5-FU = 5-fluorouracil; RDI = relative dose intensity.
Comparative dose intensity (mg/m^2/wk). Tumors weighed 2 to 3 g.
Source: Adapted from Ref. 1.

Similar calculations of dose intensity have been made for combination regimens used to treat HD and diffuse large-cell lymphomas. In both situations, there was a strong and significant correlation between cure rate and dose intensity. These calculations also show that there is probably a threshold dose below which cure is not attainable and that small increases in dose intensity might be associated with improvements in cure rate (3).

Simon and Korn (16) have described a mathematical model for rationally selecting cytotoxic drugs and dosages for combination regimens, based on the antitumor activities of the drugs, given as single agents, and their organ-specific maximum tolerated doses. Their approach does not assume that the underlying dose-response curve is steep or that maximally dose intense regimens are clinically appropriate in all situations.

3. Impact of Dose-Intense Chemotherapy on the Development of Permanent Drug Resistance

The likelihood of therapeutic success depends on the probability that there are no resistant cells and the probability that all the sensitive cells are eliminated (which may not occur if insufficient therapy is administered) (4). By extending their somatic mutation theory, Coldman and Goldie (4) have developed a model to examine the effect of different two-drug protocols, with different dose intensities, on the likelihood that doubly or singly resistant cell lines will develop (**Table 4**). Their model showed that dose-intensive therapy is superior in both reducing the likelihood that doubly resistant cell types will emerge, as well as curing the tumor. The magnitude of this reduction depends on the tumor type and the particular situation under consideration. Their model showed also that dose-intense therapy has a much greater impact on eliminating sensitive cells than it does on preventing the occurrence of doubly resistant cells. Increasing intensity may, therefore, be expected to have larger effects on response rates than on cure rates and survival time.

The model also investigated the effect of delaying treatment time, and this exercise showed that even for hypothetical tumor systems with the same variables, the use of identical regimens may not necessarily result in identical cure rates. This observation emphasizes the need for using a common time for the beginning of the period for which dose intensity is to be calculated. In clinical trials, the logical time for the commencement of the calculation of dose intensity would seem to be either the time of randomization or the time of tumor staging (4). The effect of delaying the start of treatment may have important implications in situations such as neoadjuvant chemotherapy because the probability of no double resistance is strongly influenced by the early cycles of therapy in a regimen. In other

Table 4.   Model to Examine Effect of Two-Drug Protocols.

Hypothetical regimens (maximum seven cycles)

Regimen		IRDI[a]
I	Drug A, drug B full dose, fractional kill = 99%, treatment interval = 20 day	1.0
II	Drug A, drug B half dose, fractional kill = 90%, treatment interval = 20 day	0.5
III	Drug A, drug B full dose, fractional kill = 99%, treatment interval = 40 day	0.5
IV	Drug A, drug B half dose, fractional kill = 90%, treatment interval = 40 day	0.25

Probability of no resistant cells and probability of cure (seven cycles)

Regimen	Potential curability at start of therapy (%)	Probability of no resistant cells at completion (%)	Probability of cure at completion (%)
I	77	74	74
II	77	73	56
III	77	71	71
IV	77	66	0

[a] Intended relative dose intensity compared with regimen.
Source: Adapted from Ref. 4.

words, the early dose intensity has a disproportionate effect on the probability of no double resistance and the possibility for cure. When total dose and dose intensity were examined in this model, they were found to have separate effects on cure rates, indicating that each measured important therapeutic variables (4).

4.   The Significance of Scheduling Variables

Scheduling is important because it affects dose intensity that is strongly correlated with treatment outcome; however, novel manipulation of schedules also may be used to improve current cancer chemotherapy for reasons unrelated to dose intensity (3). Alternating-chemotherapy schedules were devised as a logical approach to counteract the emergence of multidrug resistance (MDR) as predicted by the Goldie-Coldman hypothesis (17). The theory was that rapidly alternating, equally effective, non–cross-resistant combinations of drugs would be more effective than using them singly. Unfortunately, most clinical tests of the Goldie-Coldman hypothesis have not validated it. For example, in patients with HD, it appears that there is no advantage for alternating cyclical combination chemotherapy over the proper use of a four-drug combination (3).

The inability to verify the original Goldie-Coldman hypothesis in the clinic can be attributed to probable inaccuracies in its assumptions, ie, symmetry in growth characteristics in multiple tumor masses in the same patient and in different patients, symmetry in development of resistance, and equal efficacy and equal cell-kill capacity for any two regimens. A reanalysis of the hypothesis by Day (18) relaxed the requirement for symmetry and showed that in many cases sequential treatments would be superior to alternating treatments. Day formulated the "worst drug rule" whereby more or earlier doses of a treatment with less cell-kill potential are used in a strategy (19).

The worst drug rule leads to an apparently non-intuitive sequencing of two drugs with different activities. The rule predicts that earlier use of the weaker drug (active against cells resistant to the other, more potent drug) will be more effective than using the more potent drug first because delaying the use of the weaker drug would mean that its effect would be swamped by the emergence of resistant cells.

Norton also has simulated the influence of several plans of treatment on tumors growing in a Gompertzian fashion (19). Such a model also predicts an advantage for sequential administration rather than an alternating sequence that may allow overgrowth of resistant cells in between cycle intervals.

There is clinical evidence from studies in patients with breast cancer that alternating chemotherapy is no more effective than single combinations and that sequential regimens may be superior (3). In one study, patients received eight cycles of cyclophosphamide, methotrexate, fluorouracil, vincristine, and prednisone (CMFVP) followed by either six further cycles of the same regimen or a sequential vinblastine, doxorubicin, thiotepa, and fluoxymesterone (VATH) regimen. Preliminary results indicate that the relapse-free survival was significantly better for patients randomized to receive VATH (20). These results, however, may be explained by the effect of VATH rather than the superiority of sequential chemotherapy.

Clear-cut evidence for the superiority of sequential over alternating treatment was provided by a study in which patients received either doxorubicin followed by cyclophosphamide, methotrexate, and fluorouracil (CMF) or CMF alternating with doxorubicin. Both 5-year relapse-survival (58% versus 37%) and overall survival (77% versus 63%) were significantly higher with the sequential regimen (21).

Clearly, sequencing of drugs has an important effect on treatment outcome, but it is still essential to deliver drugs at an adequate dose intensity.

In a Cancer and Leukemia Group B (CALGB) study comparing mechlor-ethamine, vincristine, procarbazine, and prednisone (MOPP); Adriamycin®, bleomycin, vinblastine, and dacarbazine (ABVD); and alternating MOPP/ABVD, the alternating regimen was superior to MOPP but not ABVD (22); however, in this study doses of MOPP were reduced because of toxicity, so all that the data showed was that MOPP is less effective at lower doses than at full dose.

III. DRUG RESISTANCE

A. Mechanisms of Drug Resistance

Why is it that 90% of all cures from cancer chemotherapy occur in approxi-mately 10% of all cancers? What is it that differentiates cells that respond spectacularly to chemotherapy from other cells that demonstrate little or no response? **Table 5** shows some possible resistance mechanisms that may contribute to treatment failure. Knowledge about the mechanisms of drug resistance may lead to the development of pharmacological strategies to circumvent resistance. Alternatively, if resistance is relative, an increase in drug dose or dose intensity may be useful.

B. Multidrug Resistance

One explanation for the poor response of common visceral tumors, espe-cially to anticancer drugs derived from natural products, is the multidrug resistance phenotype. Multidrug resistance is believed to be mediated by a membrane protein with a molecular weight of 170 kD. The presence of this protein, called p-glycoprotein, has been correlated with degree of drug resistance (23,24). The p-glycoprotein actively pumps certain anticancer drugs from cells by hydrolyzing the energy-providing molecule ATP (**Figure 3**). The protein is highly conserved in nature, indicating that it plays a fundamental physiological role. Exactly what this role is remains intensely debated. An intriguing possibility is that p-glycoprotein is needed as a defense from carcinogens and xenobiotics so resistance is constantly being built to a variety of agents.

The expression of the MDR phenotype in a range of malignancies at diagnosis and relapse is shown in **Table 6**. It is apparent that in all cases (except choriocarcinoma, where the placenta has been shown to express the MDR phenotype), the normal tissue of origin of these tumors does not express p-glycoprotein, and only a minority of cells express it at the time of initial diagnosis, the exceptions being myeloid malignancies. A higher fraction of these tumors, however, express the p-glycoprotein at the time

Table 5. Mechanisms of Drug Resistance.

General mechanism	Drug	Result
Multidrug resistance	Vinca alkaloids	
	Antitumor antibiotics	Drug actively pumped out
	Etoposide	
Transport defect	Methotrexate	
	Melphalan	Low carrier-mediated uptake
	Nitrogen mustard	
	Cytosine arabinoside	Low membrane binding
Poor activation	Cytosine arabinoside	Low deoxycytidine kinase
	5-Azacytidine	Low uridine-cytidine kinase
	5-Fluorouracil	Low uridine kinase, orotic acid
		PRT, uridine phosphorylase
	6-Thioguanine	Low hypoxanthine-guanine PRT
	6-Mercaptopurine	
	Methotrexate	Low polyglutamation
	Doxorubicin	Low P-450 enzymes
Drug inactivation	Cytosine arabinoside	High cytidine deaminase
	Alkylating agents	High glutathione
	6-Thioguanine	High alkaline phosphatase
	6-Mercaptopurine	
Improved DNA repair	Alkylating agents	
	Antitumor antibiotics	High-efficiency repair of strand breaks, ligase
	Cisplatin	
Gene amplification	Methotrexate	Dihydrofolate reductase
	PALA	Aspartate transcarbamylase
	2-Deoxycoformycin	Adenosine deaminase
	5-Fluorouracil	? Thymidylate synthetase
Alternative pathways	Methotrexate	Increased thymidine salvage
	5-Fluorouracil	Increased thymidine kinase
Altered pools of competing substrate	Cytosine arabinoside	High CTP and dCTP
	5-Fluorouracil	
Target alterations	Vincristine	Tubulin
	Methotrexate	Dihydrofolate reductase
	5-Fluorouracil	Thymidylate synthetase
	Hydroxyurea	Ribonucleotide reductase
	Steroids	Receptor or receptor-DNA binding

PRT = Phosphoribosyl transferase; PALA = N-phosphomacetyl-L-aspartic acid; CTP = cytidine triphosphate; dCTP = deoxycytidine triphosphate.
Source: Adapted from Ref. 1.

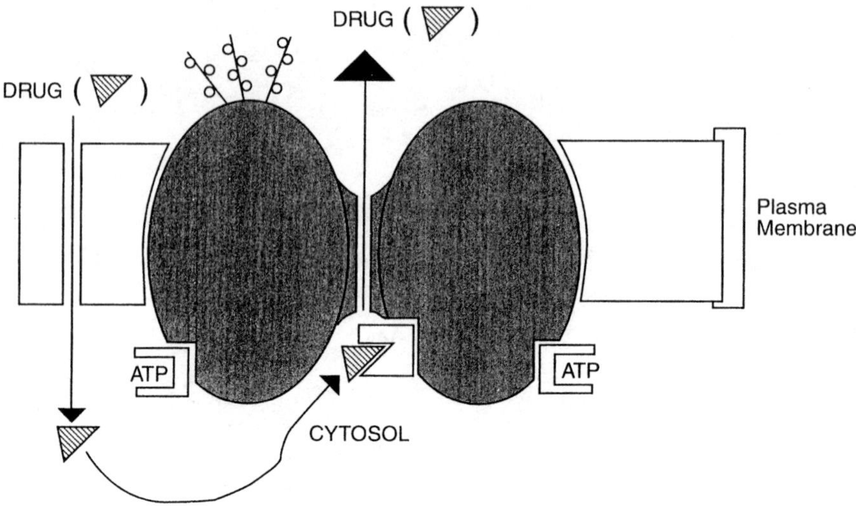

Figure 3. Schematic representation of p-glycoprotein responsible for multidrug resistance.

Table 6. Range of Expression of Multidrug Resistance (MDR) Phenotype.

		Malignant tissue	
Disease/Organ	Normal tissue	At initial diagnosis	At relapse
Choriocarcinoma	+ +	—	—
ALL	0	±	+
DLCL	0	±	+ +
AML	0	±	+
CML			
chronic phase	0	0	+ +
blast crisis	0	+ +	+ + +
Myeloma	0	+	+ +
Breast	+	±	+
Ovary	0	0	+
Head and neck	—	—	—
Esophagus	+	+	+ +
Colon	+ +	+ + +	+ + +
Pancreas	+ +	+ + +	—
Renal	+ +	+ + +	+ + +

ALL = Acute lymphoblastic leukemia; DLCL = diffuse large cell lymphoma; AML = acute myeloid leukemia; CML = chronic myelogenous leukemia.
Source: Adapted from Ref. 3.

of relapse. In contrast, in tumors responsive but not curable or poorly responsive, p-glycoprotein commonly is expressed at the time of diagnosis. The p-glycoprotein is expressed strongly in cancer of the colon and pancreas and in renal cancers, all of which are characterized by de novo resistance to chemotherapy.

An obvious strategy to overcome the MDR problem would be to develop compounds that inhibit the abnormal efflux of cytotoxic drugs from the cells and reverse the resistance. The first report of such molecules came from Tsuruo et al (25), who showed that the calcium channel blocker verapamil and the calmodulin antagonist trifluoperazine enhanced the cytotoxic effect of vincristine and increased its cellular accumulation in a resistant cell line. Since then, many compounds, with very different molecular structure and pharmacology, have been shown to partially or fully reverse drug resistance both in vitro as well as in vivo. They include chlorpromazine, quinidine, cyclosporin A, and tamoxifen. Unfortunately, for a variety of reasons, clinical trials so far have not been very successful. Problems of toxicity and the complex heterogeneous nature of MDR, seldom due uniquely to the expression of the *mdr-1* gene, are largely to be blamed for the apparent failure of this approach in the clinic. A different approach has been the selective inhibition of the *mdr-1* gene expression by antisense technology or by monoclonal antibodies, but, in spite of encouraging "in vitro" results, "in vivo" results with these technologies have so far been disappointing.

It may be possible to attempt to circumvent MDR by schedule manipulations or by pharmacological intervention with inhibitory drugs such as verapamil (3). As explained earlier, schedule manipulations may appear to be non-intuitive, for example, using effective natural product-derived antitumor agents as first treatment followed by effective combinations of alkylating agents and antimetabolites.

IV. REVIEW OF CLINICAL EVIDENCE SUPPORTING THE IMPORTANCE OF DOSE INTENSITY

A. Non-Hodgkin's Lymphoma

DeVita (1) has used the method of Hryniuk and Bush to calculate the dose intensity of each drug and the average dose intensity of the combinations of drugs used in the treatment of diffuse aggressive lymphomas. To do this, it was necessary to construct a hypothetical nine-drug combination regimen using all drugs in full doses continually as a standard (26). The analysis showed a strong correlation between the nine-drug relative dose intensity and outcome (r = 0.82, p < 0.0008) (**Figure 4**). There was no significant correlation, however, between outcome and two- or three-drug relative dose intensity. DeVita concluded that the success of third-generation regimens appears to be related to the use of more drugs, early exposure to non–cross-resistant agents, and the high dose intensity. Furthermore, he suggested that further augmentation of dose may increase the fraction of patients cured.

The importance of dose intensity also has been demonstrated for the first-generation regimen cyclophosphamide, doxorubicin, vincristine, and predisone (CHOP). A study by the South West Oncology Group (SWOG) showed that full-dose CHOP cured more patients (41%) than low-dose CHOP (12%) (**Figure 5**) (27).

A retrospective study of 73 young patients with NHL treated with CHOP showed that the dose intensity of cyclophosphamide and methotrex-

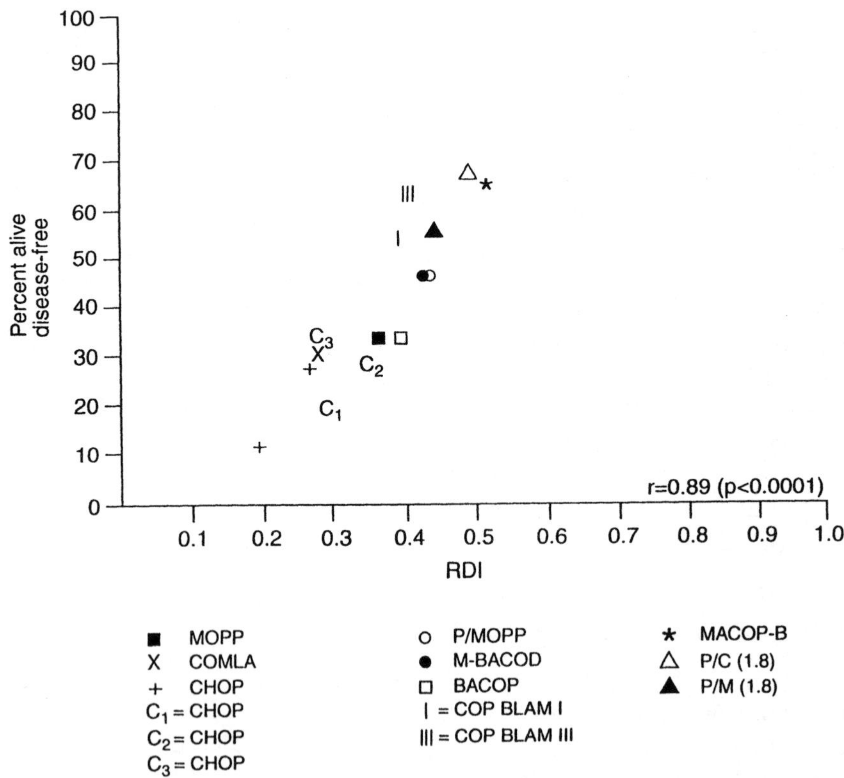

Figure 4. Relationship between nine-drug relative dose intensity and disease-free survival in patients with diffuse large-cell lymphomas. (Adapted from Ref. 1.)

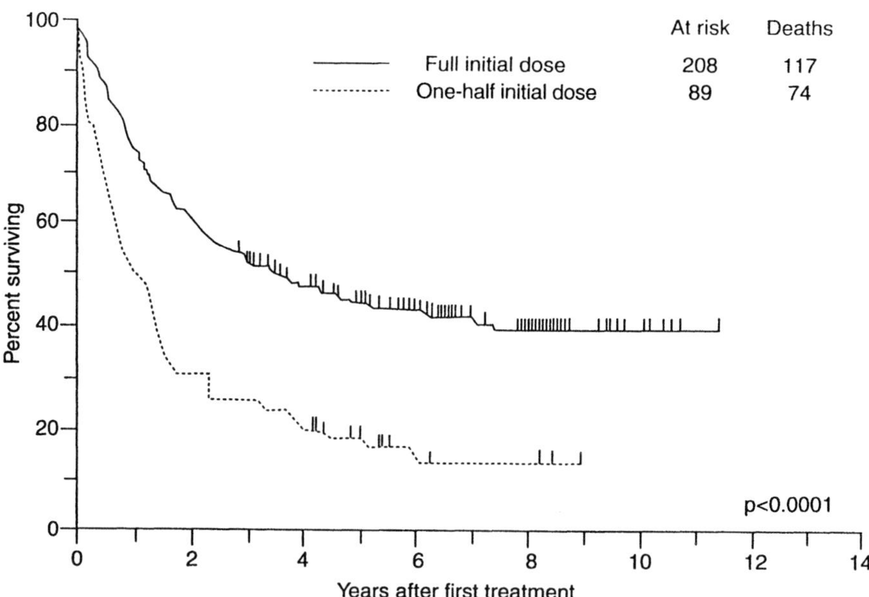

Figure 5. Effect of dose of cyclophosphamide, doxorubicin, vincristine, and predisone (CHOP) on treatment outcome in advanced diffuse large-cell lymphoma. (Adapted from Ref. 1.)

ate in the first two treatment cycles was significantly correlated with event-free survival, but that the dose intensity of these drugs in subsequent cycles and the dose intensity of other protocol drugs were not significantly associated with outcome, suggesting that the first three treatment cycles are most important (28).

In elderly patients, dose modification is common and may explain the poorer treatment outcomes observed in some studies. Dixon et al (29) observed that inferior outcomes for elderly patients in the SWOG trials may have resulted in part from less intensive chemotherapy. A subgroup of elderly patients who received full-dose chemotherapy had a higher complete response rate and higher overall survival than patients who received a low dose. Data from 266 patients treated in phase 2 SWOG studies with methotrexate, calcium leucovirin, bleomycin, doxorubicin, cyclophosphamide, vincristine, and dexamethasone (m-BACOD); predisone, methotrexate, calcium leucovirin, doxorubicin, cyclophosphamide; and etoposide (ProMACE-CytaBOM), or methotrexate, doxorubicin, cyclophosphamide, vincristine, prednisone, and bleomycin (MACOP-B), however, showed that relative dose intensity was not significantly correlated with survival after adjusting for age and performance status (30). In fact, a full analysis of the final results of 1138 patients registered for the trial (899 were eligible) suggests that CHOP remains the best available treatment for patients with this disease (31).

The weight of clinical evidence, albeit retrospective, shows that dose intensity is an important factor determining outcome in lymphoma. Excessive dose reduction or treatment delay is thus one of the most important reasons for failure to cure in patients with NHL; however, the efficacy of any new form of treatment will need to be assessed by comparing it with CHOP in a randomized trial.

B. Hodgkin's Disease

A comparison of studies of MOPP chemotherapy at different centers shows variability in treatment outcome. Although response rates were similar, the more important measurement of relapse-free survival varied markedly, being 55% at 15 years in the National Cancer Institute (NCI) (USA) series (32), but only 36% at 8 years in Milan (33) and 31% at 5 years in the SWOG studies (34). Prognostic variables in the different groups of patients do not appear to explain the different outcome. A striking difference, however, is apparent in the amount of chemotherapy given in different centers.

To try to explain the differences in survival between groups of patients treated at different centers, DeVita et al (27) calculated the relative dose intensity of chemotherapy from the different centers and correlated it with complete remission rates and disease-free survival. **Table 7** shows that there

Table 7. MOPP Regimens for Treatment of Patients with Hodgkin's Disease: Dose Intensity and Outcome.

Study	Relative dose intensity versus NCI MOPP	Actual relative dose intensity	Complete remission (%)	% Patients free of disease (yr)
NCI	1.00	0.85	84	55 (15)
Stanford	0.95	0.64	72	30 (5)
BNLI	0.82	—	52	30 (5)
SEG	0.82	0.64	46	16 (2)
CALGB	0.81	—	74	37 (5)
ECOG	0.77	0.60	73	7 (5)
Milan	0.76	0.53–0.66	74	36 (8)
SWOG	0.70	—	78	31 (5)

(*Source:* Adapted from Ref. 38.)

was a trend for a more favorable outcome for patients receiving MOPP at a greater dose intensity. The correlation was even more striking when comparisons could be made with actual relative dose intensity (amount of chemotherapy actually given rather than planned doses). Reductions of 29% and 38% in the Eastern Cooperative Oncology Group (ECOG) and Milan studies, respectively, resulted in 33% and 35% decreases in overall disease-free survival.

A retrospective analysis of experience at Stanford University (CA, USA) revealed that the dose rate and total dose of mustine, vincristine, and procarbazine were correlated significantly with complete remission rate (35). Patients receiving <65% of the projected dose of mustine had a significantly poorer survival than those receiving >65%. The NCI performed a multivariate analysis on patients treated with MOPP and found that the dose of vincristine correlated significantly with outcome (36).

Trials of MOPP or ABVD chemotherapy in comparison with alternating MOPP/ABVD failed to show a significant advantage for alternating chemotherapy, but did reveal that dose reductions in patients receiving MOPP resulted in poorer treatment outcome (22,37). For example, in the Milan trial, 35% of patients experienced a 50% reduction of the dose of vincristine, and in 9% of patients, vincristine was discontinued permanently because of toxicity. In this study, 50% of patients in the MOPP group experienced relapse in the first 24 months compared with only 34% of patients at the NCI over a 14-year follow-up (38). As well as showing the dangers of inadequate dosing, these results illustrate that dose modification is done even in major medical centers. Data clearly support the concept of dose intensity in HD and argue strongly against unnecessary dose reductions or delays. Clinical evidence thus shows that dose and dose intensity are important determinants of treatment outcome in HD.

C. Small-Cell Lung Cancer

Several randomized (**Table 8**) and non-randomized trials have studied the dose-response relationship in SCLC. In a randomized trial, the doses of CMC (cyclophosphamide, methotrexate, and CCNU) were doubled for the initial 6-week induction period, and significantly improved response rates and survival were obtained with higher drug doses (39). In a larger randomized trial when only the dose of cyclophosphamide was increased in the same regimen, there was a modest but significant increase in response rate and survival, particularly in patients with limited-stage disease (40). The dose-response effect of cyclophosphamide only has been reported for relatively low doses, and at >300 to 400 mg/m^2/week no effect has been observed, indicating that a plateau may have been reached.

Table 8. Completed Randomized Studies of Intensity and Initial Combination Chemotherapy in Small-Cell Lung Cancer.

Ref	Drug and dose (mg/m^2)	No. of patients	CR + PR rate (%)	Median survival (mo)	Comments
39	C 1000 + MTX 15 + CCNU 100[a]	23	96	10.5	High doses better: $p < 0.05$
39	C 500 + MTX 10 + CCNU 50	9	45	5.0	CR + PR, $p < 0.05$ survival
40	C 1500 + MTX 15 + CCNU[a]	175[b]	64	10.25	High dose better: $p = 0.04$
40	700 + MTX 15 + CCNU 70	174	54	9	CR + PR, $p = 0.04$ survival
81	C 2000 + VCR total 2 + MeCCNU 75 100	14	73	9	High dose better: $p < 0.05$
81	C 750 + VCR total 2 + MeCCNU 75	14	43	10.75	CR; CR + PR and survival not different
82	C 1200 + A 70 + VCR 1[c]	101	63	7	High dose better: $p = 0.04$
82	C 1000 + A 40 + VCR 1	146	53	8	CR: CR + PR and survival not different
83	C 1560 + A 59 + VCR 0.9[d]	52	71	14	No differences in CR, response, duration,
83	C 990 + A 50 + VCR 1.0	51	61	12	or survival

[a] High-dose regimen for first 6 weeks.
[b] Approximate number.
[c] High-dose regimen for first 9 weeks.
[d] Doses are actual doses given, not intended doses; high-dose regimen for first 12 weeks.

C = Cyclophosphamide; MTX = methotrexate; CCNU = carmustine; VCR = vincristine sulfate; CR = complete response; PR = partial response.

Murray retrospectively analyzed data from studies of cyclophosphamide, doxorubicin, and vincristine (CAV) and cyclophosphamide, doxorubin, and etoposide (CAE) and observed no consistent correlation between cyclophosphamide dose intensity and outcome; however, doxorubicin dose intensity in extensive stage disease was highly significantly correlated with response rates for both regimens and with median survival time for the CAE combination (41).

D. Breast Cancer

1. Retrospective Studies

The most important retrospective studies of dose and response in breast cancer have been done in the adjuvant setting. Bonadonna and Valagusse were the first to demonstrate a dose response with the CMF protocol (42). In this study, patients who received >85% of the planned total dose had significantly higher disease-free survival (56% versus 39%) and overall survival (66% versus 47%) than patients who received <65%. Although several studies have confirmed this finding, a number of others have failed to show a similar effect (**Table 9**).

Hryniuk (8) calculated the relative dose intensity of different chemotherapy regimens used in the adjuvant setting and demonstrated a significant effect of dose intensity on relapse-free survival (**Figure 6**); however, the methodology of this analysis has been questioned and Henderson et al failed to confirm the dose effect when the CMF protocol alone was analyzed (43).

In a similar retrospective analysis of chemotherapy in patients with metastatic disease, Hryniuk and Bush (15) demonstrated a relationship between dose intensity and treatment outcome (**Figure 7**); however, the methodology used in this analysis is open to criticism.

2. Randomized Studies

The results from seven randomized clinical studies evaluating the dose-response effect are summarized in **Table 10** (44). In four of these trials, there was a substantial difference in response rate in favor of the higher dose, but in the other three no difference could be detected. Of the seven trials, three showed a substantial survival difference in favor of the high-dose group.

The trial done by Tannock et al (45) showed palliation of metastatic breast cancer with a reduced-intensity CMF regimen. Response rates (30% versus 11%) and survival (15.6 months versus 12.8 months) were significantly better in patients receiving the higher dose of CMF. The higher-dose regimen also was associated with a better quality of life.

Table 9. Retrospective Analyses of Chemotherapy Dose and Disease-Free Survival or Overall Survival in Adjuvant Chemotherapy Breast Cancer Studies.

Study	Dosage comparison	Disease-free survival (%)			Overall survival (%)		
		High dose	Low dose	p-value	High dose	Low dose	p-value
Analyses confirming Bonadonna dose hypothesis							
Milan CMF	≥85 v <65	56	39	.006	66	47	.04
Mayo CFP	≥85 v <85	NA	NA	.01	—	—	—
Memorial CMF	≥85 v <65	62	49	.04	82	63	.2
Guys/Manchester CMF	≥85 v <65	74	57	.19	—	—	—
Stockholm-Gotland CMF	≥85 v <85	NA	NA	.26	—	—	—
ECOG (postmenopausal)	≥80 v <80	NA	NA	NA	—	—	—
Osakco LMF	≥90 v <90	—	—	—	74	60	.08
Analyses refuting dose hypothesis							
NSABP PAM	≥85 v <65	49	55	.005	—	—	—
NSABP PAM/5FU	≥85 v <65	60	70	.01	—	—	—
Guys/Manchester PAM	≥90 v <90	NA	NA	.07	—	—	—
Danish C	≥75 v <75	78	77	.52	—	—	—
Danish CMF	≥75 v <75	68	72	.62	—	—	—
Mayo PAM/CFP	≥85 v <65	48	62	.43	—	—	—
SEG CMF	≥85 v <85	56	63	.15	—	—	—
ECOG (premenopausal)	≥80 v <80	NA	NA	NA	—	—	—

NA = Data not available in publication; CFP = cyclophosphamide, fluorouracil, and prednisone; LMF = chlorambucil, methotrexate, and fluorouracil; PAM = melphalan; 5FU = fluorouracil; ECOG = Eastern Cooperative Oncology Group; NSABP = National Surgical Adjuvant Breast and Bowel Project; SEG = Southeastern Cancer Study Group.
Source: Adapted from Ref. 43.

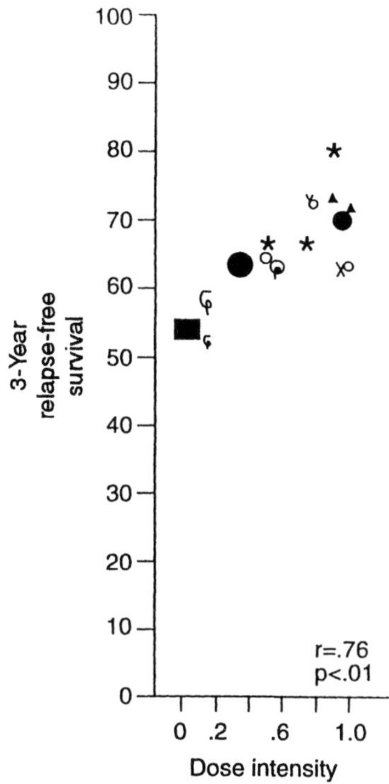

Figure 6. Effect of dose intensity on 3-year relapse-free survival in breast cancer adjuvant chemotherapy trials. Each symbol represents a different regimen. (Adapted from Ref. 13.)

In an editorial in the *Journal of Clinical Oncology*, Hryniuk (13) stated, "There can be little doubt that the higher the dose intensity of CMF, the higher the rate of remission rate of advanced breast cancer."

Probably the first solid evidence for such a dose-intensity effect in breast cancer, or at least in the adjuvant treatment of breast cancer, has come from a CALGB study that enrolled 1572 patients with stage II, node-positive breast cancer (46). All patients received adjuvant cyclophospha-mide, Adriamycin®, and 5-fluorouracil (5-FU) (CAF) chemotherapy every 4 weeks at one of three dose rates; high (600/60/600), moderate (400/40/400), or low (300/30/300). Patients in the high-dose group and the low-dose group were treated with four cycles of chemotherapy, and the dose of each drug given in the low-dose group was exactly half that given to those in the high-dose group. As a result, the cumulative dose of CAF given to patients in the low-dose group was half that given to those in the high-

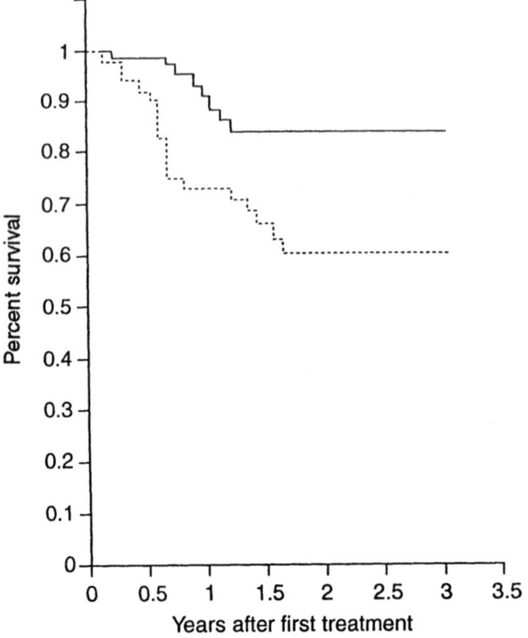

Figure 7.   Effect of dose intensity on response rate in advanced breast cancer. Solid line, high-dose patients; dotted line, low-dose patients (Adapted from Ref. 13.)

dose group. In contrast, patients randomized to the moderate-dose group of the study received six cycles of chemotherapy. As a result, the cumulative doses of each drug given to patients in the moderate-dose and the high-dose groups were identical, but the rate at which the drugs were administered was faster for patients in the high-dose group. An analysis on 1516 evaluable patients showed that after a median of 3 to 4 years, disease-free survival was significantly improved in the high-dose and moderate-dose groups compared with the low group (**Figure 8**). Overall survival also was improved significantly in the high-dose group compared with the low-dose group, but toxicity was also greater. In a sizable proportion of the patients who entered the study, prognostic factors (like S-phase fraction, ploidy, *erb*-B2, and p53) were analyzed in pathology specimens. There was only a weak correlation between S-phase fraction and dose response, but there was a marked correlation between dose response and *erb*-B2 overexpression or amplification. Those patients whose tumors showed *erb*-B2 overexpression in >50% of cells were more likely to respond to the more intensive chemotherapy regimen (47). It is therefore possible that in the not too distant future molecular prognostic markers may gain relevance as predictors of response.

Table 10. Randomized Clinical Trials to Evaluate the Effect of Dose on Response Rate and Survival.

Regimens	No. of patients	No. of responses		CR + PR (%)	Median survival times (months)
		Complete	Partial		
Cisplatin 75 mg/m^2	15	0	0	0	NA
Cisplatin 120 mg/m^2	8	0	0	0	NA
Cisplatin 60 mg/m^2	18	0	0	0	NA
Cisplatin 120 mg/m^2	19	0	4	21	NA
Doxorubicin 35 mg/m^2	24	1	5	25 } p<0.02	8 } p<0.02
Doxorubicin 70 mg/m^2	24	4	10	58	20
CMF 56–81%	79	12	33	57 } NS	14.5 } p<0.03
CMFP 76–95%	86	14	36	63	16.4
FEC 500/50 mg/m^2	39	1	17	46 } p<0.025	NA
FEc 500/100 mg/m^2	39	5	26	79	NA
CMF 300/20/300 mg/m^2	53	2	5	11 } p<0.03	12.8 } p<0.26
CMF 600/40/600 mg/m^2	53	1	15	30	15.6
FAC 500/50/500 mg/m^2	27	6	15	78 } NS	20 } NS
FAC 2500/100/1800 mg/m^2	32	8	17	78	20

CR + PR = Complete response and partial response; NA = not stated; CMF = cyclophosphamide, methotrexate, 5-fluorouracil; CMFP = cyclophosphamide, methotrexate, 5-fluorouracil, prednisone; FEC = 5-fluorouracil, epirubicin, cyclophosphamide; FAC = 5-fluorouracil, doxorubicin, cyclophosphamide.
NS = Not significant.
NA = Not available.

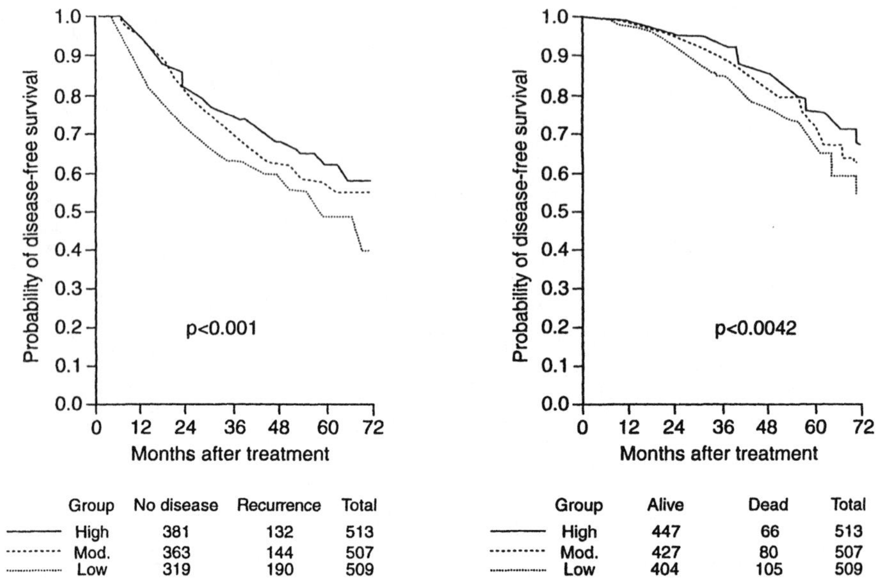

Figure 8. Actuarial survival results according to dose rate in the Cancer and Leukemia Group B (CALGB) trial of adjuvant cyclophosphamide, adriamycin, and 5-FU (CAF) chemotherapy in node-positive breast cancer. (Adapted from Ref. 46.)

In vitro tumor clonogenic tests also are being developed to assay for chemosensitivity of tumor cells from any given patient. At this time, unfortunately, the only valid test of chemosensitivity is the administration of chemotherapy to the patient and the subsequent assessment of response. (See Chapter 11 for more discussion of breast cancer.)

E. Advanced Ovarian Cancer

The effect of dose intensity of first-line chemotherapy on clinical outcome has been retrospectively analyzed by Levine and Hryniuk (48) in ovarian cancer. They calculated the average dose intensity of drugs relative to a standard cyclophosphamide, hexamethylmelamine, doxorubicin, and cisplatin (CHAP) regimen and found that the average relative dose intensity, especially of cisplatin, correlated significantly with clinical response and with the median survival time. Furthermore, there was a distinct advantage for multiagent regimens containing cisplatin over single alkylating agents.

Additional arguments in support of this important retrospective observation are provided from experimental in vitro studies which show a steep dose-response relationship for cisplatin (49). These findings emphasize the importance of maintaining chemotherapy doses and provide impetus for investigating further dose intensification in the treatment of advanced ovarian cancer.

F. Testicular Cancer

Adequate doses of first-line cisplatin combination chemotherapy will cure 70% of patients with disseminated germ cell tumors (50). Successive generations of regimens have been developed to improve the success rate while reducing toxicity by decreasing dose or substituting new drugs. The dose of vinblastine was reduced by 25%, then vinblastine maintenance was dropped, and finally vinblastine was replaced by VP-16. Despite this, dose intensity, especially for cisplatin, remains important in the success of treatment.

Several studies have shown a dose-response relationship for cisplatin in patients with testicular cancer. Samson et al (51) reported a randomized trial of 114 patients with disseminated testicular cancer which compared high-dose (120 mg/m^2, monthly) with low-dose (15 mg/m^2, daily for 5 months) cisplatin, both combined with vinblastine and bleomycin. There was a significantly higher complete response rate for high-dose therapy (63%) compared with low-dose (43%) and a survival advantage for patients randomized to the higher dose (**Figure 9**). These results show a clear relationship for dose with response and curative potential.

Ozols et al (52) reported a trial which compared a standard cisplatin, vinblastine, and bleomycin regimen with a high-dose regimen consisting of double-dose cisplatin and VP-16 combined with vinblastine and bleomycin. The high-dose regimen was associated with higher complete remission rates (88% versus 67%), lower relapse rates (17% versus 41%), and increased 5-year survival (78% versus 48%) compared with the standard regimen. The interpretation of these results is complicated because two variables were changed in the high-dose regimen. The increased effectiveness may have been due to double-dose cisplatin or the addition of VP-16. A randomized trial reported by Einhorn et al (53) failed to show any advantage for double-dose cisplatin compared with standard dose.

Data suggest that the dose of cisplatin is important in determining treatment outcome in testicular cancer, but that the value of increasing the dose above that currently accepted as standard remains to be shown. Currently, several experimental high-dose regimens are being investigated

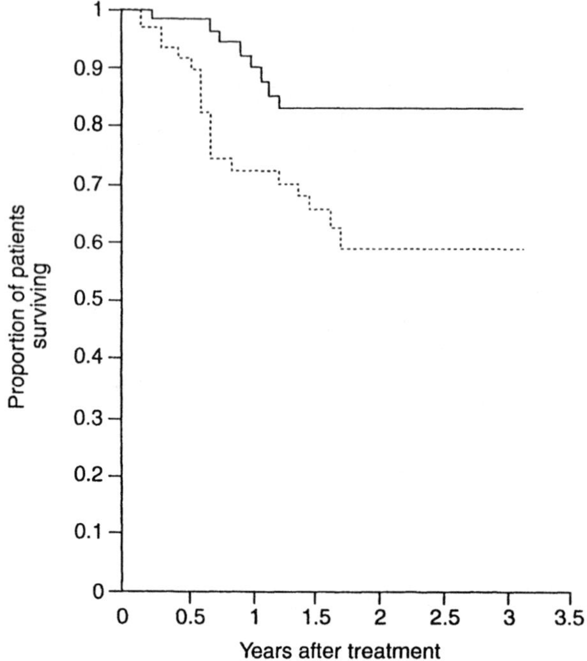

Figure 9. Survival of patients with disseminated testicular cancer following treatment with a high-dose (solid line) or low-dose (dashed line) regimen. (Adapted from Ref. 51.)

in patients with poor prognostic features or as a salvage treatment (54). (See Chapter 15 for further discussion of testicular cancer.)

G. Pediatric Malignancies

Only a minority of pediatric cancers are initially refractory to treatment. Complete remission in patients with acute leukemia ranges from 80% to 95%, and in most solid tumors complete remission is achieved at the end of chemotherapy, surgery, and, in some cases, radiation. In approximately 50% of all solid tumors, however, disease recurs due to the emergence of resistant cells. In a number of patients, an initially chemosensitive tumor recurs because of inadequate initial treatment. This may be due to initial understating and the administration of an inappropriately low-intensity regimen. Occasionally, inadequate drug doses are received because of poor patient compliance or a reluctance of physicians to accept the inevitable toxicity of high-dose therapy (55).

Clinical evidence supporting the concept of dose response comes from studies of refractory NHL in adults and children where patients resistant to conventional doses of cyclophosphamide have achieved remission when the dose was escalated. The improved results in acute myeloid leukemia (AML), high-risk leukemia, and soft tissue sarcomas have all resulted from increases in the doses of conventional chemotherapy (55).

H. Non-Curable Malignancies

A dose-response relationship can be demonstrated in some malignancies not curable by standard chemotherapy. For example, Hryniuk (8) demonstrated a steep dose-response curve for patients with advanced colorectal cancer receiving 5-FU; however, the doses in the studies analyzed were below the level considered to be the threshold for producing useful responses in advanced colorectal cancer. Unfortunately, although malignancies such as colorectal and non-small–cell lung cancer (NSCLC) may respond initially to chemotherapy, they invariably relapse due to the emergence of resistant cells. It is tempting to speculate that if there is a dose-response relationship in such conditions, it may be possible to improve outcome by significantly increasing dose. According to the model of Coldman and Goldie (4) such an approach is more likely to increase response rates, because a higher number of sensitive cells will be eliminated, than to improve overall survival because resistant cells present at diagnosis will not respond to treatment and will subsequently proliferate.

V. THE USE OF HEMATOPOIETIC GROWTH FACTORS TO OPTIMIZE DELIVERY OF CHEMOTHERAPY

A. Dose-Limiting Toxicities of Chemotherapy

Cytotoxic drugs used at curative doses produce a variety of side effects which range in severity from mild to life threatening (**Table 11**). Combinations of drugs tend to include drugs with non-overlapping toxicities, which helps to avoid additive damage to a particular organ, but increases the range of adverse effects experienced. The major dose-limiting toxicity for many combination regimens is myelosuppression.

Bone-marrow toxicity leads to the development of neutropenia and thrombocytopenia, which increase the risk of infection and requirements for platelet transfusions, respectively. The risk of infection is closely related to the degree and duration of neutropenia (56). Neutropenic infection requires hospitalization and immediate treatment with broad-spectrum intravenous (IV) antibiotics. Despite such treatment, morbidity is significant

Table 11.   Side Effects Affecting Dose of Cytotoxic Drugs.

Drug	Side effect most frequently causing dose modification	Other side effects
Doxorubicin	Myelosuppression	Cardiac toxicity Nausea and vomiting
Epirubicin	Myelosuppression	Cardiac toxicity Nausea and vomiting
Mitoxantrone	Myelosuppression	
Vincristine	Neuropathy	
Vinblastine	Myelosuppression	Neuropathy
Cyclophosphamide	Myelosuppression	Urothelial damage
Ifosfamide	Myelosuppression	Encephalopathy Urothelial damage
Etoposide	Myelosuppression	Nausea and vomiting Hypotension
Cisplatin	Nephropathy Neuropathy Nausea and vomiting	
Carboplatin	Myelosuppression	
Methotrexate	Mucositis	Renal failure Myelosuppression
Nitrosoureas	Myelosuppression	Nausea and vomiting
Paclitaxel	Myelosuppression	Neuropathy
Docetaxel	Myelosuppression	Fluid retention
Topotecan	Myelosuppression	Mucositis

and the mortality rate may be as great as 20%. Another serious consequence of neutropenia and infection, however, may be dose reduction or delay, leading to a decrease in chemotherapy dose intensity and an adverse effect on treatment outcome (1). Furthermore, myelosuppression limits attempts to further increase doses in an attempt to improve treatment outcome in poor-risk patients.

The introduction of hematopoietic growth factors offers clinicians the opportunity to reduce the impact of neutropenia on treatment strategies, which may lead to an enhanced ability to deliver regimens as planned without the need for dose modification as a result of myelotoxicity.

B. Clinical Effects of Myeloid Growth Factors After Chemotherapy

Both recombinant human granulocyte colony-stimulating factor (rHuG-CSF) and recombinant human granulocyte-macrophage colony-stimulating factor (rHuGM-CSF) have demonstrated clinical benefits in studies involving cancer patients receiving chemotherapy for a variety of malignancies (57). For example, those studies which have directly compared cycles of chemotherapy with or without Filgrastim have reported a reduction in the number of febrile days, a reduced incidence of infections, and, as a result, fewer days on antibiotics (58–65). A shorter duration of neutropenia was seen in patients receiving Filgrastim, independent of whether it was administered in the first or second treatment cycle. An additional benefit may be a reduction in the degree of mucositis experienced by patients (61).

Two randomized, double-blind, phase 3 trials involving 314 patients with SCLC receiving intensive chemotherapy (**Figure 10**) showed that Filgrastim significantly reduced the incidence of febrile neutropenia compared with placebo (60,62). The risk of being hospitalized for infection and the total duration of hospitalization also were reduced significantly by Filgrastim compared with placebo on the first cycle of chemotherapy. The benefits of Filgrastim were maintained over all six cycles of chemotherapy.

In summary, preclinical and clinical studies have shown that Filgrastim is well tolerated when administered by either the subcutaneous (SC) or IV routes. Recombinant HuGM-CSF has a much less specific action, stimulat-

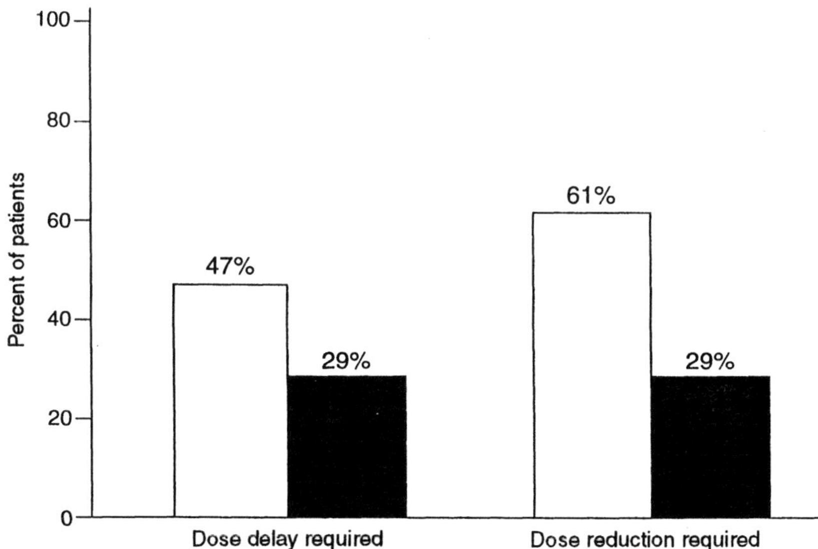

Figure 10. Filgrastim significantly reduces the need to delay chemotherapy by 2 or more days or reduces the dose by 15% or more compared with placebo. (Adapted from Ref. 62.)

ing the growth of a number of hematopoietic cell lines. This agent is considered to be well tolerated when administered by SC infusion at doses <250 $\mu g/m^2$/day for a period not exceeding 14 days, but at higher doses a range of side effects, including fever, is observed (57). Filgrastim does not appear to require the input of earlier pluripotent cells and there seems to be no experimental or clinical evidence of exhaustion of the hematopoietic system if the treatments are repeated (66). Neutrophils produced after stimulation with Filgrastim are functionally normal (67,68). This also appears to be the case with rHuGM-CSF, although one trial reported a deficit in neutrophil migration (69).

Although some studies have demonstrated the presence of hematopoietic growth factor receptors on tumor cells (70), there is no clinical evidence of enhancement of non-myeloid tumor cell growth in vivo with Filgrastim. Two randomized, double-blind, placebo-controlled trials (n = 314) have shown that Filgrastim reduces the incidence of febrile neutropenia and the duration of neutropenia (60,62). This is associated with reduced total days of hospitalization for infection, a decreased requirement for IV antibiotics, and an improved ability to deliver chemotherapy on schedule.

C. High-Dose Chemotherapy and the Use of Hematopoietic Growth Factors

1. Rationale for Dose Intensification

A dose-response relationship for chemosensitive malignancies is evident in experimental models (9) and in retrospective analyses of clinical trial results (8). In lymphoma and HD, such an analysis shows that the dose intensity of current regimens lies on the linear phase of the dose-response curve, implying that further increases in dose intensity may improve outcome (3,26). In general, only in the treatment of malignancies with a steep dose-response curve is dose intensification likely to produce significant clinical benefits. Dose intensification may be achieved by increasing the dose of one or more drugs in combination or by decreasing the interval between treatment cycles.

Coldman and Goldie (4) have predicted that increasing dose intensity has a greater impact on eliminating sensitive cells than it does in preventing the occurrence of resistant cells; however, their model assumes that resistant cells are unaffected by treatment. If drug resistance is relative and not absolute, it may be possible to increase killing of resistant cells by increasing dose. This hypothesis can be tested in the setting of minimal residual disease where clinically undetectable resistant cells remain after an initial complete response to chemotherapy.

2. Rationale for Use of Hematopoietic Growth Factors

The predominant dose-limiting toxicity of many combination chemotherapy regimens is myelosuppression, particularly neutropenia. The use of an agent such as Filgrastim significantly reduces or eliminates neutropenia after chemotherapy, so this adverse effect may cease to be the dose-limiting toxicity. The hematopoietic growth factors widen the window of opportunity for investigating a higher dose intensity of combination chemotherapy.

On the basis that many of the successes of combination chemotherapy can be attributed to the achievement of a higher equivalent cytotoxic dose by the combination of drugs with primarily non-overlapping toxic effects, Simon and Korn (16) have developed a methodology to select optimal combinations of drugs based on knowledge of their relative potencies and maximally tolerated doses. The information required can be deduced from phase 2 and phase 3 studies of the drugs in question. It is then possible to construct a graph such as the one shown in **Figure 11A**, which shows the organ-system toxicities for drugs given at doses producing a given response rate. If a particular toxicity is unique to a drug, the line will be 90° to the axis for that drug, whereas if the drugs share the toxicity, the line will slope between the two axes. The shaded area on the graph represents tolerable combinations and the upper boundary of the shaded area shows possible maximally tolerated combinations. The optimal combination occurs at the

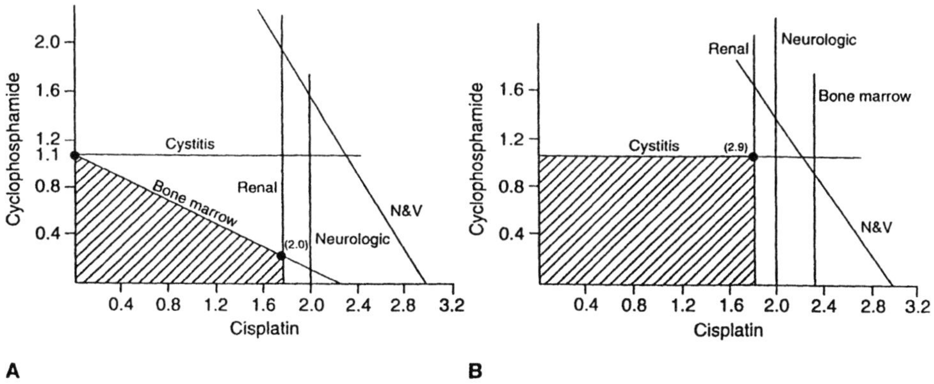

Figure 11. (A) Calculation of optimum combination of cisplatin/cyclophosphamide. Shaded areas represent tolerable combinations. Axes scaled in units of dose that gives 40% response rate. Solid lines denote organ-system toxicity constraints. (B) Same graph redrawn to eliminate bone-marrow constraint for cyclophosphamide. N & V = Nausea and vomiting. (Adapted from Ref. 16.)

intercept of the boundary lines. This technique is also applicable to combinations of more than 2 days using linear programming (16).

In the example shown in Figure 11A, the optimal combination of cisplatin and cyclophosphamide results in a total equivalent dose intensity of 2, representing an unpromising increase of only 16% over full-dose cisplatin alone; however, if a hematopoietic growth factor is used to eliminate the bone-marrow constraint for cyclophosphamide (Figure 11B), both drugs can be given in full dose, producing a total equivalent dose of 2.9, representing a 65% improvement. Such a model may be worthy of clinical evaluation. Use of a hematopoietic growth would have no effect on drug combinations limited by non-hematological toxicities.

If dose intensity is to be increased by decreasing the interval between chemotherapy cycles, it would appear to be more logical to use rHuG-CSF rather than rHuGM-CSF because of its more rapid effect (66,71).

D. High-Dose Chemotherapy Plus Hematopoietic Growth Factors

Clinical studies have already demonstrated the feasibility of using Filgrastim or rHuGM-CSF to allow delivery of high-dose chemotherapy without bone-marrow transplantation, and initial results appear promising. Neidhart et al (65,72) reported the use of rHuG-CSF and rHuGM-CSF to enable administration of high-dose cyclophosphamide, etoposide, and cisplatin to patients with chemotherapy-resistant malignancies (**Table 12**). The use of growth factor accelerated neutrophil recovery, and with Filgrastim the median duration of antibiotic therapy was not shortened.

Filgrastim was used to increase the intensity of doxorubicin in patients with breast and ovarian cancer by increasing the dose and reducing the interval between cycles (59). With this approach it was possible to escalate the doxorubicin dose intensity 4.5-fold before reaching non-hematological dose-limiting effects. A high response rate was reported, but the duration of remission was disappointingly short.

A European Organization for Research of Treatment of Cancer (EORTC) study has established that Filgrastim makes possible a shorter interval between epirubicin (100 mg/m^2)/cyclophosphamide (830 mg/m^2) treatment cycles, resulting in an increased dose intensity in patients with metastatic breast cancer (73). The interval between chemotherapy cycles was reduced from 22 days to 13 days, with Filgrastim given on days 2 to 11 in the maximally tolerated schedule. The clinical response rate was 94% (complete response, 25%).

Preliminary trials with dose-intensive chemotherapy have provided encouraging results; although high response rates have been reported, it is

Table 12. Filgrastim and rHuGM-CSF after Dose-Intensive Therapy Without Bone Marrow Transplantation.

	AGC × 10^9/L		
	<0.1	<0.5	Duration of antibiotics
Historical controls	8.5	12	9 days
Filgrastim			
40 μg/kg	7.0		
60 μg/kg	5.5	6	5
GM-CSF			
500 μg/kg day 6	8.0	NA	
day 4	6.5	NA	Significant decrease
750 μg/kg day 6	8.0	NA	
1000 μg/kg day 6	7.0	NA	

AGC = Absolute granulocyte count.
NA = Not available.
Source: Adapted from Ref. 65.

still too early to make any firm conclusions. Furthermore, these trials have already revealed some limitations of using hematopoietic growth factors to increase dose intensity, for example, an increase in non-hematological toxicity and an increased risk of thrombocytopenia associated with higher doses of chemotherapy.

The iatrogenic mortality and morbidity associated with myeloablative chemotherapy (or chemoradiotherapy) has been greatly reduced by the judicious use of recombinant hematopoietic growth factors, peripheral blood progenitor cells (PBPC), better supportive measures (eg, more potent anti-emetics), and better selection of patients. In most clinical settings, for example, procedure-related mortality is well below 5% for solid tumors (a bit higher for hematological malignancies), and well below the 10% to 20% figures observed 10 years ago. Costs are also gradually decreasing, as the total number of procedures per center increases, and more experience is gained on the out-patient management of high-dose chemotherapy protocols.

There is now good evidence to justify high-dose myeloablative chemotherapy in AML (in first or second remission). For example, in a large multicenter, randomized protocol of the EORTC and the Gruppo Italiano Malattie Ematologiche dell'Adulto (GIMEMA) cooperative groups (74) designed to compare prospectively conventional intensive chemotherapy versus consolidation with autologous bone marrow transplantation

(ABMT) or allogeneic bone marrow transplantation (allo BMT), at a median follow-up of 3 years the results appeared satisfactory, with a disease-free survival of the intensive chemotherapy group (29% at 4 years) at least equivalent to the previous studies with less intensive regimens and disease-free survival in the allo-BMT and ABMT of 54% and 49%, respectively. The two marrow groups gave significantly better results than the intensive chemotherapy group (p = 0.04). The main reason for failure is relapse in both the ABMT and allo-BMT groups, while treatment-related mortality is higher in the allo-BMT group. Nevertheless, like for most indications, there is no universal agreement on the appropriateness of bone marrow or peripheral blood transplants, and panels of experts (helped by an overall analysis of all relevant data from Medline or other sources) are being set to define clinical settings where transplants are most likely appropriate and to identify variables influencing clinical decision making (75).

There is also mounting evidence (including preliminary data from randomized studies) that ABMT should be performed in patients with relapsed chemosensitive HD and NHL, or even up front as consolidation of first remission in poor-prognosis NHL, and in some acute lymphoblastic leukemias (ALL) with poor prognosis features. For example, the 5-year survival rate for patients with intermediate and high-grade NHL is only 50% with either conventional CHOP-like regimens or with third-generation, intensive weekly regimens. The 10% or so of patients who fail to respond to initial therapy, or the 30% who relapse after complete remission, have an extremely poor prognosis, with a disease-free survival usually <10% at 3 years. Philip et al (76) have presented the final analysis of a randomized study (also known as the "Parma study") with 216 patients with relapsed intermediate- or high-grade NHL. All patients entered in this multicenter study had a previous complete response to an Adriamycin®-containing regimen, no neurological and bone marrow involvement, and no major organ dysfunction. All patients received two courses of a semi-intensive regimen (dihydroxyacetone phosphate [DHAP] chemotherapy), and 109 patients (with complete response or a very good partial response) were randomized to receive either four further courses of DHAP plus radiotherapy to involved sites or radiotherapy plus high-dose chemotherapy (BEAC [carmustine, etoposide, Ara C, and cyclophosphamide]) with ABMT rescue. No difference in terms of prognostic factors at inclusion were observed in the marrow transplantation group (of 50 patients actually transplanted) or in the chemotherapy group. All patients were included in the final analysis, and disease-free survival was 46% in the marrow transplantation group versus only 12% in the control group at 5 years follow-up. Similarly, differences were also significant in terms of overall survival: 53% versus 32% at 5 years. The authors concluded that high-dose chemotherapy and bone marrow transplantation should now be regarded as standard therapy in sensitive relapse of patients with high-grade NHL, but that at subsequent relapse is not able to salvage chemotherapy failures.

In patients with multiple myeloma (MM), the question whether early intensive therapy with autologous stem cell support offers advantage over conventional treatment has been addressed by several groups. The Myeloma French Group (77) compared in a randomized trial melphalan 140 mg/m^2 plus total body irradiation 800 cGy versus vincristine/melphalan/ cyclophosphamide/prednisone-vincristine/carmustine/doxorubicin (VMCP-VBAP) in 200 patients. The high-dose group produced a higher complete-response rate (22% versus 5%), longer progression-free survival (27 months versus 18 months), and longer overall survival (probability of 5-year survival of 52% versus 12%), respectively. Although not curative in patients with MM, high-dose chemotherapy offers good palliation and improves survival.

In numerical terms, the impact of high-dose chemotherapy is likely to be more important for some chemosensitive solid tumors and, in particular, for cancer of the breast. Several randomized studies are now ongoing to test the merits of adjuvant high-dose chemotherapy in patients with breast cancer presenting with four or more positive axillary lymph nodes at primary surgery. Five-year follow-up data are available from at least two non-randomized studies. Peters et al (78) treated 85 patients with 10 or more positive axillary lymph nodes with adjuvant high-dose cyclophosphamide, cisplatin, and carmustine (CPB) and ABMT after CAF adjuvant chemotherapy. The median age was 38 years, the median number of involved axillary nodes was 14, and therapy-related mortality was 12% (which has now decreased considerably by the use of Filgrastim and PBPC). The 5-year disease-free survival in this series was 71% (95% CI: 53% to 84%), compared with only 28% to 34% in historical control patients. In a similar study, but using a novel high-dose sequential combination chemotherapy (HDS), Gianni et al (79) from Milan reported only one procedure-related death (of 67 patients) and a disease-free survival at 5 years of 56% (overall survival, 78%) versus 33% to 41% in their best historical series (overall survival, 60%). If only patients with 10 to 15 axillary nodes were analyzed and patients with more advanced disease were excluded, then the figures became 64% for disease-free survival and 83% for overall survival in the HDS group.

A large (423 patients), prospective, randomized trial of high-dose combination alkylating agents (CPB) with ABMT rescue as consolidation for patients with metastatic breast cancer achieving a complete response after intensive induction therapy with AFM (doxorubicin/5-FU, and methotrexate) was recently presented by Peters et al (80). The disease-free survival was significantly longer for patients randomized to immediate high-dose consolidation using CPB versus observation, but, surprisingly, overall survival appears superior when high-dose therapy is administered at the time of relapse, rather than as consolidation of a complete response following induction therapy. Why the schedule of high-dose consolidation appears to be a significant variable affecting disease-free survival and overall survival remains unclear (see also Chapter 11 on breast cancer and Chapter 13 and Chapter 14 on PBPC transplantation).

VI. CONCLUSION

The dose-response relationship is a fundamental principle of chemotherapy. There is abundant experimental and clinical evidence to show that dose, whether expressed as dose intensity or total dose, is an important determinant of treatment outcome (3). Dose reduction or treatment delay reduces dose intensity and may sacrifice cure rate for a transient alleviation of toxicity. In patients with chemosensitive malignancies undergoing their first chemotherapy treatment, dose intensity should not be reduced for trivial reasons (1). Furthermore, in the design of combination regimens it is important not to compromise the dose of the most effective drugs by the addition of other agents. Such an approach may reduce the dose below the threshold of effectiveness and destroy the curative capacity of the combination. Although this may appear obvious, retrospective studies of commonly used regimens show that in many cases combination chemotherapy has resulted in significant compromises in dose intensity of the most valuable agents used in these programs (8,13,14).

In curable malignancies such as lymphoma and HD, patients receiving the highest dose intensity have higher response rates and longer survival (26,38). In the adjuvant setting, there is evidence that patients with breast cancer receiving a higher dose intensity have a better outcome than patients receiving lower doses (8,13). Furthermore, in the treatment of metastatic breast cancer there is evidence to show that better palliation is achieved with higher-dose chemotherapy (45). The message from all these observations is that chemotherapy regimens should be administered as prescribed and ad hoc reductions of dose should be avoided whenever possible.

One of the main reasons for reducing dose or delaying treatment is neutropenia. The hematopoietic growth factor Filgrastim significantly reduces neutropenia and infectious complications and consequently improves the delivery of full-dose chemotherapy on time. The use of Filgrastim to avoid dose modification improves dose intensity, and trials are ongoing to determine whether this benefit is associated with improved outcome. The clinical evidence showing that a reduction in dose intensity compromises outcome also suggests that further increases in dose intensity may improve outcome in some malignancies.

Dose intensification is frequently limited by hematological toxicity. Simon and Korn (16) have demonstrated that if a hematopoietic growth factor is used to alleviate bone marrow toxicity, it is possible to achieve significant dose escalation. Dose intensification of drugs limited by neutropenia may be achieved using Filgrastim, and it is hoped that this approach may be used to improve outcome in patients with an unfavorable prognosis. Preliminary studies have shown the feasibility of using Filgrastim to allow high dose-intensity therapy and initial results appear promising.

The clinical impact of dose intensification and the role of hematopoietic growth factors in this setting continue to be explored in controlled clinical trials. Until the results of these trials have demonstrated a clinical benefit, this approach should be regarded as experimental.

In summary, this chapter has reviewed some of the general principles underlying dose-response considerations in the chemotherapy treatment of human malignancies, and it has presented some evidence, retrospective and prospective, to verify if these theoretical principles are actually observed in the clinic. Although retrospective studies have often suggested a good correlation between dose-intensity and response rates, or even disease-free survival, the interpretation of these studies can be fraught with difficulties. For example, it is conceivable that a low dose intensity in a subgroup may be an effect of other prognostic factors (eg, poor performance status or old age), rather than being the cause of a poor outcome. Similarly, analysis using arbitrary cut-off points may be flawed because the results will change according to the dose intensity or cumulative dose chosen as the cut-off value. Furthermore, patients who do not complete the planned course of chemotherapy, for whatever reason, receive a low total dose but may appear to have received a relatively high dose intensity, and comparisons of dose delivery without reference to the period of treatment result in a bias towards the shorter regimen. In some of these retrospective studies, the inclusion of both single drug and multidrug regimens assumes that a double dose of one drug is equal to the combined dose of two drugs, ignoring drug synergy, and implying a homogeneous sensitivity of different clones within a tumor to chemotherapy (which is highly unlikely). Prospective randomized trials of dose intensity were obviously limited by myelotoxicity until the advent of recombinant hematopoietic growth factors. At least two different questions can be asked in this context: how does a reduction in standard-dose chemotherapy affect response rate and survival, and how does an increase in the dose of chemotherapy affect response rate and survival. Neither of these two fundamental questions has been unequivocally answered. Prospective studies are still ongoing and they are hampered by important difficulties in study design, including (as, for example, in metastatic breast cancer) the considerable differences that may exist in the natural history of malignant disease. Not only are there more than 200 different types of cancer, but each cancer is heterogeneous, and patients differ in their tolerance to treatment, immune status, metabolic handling of cytotoxics, etc.

In conclusion, the use of Filgrastim to alleviate chemotherapy-induced neutropenia and infection should improve the probability of delivering chemotherapy at full dose on time. Clinical evidence would suggest that facilitating chemotherapy administration may avoid some of the pitfalls of inadequate dosing.

REFERENCES

1. DeVita, V. T. (1989). Principles of chemotherapy. In *Cancer Principles and Practice of Oncology,* DeVita, V. T., Hellman, S., Rosenberg, S. A., eds., JB Lippincott, Philadelphia, pp. 276–300.
2. Skipper, H. E. (1967). Criteria associated with destruction of leukemia and solid tumor cells in animals. *Cancer Res 27:*2636–2645.
3. DeVita, V. T. (1991). The influence of information on drug resistance on protocol design. *Ann Oncol 2:*93–106.
4. Coldman, A. J., and Goldie, J. H. (1987). Impact of dose-intense chemotherapy on the development of permanent drug resistance. *Semin Oncol 14:*29–33.
5. Frei, E. III, and Canellos, G. P. (1980). Dose: a critical factor in cancer chemotherapy. *Am J Med 69:*585–594.
6. Skipper, H. Data and analyses having to do with the influence of dose intensity and duration of treatment (single drugs and combinations) on lethal toxicity and therapeutic response of experimental neoplasms. Birmingham, AL: Southern Research Institute Booklet 13, 1986.
7. Coppin, C. M. L. (1987). The description of chemotherapy delivery: options and pitfalls. *Semin Oncol 14:*34–42.
8. Hryniuk, W. M. (1988). More is better. *J Clin Oncol 6:*1365–1367.
9. Skipper, H. E. Dose intensity versus total dose of chemotherapy: an experimental basis. In: *Important Advances in Oncology,* DeVita, V. T., Hellman, S., Rosenberg, S. A., eds. Philadelphia: JB Lippincott, 1990 pp. 43–64.
10. Frei, E. III, Freireich, E. J., Gehen, E., et al (1961). Studies of sequential and combination antimetabolite therapy in acute leukaemia: 6-mercaptopurine and methotrexate. *Blood 18:*431.
11. Frei, E. III, Karon, M., Levin, R., et al (1965). The effectiveness of combinations of antileukemic agents in inducing and maintaining remissions in children with acute leukemia. *Blood 26:*642.
12. Blum, R. H., Frei, E. III, and Holland, J. F. Principles of dose, schedule and combination chemotherapy. In: *Cancer Medicine,* Holland, J. F., Frei III, E., eds., 1982, pp. 730–752.

13. Hryniuk, W. M. The importance of dose intensity in the outcome of chemotherapy. In: *Important Advances in Oncology,* DeVita, V. T., Hellman, S., Rosenberg, S. A., eds., Philadelphia: JB Lippincott, 1988, pp. 121–142.

14. Hryniuk, W., and Levine, M. N. (1986). Analysis of dose intensity for adjuvant chemotherapy trials in ovarian cancer. *J Clin Oncol 4:*1162–1170.

15. Hryniuk, W., and Bush, H. (1984). The Importance of dose intensity in chemotherapy of metastatic breast cancer. *J Clin Oncol 2:*1281–1288.

16. Simon, R., and Korn, E. L. (1990). Selecting drug combinations based on total equivalent dose (dose intensity). *JNCI 82:*1469–1476.

17. Goldie, J. H., and Coldman, A. J. (1979). A mathematic model for relating the drug sensitivity of tumors to their spontaneous mutation rate. *Cancer Treatment Rep 63:*1727–1733.

18. Day, R. S. (1986). Treatment sequencing, asymmetry and uncertainty: protocol strategies for combination chemotherapy. *Cancer Res 46:*3876–3885.

19. Norton, L., and Day, R. Potential innovations in scheduling in cancer chemotherapy. In: *Important Advances in Oncology,* DeVita, V. T., Hellman, S., Rosenberg, S. A., eds., Philadelphia: JB Lippincott, 1990, 57–77.

20. Perloff, M., Norton, L., Korzun, A., et al (1986). Advantage of an adriamycin combination plus halotestin after initial CMFVP for adjuvant therapy of node-positive Stage II breast cancer. *Proc Am Soc Clin Oncol 70:*273 (abstr).

21. Buzzoni, E., Bonadonna, G., Valagussa, P., et al (1990). Sequential versus alternating chemotherapy in the adjuvant treatment of breast cancer with more than 3 positive axillary nodes. *Proc Am Soc Clin Oncol* (abstr 67).

22. Canellos, G. P., Propert, K., Cooper, R., et al (1988). MOPP versus ABVD versus MOPP alternating with ABVD in advanced Hodgkin's disease: a prospective CALGB trial. *Proc Am Soc Clin Oncol 7:*230.

23. Deuchars, K. L., and Ling, V. (1989). p-Glycoprotein and multi-drug resistance in cancer chemotherapy. *Semin Oncol 16:*156–165.

24. Ueda, K., Carderelli, C., Gottesmann, M. N., and Pastan, I. (1987). Expression of full length cDNA for the human mdr1 (p-glycoprotein) gene confers multidrug-resistance in mouse and human cells. *Proc Natl Acad Sci 84:*3004–3008.

25. Tsuruo, T., Iida, H., Tsukagoshi, S., and Sakurai, Y. (1982). Increased accumulation of vincristine and adriamycin in drug-resistant P388 tumor cells following incubation with calcium antagonists and calmodulin inhibitors. *Cancer Res 42:*4730–4733.

26. DeVita, V. T. Jr., Jaffe, E. S., Mauch, P., and Longo, D. L. Lymphocytic lymphomas. In: *Cancer Principles and Practice of Oncology,* DeVita, V. T., Hellman, S., Rosenberg, S. A., eds., Philadelphia: JB Lippincott, 1989, pp. 1741–1798.

27. DeVita, V. T. Jr., Hubbard, S. M., and Longo, D. L. (1987). The chemotherapy of lymphomas: looking back, moving forward—The Richard and Hinda Rosenthal Foundation Award lecture. *Cancer Res 47:*5810–5824.

28. Magrath, I. T., Steinberg, S. M., Adde, M. A., and Haddy, T. B. (1989). Dose rate: an important prognostic determinant in non-Hodgkin's lymphomas in young patients. *Blood 74:*25a (abstr 80).

29. Dixon, D. O., Neilan, B., Jones, S. E., et al (1986). Effect of age on therapeutic outcome in advanced diffuse histiocytic lymphoma: the Southwest Oncology Group Experience. *J Clin Oncol 4*:295–305.

30. Dahlberg, S., Miller, T. P., Dana, B., et al (1990). Dose intensity is not associated with subsequent survival after adjustment for known prognostic factors in non-Hodgkin's lymphoma patients treated with m-BACOD, ProMACE-CytaBOM, and MACOP-B on Southwest Oncology Group Studies. *Proc Am Soc Clin Oncol 9* (abstr 986).

31. Fisher, R. I., Gaynor, E. R., Dahlberg, S., et al (1993). Comparison of a standard regimen (CHOP) with three intensive chemotherapy regimens for advanced non-Hodgkin's lymphoma. *N Engl J Med 328*:1002–1006.

32. DeVita, V. T., Simon, R. M., Hubbard, S. M., et al (1980). Curability of advanced Hodgkin's disease with chemotherapy. Long-term follow up of MOPP treated patients at the NCI. *Ann Intern Med 92*:587–595.

33. Bonadonna, G. (1987). Hodgkin's disease: the Milan Cancer Institute experience with MOPP and ABVD. *Third International Conference on Malignant Lymphoma,* June 1987 (abstr).

34. Coltman, C. A. Jr., Jones, S. E., Grozea, P. N., et al (1978). Bleomycin in combination with MOPP for management of Hodgkin's diseas: Southwest Oncology Group experience. In: *Bleomycin: Current Status and New Developments,* Carter, S. K., Crooke, S. T., eds., Orlando, FL: Academic Press, pp. 227–242.

35. Carde, P., MacKintosh, R., and Rosenberg, S. A. (1983). A dose and time response analysis of the treatment of Hodgkin's disease with MOPP therapy. *J Clin Oncol 1*:146–153.

36. Longo, D. L., Young, R. C., Wesley, M., et al (1986). Twenty years of MOPP chemotherapy for Hodgkin's disease. *J Clin Oncol 4*:1295–1306.

37. Bonadonna, G., Valagussa, P., and Santoro, A. (1986). Alternating non-cross-resistant combination chemotherapy with ABVD or MOPP in Stage IV Hodgkin's disease: a report of eight year results. *Ann Intern Med 104*:739–746.

38. Hellman, S., Jaffe, E. S., and DeVita, V. T. Jr. (1989). Hodgkin's disease. In: *Cancer Principles and Practice of Oncology,* DeVita, V. T., Hellman, S., Rosenberg, S. A., eds., Philadelphia: JB Lippincott, pp. 1696–1740.

39. Cohen, M. H., Creaven, P. J., Fossieck, B. E., et al (1977). Intensive chemotherapy of small cell bronchogenic carcinoma. *Cancer Treatment Rep 61*:349–354.

40. Mehta, C., and Vogl, S. E. (1982). High-dose cyclophosphamide in the induction chemotherapy of small cell lung cancer: minor improvements in rate of remission and survival. *Proc Am Soc Cancer Res 23*:155.

41. Murray, N. (1987). The importance of dose and dose intensity in lung cancer chemotherapy. *Semin Oncol 14*:20–28.

42. Bonadonna, G., and Valagussa, P. (1981). Dose response effect of adjuvant chemotherapy in breast cancer. *N Engl J Med 304*:10–15.

43. Henderson, I. C., Hayes, D. F., and Gelman, R. (1988). Dose-response in the treatment of breast cancer: a critical review. *J Clin Oncol 6*:1501–1515.

44. Hortobagyi, G. N. The importance of dose response in cytotoxic therapy for breast cancer. In: *Therapeutic Strategies in Oncology: Advances in Breast Can-*

cer Treatment, Henderson, I. C., Borden, E. C., eds., London: Mediscript, 1990, pp. 47–69.

45. Tannock, I. F., Boyd, N. F., DeBoer, G., et al (1988). A randomized trial of two dose levels of cyclophosphamide, methotrexate, and fluorouracil chemotherapy for patients with metastatic breast cancer. *J Clin Oncol 6:*1377–1387.

46. Wood, R., Budman, D. R., Korzun, A. H., et al. (1994). Dose and dose intensity of adjuvant chemotherapy for stage II, node-positive breast cancer. *N Engl J Med 330:*1253–1259.

47. Muss, H. B., Thor, A. D., Berry, D. A., et al (1994). C-erb-B2 expression and response to adjuvant therapy in women with node-positive early breast cancer. *N Engl J Med 330:*1260–1266.

48. Levin, L., and Hryniuk, W. M. (1987). Dose intensity analysis of chemotherapy regimens in ovarian carcinoma. *J Clin Oncol 5:*756–767.

49. Behrens, B. C., Grotzingen, K. R., Hamilton, T. C., et al (1985). Cytotoxicity of 3 cisplatin analogues in drug sensitive and a new cisplatin resistant human ovarian cancer cell line. *Proc Am Assoc Cancer Res 26:*262.

50. Einhorn, L. H., Crawford, E. D., Shipley, W. U., et al. Cancer of the testes. In: *Cancer Principles and Practice of Oncology,* DeVita, V. T., Hellman, S., Rosenberg, S. A., eds., Philadelphia: JB Lippincott, 1989, pp. 1071–1098.

51. Samson, M. K., Rivkin, S. E., Jones, S. E., et al (1984). Dose-response and dose-survival advantage for high versus low-dose cisplatin combined with vinblastine and bleomycin in disseminated testicular cancer. *Cancer 53:*1029–1035.

52. Ozols, R. F., Ihde, D. C., Linehan, W. M., et al (1988). A randomized trial of standard chemotherapy v a high-dose chemotherapy regimen in the treatment of poor prognosis nonseminomatous germ-cell tumors. *J Clin Oncol 6:*1031–1040.

53. Einhorn, L., Williams, S., Loehrer, P., et al (1990). Phase III study of cisplatin dose intensity in advanced germ cell tumors (GCT): a Southeastern and Southwest Oncology Group protocol. *Proc Am Soc Clin Oncol 9:*132 (abstr 510).

54. Bokemeyer, C., Kuczyk, M. A., Köhne, H., et al (1996). Hematopoietic growth factors and treatment of testicular cancer: biological interactions, routine use, and dose-intensive chemotherapy. *Ann Hematol 72:*1–9.

55. Pinkerton, C. R., and Philip, T. *Treatment Strategies in Pediatric Cancer: The Role of Hematopoietic Growth Factors.* Consultant Series (No.7), Macclesfield, United Kingdom: Gardiner-Caldwell Communications.

56. Bodey, G. P., Buckley, M., Sathe, Y. S., and Freireich, E. J. (1966). Quantitative relationships between circulating leukocytes and infection in patients with acute leukemia. *Ann Intern Med 64:*328–340.

57. Groopman, J. E., Molina, J. M., and Scadden, D. T. Hematopoietic growth factors. Biology and clinical applications. *N Engl J Med 321:*1449–1459.

58. Bronchud, M. H., Scarffe, J. H., Thatcher, N., et al (1987). Phase I/II study of recombinant human granulocyte colony-stimulating factor in patients receiving intensive chemotherapy for small cell lung cancer. *Br J Cancer 56:*809–813.

59. Bronchud, M. H., Howell, A., Crowther, D., et al (1989). The use of granulocyte colony-stimulating factor to increase the intensity of treatment with doxorubi-

cin in patients with advanced breast and ovarian cancer. *Br J Cancer* *60*:121–125.

60. Crawford, J., Ozer, H., Stoller, R., et al (1991). Reduction by granulocyte colony-stimulating factor of fever and neutropenia induced by chemotherapy in patients with small cell lung cancer. *N Engl J Med 325*:164–170.

61. Gabrilove, J. L., Jakubowski, A., Scher, J., et al (1988). Effect of granulocyte colony-stimulating factor on neutropenia and associated morbidity due to chemotherapy for transitional cell carcinoma of the urothelium. *N Engl J Med 318*:1414–1422.

62. Green, J. A., Trillet, V. N., and Manegold, C. (1991). R-metHuG-CSF (G-CSF) with CDE chemotherapy (CT) in small cell lung cancer (SCLC): Interim results from a randomized, placebo controlled trial. *Proc Am Soc Clin Oncol 10*:243 (abstr 832).

63. Morstyn, G., Campbell, L., Souza, L. M., et al (1988). Effect of granulocyte colony-stimulating factor on neutropenia induced by cytotoxic chemotherapy. *Lancet 1*:667–672.

64. Morstyn, G., Campbell, L., Lieschke, G. K., et al (1989). Treatment of chemotherapy-induced neutropenia by subcutaneously administered granulocyte colony-stimulating factor with optimization of dose and duration of therapy. *J Clin Oncol 7*:1554–1562.

65. Neidhart, J., Mangalik, A., Kohler, W., et al (1989). Granulocyte colony-stimulating factor stimulates recovery of granulocytes in patients receiving dose-intensive chemotherapy without bone marrow translantation. *J Clin Oncol 7*:1685–1692.

66. Lord, B. I., Bronchud, M. H., Ownes, S., et al (1989). The kinetics of human granulopoiesis following treatment with granulocyte colony-stimulating factor in vivo. *Proc Natl Acad Sci USA 86*:9499–9503.

67. Lindemann, A., Herrmann, F., Oster, W., et al (1989). Hematologic effects of recombinant human granulocyte colony-stimulating factor in patients with malignancy. *Blood 74*:2644–2651.

68. Bronchud, M. H., Potter, M. R., Morgenstern, G., et al (1988). In vitro and in vivo analysis of the effects of recombinant human granulocyte colony-stimulating factor in patients. *Br J Cancer 58*:64–69.

69. Peters, W. P., Stuart, A., Affronti, M. L., et al (1988). Neutrophil migration is defective during recombinant human granulocyte macrophage colony stimulating factor infusion after after autologous bone marrow marrow transplantation in humans. *Blood 72*:1310–1315.

70. Dedhar, S., Galloway, P., and Eaves, C. (1988). Human granulocyte macrophage colony factor is a growth factor for a variety of cell types of non hemopoietic origin. *Proc Am Soc Clin Oncol 7*:51 (abstr).

71. Mertelsmann, R., and Peters, W. *Hematopoietic Growth Factors,* Houston, TX: Educational Book, *Am Soc Clin Oncol* 1991.

72. Neidhart, J., Stidley, G., Ferguson, J., et al (1990). GM-CSF decreases duration of cytopenia in patients receiving dose intensive therapy with cyclophosphamide, etoposide and cisplatin. *Proc Am Soc Clin Oncol 9*:753A (abstr).

73. Piccart, M., van der Schueren, E., Bruning, P., et al (1991). High dose-intensity (DI) chemotherapy (CT) with epi-adriamycin (E), cyclophosphamide (C) and r-metHuG-CSF (AMGEN) in breast cancer (BC) patients. *Eur J Cancer Suppl* 2:S56 (abstract 308).

74. Zittoun, R. A., Mandelli, F., Willemze, R., et al (1995). Autologous or allogeneic bone marrow transplantation compared with intensive chemotherapy in acute myelogenous leukemia. *N Engl J Med 332*:217–223.

75. Gale, R. P., Herzig, G. P., Hocking, W. G., et al (1996). AML in 1st remission: RAND-DELPHI analysis of appropriateness of bone marrow transplants. *Proc Am Soc Clin Oncol 15*:365 (abstr 1079).

76. Philip, T., Guglielmi, C., Chauvin, F., et al (1995). Autologous bone marrow transplantation versus conventional chemotherapy (DHAP) in relapsed non Hodgkin's lymphoma: final analysis of the Parma randomised study. *Proc Am Soc Clin Oncol 14*:390 (abstr 1220).

77. Attal, M., Harousseau, J. L., Stoppa, A. M., et al (1996). A prospective, randomized trial of autologous bone marrow transplantation and chemotherapy in multiple myeloma. *N Engl J Med 335*:91–97.

78. Peters, W. P., Berry, D., Vredenburgh, J. J., et al (1995). Five year follow up of high-dose combination alkylating agents with ABMT as consolidation after standard dose of CAF for primary breast cancer involving ten or more axillary lymph nodes. *Proc Am Soc Clin Oncol 14*:317 (abstr 933).

79. Gianni, A. M., Siena, S., Bregni, M., et al (1995). Five year results of high-dose sequential adjuvant chemotherapy in breast cancer with ten or more axillary lymph nodes. *Proc Am Soc Clin Oncol 14*:90 (abstr 61).

80. Peters, W. P., Jones, R. B., Vredenburgh, J., et al (1996). A large, prospective, randomized trial of high-dose combination alkylating agents with autologous cellular support as consolidation for patients with metastatic breast cancer achieving complete remission after intensive doxorubicin-based induction therapy. *Proc Am Soc Clin Oncol 15*:121 (abstr 149).

9
Use of Filgrastim (r-metHuG-CSF) in the Treatment of Lung Cancer

Jeffrey Crawford
Duke Comprehensive Cancer Center and Duke University Medical Center, Durham, North Carolina

MaryAnn Foote
Amgen Inc., Thousand Oaks, California

I. INTRODUCTION

Lung cancer is the third most common cancer in the United States and is the most lethal form of cancer for men and women (1). Small-cell lung cancer (SCLC) represents approximately 20% of all lung cancers and has a slightly lower cure rate than that of non-small–cell lung cancer (NSCLC). Small-cell lung cancers are prone to metastases, and more than 90% of patients with SCLC present with mediastinal lymph node metastases and more than 66% with distant organ metastases (2). Non-small–cell lung cancers account for approximately 75% of all cases of lung cancer diagnosed yearly in the United States (3,4). Unfortunately, patients usually die within 1 year of diagnosis of NSCLC. Even if patients do not have extensive disease at presentation, and nearly 60% do, they will have other significant comorbid conditions, such as cardiac or pulmonary disease, that preclude surgical intervention. Approximately 20% of all patients with NSCLC will be candidates for curative surgical resection; these patients have a 5-year survival of 35% to 50% (5).

Only two decades ago, the management of patients with SCLC was limited to surgery and radiotherapy. These local methods rarely provided effective long-term control when used as a single-treatment modality (6–9). Knowledge of the frequency and extent of metastases and the sensitivity of SCLC to chemotherapy have led to the current emphasis on chemother-

Table 1. Prognostic Factors in Small-Cell Lung Cancer.

Stage (limited or extensive disease)
Performance status
Sex
Age
Histological subclassifications (presence of large-cell elements has poor
    prognosis)
Liver metastases
Central nervous system involvement
Blood biochemistry (especially lactate dehydrogenase)

apy. This new awareness resulted in a fivefold increase in survival and a long-term, disease-free survival of more than 3 years for all patients with SCLC in the intervening 20 years (2,10). Patients with limited-stage disease usually receive both chemotherapy and radiotherapy, as meta-analyses of randomized trials have shown a significant survival advantage with the combined approach (11). Patients with extensive-stage disease usually are treated with combination chemotherapy alone. However, there is no standardized treatment for SCLC.

Until recently, the majority of patients with NSCLC who were not candidates for surgical resection were not treated with further cytotoxic therapy and received only palliative radiation therapy for symptom control. The routine use of systemic chemotherapy was rare unless patients were entered into a clinical trial. Over the past 5 years, the use of systemic chemotherapy for NSCLC has become more common, primarily because several meta-analyses of randomized clinical trials compared with best-available supportive care showed a statistically significant improvement in survival (12,13).

Because therapeutic options now exist, it is important for the physician and patient to define the objective of the treatment before therapy is initiated. Palliation may be the best and most reasonable objective for elderly patients or for those with poor prognostic factors (**Table 1**). Other patients may be candidates for therapy with potential long-term survival benefits or even cure. The two key prognostic factors appear to be pretreatment performance status and extent of disease (14).

II. CHEMOTHERAPY FOR TREATMENT OF LUNG CANCER

A. Small-Cell Lung Cancer

As discussed, the main treatment modality for SCLC is chemotherapy, given with the intent to achieve the highest long-term, disease-free survival with the lowest possible morbidity. In the 1970s, median survival was prolonged by 4 to 5 times and cure was possible in 5% to 10% of the patients treated with surgery and chemotherapy (10,15). Many agents have antitumor activity in SCLC and a majority of studies have incorporated these agents: etoposide, vincristine, cyclophosphamide, doxorubicin, CCNU (lomustine), cisplatin, and methotrexate. Among the most commonly used combinations are cisplatin or carboplatin and etoposide; doxorubicin, etoposide, and cyclophosphamide; and cyclophosphamide, doxorubicin, and vincristine. There has been a trend to include etoposide in the initial treatment because results of these studies have shown that combinations with etoposide are superior to those without it (16,17). Although it is believed that three or four drugs are superior to single-drug therapy, results have suggested that epipodophyllotoxins, alone or in two-drug combinations with platinum, are as effective as multi-drug combinations in producing initial responses (15).

Although SCLC is initially chemosensitive, resistent clones present at diagnosis or emerging during therapy eventually cause treatment failure in most patients (18,19). Therefore, alternate, active, non-cross–resistant regimens are indicated to prevent the development of drug resistance and subsequent relapse. (See Chapter 8 of this volume for discussion of the Coldman-Goldie theory.) Results from large controlled trials (**Table 2**) suggest small advantages for alternating chemotherapy for limited-stage disease (14,20), extensive-stage disease (21), or both stages of disease (22), although no impressive differences were seen.

A 2-year survival rate of 10% was seen in one study in patients receiving alternating therapy, compared with patients receiving sequential therapy (23).

Dose intensification studies have shown equivocal results. Dose intensification studies usually require stem cell support (bone marrow or peripheral blood) and/or growth factors. Three randomized trials compared standard-dose and high-dose therapies (24–26). Increasing the dose of either CAV or PE increased toxicity but did not improve survival.

It remains that most patients with SCLC eventually relapse and die of progressive disease. New drugs with novel mechanisms of action are being tested to determine if they can improve cure rate and prolong survival in the patient population.

B. Non-Small–Cell Lung Cancer

As discussed in Section I, chemotherapy has not been used extensively to treat patients with NSCLC. The trend, however, has been changed, and since the early 1990s use of systemic chemotherapy has increased. Chemotherapy is not only used to treat advanced-stage disease, but has been

Table 2. Alternating Versus Sequential Chemotherapy Randomized Trials 1983–1993.

Reference	Regimen	Number of patients			Complete response (%)			Survival (months)		
		All	Ltd	Ext	All	Ltd	Ext	All	Ltd	Ext
21	S:CVLPZ × 2 Irr CVLPZ...18 mos	—	28	56	—	54	20[a]	—	10	7[b]
	A:EAM × 3 Irr CVLPZ + EAM + CVLPZ...	—	28	50	—	42	44	—	16	10
35	S:CAV × 3 + EP × 3 + Irr	—	146	144	—	45	27	—	15.5	8.0
	A:(CAV + EP) × 3 + Irr	—	154	145	—	52	29	—	15.0	9.6
22	S:CAV × 8 + Irr	—	53	99	—	33	15	—	11.1	8.9[c]
	A:EVdl + PAV × 3 + CML × 2 + Irr	—	51	99	—	52	28	—	13.4	9.9
14; 20	S:CLVE...+ P/A/Vd/M/H... 19 mos	154	—	71	44	—	21	9.5	—	7.5[c]
	A₁:CLVM + EAV...18 mos	151	—	66	37	—	18	11.5	—	7.5
	A₂:CLVM + EAV + PVdH... 18 mos	152	—	66	41	—	21	12.0	—	8.0

28	S:CAV × 6-8	—	—	294	—	—	16[d]	—	—	10.6[d]
	A:(CAV + HEM) × 3-4	—	—	283	—	—	23	—	—	11.5
23	S:CAVE × 6	30	—	18	—	44	—	—	11.8	—
	A:PE × 2 + CAVE × 2 + PE × 2	30	—	17	—	49	—	—	10	—
36	S:CAV × 16	—	—	79	—	—	24	—	—	6.9
	A:(MEP + CAV) × 8	—	—	82	—	—	25	—	—	9.2
37	S_1:CAV...12 mos	—	49	46	16	—	13	12.4	—	8.7
	S_2:PE...12 mos	—	44	51	18	—	10	11.7	—	8.3
	A:CAV + PE + CAV...12 mos	—	49	40	22	—	8	16.8	—	9.0
38	S_1:IE to max. response + CAV... 4.5 mos + Irr	—	63	99	35	—	20	12.3	—	9.1
	A:IE + CAV + IE...4.5 mos + Irr	—	72	87	40	—	15	12.5	—	8.5
39; 40	S_1:CAV...4.5 mos	—	—	159	—	—	7.1	—	—	8.3
	S_2:PE...3 mos	—	—	156	—	—	10	—	—	8.6
	A:CAV + PE + CAV... 4.5 mos	—	—	162	—	—	7.2	—	—	8.1

Probability S versus A: [a]p = 0.03; [b]p = 0.001; [c]p = 0.05; [d]p = 0.002 (after adjustment for prognostic factors).
S = Sequential; A = alternating; Irr = irradiation; Ltd = limited; Ext = extensive; mos = months.
Agents: A = doxorubicin; C = cyclophosphamide; E = etoposide; H = hexamethylmelamine; I = ifosfamide; L = lomustine; M = methotrexate; P = cisplatin; PZ = procarbazine, V = vincristine; Vd = vindesine.
Source: Adapted from Ref. 41.

shown to have benefit in patients with locally advanced disease (27). The overall improvements in cure rate and survival time have been modest, suggesting that here, too, new cytotoxic agents will offer better results.

III. NEW CHEMOTHERAPEUTIC AGENTS

As discussed above, most patients with SCLC relapse and patients with NSCLC are not routinely treated with chemotherapy. Both of these situations suggest a possible role for new drugs with novel mechanisms of action.

One of the more promising new agents for the treatment of SCLC is paclitaxel (Taxol®). Its activity was confirmed by two studies done by cooperative cancer groups. In both studies, patients received 250 mg/m^2 paclitaxel over a 24-hour infusion with Filgrastim given as routine support (28,29). Patients in the ECOG study (28) received two courses of paclitaxel, and required further treatment with cisplatin and etoposide because of limited supply of paclitaxel. Response rate, confirmed by chest radiograph after 4 weeks without a change in therapy, was 34%. In the North Central Cancer Group trial (29), patients received paclitaxel for six courses or until their disease progressed. The overall response rate in this trial was 68%.

Docetaxel (Taxotere®) is another new agent useful in the treatment of NSCLC, and is especially useful in the treatment of platinum-refractory patients. Results of a study with 80 patients were recently reported in abstract form (30). They had progressive/recurrent disease despite platinum-based therapy. Docetaxel 100 mg/m^2 every 3 weeks was administered and yielded a response rate of 15%. Median survival was 7 months, and the 1-year survival rate was 25%. Febrile neutropenia was a severe adverse event affecting 19% of the patients.

Recently, data on the use of paclitaxel in patients with SCLC were presented in abstract form (31). In this study, paclitaxel 200 mg/m^2 was added to increased doses of carboplatin and etoposide. The higher doses of paclitaxel and carboplatin resulted in higher responses and survival compared with lower doses of these chemotherapeutic agents. This suggests a dose-response relationship in the setting of SCLC, but this was not a randomized study. Increased doses of paclitaxel and carboplatin, however, did increase grade 3 and 4 leukopenia.

IV. NEUTROPENIA AND USE OF FILGRASTIM

Acquired neutropenia may result from under-production or increased destruction of neutrophils as well as from migration from the circulatory pool to the tissue pool in response to an inflammatory process. Neutrophil production may be impaired by chemotherapy or other drug administration, or by previous radiation therapy. Patients with acute forms of neutropenia (ie, chemotherapy-induced) are susceptible to fulminant, life-threatening pyogenic infections. The incidence of infection has been shown to increase with decreasing neutrophil counts (32) and there is a sharp increase in infectious complications when a patient's ANC is $< 1.0 \times 10^9/L$. Since inflammatory cells are lacking, many of the common signs of inflammation may not be present.

Until recently, the management of the infectious complications of malignancy and chemotherapy was limited to antibiotics and general supportive measures. The advent of myeloid growth factors with demonstrated ability to increase neutrophil counts are a new class of agents available to the oncologist.

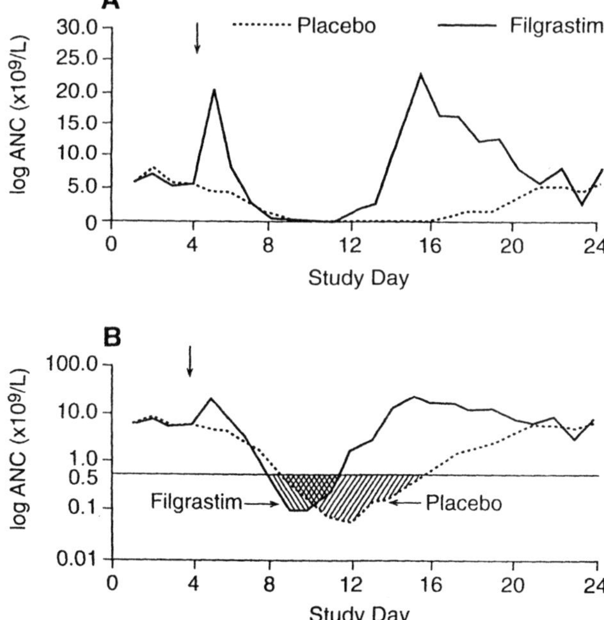

Figure 1. Mean absolute neutrophil counts (ANC) for cycle 1. Arrows denote start of study drug. Panel A, ANC given in linear scale. Panel B, ANC given in log scale. The hatched area highlights the degree and duration of neutropenia (ie, ANC $< 0.5 \times 10^9/L$). (Adapted from Ref. 26.)

Two prospective, randomized, placebo-controlled trials have shown efficacy for initiation of Filgrastim therapy with the first cycle of chemotherapy (ie, primary therapy) in reducing the incidence of febrile neutropenia by approximately 50% in patients receiving cytotoxic chemotherapy (26,33).

The Crawford et al (26) study was a multicenter, prospective, randomized, placebo-controlled trial, involving 211 patients with SCLC. Placebo or Filgrastim 230 μg/m^2/day SC was administered after CAE (cyclophosphamide, doxorubicin, etoposide) chemotherapy (up to six cycles). The duration and severity of neutropenia were significantly reduced in patients receiving Filgrastim. During cycle 1, the rate of febrile neutropenia was reduced by 50% and the difference in the cumulative event rate across all cycles was statistically significant (**Figure 1**). Antibiotic use and number of days of hospitalization were also significantly reduced by about 50% with Filgrastim. Due to the crossover design of the trial, many patients randomized to the placebo group were allowed to eventually receive open-label Filgrastim, which makes interpretation of the results more complicated.

A phase 3, multicenter, prospective, randomized, placebo-controlled trial confirmed the results of Crawford et al's pivotal trial and also showed the improved ability to deliver chemotherapy as scheduled in patients receiving Filgrastim (33). In this study, 130 patients with SCLC also received Filgrastim 230 μg/m^2/day or placebo after CAE chemotherapy. Over all

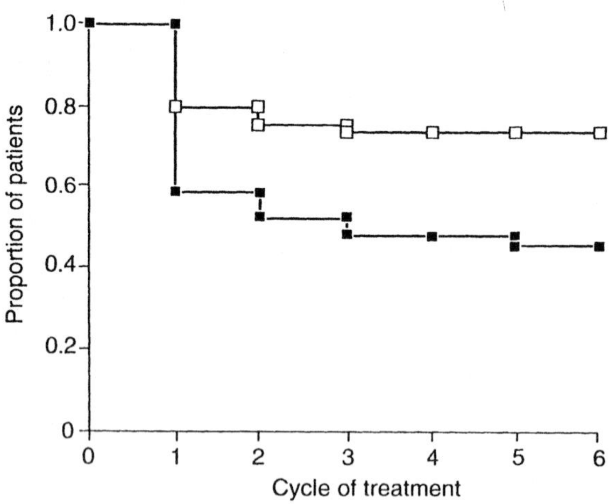

Figure 2. Proportion of patients in each treatment group who did not have at least one episode of fever with neutropenia (ie, absolute neutrophil count $< 1.0 \times 10^9$/L and fever $\geq 38.2°$C). Open squares, Filgrastim; closed squares, placebo. (Adapted from Ref. 33.)

Table 3. Use of Filgrastim in Patients with Lung Cancer.

Type of Cancer	No. Patients (Filgrastim/No)	Chemotherapy	Results	Reference
SCLC	9/0		decrease in neutropenia; decrease in documented infection	42
SCLC	32/0	doxorubicin, cyclophosphamide, etoposide	Filgrastim allowed dose intensified chemotherapy to be administered by decreased intervals between cycles	27
SCLC	103/103	cyclophosphamide, doxorubicin, etoposide	decrease in treatment delays, decrease in toxic death, increased survival	43
NSCLC	101/0	ifosfamide, mesna, cisplatin, or etoposide, cisplatin	Filgrastim allowed full doses of drugs to be delivered on schedule	44
NSCLC	130/0	vinorelbine, carboplatin	Filgrastim allowed dose intensification to 30 mg/m^2 vinorelbine	45
NSCLC	20/40	cisplatin, vindesine, ifosfamide, etoposide	improved survival in patients given intensified induction chemotherapy with subsequent tumor resection; Filgrastim supported intensified therapy	46
NSCLC	36/0	carboplatin, vinorelbine	Filgrastim is effective in assuring low incidence of febrile neutropenia across all cycles	47

cycles, Filgrastim significantly reduced the incidence of febrile neutropenia (**Figure 2**), there was a significant reduction in the requirement for parenteral antibiotics as well as a significant reduction in the number of days of hospitalization. Filgrastim significantly reduced the duration and severity of neutropenia and the need to delay or decrease the dose of chemotherapy.

Recently, results of a phase 3 trial of 63 consecutive patients with SCLC were published (34). Patients were treated with CODE (cisplatin, Oncovin®, doxorubicin, etoposide) chemotherapy with (n = 32) or without (n = 31) Filgrastim 50 μg/m^2 SC. Compared with control patients, patients receiving Filgrastim achieved a significant reduction of the duration of neutropenia and the incidence of febrile neutropenia. The use of Filgrastim in the CODE regimen was associated with an increase in delivered dose intensity, a 27-week prolongation of median survival, and about a fivefold increase in the 2-year survival rate (31.3% versus 6.5%) compared with the CODE-only group.

Other studies with Filgrastim in the treatment of patients with lung cancer are summarized in **Table 3**.

V. CONCLUSIONS

Both SCLC and NSCLC are systemic diseases requiring systemic chemotherapy. Many agents are active and some combinations of two or more drugs produce good results. In recent years, overall results have been unchanged, but this is may improve with the introduction of new agents with alternate mechanisms of action. From the clinical data on the administration of hematopoietic growth factors, the evidence suggests that the use of Filgrastim in combination with intensive chemotherapy reduces overall morbidity with less febrile neutropenic episodes.

REFERENCES

1. Wingo, P. A., Tong, T., and Bolden, S. (1995). Cancer statistics, 1995. *CA Cancer J Clin 45*:8–30.
2. Cook, R. M., Miller, Y. E., and Bunn, P. A., Jr. (1993). Small cell lung cancer: etiology, biology, clinical features, staging, and treatment. *Curr Probl Cancer 17*:69–144.
3. American Cancer Society. (1993). Cancer facts and figures—1993. Atlanta, p. 10.
4. Ihde, D. C., and Minna, J. D. (1991). Non-small cell lung cancer. Part II: treatment. *Curr Probl Cancer 15*:105–154.

5. Carney, D. N. (1996). Non-small cell lung cancer: slow but definite progress. *Semin Oncol 23*:5–6.

6. Goodman, G. E., Miller, T. P., Manning, M. M., Davis, S. L., and McMahon, L. J. (1983). Treatment of small cell lung cancer with VP-16, vincristine, doxorubicin (Adriamycin), cyclophosphamide (EVAC), and high-dose chest radiotherapy. *J Clin Oncol 1*:483–488.

7. Niiranen, A. (1988). Long-term survival in small cell carcinoma of the lung. *Eur J Cancer Clin Oncol 24*:749–752.

8. Seifter, E. J., and Ihde, D. C. (1988). Therapy of small cell lung cancer: a perspective on two decades of clinical research. *Semin Oncol 15*:278–299.

9. Vokes, E. E., and Weichselbaum, R. R. (1990). Concomitant chemoradiotherapy: rationale and clinical experience in patients with solid tumors. *J Clin Oncol 8*:911–934.

10. Hansen, H. H. (1992). Management of small-cell cancer of the lung. *Lancet 339*:846–849.

11. Pignon, J. P., Arriagada, R., Ihde, D. C., et al (1992). A meta-analysis of thoracic radiotherapy for small cell lung cancer. *N Engl J Med 327*:1618–1624.

12. Souquet, P. J., Chauvin, F., Boissel, J. P., et al (1993). Polychemotherapy in advanced non-small cell lung cancer: a meta-analysis. *Lancet 342*:19–21.

13. Non-Small Cell Lung Cancer Collaborative Group. (1995). Chemotherapy in non-small cell lung cancer: a meta-analysis using updated data on individual patients from 52 randomised clinical trials. *Br Med J 311*:899–904.

14. Østerlind, K., Pedersen, A. G., Vindeløv, L. L., et al (1986). Alternating or continuous chemotherapy of extensive stage small cell lung cancer. Results of a controlled trial of 204 patients. *Lung Cancer 2*:127–128.

15. Bunn, P. A., Cullen, M., Fukuoka, M., et al (1989). Chemotherapy in small cell lung cancer: a consensus report. *Lung Cancer 5*:127–134.

16. Hirsch, F. R., Hansen, H. H., Hansen, M., et al (1987). The superiority of combination chemotherapy including etoposide based on in vivo cell cycle analysis in the treatment of extensive small cell lung cancer. *J Clin Oncol 5*:585–591.

17. Jackson, D. V., Zekan, P. J., Caldwell, R. D., et al (1984). VP-16-213 in combination chemotherapy with chest irradiation for small cell lung cancer: a randomised trial of the Piedmont Oncology Association. *J Clin Oncol 2*:1343–1351.

18. Vindeløv, L. L., Hansen, H. H., Christensen, I. J., et al (1980). Clonal heterogeneity of small cell anaplastic carcinoma of the lung demonstrated by flow cytometric DNA analysis. *Cancer Res 40*:4295–4300.

19. Roed, H., Vindeløv, L. L., Christensen, I. J., et al (1988). The cytotoxic activity of cisplatin, carboplatin, and teniposide alone and combined determined on four human small cell lung cancer cell lines by the clonogenic assay. *Eur J Clin Oncol 24*:247–253.

20. Pedersen, A. G., Østerlind, K., Vindeløv, L. L., et al (1987). Alternating or continuous chemotherapy of small cell lung cancer. A three armed randomized trial. *Proc Am Soc Clin Oncol 6*:195 (abstr 731).

21. Daniels, J. R., Chak, L. Y., Sikic, B. I., et al (1984). Chemotherapy of small cell carcinoma of the lung: a randomized comparison of alternating and sequential combination chemotherapy programs. *J Clin Oncol 2*:1192–1199.

22. Havemann, K., Wolf, M., Holl, R., et al (1987). Alternating versus sequential chemotherapy in small cell lung cancer. A randomized German multicentre trial. *Cancer 59*:1072–1082.

23. Smith, A. P., Anderson, G., Chappell, G., et al (1991). Does the substitution of cisplatin in a standard four drug regimen improve survival in small cell carcinoma of the lung? A comparison of two chemotherapy regimens. *Thorax 46*:172–174.

24. Johnson, D. H., DeLeo, M. J., Hande, K. R., Wolff, S. N., Hainsworth, J. D., and Greco, F. A. (1987). High-dose induction chemotherapy with cyclophosphamide, etoposide, and cisplatin for extensive-stage small-cell lung cancer. *J Clin Oncol 5*:703–709.

25. Tvede, K., Ardizzoni, A., Dombernowsky, P., et al (1994). Phase II study of topotecan (TC) in patients (pts) with pretreated small-cell lung cancer (SCLC). *Lung Cancer 11*:106 (abstr 404).

26. Crawford, J., Ozer, H., Stoller, R., et al. (1991). Reduction by granulocyte colony-stimulating factor of fever and neutropenia induced by chemotherapy in patients with small-cell lung cancer. *N Engl J Med 325*:164–170.

27. Thatcher, N., Clark, P. I., Smith, D. B., et al (1995). Increasing and planned dose intensity of doxorubicin, cyclophosphamide, and etoposide (ACE) by adding recombinant human methionyl granulocyte colony-stimulating factor (G-CSF; Filgrastim) in the treatment of small cell lung cancer (SCLC). Medical Research Council Lung Cancer Working Party. *Clin Oncol 7*:293–299.

28. Ettinger, D. S., Finkel, D. M., Abeloff, M. D., et al (1990). A randomized comparison of standard chemotherapy versus alternating chemotherapy and maintenance versus no maintenance therapy for extensive-stage small cell lung cancer. A phase III study of the Eastern Cooperative Oncology Group. *J Clin Oncol 8*:230–240.

29. Kirschling, R. J., Jung, S. H., and Jett, J. R. (1994). A phase II trial of Taxol and GCSF in previously untreated patients with extensive stage small cell lung cancer (SCC). *Proc Am Soc Clin Oncol 13*:326 (abstr 1076).

30. Gandara, D. R., Yokes, E., Green, M., et al (1997). Docetaxel (Taxotere) in platinum-treated non-small cell lung cancer (NSCLC): confirmation of prolonged survival in a multicenter trial. *Proc Am Soc Clin Oncol 16*:454a (abstr 1632).

31. Hainsworth, J. D., Gray, J. R., Hopkins, L. G., et al (1997). Paclitaxel (1-hour infusion), carboplatin, and extended schedule etoposide in small cell lung cancer (SCLC): a report on 117 patients (pts) treated by the Minnie Pearl Cancer Research Network. *Proc Am Soc Clin Oncol 16*:451a (abstr 1623).

32. Bodey, G. P., Buckley, M., Sathe, Y. S., and Freireich, E. J. (1966). Quantitative relationships between circulating leukocytes and infection in patients with acute leukemia. *Ann Intern Med 64*:328–340.

33. Trillet-Lenoir, V., Green, J., Manegold, C., et al. (1993). Recombinant granulocyte colony stimulating factor reduces the infectious complications of cytotoxic chemotherapy. *Eur J Cancer 29A*:319–324.

34. Fukuoka, M., Masuda, N., Negoro, S., et al (1997). CODE chemotherapy with and without granulocyte colony-stimulating factor in small cell lung cancer. *Br J Cancer 75:*306–309.

35. Evans, W. K., Feld, R., Murray, N., et al (1987). Superiority of alternating non-cross-resistant chemotherapy in extensive small cell lung cancer. A multicenter, randomized trial by the National Cancer Institute of Canada. *Ann Intern Med 107:*451–458.

36. Wampler, G. L., Heim, W. J., Ellison, N. M., et al (1991). Comparison of cyclophosphamide, doxorubicin, and vincristine with an alternating regimen of methotrexate, etoposide, and cisplatin/cyclophosphamide, doxorubicin, and vincristine in the treatment of extensive-disease small-cell lung carcinoma: a Mid-Atlantic oncology program study. *J Clin Oncol 9:*1438–1445.

37. Fukuoka, M., Furuse, K., Saijo, N., et al (1991). Randomized trial of cyclophosphamide, doxorubicin and vincristine versus cisplatin and etoposide versus alternation of these regimens in small-cell lung cancer. *J Natl Cancer Inst 83:*855–861.

38. Wolf, M., Holle, R., Hans, K., et al (1991). Analysis of prognostic factors in 766 patients with small cell lung cancer (SCLC): the role of sex as a predictor for survival. *Br J Cancer 63:*986–992.

39. Roth, B. J., Johnson, D. H., Greco, F. A., et al (1989). A phase III trial of etoposide and cisplatin versus cyclophosphamide, doxorubicin, and vincristine versus alternation of the two therapies for patients with extensive small cell lung cancer: preliminary results. *Proc Am Soc Clin Oncol 8:*225 (abstr 875).

40. Roth, B. J., Johnson, D. H., Einhorn, L. H., et al (1992). Randomized study of cyclophosphamide, doxorubicin, and vincristine versus etoposide and cisplatin versus alternation of these two regimens in extensive small-cell lung cancer: a phase III trial of the South-Eastern Cancer Study Group. *J Clin Oncol 10:*282–291.

41. Hansen, H. H. (1994). *Small Cell Lung Cancer. The Role of Haematopoietic Growth Factors.* Gardiner-Caldwell Communications, Limited, Macclesfield, UK.

42. Bronchud, M. H., Howell, A., Crowther, D., Hopwood, P., Souza, L., and Dexter, T. M. (1989). The use of granulocyte colony-stimulating factor to increase the intensity of treatment with doxorubicin in patients with advanced breast and ovarian cancer. *Br J Cancer 60:*121–125.

43. Splinter, T. A., Kranse, R., and Wamelink, I. A. (1996). Randomized trial of chemotherapy ± G-CSF without dose-intensification in small cell lung cancer (SCLC): improved survival in the G-CSF arm. *Proc Am Soc Clin Oncol 15:*342 (abstr 993).

44. Graziano, S., Valone, F., Herndon, J., et al (1994). A randomized phase II study of ifosfamide, mesna, cisplatin plus G-CSF or etoposide/cisplatin plus G-CSF in advanced non-small cell lung cancer (NSCLC): CALGB 9132. *Proc Am Soc Clin Oncol 13:*328 (abstr 1082).

45. Crawford, J., O'Rourke, M. A., Herndon, J., Hohneker, J., and Burman, S. (1995). Sequential trials of vinorelbine (navelbine, NVB) with carboplatin in

patients with advanced non small cell lung cancer (NSCLC). *Proc Am Soc Clin Oncol 14:*376 (abstr 1162).

46. Fischer J. R., Muller, A., Bulzebruck, H., et al (1996). Induction chemotherapy with and without G-CSF in stage IIIA/B non-small cell lung cancer: 7-year follow-up. *Proc Am Soc Clin Oncol 15:*395 (abstr 1195).

47. Garst, J., O'Rourke, M. A., Herndon, J., Blackwell, S., Shoemaker, D., and Crawford, J. (1997). G-CSF titration based on the absolute neutrophil count (ANC) for use in weekly solid tumor chemotherapy regimens. *Proc Am Soc Clin Oncol 16:*119a (abstr 418).

10
Filgrastim (r-metHuG-CSF) in Lymphoma

Janice Gabrilove
Memorial Sloan-Kettering Cancer Center, New York, New York

I. INTRODUCTION

Lymphoma, cancer of the lymphatic system, is classified as either Hodgkin's lymphoma or Hodgkin's disease (HD) or non-Hodgkin's lymphoma (NHL), and they are distinguishable by cell type. Both diseases share similar symptoms, such as painless swelling of the lymph nodes, fever, and fatigue. Non-Hodgkin's lymphoma is more common and has approximately 15 subtypes; it generally occurs in individuals age 30 to 70 years. Hodgkin's disease generally occurs in individuals age 15 to 40 years; HD is considered to be more curable than NHL. Both HD and NHL are diagnosed by presence of Reed-Sternberg cells, and both diseases are classified by the same categories of grades and stages (**Table 1**).

There are several subtypes of HD. In lymphocyte predominance, the lymph nodes are composed largely of reactive lymphocytes and malignant cells that have a "popcorn" appearance, but very few Reed-Sternberg cells. This subtype accounts for 5% to 10% of all cases of HD and affects more men than women. In the nodular sclerosis subtype, the lymph nodes in the lower neck, chest, and clavicular region usually contain normal and reactive lymphocytes and Reed-Sternberg cells separated by bands of scar-like tissue. This subtype accounts for 30% to 60% of all cases of HD, and even though it is more aggressive than the lymphocyte predominance subtype, it is considered to be a limited-stage disease. In the mixed cell subtype of HD, the lymph nodes usually contain Reed-Sternberg cells and inflammatory cells; this subtype accounts for 20% to 40% of all cases. There is another

Table 1. Grades and Stages of Lymphoma.

Grade	
high grade	usually found in both B-cell and T-cell types
intermediate grade	usually found in both B-cell and T-cell types
low grade	predominantly found in B-cell types
Stages*	
I	single cancer site; no marrow involvement
II	two cancer sites; both are either above or below the diaphragm; no marrow involvement
III	cancer sites above and below the diaphragm; no marrow involvement
IV	marrow involved or other metastases have occurred

* In HD, staging is further classified as B (presence of fever, weight loss, or night sweats); A (absence of fever, weight loss, or night sweats); or E (metastases beyond lymphatic system).

subtype called lymphocyte-depleted that has two variations: in one there are sheets of differing malignant cells, and in the other, few Reed-Sternberg cells and lymphocytes with scar-like tissue.

The approximately 15 subtypes of NHL are based on the types of cancer cells present and there are several different classification systems, all using cell type and rate of cell growth to distinguish the different types. In the category of high-grade or highly aggressive NHL there are Burkitt's, non-Burkitt's, diffuse, leukemia/lymphoma, lymphoblastic, and T-cell subtypes. In the intermediate grade, there are large-cell follicular, mixed-cell diffuse, large-cell diffuse, and immunoblastic subtypes. The low-grade or indolent subtype has small lymphocytic, small cleaved-cell follicular, mixed follicular, small cleaved-cell diffuse, intermediately differentiated diffuse, and cutaneous T-cell subtypes.

It is known that environmental toxins such as pesticides can cause lymphoma, although viruses and bacteria can play a role in the disease etiology. People infected with human immunodeficiency virus (HIV) or Epstein-Barr virus (EBV) or who are taking immunosuppressive drugs after organ transplantation are at greater risk for some type of lymphoma. There is no nationwide cancer registry in the United States, but in 1996 it was estimated tha there were 85,000 new cases of lymphoid malignancies, including HD-NHL, chronic lymphocytic leukemia (CLL), acute lymphoblastic leukemia (ALL), and multiple myeloma.

With the introduction of combination chemotherapy more than 20 years ago, it has become clear that NHL is a potentially curable malignancy. New chemotherapy dosing and administration schedules have validated previously described concepts of cell tactics and have improved early treatment results and allowed successful treatment of even refractory or relapsed disease. New drugs are currently under evaluation along with other promising modalities including biologic response modifiers, immunoconjugates, and bone marrow transplantation (BMT). The use of hematopoietic growth factors, in particular, has clearly infested the field, insisting dose intensity be explored and achieved (1). Use of all high-dose chemotherapy regimens is, however, associated with the risk of neutropenia-related morbidity and mortality.

II. PREVENTION OF NEUTROPENIA DURING CHEMOTHERAPY DOSE ESCALATION

Filgrastim significantly reduced the incidence of severe neutropenia in a randomized controlled trial of patients with high-grade NHL (1). Eighty patients received high-dose VAPEC-B chemotherapy (vincristine, Adriamycin®, prednisone, etoposide, cytotaxan, and bleomycin). Of these, 39 patients were randomized to receive only their scheduled weekly chemotherapy. The remaining 41 patients were randomized to receive both chemotherapy and daily doses of Filgrastim (230 μg/m^2). Patients in both groups also received prophylactic ketoconazole and cotrimoxazole throughout the study period, as well as identical dose modification and antibiotic treatment.

The incidence of severe neutropenia (absolute neutrophil count [ANC] $<1.0 \times 10^9$/L), in patients receiving Filgrastim was significantly reduced compared with patients in the control group (37% versus 85%; p = 0.00001) (**Figure 1**). Filgrastim treatment also was associated with a reduced incidence of neutropenic fever (22% versus 44%; p = 0.04). Fewer patients who received Filgrastim required delays in their chemotherapy schedule compared with the control group (10 versus 20 patients, respectively). Delays that did occur in the Filgrastim treatment group were of shorter duration compared with those experienced by patients in the control group (p = 0.01). In general, treatment delays in the control groups were required because of the development of neutropenia. Nineteen (49%) patients in the control groups required treatment delays because of neutropenia compared with only two (5%) patients receiving Filgrastim (p < 0.00001). Chemotherapy-associated toxicities forced dose reductions in 33% of control patients compared with only 10% of Filgrastim-treated patients (p = 0.01). Lastly, those patients who received Filgrastim were able to tolerate higher doses of chemotherapy than were control patients (median, 95% versus 83%); the protocol did not, however, permit maximum dose intensification, and further dose increases may have been possible. The use

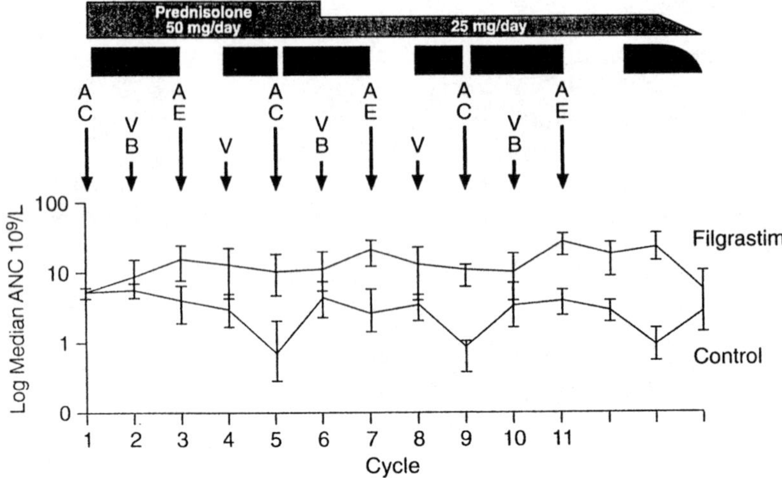

Figure 1. Filgrastim reduced the incidence of absolute neutrophil counts (ANC) $<1.0 \times 10^9$/L in NHL patients receiving weekly VAPEC-B chemotherapy. Group A received doxorubicin (A), 35 mg/m^2; cyclosphosphamide (C), 350 mg/m^2; vincristine (V) 1.4 mg/m^2; Group B received bleomycin (B), 10 mg/m^2; and group E received etoposide (E), 100 mg/m^2. Shaded areas represent the time of administration of Filgrastim (230 μg/m^2/day); Filgrastim was discontinued at week 13 or when the ANC reached 20×10^9/L (or whichever came first). (Adapted from Ref. 1.)

of intravenous (IV) antibiotics and the number of hospitalizations were comparable between the two treatment groups.

At the level of dose intensity used in this study, Filgrastim did not reduce patient mortality at a median follow-up time of 15 months, but Filgrastim was effective in reducing morbidity associated with chemotherapy-induced neutropenia.

Juliusson and Liliemark (2) showed that it was possible to use very high-dose CDE (cyclophosphamide, etoposide, and doxorubicin) as the primary therapy for high-risk NHL patients by also administering Filgrastim; additional stem cell support was not required. All patients received doxorubicin (90 mg/m^2) and Filgrastim (20 μg/kg/day). During the first course of chemotherapy, 18 patients received 1800 mg/m^2 cyclophosphamide and 450 mg/m^2 etoposide. These doses were subsequently escalated three times up to a maximum of 3900 mg/m^2 cyclophosphamide and 975 mg/m^2 etoposide. The median number of days that patients experienced ANC $<0.5 \times 10^9$/L and an untransfused platelet count $<20 \times 10^9$/L was 9 days and 16 days, respectively. The duration of fever (temperature $\geq 38.0°C$) was 5 days. Intravenous antibiotics were required for 10 days, while red cell and platelet transfusions were required for 1 day and 2 days, respectively. Six patients experienced complete remission from this first course of chemotherapy while an additional 11 patients experienced partial remissions. Toxicities experienced during this first course of chemotherapy were not related to dose level.

The 15 patients who received a second course of chemotherapy experienced fewer days with neutropenia (mean, 7.2 days), and IV antibiotics (mean, 6.3 days). The adverse events were similar to those seen with standard autologous stem cell transplantation programs; mucositis was the major non-hematologic toxicity.

III. HIGH-DOSE CHEMOTHERAPY WITH AUTOLOGOUS STEM CELL SUPPORT

A. Bone Marrow Transplantation

Continuous IV infusion of Filgrastim was administered to 54 patients with malignant lymphoma (HD or high-grade NHL) (3,4). Patients received high-dose chemotherapy and autologous bone marrow transplants (ABMT). Patients with HD received a CVB regimen (cyclophosphamide, etoposide, and carmustine [BCNU]). Patients with NHL received a BEAM regimen (carmustine [BCNU], etoposide, cytarabine, and melphan). Under both regimens, Filgrastim significantly accelerated neutrophil recovery (ANC $\geq 0.5 \times 10^9/L$) and decreased both the number and duration, of days with neutropenia. At doses of 10 $\mu g/kg/day$ and 30 $\mu g/kg/day$, Filgrastim induced neutrophil recovery after 12 days and 14 days, respectively, compared with 20 days in the group that did not receive Filgrastim (p < 0.004).

Filgrastim accelerated neutrophil recovery in a prospective, open-label, randomized trial treating patients with poor-risk NHL or relapsed HD with high-dose chemotherapy and ABMT; mean recovery time was 10 days in Filgrastim-treated compared with 18 days in control group (p $= 0.001$) (5). The high-dose chemotherapy regimen consisted of cyclophosphamide (total dose of 6 g/m^2), carmustine (BCNU) (300 mg/m^2), and etoposide (total of 900 mg/m^2 given in two equally divided doses). Patients receiving Filgrastim experienced fewer days with fever and neutropenic fever than did patients who did not receive Filgrastim (1 day versus 4 days [p $= 0.04$]; and 5 days versus 13.5 days [p $= 0.0001$] respectively). Lastly, patients in the Filgrastim-treatment group required IV antibiotics for fewer days and shorter hospital stays than patients in the control group.

B. Peripheral Blood Progenitor Cell Transplantation

Mobilized peripheral blood progenitor cell (PBPC) transplants are now used as an alternative source of stem cells (6–9). In combination with Filgrastim therapy, PBPC are becoming widely used as the only source of hematopoietic stem cells to support myeloblative therapy in advanced lymphoma (7,10–13).

Filgrastim treatment (200 μg/m^2/day) significantly increased PBPC numbers in patients with either intermediate or high-grade NHL who were receiving chemotherapy (11). Increases in the number of mobilized colony forming cells were observed, including mixed-CFC (CFC-MIX) (increased by 226-fold); granulocyte/macrophage-CFC (CFC-GM)(increased by 278-fold); and erythroid burst-forming cells (BFC-E) (increased by 29-fold). Only very modest increases in the same CFC populations (mean of 7- to 12-fold) were measured in PBPC recovered from patients who did not receive Filgrastim. Peripheral blood mononuclear cells (PBMC) were harvested from Filgrastim-treated patients either during peak PBPC induction or 2 to 4 days before. These cells effectively repopulated irradiated bone marrow stroma and restored hematopoiesis at least as effectively as bone marrow cells. In contrast, blood cells recovered before Filgrastim treatment had only a limited capacity to repopulate the bone marrow. This study showed that Filgrastim mobilized sufficient numbers of PBPC in patients with lymphoma to allow harvesting and successful engrafting without the possible complication of late hematopoietic failure.

A subsequent study by the same group (12,14) demonstrated that treatment with routine chemotherapy and Filgrastim, even in heavily pretreated patients, could mobilize enough PBPC for transplantation in a single leukapheresis. Fifty-four consecutive adult patients with either poor prognosis high-grade NHL (n = 31), HD in chemosensitive relapse (n = 12), or poor-prognosis ALL (n = 11) were included in this study. Filgrastim (300 μg/day) was administered either high-dose VAPEC-B or HiCCOM (includes methotrexate with folic acid rescue, cyclophosphamide, and cystosine arabinosine) chemotherapy to mobilize PBPC. Daily Filgrastim administration continued after PBPC autografting until an ANC $\geq 1 \times 10^9$/L was measured on two consecutive days. This protocol resulted in rapid and sustained hematopoietic engraftment in all patients. The median time to neutrophil recovery (ANC $\geq 0.5 \times 10^9$/L) was 9 days; and the median time to platelet counts of $\geq 20 \times 10^9$/L and $\geq 50 \times 10^9$/L was 10 days and 15.5 days, respectively. The recovery of neutrophils in this patient population was significantly faster than had been reported for a historical control group treated with ABMT and Filgrastim. Patients who received Filgrastim-mobilized PBPC were in the hospital for fewer days than were the historical control patients (13 days versus 19 days).

Hematologic recovery after high-dose chemotherapy in patients with high-risk NHL (n = 19) and HD (n = 4) was improved after autologous PBPC transplantation and Filgrastim treatment (10). In this study, PBPC were mobilized using cytotoxic chemotherapy before administration of Filgrastim (5 μg/kg/day). Hematopoietic reconstitution in these patients was compared with patients who received PBPC mobilized by high-dose chemotherapy in the absence of Filgrastim. Patients receiving Filgrastim-mobilized PBPC experienced a significantly faster time to neutrophil recovery (ANC > 0.5 × 10^9/L) compared with controls (median time of 10 days and 17 days respectively; p < 0.05). Recovery of self-sustaining platelet counts of >50 × 10^9/L was similar in patients receiving PBPC mobilized by either protocol (median, 17 days). A follow-up examination 6 months after PBPC transplantation revealed stable hematopoietic reconstitution.

Schmitz et al (7) also reported that infusion of Filgrastim-mobilized PBPC significantly accelerated neutrophil recovery in lymphoma patients after high-dose chemotherapy compared with patients who received ABMT (11 days versus 14 days, p = 0.005).

A novel high-dose sequential (HDS) chemotherapy program administered with Filgrastim was evaluated in patients with intermediate and high-grade NHL or with progressive HD (13). The protocol included the sequential administration of very high doses of single cytotoxic drugs (7 g/m^2 cyclophosphamide and 2 g/m^2 etoposide) with Filgrastim (5 μg/kg/day). This was followed by autografting of the Filgrastim-mobilized PBPC. Filgrastim reduced toxicities during the high-dose phase to acceptable levels and mobilized good yields of PBPC. In the final autografting phase, Filgrastim-mobilized PBPC were associated with low hematological toxicity. Overall, toxicity-related mortality in the 71 NHL patients treated with either the original or the intensified HDS chemotherapy program was 5.6%; these values are comparable to those associated with conventional chemotherapy. High-dose sequential therapy also was evaluated as first-line treatment in a series of 22 consecutive patients, presenting with advanced-stage, intermediate-grade NHL other than diffuse large cell subtype. It was associated with a complete remission rate of 82% and with a projected survival of 86% at 5 years.

High-dose chemotherapy followed by either PBPC grafting or ABMT has significantly improved survival of patients with relapsed HD or NHL. The more recent use of Filgrastim-mobilized PBPC has, at least in the short term, been shown to be the most effective regimen as well as representing a significant cost savings due to lower autograft collection costs, shorter hospital stays, and reduced requirement for supportive care (15).

IV. POTENTIAL BENEFITS OF FILGRASTIM IN ELDERLY LYMPHOMA PATIENTS

Survival of lymphoma patients decreases with age in part because elderly patients respond less well to chemotherapy than do younger patients (16,17). For example, patients who were age 60 years and older with immediate and high-grade lymphoma had a significantly poorer outcome than patients less than 60 years of age (16). Projected 5-year survival was only 36% for patients $\geq$60 years (n = 360) but was 50% for patients $\leq$60 years (n = 53) (p = 0.003).

The increased incidence of chemotherapy-related toxicities in elderly patients can often prevent the completion of standard chemotherapy regimens; elderly patients also may receive treatment regimens in which the drug doses are reduced or given less frequently, with the more severe toxicities leading to early discontinuation of chemotherapy. Consequently, the poor patient outcome of elderly lymphoma patients may, at least in part, result from the use of these lower doses at which growth of the malignancy is less well controlled (17). The inferior outcomes reported for elderly patients with advanced, diffuse large-cell (histocytic) lymphoma may be the consequence of the less intensive chemotherapy regimens tolerated by these patients. In elderly patients ($\geq$65 years) full doses of CHOP (cyclosphosphamide, doxorubicin, vincristine, and prednisone) were associated with a very high rate of febrile neutropenia (47%) and increased hospitalization rates, most of which were to treat neutropenic fevers (18). Elderly patients may, however, benefit from high-dose chemotherapy since it has been reported that older patients had similar survival rates to their younger counterparts if survival data are adjusted for age and performance status (19).

Since Filgrastim may reduce the need for dose reductions and dose delays in CHOP chemotherapy, it may have particular benefits for elderly patients. For example, Filgrastim adjunctive therapy may permit intensification of the CHOP regimen by either decreasing the cycle interval and/or increasing the doses of cyclophosphamide and doxorubicin (20). Preliminary data on Filgrastim adjunctive therapy has been reported from a small nonrandomized study of P-VEBEC (epirubicin, cyclophosphamide, eptoside, vinblastine, bleomycin, and prednisone) in 67 patients elderly patients ($\geq$65 years) with intermediate and high-grade advanced-stage NHL (21). The P-VEBEC regimen is an original chemotherapy scheme using 50 mg/m^2 epirubicin, 350 mg/m^2 cyclophosphamide, and 100 mg/m^2 etoposide (weeks 1, 3, 5, and 7, 5 mg/m^2); 5 mg/m^2 each of vinblastine and bleomycin (weeks 2, 4, 6, 8); and 50 mg/m^2 prednisone (daily in the first 2 weeks and thereafter every other day). Twenty-eight patients also received 5 μg/kg/

day Filgrastim throughout the P-VEBEC treatment starting on day 2 of every week for four consecutive days. Patients who received Filgrastim experienced both reduced mortality and a trend towards event-free survival compared with patients who did not receive Filgrastim. Patients who received Filgrastim also experienced reduced myelotoxicity compared with patients who did not receive any (89% versus 56%). Fewer patients who received Filgrastim experienced a neutrophil nadir of $<0.5 \times 10^9$/L compared with patients who did not receive Filgrastim (18% versus 56%). The rate of complete remission was influenced by the use of a relative dose intensity $>80\%$ (which may play a role in improving the outcome of patients with adverse prognostic factors such as bone marrow involvement and poor performance status). Patients who received Filgrastim received a significantly higher median relative dose intensity compared with patients not receiving Filgrastim (94% versus 79%). This study indicates that P-VEBEC can be an effective chemotherapy regimen for elderly patients because Filgrastim administration can improve the relative dose intensity received by this patient group. While interesting, additional studies are needed to confirm these conclusions and to determine if adjunctive therapy with Filgrastim has particular benefits for elderly patients with otherwise incurable lymphomas.

V. HUMAN IMMUNODEFICIENCY VIRUS–RELATED NHL

The role of high-dose chemotherapy in HIV-related NHL remains an area of investigation. In addition, the adverse side effects of high-dose chemotherapy pose particular problems for patients with HIV-related NHL. Two recent studies suggest that administration of Filgrastim can reduce the adverse side effects associated with high-dose chemotherapy to tolerable levels and, thus, can make high-dose chemotherapy a more effective therapy for these patients.

Newell et al (22) conducted a phase 1/2 trial to determine the maximum tolerated dose of CEOP (cyclophosphamide, epirubicin, vincristine, and prednisolone) for HIV-related NHL in chemotherapy-naive patients. Initially patients received Filgrastim (1.0 μg/kg/day) and antiretroviral (zidovudine [ZDV] and/or didanosine [ddI]) therapy. The dose of CEOP was subsequently introduced, with daily Filgrastim support (10 μg/kg on day 2 to day 14) and continued antiviral therapy, but it only was continued if the ANC levels remained $>1.2 \times 10^9$/L.

The maximum tolerated CEOP dosages were 500 mg/m^2 cyclophosphamide, 37.5 mg/m^2 epirubicin, 2 mg vincristine, and 75 mg/m^2 prednisolone. Higher doses of cyclophosphamide and epirubicin (750 mg/m^2 and

50 mg/m^2, respectively) were, however, administered to a second group of patients and, in combination with Filgrastim and antiretroviral therapy, were associated with encouraging responses and acceptable toxicities. The median survival of patients receiving either chemotherapy regimen was 17 months. The high-dose regimen, in combination with Filgrastim, therefore, warrants further study.

A pilot study of 25 patients with HIV-related NHL suggested that the CDE chemotherapy regimen, in combination with Filgrastim, was tolerable and effective for this patient population (23). The feasibility of combination therapy with ddl therapy also was demonstrated. Complete response occurred in 58% of patients, the median duration of which exceeded 18 months; tumor-related mortality was 20% and median survival was 18.4 months.

VI. CONCLUSIONS

Filgrastim, which has the potential to accelerate recovery from the myelosuppressive effects of chemotherapy, has proven to be an extremely beneficial adjunctive therapy in the treatment of lymphomas where it has been used successfully in various chemotherapy dose escalation regimens. As in other malignancies, Filgrastim also can be used with high-dose chemotherapy to optimize PBPC mobilization in patients with lymphoma for subsequent autologous stem-cell support after bone marrow ablation. Lastly, Filgrastim may have specific benefits for elderly patients with lymphomas also receiving high-dose chemotherapy and in the treatment of AIDS-related NHL.

REFERENCES

1. Pettengell, R., Gurney, H., Radford, J. A., et al (1992). Granulocyte colony-stimulating factor to prevent dose-limiting neutropenia in non-Hodgkin's lymphoma: a randomized controlled trial. *Blood 80:*1430–1436.
2. Juliusson, G., and Liliemark, J. (1996). Dose escalation of high-dose cyclophosphamide and etoposide with high-dose doxorubicin (CDE) and filgrastim for poor-risk non-Hodgkin's lymphoma. *Ann Oncol 7:*1037–1041.
3. Schmitz, N., Dreger, P., Zander, A., et al. (1993). A randomized, controlled, multicentre study of granulocyte colony stimulating factor (Filgrastim) in patients with Hodgkin's disease and non-Hodgkin's lymphoma undergoing autologous bone marrow transplantation. *Blood 82:*146a (abstr 568).
4. Schmitz, N., Dreger, P., Zander, A. R., et al. (1995). Results of a randomised, controlled, multicentre study of recombinant human granulocyte colony-

stimulating factor (Filgrastim) in patients with Hodgkin's disease and non-Hodgkin's lymphoma undergoing autologous bone marrow transplantation. *Bone Marrow Transplant 15:*261–266.

5. Stahel, R. A., Jost, L. M., Cerny, T. et al. (1994). Randomized study of recombinant human granulocyte colony-stimulating factor after high-dose chemotherapy and autologous bone marrow transplantation for high-risk lymphoid malignancies. *J Clin Oncol 12:*1931–1938.

6. Bensinger, W. I., Clift, R. A., Anasetti, C., et al (1996). Transplantation of allogeneic peripheral blood stem cells mobilized by recombinant human granulocyte colony-stimulating factor. *Stem Cells 14:*90–105.

7. Schmitz, N., Linch, D. C., Dreger, P., et al. (1996). Randomised trial of filgrastim-mobilized peripheral blood progenitor cell transplantation versus autologous bone-marrow transplantation in lymphoma patients. *Lancet 347:*353–357.

8. Sheridan, W. P., Lo, L. B., Brown, S. L., Grigg, A., Foote, M. A., and Juttner, C. A. (1996). Clinical use of rHuG-CSF (Filgrastim)-mobilised peripheral blood progenitor cells for transplantation. *J Drug Dev Clin Pract 8:*7–17.

9. Welte, K., Gabrilove, J., Bronchud, M. H., Platzer, E., and Morstyn, G. (1996). Filgrastim (r-metHuG-CSF): the first 10 years. *Blood 88:*1907–1929.

10. Brice, P., Divine, M., Marolleau, J. P., et al. (1994). Comparison of autografting using mobilized peripheral blood stem cells with and without granulocyte colony-stimulating factor in malignant lymphomas. *Bone Marrow Transplant 14:*51–55.

11. Demuynck H., Pettengell, R., de Campos, E., Dexter, T. M., and Testa, N. G. (1992). The capacity of peripheral blood stem cells mobilised with chemotherapy plus G-CSF to repopulate irradiated marrow stroma in vitro is similar to that of bone marrow. *Eur J Cancer 28:*381–386.

12. Pettengell, R., Morgenstern, G. R., Woll, P. J., et al (1993). Peripheral blood progenitor cell transplantation in lymphoma and leukemia using a single apheresis. *Blood 82:*3770–3777.

13. Tarella, C., Gavarotti, P., Caracciolo, D., et al (1995). Haematological support of high-dose sequential chemotherapy: clinical evidence for reduction of toxicity and high response rates in poor risk lymphomas. *Ann Oncol 6* S4:3–8.

14. Pettengell, R., Demuynck, H., Testa, N. G., and Dexter, T. M. (1992). The engraftment capacity of peripheral blood progenitor cells (PBPC) mobilised with chemotherapy elevision G-CSF. *Int J Cell Cloning 10:*59–62.

15. Smith, T. J., Hillner, B. E., Schmitz, N., et al (1997). Economic analysis of a randomized clinical trial to compare filgrastim-mobilized peripheral-blood progenitor-cell transplantation and autologous bone marrow transplantation in patients with Hodgkin's and non-Hodgkin's lymphoma. *J Clin Oncol 15:*5–10.

16. Gaynor, E. R., Dalberg, S., Fisher, R. J. (1994). Factors affecting reduced survival of the elderly with intermediate and high grade lymphoma: an analysis of SWOG-8516 (INT 0067)—the national high priority lymphoma study—a randomized comparison of CHOP vs. m-BACOD vs. ProMACE-CytaBOM vs. MACOP-B. *Proc Am Soc Clin Oncol 13:*370 (abstr 1250).

17. Dixon, D. O., Neilan B., Jones S. E., et al (1986). Effect of age on therapeutic outcome in advanced diffuse histiocytic lymphoma: the Southwest Oncology Group experience. *J Clin Oncol 4*:295–305.
18. Meyer, R. M., Browman G. P., Samosh, M. L., et al (1995). Randomized phase II comparison of standard CHOP with weekly CHOP in elderly patients with non-Hodgkin's lymphoma. *J Clin Oncol 13*:2386–2393.
19. Dahlberg, S., Miller, T. P., Dana, B., et al (1990). Dose intensity is not associated with subsequent survival after adjustment for known prognostic factors in non-Hodgkin's lymphoma patients treated with m-BACOD, ProMACE-CytaBOM and MACOP—B on Southwest Oncology Group studies. *Proc Am Soc Clin Oncol 986*:285 (abstr).
20. Blayney, D. W., Williams, S., Mortimer, J., et al. (1992). Neupogen maintenance (r-metHuG-CSF) ameliorates neutropenia during CHOP chemotherapy. *Proc Am Soc Clin Oncol 11*:320 (abstr 1088).
21. Bertini, M., Freilone, R., Vitolo, U., et al. (1994). P-VEBEC: a new 8-weekly schedule with or without rG-CSF for elderly patients with aggressive non-Hodgkin's lymphoma (NHL). *Ann Oncol 5*:895–900.
22. Newell, M., Goldstein, D., Milliken, S., et al (1996). Phase I/II trial of filgrastim (r-metHuG-CSF), CEOP chemotherapy and antiretroviral therapy in HIV-related non-Hodgkin's lymphoma. *Ann Oncol 7*:1029–1036.
23. Sparano, J. A., Wiernik, P. H., Hu, X., et al (1996). Pilot trial of infusional cyclophosphamide, doxorubicin, and etoposide plus didanosine and filgrastim in patients with human immunodeficiency virus-associated non-Hodgkin's lymphoma. *J Clin Oncol 14*:3026–3035.

11
Using Filgrastim (r-metHuG-CSF) in the Treatment of Breast Cancer

Irene Kuter
Massachusetts General Hospital Cancer Center,
Massachusetts General Hospital, Boston, Massachusetts

I. INTRODUCTION

Breast cancer is the most common malignancy affecting American women and is the leading cause of cancer-related deaths in women 34 to 54 years of age. An estimated 180,200 American women will have been diagnosed with breast cancer in 1997, and 43,900 will die of it. There is impressive variation in the incidence of breast cancer around the world. The rate is highest in "Westernized" countries such as Europe, North America, Australia, and New Zealand, where approximately 11% of women will develop the disease. In contrast, very low rates are seen in other countries such as China and Japan where the rates are approximately 20% of those observed in the United States. Intriguingly, the variation in incidence from country to country does not seem to be due to racial differences in genetic susceptibility: when individuals move from a low-risk society to a higher-risk one, their personal risk of breast cancer and the risk of their descendants rapidly increases to equal that of the new society in which they live. Observations such as these suggest that diet and/or environmental factors affect risk and offer hope that one day we will understand the etiology of breast cancer and be in a position to modify that risk. Of the other factors that influence the risk of breast cancer, the most important are family history, reproductive history, history of proliferative breast disease, and exposure to known carcinogens.

Because breast cancer is a common disease in Westernized societies, it is not uncommon to observe more than one case of the disease in a family. However, an inherited susceptibility to breast cancer is rare: only 5% to 10% of all breast cancers in the general population are thought to be due to the inheritance of a susceptibility gene (1). Chance clustering, common exposures, and polygenic inheritance are other possible reasons for other familial clustering of cases. Women with a family history of breast cancer in first-degree relatives, especially if the disease is bilateral and/or occurs in premenopausal women, are at increased risk of developing the disease. Susceptibility to breast cancer that has a genetic basis is due to the inheritance of mutations in certain critical genes. The most important of these are the BRCA1 and BRCA2 genes, mutations in which account for 80% of hereditary cases. Less common but also strongly affecting susceptibility to breast cancer are mutations in the *p53* gene found in families with the Li-Fraumeni syndrome (which confers susceptibility also to sarcomas, brain tumors, leukemia, and adrenal carcinoma) or in a recently described gene associated with Cowden's syndrome (carriers are affected by facial trichilemmomas, colonic polyps and other benign growths and may also develop thyroid or colon carcinomas). Muir's syndrome, also rare, carries with it the risk of basal cell and gastrointestinal carcinomas as well as breast cancer. In some families with Lynch syndrome type II, in which there is an inherited susceptibility to colon cancer as well as other epithelial malignancies as a result of a defective DNA mismatch repair gene, breast cancer appears to be one of the cancers to which these individuals are susceptible. Finally, carriers of a mutation in the ataxia telangiectasia gene may have a higher rate of breast cancer due to an increased spontaneous mutation rate.

Reproductive history also influences a woman's risk of breast cancer. The age at which a woman bears her first child is important. A woman who has a full-term pregnancy before the age of 18 years has one-third the risk of breast cancer as a woman who has her first child after age 30. Early menarche, late menopause, and nulliparity all increase risk. Breastfeeding has a protective effect.

It has been said that a history of previous breast biopsies increases breast cancer risk. In fact the majority of benign breast abnormalities such as fibrocystic "disease" and fibroadenomas do not confer an increased risk of breast cancer. However, women who have a diagnosis of epithelial hyperplasia do have an increased risk of breast cancer. When the hyperplasia is atypical, the risk is significantly higher. Such women are best followed in centers that can treat high-risk breast cancer.

Finally, exposure to radiation can significantly increase breast cancer risk. The age at which exposure occurred is very important. Women who were exposed to extensive fluoroscopy because of tuberculosis before age 25 years have an increased risk, whereas those exposed after that age do not. Women who had mediastinal radiation for Hodgkin's disease before age 25 years likewise have a very high risk of breast cancer beginning 8 years later. Again, such women need to be monitored carefully by physical examination and mammography.

The majority of breast cancers arise from the cells lining the ducts of the glandular tissue of the breast. A minority arise from the lobules situated at the ends of the ducts. Ductal carcinoma in situ is a preinvasive form of breast cancer, completely curable with local treatment alone, and is the precursor lesion from which invasive breast cancer arises. It is the current recommendation that all women over age 40 years have annual mammograms in an attempt to diagnose breast cancers at the earliest possible time, since the stage at diagnosis is the most important predictor of prognosis and curability. Ductal carcinoma in situ will often give rise to microcalcifications in the breast tissue even before a mass appears, and mammograms can detect these microcalcifications.

II. STAGING OF BREAST CANCER

The most common presenting sign of breast cancer is a painless lump in the breast. With improved mammographic screening, more and more cancers are being detected before they are palpable. Unfortunately, invasive breast cancer usually does not remain localized in the breast for very long. Even while the tumor is small, breast cancer cells gain access to the lymphatics or blood vessels and can disseminate to other parts of the body. The most common sites of metastases are the regional lymph nodes, bone, liver, lung, and brain. At the time of diagnosis, there is evidence for the spread of breast cancer to the axillary lymph nodes in approximately 40% to 50% of patients. Signs and symptoms of metastatic disease are seldom present at the time of diagnosis, and the few patients who present with metastatic disease have usually had a breast mass for a relatively prolonged period of time. However, with time, metastases can appear in the distant organs as microscopic deposits and grow into detectable tumor masses. The likelihood of distant metastases, usually leading to death, is related to various prognostic factors (see below). The extent to which the cancer has spread is termed the stage of the cancer. The stage at diagnosis is the most important prognostic factor. The TNM (Tumor, Node, Metastasis) classification is based on the characteristics (predominantly size) of the primary tumor, the degree of involvement of the regional (especially ipsilateral axillary) lymph nodes, and the presence or absence of detectable metastases (**Table 1**) (2). Cancers which are at stages I, IIA, IIB, and IIIA are generally considered operable. It is often difficult to distinguish inflammatory cancers (T4d) from those that are locally advanced (T4c).

Table 1. TNM Classification.

T-Tumor

T1			Tumor 2 cm or less in greatest dimension					
			T1a 0.5 cm or less in greatest dimension					
			T1b More than 0.5 cm but not more than 1 cm in greatest dimension					
			T1c More than 1 cm but not more than 2 cm in greatest dimension					
T2			Tumor more than 2 cm but not more than 5 cm in greatest dimension					
T3			Tumor more than 5 cm in greatest dimension					
T4			Tumor of any size with direct extension to chest wall or skin					

N-Regional lymph nodes

NX	Regional lymph nodes cannot be assessed (eg, previously removed)
N0	No regional lymph node metastasis
N1	Metastasis to movable ipsilateral axillary node(s)
N2	Metastasis to ipsilateral axillary node(s) fixed to one another or to other structures
N3	Metastasis to ipsilateral internal mammary lymph nodes

M-Distant metastasis

MX	Presence of distant metastasis cannot be assessed
M0	No distant metastasis
M1	Distant metastasis (includes metastasis to supraclavicular lymph nodes)

Stage Grouping

Stage 1	T1	N0	M0	Stage IIIA	T0	N2	M0
Stage IIA	T0	N1	M0		T1	N2	M0
	T1	N1	M0		T2	N2	M0
	T2	N0	M0		T3	N1,N2	M0
Stage IIB	T2	N1	M0	Stage IIIB	T4	Any N	M0
	T3	N0	M0		Any T	N3	M0
				Stage IV	Any T	Any N	M1

Source: Adapted from Ref. 2.

Traditionally, the diagnosis of inflammatory breast cancer is made on the basis of the clinical presentation of erythema and swelling (peau d'orange) covering at least one-third of the breast. There should be histological evidence of tumor invasion of dermal lymphatics (3).

III. PROGNOSTIC FACTORS

The most important prognostic factor for patients with early-stage breast cancer (primary breast cancer which is not locally advanced and which is without evidence of distant spread) is the extent of axillary lymph node involvement. Other important prognostic factors are the size of the primary tumor, the histologic subtype, the histologic grade (degree of differentiation), and the estrogen and progesterone receptor status of the tumor (2) (**Table 2**). Other prognostic factors include age at diagnosis (premenopausal patients are more likely to have a more aggressive form of the disease), the presence of lymphatic or blood vessel invasion, the mitotic index, and the overexpression of the *her-2* oncogene protein. Many other prognostic factors have been proposed, but are less useful in clinical practice. These include ploidy, cathepsin D expression, neovascularization index, and others.

Prognosis is inversely related to the number of lymph nodes involved at the time of diagnosis (**Figure 1**). On average, the 40% to 50% of patients who present with involved axillary lymph nodes have only about a 50% chance of surviving 10 years. Patients who present with four or more positive axillary nodes have a worse prognosis, and those with 10 or more involved nodes have less than a 20% chance of surviving 10 years. Women with negative axillary lymph nodes on average have a 60% to 70% chance of surviving 10 years. For these women with negative lymph nodes, the other prognostic factors become more important in assessing the risk of recurrence.

Table 2. Prognosis and Stage.

Stage	5-year Survival (%)	10-year Survival (%)
I	80	65
II	60	45
III	35–50	20–40
IV	10	5

Source: Adapted from Ref. 2.

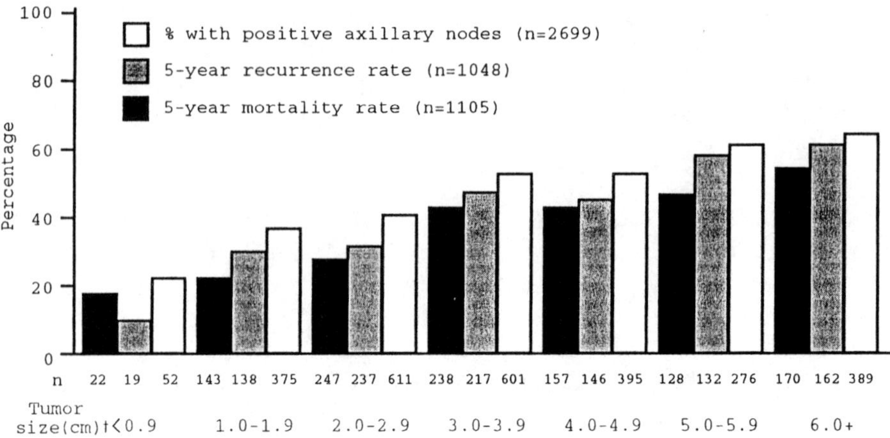

Figure 1. Relation between tumor size and axillary node involvement and recurrence and mortality rates. (Adapted from Ref. 4.)

IV. PRIMARY TREATMENT OF LOCALIZED BREAST CANCER

At one time the standard primary treatment of localized breast cancer was radical mastectomy. Later, modified radical mastectomy became the standard surgical procedure. Today breast conservation has largely replaced the mastectomy as the treatment of choice in most cases. When the breast is conserved, the primary tumor is removed with a rim of normal tissue in an attempt to achieve clean margins. After the surgery, radiation is given to eliminate residual malignant cells. When lumpectomy and radiation are both used, control of the primary tumor in the breast is satisfactory, with only about a 5% to 10% risk of recurrence in most cases. Lumpectomy with radiation is equivalent to mastectomy in terms of 10-year survival, because whether a woman lives or dies from breast cancer depends not on local control of the tumor but whether or not microscopic cells have already disseminated to distant sites before removal of the cancer. Approximately 30% of women with negative axillary lymph nodes and 60% of women with positive axillary lymph nodes at the initial staging will have a recurrence of the disease within 10 years regardless of the local therapy used. Because of this, the use of adjuvant medical therapy has become standard in the treatment of breast cancer.

V. ADJUVANT MEDICAL THERAPY

Adjuvant medical therapy can be either chemotherapy or hormonal therapy. It is usually administered after primary curative surgery and before radiation for patients who have chosen breast conservation. It is given to patients with no persisting evidence of disease, but who have a significant likelihood of recurrence. Most initial trials of adjuvant chemotherapy and hormonal therapy were done with women who had positive lymph nodes, but more recently studies have extended to women with negative lymph nodes, and it has been shown that these women benefit as well. Premenopausal women benefit more from chemotherapy than from hormonal therapy. Chemotherapy reduces recurrence risk by about 35%, whereas tamoxifen, an anti-estrogen, reduces recurrence risk by only about 12%. The reverse is true for older women: postmenopausal women benefit more from tamoxifen than from chemotherapy. The anti-estrogen reduces their recurrence risk by about 30% whereas chemotherapy reduces it by only about 20%. Hormonal therapy works better for women whose tumors carry estrogen receptors. Whether or not the combination of chemotherapy and hormonal therapy is better than chemotherapy alone in premenopausal women or better than hormonal therapy alone in postmenopausal women is controversial, although data are beginning to accumulate to support a small additive effect when both are used. The roles of oophorectomy and medical ovarian ablation are still being evaluated for premenopausal women.

VI. CHEMOTHERAPY FOR METASTATIC BREAST CANCER

Breast cancer was one of the first solid tumors to be treated with chemotherapy (5). The treatment of metastatic breast cancer has improved greatly in recent years, but there have been no objective improvements in survival in over two decades. Although good palliation and objective response rates are consistently achieved, no cures have been obtained with the standard chemotherapy regimens. If the metastatic disease has estrogen receptors, hormonal therapy is usually tried first. When hormonal therapy fails, or if the cancer is estrogen-receptor negative, chemotherapy becomes the treatment of choice. Typically, first-line conventional chemotherapy regimens can achieve objective responses in the 65% to 75% range with complete responses in the 5% to 10% range. Duration of response is usually only a few months if treatment is stopped, and median survival from initiation of chemotherapy is 18 months to 2 years. Given as second-line treatment, the rate of response is approximately half that of first-line therapy. As third-line treatment, chemotherapy typically gives inferior responses with shorter duration.

VII. DOSE INTENSIFICATION

Some early study results indicated that increasing the dose intensity of chemotherapy may result in added benefit. It has also been shown that reducing the dose or prolonging the time period between cycles is associated with reduced efficacy. (See Chapter 8 for more discussion of dose intensification.)

Since doxorubicin is considered one of the most active single agents used in treating breast cancer, and since a known dose-response relationship exists, a study was conducted to determine if it was possible to give high doses of doxorubicin every 2 weeks with the support of Filgrastim (6). Twenty-one patients entered the study, 19 with recurrent metastatic breast cancer resistant to endocrine therapy and 2 with recurrent ovarian carcinoma after chemotherapy. Seventeen of the 19 breast cancer patients had received previous radiotherapy. The median age was 53 years (range, 30 to 67). Four patients with breast cancer, two of whom received previous radiotherapy, served as the control group.

Fifteen patients received doxorubicin as a single dose of 75, 100, or 125 mg/m^2 followed by Filgrastim therapy. The cycle was repeated every 2 weeks if the patients' absolute neutrophil count (ANC) was >2.5 × 10^9/L at that time. Four control patients followed the same schedule, but received doxorubicin 75 mg/m^2 without Filgrastim. Filgrastim was administered 24 hours after chemotherapy as a continuous infusion at 10 μg/kg/day from day 1 to day 8, and 5 μg/kg/day from day 8 to day 11 of each cycle. Hematologic toxicity and recovery were compared between the Filgrastim and control groups. Absolute neutrophil counts were measured three times per week during the study. The ANC increased to normal or above normal levels by day 12 to day 14 at all dose levels of doxorubicin supported by Filgrastim. An ANC of >2.5 × 10^9/L was not reached until day 19 to 21 in the control group. The time to nadir was sooner in the Filgrastim group compared with the control groups (day 7 versus days 12 to 13). The depth of neutropenic nadir and duration of neutropenia were increased for the two higher dose levels of doxorubicin compared with the two lower dose levels whether Filgrastim was administered or not.

The nine patients who received the lower doses of doxorubicin (75 and 100 mg/m^2) were compared with the eight patients who received the higher doses (125 mg and 150 mg/m^2) in terms of tumor response and nonhematological toxicities. Of the patients with breast cancer in the study receiving <125 mg/m^2 doxorubicin, 11% achieved a complete response compared with 50%, receiving 125 or 150 mg/m^2 doxorubicin. The difference was not statistically significant. The median time to progression was 6 months.

When the nonhematological toxicities were compared, those receiving <125 mg/m^2 doxorubicin were noted to have markedly less toxicity than those receiving ≥125 mg/m^2. All patients developed mucositis, but the difference in severity between the two groups was statistically significant (p = 0.002), as was the difference in other epithelial toxicities such as erythema of the palms and soles (p = 0.005). No patient developed clinical cardiotoxicity, although two patients who received a cumulative dose of 375 mg/m^2 decreased their left ventricular ejection fraction by 15%. No toxicities attributable to Filgrastim were reported.

Patients treated with chemotherapy and Filgrastim reached normal or above-normal postnadir ANC in 12 to 14 days after doxorubicin therapy, compared with 19 to 21 days in those patients not receiving Filgrastim (**Figure 2**). This rapid ANC recovery allowed for a decrease in the dosing interval and a dose escalation of doxorubicin which resulted in 2.2- to 4.5- fold dose intensity. An overall response rate of 80% with a median time to progression of 6 months was achieved across all dose levels of doxorubicin with Filgrastim support. Filgrastim was well tolerated.

A preliminary study by Hudis et al (7) demonstrated the feasibility of adjuvant dose-intensive chemotherapy with Filgrastim support after treatment with doxorubicin in patients with high-risk stage II/III resectable breast cancer. This program delivered a projected dose intensity for doxorubicin of 16.7 mg/m^2/week and for cyclophosphamide of 400 mg/m^2/week. The group has also completed a pilot study in which post-surgical patients with node-positive primary breast cancer received three cycles of dose-intense doxorubicin followed by three of dose-intense paclitaxel and three of dose-intense cyclophosphamide, all with Filgrastim support (8).

Demetri et at (9) used an intermittent CAF (cyclophosphamide, Adriamycin®, 5-FU [5-fluorouracil]) regimen with Filgrastim given between courses to increase the delivered dose-intensity of cyclophosphamide from 248 to 471 mg/m^2/week. Hospitalization was required for neutropenic fever in 65% of patients. Ayash et al (10) used an alternative approach to increasing dose-intensity which involves concurrent administration of Filgrastim with a continuous CAF regimen. In the study, 215 patients received 5-FU 300 mg/m^2/week intravenously (IV), Adriamycin® 300 mg/m^2/week, and cyclophosphamide 60 mg/m^2/day. In comparison with standard CAF programs, the addition of Filgrastim (5 μg/kg, days 2 to 7 each week) resulted

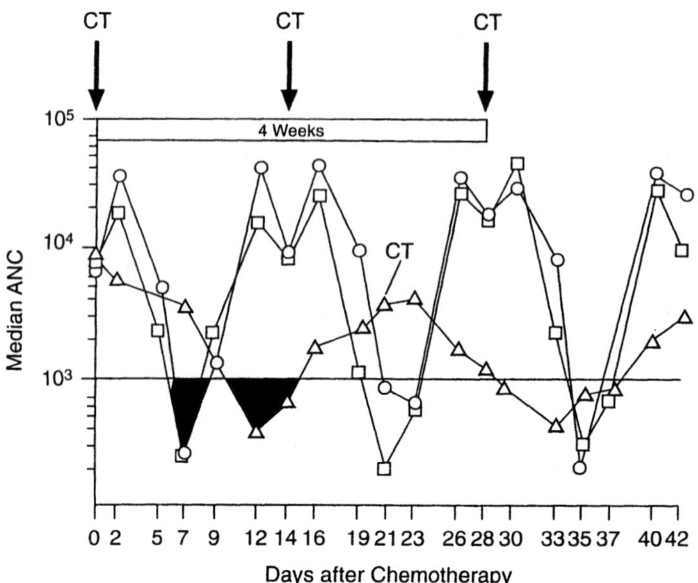

Figure 2.   Hematologic toxicity and recovery were compared between the Filgrastim and control groups. Absolute neutrophil counts (ANC) were measured three times per week during the study. The ANC increased to normal or above normal levels by day 12 to day 14 at all dose levels of doxorubicin supported by Filgrastim. An ANC >2.5 × 10^9/L was not reached until day 19 to day 21 in the control group. The time to nadir was sooner in the Filgrastim group compared with the control groups (day 7 versus days 12 to 13). CT = Chemotherapy; open circles, Filgrastim and doxorubicin 75 mg/m^2; open squares, Filgrastim and doxorubicin 100 mg/m^2; open triangles, doxorubicin 75 mg/m^2.

in an increase in delivered dose intensity of 75% (RDI = 1.76) assuming equal drug weighting.

VIII.   NEW CHEMOTHERAPEUTIC AGENTS

In recent years, studies investigating paclitaxel and docetaxel, two taxanes, have reported encouraging results for the management of breast cancer. Paclitaxel has shown similar activity in both anthracycline-sensitive and anthracycline-resistant patients, demonstrating non-cross–resistance (11–14). Paclitaxel has been shown also to have activity in patients with refractory breast cancer with responses seen at all sites of metastatic disease

(**Table 3**). Chemotherapy-naive patients appear to have a better response rate (62%) than do previously treated patients (20% to 50%). The optimal dose and schedule of paclitaxel continues to be evaluated (16). Dose-limiting toxicities of paclitaxel include neutropenia, neutropenic fever, mucositis, and nausea/vomiting (16,17).

Reichman et al (18) were the among the first to take advantage of Filgrastim's neutrophil proliferative effects to improve the therapeutic potential of the taxanes, and they administered paclitaxel (250 mg/m^2) with Filgrastim support to women who had or had not previously received adjuvant chemotherapy. Sixteen patients in this study achieved an overall response (2 complete, 14 partial) for an overall response rate of 62%. Responses occurred in all sites of metastasis and in cancers refractory to antracyclines.

Wasserheit et al (19) treated 27 patients with metastatic breast cancer who had had minimal prior chemotherapy with paclitaxel 200 mg/m^2 by 24-hour infusion. Cisplatin at 75 mg/m^2 was given on day 2 and Filgrastim was begun on day 3. Twenty-one patients were available for efficacy: the overall response rate was 52%, with 2 patients obtaining complete responses and 9 patients obtaining partial responses, with a median response duration of 3 months. The main dose-limiting toxicities were myelosuppression and cumulative neurotoxicity.

The optimal scheduling of paclitaxel has been studied in a number of trials, including one of women with untreated metastatic breast cancer (20). This study compared the use of paclitaxel given as a 3-hour or a 24-hour infusion (250 mg/m^2) with Filgrastim support.

Docetaxel also has significant activity in breast cancer (14,21–25). High response rates have been seen in metastatic visceral sites, especially the liver. Like paclitaxel, docetaxel has shown activity in anthracycline-resistant patients, demonstrating non-cross–resistance of breast cancer to these two drugs. Myelosuppression and fluid retention are the dose-limiting toxicities associated with docetaxel.

IX. INTENSIVE CHEMOTHERAPY WITH STEM CELL SUPPORT

A number of studies have suggested that dose-intensive chemotherapy with autologous stem cell support can produce high overall response rates in patients with metastatic breast cancer (26–32) (see Chapter 12 for more information concerning bone marrow transplantation and Chapters 13 and 14 for more information on peripheral blood progenitor cells). In the first randomized study to show a survival benefit for high-dose versus conven-

Table 3. Paclitaxel in Advanced Breast Cancer.

Reference	Dose (mg/m^2)	# of patients	Complete	Partial	CR + PR (%)	Toxicity	Study population	Median response duration
11	200–250 CIV 24-hour	25	3	11	56	Myelosuppression	Second-line	9 months
11	135–150 CIV 24-hour	21		6	33	Myelosuppression Stomatitis	Heavily pretreated ≥3 regimens	NA
12	135 vs 175 3-hour	113	3	27	27**	Neutropenia	Second- or third-line	NA
13	175 3-hour	15	3	4	47	Neutropenia Alopecia	Doxorubicin-refractory Anthracycline-refractory	7 months
14	250 3-hour	25		4	23.5	Myelosuppression	First-line ≥2 regimens	4 months
	175 3-hour	24		5	20.8			

* Randomized trial, overall response rate of both arms.
** No difference between two doses.
† Includes six doxorubicin-resistant patients.
CIV = Continuous intravenous infusion; NA = not available.
Source: Adapted from Ref. 15.

tional chemotherapy, Bezwoda et al (33) compared high-dose CNVp chemotherapy (cyclophosphamide 2.4 g/m^2, mitoxantrone 35 to 45 mg/m^2, etoposide 2.5 g/m^2) supported by autologous stem cells with standard CNV chemotherapy (cyclophosphamide 600 mg/m^2, mitoxantrone 12 mg/m^2, vincristine 1.4 mg/m^2) as first-line treatment in 90 patients with metastatic breast cancer. In the high-dose cohort, stem cell support consisted of autologous bone marrow transplant (ABMT) (nine patients) and autologous peripheral blood progenitor cells (PBPC) (14 mobilized by cyclophosphamide and 22 mobilized by cyclophosphamide and Filgrastim). The nine patients supported by ABMT alone received only one course of high-dose therapy, while those given PBPC received the scheduled two courses. The hematopoietic support permitted an increase in the calculated dose intensity index, compared with the conventional arm, of 3.8 times for cyclophosphamide and double for mitoxantrone. The overall response rate for the high-dose regimen was significantly greater compared with the standard-dose group (95% versus 53%, respectively), with complete responses seen in 23/45 and

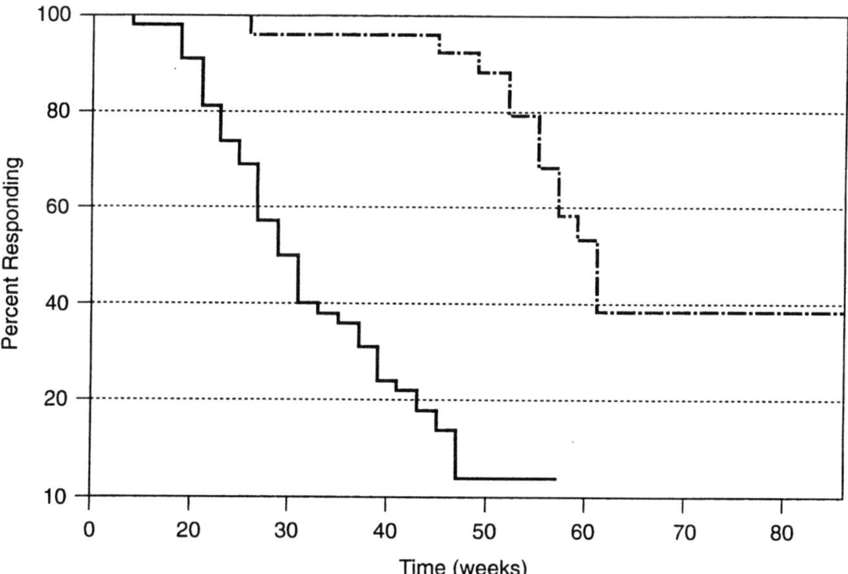

Figure 3. Duration of complete response and partial response after high-dose chemotherapy plus stem cell support in patients with metastatic breast cancer. Twenty-two patients had stem cells mobilized with Filgrastim in addition to chemotherapy. Solid line, partial response; dotted line, complete response. (Adapted from Ref. 33.)

2/45 patients, respectively. Both duration of response and survival were also significantly longer for the high-dose cohort (medians of 80 weeks versus 34 weeks and 90 weeks versus 45 weeks, respectively) (**Figure 3**). In summary, the high-dose therapy regimen produced encouraging results in terms of overall response and remission duration.

X. CONCLUSION

There is an increasing body of evidence to support a direct relationship between dose intensity and efficacy of chemotherapy for patients with high-risk or metastatic breast cancer. While it is known that breast cancer is usually sensitive to chemotherapy, the therapeutic benefit of chemotherapy dose intensification is yet to be verified. Available data, however, indicate that reduction in dose intensity is associated with reduced response and poorer prognosis for selected patients with breast cancer. The ability of Filgrastim to alleviate chemotherapy-induced neutropenia and mobilize PBPC should prove to be useful in allowing the maintenance of optimal chemotherapy intensity in situations where previously dose reduction and/ or dose delay would be necessary.

REFERENCES

1. Lindblom, A. (1995). Familial breast cancer and genes involved in breast carcinogenesis. *Breast Cancer Res Treat 34*:171–183.
2. Hermanek P., and Sobin, L. H. In: Hermanek P. and Sobin L. H., eds. *TNM Classification of Malignant Tumors. UICC International Union Against Cancer,* 4th ed. Berlin: Springer-Verlag, 1987: pp. 93–99.
3. Ellis, D. L., and Teitelbaum, S. L. (1974). Inflammatory carcinoma of the breast. A pathologic definition. *Cancer 33*:1045–1047.
4. Fisher, B. R. (1986). Prognostic and therapeutic significance of pathological features of breast cancer. *Natl Cancer Inst Monograph 1*:29–34.
5. Canellos, G. P., Pocock, S. J., Sears, M. E., et al (1976). Combination chemotherapy for metastatic breast carcinoma. Prospective comparison of multiple drug therapy with L-phenylalanine mustard. *Cancer 38*:1882–1886.
6. Bronchud, M. H., Howell, A., Crowther, D., Hopwood, P., Souza, L., and Dexter, T. M. (1989). The use of granulocyte colony-stimulating factor to increase the intensity of treatment with doxorubicin in patients with advanced breast and ovarian cancer. *Br J Cancer 60*:121–125.
7. Hudis, C., Seidman, A., Baselga, J., et al (1994). Sequential high dose adjuvant doxorubicin (A), paclitaxel (T), and cyclophosphamide (C) with G-CSF (G) is feasible for women (pts) with resected breast cancer (BC) and ≥ 4 (+) lymph nodes (LN). *Proc Am Soc Clin Oncol 13*:65 (abstr 62).

8.   Livingston, R. Adjuvant breast cancer; dose intensity concepts. In: *Filgrastim: Dose Concepts in the Treatment of Breast Cancer in the Adjuvant and Metastatic Settings.* Satellite symposium to the 5th International Conference on the Adjuvant Therapy of Primary Breast Cancer, St Gallen, Switzerland, March 2, 1995, Amgen-Rochen.

9.   Demetri, G. D., Younger, J., Shapiro, C., et al (1992). A phase I study of dose intensified CAF chemotherapy with adjunctive r-metHuG-CSF in patients with advanced breast cancer. *Proc Am Soc Clin Oncol 11:*108 (abstr 255).

10. Ayash, L. J., Elias, A., Wheeler, C., et al (1994). Double dose-intensive chemotherapy with autologous marrow and peripheral-blood progenitor-cell support for metastatic breast cancer: a feasibility study. *J Clin Oncol 12:*37–44.

11. Holmes, F. A., Valero, V., Walters, R. S., et al (1993). The M.D. Anderson Cancer Center experience with taxol in metastatic breast cancer. *Natl Cancer Inst Monograph 15:*161–169.

12. Spielmann, M. (1994). Taxol (paclitaxel) in patients with metastatic breast carcinoma who have failed prior chemotherapy: interim results of a multinational study. *Oncology 51:*25–28.

13. Gianni, L., Capri, G., Munzone, E., et al (1994). Paclitaxel (taxol) efficacy in patients with advanced breast cancer resistant to anthracycline. *Semin Oncol 5:*29–33.

14. Seidman, A. D., Hudis, C., Crown, J. P., et al (1993). Phase II evaluation of taxotere (RP56976, NSC628503) as initial chemotherapy for metastatic breast cancer. *Proc Am Soc Clin Oncol 12:*63 (abstr 52).

15. Bertheault-Cvitkovic, F., Fandi, A., and Rouëssé, J. (1991). *Breast Cancer. The Role of Haematopoietic Growth Factors.* Gardiner-Caldwell Communications, Ltd. Macclesfield, United Kingdom.

16. Sledge, G. W., Robert, N., Sparano, J. A., et al (1994). Paclitaxel (taxol)/ Doxorubicin combinations in advanced breast cancer: The Eastern Cooperative Oncology Group Experience. *Semin Oncol 21:*15–18.

17. Gelmon, K. A. (1994). Biweekly Paclitaxel (taxol) and cisplatin in breast and ovarian cancer. *Semin Oncol 21:*24–28.

18. Reichman, B. S., Seidman, A. D., Crown, J. P., et al (1993). Paclitaxel and recombinant human granulocyte colony-stimulating factor as initial chemotherapy for metastatic breast cancer. *J Clin Oncol 11:*1943–1951.

19. Wasserheit, C., Alter, R., Speyer, J., et al (1994). Phase II trial of paclitaxel and cisplatin (DDP) in women with metastatic breast cancer. *Proc Am Soc Clin Oncol 13:*100 (abstr 204).

20. Rowinsky, E. K., and Ross, C. (1995). Paclitaxel (taxol). *N Engl J Med 332:*1004–1014.

21. Ten Bokkel Huinink, W. W., Prove, A. M., Piccart, M., et al (1994). A phase II trial with docetaxel (taxotere) in second line treatment with chemotherapy for advanced breast cancer. The EORTC Study Group. *Ann Oncol 5:*527–532.

22. Valero, V., Walters, R., Thierault, R., et al (1994). Phase II study of docetaxel (Taxotere) in anthracycline-refractory metastatic breast cancer (ARMBC) (meeting abstract). *Proc Am Soc Clin Oncol 13:*470 (abstr 1636).

23. Taguchi, T., Mori, S., Abe, R., et al (1994). Late phase II clinical study of RP56976 (docetaxel) in patients with advanced/recurrent breast cancer. *Jpn J Cancer Chemother 21:*2625–2632.

24. Fumoleau, P., Chevallier, B., Dieras, V., et al (1994). Safety evaluation of two doses of taxotere (docetaxel) without routine premedication as first line in advanced breast cancer (ABC)—EORTC Clinical Screening Group (CSG) report. *Proc Am Soc Clin Oncol 13:*109 (abstr 59).

25. Trudeau, M. E., Eisenhauer, E., Lofters, W., et al (1993). Phase II study of Taxotere as first line chemotherapy for metastatic breast cancer (MBC). A National Cancer Institute of Canada Clinical Trials Group (NCIC CTG) Study. *Proc Am Soc Clin Oncol 12:*64 (abstr 59).

26. Antman, K., Bearman, S. I., Davidson, N., et al Dose intensive therapy in breast cancer: current status. In: Champlin R. E., Gale R. P. eds.; *New Strategies in Bone Marrow Transplantation.* New York: Wiley-Liss, 1991: pp. 423–436.

27. Meropol, N. J., Overmoyer, B. A., and Stadtmauer , E. A. (1992). High-dose chemotherapy with autologous stem cell support for breast cancer. *Oncology 6:*53–60.

28. Jones, R. B., Shpall, E. J., Ross, M., et al (1990). AFM induction chemotherapy followed by intensive alkylating agent consolidation with autologous bone marrow support (ABMS) for the treatment of breast cancer. Current results. *Cancer 166:*431–436.

29. Peters, W. P., Shpall, E. J., Jones, R. B., and Ross, M. (1990). High-dose combination cyclophosphamide (CPA), cisplatin (cDDP) and carmustine (BCNU) with bone marrow support as initial treatment for metastatic breast cancer: three-six year follow up. *Proc Am Soc Clin Oncol 9:*10 (abstr).

30. Dunphy, F. R., and Spitzer, G. (1992). Use of very-high dose chemotherapy with autologous bone marrow transplantation in treatment of breast cancer. *JNCI 84:*128–129.

31. Williams, S. F., Mick, R., Dresser, R., Golick, J., Beschorner, J., and Bitran, J. (1989). High-dose consolidation therapy with autologous stem cell rescue in stage IV breast cancer. *J Clin Oncol 7:*1824–1830.

32. Ghalie, R., Richman, C. M., Adler, S. S., et al (1994). Treatment of metastatic breast cancer with a split-course high-dose chemotherapy regimen and autologous bone marrow transplantation. *J Clin Oncol 12:*342–346.

33. Bezwoda, W. R., Seymour, L., and Dansey, R. D. (1995). High-dose chemotherapy with hematopoietic rescue as primary treatment for metastatic breast cancer: a randomized trial. *J Clin Oncol 13:*2477–2479.

12
Filgrastim (r-metHuG-CSF) in Bone Marrow Transplantation

William P. Sheridan
UCLA School of Medicine, Los Angeles, and Amgen Inc., Thousand Oaks, California

I. INTRODUCTION

Patients undergoing myeloablative chemotherapy and bone marrow transplantation (BMT) experience a period of pancytopenia associated with high risk of infections and other complications and a requirement for erythrocyte and platelet transfusion support. In a number of randomized clinical trials and non-randomized studies, treatment with Filgrastim after BMT has been shown to reduce the duration of severe neutropenia and antibiotic therapy. Autologous bone marrow transplantation (ABMT) is being superseded by the use of mobilized blood cells. The use of Filgrastim-mobilized blood cells obtained from normal donors for allogeneic transplantation is now under investigation. Autologous bone marrow, however, has remained an appropriate therapeutic option for many heavily pretreated patients with acute myeloid leukemia (AML), acute lymphoblastic leukemia (ALL), and chronic myelogenous leukemia (CML) (1), and most allogeneic transplant procedures are still being performed using bone marrow. The clinical experience with Filgrastim in both the autologous and allogeneic bone transplantation setting will be reviewed in this chapter.

II. CLINICAL EXPERIENCE

A. Autologous Bone Marrow Transplantation

Several nonrandomized studies in ABMT provided clinical data indicating the benefits of Filgrastim as an adjunct therapy in this setting. These studies showed that Filgrastim significantly accelerates the pace of neutrophil recovery and reduces treatment-related morbidity and overall duration of treatment. Two randomized, placebo-controlled trials have demonstrated a significant reduction in the duration of febrile neutropenia in Filgrastim-treated patients (2,3). Furthermore, Filgrastim treatment may reduce the incidence and severity of fungal infections or opportunistic infections (4). In this trial, however, mortality rates were not significantly affected. In all studies to date, Filgrastim has been well tolerated.

An early non-randomized phase 2 study involved 18 patients with relapsed or refractory Hodgkin's disease (HD) who were treated with cyclophosphamide, carmustine, and etoposide (5). Filgrastim was administered as a daily 30-minute bolus infusion at a dose of 60 μg/kg beginning 24 hours after marrow infusion. Neutrophil recovery was significantly accelerated by Filgrastim. Recovery to an absolute neutrophil count (ANC) 0.1, 0.5, and 1.0 $\times$ 10^9/L occurred 4, 9, and 14 days sooner, respectively, than was seen in historical control patients. These investigators later tested higher doses and continuous rather than short intravenous (IV) infusion, with substantially similar results (6).

In another phase 2 study, Sheridan et al (7) reported on the effects of continuous subcutaneous (SC) infusion of Filgrastim (24 μg/kg/day), commencing 24 hours after ABMT, in 15 patients with relapsed HD, non-Hodgkin's lymphoma (NHL), germ cell tumors, or ALL. Compared with 18 historical control patients, neutrophil recovery was significantly accelerated in the Filgrastim-treated patients (11 days compared with 20 days) (**Figure 1**). Accelerated neutrophil recovery was associated with many clinical benefits, including significantly fewer days of parenteral antibiotic therapy (11 days compared with 18 days) and shorter duration of isolation (10 days versus 18 days). No significant toxic effects were observed other than localized erythema at 2 of 88 infusion sites. None of the patients in this study experienced bone pain.

The study was expanded to explore lower doses and included a total of 35 patients with a variety of tumors (8). Filgrastim was administered by continuous SC infusion beginning on day 1 and continuing through to neutrophil recovery. The median days to neutrophil recovery were 12, 12, and 10 at doses of 6, 12, and 24 μg/kg, respectively.

Filgrastim (10 to 30 μg/kg/day) also significantly accelerated neutrophil recovery in 20 patients with ALL who had received total body irradiation in addition to chemotherapy (either cyclophosphamide or cytarabine) and melphalan before ABMT (9). The time to neutrophil recovery was significantly reduced from 13.5 days for the control patients (who did not receive Filgrastim) to 10.5 days in the Filgrastim-treated group. Reductions of 6 to 15 days were observed in the number of days of febrile neutropenia

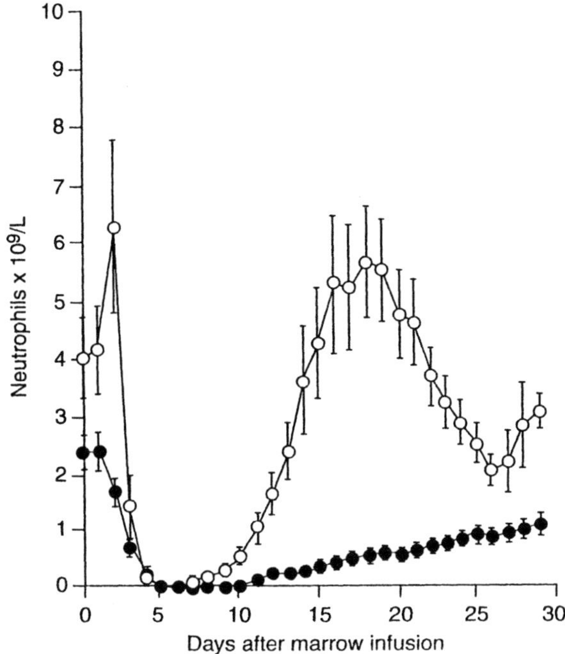

Figure 1. Acceleration of neutrophil recovery by Filgrastim after autologous bone marrow transplantation. Open circles, 15 patients treated with Filgrastim; closed circles, 18 historical control patients. (Adapted from Ref. 7.)

compared with control subjects. Furthermore, there was a significant reduction in the number of days patients had to be hospitalized (from a median of >28 to <25 days).

Passos-Coelho et al (10) reported on 19 patients with metastatic breast cancer who were treated with high-dose chemotherapy and ABMT. Treatment with Filgrastim (16 μg/kg/day) was initiated on day 0 and continued until the ANC reached 2.0×10^9/L. Twenty-four historical control subjects had received the same high-dose chemotherapy regimen and marrow transplantation without Filgrastim. Granulocyte recovery was significantly accelerated in the Filgrastim-treated patients, who also experienced a shorter duration of febrile neutropenia and hospitalization compared with control subjects. Filgrastim treatment had no effect on the number of platelet or red cell transfusions.

The effects of an alternative schedule of treatment with Filgrastim was examined in patients with advanced breast cancer (11). Compared with patients who did not receive Filgrastim, the time to engraftment and length of hospitalization was reduced in patients who received 10 μg/kg/day IV Filgrastim from 3 days before transplant until the white cell count was $>1.0 \times 10^9$/L.

These early encouraging studies have been validated in two randomized, placebo-controlled multicenter trials (3,12,13). Schmitz et al (12,13) reported on the effects of continuous IV infusion of Filgrastim in 54 patients with malignant lymphoma (HD or high-grade NHL) who had been treated with high-dose chemotherapy and ABMT. Chemotherapy regimens included cyclophosphamide, etoposide, and carmustine (BCNU) (CVB regimen) for patients with HD, or carmustine (BCNU), etoposide, cytarabine, and melphalan (BEAM regimen) for those with NHL. Administration of Filgrastim from the first day after marrow transplantation significantly accelerated neutrophil recovery and decreased the number of days with neutropenia. The median time to ANC recovery was 20, 12, and 14 days for patients treated with 0, 10, or 30 μg/kg/day of Filgrastim, respectively. Moreover, the duration of neutropenia was significantly reduced in the Filgrastim-treated groups. The control group experienced 27 days of neutropenia compared with 11 days in the group treated with 10 μg/kg/day Filgrastim, and 13 days in the group treated with 30 μg/kg/day Filgrastim. Significantly fewer days of febrile neutropenia were observed in the Filgrastim-treated groups (5 and 6 days, respectively) compared with the untreated group (10 days) (**Figure 2**). However, Filgrastim had no statistically significant effects on either the number of days with fever, the use of IV antibiotics, or duration of hospital stay. As was seen in the nonrandomized trials, Filgrastim was well tolerated without any serious side effects.

The efficacy and safety of Filgrastim after high-dose chemotherapy and ABMT was examined in a prospective, open-label, randomized trial with patients with poor-risk NHL or relapsed HD (3). Filgrastim was administered to patients by continuous SC infusion at doses of either 10 μg/kg/day (n = 19) or 20 μg/kg/day (n = 10). Fourteen control subjects did not receive Filgrastim. For purposes of analysis, the data were combined from patients treated with both doses of Filgrastim. The median time to ANC recovery was 10 days in the Filgrastim-treated patients compared with 18 days in the untreated control group. The median number of days with fever was also significantly reduced (1 versus 4) as was the number of days with neutropenic fever (5 versus 13.5). Days receiving IV antibiotics and the duration of hospitalization were shorter in the Filgrastim-treated group (although these differences were not statistically significant) (**Figure 3**).

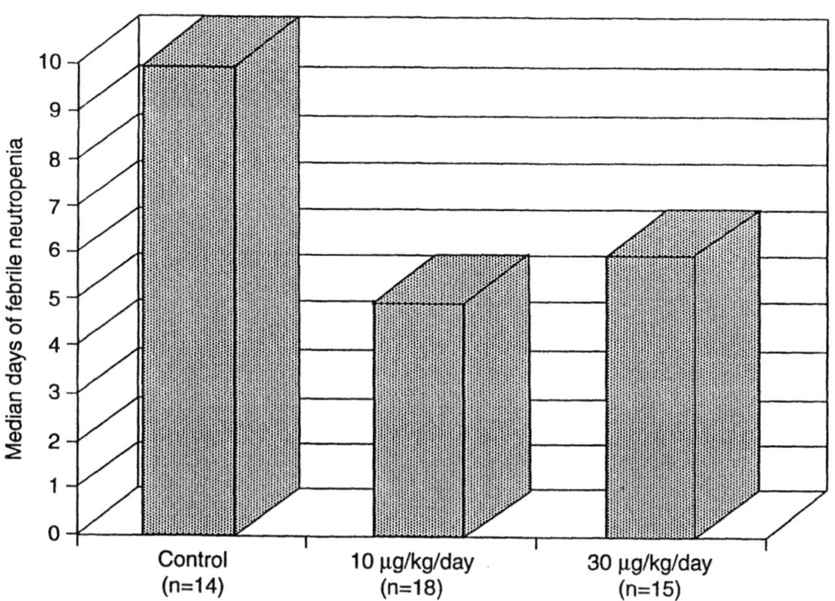

Figure 2. Reduction of duration of febrile neutropenia by Filgrastim after bone marrow transplantation for Hodgkin's disease and non-Hodgkin's lymphoma. (Adapted from Ref. 12.)

Again, only minimal toxicity was associated with the administration of Filgrastim.

B. Allogeneic Bone Marrow Transplantation

Filgrastim has also been used successfully in the setting of allogeneic bone marrow transplantation (allo-BMT). In one early non-randomized study, a short IV infusion of Filgrastim was initiated 3 to 5 days after infusion of allogeneic bone marrow from HLA-matched siblings (200 to 800 $\mu g/m^2$/ day) (14). Filgrastim significantly accelerated neutrophil recovery in 34 patients compared with 45 historical control patients (14.8 days versus 27.4 days; $p < 0.001$). There was no evidence of increased graft-versus-host disease (GVHD) in patients treated with Filgrastim.

Filgrastim was found to be effective in the phase 2 study of Ogawa et al (15). The number of days required to reach a white blood cell count $>1.0 \times 10^9$/L and an ANC $> 0.5 \times 10^9$/L were significantly less than in patients receiving either recombinant human granulocyte-macrophage colony-stimulating factor (rHuGM-CSF) or no growth factor therapy. Con-

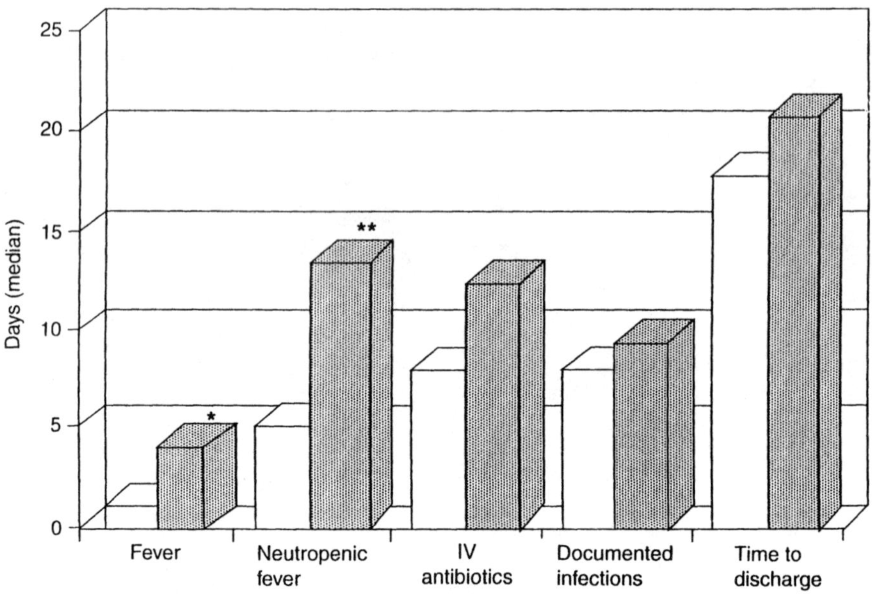

Figure 3. Improved clinical parameters in Filgrastim-treated patients with non-Hodgkin's lymphoma or relapsed Hodgkin's disease receiving autologous bone marrow transplantation. Open boxes, Filgrastim-treated patients; solid boxes, control patients; *p < 0.05; **p < 0.001. (Adapted from Ref. 3.)

sistent with data from the ABMT setting, Filgrastim significantly reduced the duration of febrile neutropenia in recipients of allogeneic marrow.

Blaise et al (9) reported that Filgrastim significantly accelerated neutrophil recovery in patients with ALL or lymphoblastic leukemia who had received total body irradiation in addition to chemotherapy, with either cyclophosphamide or cytarabine plus melphalan, before allo-BMT. Filgrastim was administered at doses of 10 to 30 μg/kg/day. Patients (n = 24) also received cyclosporine and methotrexate for the treatment of GVHD. Again, Filgrastim's effects in the allogeneic setting were similar to those seen in the autologous setting, with the exception that Filgrastim treatment did not influence the duration of hospital stay. Filgrastim treatment significantly reduced the median time to an ANC >1.0 × 10^9/L (17 days compared with 20 days) and the duration of febrile neutropenia.

A 5-year follow-up of a randomized, placebo-controlled study showed an unexpected and significant reduction in the incidence of chronic GVHD in Filgrastim-treated patients compared with untreated control subjects. Filgrastim was given at 300 μg/m^2/day for 14 days starting on day 5 after allo-BMT. Thirty-three patients were treated with Filgrastim and 37 patients received placebo. After 5 years, only 2/28 (7.1%) Filgrastim patients experienced GVHD compared with 12/34 (35.5%) control subjects (16); however, this finding should be treated with caution as the study was not primarily designed to test the possible effects of Filgrastim on GVHD. There were no statistically significant differences in survival or relapse rates between the treated and untreated groups. As in other studies, Filgrastim treatment resulted in improved neutrophil recovery; 78.1% recovery by day 19 in the Filgrastim-treated group compared with 35.3% in the placebo group (p < 0.001).

Lastly, data from 20 consecutively treated patients and 30 historical control patients who had received an allo-BMT (from an HLA-identical sibling) for acute or chronic leukemia, and eight children with acute or chronic leukemia (who received their transplant from healthy unrelated donors) were analyzed retrospectively (17). Filgrastim reduced the duration of neutropenia, without increasing the rate of relapse or the incidence and severity of GVHD in children.

III. MECHANISM OF ACTION OF FILGRASTIM IN BONE MARROW TRANSPLANTATION

The essential role of endogenous G-CSF in maintaining neutrophil production under steady-state conditions and in response to infection has been clearly demonstrated in mice genetically engineered to lack the gene for G-CSF (G-CSF $-/-$); neutrophil levels in the peripheral blood are only 20% to 30% those of wild-type mice (G-CSF $+/+$), and bone marrow granulocyte and monocytic progenitor cells are significantly reduced (18). Administration of rG-CSF reverses these granulopoietic defects. Mice lacking G-CSF also have an impaired ability to respond to infection (eg, *Listeria monocytogenes*) and show impaired neutrophil mobilization.

The beneficial consequences of Filgrastim therapy in BMT recipients most likely result from the effects of G-CSF on both neutrophil production and the functional status of mature neutrophils. G-CSF acts directly on stem cells to increase the number of neutrophil precursors (19), has a mitogenic effect on neutrophil precursors at all stages of development (20,21), and causes early release of neutrophils from the bone marrow (20–13). Filgrastim also increases the survival of mature neutrophils (24) and has other important functional effects, as evidenced by increased superoxide anion generation (25), increased binding of the chemotactic peptide N-formyl-methionyl-leucyl-phenylalanine (FMLP) (26), and the development of phagocytosis (27–30). Filgrastim improves the kinetics of recovery of neutrophils in both the peripheral blood and oral mucosa (31). In contrast, rHuGM-CSF inhibited the migration of neutrophils into experimental skin windows (32). Filgrastim has been shown to reduce post-chemotherapy mucositis (18,33), most probably by maintaining neutrophil migration into inflamed tissue sites such as the oral mucosa (31) and skin (32), and also by accelerating neutrophil recovery and reducing oral infection (33). Since much of the patient morbidity during the neutropenic period after high-dose chemotherapy and BMT is caused by mucositis and oral cavity infection, these effects of Filgrastim may help to explain its significant clinical benefits in the BMT setting.

Schmitz et al (34) have recently proposed a mathematical model of normal human granulopoiesis to quantify the most important influences of G-CSF on the regulation of white cell production. Their model showed that three effects of G-CSF act together to increase the numbers of blood neutrophils: reduction of the transit time of the postmitotic granulopoietic bone marrow cells, additional mitoses of the early granulopoietic bone marrow cells, and possibly demargination.

IV. CONCLUSIONS

Autologous bone marrow transplantation is being replaced by mobilized blood cell transplants (2,35–38). Schmitz et al (39) have recently reported on a prospective, randomized multicenter trial comparing Filgrastim-mobilized peripheral blood progenitor cells (PBPC) with autologous bone marrow cells in lymphoma patients after high-dose chemotherapy. Infusion of Filgrastim-mobilized PBPCs significantly reduced the time to ANC recovery in patients receiving PBPC (11 days) compared with patients receiving autologous bone marrow (14 days) (p = 0.005). Furthermore, the number of platelet transfusions and the time to platelet recovery also were reduced in the patients receiving PBPC transplantation (PBPCT). The median number of days with platelet transfusions after grafting was 6 days for patients receiving PBPCT and 10 days for patients receiving ABMT; time to platelet recovery $>20 \times 10^9$/L was 16 days in the PBPCT group and 23 days in the ABMT group (p = 0.02). In addition, patients randomized to receive PBPCT required fewer red blood cell transfusions (2 versus 3, p = 0.002). Patients who had received PBPCT were discharged earlier from the hospital than those who received ABMT (17 days versus 23 days).

The long-term outcome of blood cell transplantation has not yet been defined in some settings, and in the meantime many patients will receive bone marrow transplantation. For the many trials of Filgrastim in BMT described above, marked improvements in granulocytic recovery were associated with reductions in a number of clinical variables such as days of febrile neutropenia, duration of antibiotic use, and duration of hospitalization. Spitzer et al (4) provided evidence that febrile episodes are related to the duration of neutropenia with an ANC $<0.1 \times 10^9$/L. In general, the onset of fever is 4 days after patients become absolutely neutropenic. Thereafter, the proportion of patients with fever increases rapidly and remains unchanged for as long as the patient is neutropenic. Filgrastim acts by reducing the time required for full neutrophil recovery, with potentially important benefits of lowered risk of secondary infection, fungal infections and vital organ impairment resulting from prolonged infection, and toxicity resulting from antibiotic therapy. The use of Filgrastim in BMT is thus a means of providing effective pharmacologic levels of a recombinant-derived form of the primary physiological regulator of neutrophil production and function and is associated with clinically significant benefits for BMT patients.

REFERENCES

1. Boogaerts, M., Cavalli, F., Cortes-Funes, H., et al (1995). Granulocyte growth factors: achieving a consensus. *Ann Oncol 6:*237–244.
2. Schmitz, N., Linch, D. C., Dreger, P., et al (1994). A randomized phase III study of Filgrastim-mobilized peripheral blood progenitor cell transplantation (PBPCT) in comparison with autologous bone marrow transplantation (ABMT) in patients with Hodgkin's disease (HD) and non-Hodgkin's lymphoma (NHL). *Blood 84:*204a (abstr 800).
3. Stahel, R. A., Jost, L. M., Cemy, T., et al (1994). Randomized study of recombinant human granulocyte colony-stimulating factor after high-dose chemotherapy and autologous bone marrow transplantation for high-risk lymphoid malignancies. *J Clin Oncol 12:*1931–1938.
4. Spitzer, G., Dunphy, F. R., Velasquez, W. S., Petruska, P. J., and Adkins, D. R. Filgrastim (r-met-HuG-CSF) in bone-marrow and peripheral cell transplantation. In: Morstyn G. and Dexter T. M. eds. *Filgrastim (r-metHuG-CSF) in Clinical Practice.* New York: Marcel Dekker, Inc., 1994, pp. 173–195.
5. Taylor, K. M., Jagannath, S., Spitzer, G., et al (1989). Recombinant human granulocyte colony-stimulating factor hastens granulocyte recovery after high-dose chemotherapy and autologous bone marrow transplantation in Hodgkin's disease. *J Clin Oncol 17:*1791–1799.
6. Spitzer, G., Deisseroth, A., Ventura, G., et al (1990). Use of recombinant human hematopoietic growth factors and autologous bone marrow transplantation to attenuate the neutropenic trough of high-dose therapy. *Int J Cell Cloning 8:*249–259.
7. Sheridan, W. P., Morstyn, G., Wolf, M., et al (1989). Granulocyte colony-stimulating factor and neutrophil recovery after high-dose chemotherapy and autologous bone marrow transplantation. *Lancet 2:*891–895.

8. Sheridan, W., Maher, D., Wolf, M., et al (1992). Assessment of three different dosages of r-metHuG-CSF (Filgrastim) as support for marrow engraftment after autologous bone marrow transplantation (ABMT). *Blood 80:*331a (abstr 1314).

9. Blaise, D., Venant, J. P., Flere, D., et al (1992). A randomized, controlled, multicenter trial of recombinant human granulocyte colony stimulating factor (Filgrastim) in patients treated by bone marrow transplantation (BMT) with total body irradiation (TBI) for acute lymphoblastic leukemia (ALL) or lymphoblastic lymphoma (LL). *Blood 80:*248a (abstr 982).

10. Passos-Coelho, J., Davidson, N. E., Noga, S., et al (1993). G-CSF accelerates hematopoietic recovery (HR) after high-dose chemotherapy (HDC) and 4-hydroperoxycyclophosphamide (4HC) purged autologous bone marrow transplantation (PABMT) in patients with metastatic breast cancer (MBC). *Proc Am Soc Clin Oncol 12:*463 (abstr 1614).

11. Cooley, B., Lech, J., Shogan, J., et al (1993). Engraftment advantage with G-CSF in both purged and nonpurged marrow after high dose chemotherapy for advanced breast cancer. *Proc Am Soc Clin Oncol 12:*107 (abstr 230).

12. Schmitz, N., Dreger, P., Zander, A., et al (1993). A randomized, controlled, multicentre study of granulocyte colony stimulating factor (Filgrastim) in patients with Hodgkin's disease and non-Hodgkin's lymphoma undergoing autologous bone marrow transplantation. *Blood 82:*146a (abstr 568).

13. Schmitz, N., Dreger, P., Zander, A. R., et al (1995). Results of a randomized, controlled, multicentre study of recombinant human granulocyte colony-stimulating factor (Filgrastim) in patients with Hodgkin's disease and non-Hodgkin's lymphoma undergoing autologous bone marrow transplantation. *Bone Marrow Transplant 15:*261–266.

14. Masaoka, T., Takaku, F., Kato, S., et al (1989). Recombinant human granulocyte colony-stimulating factor in allogeneic bone marrow transplantation. *Exp Hematol 17:*1047–1050.

15. Ogawa, M. (1989). Phase I, II studies of recombinant granulocyte colony stimulating factor (gammaG-CSF). *Gan To Kagaku Ryoho 16:*3537–3542.

16. Hiraoaka, A., Masaoka, T., Shibata, H., et al (1995). Five years follow-up of a randomized placebo-controlled study with Filgrastim (recombinant human granulocyte colony-stimulating factor) in patients receiving allogeneic bone-marrow transplantation. *Blood 86:*222a (abstr 874).

17. Locatelli, F., Pession, A., Zecca, M., et al (1996). Use of recombinant human granulocyte colony-stimulating factor in children given allogeneic bone marrow transplantation for acute or chronic leukemia. *Bone Marrow Transplant 17:*31–37.

18. Lieschke, G. J., Grail, D., Hodgson, G., et al (1994). Mice lacking granulocyte colony-stimulating factor have chronic neutropenia, granulocyte and macrophage progenitor cell deficiency, and impaired neutrophil mobilization. *Blood 84:*1737–1746.

19. Metcalf, D., and Nicola, N. A. (1983). Proliferative effects of purified granulocyte factor (G-CSF) on normal mouse hemopoietic cells. *J Cell Physiol 116:*198–206.

20. Lord, B. I., Bronchud, M. H., Owens, S., et al (1989). The genetics of human granulopoiesis following treatment with granulocyte colony-stimulating factor in vivo. *Proc Natl Acad Sci USA 86*:9499–9503.

21. Lord, B. I. (1992). Myeloid cell kinetics in response to haemopoietic growth factors. *Baillieres Clin Haematol 5*:533–50.

22. Morstyn, G., Campbell, L., Souza, L. M., et al (1988). Effect of granulocyte colony stimulating factor on neutropenia induced by cytotoxic chemotherapy. *Lancet 1*:667–672.

23. Lord, B. I., Molineux, G., Pojda, Z., Souza, L. M., Mermod, J. J., and Dexter, T. M. (1991). Myeloid cell kinetics in mice treated with recombinant interleukin-3, granulocyte colony-stimulating factor (CSF), or granulocyte-macrophage CSF in vivo. *Blood 77*:2154–2159.

24. Begley, C. G., Lopez, A. F., Nicola, N. A., et al (1986). Purified colony-stimulating factors enhance the survival of human neutrophils and eosinophils in vitro: a rapid and sensitive microassay for colony-stimulating factors. *Blood 68*:162–166.

25. Lindemann, A., Herrmann, F., Oster, W., et al (1989). Hematologic effects of recombinant human granulocyte colony-stimulating factor in patients with malignancy. *Blood 74*:2644–2651.

26. Cohen, A. M., Hines, D. K., Korach, E. S., and Ratzkin, B. J. (1988). In vivo activation of neutrophil function in hamsters by recombinant human granulocyte colony-stimulating factor. *Infect Immun 56*:2861–2865.

27. Fabian, I., Kletter, Y., Bleiberg, I., Gadish, M., Naparsteck, E., and Slavin, S. (1991). Effect of exogenous recombinant human granulocyte and granulocyte-macrophage colony-stimulating factor on neutrophil function following allogeneic bone marrow transplantation. *Exp Hematol 19*:868–873.

28. Roilides, E., Walsh, T. J., Pizzo, P. A., and Rubin, M. (1991). Granulocyte colony-stimulating factor enhances the phagocytic and bactericidal activity of normal and defective human neutrophils. *J Infect Dis 163*:579–583.

29. Gadish, M., Kletter, Y., Flidel, O., Nagler, A., Slavin, S., and Fabian, I. (1991). Effects of recombinant human granulocyte and granulocyte-macrophage colony-stimulating factors on neutrophil function following autologous bone marrow transplantation. *Leuk Res 15*:1175–1182.

30. Weisbart, R. H., Kacena, A., Schuh, A., and Golde, D. W. (1988). GM-CSF induces human neutrophil IgA-mediated phagocytosis by an IgA Fc receptor activation mechanism. *Nature 332*:647–648.

31. Lieschke, G. J., Ramenghi, U., O'Connor, M. P., Sheridan, W., Szer, J., and Morstyn, G. (1992). Studies of oral neutrophil levels in patients receiving G-CSF after autologous marrow transplantation. *Br J Haematol 82*:589–595.

32. Peters, W. P., Kurtzberg, J., and Atwater, S. (1989). Comparative effects of rHuG-CSF and rHuGM-CSF on hematopoietic reconstitution and granulocyte function following high dose chemotherapy and autologous bone marrow transplantation (ABMT). *Proc Am Soc Clin Oncol 8*:181a (abstr 702).

33. Crawford, J., Glaspy, J., Vincent, M., Tomita, D., and Mazanet, R. (1994). Effect of Filgrastim (r-metHuG-CSF) on oral mucositis in patients with small

cell lung cancer (SCLC) receiving chemotherapy (cyclophosphamide, doxorubicin and etoposide, CAE). *Proc Am Soc Clin Oncol 13*:442 (abstr 1523).

34. Schmitz, S., Franke, H., Brusis, J., and Wichmann, H. E. (1993) Quantification of the cell kinetic effects of G-CSF using a model of human granulopoiesis. *Exp Hematol 21*:755–760.

35. Shapiro, C. L., Hurd, D., Clark, P., et al (1994). Repetitive cycles of cyclophosphamide, thiotepa and carboplatin (CTCb) intensification with peripheral blood progenitor cells (PBPC) and Filgrastim (G-CSF) in advanced breast cancer patients (PTS). *Proc Am Soc Clin Oncol 13*:66 (abstr 67).

36. Welte, K., Gabrilove, J., Bronchud, M. H., Platzer, E., and Morstyn, G. (1996). Filgrastim (r-metHuG-CSF)-the first 10 years. *Blood 88*:1907–1929.

37. Bensinger, W. I., Clift, R. S., Anasetti, C., et al (1996). Transplantation of allogeneic peripheral-blood stem-cells mobilized by recombinant human granulocyte-colony-stimulating factor. *Stem Cells 14*:90–105.

38. Sheridan, W P., Lo, L. B., Brown, S. L., Grigg, A., Foote, M. A., and Juttner, C. A. (1996). Clinical use of rHuG-CSF (filgrastim)-mobilised peripheral blood progenitor cells for transplantation. *J Drug Dev Clin Pract 8*:7–17.

39. Schmitz, N., Linch, D. C., Dreger, P., et al (1996). Randomised trial of filgrastim-mobilized peripheral blood progenitor cell transplantation versus autologous bone-marrow transplantation in lymphoma patients. *Lancet 347*:353–357.

13
Filgrastim (r-metHuG-CSF) in Autologous Peripheral Blood Cell Transplantation

Norbert Schmitz and Peter Dreger
University Hospital, Kiel, Germany

I. INTRODUCTION

Over recent years, autologous bone marrow transplantation (BMT) has largely been replaced by the transplantation of peripheral blood progenitor cells (PBPC). In 1995, 83% of a total of 6046 autologous transplants reported to the European Group for Blood and Marrow Transplantation (EBMT) (1) and approximately 70% of all autologous transplants registered with the International Bone Marrow Transplant Registry (IBMTR) were PBPC transplants; another 15% of autologous transplants reported to the latter institution consisted of a combination of blood and marrow cells. With an ongoing trend in favor of PBPC transplants, autologous marrow transplantation will virtually have disappeared by the turn of the century. A similar development is expected in the area of allogeneic stem cell transplantation.

Although exact figures which specify the use of hematopoietic growth factors for mobilization of PBPC are not available, there is good reason to believe that the vast majority of PBPC harvested for clinical use are being mobilized into the peripheral blood by the administration of recombinant human granulocyte colony-stimulating factor (rHuG-CSF). The acute side effects of rHuG-CSF are usually mild and highly predictable, and more than 1.5 million patients have now been exposed to this drug and followed for as long as 10 years (2). Thus, rHuG-CSF alone or the combination of chemotherapy followed by rHuG-CSF currently represent the safest way to mobilize PBPC. Other hematopoietic growth factors such as recombinant human forms of granulocyte-macrophage colony-stimulating factor (rHuGM-CSF), interleukin-3 (rHuIL-3), and stem cell factor (rHuSCF) or combinations thereof have been used less frequently for mobilization and usually have been investigated within study protocols or in patients with difficulties in mobilizing an adequate number of PBPC (see Chapter 14).

This chapter will focus on the use of Filgrastim in the context of autologous PBPC transplantation. Special emphasis is placed on recent achievements in the field including the publication of prospective randomized trials comparing the clinical benefits of autologous PBPC with marrow transplants and initial data addressing the economic consequences of the use of Filgrastim-mobilized PBPC, rather than autologous bone marrow, to rescue patients after high-dose chemotherapy.

II. MOBILIZATION OF PERIPHERAL BLOOD PROGENITOR CELLS

Early in the century, Maximow (3) proposed that cells capable of maintaining hematopoiesis circulated in the peripheral blood. This knowledge could not be exploited clinically, however, until blood cell separators became widely available and measures were discovered to increase the low numbers of PBPC circulating in steady state. Richman et al (4) described for the first time how myelosuppressive chemotherapy induced marked increases in progenitor cell numbers in the peripheral blood at the time of hematopoietic recovery. Some years later it was demonstrated that PBPC collected from patients with acute myeloid leukemia (AML) early during recovery from induction chemotherapy were capable of producing long-term hematopoietic reconstitution (5,6). It was only the detection that hematopoietic growth factors such as rHuGM-CSF or rHuG-CSF could induce substantial increases in the number of PBPC in the peripheral blood, however, that made the collection of PBPC more practical and attractive in a large variety of malignant disorders (7–9). The broad availability of recombinant hematopoietic growth factors caused the steep increase in PBPC transplants described previously.

The administration of Filgrastim at a dose of ≥ 10 μg/kg/day subcutaneously (SC) for 5 to 6 consecutive days followed by two to four leukapheresis procedures will allow the collection of $\geq 2 \times 10^6$ CD34$^+$ cells/kg body weight in the vast majority of patients (10). Although no general agreement exists as to what the minimum number of hematopoietic stem and progenitor cells in a PBPC collection product should be and how these cells should best be characterized, most investigators state that ≥ 2 to 6×10^6 CD34$^+$

cells/kg body weight represent sufficient numbers of hematopoietic stem and progenitor cells to guarantee prompt and durable engraftment.

There is ample evidence that the combination of aggressive chemotherapy followed by the injection of hematopoietic growth factors such as Filgrastim can mobilize higher numbers of PBPC compared with Filgrastim alone (11). When Filgrastim alone or chemotherapy followed by Filgrastim was administered to the same patients with breast cancer, Möhle et al (12) demonstrated that seven times more CD34$^+$ cells per leukapheresis could be harvested with the combined approach. An increase in the dose of rHuG-CSF up to 24 μg/kg/day (13) or combinations of early-acting hematopoietic growth factors such as rHuIL-3 and rHuSCF with rHuG-CSF or rHuGM-CSF also have been reported to improve PBPC yields (14–18). These combinations may be particularly valuable if one intends to collect high numbers of PBPC with a minimum number of apheresis procedures or if Filgrastim-only administration resulted in borderline yields of CD34$^+$ cells. There is no convincing evidence that any growth factor or combination of hematopoietic growth factors will mobilize significant PBPC numbers in patients with extremely poor yields or in patients who do not mobilize at all after chemotherapy plus Filgrastim administration.

Using a "standard" approach consisting of chemotherapy $\pm$ Filgrastim or rHuGM-CSF for PBPC mobilization, a certain fraction of patients will have low yields. The percentage of "poor mobilizers" is largely related to previous therapy and will widely differ from one disease to another depending on the extent of prior treatment given to the average patient at the time the patient becomes eligible for high-dose therapy and transplantation. Recently, Dreger et al (19) showed that the most important clinical factor influencing the yield of PBPC was the number of chemotherapy cycles containing stem cell toxic drugs administered to the patient before the harvest procedure. Low granulocyte-macrophage colony-forming cell (CFC-GM) and CD34$^+$ cell yields resulted in statistically significant delays in neutrophil and platelet recovery. Previous radiotherapy also was associated with significantly lower CFC-GM and CD34$^+$ cell yields, but had no significantly influence on engraftment (**Figure 1**). This study also included a comparison of the engraftment kinetics seen after PBPC or marrow transplantation. It was interesting to note that even in the most heavily pretreated patients, the engraftment after transplantation with Filgrastim-mobilized PBPC was faster than after marrow transplantation. Several other investigators arrived at similar conclusions and related poor progenitor cell yields and delayed engraftment to the extent of previous chemo- and/or radiotherapy (20,21).

Another interesting question regarding the mobilization of PBPC is if there is any practical possibility to predict the yield of CD34$^+$ cells in

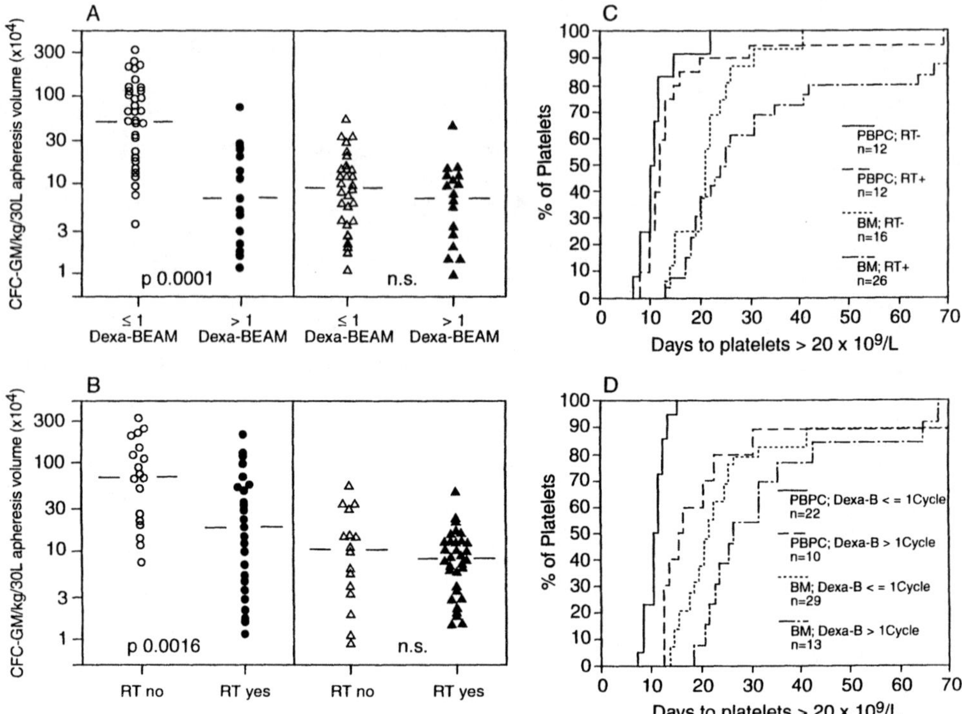

Figure 1. Influence of previous Dexa-Beam therapy (A) and previous radiotherapy (B) on GM-CFC contents of peripheral blood (*left*) and bone marrow grafts (*right*) and on platelet engraftment (C and D). Dexa-BEAM = chemotherapy with dexamethasone, carmustine, etoposide, cytarabine, and melphalan; PBPC = peripheral blood progenitor cells; BM = bone marrow; RT = radiotherapy.

the collection product before Filgrastim administration (during steady-state hematopoiesis). Schwella et al (22) calculated the preleukapheresis CD34$^+$ cell counts in the blood of patients with germ cell tumors and reported that high ($\geq 4 \times 10^4$/mL) preleukapheresis CD34$^+$ cell counts were able to predict successful collection of $\geq 2 \times 10^6$ CD34$^+$ cells/kg with a single leukapheresis. Passos-Coelho et al (23) showed that the percentage of CD34$^+$ cells in the bone marrow aspirate of a patient before PBPC mobilization correlated well with the number of CD34$^+$ cells collected during a 6-hour leukapheresis after mobilization with cyclophosphamide and rHuGM-CSF. Fruehauf et al (24) administered chemotherapy and Filgrastim to patients with lymphoma and were able to show that a concentration of 0.4×10^6/L steady-state CD34$^+$ cells in the peripheral blood before mobilization indicated a 95% probability that 2.5×10^6 CD34$^+$ cells/kg could be collected with six leukapheresis procedures. On the other hand, Roberts et al (25) were unable to correlate the number of steady-state CD34$^+$ cells in the peripheral blood of normal volunteers or patients with untreated breast cancer with the subsequent yield of CD34$^+$ cells in the collection product; however, these patients had been mobilized with Filgrastim only. Mijovic and Mufti (26) administered a single dose of Filgrastim SC (10 μg/kg) to patients with cancer and suggested that the subsequent increase of CD34$^+$ cells in the peripheral blood of these patients could predict the yield of CD34$^+$ cells in the PBPC collections. It remains to be determined if any method suggested so far will give reproducible results and if the information provided by such laboratory tests will be more predictive than taking the patient's history.

III. CHARACTERIZATION OF THE GRAFT

In the autologous setting, efforts to characterize PBPC collection products have focused on cells thought to reflect the hematopoietic potential of the graft. Traditionally, colony-forming cells committed to the myeloid (CFC-GM) or erythroid (erythroid burst-forming cells [BFC-E]) lineage have been measured to estimate the frequency of hematopoietic stem and progenitor cells. It is well established, however, that CFC-GM or BFC-E represent committed progenitor cells rather than true hematopoietic stem cells. Moreover, the respective colony assays are expensive and laborious. Therefore, more readily available and cheaper alternatives have been sought. With the detection of the CD34$^+$ antigen and the respective antibodies (reviewed in Ref. 27), immunophenotyping of harvested cells by fluorescence-activated cell sorter (FACS) analysis has become the most widely used method to enumerate CD34$^+$ stem and progenitor cells in PBPC collection products. It also has been demonstrated that subtypes of CD34$^+$ cells indicating commitment to the erythroid, myeloid, and megakaryocytic lineage (28,29) as well as immature lineage-negative cells are all contained in PBPC harvests. Recent reports give convincing evidence that the most immature stem cells detectable in humans with the plastic-adherent delta (Pδ) assay system (30), the long-term culture initiating cell (LT-CIC) assay (31,32), and the cobblestone area-forming cell (CAFC) assay (33) are present in PBPC harvests.

The latter assay systems require special expertise and are poorly standardized, costly, and labor intensive. It seems neither possible nor necessary to perform any of these assays on routine samples of PBPC collection products. With growing evidence that PBPC harvests contain all stem and progenitor cells necessary to sustain short- and long-term hematopoiesis, characterization of the graft by simply counting the CD34[+] cells seems completely adequate. Several authors have repeatedly demonstrated that the enumeration of CD34[+] cells in the collection product by FACS analysis can reliably predict hematopoietic engraftment (34,35). In the largest series reported so far, Weaver et al (1995) analyzed the engraftment kinetics of PBPC as a function of the CD34[+] content in the harvests of 692 patients treated with high-dose chemotherapy for various kinds of diseases. These patients had PBPC collected after administration of cyclophosphamide-based mobilization chemotherapy with or without Filgrastim. Using a Cox regression model, the authors showed that the CD34[+] content of the PBPC products was the most powerful predictor of both platelet and neutrophil recovery. More than 5.0×10^6 CD34 cells/kg body weight appeared to be optimal to ensure rapid neutrophil and platelet recovery, while patients receiving between 2.0 and 5.0×10^6 CD34 cells/ kg body weight had a somewhat delayed engraftment. For obvious reasons, it has never been possible to reliably define a lower threshold of CD34[+] cell numbers below which patients would carry a high risk of engraftment failure; however, it is suggested that at least 1×10^6 CD34[+] cells be collected.

IV.   TUMOR CELL CONTAMINATION

It has been claimed repeatedly that the tumor cell contamination of PBPC harvests may be less than that of bone marrow (36,37). The evidence for this always has been very tenuous, as it must be kept in mind that in most diseases treated by high-dose chemotherapy and transplantation of hematopoietic cells, tumor cells can circulate and that the number of collected cells from the blood is usually much more than from the bone marrow. Recent publications on this issue have demonstrated that using sensitive, mostly molecular techniques, it is possible to detect tumor cells in PBPC harvests of the majority of patients with multiple myeloma (38–40), low-grade non-Hodgkin's lymphoma (NHL) (32,41), mantle cell lymphoma (42), chronic lymphocytic leukemia (CLL) (43), and breast cancer (44,45). It is not known, however, if and to what extent these tumor cells often detected by polymerase chain reaction (PCR) technology are clonogenic and thus might be able to initiate recurrence of disease if infused into the patient's blood. CD34[+] selection techniques can reduce the tumor cell load of PBPC harvests, and varying numbers of PBPC collection products have tested negative for the molecular marker of the respective disease after $CD34^+$ selection (39). There are no available data proving that patients receiving $CD34^+$ selected or purged PBPC harvests clinically benefit from this procedure. Nevertheless, most investigators would be unwilling to infuse PBPC that have been demonstrated to be heavily contaminated with tumor cells to a patient. Although it has been shown that Filgrastim-mobilized PBPC harvests contain tumor cells,there is no convincing evidence that the mobilization of hematopoietic progenitors into the peripheral blood is in itself associated with mobilization of tumor cells. With new devices now entering the clinic, it may be possible to deplete tumor cells from PBPC harvests much more effectively than with the technologies used during recent years. It remains to be seen if these efforts will be able to improve the overall results of PBPC transplantation in hematologic malignancies and solid tumors.

V. CLINICAL USE OF MOBILIZED PROGENITOR CELLS

Sheridan et al (44) first demonstrated that Filgrastim-mobilized PBPC were able to accelerate hematopoietic reconstitution after high-dose therapy. When they added Filgrastim-mobilized PBPC to autologous bone marrow cells and compared the kinetics of hematopoietic recovery in patients given marrow plus PBPC or bone marrow only after high-dose therapy, platelet recovery occurred significantly faster in the PBPC transplant group. Patients reached $\geq 50 \times 10^9/L$ platelets 15 days after infusion of PBPC and marrow, whereas patients infused with only autologous marrow needed a median of 39 days to achieve $50 \times 10^9/L$ platelets (p = 0.0006). Reports by Bensinger et al (47) and Chao et al (48) using different study designs and clinical endpoints showed similar results. The infusion of Filgrastim-mobilized autologous PBPC after high-dose therapy resulted in a significant acceleration of platelet and neutrophil recovery compared with historical control patients who had been transplanted with autologous bone marrow. A reduction in the number of days of hospitalization and reduced costs was seen in some studies. In 1996, Schmitz et al (49) reported on the results of a randomized trial comparing Filgrastim-mobilized PBPC transplants with autologous bone marrow transplantation (ABMT) in heavily pretreated lymphoma patients who had received high-dose chemotherapy consisting of carmustine, etoposide, cytarabine, and melphalan (BEAM). Both patient groups received Filgrastim after transplantation. Time to platelet recovery $>20 \times 10^9/L$ was 16 (range, 8 to 52) days in the PBPC group and 23 (range, 13 to 56) days in the ABMT group (p = 0.02). The number of platelet transfusions given to individual patients participating in this study most impressively demonstrates the significant reduction of platelet transfusions needed after PBPC transplant (**Figure 2**). Time to neutrophil recovery $\geq0.5 \times 10^9$/L was reduced in the PBPC transplantation group (11 versus 14 days, p = 0.005). Patients randomized to PBPC transplantation needed fewer red cell transfusions and spent less time in hospital (17 versus 23 days, p = 0.002). Another study in patients with relapsed or refractory germ cell tumors randomized to receive either autologous bone marrow or PBPC mobilized by chemotherapy plus Filgrastim demonstrated a significantly shorter recovery time to neutrophil counts $\geq0.5 \times 10^9$/L (10.0 versus 11.0 days, p > 0.01) and platelet counts $>20 \times 10^9$/L (10.0 versus 17.0 days, p > 0.01). Fewer days to transfusion independence from red blood cells (8.0 versus 12.0 days, p > 0.05) and platelets (9.0 versus 12.0 days, p > 0.01) and fewer days of intravenous (IV) antibiotics (9.0 versus 11.0 days, p > 0.05) also were recorded (50). Neither study observed differences in overall or event-free survival for patients grafted with Filgrastim-mobilized PBPC opposed to bone marrow cells.

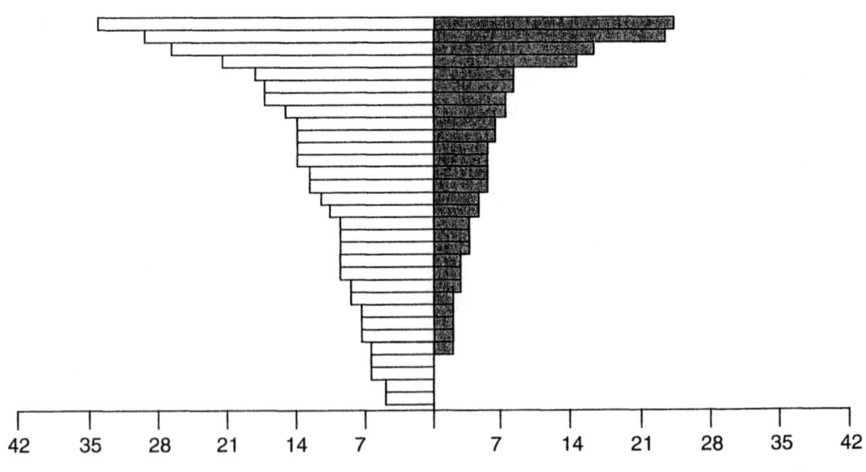

Figure 2. Number of platelet transfusions given to individual patients shows a significant reduction in the number of platelet transfusions needed by patient receiving peripheral blood progenitor cell transplantation (solid bars) compared with number needed by patients receiving autologous bone marrow transplantation. (Adapted from Ref. 49.)

VI. FILGRASTIM AFTER PERIPHERAL BLOOD PROGENITOR CELL TRANSPLANTATION

A number of studies have addressed the question of whether the administration of Filgrastim after transplantation of mobilized PBPC can further accelerate neutrophil recovery and improve the clinical outcome. The study by Spitzer et al (51) randomized patients with lymphoma and various solid tumors to receive Filgrastim plus rHuGM-CSF or no growth factor after PBPC transplantation. An absolute neutrophil count (ANC) $\geq 0.5 \times 10^9$/L was reached 10 days after PBPC infusion in the growth factor–treated group opposed to 16 days in the group without growth factor support (p = 0.0004). In addition, the duration of hospitalization was shortened (19 versus 21 days, p = 0.0112). In a second randomized study, Klumpp et al (52) administered either Filgrastim (5 μg/kg/day beginning on day +1) or standard posttransplant care to patients treated with high-dose therapy and infusion of autologous PBPC or a combination of PBPC and marrow. Administration of Filgrastim posttransplant accelerated neutrophil engraftment (11 versus 17 days, p = 0.0003, in patients who were transfused with PBPC alone; 10 versus 14 days, p = 0.02, in patients who received PBPC plus marrow). The median duration of posttransplant hospitalization (18 versus 24 days, p = 0.002) and the median number of days administered non-prophylactic antibiotics (11 versus 15 days, p = 0.03) also were reduced.

Several studies using historical control patients have addressed the question of a potential benefit of Filgrastim administration after PBPC transplant. Studies by Cortelazzo et al (53), Shimazaki et al (54), and Weaver et al (16) generally confirmed that neutrophil recovery is accelerated if Filgrastim is given after PBPC transplant. However, these studies (as well as the randomized studied) are inconclusive with regard to other important clinical endpoints such as transplant-related toxicity, overall survival, and disease-free survival.

VII. TRANSPLANTATION OF FILGRASTIM-MOBILIZED PERIPHERAL BLOOD PROGENITOR CELLS OR BONE MARROW: ECONOMIC CONSIDERATIONS

Retrospective analyses of the costs of PBPC transplantation compared with ABMT gave conflicting results. While one study reported that costs were substantially lower (55), increased costs were reported in another (56). Recently Smith et al (57) presented the results of a prospective economic analysis performed on the randomized trial by Schmitz et al (49) outlined earlier in this chapter. The total medical costs associated with the two

Table 1.   Summary Comparison of Costs and Charges of Peripheral
Blood Progenitor Cell Transplantation Versus Autologous Bone
Marrow Transplantation.

Category	PBPCT charges ($)	ABMT charges ($)	PBPCT costs ($)	ABMT costs ($)
Autograft collection	8,290	13,858	5,760	8,531
BEAM chemotherapy	13,534	13,544	10,019	10,030
PostBEAM	37,665	52,541	30,013	40,752
Total	59,489	79,943	45,792	59,314
Difference	−26%		−23%	

PBPCT = peripheral blood progenitor cell transplantation.
ABMT = autologous bone marrow transplantation.
BEAM = chemotherapy with with carmustine, etoposide, cytarabine, and melphalan.
Source: Adapted from Ref. 27.

treatments were estimated assuming that the costs for the medical resources
used in the trial were incurred by a hospital in the United States. The
authors concluded that PBPC transplantation represented a significant savings, mainly due to lower autograft collection costs, shorter hospital stays,
and less supportive care in patients grafted with Filgrastim-mobilized PBPC
compared with marrow transplant patients (**Table 1**). Although the use of
European clinical data to perform a cost analysis according to United States
standards has obvious limitations (58), cost savings of US $13,521 (−23%)
are substantial and the detailed sensitivity analysis performed by Smith et
al (57) surely corroborates their findings.

VIII.   SUMMARY AND CONCLUSIONS

Transplantation with PBPC has been shown to produce a rapid and reliable
restoration of hematopoietic activity after myelosuppressive chemotherapy.
The use of Filgrastim or other growth factors to mobilizing PBPC for
harvest has increased PBPC yields, thus allowing more patients to have
the benefit of PBPC transplantation after myelosuppressive therapy.

REFERENCES

1.   Gratwohl, A., Hermans, J., Baldomero, H., for the European Group for Blood
     and Marrow Transplantation (EBMT) (1997). Blood and marrow transplanta-
     tion activity in Europe 1995. *Bone Marrow Transplant 19:*407–419.

2. Welte, K., Gabrilove, J., Bronchud, M. H., Platzer, E., and Morstyn, G. (1996). Filgrastim (r-metHuG-CSF): the first 10 years. *Blood 88*:1907–1929.

3. Maximow, A. (1909). Der Lymphozyt als gemeinsame Stammzelle der verschiedenen Blutelemente in der embryonalen Entwicklung und im postfetaleu Leben der Säugetiere. *Verh Dtsch Ges Pathol 8*:125–134.

4. Richman, C. M., Weiner, R. S., and Yankee, R. A. (1976). Increase in circulating stem cells following chemotherapy in man. *Blood 47*:1031–1039.

5. To, L. B., Haylock, D. N., Kimber, R. J., and Juttner, C. A. (1984). High levels of circulating haemopoietic stem cells in very early remission from acute non-lymphoblastic leukaemia and their collection and cryopreservation. *Br J Haematol 58*:399–410.

6. Reiffers, J., Bernard, P. H., David B., et al (1986). Successful autologous transplantation with peripheral blood hemopoietic cells in a patient with acute leukemia. *Exp Hematol 14*:312–315.

7. Socinski, M. A., Elias, A., Schnipper, L., Cannistra, S. A., Antman, K. H., and Griffin, J. D. (1988). Granulocyte-macrophage colony stimulating factor expands the circulation haemopoietic progenitor cell compartment in man. *Lancet 1*:1194–1198.

8. Dührsen, U., Villeval, J.-L., Boyd, J., Kannourakis, G., Morstyn, G., and Metcalf, D. (1988). Effects of recombinant human granulocyte colony-stimulating factor on hematopoietic cells in cancer patients. *Blood 72*:2074–2081.

9. Gianni, A. M., Siena, S., Bregni, M., et al (1989). Granulocyte-macrophage colony-stimulating factor to harvest circulating haemopoietic stem cells for autotransplantation. *Lancet 2*:580–585.

10. Sheridan, W. Cytokine-only approaches to mobilization of progenitor cells. In: Morstyn G. and Sheridan W. eds. *Cell Therapy: Stem Cell Transplantation, Gene Therapy, and Cellular Immunotherapy,* Cambridge: Cambridge University Press, 1996, pp. 146–182.

11. Lie, A. K., Rawling, T. P., Bayly, J. L., and To, L. B. (1996). Progenitor cell yield in sequential blood stem cell mobilization in the same patients: insights into chemotherapy dose escalation and combination of haematopoietic growth factor and chemotherapy. *Br J Haematol 95*:39–44.

12. Möhle, R., Pförsich, M., Fruehauf, S., Krämer, A., and Haas, R. (1994). Filgrastim post-chemotherapy mobilizes more CD34+ cells with a different antigenic profile compared with use during steady-state hematopoiesis. *Bone Marrow Transplant 14*:827–832.

13. Zeller, W., Gutensohn, K., Stockschläder, M., et al (1996). Increase of mobilized CD34-positive peripheral blood progenitor cells in patients with Hodgkin's disease non-Hodgkin's lymphoma, and cancer of the testis. *Bone Marrow Transplant 17*:709–713.

14. Geissler, K., Peschel, C., Niederwieser, D., et al (1996). Potentiation of granulocyte colony-stimulating factor-induced mobilization of circulating progenitor cells by seven-day pretreatment with Interleukin-3. *Blood 87*:2732–2739.

15. Brugger, W., Frisch, J., Schulz, G., Pressier, K., Mertelsmann, R., and Kanz, L. (1992). Sequential administration of Interleukin-3 and granulocyte-

macrophage colony-stimulating factor following standard-dose combination chemotherapy with etoposide, ifosfamide, and cisplatin. *J Clin Oncol 10:*1452–1459.

16. Weaver, C. H., Hazelton, B., Birch, R., et al (1995). An analysis of engraftment kinetics as a function of the CD34 content of peripheral blood progenitor cell collections in 692 patients after the administration of myeloablative chemotherapy. *Blood 86:*3961–3969.

17. Tricot, G., Jagannath, S., Desikan, K. R., et al (1996). Superior mobilization of peripheral blood progenitor cells (PBPC) with r-metHuSCF (SCF) and r-metHuG-CSF (Filgrastim) in heavily pretreated multiple myeloma (MM) patients. *Blood 88:*388a (abstract 1540).

18. Glaspy, J., Davis, M. W., Parker, W. R., Foote, M., and McNiece, I. (1966). Biology and clinical potential of stem-cell factor. *Cancer Chemother Pharmacol 38:*S53–57.

19. Dreger, P., Klöss, M., Petersen, B., et al (1995). Autologous progenitor cell transplantation: Prior exposure to stem cell-toxic drugs determines yield and engraftment of peripheral blood progenitor cell but not of bone marrow grafts. *Blood 86:*3970–3978.

20. Haas, R., Möhle, R., Frühauf, S., et al (1994). Patient characteristics associated with successful mobilizing and autografting of peripheral blood progenitor cells in malignant lymphoma. *Blood 83:*3787–3794.

21. Olvieri, A., Offidani, M., Ciniero, L., et al (1994). Optimization of the yield of PBPC for autrotransplantation mobilized by high-dose chemotherapy and G-CSF: proposal for a mathematical model. *Bone Marrow Transplant 14:*273–278.

22. Schwella, N., Beyer, J., Schwaner, I., et al (1996). Impact of preleukapheresis cell counts on collection results and correlation of progenitor-cell dose with engraftment after high-dose chemotherapy in patients with germ cell cancer. *J Clin Oncol 14:*1114–1121.

23. Passos-Coelho, J. L., Braine, H. G., Davis, J. M., et al (1995). Predictive factors for peripheral-blood progenitor-cell collections using a single large-volume leukapheresis after cyclophosphamide and granulocyte-macrophage colony-stimulating factor mobilization. *J Clin Oncol 13:*705–714.

24. Fruehauf, S., Haas, R., Conradt, C., et al (1995). Peripheral blood progenitor cell (PBPC) counts during steady-state hematopoiesis allow to estimate the yield of mobilized PBPC after filgrastim (r-metHuG-GSF)-supported cytotoxic chemotherapy. *Blood 85:*2619–2626.

25. Roberts, A. W., Begley, C. G., Grigg, A. P., and Basser, R. L. (1995). Do steady-state peripheral blood progenitor cell (PBPC) counts predict the yield of PBPC mobilized by filgrastim alone? *Blood 86:*2451.

26. Mijovic, A., and Mufti, G. J. (1995). Single dose of Filgrastim (rhG-CSF) to predict mobilisation of haematopoietic progenitors in patients with haematological malignancies. *Bone Marrow Transplant 15:*813–817.

27. Krause, D. S., Fackler, M. J., Civin, C. I., and May, W. S. (1996). CD34: Structure, biology, and clinical utility. *Blood 87:*1–13.

28. To, L. B., Haylock, D. N., Dowse, T., et al (1994). A comparative study of the phenotype and proliferative capacity of peripheral blood (PB) CD34$^+$ cells mobilized by four different protocols and those of steady-phase PB and bone marrow CD34$^+$ cells. *Blood 84*:2930–2939.

29. Dercksen, M. W., Rodenhuis, S., Dirkson, M. K., et al (1995). Subsets of CD34$^+$ cells and rapid hematopoietic recovery after peripheral-blood stem-cell transplantation. *J Clin Oncol 13*:1922–1932.

30. Scott, M. A., Apperley, J. F., Jestice, H. K., Bloxham, D. M., Marcus, R. E., and Gordon, M. Y. (1995). Plastic-adherent progenitor cells in mobilized peripheral blood progenitor cell collections. *Blood 86*:4468–4473.

31. Murray, L., Chen, B., Galy, A., et al (1995). Enrichment of human hematopoietic stem cell activity in the CD34$^+$ Thy1$^+$Lin$^-$ subpopulation from mobilized peripheral blood. *Blood 85*:368–378.

32. Sutherland, H. J., Eaves, C. J., Lansdorp, P. M., Phillips, G. L., and Hogge, D. E. (1994). Kinetics of committed and primitive blood progenitor mobilization after chemotherapy and growth factor treatment and their use in autotransplants. *Blood 83*:3808–3814.

33. Breems, D. A., van Hennik, P. B., Kusadasi, N., et al (1996). Individual stem cell quality in leukapheresis products is related to the number of mobilized stem cells. *Blood 87*:5370–5378.

34. Bensinger, W., Appelbaum, F., Rowley, S., et al (1995). Factors that influence collection and engraftment of autologous peripheral-blood stem cells. *J Clin Oncol 13*:2547–2555.

35. Faucher, C., Le Corroller, A. G., Chabannon, C., et al (1996). Autologous transplantation of blood stem cells mobilized with filgrastim alone in 93 patients with malignancies: the number of CD34$^+$ cells reinfused is the only factor predicting both granulocyte and platelet recovery. *J Hematother 5*:663–670.

36. Kessinger, A., Vose, J. M., Bierman, P. J., and Armitage, J. O. (1991). High-dose therapy and autologous peripheral blood stem cell transplantation for patients with bone marrow metastases and relapsed lymphoma: An alternative to bone marrow purging. *Exp Hematol 19*:1013–1016.

37. Marit, G., Faberes, C., Pico, J. L., et al (1996). Autologous peripheral-blood progenitor-cell support following high-dose chemotherapy or chemoradiotherapy in patients with high-risk multiple myeloma. *J Clin Oncol 14*:1306–1313.

38. Schiller, G., Vescio, R., Freytes, C., et al (1995). Transplantation of CD34$^+$ peripheral blood progenitor cells after high-dose chemotherapy for patients with advanced multiple myeloma. *Blood 86*:390–397.

39. Lemoli, R. M., Fortuna, A., Motta, M. R., et al (1996). Concomitant mobilization of plasma cells and hematopoietic progenitors into peripheral blood of multiple myeloma patients: positive selection and transplantation of enriched CD34$^+$ cells to remove circulating tumor cells. *Blood 87*:1625–1634.

40. Dreyfus, F., Ribrag, V., Leblond, V., et al (1995). Detection of malignant B cells in peripheral blood stem cell collections after chemotherapy in patients with multiple myeloma. *Bone Marrow Transplant 15*:707–711.

41. Hardingham, J. E., Kotasek, D., Sage, R. E., et al (1995). Significance of molecular marker-positive cells after autologous peripheral-blood stem-cell transplantation for Non-Hodgkin's lymphoma. *J Clin Oncol 13*:1073–1079.

42. Corradini, P., Astolfi, M., Cherasco, C., et al (1997). Molecular monitoring of minimal residual disease in follicular and mantle cell non-Hodgkin's lymphomas treated with high-dose chemotherapy and peripheral blood progenitor autografting. *Blood 89*:724–731.

43. von Neuhoff, N., Suttorp, M., Dreger, P., Löffler, H., and Schmitz, N. (1996). Clinical implications of molecular monitoring in patients (pts) with low-grade non-Hodgkin lymphoma (NHL) after transplantation of autologous bone marrow (BM) or peripheral blood progenitor cells (PBPC). *Br J Haematol 93*:303.

44. Sharp, J., Kessinger, A., Mann, S., et al (1992). Detection and clinical significance of minimal tumor cell contamination of peripheral blood stem cell harvests. *Int J Cell Cloning 10*:92–94.

45. Brugger, W., Bross, K. J., Glatt, M., Weber, F., Mertelsmann, R., and Kanz, L. (1994). Mobilization of tumor cells and hematopoietic progenitor cells into peripheral blood of patients with solid tumors. *Blood 83*:636–640.

46. Sheridan, W. P., Begley, C. G., Juttner, C. A., et al (1992). Effect of peripheral-blood progenitor cells mobilised by Filgrastim (G-CSF) on platelet recovery after high-dose chemotherapy. *Lancet 339*:640–644.

47. Bensinger, W., Singer, J., Appelbaum, F., et al (1993). Autologous transplantation with peripheral blood mononuclear cells collected after administration of recombinant granulocyte stimulating factor. *Blood 81*:3158–3163.

48. Chao, N. J., Schriber, J. R., Grimes, K., et al (1993). Granulocyte colony-stimulating factor "mobilized" peripheral blood progenitor cells accelerate granulocyte and platelet recovery after high-dose chemotherapy. *Blood 81*:2031–2035.

49. Schmitz, N., Linch, D. C., Dreger, P., et al (1996). Randomised trial of Filgrastim-mobilised peripheral blood progenitor cell transplantation versus autologous bone-marrow transplantation in lymphoma patients. *Lancet 347*:353–357.

50. Beyer, J., Schwella, N., Zingsem, J., et al (1995). Hematopoietic rescue after high-dose chemotherapy using autologous peripheral-blood progenitor cells or bone marrow: a randomized comparison. *J Clin Oncol 13*:1328–1335.

51. Spitzer, G., Adkins, D. R., Spencer, V., et al (1994). Randomized study of growth factors post-peripheral-blood stem cell transplant: neutrophil recovery is improved with modest clinical benefit. *J Clin Oncol 12*:661–670.

52. Klumpp, T. R., Mangan, K. F., Goldberg, S. L., Pearlman, E. S., and Macdonald, J. S. (1995). Granulocyte colony-stimulating factor accelerates neutrophil engraftment following peripheral-blood stem-cell transplantation: A prospective, randomized trial. *J Clin Oncol 13*:1323–1327.

53. Cortelazzo, S., Viero, P., Bellavita, P., et al (1995). Granulocyte colony-stimulating factor following peripheral-blood progenitor-cell transplant in non-Hodgkin's lymphoma. *J Clin Oncol 13*:935–941.

54. Shimazaki, C., Oku, N., Uchiyama, H., et al (1994). Effect of granulocyte colony-stimulating factor on hematopoietic recovery after peripheral blood progenitor cell transplantation. *Bone Marrow Transplant 13*:271–275.

55. Uyl-de Groot, C. A., Ossenkoppele, G. J., van Riet, A. A., and Rutten, F. F. (1994). The costs of peripheral blood progenitor cell reinfusion mobilised

by granulocyte colony-stimulating factor following high dose melphalan as compared with conventional therapy in multiple myeloma. *Eur J Cancer 30A*:457–459.

56. Bennett, C. L., Armitage, J. L., Armitage, G. O., et al (1995). Costs of care and outcome for high-dose therapy and autologous transplantation for lymphoid malignancies: results from the University of Nebraska 1987 through 1991. *J Clin Oncol 13*:969–973.

57. Smith, T. J., Hillner, B. E., Schmitz, N., et al (1997). Economic analysis of a randomized clinical trial to compare filgrastim-mobilized peripheral blood progenitor cell transplantation and autologous bone marrow transplantation in patients with Hodgkin's and non-Hodgkin's lymphoma. *J Clin Oncol 15*:5–10.

58. Waters, T. M., Bennett, C. L., and Vose, J. M. (1997). Economic analyses of new technologies: the case of stem-cell transplantation. *J Clin Oncol 15*:2–4.

14
Allogeneic Peripheral Blood Progenitor Cell Transplantation

William I. Bensinger
Fred Hutchinson Cancer Research Center and University of Washington, Seattle, Washington

C. Dean Buckner
Response Oncology, Inc., Seattle, Washington

I. INTRODUCTION

It has been known for more than 30 years that hematopoietic precursors circulate in the peripheral blood and could potentially be utilized for autologous or allogeneic transplantation (1,2). Low steady-state concentration of these cells, however, precludes harvesting of useful numbers within a practical number of apheresis procedures, given the current technology. In an early attempt at allogeneic peripheral blood progenitor cell (PBPC) transplantation, a patient received a series of 10 buffy coat infusions that were depleted of T cells after collection from a donor who did not receive a growth factor (3). Although neutrophils recovered by day 11 and a bone marrow examination demonstrated trilineage engraftment on day 27, death from infection occurred on day 32 which prevented evaluation of platelet engraftment and graft-versus-host disease (GVHD). Clearly, ten 2- to 4-hour apheresis procedures are neither practical nor cost-effective for procuring hematopoietic stem cells from normal donors on a routine basis; however, PBPC collected in one to three apheresis procedures from patients with malignant disease after the administration of a recombinant growth factor can produce rapid and durable autologous hematopoietic reconstitution when infused after myeloablative therapy (4–6). These observations led to the demonstration that the administration of recombinant human granulocyte colony-stimulating factor (rHuG-CSF) to normal individuals increased the number of hematopoietic precursors in the peripheral blood where they could be collected in large quantities and used for syngeneic transplantation, and potentially for allogeneic transplantation (7–9). Syngeneic transplants were ideal for initial trials of PBPC collected from normal donors as the infusion of large numbers of syngeneic T cells should not cause severe GVHD. Several successful syngeneic PBPC transplants have been performed with durable engraftment and follow-up extending beyond 3 years (9). The administration of relatively high doses of rHuG-CSF to normal syngeneic PBPC and allogeneic granulocyte donors has been well tolerated and without measurable effects over 5 years (9–11).

The first successful allogeneic transplant using unmodified PBPC collected from a normal donor after the administration of rHuG-CSF was performed in 1991 for the treatment of graft failure unresponsive to two bone marrow infusions (12). In this case, the donor did not wish to undergo a third bone marrow aspiration. This patient is currently alive and well more than 4 years after PBPC infusion. The first primary allogeneic transplant using PBPC collected after rHuG-CSF was reported by Russell et al (13). This patient had acute lymphoblastic leukemia (ALL) in second remission and was treated with cyclophosphamide and total body irradiation followed by unmodified PBPC and GVHD prophylaxis with cyclosporine and methotrexate. He engrafted promptly and did not develop acute or chronic GVHD, but relapsed 36 months after transplantation. Peripheral blood progenitor cells were used in this case because the donor was at high risk for anesthetic complications because of morbid obesity. These initial observations led to rapid application of this technology, and several recent pilot studies, editorials, and review articles have been published (14–25).

Results of allogeneic PBPC transplantation to date suggest that this technique can produce substantially more rapid engraftment than observed with bone marrow. Further, contrary to widespread expectations, acute GVHD has not been intolerable, even with unmanipulated PBPC containing many more T cells than are present in a normal bone marrow graft. There also have been successful reports of using allogeneic PBPC after second transplants for graft rejection, graft failure, or relapse.

II. CHARACTERISTICS OF GROWTH FACTOR–MOBILIZED PERIPHERAL BLOOD PROGENITOR CELLS

A. Phenotypic and Functional Characteristics

The phenotypic characteristics of PBPC collected after growth factor administration have been reviewed by Rice and Reiffers (26), who concluded that a mixture of primitive and committed hematopoietic precursors were present that were not remarkably different from bone marrow. Antigenic analysis of CD34$^+$ PBPC revealed that these cells co-expressed a variety of differentiation antigens with properties that were similar to bone marrow–derived CD34$^+$ cells including subpopulations identified as CD34$^+$/CD33$^-$, CD34$^+$/HLADR$^-$, CD34$^+$/HLADR$^+$, and CD34$^+$/CD33$^+$ (27,28).

In humans, rHuG-CSF mobilizes cells which are resistant to 5-fluorouracil (5-FU) in culture and contain a mixture of hematopoietic progenitor cells similar in characteristics to bone marrow–derived stem cells (29). These 5-FU–resistant PBPC are capable of cytokine-mediated expansion in vitro (30). These data provide circumstantial evidence that Filgrastim-mobilized PBPC contain primitive hematopoietic precursors.

Published reports indicate that CD34$^+$ cells collected from the peripheral blood after rHuG-CSF differ from those in bone marrow in terms of adhesion molecule expression, proliferative potential, and cytokine response (31,32). Any and all of these characteristics may provide some advantage to Filgrastim-mobilized CD34$^+$ cells in terms of being more efficient in homing to the microenvironment and being more responsive to the signals produced there. Additional studies are needed to define qualitative or quantitative differences between bone marrow and PBPC that influence engraftment kinetics and determine if these differences are dependent on the specific growth factor or combination of growth factors used for mobilization.

B. Accessory Cell Content of Peripheral Blood Progenitor Cell Harvests

Rapid engraftment of autologous or allogeneic PBPC may not be due entirely to increased numbers or altered function of CD34$^+$ cells, but may be due to the presence of accessory cells that are increased in number or are qualitatively more capable of enhancing the milieu of the hematopoietic microenvironment. Initial studies designed to test this hypothesis evaluated the generation of cytokines by stromal cells co-cultured with bone marrow mononuclear cells or mononuclear cells collected from the blood after the administration of rHuG-CSF or with bone marrow mononuclear cells. Interleukin (IL)-6 and G-CSF levels were increased when bone marrow or rHuG-CSF–stimulated peripheral blood mononuclear cells were added to stroma, but the increase was tenfold greater with the latter (33).

C. T-Cell Content of Peripheral Blood Progenitor Cell Harvests

The T-cell content of normal PBPC grafts, compared with bone marrow grafts, is shown in **Table 1**. There are 1 to 2 logs more T cells infused with

Table 1. Cellular Composition of Marrow and Peripheral Blood
Progenitor Cell Grafts. Median and Range Are Given.

Cell phenotype	Marrow (N = 18) $\times 10^6$/kg	PBPC (N = 48) $\times 10^6$/kg
CD34$^+$	4.1 (0.6–5.7)	12 (1.74–23.6)
CD3$^+$	31 (10–65)	402 (81–974)
CD4$^+$	19 (1–47)	224 (45–578)
CD8$^+$	10 (1–24)	122 (33–354)

PBPC compared with bone marrow. Although initial studies did not report
an increase in T cells in the peripheral blood after the administration of
16 μg/kg Filgrastim (34), in another study of doses of 2.5 μg/kg for 6
consecutive days followed by 5 μg/kg for 4 days, investigators found a
significant increase in circulating T cells by day 8 (35). Dreger et al (7)
found that PBPC contained seven times more T cells and 20 times more
natural killer (NK) cells than bone marrow. Körbling et al (20) found a
1.0- to 1.4-log increase in T-cell subsets in PBPC harvests compared with
bone marrow.

   These data show that the numbers of T cells infused will be directly
related to the number of apheresis collections infused. Efforts to increase
the CD34$^+$ content of each collection to achieve the desired number with
the fewest collections will reduce the number of T cells infused in the graft.

D. Immunologic Characteristics of Peripheral Blood Progenitor Cells

Based on preliminary data from several studies, it can be concluded with
some assurance that the risk of acute GVHD is not increased after PBPC
transplantation, despite the infusion of 1 to 2 logs more T cells compared
with bone marrow. There are at least two possible explanations for this
observation. First, it is possible that the number of T cells in a bone marrow
graft is sufficient to produce the maximum possible severity of GVHD for
the degree of genetic disparity between donor and recipient, and that further
increases in T-cell dose have no additive effect. Support for this hypothesis
comes from clinical studies of T-cell–depleted allogeneic bone marrow in
which a threshold dose of approximately 10^5 residual CD3 cells/kg was
identified below which acute GVHD was significantly reduced (36,37). The
second explanation is that the T cells are functionally altered in rHuG-
CSF–mobilized PBPC because rHuG-CSF could alter the function of the

lymphocytes infused or could alter cytokine production by infused accessory cells. Support for this hypothesis comes from studies in mice, where rHuG-CSF pretreatment was found to polarize $CD4^+$ cells toward a T helper type-2 (Th2) response with a decrease in IL-2 and interferon production and a corresponding reduction in acute GVHD and improved survival (38). In vitro studies of patient samples have identified increased suppressor cell activity of $CD14^+$ cells in rHuG-CSF–mobilized PBPC from patients (39).

Neubauer et al (40) demonstrated that precursor cells capable of acquiring IL-2–inducible lymphokine-activated killer (LAK) activity were present in rHuG-CSF–mobilized PBPC and that LAK activity could be detected in the blood of all patients post-autologous PBPC transplant. Although such observations have little importance at the present time for allogeneic PBPC transplantation, they could become important if the cell populations that cause GVHD could be separated from the cells that are responsible for a graft-versus-tumor effect.

E. Allogeneic Peripheral Blood Progenitor Cell Transplants in Animals

Molineux et al (41) performed the first study of Filgrastim-mobilized PBPC in an animal model where syngeneic transplants were done in mice using sex-mismatched donors to evaluate donor/host chimerism. They were able to demonstrate durable bone marrow repopulation with donor cells as long as 192 days after transplant.

Successful allogeneic PBPC transplants have been performed in the canine model using peripheral blood mononuclear cells collected without chemotherapy or growth factor administration (42,43). Recent studies in dogs showed recombinant canine (rc)G-CSF and rc stem cell factor (SCF) to be synergistic for mobilizing transplantable PBPC (44,45). Allogeneic PBPC transplants have been performed in dogs conditioned with 9.2 Gy total body irradiation and no post-transplant immunosuppression. In both the DLA-identical and the DLA-haploidentical settings, all animals engrafted promptly, which was documented by variable number tandem repeat probes and, in the cases of sex-mismatched transplants, by cytogenetics. The rate of engraftment of DLA haploidentical allogeneic PBPC was significantly greater than observed with bone marrow alone, where 50% of dogs historically failed to engraft or rejected; however, all nine dogs that received haploidentical transplants were euthanized because of hyperacute GVHD. In the nine dogs that received DLA-identical transplants, eight dogs developed GVHD which was transient in five and fatal in three (45). The incidence and severity of acute GVHD was no greater, but not less, than would have been expected using bone marrow from DLA-identical donors where no post-transplant immunosuppression was used. In several animals, the cytokine-mobilized donor cells failed to proliferate in response to mismatched allogeneic cells in mixed leukocyte culture and did not stimulate mismatched allogeneic lymphocytes as well as before cytokine treatment. These findings are consistent with those made with human cells after in vitro treatment with rHuG-CSF (46).

In an outbred animal model, red burgundy rabbits were used as donors for New Zealand white rabbits of the opposite sex who served as recipients (47). It was demonstrated that PBPC mobilized with rHuG-CSF can engraft across major histocompatibility barriers with an incidence of GVHD that was no different from unmanipulated bone marrow.

Taken together, these limited data from animal models suggest that growth factor–mobilized allogeneic PBPC engraft better than bone marrow and cause no more acute GVHD despite the infusion of large numbers of lymphocytes. The increased rate of engraftment observed in the mismatched situation could be due to the added stem cells or to the increased number of lymphocytes infused. There is increasing evidence that G-CSF directly or indirectly alters the immunologic function of the PBPC graft, which probably modifies the manifestations of GVHD in the recipient; however, much work needs to be done in animal models to clearly define the effect of growth factors on the immunological function of the graft.

III. ALLOGENEIC PERIPHERAL BLOOD PROGENITOR CELL TRANSPLANT IN HUMANS

A. Collection of Peripheral Blood Progenitor Cells After the Administration of Filgrastim in Normal Donors

There are extensive data from the autologous PBPC transplant experience to predict toxicities likely to be observed in normal donors receiving recombinant growth factors followed by one or more apheresis procedures (4–6). There also are accumulating data on engraftment after autologous and syngeneic PBPC transplants that, by extrapolation, may assist in making decisions about cell dose requirements for allogeneic PBPC transplantation (4,5,9).

B. Toxicities of Filgrastim

Although other growth factors have been evaluated in animals and in humans undergoing autologous PBPC transplantation, rHuG-CSF has been the predominant drug evaluated in normal donors for granulocyte and PBPC harvesting. This hematopoietic factor was chosen for evaluation in normal donors because of its low toxicity profile in patients receiving autologous PBPC transplants. Administration of Filgrastim in doses as great as 16 μg/kg/day has been well tolerated in normal donors, with bone pain, flu-like symptoms, and myalgias being the only major side effects (16). There are no reports of discontinuation of Filgrastim because of immediate side effects in normal PBPC donors. Symptoms and the granulocytosis induced are, in general, reversed within 48 hours of discontinuing the drug.

In donors undergoing daily administration of Filgrastim and daily granulocyte collections for 7 to 14 days, there was a significant decrement in platelet levels when compared with patients undergoing the same collection procedures without the administration of Filgrastim (10). In patients undergoing one or two apheresis procedures for the collection of PBPC, platelet decrements to levels $<100 \times 10^9$/L are usually not observed. It is known from phase 1 studies of Filgrastim administration to patients who were not apheresed that platelet levels decrease (48). Similar findings of platelet decrements have also been observed after human or canine rHuGM-CSF administration in a dog model (49). The exact mechanisms of thrombocytopenia following cytokine-facilitated PBPC or granulocyte harvest are unknown.

C. Comparison of the Risks of Bone Marrow Versus Peripheral Blood Harvests

There is concern that the administration of rHuG-CSF to normal individuals could result in future disorders of hematopoiesis including leukemic transformation. These potential complications of rHuG-CSF administration would be difficult or impossible to detect after autologous transplantation where myelodysplastic syndromes are relatively common (50). Filgrastim has been given to more than 100 normal individuals to facilitate the collection of granulocytes or PBPC with the longest follow-up being 5 years (9,10,15,17,18). These donors have not had routine follow-up examinations beyond the immediate period of PBPC or granulocyte harvests, but there have been no known instances of hematologic problems. Although there is genuine concern about the long-term effects of administering rHuG-CSF to normal individuals, the existence and frequency of such effects will not be confidently known until very large numbers of donors are evaluated over a long period of time. It is relevant to reflect that it took many donations and many years for the risks of marrow harvesting to be assessed. Presently, it is difficult to decide if these unknown risks of rHuG-CSF administration are outweighed by potential benefits to the donor.

In addition to the unknown long-term effects of rHuG-CSF, donors of PBPC are subjected to one or more apheresis procedures that are generally considered to be low risk for morbidity and mortality; however, one donor with a history of cardiac disease who was deemed to be at high risk for anesthetic complications because of cardiac disease developed a myocardial infarction after the first apheresis procedure for PBPC harvest (15). Some donors with inadequate peripheral veins may require the placement of large-bore double-lumen catheters for vascular access which carries the risk of pneumothorax, bleeding, and thrombophlebitis. In a series of 90 normal donors undergoing apheresis for PBPC collections at the Fred Hutchinson Cancer Research Center (FHCRC), 14 required placement of a large-bore double-lumen subclavian or jugular catheter while the remainder were able to undergo the procedure with a peripheral vein-to-vein technique. Less invasive venous access techniques in individuals with inadequate peripheral veins are needed to decrease morbidity related to PBPC collections.

The advantages to the donor of PBPC rather than bone marrow harvesting include avoidance of general anesthesia and other complications. In a review of 1549 bone marrow harvests at the FHCRC, 27% had "significant" complications, including 3% that were considered major (more than five units of blood administered, more than 21 days of hip pain requiring hospitalization, or severe hypotension) and 0.4% that were considered life-threatening complications (cardiac arrest, severe hypotension, septicemia, and osteomyelitis) (51). Elsewhere there have been two unreported postoperative deaths after marrow harvest, both occurring in elderly donors. Any consideration of the risks to normal individuals from rHuG-CSF administration and PBPC harvest needs to be weighed against the relatively well-defined and significant risks of marrow harvest.

D. Peripheral Blood Progenitor Cell Dose for Allogeneic Transplantation

The first consideration in evaluating the dose of PBPC is to determine what benchmark should be used for judging the adequacy of harvests for allogeneic transplantation. Numerous autologous studies have not found total nucleated or mononuclear cell counts to be meaningful, while the results of colony-forming assays vary widely and, in any case, cannot be used for concurrent decision making. The number of $CD34^+$ cells infused per kg of recipient body weight appears to be the most reliable indicator of hematopoietic adequacy for autologous transplantation (4–6). For autologous transplants, 2.5×10^6 $CD34^+$ cells/kg is considered by some to be a minimum cell dose for consistent prompt recovery of granulocytes and platelets. At CD34$^+$ cell doses of 2.5 to 5.0 $\times$ 10^6/kg, compared with doses >5.0 $\times$ 10^6/kg, a significant proportion of patients receiving an autologous transplant will have a delay in platelet recovery that is exacerbated by the administration of rHuG-CSF or rHuGM-CSF after PBPC infusion (6). This phenomenon of "lineage diversion" observed after autologous PBPC transplantation should engender caution in the routine use of post-transplant myeloid growth factors following allogeneic PBPC transplantation as prolonged thrombocytopenia could occur. A dose of 5.0 $\times$ 10^6/kg is probably the best target CD34$^+$ cell dose for consistent and prompt engraftment of autologous PBPC without the administration of a post-transplant growth factor. Therefore, this should be a minimum CD34$^+$ cell dose for attempts of allogeneic PBPC transplantation.

Successful allogeneic PBPC transplants have been achieved with less than 5 $\times$ 10^6 CD34$^+$ cells/kg in several patients, but the minimum CD34$^+$ cell dose needed for rapid and complete allogeneic engraftment of all patients will remain unknown until more patients are evaluated.

Although the median CD34$^+$ cell dose collected at the FHCRC after administration of 16 μg/kg Filgrastim was high, 4 of 53 donors yielded only 0.6, 1.49, 1.55, and 1.74 CD34$^+$ cells/kg in two to four apheresis procedures. All other donors yielded 5.0 $\times$ 10^6 or more CD34$^+$ cells/kg. The low cell doses collected in these four individuals would, in general, be deemed inadequate for autologous engraftment; however, despite the low cell doses, successful engraftment was achieved with all four collections. Two of the four low CD34$^+$ cell doses were infused into syngeneic recipients and resulted in prompt engraftment of granulocytes and platelets. In another instance, a patient with graft failure received a CD34$^+$ cell dose of 0.6 $\times$ 10^6/kg that resulted in prompt recovery of granulocytes while platelets took 31 days to recover. In the fourth instance, where the dose was 1.74 $\times$ 10^6 CD34$^+$ cells/kg, bone marrow was harvested and infused after the PBPC. The bone marrow harvest in this donor yielded only 1.5 $\times$ 10^8 total nucleated cells/kg. This patient had recovery of granulocytes and platelets by day 20 after administration of both bone marrow and PBPC.

The administration of myelosuppressive agents after PBPC infusion could also affect CD34$^+$ cell-dose requirements. To achieve the benefit of earlier engraftment, larger CD34$^+$ cell doses may be required in patients receiving methotrexate than for patients receiving immunosuppressive anti-GVHD prophylaxis regimens that are not myelosuppressive.

Table 1 shows the cellular composition of 18 marrows harvested from normal unrelated donors from multiple centers compared with 48 unmodified allogeneic PBPC grafts. The goal in this study was to collect 15 $\times$ 10^6 CD34$^+$ cells/kg of recipient body weight in a maximum of two apheresis procedures for allogeneic PBPC transplantation.

Peripheral blood stem cells have been collected from normal donors after the administration of Filgrastim in doses of 2 to 16 μg/kg/day, and this subject has recently been reviewed (16). There is, in general, an increase in the number of CD34$^+$ cells collected with increasing doses of Filgrastim, but variations in CD34$^+$ cell yield also could be influenced by the timing of apheresis, centrifugation characteristics, or the duration of apheresis. A recently published, small study compared mobilizing doses of 3, 5, or 10 μg/kg of Filgrastim in normal donors (52). Although there was considerable overlap of CD34$^+$ cells and granulocyte-macrophage colony-forming cells (CFC-GM) in peripheral blood among the different dose cohorts, significantly higher median levels of CD34$^+$ cells and CFC-GM were found in the 10-μg dose level.

Doses of Filgrastim greater than 16 μg/kg/day for PBPC mobilization in normal individuals have not been reported, but it would be of interest to determine if a dose-response relationship persists at higher dose levels. For autologous patients, the optimal dose of Filgrastim for mobilization and collection of CD34$^+$ cells appears to be approximately 30 μg/kg/day (53). It will be important to determine the optimal dose and schedule of Filgrastim to maximize collections of CD34$^+$ cells and to provide background data for future studies of combinations of growth factors. In all the series of PBPC mobilization in normal donors, there is an approximate 1 log variation in PBPC yield among a particular group. One study of fixed-dose Filgrastim found a twofold difference in day 5 peripheral blood CFC-GM between young (20 to 30 years) and elderly (70 to 80 years) normal volunteers (54). Other reports have found an inverse relationship between peak CD34 peripheral blood cell concentration and age (7). This variation, however, is only partially explained by donor size and age. Thus any study designed to evaluate the optimum dose of Filgrastim for mobilization will require relatively large numbers of subjects to account for the intersubject variability.

Most investigators agree that the optimal collection days for PBPC are on the fourth and fifth days of rHuG-CSF administration, and 10 to 20 L of blood are processed in most studies (8,54). In a recent study, 77 normal donors mobilized with 12 μg/kg/day of rHuG-CSF were collected beginning on day 4 or day 5. Day-5 collections yielded a higher number of CD34$^+$ cells in one apheresis (55). In a detailed time course study of low-dose (2 μg/kg) rHuG-CSF given to normal donors, an additional increase in CFC-GM, mixed colony-forming cells (CFC-Mix), and erythroid burst-forming cells (BFC-E) were noted 4 to 6 hours after the day-3 and day-5 doses of cytokine (56). If this observation holds true for higher doses of rHuG-CSF, optimum scheduling for mobilization would include administration of the day 4 and day 5 rHuG-CSF 4 to 6 hours before collection.

The only other growth factor evaluated in normal individuals for PBPC harvesting has been rHuGM-CSF which was given to five normal donors resulting in relatively poor collections (57). The same investigators evaluated the combination of rHuG-CSF and rHuGM-CSF in five normal donors with collection of CD34$^+$ cell numbers similar to that achieved with rHuG-CSF alone. Whether or not there will be advantages to different doses and schedules of rHuG-CSF and rHuGM-CSF given together remains to be determined.

E. Cryopreservation of Allogeneic Peripheral Blood Progenitor Cells

Cryopreserved allogeneic PBPC can be used successfully for allografting (20,22,24). The recovery of CFC-GM and CD34$^+$ cells was good after thawing and there were no apparent problems with engraftment. Cryopreservation of allogeneic PBPC provides flexibility in the scheduling of the apheresis procedures and the transplant, and all the cells can be thawed and infused on 1 day. For the majority of patients receiving PBPC transplants, however, fresh cells can be administered without major difficulties with avoidance of cell loses and the expense of freezing and thawing.

F. T-Cell–Depletion Studies

Aversa et al (58) reduced the number of CD3$^+$ cells in PBPC collections to 1.24×10^5/kg with good recovery of CD34$^+$ cells using a soybean agglutination and E-rosetting technique. Dreger et al (77) compared three other approaches to T-cell depletion of PBPC: CAMPATH-1 plus autologous complement, immunomagnetic CD34$^+$ selection, and biotin-avidin-mediated CD34$^+$ selection. They found the immunomagnetic CD34$^+$ selection technique to be the most effective with the elimination of 4 logs of T cells. Suzue et al (59) reported three different techniques for T-cell depletion of PBPC, all of which were associated with a large loss of hematopoietic progenitor cells.

In a recent report, 10 patients received CD34$^+$ cell-enriched PBPC from HLA-matched, allogeneic, sibling donors (60). A median of 1.2×10^6 CD3$^+$ cells/kg were infused. Engraftment was rapid but grade 3 or 4 GVHD occurred in four of five patients receiving cyclosporine alone as prophylaxis and one of five patients receiving cyclosporine and methotrexate. These results were confirmed in another study of 16 patients with advanced hematologic malignancies transplanted with HLA-identical, allogeneic PBPC that were enriched for CD34$^+$ cells by the same avidin-biotin immunoadsorption technique (18). A median of 18.64 (range, 6.74

to 34.97) $\times$ 10^6/kg were collected and a median of 8.96 (range, 2.62 to 17.34) $\times$ 10^6 CD34$^+$ cells/kg were recovered after avidin-biotin adsorption which represented a median CD34$^+$ cell yield of 53% with a median purity of 62%. There was a 2.8-log reduction in T cells resulting in the infusion of 0.73, 0.40, and 0.32 $\times$ 10^6 CD3$^+$, CD4$^+$, and CD8$^+$ cells/kg, respectively. The median day to achieve neutrophils of 0.5 $\times$ 10^9/L and platelet of 20 $\times$ 10^9/L were 15 and 11 days, respectively. Grade 2 to 4 acute GVHD occurred in 87% and grade 3 or 4 in 40% of evaluable patients. These studies demonstrated the feasibility of allogeneic transplantation using CD34$^+$ cell-selected PBPC that result in prompt and sustained engraftment of all patients. The high incidence of acute and chronic GVHD that occurred despite significant T-cell depletion may partly be explained by the older age of patients (median, 48 years; range, 37 to 66), but it is also possible that populations of cells capable of modulating GVHD are removed by this technique of CD34$^+$ selection (18). In a more recent study of 20 older allograft recipients of HLA-identical sibling, CD34$^+$ cell-enriched PBPC, there was only a 10% incidence of grade 3 or 4 acute GVHD (61). CD34$^+$ cell selection in this study was accomplished using the immunomagnetic bead technique which resulted in a 3.9 log T-cell reduction. The lower incidence of severe GVHD in that study may be due to fewer infused T cells that were on the order of 10^5 CD3$^+$ cells/kg.

One theoretical advantage of CD34$^+$ cell enrichment over complement lysis techniques is that a second-stage selection technique could be used to select cell populations such as CD56$^+$ or CD8$^+$ cells that could be added back to the graft.

G. Addition of Peripheral Blood Progenitor Cells to Marrow for Allografting in HLA-Mismatched Allografts

Resistance to allogeneic grafts is affected by cell dose, the degree of genetic disparity between host and donor (62), transfusion-induced immunity in the patient (63), and depletion of T cells from the graft (64). Experimental animal data have indicated that increasing the number of donor hematopoietic progenitor cells in the graft improves the probability of engraftment across an allogeneic mismatched barrier (64). There are data from patients with aplastic anemia receiving HLA-identical transplants demonstrating a decrease in rejection and an improved survival with the infusion of a large number of cells from bone marrow (65). The number of donor hematopoietic cells available from marrow harvesting is limited and often suboptimal. The ability to collect PBPC after the administration of Filgrastim has made it possible to test the hypothesis that increasing the dose of donor hematopoietic cells will result in more consistent engraftment.

The incidence of graft rejection has exceeded 50% in recipients of T-cell–depleted marrow transplants from donors incompatible for two or three antigens using a soybean-lectin technique (66). Using the same technique for T-cell depletion not only of marrow but also of PBPC collected after rHuG-CSF, Aversa et al (58) have reported engraftment in 16 of 17 patients transplanted from HLA-haploidentical donors. It is remarkable to note that despite infusion of up to 620,000 T cells/kg, only one patient developed acute GVHD; however, survival was only 25% at 6 months, with deaths occurring predominantly from toxicity related to the very intense conditioning regimen that included cyclophosphamide, thiotepa, antithymocyte globulin, and single-dose total body irradiation given at a high dose rate. Results of this study suggest that the addition of T-cell–depleted PBPC assisted the establishment of allogeneic engraftment; however, T-cell depletion of both bone marrow and PBPC requires a complicated, labor intensive, and error-prone laboratory separation procedure that may well be simplified in the future by CD34$^+$ cell selection. Based on the emerging data from HLA-matched transplants with allogeneic PBPC, it should be possible to omit the bone marrow harvest in future studies of T-cell depletion and still infuse large quantities of CD34$^+$ cells.

H. Allogeneic Transplants Using Peripheral Blood Progenitor Cells Alone

1. Engraftment

Table 2 summarizes published data on engraftment after allogeneic PBPC transplants, without bone marrow, from HLA-matched siblings. Most pa-

Table 2. Engraftment After Allogeneic Peripheral Blood Progenitor Cell Transplantation. Median Day and Range for Engraftment Is Given.

Reference	N patients	Median day ANC $>5 \times 10^9$/L	Median day platelets $>20 \times 10^9$/L
17	37	14 (9–33)	11 (1–46)
20	16	9 (8–10)	15 (8–84$^+$)
67	33	14 (10–24)	14 (5–53)
24	24	16 (11–28)	14 (8–32)
76	11	15	13
25	8	15 (12–23)	19 (11–39)
22	25	10 (8–13)	14 (9–86)
14	17	14 (10–20)	14 (9–88)
61	19	13 (10–17)	10 (7–22$^+$)

tients had rapid engraftment of both neutrophils and platelets. In the Seattle patients, the median time to reach a peripheral neutrophil count $>0.5 \times 10^9$/L was day 14 for the PBPC group and 16 for the historical bone marrow group while the median time to achieve a platelet count $>20 \times 10^9$/L was day 11 in the PBPC group and 15 in the bone marrow group (p = 0.00063 and p = 0.00014, respectively).

The role of post-transplant growth factors in allogeneic PBPC transplants has yet to be determined. We have not routinely administered growth factors after allogeneic PBPC transplantation because of the marginal benefit on granulocyte recovery observed following autologous PBPC transplantation and the potential detrimental effect on platelet recovery (4–6).

In the Seattle series of allogeneic PBPC transplants, 19 of 37 patients received methotrexate as part of their prophylaxis for GVHD, and there was a 2- to 3-day delay in recovery of granulocytes and platelets for these patients compared with the 18 patients not receiving methotrexate. Other reports have confirmed that methotrexate delays the recovery of neutrophils and platelets (67). A delay in early platelet recovery was noted in patients given rHuG-CSF post-transplant when compared with patients not given the growth factor.

In patients with informative cytogenetic or molecular markers, chimerism was demonstrated that was not obviously different than that observed after marrow transplantation. Although the data are limited, with the longest follow-up being 3 years, there have been no late-graft failures after primary transplantation, suggesting that PBPC are capable of long-term engraftment.

2. Graft-Versus-Host Disease

Table 3 summarizes the incidence and severity of acute and chronic GVHD in recipients of HLA-matched allogeneic PBPC transplants. Most patients received cyclosporine with either methotrexate or prednisone for prophylaxis of GVHD. In the Seattle series, the estimated risks of developing grades 2 to 4 acute GVHD were 37% for the PBPC group and 56% for the historical bone marrow group (p = 0.18), while the estimated risks of grades 3 or 4 acute GVHD were 14% for the PBPC group and 33% for the bone marrow group (p = 0.05). In general, the estimated risks for developing acute GVHD after PBPC are not greater than would be expected after bone marrow grafts. It is of interest to note that the lowest incidence of acute GVHD (13%) was observed in a group of 16 patients who received cyclosporine and prednisone for prophylaxis (6).

In the Seattle series, the 7 of 18 evaluable patients receiving PBPC developed clinical chronic GVHD compared with 6 of 23 historical control

Table 3. Graft-Versus-Host Disease After HLA-Matched Allogeneic Peripheral Blood Progenitor Cell Transplantation.

Reference	N patients	Acute GVHD		Chronic GVHD all grades
		≥2	≥3	
61	37	37%	14%	40%
20	16	47%	NA	60%
67	33	37%	21%	36%
24	24	37%	NA	24%
76	8	50%	13%	16%
25	8	25%	13%	NA
22	25	42%	22%	NA
14	17	50%	23%	NA
17	19	13%	6%	33%

GVHD = graft-versus-host disease.
NA = not available.

patients receiving bone marrow. Similar findings have been reported by others; however, the incidence of chronic GVHD cannot yet be determined with accuracy due to the relatively small number and heterogeneity of the patients evaluated and the limited follow-up.

3. Outcome and Survival

The majority of patients given allogeneic PBPC have had advanced, and often refractory, hematologic malignancies, making evaluation of outcome difficult; however, there is the general impression that allogeneic PBPC transplants are associated with less morbidity than marrow transplants. In most studies, early survival is better for recipients of PBPC compared with bone marrow, but the numbers are small and the studies were not randomized. In the Seattle series, the estimated transplant-related mortality at 200 days was 27% for the PBPC group and 45% for the bone marrow group (p = 0.33), and overall survival was 50% and 41%, respectively (p = 0.39).

In the Seattle series, the probabilities of relapse were 70% for the PBPC group and 53% for the bone marrow group (p = 0.27). At the present time, there is no indication of a higher or lower relapse rate after infusion of PBPC; however, this needs to be carefully evaluated because of data suggesting that growth-factor–mobilized PBPC may be less alloreactive than bone marrow, which could have an adverse effect on relapses.

I. Peripheral Blood Progenitor Cell Transplantation for Relapse After Marrow Transplantation

Peripheral blood progenitor cells are being increasingly used instead of bone marrow for patients undergoing high-dose therapy and a second transplant for relapse after a first marrow transplant. Körbling et al (20) reported four patients who underwent high-dose therapy and allogeneic PBPC transplants from the same donors after relapse following an allogeneic marrow transplant. In three patients, acute GVHD was less than observed after the original marrow transplant and in one case it was more severe. In one patient, the remission after PBPC transplant was 8 months longer than the remission occurring after the first marrow transplant. Russell et al (68) reported second transplants in 10 patients using PBPC from the original bone marrow donor in eight instances and from a second donor in two. Four patients survived and all were transplanted more than 1 year after the first marrow transplant. Martinez et al (69) have reported a successful second transplant using allogeneic PBPC in a patient with acute myeloid leukemia (AML) who had relapsed 21 months after an allogeneic bone marrow transplant and is alive and without morbidity 7 months after the second transplant.

J. Peripheral Blood Progenitor Cell Infusion for Graft Failure

A small but significant fraction of patients have poor graft function or graft failure, usually of unknown etiology, after allogeneic marrow transplantation despite the presence of detectable donor cells in the bone marrow and peripheral blood. Previous studies of infusing additional unmodified bone marrow have been disappointing, with no improvement in the aplasia and resulting in many patients in increased acute GVHD (70). Dreger et al (12) reported recovery of blood counts after the infusion of unmodified PBPC after two unsuccessful attempts with bone marrow from the same HLA-identical donor. Recently Molina et al (71) reported two cases of graft failure that were reversed with PBPC from the original HLA-identical sibling bone marrow donor. In both instances, donor cells were detected in the bone marrow of the recipients before the infusion of PBPC. Both patients responded and had normal blood counts 280 and 324 days after the infusion of PBPC. Kook et al (72) reported the correction of graft failure, associated with the persistence of donor cells, in a patient with aplastic anemia who had received an HLA-identical bone marrow graft. This patient received antithymocyte globulin and PBPC from the original donor. These results achieved in four patients with graft failure of unknown etiology after HLA-identical sibling marrow transplants are of major interest. If the findings are reproducible, the use of PBPC would help solve a major problem as second bone marrow harvests within 1 to 2 months of the first produce very low cell yields, cause major discomfort, and are rarely successful (70).

K. Peripheral Blood Progenitor Cell Infusions for Graft Rejection

Immunologic graft rejection, defined as graft failure without the persistence of donor cells in bone marrow and blood, is difficult to treat. Further immunosuppression is probably required for successful engraftment. In the past, most patients died of complications from the toxicities of the reconditioning regimens unless the second transplant was performed for late graft failure. We have recently reported engraftment after second bone marrow infusions in four of seven patients conditioned with high-dose methylprednisolone and an anti-CD3-specific murine monoclonal antibody followed by cyclosporine. Two patients died of acute GVHD and two survive with normal hematopoietic function beyond 2 years from the second bone marrow infusion (73). Nine similar patients with graft rejection have received methylprednisolone and the same anti-CD3-specific murine monoclonal antibody followed by allogeneic PBPC from the original bone marrow donor. Engraftment occurred in six of nine patients, four of whom survive with good grafts more than 100 days after allogeneic PBPC infusion. Acute GVHD, grade 2 or 3, occurred in two of six patients with engraftment despite the infusion of an average of 6.5×10^9/kg CD3$^+$ T cells obtained in all but one case from HLA-mismatched or unrelated donors. It remains to be determined if there are advantages to using PBPC rather than bone marrow in this situation; however, all the donors who have donated bone marrow and PBPC have voiced a preference for the PBPC harvest if given a choice.

L. Allogeneic Peripheral Blood Progenitor Cell Infusions for Relapse

Lymphocyte infusions from the original marrow donor have been used as immunotherapy for post-transplant relapses with some success, especially in patients with chronic myeloid leukemia (CML) (74); however, a significant fraction of patients with CML who receive infusions of buffy coat cells from their bone marrow donor develop pancytopenia. Allogeneic PBPC collected after rHuG-CSF potentially no only provide lymphocytes for a graft-versus-leukemia effect but also provide hematopoietic precursors that should prevent aplasia. Majolino et al (78) have given allogeneic PBPC without prior ablative therapy to two patients who relapsed after an allogeneic marrow transplant. One patient with CML with a cytogenetic relapse achieved a remission without aplasia. The second patient had acute lymphocytic leukemia (ALL) in remission and was not evaluable for response. Grade 2 acute GVHD occurred in one of the two patients. In another study, however, there was no difference in the incidence of aplasia associated with buffy coat infusion or with rHuG-CSF–mobilized PBPC given for cytogenetic or clinical relapse after an allogeneic marrow transplant (75). Thus, it remains to be determined whether or not growth factor–mobilized PBPC will have an advantage over infusion of lymphocytes collected in the steady state.

M. Current and Future Studies

Animal and laboratory studies are being done to determine the qualitative differences between bone marrow and PBPC. The immediate goal for improvements in procurement of PBPC is to obtain enough cells for engraftment in one apheresis procedure. While this can be accomplished when unmodified PBPC are used, this may be difficult if T-cell–depletion techniques are used because of CD34$^+$ cell losses from the procedures. It will be extremely important to determine the optimal cellular composition of the graft to achieve rapid engraftment without severe GVHD while retaining a graft-versus-leukemia effect when needed. At the present time, there are no contraindications to pursuing the goal of optimizing the cellular composition of the PBPC graft. Further long-term evaluation of normal donors who have received growth factors for PBPC mobilization will help answer questions about the safety of this approach. A future goal will be to develop safe and effective growth factor regimens that will facilitate collection of adequate numbers of PBPC from 1 to 2 units of blood without apheresis altogether. Two randomized, controlled trials of allogeneic marrow versus PBPC are currently being done in Europe and in the United States that should answer questions related to engraftment, acute and chronic GVHD, relapse, and survival differences; however, the data reviewed in this chapter strongly suggest that future use of bone marrow will be limited and growth factor–mobilized PBPC will be the preferred source of hematopoietic stem cells.

ACKNOWLEDGMENTS

This study was supported in part by research grants CA18029, CA47748, CA18221, CA15704, and K08 CA01483 from the National Cancer Institute, the National Institutes of Health (Bethesda, MD), the José Carreras Foundation Against Leukemia (Barcelona, Spain), and the Joseph Steiner Krebsstiftung (Bern, Switzerland).

REFERENCES

1. Goodman, J. W., and Hodgson, G. S. (1962). Evidence for stem cells in the peripheral blood of mice. *Blood 19:*702–714.
2. McCredie, K. B., Hersh, E. M., and Freireich, E. J. (1971). Cells capable of colony formation in the peripheral blood of man. *Science 171:*293–294.
3. Kessinger, A., Smith, D. M., Strandjord, S. E., et al (1989). Allogeneic transplantation of blood-derived, T cell-depleted hemopoietic stem cells after myeloablative treatment in a patient with acute lymphoblastic leukemia. *Bone Marrow Transplant 4:*643–646.
4. Bensinger, W., Singer, J., Appelbaum, F., et al (1993). Autologous transplantation with peripheral blood mononuclear cells collected after administration of recombinant granulocyte stimulating factor. *Blood 81:*3158–3163.
5. Bensinger, W. I., Longin, K., Appelbaum, F., et al (1994). Peripheral blood stem cells (PBPCs) collected after recombinant granulocyte colony stimulating factor (rhG-CSF): an analysis of factors correlating with the tempo of engraftment after transplantation. *Br J Haematol 87:*825–831.
6. Bensinger, W., Appelbaum, F., Rowley, S., et al (1995). Factors that influence collection and engraftment of autologous peripheral-blood stem cells. *J Clin Oncol 13:*2547–2555.
7. Dreger, P., Haferlach, T., Eckstein, V., et al (1994). G-CSF-mobilized peripheral blood progenitor cells for allogeneic transplantation: safety, kinetics of mobilization, and composition of the graft. *Br J Haematol 87:*609–613.
8. Schwinger, W. (1993). Single dose of filgrastim (rhG-CSF) increases the number of hematopoietic progenitors in the peripheral blood of adult volunteers. *Bone Marrow Transplant 11:*489–492.
9. Weaver, C. H., Buckner, C. D., Longin, K., et al (1993). Syngeneic transplantation with peripheral blood mononuclear cells collected after the administration of recombinant human granulocyte colony-stimulating factor. *Blood 82:*1981–1984.
10. Bensinger, W. I., Price, T. H., Dale, D. C., et al (1993). The effects of daily recombinant human granulocyte colony stimulating factor administration on normal granulocyte donors undergoing leukapheresis. *Blood 81:*1883–1888.
11. Casper, C. B., Seger, R. A., and Berger, J. (1993). Effective stimulation of donors in granulocyte transfusions with recombinant methionyl granulocyte colony-stimulating factor. *Blood 81:*2866–2871.
12. Dreger, P., Suttorp, M., Haferlach, T., Löffler, H., Schmitz, N., and Schroyens, W. (1993). Allogeneic granulocyte colony-stimulating factor-mobilized peripheral blood progenitor cells for treatment of engraftment failure after bone marrow transplantation. *Blood 81:*1404–1407.

13. Russell, N. H., Hunter, A., Rogers, S., Hanley, J., and Anderson, D. (1993). Peripheral blood stem cells as an alternative to marrow for allogeneic transplantation. *Lancet 341*:1482.

14. Azevedo, W. M., Aranha, F. J., Gouvea, J. V., et al (1995). Allogeneic transplantation with blood stem cells mobilized by rhG-CSF for hematologic malignancies. *Bone Marrow Transplant 16*:647–653.

15. Bensinger, W. I., Weaver, C. H., Appelbaum, F. R., et al (1995). Transplantation of allogeneic peripheral blood stem cells mobilized by recombinant human granulocyte colony-stimulating factor. *Blood 85*:1655–1658.

16. Bensinger, W. I., Clift, R. A., Anasetti, C., et al (1996). Transplantation of allogeneic peripheral blood stem cells mobilized by recombinant human granulocyte colony-stimulating factor. *Stem Cells 14*:90–105.

17. Bensinger, W. I., Clift, R., Martin, P., et al (1996). Allogeneic peripheral blood stem cell transplantation in patients with advanced hematologic malignancies: A retrospective comparison with marrow transplantation. *Blood 88*:2794–2800.

18. Bensinger, W. I., Buckner, C. D., Shannon-Dorcy, K., et al (1996). Transplantation of allogeneic CD34$^+$ peripheral blood stem cells (PBPC) in patients with advanced hematologic malignancy. *Blood 88*:4132–4138.

19. Goldman, J. (1995). Peripheral blood stem cells for allografting. *Blood 85*:1413–1415.

20. Körbling, M., Przepiorka, D., Huh, Y. O., et al (1995). Allogeneic blood stem cell transplantation for refractory leukemia and lymphoma: Potential advantage of blood over marrow allografts. *Blood 85*:1659–1665.

21. Majolino, I., Aversa, F., Bacigalupo, A., Bandini, G., Arcese, W., and Reali, G. (1995). Allogeneic transplants of rhG-CSF-mobilized peripheral blood stem cells (PBPC) from normal donors. *Haematologica 80*:40–43.

22. Przepiorka, D., Anderlini, P., Ippoliti, C., et al (1997). Allogeneic blood stem cell transplantation: reduction in early treatment-related morbidity and mortality for patients with advanced hematologic malignancies. *Bone Marrow Transplant* (in press).

23. Russell, N. H., and Hunter, A. E. (1994). Peripheral blood stem cells for allogeneic transplantation. *Bone Marrow Transplant 13*:353–355.

24. Russell, J. A., Bowen, T., Brown, C., et al (1996). Allogeneic blood cell transplants for haematological malignancy: Comparison of engraftment and graft-versus-host disease with bone marrow transplantation. *Bone Marrow Transplant 17*:703–708.

25. Schmitz, N., Dreger, P., Suttorp, M., et al (1995). Primary transplantation of allogeneic peripheral blood progenitor cells mobilised by Filgrastim (G-CSF). *Blood 85*:1666–1672.

26. Rice, A., and Reiffers, J. (1992). Mobilized blood stem cells: Immunophenotyping and functional characteristics. *J Hematother 1*:19–26.

27. Bender, J. G., Unverzagt, K. L., Walker, D. E., et al (1991). Identification and comparison of CD34-positive cells and their subpopulations from normal peripheral blood and bone marrow using multicolor flow cytometry. *Blood 77*:2591–2596.

28. To, L. B., Haylock, D. N., Dowse, T., et al (1994). A comparative study of the phenotype and proliferative capacity of peripheral blood (PB) CD34$^+$ cells mobilized by four different protocols and those of steady-phase PB and bone marrow CD34$^+$ cells. *Blood 84:*2930–2939.

29. Rice, A., Barbot, C., Lacombe, F., et al (1993). 5-Fluorouracil permits access to a primitive subpopulation of peripheral blood stem cells. *Stem Cells 11:*326–335.

30. Rice, A., Boiron, J. M., Barbot, C., et al (1995). Cytokine-mediated expansion of 5-FU-resistant peripheral blood stem cells. *Exp Hematol 23:*303–308.

31. McQuaker, G., Haynes, A. P., Long, S. G., Hunter, A. E., and Russell, N. H. (1995). Study of the release of CD34 cells and colony forming cells in healthy subjects using G-CSF. *Bone Marrow Transplant 15:*S27.

32. Scott, M. A., Jestice, H. K., and Apperley, J. F. (1994). Quality and quantity of progenitor cells in mobilized peripheral blood. *Bone Marrow Transplant 2:*172.

33. Mielcarek, M., Roecklein, B. A., and Torok-Storb, B. (1996). CD14$^+$ cells in granulocyte colony-stimulating factor (G-CSF)-mobilized peripheral blood mononuclear cells induce secretion of interleukin-6 and G-CSF by marrow stroma. *Blood 87:*574–580.

34. Weaver, C. H., Longin, K., Buckner, C. D., and Bensinger, W. (1994). Lymphocyte content in peripheral blood mononuclear cells collected after the administration of recombinant human granulocyte colony-stimulating factor. *Bone Marrow Transplant 13:*411–415.

35. Matsunaga, T., Sakamaki, S., Kohogo, Y., Ohi, S., Hirayama, Y., and Niitsu, Y. (1993). Recombinant human granulocyte colony-stimulating factor can mobilize sufficient amounts of peripheral blood stem cells in healthy volunteers for allogeneic transplantation. *Bone Marrow Transplant 11:*103–108.

36. Kernan, N. A., Collins, N. M., Juliano, L., Cartagenia, T., Dupont, B., and O'Reilly, R. J. (1986). Clonable T lymphocytes in T cell-depleted bone marrow transplants correlate with development of graft-v-host disease. *Blood 68:*770–773.

37. Verdonck, L. F., Dekker, A. W., de Gast, G. C., van Kempen, M. L., Lokhorst, H. M., and Nieuwenhuis, H. K. (1994). Allogeneic bone marrow transplantation with a fixed low number of T cells in the marrow graft. *Blood 83:*3090–3096.

38. Pan, L., Delmonte, J. Jr., Jalonen, C. K., and Ferrara, J. L. (1995). Pretreatment of donor mice with granulocyte colony-stimulating factor polarizes donor T lymphocytes toward type-2 cytokine production and reduces severity of experimental graft versus host disease. *Blood 86:*4422–4429.

39. Mielcarek, M., Martin, P., and Torok-Storb, B. (1997). Suppression of alloantigen-induced T-cell proliferation by CD14$^+$ cells derived from granulocyte colony-stimulating factor–mobilized peripheral blood mononuclear cells. *Blood 89:*1629–1634.

40. Neubauer, M. A., Benyunes, M. C., Thompson, J. A., et al (1994). Lymphokine-activated killer (LAK) precursor cell activity is present in infused peripheral blood stem cells and in the blood after autologous peripheral blood stem cell transplantation. *Bone Marrow Transplant 13:*311–316.

41. Molineux, G., Pojda, Z., Hampson, I. N., Lord, B. I., and Dexter, T. M. (1990). Transplantation potential of peripheral blood stem cells induced by granulocyte colony-stimulating factor. *Blood 76*:2153–2158.

42. Carbonell, F., Calvo, W., Fliedner, T. M., et al (1984). Cytogenetic studies in dogs after total body irradiation and allogeneic transfusion with cryopreserved blood mononuclear cells: observations in long-term chimeras. *Int J Cell Cloning 2*:81–88.

43. Storb, R., Epstein, R. B., Ragde, H., and Thomas, E. D. (1967). Marrow engraftment by allogeneic leukocytes in lethally irradiated dogs. *Blood 30*:805–811.

44. de Revel, T., Appelbaum, F. R., Storb, R., et al (1994). Effects of granulocyte colony stimulating factor and stem cell factor, alone and in combination, on the mobilization of peripheral blood cells that engraft lethally irradiated dogs. *Blood 83*:3795–3799.

45. Sandmaier, B. M., Storb, R., Santos, E. B., et al (1996). Allogeneic transplants of canine peripheral blood stem cells mobilized by recombinant canine-hematopoietic growth factors. *Blood 87*:3508–3513.

46. Gerritsen, W. R., and O'Reilly, R. J. (1994). Granulocyte colony-stimulating factor (CSF) but not interleukin-1 (IL-1), IL-3, and granulocyte-macrophage CSF protect bone marrow progenitor cells from suppression by allosensitized cytotoxic T cells. *Blood 84*:1906–1912.

47. Gratwohl, A., Baldomero, H., John, L., et al (1995). Allogeneic G-CSF mobilised peripheral blood stem cell transplants (PBPC) in rabbits. *Bone Marrow Transplant 16*:63–68.

48. Lindemann, A., Herrmann, F., Oster, W., et al (1989). Hematologic effects of recombinant human granulocyte colony-stimulating factor in patients with malignancy. *Blood 74*:2644–2651.

49. Nash, R. A., Burstein, S. A., Storb, R., et al (1995). Thrombocytopenia in dogs induced by granulocyte-macrophage colony-stimulating factor: Increased destruction of circulating platelets. *Blood 86*:1765–1775.

50. Stone, R. M., Neuberg, D., Soiffer, R., et al (1994). Myelodysplastic syndrome as a late complication following autologous bone marrow transplantation for non-Hodgkin's lymphoma. *J Clin Oncol 12*:2535–2542.

51. Buckner, C. D., Petersen, F. B., and Bolonesi, B. A. Bone marrow donors. In: Forman S. J., Blume K. G., and Thomas E. D. eds. *Bone Marrow Transplantation,* Boston, MA: Blackwell Scientific Publications, 1994, pp. 259–270.

52. Grigg, A. P., Roberts, A. W., Raunow, H., et al (1995). Optimizing dose and scheduling of filgrastim (granulocyte colony-stimulating factor) for mobilization and collection of peripheral blood progenitor cells in normal volunteers. *Blood 86*:4437–4445.

53. Weaver, C. H., Hazelton, B., Palmer, P. A., et al (1996). A randomized dose finding study of filgrastim for mobilization of peripheral blood progenitor cells (PBPCs). *Proc Am Soc Clin Oncol 15*:341 (abstr 990).

54. Anderlini, P., Przepiorka, D., Huh, Y., et al (1996). Duration of filgrastim mobilization and apheresis yield of CD34$^+$ progenitor cells and lymphoid

subsets in normal donors for allogeneic transplantation. *Br J Haematol* 93:940–942.

55. Chatta, G. S., Price, T. H., Allen, R. C., and Dale, D. C. (1994). Effects of in vivo recombinant methionyl human granulocyte colony-stimulating factor on the neutrophil response and peripheral blood colony-forming cells in healthy young and elderly adult volunteers. *Blood* 84:2923–2929.

56. Sato, N., Sawada, K., Takahashi, T. A., et al (1994). A time course study for optimal harvest of peripheral blood progenitor cells by granulocyte colony-stimulating factor in healthy volunteers. *Exp Hematol* 22:973–978.

57. Lane, T. A., Law, P., Maruyama, M., et al (1995). Harvesting and enrichment of hematopoietic progenitor cells mobilized into the peripheral blood of normal donors by granulocyte-macrophage colony-stimulating factor (GM-CSF) or G-CSF; potential role in allogeneic marrow transplantation. *Blood* 85:275–282.

58. Aversa, F., Tabilio, A., Terenzi, A., et al (1994). Successful engraftment of T-cell-depleted halpoidentical "Three-Loci" incompatible transplants in leukemia patients by addition of recombinant human granulocyte colony-stimulating factor-mobilized peripheral blood progenitor cells to bone marrow inoculum. *Blood* 84:3948–3955.

59. Suzue, T., Kawano, Y., Takaue, Y., and Kuroda, Y. (1994). Cell processing protocol for allogeneic peripheral blood stem cells mobilized by granulocyte colony-stimulating factor. *Exp Hematol* 22:888–892.

60. Link, H., Arseniev, L., Bahre, O., Kadar, J. G., Diedrich, H., and Poliwoda, H. (1996). Transplantation of allogeneic CD34$^+$ blood cells. *Blood* 87:4903–4909.

61. Bensinger, W. I., Rowley, S., CD34 enriched, allogeneic peripheral blood stem cells used for transplantation. Proceedings of the 2nd International Symposium on Allogeneic Peripheral Blood and Cord Blood Transplantation. Oct 30–Nov 1, 1997:15.

62. Anasetti, C., Amos, D., Beatty, P. G., et al (1989). Effect of HLA compatibility on engraftment of bone marrow transplants in patients with leukemia or lymphoma. *N Engl J Med* 320:197–204.

63. Storb, R., Etzioni, R., Anasetti, C., et al (1994). Cyclophosphamide combined with antithymocyte globulin in preparation for allogeneic marrow transplants in patients with aplastic anemia. *Blood* 84:941–949.

64. Kernan, N. A. T-cell depletion for prevention of graft-versus-host-disease. In: Forman S. J., Blume K. G., and Thomas E. D. eds. *Bone Marrow Transplantation*, Boston, MA: Blackwell Scientific Publications, 1994, pp. 124–135.

65. Niederwieser, D., Pepe, M., Storb, R., Loughran, T. P. Jr., and Longton, G., for the Seattle Marrow Transplant Team (1988). Improvement in rejection, engraftment rate and survival without increase in graft-versus-host disease by high marrow cell dose in patients transplanted for aplastic anemia. *Br J Haematol* 69:23–28.

66. O'Reilly, R. J., Kernan, N., Cunningham, I., et al. Soybean lectin agglutination and E-rosette depletion for removal of T-cells from HLA-identical and non-identical marrow grafts administered for the treatment of leukemia. In: Martelli M. F., Grignani F., and Reisner Y. eds. *T-Cell Depletion in Allogeneic Bone-Marrow Transplantation,* Rome: Ares-Serono Symposia, 1988, pp. 123–129.

67. Urbano-Ispizua, A., Solano, C., Brunet, S., et al (1996). Allogeneic peripheral blood progenitor cell transplantation: analysis of short-term engraftment and acute GVHD incidence in 33 cases. *Bone Marrow Transplant 18*:35–40.

68. Russell, J. A., Bowen, T., Brown, C., et al (1996). Second allogeneic transplants for leukemia using blood instead of bone marrow as a source of hemopoietic cells. *Bone Marrow Transplant 18*:501–505.

69. Martinez, J. A., Picón, I., Carral, A., de la Rubia, J., Sanz, G. F., and Sanz, M. A. (1995). Allogeneic peripheral blood progenitor cells mobilized by G-CSF (filgrastim) for a second transplant in a patient with acute myeloid leukemia in relapse. *Bone Marrow Transplant 15*:149–151.

70. Bolger, G. B., Sullivan, K. M., Storb, R., et al (1986). Second marrow infusion for poor graft function after allogeneic marrow transplantation. *Bone Marrow Transplant 1*:21–30.

71. Molina, L., Chabannon, C., Viret, F., et al (1995). Granulocyte colony-stimulating factor-mobilized allogeneic peripheral blood stem cells for rescue graft failure after allogeneic bone marrow transplantation in two patients with acute myeloblastic leukemia in first complete remission. *Blood 85*:1678–1679.

72. Kook, H., Kim, H. J., Hwang, T. J., Suh, J. P., and Chung, I. J. (1995). The use of allogeneic peripheral blood stem cells (PBPC) boost in the treatment of late graft failure after allogeneic bone marrow transplant (BMT) in a patient with severe aplastic anemia. *Proc Am Soc Clin Oncol 14*:77 (abstr 15).

73. Anasetti, C., Martin, P. J., Storb, R., and Hansen, J. A. (1994). Engraftment of allogeneic hematopoietic stem cells in patients conditioned only with anti-CD3 monoclonal antibody BC3 plus methylprednisolone. *Blood 84*:249a (abstr 980).

74. Kolb, H. J., Mittermüller, J., Clemm, C., et al (1990). Donor leukocyte transfusions for treatment of recurrent chronic myelogenous leukemia in marrow transplant patients. *Blood 76*:2462–2465.

75. Flowers, M. E., Sullivan, K. M., Martin, P., et al (1995). G-CSF stimulated donor peripheral blood infusion(s) as immunotherapy in patients with hematologic malignancies relapsing after allogeneic transplantation. *Blood 86*:564a (abstr 2242).

76. Bacigalupo, A., Majolino, I., Van Lint, M. T., et al (1995). Transplantation of rh-G-CSF mobilized allogeneic peripheral blood cells from HLA identical sibling donors. *Bone Marrow Transplant 15*:S5 (abstr).

77. Dreger, P., Viehmann, K., Steinmann, J., et al (1995). G-CSF-mobilized peripheral blood progenitor cells for allogeneic transplantation: Comparison of T cell depletion strategies using different CD34$^+$ selection systems or CAMPATH-1. *Exp Hematol 23*:147–154.

78. Majolino, I, Buscemi, F., Scime, R., et al (1994). G-CSF mobilized PBPC for allogeneic transplantation. *Bone Marrow Transplant 84*:92a (abstr).

15
The Role of Filgrastim (r-metHuG-CSF) in the Treatment of Testicular Cancer

Craig Nichols and Scott Saxman
Indiana University School of Medicine,
Indianapolis, Indiana

I. OVERVIEW

In 1996, there were approximately 7300 new cases of germ-cell tumor in the United States. While still a rare disease, it is a disease of increasing importance since there appears to be a near doubling of the incidence of the disease since the 1940s and the disease strikes young men at the very beginning of productive lives. It is estimated that successful treatment of germ-cell cancer results in a net savings of $150,000,000 annually in the United States by returning these patients to the work force (1).

Treatment of testicular cancer has become a model in clinical oncology in the rational planning and successful completion of serial clinical studies that carefully defined therapy in all stages of the disease (2). Through these investigations, safe and tolerable therapies have been developed that cure all but 3% to 5% of patients presenting with germ-cell tumor and 70% to 80% of patients with disseminated germ-cell tumors.

II. DEVELOPMENT OF CHEMOTHERAPY FOR GERM-CELL TUMORS AND CURRENT STANDARDS OF TREATMENT

The breakthrough in the treatment of germ-cell cancer was the addition of cisplatin to regimens containing vinblastine and bleomycin (3). After this initial advance, subsequent studies throughout the 1970s confirmed the efficacy of this combination and refined the dose and schedule. By the early 1980s, there was sufficient experience to accurately assign prognosis and identify patients who were highly likely (>90%) to be cured with standard treatments and patients who by virtue of the extent and bulk of disease were at a significant risk of failing treatment (50%) (4). It was this seemingly simple information set that allowed for a divergence in development of clinical trials. Patients in whom nearly universally favorable outcome could be expected were entered into trials designed to address the issues of maintaining therapeutic efficacy while minimizing toxicity. Patients with advanced disease and a lesser chance of remission with standard chemotherapy trials were enrolled in trials designed to examine higher doses of chemotherapy and new agents in an effort to enhance the therapeutic effect of treatment.

Based on these premises, randomized studies have been able to define effective therapy with minimal toxicity for patients with favorable presenting features. For those patients with testis and retroperitoneal primaries, favorable markers, and no non-pulmonary visceral metastases, the standard of care is three cycles of cisplatin, etoposide, and bleomycin (BEP) or four cycles of the two-drug combination cisplatin and etoposide (EP). Clinical trials have addressed issues of substitution of carboplatin for cisplatin in these regimens and deletion of bleomycin from the combination of BEP (5,6). Both these attempts at minimizing toxicity have resulted in significant decrements of the cure rates. It is unlikely that major alterations of these well-tolerated, 9- to 12-week programs that cure 90% of patients will be made. Indeed, there are few ongoing clinical trials in this setting as the remaining questions are likely trivial.

In patients with poor-risk features, such as a mediastinal nonseminomatous primary, the presence of non-pulmonary visceral metastases, or unfavorable marker elevations, the anticipated therapeutic effect is diminished significantly. It is in these patients in whom newer therapies with higher doses and novel approaches are being tested to define a more effective treatment. Until 1985, the standard of care for patients with poor-risk features was four cycles of cisplatin, vinblastine, and bleomycin. At this time, a randomized comparison of cisplatin, vinblastine, and bleomycin (PVB) versus BEP demonstrated that the substitution of etopside for vinblastine was less toxic with respect to neuromuscular complications and more effective in the poor-risk subset (7). Accordingly, four cycles of BEP became standard therapy for patients with poor-risk disseminated germ-cell tumor. Subsequently, randomized trials in patients with poor-risk disease demonstrated that doubling of the cisplatin dose or the substitution of ifosfamide for bleomycin in the template of BEP resulted in increased toxicity without corresponding therapeutic benefit (8,9). The current intergroup trial in the United States in patients with poor-risk germ-cell tumors compares standard therapy with four cycles of BEP to two cycles of BEP followed by two cycles of high-dose chemotherapy with carboplatin, etoposide, and cyclophosphamide with stem-cell support.

Germ-cell cancer is unique in that effective, curative salvage therapy exists for patients failing primary chemotherapy. The current standard salvage chemotherapy that serves as a basis for comparison is the regimen reported at Indiana University of vinblastine, ifosfamide, and cisplatin (VeIP) (10). One hundred and twenty-four patients received cisplatin 20 mg/m^2 daily for 5 days, ifosfamide 1.2 g/m^2 daily for 5 days, and vinblastine 0.11 mg/kg on days 1 and 2. Patient characteristics reflected this poor-risk population. Advanced disease by the Indiana Classification system was present in 81 patients at initial presentation and 59 (48%) at the time of VeIP. Thirty-one patients had an extragonadal primary site. The toxicity of the regimen in this pretreated population was significant, with 73% developing granulocytopenic fever. This trial, however, was conducted in an era before the availablity of hematopoietic growth factors. Transfusions of platelets (28%) and red blood cells (48%) were common. Three patients died of treatment-related causes. Despite the formidable toxicity, the therapeutic results were gratifying. Fifty-six patients (45%) achieved a disease-free status with either chemotherapy alone (34 patients) or by resection of teratoma (15 patients, 12%) or viable carcinoma (7 patients, 6%). Twenty-nine of these patients are continuously disease-free and 37 (30%) are currently disease-free (minimum follow-up 27 months). Of the 77 patients who never obtained a disease-free status with primary BEP, 17 became disease-free with VeIP and 11 (14%) are continuously free of disease. These results have been confirmed by other investigators (11).

III. DOSE INTENSITY IN THE TREATMENT OF GERM-CELL CANCER

The concept that the dose of chemotherapy delivered is a significant determinant of therapeutic results in cancer therapy is logical and forms a mainstay of oncologic thought (see Chapter 8). It is this concept that underlies the development of new technologies to maximize dose of chemotherapy given such as hematopoietic growth factors, chemoprotectants, and bone marrow transplantation.

The dictum that "more is better" has arisen from clear-cut preclinical evidence of dose response and, largely, retrospective analysis of pooled clinical studies. Unfortunately, most of the clinical evidence supporting this dictum is flawed. In a single-arm study, it is impossible to determine whether superior outcome of more intensively treated patients is a consequence of better therapy or that the ability to tolerate full-dose therapy is a subtle reflection of better performance status, less bulky disease, younger age, and fewer concommitant health problems, factors that independently contribute to outcome regardless of dose received. Likewise, willingness to treat patients aggressively may be a reflection of the investigator's experience with the disease, better ancillary support, and other factors that may lessen therapy-related morbidity and mortality. Finally, different studies enter patients with diverse prognositic features, and different endpoints are used.

Additional difficulty arises in the simple calculation of dose intensity across studies. In most studies, dose actually received is rarely tabulated, making true comparisons of dose impossible. Assumptions regarding the equivalence of certain drugs are often empirical and not based on randomized trials. Such an approach almost always scores regimens with more drugs as more dose intense even if the additional drugs have marginal activity in that disease type. Total dose is usually ignored in this setting and the dose rate is the primary focus of most analyses of dose intensity. Such analyses do not consider that the addition of drugs with similar mechanisms of action are not necessarily additive in that they will likely act on similar sensitive populations of cells. Conclusions regarding dose intensity also reflect the standard chosen for comparison, which is again arbitrary.

With all these limitations, retrospective reviews of dose intensity in clinical trials of cancer therapy are best viewed as important for the generation of therapeutic hypotheses, but do not constitute proof that more drug delivered in a shorter period of time results in improved outcome. Such proof can only come from randomized clinical trials in which dose is assigned prospectively, prognostic factors are randomly allocated, and there is strict accounting of the dose actually delivered. Preferably, these trials should test the contribution of one factor at a time without altering other components of the regimen.

Clinical trials in testis cancer have contributed to the development of highly effective therapy in this rare neoplasm. Of interest, many of the straightforward randomized comparisons of chemotherapy dose intensity have been performed in patients with testicular cancer. Review of these trials provides an opportunity to explore the impact of dose intensity on treatment outcome in this rare but important tumor.

IV. PHASE 2 TRIALS OF INTENSIVE CHEMOTHERAPY IN POOR-RISK GERM-CELL CANCER

A variety of phase 2 clinical trials have been performed with the goal of intensifying the dose of available agents in patients with poor-risk germ-cell cancer. Most of these protocols have used high-dose cisplatin with etoposide, bleomycin, and, frequently, with the addition of agents of unknown efficacy in germ-cell cancer, such as vincristine and methotrexate. These trials have reported superior outcome in this patient population and such trials have been held as validation of the concept of dose intensity in germ-cell cancer. These high-dose phase 2 trials and relative dose intensity compared with BEP are summarized in **Table 1.**

These trials illustrate some of the difficulties in drawing conclusions regarding dose intensity from single-arm, pooled clinical studies. The investigators used a myriad of classification systems to assign poor risk. Many trials include patients who by other classification systems would be considered good risk (particularly patients with bulky abdominal disease only).

Table 1. Cisplatin/Etoposide-Based Therapy in Patients with Poor-Risk Germ-Cell Cancer.

Regimen	Dose intensity (relative to BEP)	% NED	References
VAB-6/EP	0.57 + ACT-D + VLB + CTX	54	26
BOP/VIP	0.72 + IFX + VNC	65	27
POMB-ACE	0.80 + others	80	12
BOP/BEP	0.82 + vincristine	66	28
BEP	1.00	88	29
BEP	1.00	63	7
BEP	1.00	73	9
PVeBV X 2 + PEC	1.13 + cytoxan	41	30
BEP^2	1.33	68	31
PVeBV	1.33 + vinblastine	88	14
BEP	1.61	86	32

ACT-D = dactinomycin.
BEP = bleomycin, etoposide, and cisplatin.
BOP/BEP = bleomycin, vincristine, cisplatin followed by bleomycin, etoposide, cisplatin.
BOP/VIP = bleomycin, vincristine, cisplatin followed by etoposide, ifosfamide, cisplatin.
CTX = cyclophosphamide (Cytoxan®).
IFX = ifosfamide.
NED = no evidence of disease.
PEC = cisplatin, epidoxorubicin and cyclophosphamide.
POMB-ACE = cisplatin, vincristine, methotrexate, bleomycin, actinomycin D, cyclophosphamide, and etoposide.
PVeBV = high dose CDDP + VBL + VP-16 + BLM (high dose cisplatin, vinblastine, etoposide, and bleomycin sulfate).
VAB-6/EP = cisplatin, vinblastine, bleomycin, cyclophosphamide, and dactinomycin followed by etoposide and cisplatin.
VLB = vinblastine.

Inclusion of such patients into poor-risk trials will seem to validate the newer regimen since such patients would have excellent outcome with standard therapy. When reclassified by other more stringent classification systems, response rates diminish and often are comparable to standard therapy. These trials are performed over a long period of time allowing for other non-treatment–related factors to influence the comparison with historical control groups.

The inclusion of more drugs into the combination will, by most methodologies for dose intensity calculation, increase the dose intensity. Examples of such an approach would be POMB-ACE, which incorporates vincristine and methotrexate along with agents of demonstrated activity in testicular cancer, cisplatin, etoposide, bleomycin, and actinomycin-D (12).

When all these limitations are considered, it is clear that the most proper (and perhaps only) way to truly test the concept of dose intensity in the clinical setting is by randomized comparisons. Again, in germ-cell cancer we have a number of trials that both illustrate some methodologic problems in trial design, some misinterpretations of properly conducted randomized trials, and, finally, several trials that allow for reasonable conclusions to be drawn. The number of randomized trials of dose intensity in germ-cell cancer exceeds the number in any other malignancy. Results of these trials are given in **Table 2**.

Table 2. Randomized Clinical Trials of Dose Intensity in Germ-Cell Cancer.

Study	Results	Reference
PV(0.4)B versus PV(0.3)B	No difference	33
low-dose PVB versus standard dose PVB	Favors standard dose arm	13
BEP versus EP	Favors BEP	6
PVeBV versus PVB	Favors four-drug, etoposide arm	14
BEP^2 versus BEP	No difference	9
PveVB versus PVeBV + PEC	Favors standard arm	30
BOP/VIP versus BEP	Favors standard arm	27

BEP = bleomycin, etoposide, and cisplatin.
BOP/VIP = bleomycin, vincristine, cisplatin followed by etoposide, ifosfamide, cisplatin.
EP = etoposide and cisplatin.
PEC = cisplatin, epidoxorubicin, and cyclophosphamide.
PV(0.3)B = cisplatin, vinblastine (0.3 mg/kg), bleomycin.
PV(0.4)B = cisplatin, vinblastine (0.4 mg/kg), bleomycin.
PVB = cisplatin, vincristine, and bleomycin.
PVeBV = high-dose CDDP + VBL + VP-16 + BLM (high-dose cisplatin, vinblastine, etoposide, and bleomycin sulfate).

V. RANDOMIZED CLINICAL TRIALS OF DOSE INTENSITY IN GERM-CELL CANCER

The common feature of most trials of intensive chemotherapy in germ-cell cancer is cisplatin dose intensity. Several early trials seem to suggest a steep dose–response curve for cisplatin in germ-cell cancer. The first trial in this regard, and a trial frequently cited as validation of the concept of dose intensity in cancer therapy is the study of the Southwest Oncology Group (SWOG) (13). Patients with disseminated germ-cell cancer were randomly assigned to receive vinblastine, bleomycin, and either cisplatin 120 mg/m^2 every 4 weeks (30 mg/m^2/week) or cisplatin 75 mg/m^2 every 4 weeks (19 mg/m^2/week). There was a statistically significant increase in the complete remission rate and survival for patients receiving the higher-dose therapy. This advantage was most apparent when the group of patients with maximal disease was considered (57% complete response compared with 34% complete response; p = 0.02). It clearly demonstrates that the 50% dose reduction of what is now considered standard cisplatin dose intensity does not maintain therapeutic outcome. As a point of reference, the low-dose cisplatin arm of this trial has a relative-dose cisplatin dose intensity of 0.58 compared with standard BEP.

The second important and frequently cited trial in this regard is the study conducted at the National Cancer Institute (NCI) (14). The clinical question posed was whether maximum escalation of chemotherapy doses resulted in superior outcome relative to standard therapy. Fifty-two patients with poor-risk germ-cell cancer were randomized to receive high-dose therapy with high-dose cisplatin (40 mg/m^2 daily for 5 days), etoposide (100 mg/m^2 daily for 5 days), vinblastine (0.2 mg/kg day 1), and bleomycin (30 units day 1, 8, and 15); or standard therapy with cisplatin (20 mg/m^2 daily for 5 days), vinblastine (0.3 mg/kg day 1), and bleomycin (30 units day 1, 8, and 15). Patients were randomized in a 2:1 ratio. Of 34 patients randomized to the high-dose therapy, 30 (88%) obtained disease-free status compared with 12 of the 18 patients (67%) who were randomized to the standard-dose therapy (p = 0.14). Sixty-eight percent of patients (23/34) randomized to the high-dose therapy arm remained alive and continuously free of disease compared with 33% (6/18) of patients randomized to the standard-dose therapy (p = 0.02). This small study with this particular design illustrates some of the difficulty in dose-intensity analysis of regimens containing different drugs. In this study, it is difficult to ascribe the apparent improvement in outcome solely to the high-dose cisplatin component of the regimen since full-dose etoposide was the other addition to the high-dose regimen. In particular, since the randomized comparison of cisplatin, bleomycin, and either etoposide or vinblastine in advanced germ-cell cancer performed by the Southeastern Cancer Study Group (SECSG) showed an advantage for the etoposide-containing group, one might argue that the seeming improvement in outcome is due to the inclusion of etoposide rather than the doubling of the cisplatin dose (7). Of note, this trial of the SECSG demonstrated a complete remission rate of 63% in patients with advanced disease receiving standard-dose BEP. We and other investigators have noted continued improvements in remission rate and survival rates using identical doses of chemotherapy.

To define the role of maximum-tolerated cisplatin therapy in advanced germ-cell cancer, Indiana University in conjunction with participating institutions in the SECSG and SWOG designed a straightforward comparison of etoposide and bleomycin and either standard-dose cisplatin (20 mg/m^2 days 1 to 5) or high-dose cisplatin (40 mg/m^2 days 1 to 5). Over a period of 4 years, 159 patients were enrolled in this trial. On the standard-dose cisplatin arm, 77 patients were eligible and evaluable for survival toxicity and response. Thirty-six patients (47%) had complete responses to chemotherapy, and an additional 20 patients (26%) became free of disease by total resection of teratoma (15 patients) or resection of residual cancer (5 patients). In the high-dose cisplatin group, 76 patients were eligible and evaluable for survival, toxicity, and response. Thirty-five high-dose patients (45%) obtained complete remission with chemotherapy alone while 16 patients (21%) required surgery to obtain disease-free status (15 teratoma and 1 cancer). Overall, 73% of patients receiving standard-dose therapy became disease-free while 67% of patients receiving high-dose therapy became disease-free.

Dose-intensity analysis was performed for each drug in the treatment regimen. Treatment courses in 72 patients receiving high-dose therapy could be analyzed for dose intensity. Three patients with early deaths and one patient who dropped out of treatment after 5 days were excluded from this analysis. The median cisplatin dose intensity for this group of patients was 60 mg/m^2/week with 47% of patients receiving >90% of projected dose and 79% receiving ≥80% of the projected dose of cisplatin. The median dose intensity of etoposide was 152 mg/m^2/week. Of patients in the high-dose group, 49% received ≥90% of the projected etoposide dose intensity and 79% received ≥80% of the projected etoposide dose intensity. Median dose intensity for bleomycin was 14 units/m^2/week with 44% of patients receiving ≥90% and 75% receiving ≥80% of projected bleomycin dose intensity.

Of patients receiving standard-dose therapy, 76 records could be analyzed for dose intensity. One patient died early related to treatment and was excluded from this analysis. The median dose intensity for cisplatin in this treatment group was 33 mg/m^2/week with 91% of patients receiving ≥90% of the projected dose intensity and 95% receiving ≥80%. The median dose intensity of etoposide in the standard-dose group was 166 mg/m^2/week with 91% of patients receiving $\geq$90% of projected etoposide dose intensity, and 95% received $\geq$80%. The median bleomycin dose intensity for this treatment group was 14 mg/m^2/week with 48% of patients receiving $\geq$90% of the projected bleomycin dose-intensity and 75% receiving $\geq$80%.

Overall dose intensity was calculated for each group and comparisons made for the most important single agent, cisplatin; the two most important agents, cisplatin and etoposide; and the overall dose intensity of all three agents. In all these analyses, the agents were given equal weight in consideration of dose intensity. The three univariate logistic regression analyses investigating the relationship between dose intensity and treatment were unable to conclude that dose intensity had an effect on response: cisplatin (p = 0.42), cisplatin and etoposide (p = 0.33), and cisplatin, etoposide, and bleomycin (p = 0.15).

This trial provides one of the best examples of a prospective analysis of dose intensity. The straightforward comparison of one variable (cisplatin dose intensity) simplifies the analysis. The randomized format assured that poor prognostic features (and inherent inability to tolerate aggressive treatment) were distributed equally between the two treatment groups. There was a full accounting of the dose actually delivered. The young age of the patients and the experience of the investigators allowed for maintenance of dose intensity in both treatment groups. Indeed, the median dose intensity for the entire combination in the standard-dose group was 0.97. The difference in dose intensity between the two groups (0.97 to 1.21) was substantially greater than the usual comparisons of dose intensity which cluster in the range of 0.6 to 0.8 relative dose intensity. This difference is even more striking when one considers the most important component in the BEP combination, cisplatin, for which the relative dose intensity was 1.82 to 1.00 in the high-dose and standard-dose groups, respectively, or even when the comparison is made with the two most important drugs, cisplatin and etoposide, for which the difference in relative dose intensity was 1.35 to 1.00.

The impact of achieving high cisplatin dose intensity in this study is clear. There was a marked increase in ototoxicity, neurotoxicity, gastrointestinal complications, and severe myelosuppression. Therapy-related deaths, while rare, occurred more commonly in the high-dose group. This increase in therapy-induced toxicity was not offset by an increase in complete remission rate or survival. With identical therapeutic results achieved with much less toxicity, cisplatin (100 to 120 mg/m^2 per course), etoposide, and bleomycin remains the standard for patients with poor-risk germ-cell cancer. Higher doses of cisplatin should not be used outside the context of a controlled clinical trial.

Since the mid-1980s, very-high-dose chemotherapy has been explored in germ-cell tumors. Initially studies were conducted in patients with highly refractory disease to define tolerable doses and to discern whether high-dose chemotherapy could cure patients with disease resistant to standard-dose chemotherapy. Single-institution and cooperative-group studies demonstrated that such attempts were warranted (11,15,16). Despite initial high therapy-related mortality rates (15% to 20%), a significant fraction of patients were cured (10% to 20%). Efforts began to improve the preparative regimen and improve patient selection for these procedures. Also, investigations have begun to evaluate the role of high-dose chemotherapy in earlier phases of treatment such as use with initial salvage treatment or in untreated poor-risk patients.

VI. NEUTROPENIC COMPLICATIONS IN THE TREATMENT OF GERM-CELL TUMORS

A. Neutropenic Fever

Chemotherapy with BEP induces significant neutropenia in the majority of patients receiving such treatment. Despite this, in patients with favorable prognostic features, neutropenic fever, sepsis, and infectious death are quite rare. In serial trials of chemotherapy in patients with favorable prognosis, the incidence of neutropenic fever is <15% and infectious death <1% (6,17).

Of interest, despite identical doses of chemotherapy, BEP results in a substantially higher incidence of neutropenic complications in patients with poor-risk features. These patients often present with symptomatic disease, low performance status, bulky tumors with obstruction, and poor nutritional status. In this group of patients, neutropenic fever occurs in approximately 20% to 30% of patients receiving BEP, and there is a 2% to 3% incidence of septic death. The substitution of ifosfamide for bleomycin in first-line treatment increases the incidence of neutropenic complications.

Management of patients with germ-cell tumors who develop neutropenic fever is controversial. Before the availability of hematopoietic growth factors, most investigators would institute a 20% to 25% dose reduction of the myelosuppressive components of chemotherapy (etoposide, vinblastine, or ifosfamide) with no modifications of cisplatin or bleomycin for subsequent courses of chemotherapy. There is no evidence that such modest decrements in dose intensity of the whole program has any adverse effect on therapeutic outcome.

There are very little data to guide the clinician regarding the use of growth factors in patients receiving primary chemotherapy for germ-cell tumors. There are no studies in good-risk patients receiving three cycles of BEP or four cycles of EP supporting or refuting the value of growth factors in this clinical setting. With the low anticipated rate of neutropenic fever and negligble impact on dose intensity of modest dose reduction, most investigators do not use hematopoietic growth factors in patients with good-risk disseminated germ cell tumors and, in those rare patients who develop neutropenic fever, institute dose reductions rather than growth factors.

In patients with poor-risk germ-cell tumors, there are some data suggesting that primary prophylaxis is appropriate. In a randomized trial in patients with poor-risk, germ-cell tumor, patients were randomized to receive standard BEP/EP for six cycles or a very schedule-intensive and dose-intensive regimen of bleomycin, vincristine, and cisplatin followed by etoposide, ifosfamide, and cisplatin (BOP/VIP) (18). Patients also were randomized to receive either Filgrastim or no hematopoietic support. Preliminary results in 263 randomized patients show that cycle delay (34% versus 21%), dose reduction (61% versus 24%), and granulocytopenic fever (29% versus 17%) all were significantly less in patients receiving Filgrastim. Perhaps more importantly, infectious deaths were more common (six patients compared with one) in patients not receiving Filgrastim; however, there was no difference in complete response rates or total deaths in either group. These results suggest that primary prophylaxis is appropriate in this group of ill patients who have as high as 30% incidence of neutropenic fever. The current intergroup trial in the United States includes Filgrastim in both treatment groups (standard BEP versus BEP + transplant).

B. Neutrophil Recovery

The current recommendations for treatment of disseminated germ-cell tumors with BEP or EP-based treatments call for repeating cycles of treatment every 21 days. The recommendation is to restart therapy on day 22 irrespective of the neutrophil count. This engenders some concern by clinicians accustomed to dose reductions and delays based on neutrophil recovery. In practical terms, current clinical practice is to initiate therapy on day 22 and, for those patients with very low granulocyte counts (absolute neutrophil count [ANC] $<0.5 \times 10^9$/L), simply obtain a blood count on day 4 of the 5-day cycle of treatment. In almost all patients, incipient recovery will be demonstrated and treatment proceeds. In those rare patients with ANC $<1.0 \times 10^9$/L on day 4, the last day of the myelosuppressive portions of the combination is deleted (etoposide and/or ifosfamide). There is no evidence that such a practice diminishes therapeutic efficacy, and sluggish neutrophil recovery is not an indication for the initiation of growth factor support.

VII. DOES G-CSF INTERACT WITH BLEOMYCIN?

The most serious complication of bleomycin is pulmonary toxicity, which most often consists of pneumonitis or pulmonary fibrosis. The reported incidence of pulmonary toxicity ranges from 3% to 40% in various series, with fatal toxicity occurring in 1% to 15% of patients (19). Several clinical characteristics are known to increase the risk of bleomycin lung toxicity. These include increasing age, cumulative bleomycin dose, concurrent chest radiation therapy, presence of renal insufficiency, and oxygen administration (20).

Recent anecdotal reports have suggested that concurrent administration of recombinant G-CSF may enhance bleomycin-induced pulmonary toxicity. Matthews reported five patients with Hodgkin's disease (HD) who received ABVD (Adriamycin®, bleomycin, vinblastine, dacarbazine) chemotherapy with rHuG-CSF given concurrently for neutrophil support (21). Four of the five patients developed clinical symptoms of pulmonary toxicity with cumulative doses of bleomycin of 70 to 80 U/m^2.

A larger series has been reported by Lei et al in which 12 Chinese patients with non-Hodgkin's lymphoma (NHL) were treated with BACOP (bleomycin, Adriamycin®, cyclophosphamide, vincristine, prednisone) with rHuG-CSF given between cycles for neutrophil support (22). The median cumulative dose of bleomycin was 54.5 U (range, 28 to 108 U). Four of these patients developed a rapidly progressive pulmonary syndrome that was characterized by diffuse pulmonary infiltrates and hypoxemia, with no identifiable infectious etiology. The investigators compared this group retrospectively with a group of 24 patients receiving BACOP without rHuG-CSF in which there was only one incidence of pulmonary toxicity and concluded that rHuG-CSF may have enhanced the bleomycin-induced pulmonary toxicity.

Bastion et al (23) have questioned the purported relationship between rHuG-CSF and bleomycin pulmonary toxicity in their analysis of two prospective studies in patients NHL. In these studies, 278 patients were randomized to receive a combination chemotherapy regimen that included bleomycin with or without the addition of rHuG-CSF. Overall the frequency of serious pulmonary toxicity that was felt to be due to bleomycin was nearly identical in the two groups (5.0% versus 6.5%); however, the total cumulative dose of bleomycin was much lower in these studies (80 U) than in either the previous case reports (70 to 80 U/m^2).

At Indiana University, we conducted a retrospective analysis of bleomycin toxicity in two groups of patients with advanced-stage germ-cell tumors (24). Group A (bleomycin with Filgrastim) consisted of 29 patients with advanced-stage germ-cell tumors. Twenty-three of these patients were treated on an Indiana University phase 2 study evaluating a five-drug chemotherapy regimen consisting of four cycles of etoposide, ifosfamide, cisplatin, vinblastine, and bleomycin (25). All patients received concurrent prophylactic Filgrastim. Six additional patients included in group A were treated with four cycles of bleomycin, cisplatin, and either vinblastine or etoposide. These patients developed granulocytopenic fever after the first chemotherapy cycle and received Filgrastim during cycles 2 through 4. Group B (bleomycin without Filgrastim) consisted of 57 patients who were treated on a phase 3 study comparing standard BEP to BEP with twice the cisplatin dose (9). At the time this study was conducted, Filgrastim was not commercially available, therefore none of these patients received growth factor. In both groups, bleomycin was given as a weekly bolus intravenous (IV) injection in a dose of 30 U for 12 weeks. Bleomycin pulmonary toxicity requiring discontinuation of the drug was based on either radiographic findings consistent with an active interstitial process, clinical exam findings suggestive of an interstitial process, or symptoms suggestive of pneumonitis including dyspnea or development of a persistent dry cough. The median dose of bleomycin administered in both groups was 300 U (range, 180 to 360 U). Filgrastim was administered in a dose of 5 μg/kg subcutaneously (SC) for 10 days starting on day 7 (24 to 48 hours after chemotherapy). Bleomycin was administered concurrently with Filgrastim on days 8 and 15.

Of the 29 patients who received concurrent chemotherapy and Filgrastim, 10 (34%; 95% CI 17.9% to 54.3%) were felt to have clinically significant bleomycin toxicity. Two of these patients subsequently died of bleomycin pulmonary toxicity after undergoing surgical resection of residual disease soon after completion of chemotherapy. Of the 57 patients who did not receive growth factor, 19 (33%; 95% CI, 21.4% to 47.1%) had bleomycin-related toxicity. One of these patients subsequently died of bleomycin-related pulmonary complications. There was no difference in the incidence of pulmonary toxicity between the groups (p = 1.00 by Fisher's exact test). Overall 33.7% (95% CI, 23.5% to 43.9%) of patients with advanced-stage germ-cell tumors receiving bleomycin had clinical evidence of pulmonary toxicity.

Although this was a retrospective study, we found that there was no increase in pulmonary toxicity with co-administration of Filgrastim and bleomycin compared with bleomycin alone in two populations of nearly identical patients with advanced germ-cell tumors. These results are consistent with the findings of the randomized European Organization for Research of Treatment of Cancer (EORTC) study which found three bleomycin deaths in the group of 130 patients not receiving rHuG-CSF and no bleomycin-related deaths in the 133 patients receiving rHuG-CSF concomitant with bleomycin. In any case, frequent clinical evaluation for signs of pulmonary fibrosis is necessary and bleomycin should be discontinued as soon as any sign of pulmonary toxicity is evident.

VIII. USE OF GROWTH FACTORS TO SUPPORT HIGH-DOSE CHEMOTHERAPY

Currently, the vast majority of high-dose chemotherapy programs for treatment of germ-cell tumors use emerging peripheral stem cell technology to supply hematopoietic support. In most settings, growth factor mobilization rather than chemotherapy mobilization is used before stem cell harvest. In almost all patients, sufficient yields are obtained for double autologous transplants that are commonly used in the United States. Most patients receive growth factors after high-dose chemotherapy in attempts to hasten recovery. At Indiana University, patients receive 10 μg/kg of Filgrastim for 4 days as mobilization, followed by peripheral stem cell harvest. This dose of Filgrastim begins as the patient is exhibiting granulocyte recovery from prior chemotherapy. After infusion of stem cells, patients receive 5 μg/kg/day until granulocyte recovery, which usually occurs between days 8 and 10.

IX. SUMMARY

Hematopoietic growth factors have been used extensively in patients with disseminated germ-cell tumor to support both standard chemotherapy and those patients receiving high-dose chemotherapy. Currently, we recommend primary prophylactic use of Filgrastim for those patients presenting with poor-risk features who are receiving four cycles of BEP, as well as for all patients receiving salvage chemotherapy. We do not believe primary prophylaxis with Filgrastim is appropriate for patients presenting with good-risk features or that hematopoietic growth factors should be used to hasten recovery of neutrophil counts. Filgrastim plays a critical role in mobilizing for stem cell harvest for those patients undergoing high-dose chemotherapy and is frequently used after high-dose chemotherapy to hasten neutrophil recovery.

REFERENCES

1. Shibley, L., Brown, M., Schuttinga, J., Rothenberg, M., and Whalen, J. (1990). Cisplatin-based combination chemotherapy in the treatment of advanced-stage testicular cancer: cost-benefit analysis. *J Natl Cancer Inst 82*:186–192.

2. Einhorn, L. H. (1990). Treatment of testicular cancer: a new and improved model. *J Clin Oncol 8:*1777–1781.
3. Einhorn, L., and Donohue, J. (1977). Cis-diaminedichloroplatinum, vinblastine, and bleomycin combination chemotherapy in disseminated testicular cancer. *Ann Int Med 87:*293–298.
4. Birch, R., Wiliams, S., Cone, A., et al (1986). Prognostic factors for favorable outcome in disseminated germ cell tumors. *J Clin Oncol 4:*400–407.
5. Bajorin, D. F, Sarosdy, M. F, Pfister, D. G., et al (1993). A randomized trial of etoposide and cisplatin versus etoposide and carboplatin in patients with good-risk germ cell tumors: a multiinstitutional study. *J Clin Oncol 11:*598–606.
6. Loehrer, P. J., Sr., Johnson, D., Elson, P., Einhorn, L. H., and Trump, D. (1995). Importance of bleomycin in favorable-prognosis disseminated germ cell tumors: an Eastern Cooperative Oncology Group trial. *J Clin Oncol 13:*470–476.
7. Williams, S. D., Birch, R., Einhorn, L. H., Irwin, L., Greco, F. A., and Loehrer, P. J. (1987). Treatment of disseminated germ cell tumors with cisplatin, bleomycin, and either vinblastine or etoposide. *N Engl J Med 316:*1435–1440.
8. Loehrer, P. J., Sr., Einhorn, L., Elson, P., et al (1993). Phase III study of cisplatin (p) plus etoposide (VP-16) with either bleomycin (B) or ifosfamide (I) in advanced stage germ cell tumors (GCT): an intergroup trial. *Proc Am Soc Clin Oncol 12:*261 (abstr 831).
9. Nichols, C., Williams, S., Loehrer, P., et al (1991). Randomized study of cisplatin dose intensity in advanced germ cell tumors: a Southeastern Cancer Study Group and Southwest Oncology Group protocol. *J Clin Oncol 9:*1163–1172.
10. Einhorn, L., Weathers, T., Loehrer, P., and Nichols, C. (1992). Second line chemotherapy with vinblastine, ifosfamide, and cisplatin after initial chemotherapy with cisplatin, VP-16 and Bleomycin (PVP-16B) in disseminated germ cell tumors (GCT). *Proc Am Soc Clin Oncol 11:*196 (abstr 599).
11. Motzer, R., Cooper, K., Geller, N., et al (1992). The role of cisplatin + fosfamide based chemotherapy as salvage therapy for patients with refractory germ cell tumors. *Cancer 66:*2476–2481.
12. Newlands, E. S., Begent, R. H., Rustin, G. J., Parker, D., and Bagshawe, K. D. (1983). Further advances in the management of malignant teratomas of the testis and other sites. *Lancet 1:*948–951.
13. Sampson, M. K., Rivkin, S. E., Jones, S. E., et al (1984). Dose-response and dose-survival advantage for high versus low-dose cisplatin combined with vinblastine and bleomycin in disseminated testicular cancer: a Southwest Oncology Group Study. *Cancer 53:*1029–1035.
14. Ozols, R. F., Ihde, D. C., Linehan, W. M., Jacob, J., Ostchega, Y., and Young, R. C. (1988). A randomized trial of standard chemotherapy v a high-dose chemotherapy regimen in the treatment of poor prognosis nonseminomatous germ-cell tumors. *J Clin Oncol 6:*1031–1040.
15. Nichols, C. R., Tricot, G., Williams, S. D., et al (1989). Dose-intensive chemotherapy in refractory germ cell cancer—a phase I/II trial of high dose carboplatin and etoposide with autologous bone marrow transplantation. *J Clin Oncol 7:*932–939.

16. Nichols, C. R, Williams, S. D, Loehrer, P. J., et al (1991). Randomized study of cisplatin dose intensity in poor-risk germ cell tumors: a Southeastern Cancer Study Group and Southwest Oncology Group protocol. *J Clin Oncol 9*:1163–1172.

17. Einborn, L. H., Williams, S. D., Loehrer, P. J., et al (1989). Evaluation of optimal duration of chemotherapy in favorable-prognosis disseminated germ cell tumors: a Southeastern Cancer Study Group protocol. *J Clin Oncol 7*:387–391.

18. Fossa, S., Kaye, S., Mead, G., et al (1995). An MRC/EORTC randomised trial in poor prognosis metastatic teratoma comparing treatment with/without filgastrim (G-CSF). *Proc Am Soc Clin Oncol 14*:245.

19. Jules-Elysee, K., and White, D. A. (1990). Bleomycin-induced pulmonary toxicity. *Clin Chest Med 11*:1–20.

20. Comis, R. (1992). Bleomycin pulmonary toxicity: current status and future directions. *Semin Oncol 19*:64–70.

21. Matthews, J. H. (1993). Pulmonary toxicity of ABVD chemotherapy and G-CSF in Hodgkin's disease: possible synergy (letter). *Lancet 342*:988.

22. Lei, K. L., Leung, W. T., and Johnson, P. J. (1994). Serious pulmonary complications in patients receiving recombinant granulocyte colony-stimulating factor during BACOP chemotherapy for aggressive non-Hodgkin's lymphoma. *Br J Cancer 70*:1009–1013.

23. Bastion, Y., Reyes, F., Bosly, A., et al (1994). Possible toxicity with the association of G-CSF and bleomycin. *Lancet 343*:1221–1222.

24. Saxman, S., Nichols, C., and Einhorn, L. (1997). Pulmonary toxicity in patients with advanced-stage germ cell tumors receiving bleomycin with and without granulocyte colony stimulating factor. *Chest 111*:657–660.

25. Blanke, C., Loehrer, P., Einhorn, L., and Nichols, C. (1994). A phase II study of VP-16 plus ifosfamide plus cisplatin plus vinblastine plus bleomycin (VIP/VB) with filgrastim for advanced stage testicular cancer. *Proc Am Soc Clin Onc 13*:234 (abstr 723).

26. Bosl, G. J., Geller, N. L., Vogelzang, N. S., et al (1987). Alternating cycles of etoposide plus cisplatin and VAB-6 in the treatment of poor-risk patients with germ cell tumors. *J Clin Oncol 5*:436–440.

27. Kaye, S., Mead, G., Fossa, S., et al (1995). An MRC/EORTC randomised trial in poor-prognosis metatatic teratoma, comparing BEP to BOP/VIP. *Proc Am Soc Clin Oncol 14*:246 (abstr 1525).

28. Horwich, A., Brada, M., and Nicholls, J. (1989). Intensive induction chemotherapy for poor risk non-seminomatous germ cell tumors. *Eur J Cancer Clin Oncol 25*: 177–184.

29. Pizzocaro, G., Piva, L., Salvioni, R., Zanoni, F., and Milani, A., (1985). Cisplatin etoposide, bleomycin first-line therapy and early resection of residual tumor in far-advanced germinal testis cancer. *Cancer 56*:2411–2415.

30. Droz, J., Pico, J., Biron, P., et al (1992). No evidence of a benefit of early intensified chemotherapy (HDCT) with autologous bone marrow transplantation (ABMT) in first line treatment of poor risk non seminomatous germ cell tumors. *Proc Am Soc Clin Oncol 11*:197.

31. Nichols, C. R., Andersen, J., Lazarus, H. M., et al (1992). High dose carboplatin and etoposide with autologous bone marrow transplantation in refractory germ cell cancer: an Eastern Cooperative Oncology Group protocol. *J Clin Oncol 10:*558–563.

32. Davguard G, Rorth M. (1986). High-dose cysplatin and VP-16 with bleomycin in the management of advanced metastatic germ-cell tumors. *Eur J Cancer Clin Oncol 22:*477–485.

33. Einhorn, L. H., and Williams, S. D. (1980). Chemotherapy of disseminated testicular cancer. A random prospective study. *Cancer 46:*1339–1344.

16
Use of Myeloid Hematopoietic Growth Factors in AIDS-Related Malignancies

Fa-Chyi Lee and Ronald T. Mitsuyasu
UCLA, Los Angeles, California

I. INTRODUCTION

Neutropenia in patients with human immunodeficiency virus (HIV) infection can be caused by a variety of mechanisms. These include direct or indirect effects of HIV in inhibiting hematopoiesis, generation of myelosuppressive inflammatory cytokines by HIV or opportunistic infections, and myelosuppression due to a variety of medications, infections, or tumors. Available data do not point to any specific findings in bone marrow that are uniformly associated with HIV infection and neutropenia (1). Several Food and Drug Administration (FDA)–licensed antiretroviral drugs can induce neutropenia, especially zidovudine (ZDV) (2). Hematopoietic growth factors have been used as adjunctive treatments to antiretroviral therapy (3), and the use of Filgrastim in the HIV-infected patient without malignancies is reviewed in another chapter of this volume (see Chapter 21).

Treating malignancies with cytotoxic drugs in patients with compromised marrow function can be challenging, and this is particularly true in patients with HIV infection and multiple causes of neutropenia. Since approval of Filgrastim in February 1991, it has been used to support hematopoiesis in just about all malignant diseases that require systemic chemotherapy (4). This chapter will discuss the use of Filgrastim in HIV-associated malignancies. In some cases, the clinical trial data are quite compelling in support of the routine use of myeloid hematopoietic growth factors, while in others its use is suggested based on clinical experiences, without many data from controlled clinical studies. It should be noted that Filgrastim is approved for the treatment of neutropenia of HIV infection only in Australia, India, and Japan.

II. HIV-ASSOCIATED MALIGNANCIES

Kaposi's sarcoma, non-Hodgkin's lymphoma (NHL), including primary central nervous system lymphoma, and invasive cervical carcinoma are the three malignancies classified by the U.S. Centers for Disease Control (CDC) as defining acquired immunodeficiency syndrome (AIDS). Other cancers such as Hodgkin's disease (HD), anal carcinoma, lung cancer, and non-melanomatous skin cancers can be found in the literature identified in HIV-infected individuals (5,6). A study involving 15,656 homosexual men in New York City and in San Francisco confirmed an increased incidence of anal cancer, HD, Kaposi's sarcoma, NHL, and HD in association with HIV infection (7). A high incidence of abnormal cervical cytology (8) and advanced-stage and aggressive invasive cervical carcinoma also has been seen in HIV-infected women (9).

III. KAPOSI'S SARCOMA

A. Clinical Presentation

Kaposi's sarcoma is the most common HIV-associated malignancy (10) and was one of the first described manifestations of AIDS (11). Based on the epidemiologic data, an infectious agent has long been suspected to play an important role in the etiology of Kaposi's sarcoma. Recently, sequences of a γ herpesvirus have been identified in tissue. This Kaposi's sarcoma-associated herpesvirus (KSHV), also termed human herpesvirus type-8 (HHV-8), is present in individuals with Kaposi's sarcoma with or without HIV infection (12,13).

Clinical manifestations of Kaposi's sarcoma are variable (14), as is its disease course. Some HIV-infected patients with Kaposi's sarcoma have indolent disease initially and may be observed or treated simply for control of local symptoms (15). As skin lesions, Kaposi's sarcoma starts as a brownish, hyperpigmented, macular lesion, and progresses to become raised and sometimes nodular, deep purplish-blue tumors. In some cases, the raised lesions become confluent and may form large plaques or a mass that may become ulcerated. Visceral Kaposi's sarcoma may involve any organ, and multiple organ involvement within an individual is not uncommon (16). Uniform treatment recommendations for the Kaposi's sarcoma of AIDS have been difficult to establish due to its varied presentation, other AIDS-associated diseases, and the different rates of disease progression. Current treatment usually does not cure Kaposi's sarcoma associated with AIDS; however, many patients can achieve significant tumor shrinkage and even complete remission with effective treatment. Most AIDS patients with Kaposi's sarcoma do not die as a result of their tumors, but may succumb to opportunistic infections. When Kaposi's sarcoma is the proximate cause of death, it is most commonly due to either pulmonary or other visceral organ compromise. The primary goal of treatment should be to relieve tumor-induced physical or psychological symptoms and should be a part of an overall HIV management plan.

B. Treatment of Kaposi's Sarcoma

For patients with limited cutaneous disease without evidence of visceral involvement, local treatment either with radiation or intralesional injection of sclerosing agents such as sotradecol (3% sodium tetradecyl) (17) or low concentrations of vinblastine can be effective and does not cause bone marrow suppression (18). Cryotherapy with liquid nitrogen (19), palliative local radiation therapy, and laser therapy also are effective and have few systemic side effects. Filgrastim use in these patients is typically not necessary.

Recombinant interferon-α (IFN-α) is the only biologic agent approved by the FDA for treatment of Kaposi's sarcoma in AIDS patients. The label recommendation for single-agent use of IFN-α is to administer 10 to 18 million international units (MIU) subcutaneously (SC) daily for the first 8 to 10 weeks followed by the same dose given three times weekly (20,21). The overall response rate is 35%, with better response rates seen in patients with CD4 counts >300/μL and who lack systemic symptoms of fever or night sweats (22). Myelosuppression can occur with IFN-α, although it appears to be dose related and generally is reversible upon stopping treatment.

The combination of ZDV and IFN-α has been shown to increase overall tumor response rates; however, neutropenia and anemia, which are the major dose-limiting toxicities of IFN-α, are more profound and occur more frequently (23). When 9 MIU/day IFN-α are combined with ZDV 1200 mg/day, the incidence of neutropenia increases to >47%. Low doses of recombinant human granulocyte-macrophage colony-stimulating factor (rHuGM-CSF) at 1.25 μg/kg/day have been shown to be effective in alleviating neutropenia in this setting (24,25). Clinical trials using Filgrastim to counteract neutropenia induced by the combination of IFN-α and ZDV have not been reported; however, case reports have demonstrated that rHuG-CSF given at 150 μg subcutaneously three times a week is effective in preventing neutropenia induced by IFN-α given with either vincristine or ZDV (26). This would suggest that Filgrastim at doses of 150 or 300 μg per day three or more times a week may be useful in preventing the development of neutropenia with regimens that combine IFN-α and ZDV or other myelosuppressive medications.

Chemotherapy drugs with activity against Kaposi's sarcoma include teniposide, vinblastine, vincristine, etoposide, bleomycin, doxorubicin, vinorelbine, liposomal doxorubicin, liposomal daunorubicin, and paclitaxel. Listed in **Table 1** are the dosing and major toxicities of the commonly used agents. As can be seen from Table 1, the most commonly encountered chemotherapy toxicity in AIDS patients with Kaposi's sarcoma is neutropenia, although skin rash, neuropathy, anemia, and palmar-plantar syndrome also can occur with certain medications.

1. Combination Chemotherapy

Most of the multidrug chemotherapy regimens combine two or more of the effective single agents listed in Table 1. Before the availability of the liposomal antracyclines, combination chemotherapy regimens such as doxorubicin, bleomycin, and vincristine (ABV) were felt to induce more responses than single-agent treatment, although prospective comparison studies are few. In one study, three regimens, doxorubicin alone, bleomycin with vincristine (BV), and the triple combination ABV regimen, were compared in AIDS patients with pulmonary Kaposi's sarcoma. The response rates for the combination regimens of either BV or ABV were superior to doxorubicin alone (27). In a larger study, major objective tumor responses were higher with the combination of ABV than with doxorubicin alone, 88% versus 48% (28). In both studies, the major toxicity was hematologic; neutropenia occurred commonly. For patients with symptomatic pulmonary Kaposi's sarcoma, the ABV regimen was effective in improving dyspnea is >90% of patients and induced radiologic responses in 64% (29). Certain anti-HIV medications, eg, ZDV, when combined with doxorubicin or ABV can cause profound neutropenia that may be overcome or prevented with myeloid hematopoietic growth factors (ACTG 075) (30).

2. Newer Single-Agent Chemotherapy for the Kaposi's Sarcoma of AIDS

As a result of recently reported clinical trials comparing liposomal doxorubicin or liposomal daunorubicin with BV or ABV, single-agent liposomal antracycline therapy is rapidly becoming the initial treatment of choice for patients with advanced Kaposi's sarcoma requiring treatment with cytotoxic chemotherapy. A multicenter, randomized trial comparing single-agent li-

Table 1. Single-Agent Chemotherapy for the Kaposi's Sarcoma of AIDS.

Agent	Dosage	Response rate (%)	Dose-limiting toxicity	Reference
Bleomycin	5 mg/day IM × 3 days or 6 mg/m^2/ day CIV × 4 day	48	Neutropenia, rash	76
Vincristine	1–2 mg IV qw for 2–5 weeks then q2w	61	Neuropathy	77
Vinblastine	4 mg IV qw, increase dose as tolerated	26	Neutropenia	78
Teniposide	360 mg/m^2 IV q3w	40	Neutropenia	79
Etoposide	150 mg/m^2 IV × 3 day q 28 day	76	Neutropenia	80
Doxorubicin	20 mg/m^2 IV q2w	48	Neutropenia, cardiomyopathy	28
Paclitaxel	100–135 mg/m^2 IV over 3 hours, q3w	65	Neutropenia, neuropathy	81
Liposomal doxorubicin	20 mg IV q2–3w	74	Neutropenia, palmar-plantar rash	82
Liposomal daunorubicin	40 mg/m^2 IV q2w	63	Neutropenia	83
Vinorelbine	30 mg/m^2 IV q2w	52	Neutropenia	84

CIV = continuous intravenous.
IV = intravenous.
IM = intramuscular.
qw = every week.
q2w = every 2 weeks.
q3w = every 3 weeks.

posomal doxorubicin with ABV showed a higher response rate with liposomal doxorubicin than with ABV, 43.2% vs 25.5%, respectively, in previously treated advanced-stage AIDS patients with Kaposi's sarcoma (31). A trial comparing single-agent liposomal daunorubicin with ABV in chemotherapy-naive patients showed that the response rates were equivalent, 28% and 25%, respectively; however, overall toxicity was less with liposomal daunorubicin (32). Complications from chemotherapy, eg, alopecia and neuropathy, were significantly more frequent in the combination group, whereas the incidence of severe neutropenia requiring Filgrastim treatment was more frequent with single-agent liposomal daunorubicin (32). In a study comparing liposomal doxorubicin with liposomal doxorubicin and bleomycin and vincristine (ACTG 286), a somewhat higher mortality rate was seen with the combination therapy than with the single-agent treatment at the interim analysis (33).

Paclitaxel (Taxol®) has also been shown to have significant antitumor activity against Kaposi's sarcoma at 135 mg/m^2 infused over 3 hours in a study by Saville et al (34). Good responses also were seen at a lower dose of paclitaxel 100 mg/m^2 by Gill et al (32). Major toxicity with this drug is neutropenia which can be overcome with Filgrastim.

3. Use of Myeloid Growth Factors in the Treatment of Kaposi's Sarcoma

A number of published studies assessing the usefulness of myeloid hematopoietic growth factors in treatment of AIDS patients with Kaposi's sarcoma have involved the use of rHuGM-CSF. A few of these studies specifically looked at the use of growth factor support with chemotherapy. Gill et al (35) studied the use of rHuGM-CSF to support the combination chemotherapy regimen ABV. The dose of rHuGM-CSF used was 250 μg/m^2 every 12 hours given on days 2 to 12 of each chemotherapy cycle. Antiretroviral medications were not allowed while the patient was undergoing chemotherapy. Patients were treated for a median of six cycles, and transient neutropenia was the most common toxicity. No patient experienced bacterial infection, and all patients remained p24 antigen negative throughout the treatment period. Higher doses of rHuGM-CSF were associated with systemic toxicities of fever, fatigue, bone pain, and diarrhea. Bakker et al (36) reported their experience of treating Kaposi's sarcoma with ABV and rHuGM-CSF given SC at a dose of 5 μg/kg/day from days 2 to 12, when the absolute neutrophil count (ANC) became <0.5 × 10^9/L. Treatment-induced neutropenia was reduced by using rHuGM-CSF.

Sloand et al (37) published their results using a six-drug combination, (doxorubicin, bleomycin, vincristine, vinblastine, actinomycin D, and dacarbazine) in 18 patients with pulmonary Kaposi's sarcoma. Antiretroviral medications, including ZDV with zalcitabine or didanosine, were continued throughout chemotherapy. Filgrastim was given at 300 μg/day to all patients with ANC <1.0 × 10^9/L until their neutropenia resolved. It was found that use of Filgrastim shortened the duration of neutropenia and made possible the delivery of full doses of this multidrug regimen. No patients receiving Filgrastim required dose reduction. Preliminary results of the study of BV plus etoposide with or without Filgrastim showed that the group receiving Filgrastim had a lower incidence of neutropenia and required fewer dose reductions (38).

IV. LYMPHOMAS

A. Clinical Presentation

Non-Hodgkin's lymphoma is the second most common malignancy associated with HIV infection. Incidence studies show that HIV-infected individuals have a 60-fold higher risk of developing NHL than the general population (39). In the United States, as well as in Europe, NHL constitutes 3% to 5% of AIDS-presenting illness (40). Patients with lower CD4 counts generally have a worse prognosis and more advanced HIV disease. This is especially true of the subgroup of AIDS patients with primary central nervous system lymphoma in whom the CD4 counts are almost invariably <50/μL (41). Most cases of NHL in AIDS patients are B-cell immunoblastic, small non-cleaved, lymphoblastic, or large-cell lymphomas (42,43). Similar to NHL in non-HIV–infected patients, the AIDS-associated ones are a heterogeneous group of diseases (44,45).

A large majority (80% to 90%) of the AIDS patients with NHL present with "B" symptoms, and many have extranodal disease (46). Usually, patients present with a rapidly growing bulky tumor. Even patients who present with an initial diagnosis of gastrointestinal lymphoma have been found to have a high prevalence of extranodal involvement, 73% (47). For treatment purposes, there may be no truly localized disease in NHL in AIDS patients, and all cases should be regarding as disseminated at diagnosis. For this reason, systemic chemotherapy with or without local radiation is the primary treatment for all patients with non-primary central nervous system lymphomas.

Primary central nervous system lymphoma in HIV-infected individuals is generally of the aggressive, large-cell type, and all have integrated Epstein-Barr virus (EBV) genome. Most patients also have detectable EBV DNA in their cerebrospinal fluid (48,49). Primary central nervous system lymphoma is rare, 0.4% of AIDS cases, although its absolute incidence in persons with AIDS is 3600-fold higher than in the general population (50).

A possible association of HD and HIV infection was first reported in 1984 (51). An increased incidence of HD in HIV-infected individuals subsequently was reported among young men in San Francisco (7) and in Italy (52). The French registry experience showed the following distinctive clinical characteristics of patients with the disease: 1) low rate, 13% of mediastinal involvement; 2) 75% presented with clinical stage 3 or 4 disease; 3) 80% had B symptoms within 3 months of diagnosis; 4) higher CD4 counts than patients with NHL, median of $306/\mu L$; and 5) 11% had antecedent AIDS diagnosis (53).

B. Treatment of AIDS Lymphoma

1. Non-Hodgkin's Lymphoma

The major concern of treating patients with AIDS lymphoma has been whether dose-intensive chemotherapy or even standard-dose chemotherapy would further compromise the immune system and induce unacceptably high rates of infection. One small study with 22 patients compared standard dose m-BACOD (methotrexate, calcium leucovorin, bleomycin, doxorubicin, cyclophosphamide, vincristine, and dexamethasone) with a novel intensive regimen consisting of L-asparaginase vincristine and high-dose cytosine arabinoside 3 g/m^2 on day 1, cyclophosphamide 1.5 g/m^2 on day 22, and methotrexate 2 g/m^2 on day 36 (54). The aim of the study was to evaluate if high-dose-intensive chemotherapy would improve tumor response rates, reduce tumor relapses, and prolong survival. The opposite was demonstrated, however. The group treated with high-dose chemotherapy experienced more opportunistic and other infections and had lower survival than patients receiving the standard-dose treatment. Central nervous system relapse also occurred despite intensive treatment. This finding was confirmed by another study using an intensive MACOP-B (methotrexate, doxorubicin, cyclophosphamide, vincristine, prednisone, and bleomycin) regimen (55). High-dose chemotherapy using cyclophosphamide 120 mg/kg with either bulsulfan 16 mg/kg or total body irradiation of 10 Gy followed by peripheral blood stem cell and Filgrastim support in HIV-infected patients with hematologic malignancies and CD4 counts $>300/\mu L$ found that treatment was well tolerated; however, relapses continued to occur (56).

Retrospective studies of several standard lymphoma chemotherapy regimens with Filgrastim support showed that HIV-infected individuals tolerated CHOP (cyclophosphamide, doxurubicin, vincristine, and prednisone). COP (cyclophosphamide, vincristine, and prednisone), or MACOP-

B as well as non-HIV–infected patients, except for the subgroup of patients with prior opportunistic infections and lower CD4 counts in whom mucositis and myelosuppression were more severe (57). Dose-intensive chemotherapy appears to be best tolerated and response rates higher in patients with CD4 counts >100/μL and those with better performance status and no prior AIDS diagnoses (58).

Filgrastim has been used to prevent neutropenia in a regimen using methotrexate 1 g/m^2 every week with increasing dose of ZDV at either 2, 4, or 6 g/m^2 given every 2 weeks (59). Another study showed that using Filgrastim allowed dose intensification of cyclophosamide and epirubicin with acceptable hematologic toxicity in a regimen consisting of cyclophosphamide, epirubicin, vincristine, and prednisone (CEOP) in AIDS patients with lymphoma (60).

It is recommended that the majority of patients with AIDS-related lymphoma receive moderately dose-intensive first- or second-generation chemotherapy regimens (such as CHOP or m-BACOD) along with hematopoietic growth factor support, and that more dose-intensive regimens should be reserved for selected populations of patients, such as those with small non-cleaved lymphomas. It also is recommended that patients receive effective antiretroviral treatment with their chemotherapy whenever possible.

2. Primary Central Nervous System Lymphoma

The major treatment for primary central nervous system lymphoma is cranial irradiation. Whole brain radiation therapy improves survival in HIV patients with primary central nervous system lymphoma, 134 days compared with 48 days in patients not receiving radiation (61). Radiation followed by three cycles of procarbazine, lomustine, and vincristine (PCV) prolonged median survival to 13.5 months in four patients (62). Another combined modality therapy using methotrexate 3 g/m^2, thiotepa 30 mg/m^2, and procarbazine 50 mg/m^2 showed that patients who achieved complete response survived for an average of 7 months (63). All patients treated with chemotherapy developed neutropenia, and two died of infections before the availability of Filgrastim. Use of Filgrastim is highly recommended in the setting of combined modality therapy for primary central nervous system lymphoma.

3. Hodgkin's Disease

Hodgkin's disease in HIV-infected individuals does respond to chemotherapy. Treatment approaches differ between different centers; however, all regimens used to date are the same as those used in non-HIV–infected individuals, ie, MOPP, ABVD, or alternating MOPP-ABVD. The complete response rate, however, is approximately 44% as reported by the Cooperative Study Group of Malignancies Associated with HIV Infection, much lower than the reported complete response rate in non-HIV–infected patients (64). A French multicenter, retrospective, case-cohort study showed that with MOPP, ABVD, or MOPP-ABVD, a higher complete response rate of 79% could be seen in HIV-infected patients with HD, equivalent to that seen in their non-HIV–infected patients (65). Another multicenter, retrospective study from the Italian Cooperative Group on AIDS and Tumors (GICAT) reported that patients treated with MOPP-ABVD seemed to have higher complete response rates and lower rates of opportunistic infections than patients treated with MOPP alone (66). A prospective trial using epirubicin, vinblastine, and bleomycin (EVB) compared with full-dose bleomycin, but with a 50% dose reduction in epirubicin and vinblastine, showed an overall complete response of 82% (67). The dose-limiting toxicity in both groups was bone marrow suppression. Instead of using myeloid growth factor, however, the study reduced doses of chemotherapy based on neutrophil count before its scheduled administration. In this study, most of the complete responders were those treated with full-dose EVB. Therefore, patients with better performance status and less opportunistic infections should receive chemotherapy with Filgrastim support to allow chemotherapy to be given in full doses and on schedule.

V. TREATMENT OF OTHER TUMORS IN HIV-INFECTED PATIENTS

A. Cervical/Anal Carcinoma

Cervical carcinoma is the only AIDS-defining malignancy where an effective screening procedure for early detection and intervention is available. Treatment should focus on treating premalignant conditions (eg, cervical intraepithelial neoplasia) with either loop excision, cryoablation, conization, or laser surgery. An AIDS Clinical Trials Group study (ACTG 200) is investigating the role of topical 5-fluorouracil (5-FU) in the treatment of premalignant lesions in HIV-infected women. Invasive cervical carcinoma, once developed, is best treated with surgery. Radiation and chemotherapy have more limited effectiveness in treating invasive cervical carcinoma. Filgrastim use would be helpful in supporting chemotherapy for metastatic disease.

Cytologic screening for early preinvasive malignant anal lesions and anal intraepithelial neoplasia is currently under investigation (68). Definitive treatment for anal carcinoma is surgery. Concurrent chemotherapy with 5-FU and mitomycin C has been shown to achieve better results than radiation alone (69). HIV-infected individuals treated with concurrent radiation and chemotherapy experienced more toxicities such as local skin reaction and severe bone marrow suppression that may require withholding therapy or reducing dosage (70). Long-term disease-free survival may be achieved with this combination modality (71). Use of Filgrastim to support full-dose treatment in this situation is justified.

B. Testicular Cancer

An increased incidence of testicular cancer was found in the Multicenter AIDS Cohort Study (MACS) (6). A retrospective study from the University of California at San Francisco found that six of seven HIV-infected individuals developed neutropenia with ANC $<0.5 \times 10^9$/L when standard doses of cisplatinum, etoposide, and bleomycin (BEP) or cyclophosphamide, etoposide, and carmustine (CVB) were used to treat this tumor (72). Use of Filgrastim in this setting is highly recommended.

C. Other Malignant Tumors

Lung cancer, head and neck cancer, basal cell carcinoma of the skin, soft tissue sarcoma in pediatric patients, and leukemia have all been reported to develop in HIV-infected individuals (73,74). Until more direct experience is reported, these tumors in HIV-infected patients should be treated the same way as in non-HIV–infected patients. The use of Filgrastim should be determined based on the probability of severe neutropenia (ANC $<0.5 \times 10^9$/L) with the specific therapy contemplated.

VI. USE OF MYELOID GROWTH FACTOR TO SUPPORT CHEMOTHERAPY IN AIDS

Chemotherapy clearly can induce tumor regression and improve survival in patients with advanced-stage AIDS-related malignancies. Most chemotherapy, when used in full or even modified doses, can induce significant neutropenia in patients with AIDS. Filgrastim has been shown in small studies to correct or reduce the incidence of neutropenia in patients receiving chemotherapy. Use of Filgrastim to prevent chemotherapy-induced neutropenia conforms to guidelines already established in other chemotherapy-treated malignancies (75). In addition, the cost benefit of using Filgrastim to prevent neutropenia in cancer patients has been reported. For patients at significant risk of neutropenia, ie, those with low CD4 counts, history of prior opportunistic infections, or marrow involvement with infections or tumors, and those receiving other myelosuppressive medications, the use of Filgrastim with myelosuppressive chemotherapy is clearly warranted.

The optimal dose and schedule of Filgrastim for patients with AIDS-related malignancies has not been fully established and may vary with the chemotherapy regimen and schedule and dose. Filgrastim administered at 5 μg/kg/day may be started 24 to 72 hours after administering chemotherapy and continued until ANC $>10 \times 10^9$/L. For patients receiving myelosuppressive regimens, the duration of Filgrastim administration may be shorter or its initiation delayed to parallel the peak time for myelosuppression. An example of this latter situation may be the initiation of Filgrastim 5 to 7 days after liposomal antracycline for Kaposi's sarcoma, as the peak myelosuppressive effects of these agents occur considerably later than after other chemotherapy drugs. For patients receiving Filgrastim for reasons other than chemotherapy, it should be discontinued 24 hours before administration of cytotoxic agents and resumed 24 hours after treatment.

VII. CONCLUSIONS

Filgrastim is licensed for use in the treatment and prevention of chemotherapy-induced neutropenia and can equally be applied in the HIV-infection as in the non–HIV-infection setting. Published clinical trials experience with Filgrastim in the AIDS-malignancy setting, however, is scanty. Studies in HIV-infected patients without malignancies have demonstrated good neutrophil response to administration of Filgrastim, and several small trials have been reported demonstrating its usefulness in overcoming or preventing severe neutropenia in patients with AIDS and Kaposi's sarcoma or NHL. Whether chemotherapy dose intensification using Filgrastim with or without hematopoietic stem cell support will result in longer remissions or survival in patients with AIDS-associated malignancies has not as yet been answered. In the case of advanced-stage NHL in AIDS patients, it would appear that dose intensification in and of itself even with hematopoietic growth factor support is not the answer for the majority of patients, and novel and more effective drugs or treatment approaches will be needed to control this tumor. In the meantime, continued treatment with chemotherapy and other myelosuppressive treatments will be necessary. Myeloid hematopoietic growth factor support can clearly allow these treatments to be given safely and effectively.

ACKNOWLEDGMENTS

This manuscript was supported in part by grants from the State of California, Universitywide Task Force on AIDS to the UCLA AIDS Clinical Research

Center (CC 96-LA-175) and U.S. Public Health Service National Institutes of Health grants Al-27660, CA-70080, and RR-00865.

REFERENCES

1. Harbol, A. W., Liesveld, J. L., Simpson-Haidaris, P. J., and Abboud, C. N. (1994). Mechanisms of cytopenia in human immunodeficiency virus infection. *Blood Rev 8*:241–251.
2. Koch, M. A., Volberding, P. A., Lagakos, S. W., Booth, D. K., Pettinelli, C., and Myers, M. W. (1992). Toxic effects of zidovudine in asymptomatic human immunodeficiency virus-infected individuals with CD4$^+$ cells counts of 0.5×10^9/L or less: detailed and updated results from protocol 019 of the AIDS Clinical Trials Group. *Arch Intern Med 152*:2286–2292.
3. Miles, S. A. (1992). Hematopoietic growth factors as adjuncts to antiretroviral therapy. *AIDS Res Human Retroviruses 8*:1073–1080.
4. Welte, K., Gabrilove, J., Bronchud, M. H., Platzer, E., and Morstyn, G. (1996). Filgrastim (r-metHuG-CSF): The first 10 years. *Blood 88*:1907–1929.
5. Tirelli, U., Vaccher, E., and Spina, M. (1994). Other cancers in HIV-infected patients. *Curr Opinion Oncol 6*:508–511.
6. Lyter, D. W., Bryant, J., Thackeray, R., Rinaldo, C. R., and Kingsley, L. A. (1995). Incidence of human immunodeficiency virus-related and nonrelated malignancies in a large cohort of homosexual men. *J Clin Oncol 13*:2540–2546.
7. Hessol, N. A., Koblin, B. A., et al (1996). Increased incidence of cancer among homosexual men, New York City and San Francisco, 1978–1990. XI Int Conf AIDS (abstr C.433).
8. Provencher, D., Valme, S., Averette, H. E., et al (1988). HIV status and positive Papanicolaou screening: identification of a high-risk population. *Gynecol Oncol 31*:184–190.
9. Maiman, M. (1994). Cervical neoplasia in women with HIV infection. *Oncol 8*:83–94.
10. Bigger, R. J., and Rabkin, C. S. (1996). The epidemiology of AIDS-related neoplasms. *Hematol/Onc Clin NA 10*:997–1010.
11. Durack, D. T. (1981). Opportunistic infection and Kaposi's sarcoma in homosexual men. *N Engl J Med 305*:1465–1467.
12. Chang, Y., Cesarman, E., Pessin, M. S., et al (1994). Identification of herpesvirus-like DNA sequences in AIDS-associated Kaposi's sarcoma. *Science 266*:1865–1869.
13. Moore, P. S., and Chang, Y. (1995). Detection of herpesvirus-like DNA sequences in Kaposi's sarcoma in patients with and those without HIV infection. *N Engl J Med 332*:1181–1185.
14. Safai, B. (1996). Clinical manifestations of Kaposi's sarcoma. *Oncology 10*:S13–S23.
15. Mitsuyasu, R. T., and Groopman, J. E. (1984). Biology and therapy of Kaposi's sarcoma. *Sem Oncol 11*:53–59.

16. Ioachim, H. L., Adsay, V., Giancotti, F. R., et al (1995). Kaposi's sarcoma of internal organs. *Cancer 75*:1376–1385.

17. Lucatorto, F. M., and Sapp, J. P. (1993). Treatment of oral Kaposi's sarcoma with sclerosing agent in AIDS patients. A preliminary study. *Oral Surg Oral Med Oral Pathol 75*:192–198.

18. Wang, C. Y., Schroeter, A. L., and Su, W. P. (1995). Acquired immunodeficiency syndrome-related Kaposi's sarcoma. *Mayo Clin Proc 70*:869–879.

19. Tappero, J. W., Berger, T. G., Kaplan, L. D., Volberding, P. A., and Kahn, J. O. (1991). Cryotherapy for cutaneous Kaposi's sarcoma (KS) associated with acquired immunodeficiency syndrome (AIDS): a phase II trial. *J AIDS 4*:839–846.

20. Schaart, F. M., Bratzke, B., Ruszczak, Z., Stadler, R., and Orfanos, C. E. (1991). Long-term therapy of HIV-associated Kaposi's sarcoma with recombinant interferon alpha-2A. *Br J Dermatol 124*:62–68.

21. Evans, L. M., Itri, L. M., Campion, M., et al (1991). Interferon-alpha 2a in the treatment of acquired immunodeficiency syndrome-related Kaposi's sarcoma. *J Immunol 10*:39–50.

22. Rozenbaum, W., Gharakhanian, S., Navarette, M. S., De Sahb, R., Cardon, B., and Rouziouz, C. (1990). Long-term follow-up of 120 patients with AIDS-related Kaposi's sarcoma treated with interferon-alpha 2A. *J Invet Dermatol 95*:161S–165S.

23. Fischi, M. A., Uttamchandani, R. B., Resnick, L., et al (1991). A phase I study of recombinant human interferon-alpha 2a or human lymphoblastoid interferon-alpha n1 and concomitant zidovudine in patients with AIDS-related Kaposi's sarcoma. *J AIDS 4*:1–10.

24. Scadden, D. T., Bering, H. A., Levine, J. D., et al (1991). Granulocyte-macrophage colony-stimulating factor mitigates the neutropenia of combined interferon alpha and zidovudine treatment of acquired immune deficiency syndrome-associated Kaposi's sarcoma. *J Clin Oncol 9*:802–808.

25. Krown, S. E., Paredes, J., Bundow, D., et al (1992). Interferon-alpha, zidovudine, and granulocyte-macrophage colony-stimulating factor: a phase I AIDS Clinical Trial Group study in patients with Kaposi's sarcoma associated with AIDS. *J Clin Oncol 10*:1344–1351.

26. Schroder, K., Garbe, C., Waibel, M., et al (1992). Granulocyte colony stimulating factor (G-CSF) in treatment of patients with HIV-associated mucocutaneous Kaposi's sarcoma. Successful use in virus and drug-induced leukopenia. *Hautarzt 43*:700–706.

27. Gill, P. S., Akil, B., Colletti, P., et al (1989). Pulmonary Kaposi's sarcoma: clinical findings and results of therapy. *Am J Med 87*:57–61.

28. Gill, P. S., Rarick, M., McCutchan, J. A., et al (1991). Systemic treatment of AIDS-related Kaposi's sarcoma: results of a randomized trial. *Am J Med 90*:427–433.

29. Cadranel, J. L., Kammoun, S., Chevret, S., et al (1994). Results of chemotherapy in 30 AIDS patients with symptomatic pulmonary Kaposi's sarcoma. *Thorax 49*:958–960.

30. Gill, P. S., Miles, S. A., Mitsuyasu, R., et al (1994). Phase I study of ABV chemotherapy with zidovudine in the treatment of AIDS-related Kaposi's sarcoma: Results of a phase I ACTG study. *AIDS 8:*1695–1699.

31. Northfelt, D. W., Dezube, B., Miller, B., et al (1995). Randomized comparative trial of Doxil vs adriamycin, bleomycin, and vincristine (ABV) in the treatment of severe AIDS-related Kaposi's sarcoma (AIDS-KS). *Blood 86:*882a (abstr 1515).

32. Gill, P. S., Wernz, J., Scadden, D. T., et al (1996). Randomized phase III trial of liposomal daunorubicin versus doxorubicin, bleomycin, and vincristine in AIDS-related Kaposi's sarcoma. *J Clin Oncol 14:*2353–2364.

33. Mitsuyasu, R., Von Roenn, J., Krown, S. E., et al (1997). Comparison study of liposomal doxorubicin alone or with bleomycin and vincristine for treatment of advanced AIDS-associated Kaposi's sarcoma: ACTG protocol 286. *Proc Am Soc Clin Oncol* 55a (abstr 191).

34. Saville, M. W., Lietzau, J., Wilson, W., et al (1994). A trial of Paclitaxel (Taxol) in patients with HIV-associated Kaposi's sarcoma. *Proc Ann Meet Am Soc Clin Oncol 13:*5A (abstr 20).

35. Gill, P. S., Bernstein-Singer, M., Espina, B. M., et al (1992). Adriamycin, bleomycin and vincristine chemotherapy with recombinant granulocyte-macrophage colony-stimulating factor in the treatment of AIDS-related Kaposi's sarcoma. *AIDS 6:*1477–1481.

36. Bakker, P. J., Danner, S. A., ten Napel, C. H., et al (1995). Treatment of poor prognosis epidemic Kaposi's sarcoma with doxorubicin, bleomycin, vindesine and recombinant human granulocyte-monocyte colony stimulating factor (rh GM-CSF). *Eur J Cancer 31A:*188–192.

37. Sloand, E., Kumar, P. N., and Pierce, P. F. (1993). Chemotherapy for patients with pulmonary Kaposi's sarcoma: benefit of filgrastim (G-CSF) in supporting dose administration. *South Med J 86:*1219–1224.

38. Routy, J. P., Macleod, J., and Urbanek, A. (1996). Bleomycin + Vincristine/VP16 with or without G-CSF in AIDS patients with Kaposi's sarcoma. XI Int Conf AIDS (abstr B.1249).

39. Beral, V., Peterman, T., Berkelman, R., and Jaffe, H. (1991). AIDS-associated non-Hodgkin lymphoma. *Lancet 337:*805–809.

40. Pedersen, C., Barton, S. E., Chiesi, A., et al (1995). HIV related non-Hodgkin's lymphoma among European AIDS patients. *Eur J Haematol 55:*245–250.

41. Pluda, J. M., Venzon, D. J., Tosato, G., et al (1993). Parameters affecting the development of non-Hodgkin's lymphoma in patients with severe human immunodeficiency virus infection receiving antiretroviral therapy. *J Clin Oncol 11:*1099–1107.

42. Hamilton-Dutoit, S. J., Pallesen, G., Franzmann, M. B., et al (1991). AIDS-related lymphoma: Histopathology, immunophenotype, and association with Epstein-Barr virus as demonstrated by in situ nucleic acid hybridization. *Am J Pathol 138:*149–163.

43. Lucas, S. B., Diomande, M., Hounnou, A., et al (1994). HIV-associated lymphoma in Africa: an autopsy study in Cote d'Ivoire. *Int J Cancer 59:*20–24.

44. Shiramizu, B., Herndier, B., Meeker, T., Kaplan, L., and McGrath, M. (1992). Molecular and immunophenotypic characterization of AIDS-associated, Epstein-Barr virus-negative, polyclonal lymphoma. *J Clin Oncol 10:*383–389.

45. Ballerini, P., Gaidano, G., Gong, J. Z., et al (1993). Multiple genetic lesions in acquired immunodeficiency syndrome-related non-Hodgkin's lymphoma. *Blood 81:*166–176.

46. Stein, M. E., Spencer, D., Dansey, R., Bezwoda, W. R. (1994). Biology of disease and clinical aspects of AIDS-associated lymphoma: a review. *E Africa Med J 71:*219–222.

47. Cappell, M. S., and Botros, N. (1994). Predominantly gastrointestinal symptoms and signs in 11 consecutive AIDS patients with gastrointestinal lymphoma: a multicenter, multiyear study including 763 HIV-seropositive patients. *Am J Gastroenterol 89:*545–549.

48. Cinque, P., Brytting, M., Vago, L., et al (1993). Epstein-Barr virus DNA in cerebrospinal fluid from patients with AIDS-related primary lymphoma of the central nervous system. *Lancet 342:*398–401.

49. Ciardi, M., Borgese, L., De Carlo, A., et al (1996). Diagnostic significance of EBV-DNA detection by PCR in CSF samples of HIV-infected patients. *XI Int Conf AIDS* (abstr A4078).

50. Cote, T. R., Manns, A., Hardy, C. R., Yellin, F. J., Hartge, P. (1996). Epidemiology of brain lymphoma among people with or without acquired immunodeficiency syndrome. AIDS/Cancer Study Group. *J Natl Cancer Inst 88:*675–679.

51. Robert, N. J., and Schneiderman, H. (1984). Hodgkin's disease and the acquired immunodeficiency syndrome (letter). *Ann Intern Med 101:*142–143.

52. Serraino, D., Pezzotti, P., Cozzi, L. A., et al (1996). Incidence of Hodgkin's disease and other cancers in a cohort of HIV serconverters. *IX Int Conf AIDS* (abstr B.4235).

53. Andrieu, J. M., Roithmann, S., Tourani, J. M., et al (1993). Hodgkin's disease during HIV1 infection: the French registry experience. French Registry of HIV-associated Tumors. *Annals of Oncology,* Sep, 4(8):635–641.

54. Gill, P. S., Levine, A. M., Krailo, M., et al (1987). AIDS-related malignant lymphoma: results of prospective treatment trials. *J. Clin Oncol 5:*1322–1328.

55. Schurmann, D., Grunewald, T., Weiss, R., Jautzke, G., Pohle, H. D., and Ruf, D. (1995). Intensive treatment of AIDS-related non-Hodgkin's lymphoma with the MACOP-B protocol. *Eur J Haematol 54:*73–77.

56. Rio, B., Compagnucci, A., Rochemaure, J., et al (1996). High dose chemotherapy (CT) and autologous bone marrow transplantation (ABMT) for malignant hemopathies in HIV-infected patients. *IX Int Conf AIDS* (abstr B.1292).

57. Bermudez, M. A., Grant, K. M., Rodvien, R., and Mendes, F. (1989). Non-Hodgkin's lymphoma in a population with or at risk for acquired immunodeficiency syndrome: indications for intensive chemotherapy. *Am J Med 86:*71–76.

58. Gisselbrecht, C., Oksenhendler, E., Tirelli, U., et al (1993). Human immunodeficiency virus-related lymphoma treatment with intensive combination chemotherapy. *Am J Med 95:*188–196.

59. Tosi, P., Gherlinzoni, F., Visani, G., Coronado, O., Chiodo, F., and Tura, S. (1995). AZT+ Methotrexate in high-grade HIV-related non-Hodgkin lymphomas: interim report on feasibility and tolerance. *Haematologica 80:*31–34.

60. Goldstein, D., Newell, M., Milliken, S., et al (1994). Filgrastim, CEOP and antiretroviral therapy in HIV-related non-Hodgkin's lymphoma (NHL). *Ann Conf Australia Soc HIV Med 6:*167.

61. Baumgartner, J. E., Rachlin, J. R., Beckstead, J. H., et al (1990). Primary central nervous system lymphomas: natural history and response to radiation in 55 patients with acquired immunodeficiency syndrome. *J Neurosurg 73:*206–211.

62. Chamberlain, M. C. (1994). Long survival in patients with acquired immune deficiency syndrome-related primary central nervous system lymphoma. *Cancer 73:*1728–1730.

63. Forsyth, P. A., Yahalom, J., and De Angelis, L. M. (1994). Combined-modality therapy in the treatment of primary central nervous system lymphoma in AIDS. *Neurology 44:*1473–1479.

64. Rubio, R. (1994). Hodgkin's disease associated with human immunodeficiency virus infection. A Clinical Study of 46 cases. Cooperative Study Group of Malignancies Association with HIV infection of Madrid. *Cancer 73:*2400–2407.

65. Levy, R., Colonna, P., Tourani, J. M., et al (1995). Human immunodeficiency virus associated Hodgkin's disease: report of 45 cases from the French Registry of HIV-Associated Tumors. *Leuk Lymphoma 16:*451–456.

66. Tirelli, U., Errante, D., Vaccher, E., et al (1992). Hodgkin's disease in 92 patients with HIV infection: the Italian experience. *Ann Oncol 3:*69–72.

67. Errante, D., Tirelli, U., Gastaldi, R., et al (1994). Combined antineoplastic and antiretroviral therapy for patients with Hodgkin's disease and human immunodeficiency virus infection. A prospective study of 17 patients. The Italian Cooperative Group on AIDS and Tumors. *Cancer 73:*437–444.

68. Surawicz, C. M., Kirby, P., Critchlow, C., Sayer, J., Dunphy, C., and Kiviat, N. (1993). Anal dysplasia in homosexual men: role of anoscopy and biopsy. *Gastroenterology 105:*658–666.

69. Cummings, B. J., Keane, T. J., O'Sullivan, B., Wong, C. S., and Catton, C. T. (1991). Epidermoid anal cancer: treatment by radiation and 5-fluorouracil with and without mitomycin C. *Int J Radiation Oncol Biol Phys 21:*1115–1125.

70. Holland, J. M., and Swift, P. S. (1994). Tolerance of patients with human immunodeficiency virus and anal carcinoma to treatment with combined chemotherapy and radiation therapy. *Radiology 193:*251–254.

71. Chadha, M., Rosenblatt, E. A., Malamud, S., Pisch, J., and Berson, A. (1994). Squamous-cell carcinoma of the anus in HIV-positive patients. *Dis Colon Rectum 37:*861–865.

72. Timmerman, J. M., Northfelt, D. W., and Small, E. J. (1995). Malignant germ cell tumors in men infected with the human immunodeficiency virus: natural history and results of therapy. *J Clin Oncol 13:*1391–1397.

73. McClain, K. L., Joshi, V. V., and Murphy, S. B. (1996). Cancers in children with HIV infection. *Hematol Oncol Clin North Am 10:*1189–1201.

74. Remick, S. C. (1996). Non-AIDS-defining cancers. *Hematol Oncol Clin North Am 10:*1203–1213.

75. American Society of Clinical Oncology (1994). Recommendations for the use of hematopoietic colony-stimulating factors: evidence-based, clinical practice guidelines. *J Clin Oncol 12:*2471–2508.

76. Lassoued, K., Clauvel, J. P., Katlama, C., Janier, M., Picard, C., and Matherson, S. (1990). Treatment of the acquired immune deficiency syndrome-related Kaposi's sarcoma with bleomycin as a single agent. *Cancer 66:*1869–1872.

77. Mintzer, D. M., Real, F. X., Jovino, L., and Krown, S. E. (1985). Treatment of Kaposi's sarcoma and thrombocytopenia with vincristine in patients with the acquired immunodeficiency syndrome. *Ann Intern Med 102:*200–202.

78. Volberding, P. A., Abrams, D. I., Conant, M., Kaslow, K., Vranizan, K., and Ziegler, S. (1985). Vinblastine therapy for Kaposi's sarcoma in the acquired immunodeficiency syndrome. *Ann Intern Med 103:*335–338.

79. Schwartsmann, G., Sprinz, E., Kronfeld, M., et al (1991). Phase II study of teniposide in patients with AIDS-related Kaposi's sarcoma. *Eur J Cancer 27:*1637–1639.

80. Laubenstein, L. J., Krigel, R. L., Odajnyk, C. M., et al (1984). Treatment of epidemic Kaposi's sarcoma with etoposide or a combination of doxorubicin, bleomycin, and vinblastine. *J Clin Oncol 2:*1115–1120.

81. Saville, M. W., Lietzau, J., Pluda, J. M., et al (1995). Treatment of HIV-associated Kaposi's sarcoma with paclitaxol. *Lancet 346:*26–28.

82. Harrison, M., Tomlinson, D., and Stewart, S. (1995). Liposomal-entrapped doxorubicin: An active agents in AIDS-related Kaposi's sarcoma. *J Clin Oncol 13:*914–920.

83. Presant, C. A., Scolaro, M., Kennedy, P., et al (1993). Liposomal daunorubicin treatment of HIV-associated Kaposi's sarcoma. *Lancet 341:*1242–1243.

84. Nasti, G., Errante, M., Fasan, G., et al (1996). Evidence of activity of vinorelbine in patients with previously untreated AIDS-associated Kaposi's sarcoma. Third International Congress on Drug Therapy in HIV Infection, Birmingham. *AIDS 10*(2 suppl): S13.

17
Use of Hematopoietic Colony-Stimulating Factors in the Early Clinical Development of New Anticancer Drugs

Eric Keith Rowinsky
University of Texas Health Sciences Center at San Antonio and Institute for Drug Development, Cancer Therapy and Research Center, San Antonio, Texas

I. INTRODUCTION

The principal objective of early developmental trials of new cytotoxic agents, particularly phase 1 and 2 evaluations, is to determine if the new agents are capable of producing significant antitumor activity with acceptable therapeutic indices. The incorporation of hematopoietic colony-stimulating factors (CSF) into early developmental studies to facilitate the achievement of this goal is somewhat controversial. The crux of the controversy is that there is insufficient evidence demonstrating that the magnitude of dose escalation that can be safely achieved with hematopoietic CSF results in drug doses that are capable of producing significantly higher response rates and longer survival compared with conventional doses. It also can be argued that conventional methodological approaches used in the majority of developmental trials of new agents with hematopoietic CSF have largely obstructed optimal evaluations of the utility of hematopoietic CSF in early developmental trials of new cytotoxic agents. Methodological flaws in study design and patient eligibility criteria may result in missed opportunities to appreciate the usefulness of hematopoietic CSF in facilitating the detection of significant antitumor activity in early clinical evaluations of new cytotoxic agents. This chapter summarizes the rationale for incorporating hematopoietic CSF into early developmental studies of new agents and describes circumstances when hematopoietic CSF may be particularly helpful, as well as several methodological approaches used to date. This chapter principally focuses on the use of hematopoietic CSF to ameliorate neutropenia; however, the discussion is also applicable to factors that are capable of stimulating megakaryopoeisis and erythropoiesis.

II. OBJECTIVES

Overall efforts to incorporate hematopoietic CSF into early clinical trials of new anticancer agents have generally been aimed at accomplishing at least one of the following goals: substantial dose escalation of new myelosuppressive agents used alone or in combination with other cytotoxic agents or therapeutic modalities and/or amelioration of the depth, duration, and complications of severe myelosuppression, particularly neutropenia.

III. DOSE ESCALATION OF NEW AGENTS WITH HEMATOPOIETIC COLONY-STIMULATING FACTORS: SELECTION OF NEW AGENTS

The magnitude of dose escalation possible during phase 2 "disease-directed screening" evaluations should not be an important consideration for all new agents, and it is not reasonable to perform phase 1 studies aimed at determining the magnitude of dose escalation achievable using hematopoietic CSF with all cytotoxic agents. There are several situations in which hematopoietic CSF may be particularly warranted to enable optimal screening and development of new cytotoxic drugs. Several situations are listed in **Table 1.**

The availability of preclinical studies that demonstrate positive relationships between antitumor activity and either drug dose or drug concentration is perhaps the most compelling argument to support the use of high doses of new cytotoxic agents with hematopoietic CSF support during phase 2 evaluations. As a corollary to this, if such information is not available for the new agent itself, it still may be reasonable to pursue phase 2 evaluations of new agents at maximal doses using hematopoietic CSF support if positive dose-response relationships have been identified for other agents with similar characteristics and mechanisms of action. For example, it may be logical to consider dose escalation of new alkylating agents or topoisomerase I- or topoisomerase II-targeting agents with hematopoietic CSF since steep dose–response relationships have been previously demonstrated for these types of compounds (1–4). An overriding concern in attempting to

Table 1. Situations in New Drug Development When Dose Escalation of New Agents Using Hematopoietic Colony-Stimulating Factors May Be Reasonable.

- Well-characterized dose–response relationship for the new agent or agents with similar mechanisms of action.

- Isolated neutropenia or thrombocytopenia is the principal dose-limiting toxicity of the new agent at conventional doses without hematopoietic colony-stimulating factors (CSF).

- Non-hematological toxicities at the new agent's maximum tolerated dose without hematopoietic CSF are infrequent or modest in severity.

- The maximum-tolerated dose of the new agent used alone or in a combination regimen is substantially lower than clinically relevant single-agent doses that have been demonstrated to be important in phase 2 studies.

- Relevant pharmacological parameters (C_{max}, AUC, duration of exposure above biologically relevant concentrations) are not achieved at the maximum tolerated dose of the new agent used alone or in a combination regimen.

escalate drug doses above conventional doses using hematopoietic CSF relates to the severity of nonhematological toxicities relative to hematological toxicities at the maximum-tolerated dose (MTD) level, and whether or not the agent belongs to a class of compounds in which there is a significant disparity between doses that portend dose-limiting hematological and non-hematological toxicities (eg, alkylating agents and topoisomerase I- and topoisomerase II-directed agents). There still may be significant variability within any particular class of cytotoxic agents with respect to the differences in doses that produce dose-limiting hematological and nonhematological effects. For example, the topoisomerase I-targeting agents topotecan and irinotecan (CPT-11) are quite different in this regard (1–7). For topotecan, nonhematological toxicities are uncommon at conventional doses that can be administered without hematopoietic CSF, and significant dose escalation of topotecan (approximately 2.5-fold) with Filgrastim is feasible in un-treated or minimally pretreated patients (1–3,5). Dose-limiting gastrointestinal toxicities are relatively common at conventional doses of irinotecan, precluding dose escalation of irinotecan with Filgrastim above approximately 1.5-fold (1,2,6,7).

Irrespective of the potential for significant dose escalation of new agents using hematopoietic CSF, hematopoietic CSF may enable dose escalation of new cytotoxic agents in combination-chemotherapy regimens when clinically relevant doses of the new agent are not readily achievable because of severe, noncumulative hematological toxicities. For example, severe neutropenia initially thwarted attempts to administer clinically relevant single-agent doses of paclitaxel (>135 mg/m^2) as a 24-hour infusion in combination with topotecan in women with recurrent or refractory advanced ovarian carcinoma (8). Although the use of Filgrastim did not permit substantial dose escalation of either drug in the regimen, it did enable the administration of both agents in combination at clinically relevant single-agent doses (paclitaxel 135 mg/m^2 on a 24-hour schedule and topotecan 0.75 mg/m^2 over 30 minutes daily for 5 days), permitting subsequent evaluations of the regimen in this clinical setting.

The decision to use hematopoietic CSF to enable dose escalation of new cytotoxic agents administered as single agents or in combination-chemotherapy regimens may also be reasonable if pertinent pharmacological variable determinants of antitumor activity as determined from preclinical studies are not readily achieved at conventional drug doses due to severe hematological toxicities. Relevant pharmacokinetic determinants may include peak plasma concentration (C_{max}), area under the concentration-versus-time curve (AUC), and duration of drug exposure above a critical threshold concentration.

IV. RATIONALE FOR PHASE 2 STUDIES WITH HEMATOPOIETIC COLONY-STIMULATING FACTORS

It may reasonable to perform phase 2 trials of new agents at high doses with hematopoietic CSF support in the clinical situations listed in **Table 2.** Several of these situations are discussed in this section.

A. Limited Drug Supply—The Case with Paclitaxel

Assuming that myelosuppression is the principal dose-limiting toxicity of a new cytotoxic agent and that nonhematological effects are relatively uncommon at the MTD without hematopoietic CSF, there are few situations when it is reasonable to perform broad phase 2 studies with hematopoietic CSF before the completion of phase 2 studies at lower drug doses without hematopoietic CSF. One situation is when the supply of the new agent is limited at the time of phase 2 screening studies and there is sufficient reason to indicate that antitumor activity may be overlooked if phase 2 disease-directed screening trials use drug doses that are substantially lower than the MTD. A concern during the early development of paclitaxel was that its limited supply would preclude the scope and number of phase 2 evaluations (9,10). Since heavily pretreated women with advanced ovarian cancer were able to tolerate substantially lower paclitaxel doses than untreated or minimally pretreated patients, investigators initially opted to perform phase 2 evaluations of paclitaxel at its MTD with Filgrastim support, in addition

Table 2. Clinical Situations That May Support a Rationale for Phase 2 Trials of New Agents with Hematopoietic Colony-Stimulating Factors.

- Well-characterized dose–response relationships.
- No or minimal antitumor activity at clinically relevant doses in phase 1 trials involving appropriate patient populations.
- Neutropenia is the principal dose-limiting toxicity when the new agent is used alone or in a combination regimen at the maximum-tolerated dose without hematopoietic colony-stimulating factors (CSF).
- A low incidence of nonhematological toxicities when the new agent is used alone or in a combination regimen at the maximum-tolerated dose without hematopoietic CSF.
- Patient population or clinical setting predisposes patients to the serious complications of neutropenia.
- Mucosal barrier breakdown associated with diarrhea and/or mucositis.
- Multiple prior surgeries involving gastrointestinal mucosa (eg, patients with gynecological malignancies and aerodigestive tract cancers).
- Limited drug supply.
- Relevant pharmacological objectives are not met using maximum-tolerated doses without hematopoietic CSF.
- The targeted patient population has inherently limited bone marrow reserves or is heavily pretreated:
    —Elderly adults
    —Heavily pretreated children.

to phase 2 studies at lower paclitaxel doses without Filgrastim support (11–27). In several clinical situations, such as evaluations of paclitaxel in untreated or minimally pretreated women with advanced breast cancer, phase 2 studies using hematopoietic CSF were performed from the outset (17). In essence, the decision to evaluate paclitaxel at its MTD with Filgrastim was based on the possibility that the demonstration of unimpressive activity at higher paclitaxel doses would obviate the need to perform additional phase 2 trials of the agent at conventional doses. In contrast, the demonstration of negligible activity at submaximal clinical doses would not conclusively prove that paclitaxel was inactive across its entire dosing spectrum, and additional studies evaluating higher drug doses might be required. In studies of paclitaxel performed at its MTD, negligible activity was demonstrated in patients with advanced melanoma and colorectal, renal, and gastric cancers (9,10), which argued strongly against the need to perform additional trials in these tumor types at the lower doses. However, the potential benefits of this approach, which results in unequivocally negative conclusions when phase 2 disease-directed screening studies of high doses are negative, are offset by the dilemma that is often faced when significant antitumor activity is observed in nonrandomized phase 2 studies using maximal drug doses with hematopoietic CSF support. In this situation, as was the case demonstrated for paclitaxel in phase 2 trials in lymphoma and carcinomas of the ovary, breast, lung, cervix, and head and neck, further clinical investigations may focus solely on using the most intensive dose schedules when antitumor activity at lower drug doses and/or dose intensity is equivocal (9–23).

B. Rationale Based on Dose–Response Relationships

In the initial five phase 2 studies of paclitaxel using a 24-hour infusion schedule in women with recurrent or refractory ovarian cancer, paclitaxel doses ranged from 110 to 300 mg/m^2 (9–15). In individual study reports, each using somewhat different criteria to define patient eligibility and evaluability for response, response rates were seemingly higher (36% to 48%) in trials evaluating higher paclitaxel doses (170 to 300 mg/m^2) and dose intensity compared with those in which most patients received lower paclitaxel doses (response rate, 30% to 37%) or higher doses, albeit less dose-intensive treatment (response rate, 20%; 250 mg/m^2) with treatment delays (11–15). Filgrastim also was used from the outset to ameliorate severe neutropenia and maintain dose intensity in trials that administered high paclitaxel doses or dose intensity (14,15). Using the correlative methods of Hryniuk and Levin, Reed et al (28) retrospectively studied the relationships between paclitaxel dose intensity and response in phase 2 trials of paclitaxel in women with advanced ovarian and breast cancers and identified strong positive relationships in both clinical situations (p = 0.022 and 0.004, respectively).

Nevertheless, one potential flaw of retrospective analyses of this type is the likelihood that there may be significant interstudy differences in both study design and patient eligibility requirements, particularly with respect to relevant prognostic variables. A recent meta-analysis of the effects of paclitaxel dose and dose intensity using audited data from the initial five phase 2 trials of paclitaxel in women with refractory or recurrent ovarian cancer that controlled for interstudy differences in several potentially relevant determinants of response, including age, performance status, number of prior regimens, response to prior therapy, and platinum sensitivity, illustrates this point. This analysis demonstrated that the probability of achieving an objective response was not related to the average dose of paclitaxel (odds ratio = 1.0/10 mg/m^2, p = 0.60) (29). In fact, the probability of responding actually decreased with increasing dose intensity (odds ratio =

0.77/10 mg/m^2/week, p = 0.060). A strong negative relationship between the maximum percent reduction in tumor size and dose intensity was also apparent in both the pooled data and in each study individually (odds ratio = −6.1/10 mg/m^2/week, p = 0.0077). In addition, there was moderate evidence for a negative relationship between average paclitaxel dose and survival (Hazard ratio = 1.06/10 mg/m^2, p = 0.08) and strong evidence for a negative relationship between dose intensity and survival (Hazard ratio = 1.3/10 mg/m^2/week, p = 0.0001).

The paclitaxel dose-response issue in this clinical setting has been evaluated prospectively in a randomized intergroup study in which women with refractory or recurrent ovarian cancer were treated with paclitaxel on a 24-hour infusion schedule at one of three doses: either 135 or 175 mg/m^2 or 250 mg/m^2 plus Filgrastim 5 or 10 μg/kg/day starting 24 hours after treatment with paclitaxel (30). Although accrual into the lowest dose arm was terminated after the regulatory approval of paclitaxel, which resulted in reduced patient accrual, a preliminary review of the study results indicates only modest differences in response, 36% versus 28%, between the 250 mg/m^2 plus Filgrastim and 175 mg/m^2 treatment arms, respectively, but no differences in either progression-free or overall survival (30). In addition, a retrospective subset analysis revealed that the propensity of responding was much lower at the higher paclitaxel doses for women projected to be sensitive to platinum-based therapies, which is counterintuitive.

Hematopoietic CSF are being incorporated into prospective randomized evaluations of new agents to determine if significant dose-response relationships exist. For example, a Cancer and Leukemia Group B (CALGB) phase 3 trial is addressing the paclitaxel dose–response issue in breast cancer by randomizing women who have received one prior therapy for metastatic disease or have relapsed within 6 months of completing adjuvant therapy to treatment with paclitaxel as a 3-hour infusion at one of three paclitaxel doses encompassing a relatively broad dosing range: 175 mg/m^2, 210 mg/m^2, or 250 mg/m^2 with Filgrastim added as secondary prophylaxis in the 210 and 250 mg/m^2 cohorts. Filgrastim also is being used to determine if there is an optimal paclitaxel schedule of administration (9,10). The National Surgical Adjuvant Breast and Bowel Project (NSABP) is prospectively randomizing women with metastatic or locally advanced breast cancer who have not received prior chemotherapy for metastatic disease to treatment with paclitaxel at a dose of 250 mg/m^2 given by either a 3- or 24-hour infusion with Filgrastim since these dose schedules are essentially equitoxic (9,10). In another study with similar principal objectives, women with metastatic or locally advanced breast cancer were randomized to treatment with paclitaxel 175 mg/m^2 given as either a 3- or 24-hour infusion (31). In contrast to the equitoxic paclitaxel dose schedules used in the NSABP study, these paclitaxel dose schedules resulted in substantially different degrees of hematological toxicities. Although the overall response rates between the treatment groups were not significantly different, treatment with paclitaxel over 24 hours resulted in superior activity in women who received adjuvant therapy only, and patients treated in the 24-hour paclitaxel group had substantially more myelotoxicity than those who were treated with paclitaxel on a 24-hour schedule. Therefore, it is possible that the difference in response rates in the less heavily pretreated patient group reflects the fact that the dose schedules are not equitoxic, and not the fact that the 24-hour infusion schedule is an inherently superior dose schedule.

C. Pharmacologic Rationale

The performance of phase 2 evaluations of new agents at high doses with hematopoietic CSF support may be proposed as an alternative to broad phase 2 trials using conventional doses when pharmacologic conditions that may portend optimal antitumor activity are not likely to be achieved at the MTD established without hematopoietic CSF. However, the use of hematopoietic CSF in this situation, as well as all other situations in which hematopoietic CSF are being used to evaluate the potential for dose escalation of new cytotoxic agents, should be considered only when neutropenia is the principal dose-limiting toxicity of the new agent administered alone or in a combination regimen. In some situations, critical pharmacologic determinents of antitumor activity (eg, C_{max}, AUC, duration of exposure above a critical threshold concentration) may have already been established in preclinical studies. If it is not known which pharmacologic parameters are relevant for cytotoxicity, a general notion of potentially important pharmacodynamic relationships may be ascertained from preclinical data. However, the use of preclinical pharmacologic data from which to gain a notion of the magnitude of the relevant pharmacologic variable to target in clinical trials may be to no avail if there are significant interspecies differences in the pharmacologic behavior of the new agent.

D. Mechanistic Rationale—Overcoming Drug Resistance: The Case with Topotecan

It may be rational to perform phase 2 screening trials of high doses of new agents with hematopoietic CSF support in situations where clear dose-response relationships have been identified in preclinical studies, and it is possible that the lack of clinical antitumor activity observed at conventional doses without hematopoietic CSF is due to a mechanism of drug resistance that may be overcome by using higher doses. Despite encouraging activity in preclinical models of colon cancer and other gastrointestinal cancers with topotecan, a semisynthetic hydrophilic analog of camptothecin that targets the nuclear enzyme topoisomerase I, conventional doses of topotecan without hematopoietic CSF demonstrated negligible antitumor activity in patients with colorectal carcinoma in phase 2 trials (32,33). Although several mechanisms of drug resistance are shared by topoisomerase I-targeting agents, P-glycoprotein (Pgp) overexpression associated with the multidrug resistance (MDR) phenotype was demonstrated to confer a low level of cross-resistance to topotecan in vitro (approximately two- to five-fold), and a low level of resistance to topotecan relative to the prototypic topoisomerase I inhibitor, camptothecin, was observed with certain Pgp-expressing cell lines in vivo (34,35). As it was hypothesized that these observations may account for the poor activity achieved with conventional topotecan doses in colorectal and other cancers that are derived from tissues that constitutively overexpress the Pgp multidrug transporter, the feasibility of administering high-dose topotecan with Filgrastim was evaluated in patients with solid tumors (5,36,37). In one phase 1 and pharmacologic study of topotecan and Filgrastim in untreated or minimally pretreated patients, severe neutropenia and thrombocytopenia precluded escalation of topotecan doses >3.5 mg/m^2/day given over 30 minutes daily for 5 days every 3 weeks with Filgrastim 5 μg/kg/day starting on day 6 and continued until the resolution of neutropenia (5). This study demonstrated that a 2.3-fold dose escalation of topotecan on this schedule with post-treatment Filgrastim was feasible in such patients. At the recommended phase 2 dose, 3.5 mg/m^2/day, respective mean AUC values for total topotecan and lactone were also 2.2- and 2.3-fold higher than those achieved at recommended phase 2 doses with Filgrastim. As this low level of cross-resistance conferred by MDR in vitro is able to be overcome by treating cells with higher topotecan concentrations, longer treatment durations, or greater overall drug exposures (ie, AUC) (34,35), the evaluation of higher doses of topotecan with Filgrastim support was thought to be a rational pursuit in clinical situations in which resistance may be conferred by low-level MDR that may be readily overcome in vivo. However, although it was encouraging to observe a major antitumor response in a patient with 5-fluorouracil (5-FU)-refractory colorectal cancer, a tumor type that is derived from tissues that constitutively overexpress Pgp and has a low response rate to lower doses of topotecan, in a phase 1 trial of topotecan doses ranging from 2.5 to 4.2 mg/m^2/day with Filgrastim, no antitumor effect was observed in a phase 2 trial of high-dose topotecan 3.5 mg/m^2/day administered for 5 days every 3 weeks with Filgrastim 5 μg/kg/day beginning on day 6 in patients with fluoropyrimidine-refractory colorectal carcinoma (38).

E. No Known Mechanistic Rationale to Determine the Toxicities at High Doses and Maximum-Tolerated Doses for Subsequent Trials

For new agents that predominately induce myelosuppression, particularly isolated neutropenia, some investigators routinely attempt to determine the MTD without and with hematopoietic CSF support during early phase 1 evaluations. This approach is generally used for agents in which clear dose-response relationships have been identified in preclinical studies, even when there are no immediate plans for phase 2 development of the new agent at high doses with hematopoietic CSF support. The incorporation of dose-finding studies using hematopoietic CSF support into conventional phase 1 evaluations may obviate the need for additional phase 1 studies if subsequent preclinical or clinical data emerge to support a rationale for performing phase 2 studies of the new agent at high doses with hematopoietic CSF support. This approach was undertaken by investigators at the National Cancer Institute Medicine and Pediatric Branches during phase 1 studies of the anthrapyrazole piroxantrone (oxantrazole, CI-942) (39). Piroxantrone was originally developed to decrease the formation of semi-quinone free radicals presumed to be responsible for anthracycline-related cardiac toxicity. Severe neutropenia precluded dose escalation of piroxantrone >150 mg/m^2 as a single infusion every 3 weeks without hematopoietic CSF support, whereas thrombocytopenia precluded dose escalation of the agent >355 mg/m^2 on an identical dosing schedule with Filgrastim, which represented a greater than 2-fold increase in the MTD. However, piroxantrone's potential to induce severe cumulative cardiotoxicity was manifested in this high-dose study in that seven patients developed severe symptomatic congestive heart failure at cumulative piroxantrone doses ranging from 855 to 2475 mg/m^2, and two patients died of cardiotoxicity.

Investigators at the Cancer Therapy and Research Center in San Antonio have evaluated the feasibility of escalating doses of the antimicro-tubule agent rhizoxin on a 72-hour infusion schedule as part of a comprehensive phase 1 evaluation (40). In the initial stage of the study, an unacceptably high rate of severe neutropenia associated with fever and infection precluded dose escalation of rhizoxin >1.2 to 2.4 mg/m^2 as a 72-hour continuous infusion every 3 weeks. The rationale for further dose escalation of rhizoxin with Filgrastim was based on preclinical studies demonstrating a positive relationship between rhizoxin dose and antitumor response. A similar approach was undertaken during the development of paclitaxel on several dosing schedules. For example, Schiller et al (41) determined the principal toxicities and MTD for paclitaxel administered on a 3-hour schedule without and with Filgrastim in an early phase 1 study, and both dosing schedules were subsequently evaluated in several disease settings.

F. Amelioration of Neutropenic Complications During Phase 2 and Phase 3 Development

It may be reasonable to consider incorporating hematopoietic CSF into phase 2 studies of new agents if the new agent produces an unacceptable degree of complications due to severe neutropenia when administered either alone or in a combination-chemotherapy regimen, regardless of whether high doses are being evaluated. In contrast to the situation with topotecan, the use of Filgrastim did not permit doses of 9-aminocampthotecin (9AC), a lipophilic topoisomerase I-targeting agent, to be substantially increased in phase 1 trials (42,43). Both severe neutropenia and thrombocytopenia precluded dose escalation of 9AC doses >35 to 45 μg/m^2/hour as a 72-hour infusion given every 2 weeks, whereas the MTD of 9AC with Filgrastim was 47 μg/m^2/hour on an identical dosing schedule (42,43). However, the patient eligibility requirements pertaining to the extent of prior therapy in phase 1 studies of 9AC with Filgrastim were less conservative compared with those used in phase 1 trials of topotecan with Filgrastim (5), and it is possible that the disparate results regarding the potential for dose escalation of topotecan and 9AC may reflect interstudy differences in the hematopoietic reserves of the patient populations. Despite the fact that the MTD of 9AC achievable without and with Filgrastim are not significantly different, phase 2 studies of 9AC were performed both without and with Filgrastim to determine if the use of Filgrastim reduces the complications of severe neutropenia, minimizes the need for dose reduction, and increases the overall therapeutic index of 9AC (44–46).

Hematopoietic CSF support has been used during several pivotal phase 2 and 3 trials of topotecan in patients who might have otherwise required substantial dosage reductions, resulting in less favorable outcomes. In an Eastern Cooperative Oncology Group (ECOG) phase 2 study of topotecan in extensive small-cell lung cancer (SCLC), 48 previously untreated patients received 2.0 mg/m^2/day topotecan for 5 days every 3 weeks (47). This dose encompassed the high end of the range of doses without hematopoietic CSF support that were recommended for phase 2 trials. The first 13 patients were treated without CSF support, but the next 35 patients received Filgrastim 5 μg/kg/day for 10 to 14 days starting on day 6. Ninety-two percent of patients treated without Filgrastim developed grade 3 or 4 neutropenia compared with 29% who received Filgrastim. However, the incidence of neutropenia with fever was similar between the two groups (8% and 11%, respectively) and one patient in each group died of infection associated with severe neutropenia. In addition, there were no differences in objective tumor response, duration of response, time to treatment failure, and survival between the 13 patients who entered the study before Filgrastim administration was mandated and the 35 patients who received Filgrastim. Several phase 2 and phase 3 evaluations of topotecan in heavily pretreated women with refractory or recurrent ovarian cancer also have permitted secondary prophylaxis with Filgrastim to minimize the number of dosage reductions and the use of suboptimal topotecan doses (48–51). In the three phase 2 and one phase 3 study of topotecan that formed the basis for approval for use in women with refractory or recurrent ovarian cancer, 445 patients received a total of 2019 topotecan courses, of which 1673 courses (83%) were administered at the starting dose of 1.5 mg/m^2/day for 5 days every 3 weeks (52). The administration of Filgrastim was permitted after a patient's first course, before dose level reduction to maintain dose intensity, if neutropenia or sequelae of neutropenia would have been the sole reason for dose reduction. Overall, Filgrastim was used prophylactially in 388 of the 2019 (19%) treatment courses. Retreatment commenced within 5 days of the scheduled start dates in 1206 (77%) of 1574 courses after the first treatment course. Overall, the patients received a median of four courses (range, 1 to 33 courses), a median dose intensity of 2.39 mg/m^2/day, and a median cumulative dose of 30 mg/m^2.

The incorporation of hematopoietic CSF into early developmental trials of new cytotoxic agents may be particularly useful in situations when the new agent has the capacity to disrupt gastrointestinal mucosal barriers. The propensity for developing fever and infection associated with severe neutropenia appears to be increased if there is mucous membrane barrier breakdown and is often associated with the concurrent development of diarrhea and mucositis involving perioral, perivaginal, or perianal tissues. In addition, there are special clinical circumstances when patients may be especially prone to the development of fever and infection during prolonged periods of severe neutropenia. For example, patients with gynecological, head and neck, and upper aerodigestive tract malignancies, particularly patients who have had multiple prior intraabdominal surgical procedures (eg, ovarian carcinoma), appear to be predisposed to developing fever and infection during periods of severe neutropenia that may be well tolerated by other types of patients. Although Filgrastim was used to facilitate the development of combination-chemotherapy regimens consisting of paclitaxel (24-hour infusion) and topotecan (30-minute infusion daily for 5 days) (8), its use did not enable investigators to increase doses of either paclitaxel or topotecan above minimally relevant single-agent dose levels (paclitaxel 135 mg/m^2 as a 24-hour infusion combined with topotecan 0.75 mg/m^2 as a 30-minute infusion daily for 5 days) nor reduce the depth of severe neutropenia in women with refractory or recurrent ovarian cancer compared with drug doses in the regimen without Filgrastim. However, the incorporation of Filgrastim substantially reduced the duration of severe neutropenia and the rates of fever and infection, which enabled the development of the new regimen in this disease setting. Hematopoietic CSF may be particularly useful in reducing the complications of severe neutropenia during developmental studies of multiagent-chemotherapy regimens consisting of combinations of new cytotoxic agents with known mucosal toxins like doxorubicin, etoposide, and 5-FU (53–55).

G. Special Patient Populations

Hematopoietic CSF may be used as adjuncts in phase 2 trials of new cytotoxic agents in patient populations that may be especially prone to developing severe neutropenia at clinically relevant doses, and who might otherwise require significant reduction of doses, possibly to subtherapeutic levels. Hematopoietic CSF support might be particularly useful during phase 2 "screening" trials of new agents in pediatric patients who are usually more heavily pretreated than adult patients at the time of phase 2 study entry. Therefore, adoption of overly restrictive eligibility requirements that exclude heavily pretreated pediatric patients from participating in phase 1 and phase 2 studies may not be feasible since the majority of pediatric neoplasms are chemosensitive, and conventional treatment practices often dictate the use of intensive combination chemotherapy in both first-line and salvage settings. Filgrastim was required to administer clinically relevant doses of topotecan as a 72-hour infusion in a phase 1 study that involved 27 heavily pretreated pediatric patients, of whom 26 received a median of three prior chemotherapy regimens (56). The utility of hematopoietic CSF in permitting the administration of new agents at clinically relevant doses during phase 2 disease-directed screening trials in children is illustrated by the fact that the MTD of topotecan with Filgrastim in heavily pretreated children is equivalent to topotecan doses that are readily tolerated by untreated or minimally pretreated adults who do not receive hematopoietic CSF support.

It is commonly felt that elderly individuals are particularly prone to the myelosuppressive effects of chemotherapy due to age-related deficits in clearance-organ function and because hematopoietic function generally decreases with increasing age. For these reasons, age restrictions are often incorporated into early clinical investigations, particularly phase 1 studies. However, elderly patients are also more prone to the majority of malignancies for which new drug development is often targeted. By excluding elderly individuals, who are the ultimate target population for new therapeutic regimens, maximally tolerated and recommended phase 2 doses may be defined for atypical patients who have lower rates of comorbid medical illnesses. Thus, clinical observations and recommended phase 2 doses based on the results of such trials may not be applicable to most clinically relevant patient populations, and hematopoietic CSF may be used in the phase 1 setting to define maximum-tolerated and recommended phase 2 doses for the elderly and other types of patients who may be particularly prone to developing severe hematological toxicities. This approach is likely to result in the development of more clinically applicable therapeutic regimens that require fewer dosage reductions and treatment delays in elderly individuals during subsequent phase 2 and phase 3 evaluations.

V. HEMATOPOIETIC COLONY-STIMULATING FACTORS AND THE DEVELOPMENT OF COMBINATION CHEMOTHERAPY REGIMENS CONTAINING NEW CYTOTOXIC AGENTS

Hematopoietic CSF are being used more frequently in the early phases of new drug evaluations to facilitate the development of combination chemotherapy regimens consisting of clinically relevant doses of new cytotoxic agents. A major concern during the development of the chemotherapy regimen consisting of cisplatin and paclitaxel was that the MTD of paclitaxel 135 mg/m^2 on a 24-hour schedule that could be administered safely before cisplatin 75 mg/m^2 was significantly lower than paclitaxel doses (250 mg/m^2) that were determined to be active in many tumor types in phase 2 studies (57). Since neutropenia was the principal toxicity that precluded further escalation of paclitaxel doses in the combination regimen, phase 1 studies of the regimen subsequently focused on using Filgrastim to enable further dose escalation of paclitaxel combined with cisplatin (58). In these trials, peripheral neuroxicity precluded repetitive administration of paclitaxel doses >250 mg/m^2 (day 1) on a 24-hour infusion schedule followed by cisplatin 75 mg/m^2 (day 2) and Filgrastim 5 μg/kg/day (starting on day 3 and continuing until the resolution of neutropenia). In view of the acceptable toxicity profile demonstrated for cisplatin 75 mg/m^2 combined with higher doses of paclitaxel 250 mg/m^2 with Filgrastim, ECOG randomized chemotherapy-naive patients with metastatic non-small–cell lung cancer (NSCLC) to treatment with either "standard" therapy, consisting of etoposide 100 mg/m^2 on days 1 to 3 and cisplatin 75 mg/m^2, or cisplatin 75 mg/m^2 combined with either low doses of paclitaxel 135 mg/m^2 (24-hour infusion) or higher doses of paclitaxel 250 mg/m^2 (24-hour infusion) with Filgrastim 5 μg/kg/day (59). The preliminary results of this study demonstrated that the response rates in both the high- and low-dose paclitaxel arms (26.5% and 32.1%, respectively) and median survival times (9.56

months and 9.99 months, respectively) were superior to the etoposide treatment (12% and 7.69 months, respectively). Although the rates of severe neutropenia, fever, and infection were similar in both paclitaxel groups, the overall impact of high-dose paclitaxel with Filgrastim support in this disease setting was negligible. A similar ECOG study designed to assess the benefit of cisplatin combined with high doses of paclitaxel is being performed in patients with locoregional recurrent or metastatic head and neck carcinoma who are generally more susceptible to the complications of severe neutropenia becaue of a high incidence of comorbid medical illnesses and upper aerodigestive tract mucosal barrier breakdown related to tumor and radiation therapy. Preliminary results of this trial have demonstrated that response rates were similar in patients randomized to treatment with cisplatin 75 mg/m^2 combined with either high-dose paclitaxel 200 mg/m^2 (24-hour infusion) with Filgrastim or low-dose paclitaxel 135 mg/m^2 (24-hour infusion) (60).

Although investigators have consistently demonstrated that clinically relevant single-agent doses of paclitaxel can be combined with cisplatin and the principal issue regarding the development of this combination regimen pertains to the relative benefits of high- versus low-dose paclitaxel, it has been much more difficult to administer clinically relevant single-agent doses of paclitaxel in combination with other cytotoxic agents, particularly agents that are more myelosuppressive than cisplatin, without using hematopoietic CSF.

The incorporation of hematopoietic CSF into phase 1 evaluations has made it possible to develop chemotherapy regimens consisting of paclitaxel and carboplatin (61–63), doxorubicin (64–67), epirubicin (68), liposomal doxorubicin (69,70), cyclophosphamide (71–73), edatrexate (74), vinorelbine (75), and ifosfamide (76). In addition, the development of almost all paclitaxel-based combinations involving three chemotherapeutic agents has required hematopoietic CSF support due to profound neutropenia at modest single-agent drug doses (77–79). Although the administration of paclitaxel combined with carboplatin at clinically relevant single-agent doses without hematopoietic CSF was demonstrated to be feasible in women with metastatic ovarian cancer (73), repetitive treatment of patients with advanced NSCLC with similar paclitaxel-carboplatin regimens without hematopoietic CSF support resulted in high rates of severe hematological toxicities and treatment delay (61,62). Repetitive treatment of patients with advanced NSCLC with carboplatin doses targeting AUC ranging from 6 to 7 mg/mL.min combined with paclitaxel 135 to 175 mg/m^2 (24-hour infusion) and Filgrastim was subsequently demonstrated to be feasible. In fact, this regimen has resulted in impressive antitumor activity in chemotherapy-naive patients (61,62). In one study, Langer et al (61) reported a 62%

response rate, a median survival of 53 weeks, and a 1-year survival of 54% in patients with metastatic NSCLC.

In the absence of hematopoietic CSF support, it has been very difficult to combine some of the most important new cytotoxic agents, such as topotecan and irinotecan, with most other myelosuppressive agents. Multi-agent regimens consisting of either topotecan or irinotecan combined with carboplatin (80,81), doxorubicin (82), etoposide (83,84), paclitaxel (8,85,86), or cyclophosphamide (87) have required Filgrastim (88). For regimens such as the combination of irinotecan and etoposide without hematopoietic CSF support, high rates of severe neutropenia, fever, and infection preclude administering even low single-agent doses of irinotecan in combination with other chemotherapy agents (83). The recommended doses of irinotecan and etoposide are 60 and 60 mg/m^2 on days 1 through 3 every 3 to 4 weeks with Filgrastim 50 μg/m^2/day on days 4 to 17. Although it has been less difficult to combine topoisomerase I-targeting agents with cisplatin, severe neutropenia associated with fever and infection has been the principal dose-limiting toxicity of such combinations, and Filgrastim often has been incorporated into these regimens to enable further development (88–95). This has been particularly true for the combination of irinotecan and cis-platin, which produces a high rate of severe neutropenia associated with fever and diarrhea (93). The recommended doses of drugs in the regimen for phase 2 studies were cisplatin 80 mg/m^2 on day 1 and irinotecan 60 mg/m^2 on days 1, 8, and 15. The incorporation of Filgrastim into the regimen at a dose of 2 μg/kg/day on days 4 to 21, except on the days of irinotecan treatment, enabled the dose of CPT-11 to be increased by 33% to 80 mg/m^2, with further CPT-11 dose escalation precluded by a high rate of severe diarrhea (94). The development of irinotecan-cisplatin regimens on alter-nate administration schedules, such as irinotecan combined with infusional cisplatin, has also required hematopoietic CSF support (95).

In contrast, hematopoietic CSF support has not been used to develop docetaxel-based combination chemotherapy regimens (96–98). The defini-tions of dose-limiting toxicity used during the development of docetaxel have generally been much less conservative than in studies of paclitaxel- and topotecan-based combination chemotherapy regimens. In phase 1 and phase 2 of docetaxel-based regimens, absolute neutrophil counts (ANC) <0.5 × 10^9/L lasting as long as 7 days and/or fever associated with severe neutropenia for as long as 3 days has been considered to be acceptable. Nevertheless, these criteria may not be acceptable in standard clinical prac-tice. Therefore, dosing recommendations based on these criteria may not be applicable, and it is possible that patients participating in phase 2 and phase 3 studies of docetaxel-based regimens, such as the combination of docetaxel and doxorubicin, may have higher rates of unacceptable toxicity than patients in phase 1 trials who generally have better performance capabilities.

Although the CALGB demonstrated that combinations of topotecan-cisplatin and topotecan-paclitaxel produce acceptable toxicity profiles in multicenter phase 1 evaluations (85,91), substantial neutropenia was observed in phase 2 studies. In a CALGB phase 1 study of cisplatin-topotecan, 38 patients with advanced solid tumors who had not been previously treated with platinum compounds were treated with topotecan on days 1 to 5 and cisplatin on day 1 of a 21-day cycle (99). The topotecan dose was fixed at 1.0 mg/m^2/day in the first four dose levels, and cisplatin was escalated in 25 mg/m^2 increments from 25 to 100 mg/m^2. The principal toxicity was neutropenia and the combination of topotecan and cisplatin induced more severe neutropenia than either drug alone at the same doses. Based on the definition of dose-limiting toxicity as grade 4 neutropenia lasting longer than 7 days, the CALGB recommended that phase 2 evaluations should be performed using topotecan 1.0 mg/m^2 for 5 days in combination with cisplatin 50 mg/m^2 on day 1 without Filgrastim, or alternatively cisplatin 75 mg/m^2 on day 1 with Filgrastim support. In addition, neutropenia was determined to be the dose-limiting toxicity of the regimen in a multicenter CALGB phase 1 trial involving 46 patients with advanced solid tumors who had received one prior chemotherapy regimen (85). Topotecan was infused over 30 minutes at a fixed dose of 1.0 mg/m^2/day for 5 days every 3 weeks. Paclitaxel at the starting dose of 50 mg/m^2 was infused over 3 hours on day 1 before the administration of topotecan. As projected based on the myelosuppressive effects of both paclitaxel and topotecan as single agents, neutropenia was the principal dose-limiting toxicity of the combination. Without Filgrastim, the MTD of paclitaxel in this combination was 80 mg/m^2. However, the dose of paclitaxel could be escalated to 230 mg/m^2 when Filgrastim was administered on days 6 to 14. The recommended doses for phase 2 studies were paclitaxel 230 mg/m^2 on day 1 and topotecan 1.0 mg/m^2/day for 5 days with Filgrastim on days 6 to 10 (85).

Similar to the situation discussed for the development of docetaxel, it should be noted that the MTD and recommended doses for the CALGB phase 2 studies were based on a definition of dose-limiting neutropenia consisting of an ANC $<0.5 \times 10^9$/L for longer than 7 days, which may be inappropriately long, particularly for patients prone to developing severe neutropenia complicated by fever and infection. This is illustrated by the preliminary results of a subsequent CALGB report of an unexpectedly high rate of treatment-related fatal sepsis from both topotecan-cisplatin and topotecan-paclitaxel combinations in patients with extensive SCLC (91). Three of the 12 patients treated with topotecan-cisplatin at the doses cited previously and 3 of 13 patients treated with topotecan-paclitaxel with

Filgrastim developed fatal sepsis, leading to the suspension of patient accrual in these two arms of a three-arm randomized phase 2 trial, which also included a paclitaxel/cisplatin treatment arm. Based on the findings of Blackwell and Crawford (100), who reported that the rate of severe neutropenia complicated by fever and infection was directly related to the duration of grade 4 neutropenia in patients with SCLC and based on a retrospective analysis of infection rate as a function of the duration of grade 4 neutropenia in their prior phase 1 trials, the CALGB recommended that investigators adjust doses according to the specific patient population and the risk of prolonged or febrile neutropenia that is considered acceptable (91).

VI. PHASE 1 TRIALS OF NEW AGENTS WITH HEMATOPOIETIC COLONY-STIMULATING FACTORS—METHODOLOGY

A. Patient Selection

In evaluating the use of hematopoietic CSF as an adjunctive supportive measure to escalate doses of new cytotoxic agents alone or in combination-chemotherapy regimens, it is essential that patient eligibility be restricted to the type of patients who will be eventually targeted in phase 2 and phase 3 evaluations of the high-dose regimen. Since the principal objective of phase 1 studies involving high doses of new agents given either alone or in combination with hematopoietic CSF is to devise feasible high-dose regimens intended for front-line therapy, patient eligibility should be restricted to subjects who have received no or minimal prior myelosuppressive therapy. In addition, heavily pretreated individuals constitute a patient population that is heterogeneous with respect to hematopoietic function and bone marrow reserve, and, therefore, these patients are more likely to experience significant intersubject and intrasubject variability with respect to hematological toxicity. The following criteria, albeit somewhat arbitrary, have been used in previous studies to define "heavily pretreated" patients with respect to prior myelosuppressive therapy (5,57,58):

- Prior treatment with more than four to six courses of combination chemotherapy containing alkylating agents except for regimens containing low or moderate doses of cisplatin.
- Prior treatment with irreversible hematopoietic stem cell toxins such as mitomycin C or nitrosourea.
- Radiation to greater than 25% of bone marrow–bearing areas (eg, pelvic radiation).

- Widespread bone metastases or bone marrow involvement with bone marrow biopsies required for patients with tumor types that have a high likelihood of bone marrow involvement.

If the new agent given alone or in combination with hematopoietic CSF is ultimately intended for evaluation in heavily pretreated patients as "salvage" therapy, the patients who are enrolled in the phase 1 dose-finding study should be nearly identical to the intended target population with respect to prior myelosuppressive chemotherapy from the outset. A second stage of the phase 1 trial may be initiated for patients who are heavily pretreated after MTD, and recommended phase 2 doses are defined for patients who have had no or minimal prior myelosuppressive therapy. As discussed previously, a similar approach using hematopoietic CSF support may be used to define doses for special patients such as elderly individuals who may be more prone to the myelosuppressive effects of chemotherapy due to inherently low hematopoietic function and drug clearance.

In addition, the use of variable patient eligibility criteria may result in profoundly different study conclusions regarding the feasibility and magnitude of drug dose escalation with hematopoietic CSF support. For example, substantial topotecan dose escalation was demonstrated to be feasible in a phase 1 study that used conservative patient eligibility criteria similar to those discussed in the previous section, whereas topotecan doses were not able to be escalated above conventional doses without Filgrastim in studies using more liberal patient eligibility criteria pertaining to the extent of prior myelosuppressive therapy (5,36).

B. Dose Escalation

In phase 1 studies of new agents given alone or in combination-chemotherapy regimens with hematopoietic CSF support, three to four patients are conventionally treated at each dose level in which consistent dose-limiting toxicity (ie, two or more dose-limiting events defined according to the projected level of tolerance of the targeted patient population) does not occur. If one patient develops dose-limiting toxicity, a total of six patients are then treated at that particular dose level. The MTD generally has been defined as the highest dose level in which less than one of the first six patients develops a dose-limiting event. However, at least eight total patients have usually been treated at the MTD to yield ample toxicological and pharmacological data at the clinically relevant dose level.

C. Scheduling of Hematopoietic Colony-Stimulating Factors

The scheduling of hematopoietic CSF with chemotherapy may be critical in enabling their optimal use to ameliorate hematological toxicity and maximal dose escalation. This is illustrated by the results of a phase 1 and pharmaco-kinetic study in which the principal objective was to prospectively determine whether the scheduling of Filgrastim variably affected the ability to escalate doses of topotecan (5). In this study, untreated or minimally pretreated patients received escalating doses of topotecan as a 30-minute infusion daily for 5 days every 3 weeks with Filgrastim initially given concurrently with topotecan (starting on day 1). Severe neutropenia and thrombocyto-penia precluded the administration of topotecan at even the lowest dose level evaluated (2.0 mg/m^2). In fact, the myelosuppressive effects of topo-tecan and concurrent Filgrastim were more severe than the myelosuppres-sive effects due to topotecan without hematopoietic CSF support (101). Therefore, an alternate post-treatment schedule, in which treatment with Filgrastim began immediately after topotecan (starting on day 6), was subse-quently studied. Unlike the concurrent schedule, dose escalation of topo-tecan with Filgrastim on the post-treatment schedule proceeded to 4.2 mg/m^2/day (5). In addition, the pharmacologic behavior of high-dose topotecan was not significantly altered by the scheduling of Filgrastim.

The enhancement of myelosuppression due to the concurrent adminis-tration of cytotoxic agents and hematopoietic CSF also has been observed in preclinical and clinical studies of other antineoplastic agents, particularly inhibitors of DNA synthesis (53,102–105). The profound myelosuppressive effects of combinations of cytotoxic agents and hematopoietic CSF on a concurrent schedule may be attributed to the stimulatory effects of CSF on hematopoietic precursors, possibly increasing the susceptibility of these cells to cytotoxic agents, particularly antimetabolites and inhibitors of DNA synthesis. This hypothesis most likely explains the profound hematologic toxicity resulting from the concurrent administration of Filgrastim and topotecan since DNA replication processes during the S phase of the cell cycle convert drug-induced reversible single-strand DNA breaks into irre-versible double-strand DNA lesions (1). Specifically with regard to topo-isomerase I-targeting agents, intracellular levels of topoisomerase I have been demonstrated to decrease during myeloid maturation which indicates that early myeloid cells with proliferative potential may be particularly sensitive to the effects of topoisomerase I-targeting agents, especially after concurrent treatment with topoisomerase I-targeting agents and hemato-poietic CSF (106).

The thrombocytopenic effects of the concurrent schedule of topotecan and Filgrastim were also somewhat surprising since Filgrastim is generally considered to specifically affect the myeloid compartment. However, this observation is not incongruent with the known stimulatory effects of G-CSF which include the proliferation of multipotent hematopoietic progenitors (107–109) in addition to its known proliferative and differentiating effects on committed myeloid progenitors. Furthermore, the collection and infusion of these multipotent progenitor cells after stimulation with Filgrastim results in the accelerated recovery of both platelet and neutrophils after myeloablative chemotherapy (107–110). Although Filgrastim induces the proliferation of multipotent progenitors inherently capable of differentiating into either committed myeloid precursors or megakaryocytes, it specifically induces myeloid differentiation, and, therefore, neutrophil counts increase most dramatically after treatment with Filgrastim. It is likely that the concurrent topotecan-Filgrastim treatment scheme stimulates the proliferation of multipotent progenitor cells (ie, capable of differentiating into myeloid, erythroid, and megakaryoctic precursors), which, in turn, are cytokinetically susceptible to the effects of topotecan, thereby explaining the substantial decrements in both neutrophil and platelet counts with concurrent treatment. In addition to severe neutropenia, severe thrombocytopenia also has been observed when hematopoietic CSF are administered concurrently with other cytotoxic therapies, including chemotherapy and/or radiotherapy (53,102–105).

The concerns regarding accentuation of myelosuppression due to the concurrent administration of cytotoxic agents and hematopoietic CSF may be particularly important with cytotoxic agents administered on frequent or prolonged dosing schedules. In developing cytotoxic agents that are administered on weekly schedules with hematopoietic CSF, investigators often have administered hematopoietic CSF for short periods in the intervals between treatments with the cytotoxic agents. For example, in developing a clinically relevant high-dose regimen of vinorelbine, a novel *Vinca* alkaloid that is active in NSCLC and breast carcinoma and is most commonly administered on a weekly schedule, for heavily pretreated women with metastatic breast cancer, one group of investigators successfully used Filgrastim 300 μg/day twice weekly in the interval between vinorelbine doses (111). A similar approach has been undertaken with paclitaxel in that the use of hematopoietic CSF support has permitted the development of schedules with short intervals between treatment (eg, 3-hour infusion every 2 weeks), resulting in increased paclitaxel dose intensity compared with conventional dosing schedules (112).

Even if sufficient time elapses between the administration of myelosuppressive cytotoxic agents and hematopoietic CSF, antitumor agents may possess distinct pharmacologic attributes that may increase the likelihood of cytokinetic interactions, resulting in suboptimal hematopoietic stimulation. If hematopoietic CSF are administered immediately after treatment with cytotoxic agents, such as the nitrosoureas that have relatively low plasma and/or tissue clearance rates, the combination may hypothetically affect hematopoietic progenitor cells similar to that of a "concurrent" administration schedule, resulting in suboptimal hematopoietic stimulation and possibly cytopenias.

VII. SUMMARY

Hematopoietic CSF are being used to enable evaluations of new cytotoxic agents at a full range of doses either alone or in combination-chemotherapy regimens. To date, phase 2 studies of new cytotoxic agents performed at high doses with hematopoietic CSF support have not demonstrated that hematopoietic CSF broaden the clinical antitumor spectra for the majority of novel compounds. However, the use of hematopoietic CSF may increase the efficiency of phase 2 evaluations of new agents at clinically relevant doses in special patient populations such as elderly individuals who tolerate cytotoxic chemotherapy less well than adults due to reduced hematopoietic function and drug clearance. Hematopoietic CSF support also may play a role in facilitating the development of chemotherapy regimens incorporating optimal single-agent doses of myelosuppressive drugs. Initially, the use of hematopoietic CSF support during new drug evaluations was viewed cynically as an unnecessary component of an extraordinarily complex developmental process. However, the incorporation of hematopoietic CSF into early drug investigations has enabled the development of several novel classes of cytotoxic agents such as the taxanes and topoisomerase-I inhibitors in the salvage setting, as well as in the front-line setting where any new agent is likely to make its greatest therapeutic impact.

REFERENCES

1. Slichenmyer, W. J., Rowinsky E. K., Donehower, R. C., and Kaufman, S. H. (1993). The current status of camptothecin analogues as antitumor agents. *J Natl Cancer Inst* 85:271–291.
2. Takimoto, C., and Arbuck, S. Camptothecins. In: Chabner, B. A. and Longo, D. L. eds. *Cancer Chemotherapy and Biotherapy,* Philadelphia: Lippincott Co., 1996, pp. 463–484.
3. Pommier, Y. G., Fesen, M. R., and Goldwassar, F. Topoisomerase II inhibitors: The epidophylolotoxins, m-AMSA, and ellipticine dervatives. In: Chabner B. A. and Longo D. L. eds. *Cancer Chemotherapy and Biotherapy,* Philadelphia: J.B. Lippincott Co., 1996, pp. 463–484.
4. Tew, K., Colvin, O. M., and Chabner, B. A. Alkylating agents. In: Chabner, B. A. and Longo, D. L. eds. *Cancer Chemotherapy and Biotherapy,* Philadelphia: J.B. Lippincott Co., 1996, pp. 297–332.

5. Rowinsky, E. K., Grochow, L. B., Sartorious, S. E., et al (1996). A phase I and pharmacologic study of high doses of the topolsomerase I inhibitor topotecan with granulocyte colony-stimulating factor. *J Clin Oncol 14*:1224–1235.

6. O'Reilly, S., and Rowinsky, E. K. (1996). Irinotecan (CPT-11). *Crit Rev Oncol 24*:47–70.

7. Lestingi, T. M., Vokes, E. E., Gray, W., et al (1995). CPT-11 in solid tumors with G-CSF and antidiarrheal support. *Proc Am Soc Clin Oncol 14*:480 (abstr 1563).

8. O'Reilly, S., Baker, S., Fleming, G. F., et al (1997). A phase I and pharmacologic trial of sequences of paclitaxel and topotecan in patients with advanced ovarian epithelial malignancies: a Gynecologic Oncology Group Study. *J Clin Onçol 15*:177–186.

9. Rowinsky, E. K. (1996). The development and clinical utility of the taxane class of antimicrotubule chemotherapy agents. *Ann Rev Med 48*:353–374.

10. Rowinsky, E. K., and Donehower, R. C. (1995). Paclitaxel (Taxol). *N Engl J Med 332*:1004–1114.

11. McGuire, W. P., Rowinsky, E. K., Rosenshein, N. B., et al (1989). Taxol: a unique antineoplastic agent with significant activity in advanced ovarian epithelial neoplasms. *Ann Intern Med 111*:273–279.

12. Einzig, A. I., Wiernik, P., Sasloff, J., Runowicz, C. D., and Goldberg, G. L. (1992). Phase II study and long-term follow up of patients treated with taxol for advanced ovarian adenocarcinoma. *J Clin Oncol 10*:1748–1753.

13. Thigpen, T., Blessing, J., Ball, H., Hummel, S., and Barret R. (1994). Phase II trial of paclitaxel in patients with progressive ovarian carcinoma after platinum-based chemotherapy: a Gynecological Oncology Group study. *J Clin Oncol 12*:1748–1753.

14. Kohn, E. C., Sarosy, G., Bicher, A., et al (1994). Dose-intense taxol: high response rate in patients with platinum-resistant recurrent ovarian cancer. *J Natl Cancer Inst 86*:18–24.

15. Sarosy, G., Kohn, E., Stone, D. A., et al (1992). Phase I study of taxol and granulocyte colony-stimulating factor in patients with refractory ovarian cancer. *J Clin Oncol 10*:1165–1170.

16. Holmes, F. A., Walters, R. S., Theriault, R. L., et al (1991). Phase II trial of taxol, an active drug in metastatic breast cancer. *J Natl Cancer Inst 83*:1797–1805.

17. Reichman, B., Seidman, A., Crown, J., et al (1993). Paclitaxel and recombinant human granulocyte colony-stimulating factor as initial chemotherapy for metastatic breast cancer. *J Clin Oncol 11*:1943–1951.

18. Murphy, W. K., Fossella, F. V., Winn, R. J., et al (1993). Phase II study of taxol in patients with untreated non-small cell lung cancer. *J Natl Cancer Inst 85*:384–388.

19. Chang, A., Kim, K., Glick, J., et al (1993). Phase II study of taxol, merbarone, and piroxantrone in stage IV non-small cell lung cancer; the Eastern Cooperative Oncology Group (ECOG) results. *J Natl Cancer Inst 85*:388–394.

20. Ettinger, D. S., Finkelstein, D. M., Sarma, R., and Johnson, D. H. (1993). Phase II study of taxol in patients with extensive-stage small cell lung cancer. *Proc Am Soc Clin Oncol 12:*329 (abstr).

21. Kirschling, R. J., Jung, S. H., and Jett, J. R. (1994). A phase II trial of taxol and G-CSF in previously untreated patients with extensive stage small cell lung cancer. *Proc Am Soc Clin Oncol 13:*326 (abstr 1076).

22. Forastiere, A. A., Neuberg, D., Taylor, S. G., et al (1993). Phase II evaluation of taxol in advanced head and neck cancer: An Eastern Cooperative Oncology Group Trial. *Monogr Natl Cancer Inst 15:*181–184.

23. Wilson, H. W., Chabner, B. A., Bryant, G., et al (1995). Phase II study of paclitaxel in relapsed non-Hodgkin's lymphoma. *J Clin Oncol 13:*381–386.

24. Roth, B. J., Dreicer, R., Einhorn, L. H., et al (1994). Significant activity of paclitaxel in advanced transistional-cell carcinoma of the urothelium: a phase II study of the Eastern Cooperative Oncology Group. *J Clin Oncol 12:*2264–2270.

25. Ajani, J. A., Ilson, D., Daugherty, L., et al (1994). Activity of taxol in patients with squamous cell carcinoma and adenocarcinoma of the esophagus. *J Natl Can Inst 86:*1086–1091.

26. Motzer, R., Bajorin, D., and Schwartz, L. (1994). Phase II trial of paclitaxel shows antitumor activity in patients with previously treated germ cell tumors. *J Clin Oncol 12:*2277–2283.

27. Gill, P. S., Tulpule, A., Reynolds, T. (1996). Paclitaxel (Taxol) in the treatment of relapsed or refractory AIDs-related Kaposi's sarcoma. *Proc Am Soc Clin Oncol 15:*306 (abstr 854).

28. Reed, E., Bitton, R., Sarosy, G., and Kohn, E. (1996). Paclitaxel dose intensity. *J Infusion Chemother 6:*59–63.

29. Rowinsky, E. K., Mackey, M. K., and Goodman, S. N. (1995). Meta analysis of paclitaxel dose-response and dose-intensity in recurrent or refractory ovarian cancer. *Proc Am Soc Clin Oncol 15:*284 (abstr 770).

30. Omura, G. A., Brady, M. F., Delmore, J. E., et al (1995). A randomized trial of paclitaxel at 2 dose levels and filgastrim (G-CSF) in platinum pretreated epithelial ovarian cancer: a Gynecologic Oncology Group, SWOG, NCCTG, and ECOG study. *Proc Am Soc Clin Oncol 15:*280 (abstr 755).

31. Peretz, T., Sulkes, A., and Chollet, P. (1995). A multicenter randomized study of two schedules of paclitaxel in patients with advanced breast cancer. *Eur J Cancer 31A:*(suppl 5):S75.

32. Creemers, G. J., Gerrits, C. J. M., Schellens, J. H. M., et al (1996). Phase II and pharmacologic study of topotecan administered as a 21-day continuous infusion to patients with colorectal cancer. *J Clin Oncol 14:*2540–2545.

33. Sugarman, S. M., Ajani, J. A., Daugherty, K., et al (1994). A phase II trial of topotecan (TPT) for the treatment of advanced, measurable colorectal cancer. *Proc Am Soc Clin Oncol 13:*225 (abstr 687).

34. Hendricks, C. B., Rowinsky, E. K., Growchow, L. B., and Kaufman, S. H. (1992). Effect of P-glycoprotein expression on accumulation and cytotoxicity of topotecan (SK&F 104864), a new camptothecin analog. *Cancer Res 52:*2268–2278.

35. Mattern, M. R., Hofmann, G. A., and Polsky, R. M. (1993). In vitro and in vivo effects of clinically important campothecin analogues in multidrug-resistant cells. *Oncol Res 5:*467–474.

36. Saltz, L., Sirott, M., Young C., et al (1993). Phase I clinical and pharmacology study of topotecan given daily for 5 consecutive days to patients with advanced solid tumors, with attempt at dose intensification using recombinant granulocyte colony-stimulating factor. *J Natl Cancer Inst 85:*1499–1507.

37. Abbruzzese, J. L., Madden, T., Sugarman, S. M., et al (1996). Phase I clinical and plasma and cellular pharmacological study of topotecan without and with granulocyte colony-stimulating factor. *Clin Cancer Res 2:*1489–1497.

38. Rowinsky, E. K., O'Reilly, S., Burks, K., Donehower, R. C., and Growchow, L. B. (1998). Phase III trial of high-dose topotecan and granulocyte-colony-stimulating factor. *Ann Oncol* (in press).

39. Savarese, D. M., Denicoff, A. M., Berg, S. L., et al (1993). Phase I study of high-dose piroxantrone with granulocyte colony-stimulating factor. *J Clin Oncol 11:*1795–1803.

40. Aylesworth, C., Von Hoff, D., Eckardt, G., et al (1997). A phase I trial of rhizoxin (NSC 332598) administered as a 72-hour continuous intravenous infusion every 3 weeks. *Proc Am Soc Clin Oncol 16:*230a (abstr 809).

41. Schiller, J. H., Storer, B., Tutsch, K., et al (1994). Phase I trial of 3-hour infusions of paclitaxel with or without granulocyte colony-stimulating factor in patients with advanced cancer. *J Clin Oncol 12:*241–248.

42. Rubin, E., Wood, V., Bharti, A., et al (1995). A phase I and pharmacokinetic study of a new camptothecain derivative, 9-aminocamptothecin. *Clin Cancer Res 1:*269–276.

43. Dahut, W., Harold, N., Takimoto, C., et al (1996). Phase I and pharmacologic study of 9-aminocamptothecin given by 72-hour infusion in adult cancer patients. *J Clin Oncol 14:*1236–1244.

44. Saltz, L., Kenemy, N., Soignet, S., et al (1996). A phase II study of 9-aminocamptothecin (9AC) in patients with fluorouracil(5U)-refractory colorectal cancer. *Proc Am Soc Clin Oncol 15:*204 (abstr).

45. Ansari, R. H., Masters, G. A., Hoffman, P. C., et al (1996). A phase II trial of 9-aminocamptothecin (9-AC) in advanced non-small cell lung cancer (NSCLC). *Proc Am Soc Clin Oncol 15:*408 (abstr 1247).

46. Takimoto, C. H., Dahut, W., Harold, N., et al (1996). A phase I trial of 9-aminocamptothecin (9-AC) in adult patients with solid tumors. *Proc Am Soc Clin Oncol 15:*488 (abstr 1554).

47. Schiller, J. H., Kim, K., Hutson, P., et al (1996). Phase II study of topotecan in patients with extensive stage small-cell lung carcinoma of the lung: an Eastern Cooperative Oncology Group trial. *J Clin Oncol 14:*2345–2352.

48. Gordon, A., Bookman, M., Malmstrom, H., et al (1996). Efficacy of topotecan in advanced epithelial ovarian cancer after failure of platinum and paclitaxel: International Topotecan Study Group trial. International phase II study. *Proc Am Soc Clin Oncol 15:*282 (abstr 763).

49. Carmicheal, J., Gordon, A., Malfetano, J., et al (1996). Topotecan, a new active drug, versus paclitaxel in advanced epithelial ovarian carcinoma: Inter-

national Topotecan Study Group trial. *Proc Am Soc Clin Oncol 15*:283 (abstr 765).

50. Kudelka, A. P., Treusukosol, D., Edwards, C. L., et al (1996). Phase II study of intravenous topotecan as a 5-day infusion for refractory epithelial ovarian carcinoma. *J Clin Oncol 14*:1552–1557.

51. Creemers, G. J., Bolis, G., Gore, M., et al (1996). Topotecan, an active drug in the second-line treatment of epithelial ovarian cancer: results of a large European phase II study. *J Clin Oncol 14*:3056–3061.

52. Smith Kline Beecham (1996). New Drug Application 20–671. Hycamptin (topotecan hydrochloride) for injection. Briefing Document for FDA Oncology Drug Products Advisory Committee.

53. Meropol, N. J., Miller, L. L., Korn, E. L., et al (1992). Severe myelosuppression resulting from concurrent administration of granulocyte colony-stimulating factor and cytotoxic chemotherapy. *J Natl Cancer Inst 84*:1201–1203.

54. Fisherman, J. S., Cowan, K. H., Noone, M., et al (1996). Phase I/II study of 72-hour infusional paclitaxel and doxorubicin with granulocyte colony-stimulating factor in patients with metastatic breast cancer. *J Clin Oncol 14*:774–782.

55. Karato, A., Sasaki, Y., Shinkai, T., et al (1993). Phase I study of CPT-11 and etoposide in patients with refractory solid tumors. *J Clin Oncol 10*:2030–2035.

56. Pratt, C. B., Stewart, C., Santana, V. M., et al (1994). Phase I study of topotecan for pediatric patients with malignant solid tumors. *J Clin Oncol 12*:539–543.

57. Rowinsky, E. K., Gilbert, M., McGuire, W. P., et al (1991). Sequences of taxol and cisplatin: a phase I and pharmacologic study. *J Clin Oncol 9*:1692–1703.

58. Rowinsky, E. K., Chaudhry, V., Forastiere, A. A., et al (1993). A phase I and pharmacologic study of taxol and cisplatin with granulocyte colony-stimulating factor: neuromuscular toxicity is dose-limiting. *J Clin Oncol 11*:2010–2020.

59. Bonomi, P., Kim, K., Chang, D., and Johnson, D. (1996). Phase III trial comparing etoposide, cisplatin, versus taxol with cisplatin-G-CSF versus taxol-cisplatin in advanced non-small cell lung cancer. An Eastern Cooperative Oncology Group (ECOG) trial. *Proc Am Soc Clin Oncol 15*:382 (abstr 1145).

60. Forastiere, A. A., Leong, T., Murphy, B., Rowinsky, E., De Conti, R., Karp, D., and Adams, G. (1997). A phase III trial of high dose paclitaxel and cisplatin and G-CSF versus low dose paclitaxel and cisplatin in patients with advanced squamous cell carcinoma of the head and neck: an Eastern Cooperative Oncology Group trial. *Proc Am Soc Clin Oncol 16*:384 (abstr).

61. Langer, C. J., Leighton, J. C., Comis, R. L., et al (1995). Paclitaxel and carboplatin in combination in the treatment of advanced non-small-cell lung cancer: a phase II toxicity, response, and survival analysis. *J Clin Oncol 13*:1860–1870.

62. Johnson, D. H., Paul, D. M., Hande, K. R., et al (1996). Paclitaxel plus carboplatin in advanced non-small cell lung cancer: a phase II trial. *J Clin Oncol 14*:2054–2060.

63. Bookman, M. A., McGuire, W. P. III, et al (1996). Carboplatin and paclitaxel in ovarian and peritoneal carcinoma: a phase I study of the Gynecologic Oncology Group. *J Clin Oncol 14:*1895–1902.
64. Jacobs, S. A., Stoller, R. G., Earle, M. F., et al (1996). Phase I study of sequential adriamycin and taxol with neurogen support in advanced breast cancer. *Proc Am Soc Clin Oncol 15:*97 (abstr 55).
65. Gianni, L., Munzone, E., Capri, G., et al (1995). Paclitaxel by 3-hour infusion in combination with bolus doxorubicin in women with untreated metastatic breast cancer: high antitumor efficacy and cardiac effects in a dose-finding and sequence-finding study. *J Clin Oncol 13:*2688–2699.
66. Sledge, G. W., Robert, N., Sparano, J., et al (1994). Paclitaxel (Taxol)/doxorubicin combinations in advanced breast cancer. The Eastern Cooperative Oncology Group Experience. *Sem Oncol 21*(suppl 8):15–18.
67. Sledge, G. W., Jr., Robert, N., Sparano, J. A., et al (1995). Eastern Cooperative Oncology Group studies of paclitaxel and doxorubicin in advanced breast cancer. *Sem Oncol 3*(suppl 6):105–108.
68. Lalisang, R., Wils, J., Nortier, J., et al (1996). Dose intensification of epirubicin and paclitaxel with G-CSF (Filgastrim) in metastatic breast cancer. *Proc Am Soc Clin Oncol 15:*99 (abstr 62).
69. Early, E., Shorter, S., Sugarman, A., et al (1996). Phase I trial of dose-intense liposomal encapsulated doxorubicin (LED) with G-CSF in patients with advanced soft tissue sarcoma. *Proc Am Soc Clin Oncol 15:*524 (abstr 1690).
70. Shapiro, C. L., Ervin, T., Azarnia, N., et al (1996). Phase II trial of high dose liposome-encapsulated doxorubicin (D-99) with G-CSF in metastatic breast cancer. *Proc Am Soc Clin Oncol 15:*112 (abstr 115).
71. Tolcher, A. W., Cowan, K. H., Noone, M. H., et al (1996). Phase I study of paclitaxel in combination with cyclophosphamide and granulocyte colony-stimulating factor in metastatic breast cancer patients. *J Clin Oncol 14:*95–102.
72. Kennedy, M. J., Donehower, R. C., and Rowinsky, E. K. (1995). Treatment of metastatic breast cancer with the paclitaxel/cyclophosphamide combination. *Sem Oncol 22*(suppl 8):23–27.
73. Kennedy, M. J., Zahurak, M. L., Donehower, R. C., et al (1996). Phase I and pharmacologic study of sequences of paclitaxel and cyclophosphamide supported by granulocyte colony-stimulating factor in women with previously-treated metastatic breast cancer. *J Clin Oncol 14:*783–791.
74. Diamandidis, D. T., Lee, J. S., Shin, D. M., et al (1995). Phase I study of taxol and edatrexate (EDAM) combination with G-CSF support in solid tumors. *Proc Am Soc Clin Oncol 14:*470 (abstr 1523).
75. Weiselberg, L., Budman, D. R., O'Mara, V., et al (1996). Phase I trial of sequential vinorelbine-paclitaxel in patients with metastatic breast cancer: early evidence of tolerability and efficacy. *Proc Am Soc Clin Oncol 15:*97 (abstr 54).
76. Shepherd, F., Latreille, E. Eisenhauer, E., et al (1996). Phase I trial of paclitaxel (Taxol) and ifosfamide in previously untreated patients with nonsmall cell lung cancer. *Proc Am Soc Clin Oncol 14:*373 (abstr 1153).

77. Kohn, E. C., Sarosy, G. A., Davis, C., et al (1996). A pilot study of cyclophosphamide (CTX), paclitaxel (T), and cisplatin (P) with G-CSF for newly diagnosed ovarian cancer patients. *Proc Am Soc Clin Oncol 15*:281 (abstr 757).

78. Stemmer, S. M., Cagnoni, P. J., Shpall, E. J., et al (1996). High-dose paclitaxel, cyclophosphamide, and cisplatin with autologous hematopoietic progenitor-cell support: a phase I trial. *J Clin Oncol 14*:1463–1472.

79. Javed, T., Reed, C., Walle, T., et al (1995). A regimen of paclitaxel, cisplatin, and 5-fluorouracil followed by G-CSF is highly active against epidermoid and adenocarcinoma of esophagus. *Proc Am Soc Clin Oncol 14*:195 (abstr 456).

80. Okamoto, H., Nagatomo, A., Kunitoh, H., and Watanabe, K. (1995). A combination phase I-II study of carboplatin/irinotecan plus G-CSF with the calvaert equation in patients with non-small cell lung cancer (NSCLC). *Proc Am Soc Clin Oncol 14*:373 (abstr 1150).

81. Heideman, R., Kuttesch, J., Stewart, C., et al (1995). A phase I trial of a fixed systemic exposure of carboplatin with continuous infusion topotecan in pediatric solid tumors. *Proc Am Soc Clin Oncol 14*:447 (abstr 1430).

82. Tolcher, A. W., O'Shaughnessy, J. A., Weiss, R. B., et al (1994). A phase I study of topotecan (a topoisomerase I inhibitor) and doxorubicin (a topoisomerase II inhibitor). *Proc Am Soc Clin Oncol 13*:157 (abstr 422).

83. Karato, A., Sasaki, Y., Shinkai, T., et al (1993). Phase I study of CPT-11 and etoposide in patients with refractory solid tumors. *J Clin Oncol 11*:2030–2035.

84. Eckardt, J. R., Burris, H. A., Von Hoff, D. D., et al (1994). Measurement of tumor topoisomerase I and II levels during the sequential administration of topotecan and etoposide. *Proc Am Soc Clin Oncol 13*:141 (abstr 358).

85. Lilenbaum, R. C., Ratain, M. J., Miller, A. A., et al (1995). Phase I study of paclitaxel and topotecan in patients with advanced tumors: a Cancer and Leukemia Group B study. *J Clin Oncol 13*:2230–2237.

86. Hochster, H., Speyer, J., Oratz, R., et al (1995). Phase I study of taxol with 14-day topotecan continuous low-dose infusion. *Proc Am Soc Clin Oncol 14*:486 (abstr 1586).

87. Murren, J. R., Fedele, J., Anderson, S., et al (1995). Phase I trial of cyclophosphamide and topotecan in refractory cancer. *Proc Am Soc Clin Oncol 14*:475 (abstr 1542).

88. Rowinsky, E. K., and Kaufmann, S. H. (1998). Topotecan in combination chemotherapy. *Sem Oncol* (in press).

89. Rothenberg, M. L., Burris, H. A., III., Eckardt, J. R., et al (1993). Phase I/II study of topotecan + cisplatin in patients with non-small cell lung cancer (NSCLC). *Proc Am Soc Clin Oncol 12*:156 (abstr 423).

90. Rowinsky, E. K., Kaufmann, S. H., Baker, S. D., et al (1996). Sequences of topotecan and cisplatin: phase I, pharmacologic and in vitro studies examining sequence-dependence. *J Clin Oncol 14*:3074–3084.

91. Miller, A. A., Hargis, J. B., Lilenbaum, R. C., et al (1994). Phase I study of topotecan and cisplatin in patients with dvanced solid tumors: a cancer and leukemia group B study. *J Clin Oncol 12*:2743–2750.

92. Saltz, L., Kanowitz, J., Schwartz, G., et al (1995). Phase I trial of cisplatin (DDP) plus topotecan on a daily × 5 schedule in patients with solid tumors. *Proc Am Soc Clin Oncol 14*:475 (abstr).

93. Masuda, N., Fukuoka, M., Takada, M., et al (1992). CPT-11 in combination with cisplatin for advanced non-small cell lung cancer. *J Clin Oncol 10:*1775–1780.

94. Masuda, N., Fukuoka, M., Kudoh, S., et al (1994). Phase I study of irinotecan and cisplatin with granulocyte colony-stimulating factor support for advanced non-small-cell lung cancer. *J Clin Oncol 12:*90–96.

95. Mori, K., Ohnishi, T., Yokoyama, M., and Tominaga, K. (1996). A phase I study of irinotecan and infusional cisplatin with recombinant human granulocyte colony-stimulating factor support in the treatment of advanced non-small cell lung cancer. *Proc Am Soc Clin Oncol 15:*384 (abstr 1153).

96. Cole, J. T., Gralla, R. V., Marques, C. B., and Rittenberg, C. N. (1995). Phase I-II study of cisplatin + docetaxel (Taxotere) in non-small cell lung cancer. *Proc Am Soc Clin Oncol 14:*357 (abstr 1087).

97. Zalcberg, J. R., Bishop, J. F., Millward, M. J., et al (1995). Preliminary results of the first phase II trial of docetaxel in combination with cisplatin in patients with metastatic or locally advanced non-small cell lung cancer. *Proc Am Soc Clin Oncol 14:*351 (abstr 1062).

98. Bourgeois, H., Gruia, G., Dieras, V., et al (1996). Docetaxel in combination with doxorubicin as first line CT of metastatic breast cancer. *Proc Am Soc Clin Oncol 15:*148 (abstr 259).

99. Miller, A. A., Lilenbaum, R. C., Lynch, T. J., et al (1996). Treatment related fatal sepsis from topotecan/cisplatin and topotecan/paclitaxel. *J Clin Oncol 14:*1964–1965.

100. Blackwell, S., and Crawford, J. Filgrastim (r-metHuG-CSF) in the chemotherapy setting. In: Morstyn G., and Dexter T. M. eds. *Filgrastim in Clinical Practice,* New York: Marcel Dekker, 1994, pp. 103–116.

101. Rowinsky, E. K., Grochow, L. B., Hendricks, C., et al (1992). Phase I and pharmacologic study of topotecan (SK&F 104864): A novel topoisomerase I inhibitor. *J Clin Oncol 10:*647–656.

102. Moore, M. A., Stolfi, R. L., Martin, D. S., et al (1990). Hematologic effects of interleukin 1-β, granulocyte colony-stimulating factor, and granulocyte-macrophage colony-stimulating factor in tumor-bearing mice treated with fluoruracil. *J Natl Cancer Inst 82:*1031–1037.

103. Kaplan, L. D., Kahn, J. O., Crowe, S., et al (1991). Clinical and virologic effects of recombinant human granulocyte-macrophage colony-stimulating factor in patients receiving chemotherapy for human immunodeficiency virus-associated non-Hodgkins lymphoma: results of a randomized trial. *J Clin Oncol 9:*929–940.

104. Osborne, C. K., Sutherland, M. C., Neidhart, J. A., et al (1994). Failure of GM-CSF to permit dose-escalation in an every other week dose-intensive regimen for advanced breast cancer. *Ann Oncol 5:*43–47.

105. Clark, D. A. and Neidhart, J. A. (1992). Granulocyte-macrophage colony-stimulating factor with dose-intensified treatment of cancer. *Semin Hematol 29*(suppl 3):27–32.

106. Kaufmann, S. H., Charron, M., Burke, P. J., and Karp, J. E. (1995). Changes in topoisomerase I levels and localization during myeloid maturation in vitro and in vivo. *Cancer Res 55:*1255–1260.

107. Pettengell, R., Morgenstern, G. R., Woll, P. J., et al (1993). Peripheral blood progenitor cell transplantation in lymphoma and leukemia using a single apheresis. *Blood 82*:3770–3777.

108. Gale, R. P., Henon, P., and Juttner, C. (1992). Blood stem cell transplants come of age. *Bone Marrow Transplant 9*:151–155.

109. Moore, M. A. S. (1991). Clinical implications of positive and negative hematopoietic stem cell regulators. *Blood 78*:1–9.

110. Moore, M. A. S. (1992). Does stem cell exhaustion result from combining hematopoietic growth factors with chemotherapy? If so, how do we prevent it? *Blood 80*:3–7.

111. Ranuzzi, M., Nistico, C., Garufi, C., et al (1996). Vinorelbine (VNR) in weekly schedule with G-CSF in patients with advanced breast cancer. *Proc Am Soc Clin Oncol 15*:125 (abstr 166).

112. Parimoo, D., Garcia, A., Muggia, F., et al (1996). Tolerance of paclitaxel (Taxol) 3-hr infusion with G-CSF on a biweekly schedule. *Proc Am Soc Clin Oncol 15*:181 (abstr 376).

18
Use of Combinations of Cytokines with Filgrastim (r-metHuG-CSF)

Juan W. Valle and J. Howard Scarffe
*University of Manchester, Christie Hospital NHS Trust,
Manchester, England*

I. INTRODUCTION

Hematopoiesis is a function regulated not by a single factor, but by the complex interaction of a variety of cytokines acting on hematopoietic cells at various stages of maturation. There is increasing evidence of in vitro and in vivo interactions between these individual cytokines. The potential of these interactions is being explored to improve the production of blood cells, which is impaired in a variety of diseases and during myelosuppressive treatment, and to optimize the yield of peripheral blood hematopoietic progenitor cell collection. In this chapter we cover some of the current knowledge regarding the synergy between granulocyte colony-stimulating factor (G-CSF) and other cytokines. While the focus is a predominantly clinical one, it is necessary to include the results of a modest amount of preclinical laboratory work.

Sections II to VII present the results of a variety of in vitro, in vivo, and clinical studies examining the combination of various cytokines with Filgrastim. Section VIII discusses the ex vivo expansion of peripheral blood progenitor cells (PBPC) after myelosuppressive treatment, also using a variety of combinations of cytokines with Filgrastim. The presentation of the research in this chapter is not intended to be comprehensive.

II. G-CSF AND GM-CSF

Synergy between G-CSF and granulocyte-macrophage colony-stimulating factor (GM-CSF) has been demonstrated in in vitro, in vivo, and clinical studies.

A. In Vitro Studies

In agar cultures of murine bone marrow cells, the addition of G-CSF and GM-CSF results in greater numbers of granulocyte-macrophage colony-forming cells (CFC-GM) than with the addition of each factor alone. Their addition also results in increased colony size (1). In human bone marrow cultures, the addition of G-CSF and GM-CSF increases colony size and numbers (2), also to supradditive levels (3), an effect that appears to be dependent on the dose of each cytokine (4). This cytokine combination appears to increase the number of granulocyte colony-forming cells (CFC-G), as well as increase the size and survival of the colonies. At the same time, the combination of GM-CSF and macrophage colony-stimulating factor (M-CSF) has its predominant effect on macrophage colony-forming cells (CFC-M) (5).

The addition of doses of G-CSF and GM-CSF in combination also shows strong synergy in increasing the numbers of CFC-GM and erythroid burst-forming cells (BFC-E) produced in long-term culture (6); G-CSF and GM-CSF are individually inactive in this context. Moreover, G-CSF and GM-CSF individually induce a degree of maturation of monocytes in the early stages of differentiation, as measured by phenotypic changes. When G-CSF and GM-CSF are added in combination, this process appears to be maximally induced (7).

To assess the influence of other combinations of cytokines, Heyworth et al (8) used a non-leukemic, multipotent, hematopoietic cell line derived from long-term bone marrow cultures. The addition of G-CSF to GM-CSF and interleukin (IL)-3 has been shown to promote differentiation of the cell line into mature neutrophils in favor of self-renewal, in contrast to high IL-3 concentrations which favor the latter. Also, G-CSF and GM-CSF act synergistically in 14-day cultures of purified CD34$^+$ bone marrow cells in terms of the numbers of cells cycling (a subadditive effect suggesting partial overlap of the different target cells), and they accelerate transit through the cell cycle (suggesting an overlapping signaling pathway of G-CSF and GM-CSF) (9).

In vitro studies also have been performed with progenitor cells from myelodysplastic and leukemic cell lines. The combination of M-CSF, G-CSF, and GM-CSF in culture was the most effective stimulus for non-erythroid progenitor cell proliferation, but not differentiation, in highly purified marrow-progenitor cells from patients with myelodysplasia (10). However, there is marked heterogeneity in the response of individual leukemic cell lines to G-CSF and GM-CSF. In some leukemic cell lines, for example, G-CSF antagonizes the effect of GM-CSF on proliferation, while acting synergistically on others (11). Blast progenitors from patients with acute myeloid leukemia (AML) also show marked heterogeneity in their response to growth factors; they show strong synergy between G-CSF and GM-CSF (measured between studies as increased colony formation, growth, or differentiation) in approximately half of the samples (12–14). Moreover, the potential effect of administered cytokines on the resistance of AML blast cells to antileukemia drugs was examined by Norgaard et al (15). Cells stimulated by the combination of G-CSF, GM-CSF, and IL-3 showed non-significant changes in in vitro drug resistance to aclarubicin and mitoxantrone, significantly increased sensitivity to Ara-C, but significantly increased resistance to daunorubicin.

B. In Vivo Studies

In vivo studies include the evaluation of the effects of pre–total body irradiation (TBI) treatment of mice with G-CSF and GM-CSF alone and in combination (either sequentially or simultaneously). Both cytokines alone protected a significant fraction of mice from the lethal effects of TBI, with G-CSF displaying the more potent activity. Administration of G-CSF followed by GM-CSF was slightly more effective than G-CSF alone in supralethally irradiated mice but not in lethally irradiated mice (16). Using an infective model, Abramson et al (17) showed that the polymorphonuclear leukocytes (PMN) of chinchillas could be primed by pretreatment with G-CSF and/or GM-CSF, and that this resulted in a reversal of influenza-A virus–induced PMN dysfunction; however, this did not result in a decrease in the incidence of secondary pneumococcal illness.

C. Clinical Studies

Clinical studies have established that the number of circulating hematopoietic progenitor cells in peripheral blood may be markedly increased by the administration of Filgrastim and recombinant human (rHu)GM-CSF. These progenitor cells, collected by apheresis, are able to induce bone marrow reconstitution after myeloablative treatments, resulting in a more rapid recovery of myeloid and megakaryocytic cell lines. Given the results of these studies, clinical trials have been performed to assess the safety, tolerability, and efficacy of the combination of Filgrastim and rHuGM-CSF administration. These studies have also been performed in patients with bone marrow failure states to evaluate the optimal mobilization of PBPC.

A phase 1–2 study by Winter et al (18) looked at the mobilization of hematopoietic cells in five cohorts of patients with the following dosage regimens: cohorts 1, 2, and 3 received Filgrastim 5 μg/kg/day on days 1 to 12 and rHuGM-CSF 0.5, 1, or 5 μg/kg/day, respectively, from days 7 to 12. To assess any possible priming effect by one cytokine before the introduction of another, cohort 4 received rHuGM-CSF 5 μg/kg/day on days 1 to 12 and Filgrastim 5 μg/kg/day from days 7 to 12 (ie, the reverse order of cohort 3). The final cohort received both Filgrastim and rHuGM-CSF, each at 5 μg/kg/day, from days 1 to 12. Ten-liter aphereses were performed on day 1 (after pretreatment with growth factors) and on days 5, 7, 11, and 13. The toxicity profile included fever, bone pain, myalgia, mild thrombocytopenia, nausea, and localized injection-site reactions. There were no dose-limiting toxicities in any of the treatment groups, and the toxicity profile from the combined cytokine administration was no different from that of patients who received a single growth factor. The results of this study showed that Filgrastim and rHuGM-CSF (at 5 μg/kg/day) were equally effective in enhancing hematopoietic progenitor cell collections (CFC-GM and BFC-E) on days 5 and 7. When Filgrastim was added to rHuGM-CSF (cohort 4), there was a significantly higher yield of CFC-GM/kg and E-BFC/kg on days 11 and 13, versus each factor alone. This effect did not occur in the reverse case with the addition of rHuGM-CSF to Filgrastim (cohort 3). In patients who received both cytokines from day 1 (cohort 5), there was a higher CFC-GM yield on day 5 that became statistically significant by day 7. The effect on cell yield declined over the 12-day treatment period. After PBPC mobilization, patients went on to receive myeloablative treatment and the harvested progenitor cells were then infused. After the transplant, patients received either Filgrastim or both Filgrastim and rHuGM-CSF; because of infusion of supplemental aphereses or bone marrow collections in some cases, the engraftment data are limited.

In another study, Spitzer et al (19) evaluated the use of Filgrastim and rHuGM-CSF in bone marrow reconstitution PBPC transplantation. Patients were apheresed after receiving a combination of Filgrastim 10 μg/kg/day and rHuGM-CSF 5 μg/kg/day for 6 days. High-dose chemotherapy was given according to the diseases of the patients, who were randomized to receive either no growth factors or Filgrastim 7.5 μg/kg and rHuGM-CSF 2.5 μg/kg twice daily as a 2-hour infusion until absolute neutrophil count (ANC) recovery (1.5×10^9/L for two consecutive days). Although neutrophil recovery occurred significantly faster and the duration of hospitalization was shortened by 2 days in the treatment group, the duration of fever, the number of documented septic episodes, and the number of blood products transfused were not statistically different between the two cohorts.

In a phase 2 pilot study, Weinthal et al (20) addressed the effect of the administration of rHuGM-CSF (500 mg/m^2) followed by Filgrastim (10 μg/kg/day) (when the ANC reached 0.5×10^9/L), versus rHuGM-CSF alone, after bone marrow transplantation. There was enhanced myelopoiesis with a shorter time to neutrophil recovery and decreased length of hospitalization in the sequentially treated cohort (3.2 versus 9.1 days, respectively) (p < 0.0001); however, there was no statistical difference in the number of blood products transfused, documented infections, or toxicity. Weisdorf et al (21) looked at the sequential administration of rHuGM-CSF (250 mg/m^2/day) on days 1 to 7 plus Filgrastim 5 μg/kg/day from days 8 to 14 versus rHuGM-CSF 250 mg/m^2/day alone from days 1 to 14 in graft failure after bone marrow transplantation. There was no significant difference in the recovery times of neutrophils or red blood cell and platelet transfusion independence between the two cohorts.

It appears, therefore, that the combination of Filgrastim and rHuGM-CSF can improve the yield of PBPC harvests that may reduce the numbers of aphereses and bone marrow harvests (because of a low PBPC yield) required. The evidence so far, however, appears to show modest benefits with the co-administration of these growth factors on recovery after PBPC transplantation compared with Filgrastim alone.

III. FILGRASTIM AND STEM CELL FACTOR

Preclinical studies have shown that stem cell factor (SCF), a ligand for the receptor encoded by the *c-kit* proto-oncogene, shows synergistic properties with other cytokines, including Filgrastim. This synergy has clinical implications that are under investigation in ongoing studies.

A. In Vitro Studies

McNiece et al (22), using human bone marrow-progenitor cells, showed that rHuSCF alone resulted in no significant colony formation in vitro; however, there was a synergistic increase in CFC-GM colony size and numbers with rHuSCF in combination with Filgrastim. In another study, by Bernstein et al (23), rHuSCF and Filgrastim in combination was not only shown to increase the numbers of CFC-GM colonies from human bone marrow CD34$^+$ cells, but that the most dramatic growth was within a small subpopulation of primitive, lineage-uncommitted stem cells. Similar results were seen both in serum-containing and serum-deprived conditions (24,25). Heyworth et al (24), also demonstrated that the addition of rHuSCF to colony-forming assays reduced the concentration of Filgrastim required for optimal colony formation.

A study using purified progenitor CD34$^+$ soybean agglutinin-negative PBPC from human cord blood showed a tenfold increase in CFC, using both rHuSCF and Filgrastim. When rHuSCF was used alone, it was unable to maintain cell numbers or CFC (26), an effect related to the feeding regimen of the cultures (27).

In murine studies, rSCF in combination with Filgrastim increases the total number of primitive progenitor cells, defined as high–proliferative potential colony-forming cells (HPP-CFC) and GM-CFC (28), as well as the number of committed progenitor cells unable to proliferate further (29).

B. In Vivo Studies

Synergy between rHuSCF and Filgrastim has also been reflected in in vivo studies. Poly[ethylene glycol]-derivitized (PEG) recombinant rodent (rr) SCF and rHuG-CSF administered to normal mice resulted in a supradditive increase in blood neutrophils and circulating spleen colony-forming cells (CFC-S). This effect was dependent on the doses of cytokines used such that in combination with rrSCF, subefficacious doses of Filgrastim produced an increase in nucleated cells per mL blood. The co-administration of these growth factors to splenectomized mice in the same study resulted in an earlier, and higher, peak of peripheral blood neutrophil numbers (30) (**Figure 1**).

Recombinant rSCF and Filgrastim injected for 1 week in rats resulted in an increase in the number of mature bone marrow neutrophils and a greater than additive increase (4- to 17-fold) in the rate of release of these neutrophils, compared with each cytokine alone, thus causing a peripheral neutrophilia. This was accompanied by a decrease in marrow erythroid and lymphoid elements and an increase in splenic granulopoiesis and erythropoiesis (the latter may be a compensatory effect due to decreased marrow erythropoiesis) (31).

In splenectomized mice, the combined administration of rrSCF (25 or 200 μg/kg/day) and Filgrastim (200 μg/kg/day) resulted in a greater than additive increase in peripheral blood low-density mononuclear cells, CFC-GM, and HPP-CFC versus each factor alone (32). In another study using splenectomized mice, Filgrastim was shown to inhibit erythropoietin (EPO)–induced erythropoiesis and, conversely, EPO-inhibited, G-CSF–induced granulopoiesis. The inhibitory effect appeared to take place at an advanced stage of lineage differentiation. The co-administration of rHuSCF with the other two factors, however, resulted in an increase in both erythroid and myeloid cell lines with a reduction of the effect of the mutual inhibition (33).

Cairo et al (34) demonstrated that the administration of rHuSCF and Filgrastim to neonatal rats, either simultaneously for 14 days or sequentially (7 days of rHuSCF followed by 7 days of Filgrastim), resulted in a significant increase in the peripheral ANC. The rHuSCF + Filgrastim group also

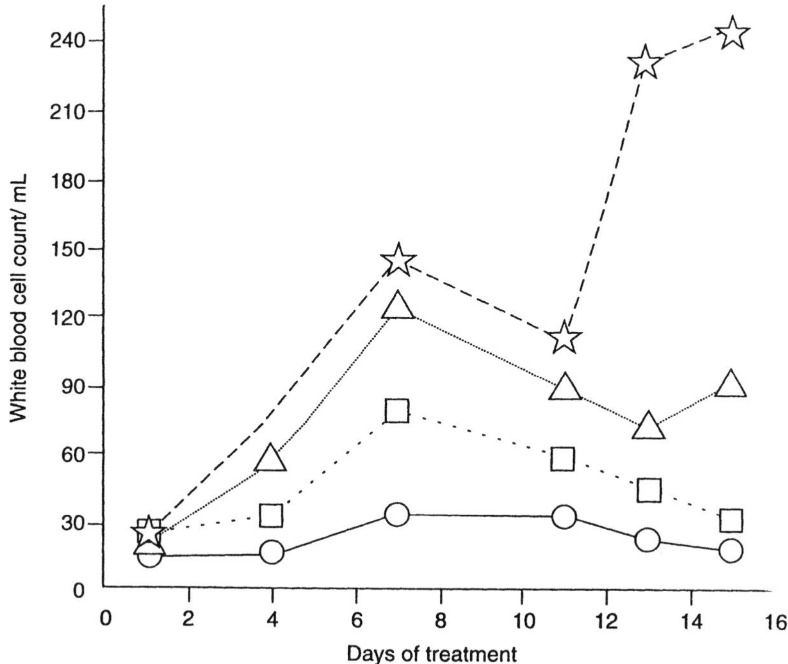

Figure 1. The changes in peripheral blood nucleated cell count in mice during treatment with recombinant rodent stem cell factor (rrSCF) (100 μg/kg/day) either alone or in combination with Filgrastim at doses between 0.1 and 2.5 μg/kg/day. Circles = rrSCF; squares = rrSCF + Filgrastim 0.1 μg/kg/day; triangles = rrSCF + Filgrastim 1 μg/kg/day; stars = rrSCF + Filgrastim 25 μg/kg/day.

showed a significant increase in the bone marrow–myeloid pools and proliferative rates and a reduction in mortality rate after experimental group B streptococcal sepsis.

McNiece et al (35) showed that the co-administration of rHuSCF and Filgrastim in mice resulted in a synergistic increase in the peripheral white blood cell count, a threefold increase in mature progenitors (CFC-GM) versus Filgrastim alone, and an increase in the number of early progenitors (HPP-CFC) and BFC-E numbers. Survival of these mice was equivalent or greater than mice treated with Filgrastim alone. Non-human primates that received both growth factors showed a faster increase in white blood cell count, higher than with Filgrastim alone, and a greater than 5-fold increase in CFC-GM. The latter was sustained for the period of cytokine administration, unlike Filgrastim alone, which caused a single peak. The sequential administration of rHuSCF followed by Filgrastim did not result in an enhancement of the bone marrow response.

In a follow-up to their earlier work, Andrews et al (36) administered low-dose rHuSCF (25 μg/kg/day) plus Filgrastim (100 μg/kg/day) or Filgrastim alone to baboons. The cohort receiving the two cytokines had a greater number of progenitor cells in the leukapheresis product (14-fold), and the blood cells mobilized from this cohort engrafted lethally irradiated recipients more rapidly (shorter time to platelets $>20 \times 10^9$/L, white blood cells $>1 \times 10^9$/L, and ANC $>0.5 \times 10^9$/L), compared with the group receiving Filgrastim alone.

De Revel et al (37) used a canine model of PBPC rescue after lethal irradiation. Dogs were treated for 7 days with the following cytokine regimens: recombinant canine (rc) SCF alone at two dose levels (25 and 200 μg/kg, respectively), rcG-CSF (10 μg/kg) alone, low-dose rcSCF (25 μg/kg) + rcG-CSF (10 μg/kg), or no cytokines at all. A single leukapheresis was then performed. Dogs treated with rcSCF + rcG-CSF were found to have a greater than additive increase in circulating CFC-GM per mL of blood. One month after the collection and cryopreservation of PBPC, the dogs were exposed to a lethal dose of TBI, then all were infused with a standard number of mononuclear cells (1 $\times$ 10^8/kg). Control animals (treated with PBPC harvested with no growth factors) all died, as did animals who received low-dose rcSCF. All animals who had received either high-dose rcSCF alone, rcG-CSF alone, or low-dose rcSCF + rcG-CSF recovered granulocyte function. In the combination group, all animals became long-term survivors. In each single cytokine group, four of five animals became long-term survivors.

Studies with leukemic blast cells have shown that, in combination, Filgrastim and rHuSCF stimulate the proliferation of acute myeloid leukemia (AML) cells, expanding the leukemia colony-forming cell (CFC-L) pool (38). Filgrastim at optimal concentrations also increases the sensitivity of blast cells to rHuSCF (39). This effect, both on cell lines and fresh cells from patients with AML, is heterogeneous (40), and it appears that costimulation by the two factors probably stimulates subsets of AML cells that do not respond to the individual factors in isolation (41).

C. Clinical Studies

Given these preclinical data, although prospective randomized studies have not been published, preliminary results of clinical trials are appearing as abstracts. In a phase 1–2 study, McNiece et al (42) administered Filgrastim alone (10 μg/kg/day) or in combination with rHuSCF (5 or 10 μg/kg/day) to patients with high-risk breast cancer. They reported an increase in CD34$^+$

cells and GM-CFC in the leukapheresis product of patients receiving both cytokines. There was also an increase in the median white blood cell count in the co-administration group. This increase was dose dependent: twofold for the low-rHuSCF dose and fourfold for the higher rHuSCF dose.

Briddell et al (43) performed leukapheresis harvests on patients with high-risk breast cancer after mobilization with rHuSCF (5 μg/kg/day) alone, Filgrastim (10 μg/kg/day) alone, or Filgrastim with rHuSCF at either 5 or 10 μg/kg/day. In the leukapheresis harvests, the cohorts treated with Filgrastim + rHuSCF showed an increase in CFC-GM and BFC-E (three- to fourfold) and CD34$^+$ cells (threefold), compared with treatment with Filgrastim alone; this effect was once again related to the dose of rHuSCF. In the cohort treated with Filgrastim + rHuSCF (10 μg/kg/day), there was also a 13-fold increase in mature megakaryocyte colony-forming cells (CFC-MK) and 70-fold increase in primitive megakaryocyte burst-forming cells (BFC-MK). Mixed colony-forming cells (CFC-Mix) also were significantly increased (**Table 1**). After PBPC transplantation, engraftment data appear to show more rapid platelet recovery in patients mobilized with Filgrastim + rHuSCF compared with Filgrastim alone.

In a dose-escalating, phase 1 study (44), patients were randomized to receive rHuSCF in doses of 10, 25, and 50 μg/kg/day. At the 50-μg/kg/day dose level, there was toxicity suggestive of mast cell activation (urticaria, dyspnea, and throat tightness), and this was considered to be the maximum tolerated dose. Antihistamines and β-adrenergic agonists are now administered prophylactically.

In a phase 1–2 study by Glaspy et al (45), also in patients with advanced breast cancer, patients received either rHuSCF (5 μg/kg/day, for 13 days) alone, Filgrastim (10 μg/kg/day, for 7 days) alone, or the combina-

Table 1. Leukapheresis Harvest Results After the Administration of rHuSCF and/or Filgrastim.

Colony-forming cells/kg	CFC-MK	BFC-MK	CFC-MIX
rHuSCF (5 μg/kg)	1125	0	2531
Filgrastim (10 μg/kg)	7202	181	13,178
rHuSCF (5 μg/kg) + Filgrastim	21,459	4239	12,437
rHuSCF (10 μg/kg) + Filgrastim	93,834	13,924	28,230

BFC-MK = megakaryocyte burst-forming cell.
CFC-MIX = mixed colony-forming cell.
CFC-MK = megakaryocyte colony-forming cell.
rHuSCF = recombinant human stem cell factor.
Source: Adapted from Ref. 53.

tion of Filgrastim (10 μg/kg/day) and one of four doses of rHuSCF (5, 10, 15, or 20 μg/kg/day) for 13 days. All patients receiving rHuSCF were premedicated using antihistamines and β-adrenergic agonists. The optimal dose of rHuSCF plus Filgrastim produced a twofold increase in the number of circulating mononuclear cells compared with Filgrastim alone (unchanged in the rHuSCF alone group), and there was a significant reduction in the time to platelet recovery after transplantation, by up to 4 days. No generalized adverse events were reported, although localized skin reactions at the injection site were common.

After these studies, which used cytokine-only mobilization, a study by Weaver et al (46) looked at PBPC mobilization using the cytokines in combination with chemotherapy. Thirty-six patients with ovarian cancer at various tumor stages were treated with cyclophosphamide (3 g/m^2) and were randomized to receive one of four cytokine combinations: Filgrastim 5 μg/kg/day alone or with rHuSCF at one of three dose levels (5, 10, or 15 μg/kg/day). Growth factors were started 3 days after chemotherapy, and a single apheresis was performed when the peripheral white blood count reached $\geq 4 \times 10^9$/L. There was a statistically significant increase in both colony-forming cells (CFC-GM and BFC-E) and CD34$^+$ cells with increasing doses of rHuSCF (**Table 2**). The progenitor cell harvest was divided into four aliquots and infused after each of four subsequent cycles of chemotherapy (cyclophosphamide 900 mg/m^2 and carboplatin at an area under the curve [AUC] of 7.5 mg/mL/minute). The engraftment data are yet to be published. Follow-up data from these and other ongoing studies are awaited to further analyze the benefits of the combination of rHuSCF and Filgrastim in this setting. In any event, the results outlined are encouraging.

IV. FILGRASTIM AND INTERLEUKIN-3

A. In Vitro Studies

The combination of Filgrastim and rHuIL-3 has been shown to increase proliferation (decreased time to colony formation) of multilineage and total colony formation by murine spleen cells compared with either factor alone (47). The combination has also been shown to increase the percentage of progenitors in S-phase of the cell cycle in murine bone marrow and spleen (48,49) and increase the mobilization of murine CFC-S into the peripheral blood (50). This may be in part because of upmodulation of IL-3 receptors on highly enriched bone marrow-progenitor cells (51). The addition of Filgrastim (with no direct effect on megakaryocyte colony formation) to rHuIL-3 in normal murine bone marrow cultures increased the number

Table 2. Harvest from Single Apheresis (Median and Range) (All Patients Received Cyclophosphamide Before Growth Factor Administration).

	Filgrastim 5 μg/kg/day (n = 9)	Filgrastim 5 μg/kg/day + rHuSCF 5 μg/kg/day (n = 9)	Filgrastim 5 μg/kg/day + rHuSCF 10 μg/kg/day (n = 8*)	Filgrastim 5 μg/kg/day + rHuSCF 15 μg/kg/day (n = 9)	p-value for dose response
MNC $\times$ 10^8/kg	3.5 (1.3–1.3)	6.0 (2.8–10.2)	5.5 (3.7–8.7)	4.5 (2.6–8.6)	p = 0.4
CFC-GM $\times$ 10^4/kg	124 (23–653)	107 (83–138)	151 (119–373)	423 (167–706)	p < 0.001
BFC-E $\times$ 10^4/kg	150 (24–909)	143 (100–308)	235 (134–656)	346 (196–667)	p = 0.003
CD34$^+$ $\times$ 10^6/kg	5.0 (1.1–17.0)	4.6 (1.8–25.0)	6.6 (1.2–15.8)	9.9 (6.3–18.3)	p = 0.001

* Data unavailable for one patient
BFC-E = erythroid burst-forming cell
CFC-GM = granulocyte-macrophage colony-forming cell
MNC = mononuclear cell
rHuSCF = recombinant human stem cell factor
Source: Adapted from Ref. 46.

and size of megakaryocyte colonies compared with rHuIL-3 alone (52); however, Heyworth et al (58) found an increase in size of murine CFC-GM occurred with rHuIL-3 plus rHuGM-CSF, rHuIL-3 plus GM-CSF plus Filgrastim, or rHuIL-3 plus rHuM-CSF, but not with rHuIL-3 plus Filgrastim.

Using normal human bone marrow, rHuIL-3 and Filgrastim in combination also has been shown to increase the number and size of granulocytic colonies compared with Filgrastim alone with a higher percentage of immature cells from normal human bone marrow (54,55). In vitro addition of Filgrastim and rHuIL-3 to purified hematopoietic progenitor cells from PBPC harvests resulted in a synergistic increase in GM-CFC colony numbers in a pediatric study (56).

B. In Vivo Studies

In a non-human primate model, administration of rHuIL-3 alone for up to 4 days to cynomolgus monkeys resulted in platelet production only; rHuIL-3 given for 8 days followed by Filgrastim resulted in myelocytic cell development; and the reverse schedule (Filgrastim followed by rHuIL-3) induced a twofold increase in platelets (57).

C. Clinical Studies

Clinical studies with both of these cytokines are few and most publications are pilot studies to assess tolerability, mobilization of PBPC, and bone marrow reconstitution after transplantation. Hocker et al (58) compared the effects of mobilization of circulating progenitor cells in seven patients after treatment with Filgrastim (5 μg/kg/day) or rHuGM-CSF (5 μg/kg/day) alone and, after an interval, in the same patients, after priming with rHuIL-3 (5 μg/kg/day) for 7 days. Recombinant HuIL-3 alone did not mobilize PBPC; however, it significantly potentiated concurrent rHuGM-CSF–induced or Filgrastim-induced mobilization of PBPC to levels greater than each growth factor alone. This effect was more marked with Filgrastim.

Another pilot study (59) administered the cytokines according to three schedules: rHuIL-3 (7.5 μg/kg/day) and Filgrastim (5 μg/kg/day) concurrently for 7 days; rHuIL-3 (7.5 μg/kg/day) for 7 days followed by Filgrastim (12 μg/kg/day) for 7 days; and the third schedule had a partial overlap (3 days) of the two cytokines (doses as for the sequential regimen). This study was primarily a safety and toxicity defining study, although PBPC were successfully collected by leukaphereses. Side effects appeared similar to the administration of each growth factor alone, namely fever, flu-like symptoms, headaches, and nausea/vomiting. All patients had WHO grades of I or II and were easily manageable.

In a study measuring endogenous cytokine levels by enzyme-linked immunosorbent assay (ELISA) in patients after PBPC transplantation, G-CSF levels were found to increase immediately after transplantation and decrease with an increasing neutrophil count. Interleukin-3 also showed a transient increase suggesting that these two cytokines contribute to the early hematopoietic reconstitution after PBPC transplant (60). Lemoli et al (61) administered rHuIL-3 and Filgrastim to patients with Hodgkin's disease (HD) and non-Hodgkin's lymphoma (NHL) who had received the same pretransplant conditioning. Two cohorts were studied: Filgrastim (5 μg/kg/day) from day 1 after the infusion of autologous stem cells with the addition of rHuIL-3 (10 μg/kg/day) concurrently from day 6 versus Filgrastim alone. Actual numbers of bone marrow CFC-GM and BFC-E per given number of cells plated were the same or slightly reduced in both cohorts; however, because of an increase in bone marrow cellularity in the Filgrastim and rHuIL-3 cohort, the total numbers of these colonies were significantly greater. In addition, the Filgrastim + rHuIL-3 cohort showed a significantly greater percentage of CFC-GM, BFC-E, CFC-MK, and BFC-MK in the S-phase of the cell cycle compared with baseline. In the Filgrastim-alone cohort, there was an increase in the cell cycling of CFC-GM (although this was significantly lower than the Filgrastim + rHuIL-3 cohort) but not of BFC-E, CFC-MK, or BFC-MK.

The exact dosing schedule using these cytokines may alter the outcome of hematopoietic recovery after myeloablative treatment: rHuIL-3 (5 μg/kg/day) from days 7 to 11 after high-dose melphalan for multiple myeloma followed by Filgrastim (5 μg/kg/day) from days 12 to 20 did not shorten the duration of aplasia, whereas this was significantly shorter when rHuIL-3 was administered from days 1 to 7 and Filgrastim from days 4 until recovery, ie, an overlapping period of 4 days (62). Since this was a small study (n = 5 patients), further studies will be needed to elaborate these findings. Vannucchi et al (63) used the combination of rHuIL-3 and rHuGM-CSF in delayed failure in bone marrow recovery after autologous bone marrow transplantation (ABMT) after a 5-day course of Filgrastim. This sequential treatment induced neutrophil recovery within 2 days of administering all three cytokines, suggesting the possibility of a priming effect. This also would need follow-up in larger studies.

V. FILGRASTIM AND ERYTHROPOIETIN

The combination of Filgrastim and rHuEPO has been used after marrow and PBPC transplantation and in the treatment of hematological malignancies: aplastic anemia and myelodysplasia.

In a pediatric pilot study, Locatelli et al (64) treated 15 consecutive patients undergoing allogeneic bone marrow transplantation (allo-BMT) with Filgrastim and rHuEPO. They then compared variables for marrow engraftment with 15 patients previously treated with rHuEPO alone and 16 historical control patients. There was no toxicity attributable to the co-administration of Filgrastim and rHuEPO. Erythroid repopulation, measured by assaying serum transferrin receptor and reticulocyte count, was faster than historical controls in the combined cytokine group and the rHuEPO alone group, although there was no statistical difference between these two cohorts. Red blood cell transfusion requirements, however, were reduced in the rHuEPO-alone group and significantly more so in the Filgrastim plus rHuEPO group compared with controls. There also was a significantly shorter time to neutrophil recovery with the cytokine combination with fewer infective episodes, fewer days of fever, faster time to platelet recovery, and fewer platelet transfusions. There was no Filgrastim-only arm in the study.

In another study, this time after autologous bone marrow transplantation (ABMT) for 35 patients with HD and NHL (65), cohorts were randomized to receive either rHuEPO (600 U/kg three times per week) or placebo for 3 weeks before high-dose chemotherapy. This was then stopped for the week of conditioning chemotherapy and restarted on the day of bone marrow transplantation along with Filgrastim (10 μg/kg/day, intravenously [IV]). There was no difference in the time to neutrophil recovery, platelet count recovery, and transfusion requirements of either red blood cells or platelets when rHuEPO was added to Filgrastim compared with the Filgrastim and placebo group. Once again, the combination was safe and well tolerated with no directly attributable grade 4 toxicities.

With the increasing use of PBPC transplants, a phase 1–2 study was reported by Pierelli et al (66) in which 15 patients were given Filgrastim (5 μg/kg) daily from the day of infusion of the stem cells until day 12 plus rHuEPO (150 U/kg) every 48 hours until day 11. Their recovery was compared with that of eight historic and control patients who did not receive cytokine support, and there was no Filgrastim-only arm. The combination of cytokines was well tolerated and resulted in significantly faster white blood cell, neutrophil, and platelet count recovery; significantly lower platelet transfusion requirements; and shorter hospitalization. There were no episodes of neutropenic fever in the cytokine-treated group, but the control group experienced fever >38°C for a median of 4 days. Erythroid recovery, however, was not accelerated. Without a Filgrastim-only cohort, it is difficult to wholly attribute these findings to the combined administration of Filgrastim and rHuEPO.

The use of Filgrastim and rHuEPO in patients with aplastic anemia is limited mainly to case reports. Bessho et al (67) describe two cases of severe aplastic anemia responding to rHuG-CSF and rHuEPO. The first patient was treated initially with antilymphocyte globulin, high-dose steroids, and rHuG-CSF. Recombinant HuEPO was added 2 months later and resulted in trilineage recovery. On stopping rHuG-CSF and rHuEPO, his aplasia returned. Trilineage recovery was later reinduced with the same combination of growth factors. A second patient was treated with rHuG-CSF and rHuEPO from the onset; 11 months later he showed trilineage recovery. It may be that the antilymphocyte globulin and high-dose steroids induced the recovery in the first patient. However, the relapse on stopping the growth factors and subsequent recovery with their reintroduction, along with the outcome of the second case, support the hypothesis that recovery may be induced and maintained with rHuG-CSF and rHuEPO. Weide et al (68) reported the case of a patient with severe aplastic anemia who had become resistant to antilymphocyte globulin and steroids and who had failed to respond to the combination of rHuEPO and cyclosporin-A after 10 weeks of treatment. The subsequent addition of rHuG-CSF resulted in trilineage recovery which was sustained after 6 weeks. Interestingly, in another case of severe aplastic anemia, Filgrastim, rHuEPO, and high-dose methyl prednisolone resulted initially in a trilineage response; however, after 10 weeks of treatment, the patient developed cytogenetically confirmed myelodysplasia (69).

Given this background, Imamura et al (70) used Filgrastim (as a 30-minute IV infusion) and rHuEPO in a dose-finding study for efficacy in patients with moderate and severe aplastic anemia. There was an effect on erythrocytes in 44.4% of patients overall, although the majority were patients who had moderate (n = 9/14) rather than severe (n = 3/13) disease. In 6 of 27 patients, a trilineage response was seen and a delayed and long-lasting effect was seen in 5 of 27 patients. Optimum doses appeared to be Filgrastim 400 μg/m^2/day and rHuEPO 100 U/kg/day, subcutaneously (SC).

The synergistic effects of Filgrastim and rHuEPO also have been explored in myelodysplastic syndromes (MDS). In a phase 2 trial of 22 patients, Hallstrom-Lindberg et al (71) used rHuEPO and Filgrastim for a 12-week period after Filgrastim alone had been given for 6 weeks. A significant increase in hemoglobin value was seen in 38% of the evaluable patients, including a patient who had become unresponsive to rHuEPO alone. Responses were more likely if there was less advanced cytopenia and if there were ring sideroblasts in the bone marrow and lower endogenous serum levels of EPO. This latter observation was also seen by Negrin et al (72). In their study, Filgrastim was administered to normalize the white cell count and rHuEPO was added at a dose of 100 U/kg/day and escalated to 150 and 300 U/kg/day every 4 weeks. Forty-two percent of the evaluable patients had erythroid responses, all patients had neutrophil responses, and 25% of patients became transfusion-independent during therapy. In contrast to pretreatment endogenous serum EPO levels, erythroid responses were found to be independent of patient age, French-American-British (FAB) subtype, duration of disease, prior transfusion requirements, or cytogenetic abnormalities.

Combination therapy of 1) all-*trans* retinoic acid (ATRA) and Filgrastim and 2) ATRA, Filgrastim, rHuEPO, and tocopherol in myelodysplasia has been reported in two studies by Maurer et al (73). They found that in both groups the automated nucleated count increased in 95% of cases. An increase in hemoglobin concentration and platelet count, and a reduction in transfusion requirements occurred in 40% of cases overall. Moreover, this response correlated strongly with good in vitro BFC-E growth.

VI. FILGRASTIM AND INTERFERON

The use of the combination of Filgrastim and interferon (IFN) has been investigated for the treatment of hairy cell leukemia with neutropenia. Interferon has a recognized role in the treatment of hairy cell leukemia; however, the consequent myelosuppression that may result from this therapy makes the treatment of neutropenic patients difficult. Glaspy et al (74) treated 10 patients with hairy cell leukemia with neutropenia with daily Filgrastim (at 3.6 or 7.2 μg/kg) until neutrophil counts had normalized; this normalization occurred in all cases. Interferon (10 mg/m^2 SC, three times a week) was started concurrently with the Filgrastim. After 3 months, Filgrastim was stopped and the IFN continued for 1 year. No patients developed neutropenia on starting IFN therapy, and of nine evaluable patients, eight showed hematologic improvement and three attained complete responses.

In a report of six patients (75), Filgrastim (5 μg/kg) and IFN were administered concurrently from the onset of treatment. In five of these cases, the neutrophil counts normalized within 2 to 11 days, an effect which reversed on stopping the Filgrastim support after 2 to 5 weeks of therapy. A sustained effect may have been attained with longer treatment.

Both of these studies support the further study of the use of Filgrastim as an adjunct to IFN therapy in the early treatment phase of hairy cell leukemia with neutropenia.

VII. OTHER COMBINATIONS

A. Filgrastim and IL-6

Two abstracts report the co-administration of Filgrastim and rHuIL-6. In the first by Crawford et al (76), patients with non-small–cell lung cancer (NSCLC) receiving chemotherapy were administered Filgrastim (5 μg/kg/day) either alone or with rHuIL-6 at one of three doses (1, 2.5, or 5 μg/kg/day). No severe toxicities were attributable to the combination, and full data on bone marrow recovery are awaited. In a phase 2, randomized study of patients undergoing ABMT for breast cancer, Devine et al (77) administered Filgrastim alone or with one of two doses of rHuIL-6 post-transplant. There was no difference between the groups in the time to ANC or platelet recovery or the number of platelet transfusions given; however, there was a significantly increased incidence of venocclusive disease in the dual-cytokine–treated patients when compared with historical control patients. The study was consequently halted.

B. Filgrastim and IL-11

Women undergoing ABMT for high-risk breast cancer were administered Filgrastim (5 μg/kg/day) and rHuIL-11 (10 or 25 μg/kg/day) in a study reported by Champlin et al (78). The combination was well tolerated with no grade 3 toxicity at either rHuIL-11 dose level. The median time to platelet recovery (to $>20 \times 10^9$/L) was 21 days, compared with 24 days in a historical control group.

VII. EX VIVO EXPANSION

Peripheral blood progenitor cells are used with increasing frequency for bone marrow repopulation after treatments that are significantly myelosuppressive. The collection of the PBPC requires the processing of large volumes of blood by leukapheresis. One of the more recent developments in this area has been the successful mobilization, isolation, and ex vivo expansion of CD34$^+$ cells.

Haylock et al (79) investigated the ability of various hematopoietic growth factor combinations to generate CFC-GM colonies in vitro. Cryopreserved cells were used from three patients who were apheresed during hematopoietic recovery after chemotherapy, one of whom, in addition, had received Filgrastim 12 μg/kg/day for 7 days. CD34$^+$ cells were isolated by fluorescence-activated cell sorting (FACS) and cultured with growth factors, either singly or in combination. The optimum combination of Filgrastim, rHuIL-1, rHuIL-3, rHuIL-6, rHuG-CSF, and rHuSCF produced a 29- to 39-fold increase in CFC-GM in a 7-day suspension culture of peripheral blood CD34$^+$ cells. Using the same combination of growth factors in large-volume suspension cultures, there was a significant increase in the number of CFC-GM (66-fold increase at day 14) and a 3-log increase in nucleated cells by day 21. These cells were myeloid progenitor cells and normally maturing neutrophil precursors. Using a system of weekly exchange of media and cytokines, Shapiro et al (80) found the optimum combination to be rHuSCF, rHuIL-3, rHuIL-6, rHuEPO, and Filgrastim. This may reflect the different method used for ex vivo expansion.

Brugger et al (81), in a similar study using PBPC harvested from patients after chemotherapy and Filgrastim mobilization, found a median 190-fold increase in total clonogenic progenitor cells (CFC-GM, BFC-E, and granulocyte-erythroid-macrophage-monocyte-CFC [CFC-GEMM]) at day 14 with the combination of rHuSCF, rHuEPO, rHuIL-1, rHuIL-3, and rHuIL-6. There was, however, only a 12-fold increase in CD34$^+$ cells during ex vivo expansion. The addition of Filgrastim resulted in reduced colony formation possibly due to an increase in terminal differentiation resulting in neutrophils with no clonogenic activity. In this and subsequent studies, CD34$^+$ cells were separated by immunoaffinity column separation, as FACS sorting would have been excessively time consuming. Henschler et al (82), using the same five-cytokine combination of rHuSCF, rHuEPO, rHuIL-1, rHuIL-3, and rHuIL-6, demonstrated that long-term culture-initiating cells (LT-CIC) are mobilized into the peripheral blood during apheresis after treatment with chemotherapy and Filgrastim. They also demonstrated that the LT-CIC survive (functionally and numerically) the processes of CD34$^+$ selection by immunoaffinity columns as well as ex vivo expansion; LTCIC themselves were preserved but not expanded during the culture period.

In a study by Urashima et al (83), umbilical cord blood cells were used for ex vivo expansion. After 4 weeks of culture, CD34$^+$CD38$^-$ cells (non-lineage–committed progenitor cells) disappeared in all cytokine combinations, although rHuSCF + rHuIL-6 and rHuSCF + rHuIL-6 + rHuIL-3 + rHuIFN were more effective in maintaining CD34$^+$CD38$^-$ cells. Maximal expansion of CD34$^+$CD38$^+$ (lineage-committed progenitor cells) was seen with the combination of rHuSCF + rHuIL-3 + rHuIL-6, and either Filgrastim or rHuEPO; however, this expansion was transient, reduced to zero in 3 weeks.

Given these data, studies are emerging to determine the safety, tolerability, and efficacy of the infusion into patients of hematopoietic ex vivo expansion products. In a small phase 1 study, Chang et al (840) used the combination of rHuIL-3 + rHuGM-CSF + Filgrastim to culture PBPC for

3 days in LifeCell bags. They had found this combination to be better than any of the growth factors alone, with respect to the number of CFC-GM and granulocytes retrieved. The infusions were well tolerated in five of six patients, with only a minor side effect in the remaining patient. There was no significant decrease in the duration of neutropenia.

IX. CONCLUSION

Hematopoietic growth factors are endogenous compounds present in the body in small amounts. The advent of recombinant growth factors has made possible the study of the physiology of these cytokines and their pharmacodynamic effects, singly or in combination. Given the evidence of in vitro and in vivo synergy between Filgrastim and other cytokines, clinical trials have been performed and are currently ongoing to see if this synergy can be used to produce a significant clinical effect.

Most clinical studies to date are limited to phase 1–2 safety and toxicity-defining studies. Prospective, randomized studies are needed to further assess the effects of the coadministration of Filgrastim and other cytokines, whether they are on optimizing PBPC collection, rescue of hematopoietic activity after myelosuppressive treatment, or in different disease states. The possibility of ex vivo expansion of hematopoietic progenitor cells is an interesting new development that may result in smaller and fewer leukaphereses, and it even may be possible to limit cell collection to a single venisection product.

REFERENCES

1. McNiece, I. K., Stewart, F. M., Deacon, D. M., and Quensenberry, P. J. (1988b). Synergistic interactions between haematopoietic growth factors as detected by in vitro mouse bone marrow colony formation. *Exp Haematol 16:*383–388.
2. Hassan, H. T., Zyada, L. E., Ragab, M. H., and Rees, J. K. (1991). Synergistic interactions between recombinant human interleukin-3, GM-CSF and G-CSF in normal human marrow granulocyte-macrophage colony formation. *Cell Biol Int Rep 15:*211–219.
3. McNiece, I. K., Andrews, R., Stewart, M., et al (1989). Action of interleukin-3, G-CSF, and GM-CSF on highly enriched human haematopoietic progenitor cells: synergistic interaction of GM-CSF plus G-CSF. *Blood 74:*110–114.
4. Hara, H., and Namiki, M. (1989). Mechanism of synergy between granulocyte-macrophage colony stimulating factor and granulocyte colony-stimulating factor in colony formation from human marrow cells in vitro. *Exp Haematol 17:*816–821.

5. Bot, F. J., van Eijk, L., Schipper, P., Backx, B., and Lowenberg, B. (1990). Synergistic effects between GM-CSF and G-CSF or M-CSF on highly enriched human marrow progenitor cells. *Leukemia 4*:325–328.

6. Hogge, D. E., Cashman, J. D., Humphries, R. K., and Eaves, C. J. (1991). Differential and synergistic effects of human granulocyte-macrophage colony-stimulating factor and human granulocyte colony stimulating factor on hematopoiesis in human long-term marrow cultures. *Blood 77*:493–499.

7. Geissler, K., Harrington, M., Srivastava, C., et al (1989). Effects of recombinant human colony-stimulating factors (CSF) (granulocyte-macrophage CSF, granulocyte CSF, and CSF-1) on human monocyte/macrophage differentiation. *J Immunol 143*:140–146.

8. Heyworth, C. M., Dexter, T. M., Kan, O., and Whetton, A. D. (1990). The role of hematopoietic growth factors in self-renewal and differentiation of IL-3 dependent multipotential stem cells. *Growth Factors 2*:197–211.

9. Lardon, F., Van Bockstaele, D. R., Snoeck, H. W., and Peetermans, M. E. (1993). Quantitative cell cycle progression analysis of the first three successive cell cycles of granulocyte colony stimulating factor and/or granulocyte macrophage colony stimulating factor stimulated human CD34$^+$ bone marrow cells in relation to their colony formation. *Blood 81*:3211–3216.

10. Sawada, K., Sato, N., Notoya, A., et al (1995). Proliferation and differentiation of myelodyplastic CD34$^+$ cells in serum free medium. II. Response to combined colony stimulating factors. *Eur J Haematol 54*:85–94.

11. Santoli, D., Yang, Y. C., Clark, S. C., et al (1987). Synergistic and antagonistic effects of recombinant human interleukin (IL) 3, IL-1 alpha, granulocyte and macrophage colony stimulating factors (G-CSF and M-CSF) on the growth of GM-CSF dependent leukaemic cell lines. *J Immunol 139*:3348–3354.

12. Kelleher, C., Miyauchi, J., Wong, G. (1987). Synergism between recombinant growth factors, GM-CSF and G-CSF, acting on the blast cells of acute myeloblastic leukemia. *Blood 69*:1498–1503.

13. Murohashi, I., Nagata, K., Suzuki, T., et al (1988). Effects of recombinant G-CSF and GM-CSF on the growth in methylcellulose and suspension of the blast cells in acute myeloblastic leukaemia. *Leuk Res 12*:433–440.

14. Vellenga, E., Young, D. C., Wagner, K., et al (1987). The effects of GM-CSF and G-CSF in promoting growth of clonogenic cells in acute myeloblastic leukaemia. *Blood 69*:1771–1776.

15. Norgaard, J. M., Langkjer, S. T., Ellegaard, J., et al (1993). Synergistic and antagonistic effects of myeloid growth factors on in vitro cellular killing by cytotoxic drugs. *Leuk Res 17*:689–694.

16. Waddick, K. G., Song, C. W., Souza, L., and Uckun, F. M. (1991). Comparative analysis of the in vivo radioprotective effects of recombinant granulocyte colony stimulating factor (G-CSF), recombinant granulocyte macrophage CSF, and their combination. *Blood 77*:2364–2371.

17. Abramson, J. S., and Hudnor, H. R. (1994). Effect of priming polymorphonuclear leukocytes with cytokines (granulocyte-macrophage colony-stimulating factor [GM-CSF] and G-CSF) on the host resistance to Streptococcus pneumoniae in chinchillas infected with influenza A virus. *Blood 83*:1929–1934.

18. Winter, J. N., Lazarus, H. M., Rademaker, A., et al (1996). Phase I/II study of combined granulocyte colony stimulating factor and granulocyte macrophage colony stimulating factor administration for the mobilization of haematopoietic progenitor cells. *J Clin Oncol 14:*277–286.

19. Spitzer, G., Adkins, D. R., Spencer, V., et al (1994). Randomized study of growth factors post peripheral blood stem cell transplant: neutrophil recovery is improved with modest clinical benefit. *J Clin Oncol 12:*661–670.

20. Weinthal, J., Sender, L., Briones, G., et al (1993). The sequential combination of GM-CSF followed by G-CSF significantly enhances myeloid engraftment compared to GM-CSF alone following bone marrow transplantation. *Blood 82:*286a (abstr 1129).

21. Weisdorf, D. J., Verfaille, C. M., Davies, S. M., et al (1995). Hematopoietic growth factors for graft failure after bone marrow transplantation: a randomised trial of granulocyte macrophage colony stimulating factor (GM-CSF) versus sequential GM-CSF plus granulocyte-CSF. *Blood 85:*3452–3456.

22. McNiece, I. K., Langley, K. E., and Zsebo, K. M. (1991). Recombinant human stem cell factor synergises with GM-CSF, G-CSF, IL-3 and epo to stimulate human progenitor cells of the myeloid and erythroid lineages. *Exp Haematol 19:*226–231.

23. Bernstein, I. D., Andrews, R. G., and Zsebo, K. M. (1991). Recombinant human stem cell factor enhances the formation of colonies by $CD34^+$ and $CD34^+lin^-$ cells, and the generation of colony-forming cell progeny from $CD34^+lin^-$ cells cultured with interleukin-3, granulocyte colony-stimulating factor, or granulocyte-macrophage colony-stimulating factor. *Blood 77:*2316–2321.

24. Heyworth, C. M., Whetton, A. D., Nicholls, S., Zsebo, K., and Dexter, T. M. (1992). Stem cell factor directly stimulates the development of enriched granulocyte-macrophage colony-forming cells and promotes the effect of other colony-stimulating factors. *Blood 80:*2230–2236.

25. Tsuji, K., Lyman, S. D., Sudo, T., et al (1992). Enhancement of murine hematopoiesis by synergistic interactions between steel factor (ligand for c-kit), interleukin-11, and other early acting factors in culture. *Blood 79:*2855–2860.

26. Migliaccio, G., Migliaccio, A. R., Druzin, M. L., et al (1992). Long term generation of colony forming cells in liquid culture of CD34+ cord blood cells in the presence of recombinant human stem cell factor. *Blood 79:*2620–2627.

27. Migliaccio, A. R., Migliaccio, G., Durand, B., et al (1993). The generation of colony forming cells (CFC) and the expansion of haematopoiesis in cultures of human cord blood cells is dependent on the presence of stem cell factor (SCF). *Cytotechnology 11:*107–113.

28. Williams, N., Bertoncello, I., Kavnoudias, H., et al (1992). Recombinant rat stem cell factor stimulates the amplification and differentiation of fractionated mouse stem cell populations. *Blood 79:*58–64.

29. Metcalf, D. (1993). The cellular basis for enhancement interactions between stem cell factor and the colony stimulating factors. *Stem Cells 11:*1–11.

30. Molineux, G., Migdalska, A., Szmitkowski, M., et al (1991). The effects on haematopoiesis of recombinant stem cell factor (ligand for c-kit) administered

in vivo to mice either alone or in combination with granulocyte colony-stimulating factor. *Blood 78:*961–966.

31. Ulich, T. R., del Castillo, J., McNiece, I. K., et al (1991). Stem cell factor in combination with granulocyte colony stimulating factor (CSF) or granulocyte macrophage CSF synergistically increases granulopoiesis in vivo. *Blood 78:*1954–1962.

32. Briddell, R., Hartley, C., Moss, S., and McNiece, I. (1992). The role of stem cell factor in mobilization of peripheral blood progenitor cells (PBPC) with marrow repopulating ability in mice. *Blood 80:*12a (abstr 38).

33. de Haan, G., Engel, C., Dontje, B., Nijof, W., and Loeffler, M. (1994). Mutual inhibition of murine erythropoiesis and granulopoiesis during combined erythropoietin, granulocyte colony-stimulating factor, and stem cell factor administration: in vivo interactions and dose-response surfaces. *Blood 84:*4157–4163.

34. Cairo, M. S., Plunkett, J. M., Nguyen, A., and van de Ven, C. (1992). Effect of stem cell factor with and without granulocyte colony-stimulating factor on neonatal hematopoiesis: in vivo induction of newborn myelopoiesis and reduction of mortality during experimental group B streptococcal sepsis. *Blood 80:*96–101.

35. McNiece I. K., Briddell, R. A., Hartley, C. A., et al (1993). Stem cell factor enhances in vivo effects of granulocyte colony stimulating factor for stimulating mobilisation of peripheral blood progenitor cells. *Stem Cells 11:*36–41.

36. Andrews, R. G., Briddell, R. A., Knitter, G. H., Rowley, S. D., Appelbaum, F. R., McNiece, I. K., et al. (1995). Rapid engraftment by peripheral blood progenitor cells mobilized by recombinant human stem cell factor and recombinant human granulocyte colony-stimulating factor in nonhuman primates. *Blood 85:*15–20.

37. de Revel, T., Appelbaum, F. R., Storb, R., et al (1994). Effects of granulocyte colony stimulating factor and stem cell factor, alone and in combination, on the mobilization of peripheral blood cells that engraft lethally irradiated dogs. *Blood 83:*3795–3799.

38. Carlesso, N., Pregno, P., Bresso, P., et al (1992). Human recombinant stem cell factor stimulates in vitro proliferation of acute myeloid leukemia cells and expands the clonogenic pool. *Leukaemia 6:*642–648.

39. Murohashi, I., Endoh, K., Feng, M., et al (1995). Roles of stem cell factor in the in vitro growth of blast clonogenic cells from patients with acute myeloblastic leukaemia. *J Interferon Cytokine Res 15:*829–835.

40. Pietsch, T., Kyas, U., Steffens, U., et al (1992). Effects of human stem cell factor (c-kit ligand) on proliferation of myeloid leukaemia cells: heterogeneity in response and synergy with other haematopoietic growth factors. *Blood 80:*1199–1206.

41. Budel, L. M., Delwel, R., van Buitenen, C., et al (1993). Effects of mast cell growth factor on acute myeloid leukaemia cells in vitro: effects of combinations with other cytokines. *Leukaemia 7:*426–434.

42. McNiece, I. K., Glaspy, J., LeMaistre, F., et al (1993). Effects of recombinant methionyl human stem cell factor (rhSCF) and filgrastim (rhG-CSF) on mobili-

sation of peripheral blood progenitor cells: preliminary laboratory results from a Phase I/II study. *Blood 82:*84a (abstr 325).

43. Briddell, R., Glaspy, J., Sphall, E. J., LeMaistre, F., Menchaca, D., and Mc-Niece, I. (1994). Mobilization of myeloid, erythroid and megakaryocyte progenitors by recombinant human stem cell factor (rhSCF) plus Filgrastim (rhG-CSF) in patients with breast cancer. *Proc Am Soc Clin Oncol 13:*77 (abstr 109).

44. Demetri, G., Costa, J., Hayes, D., et al (1993). A phase I trial of recombinant methionyl human stem cell factor (SCF) in patients with advanced breast carcinoma pre- and post-chemotherapy (CHEMO) with cyclophosphamide (C) and doxorubicin (A). *Proc ASCO 12:*142 (abstr 367).

45. Glaspy, J., McNiece, I., LeMaistre, F., et al (1994). Effects of stem cell factor (rhSCF) and Filgrastim (rhG-CSF) on mobilization of peripheral blood progenitor cells and on hematological recovery posttransplant: early results from a phase I/II study. *Proc Am Soc Clin Oncol 13:*68 (abstr 76).

46. Weaver, A., Slowley, C., Woll, P. J., et al (1995). A randomized comparison of progenitor cell mobilisation using chemotherapy plus stem cell factor (r-metHuSCF) and filgrastim (r-metHuG-CSF) or chemotherapy plus filgrastim. *Proc Am Soc Clin Oncol 15:*269 (abstr 712).

47. Ikebuchi, K., Clark, S. C., Ihle, J. N., Souza, L. M., and Ogawa, M. (1988). Granulocyte colony-stimulating factor enhances interleukin 3-dependent proliferation on multipotential hemopoietic progenitors. *Proc Natl Acad Sci USA 85:*3445–3449.

48. Ikebuchi, K., Ihle, J. N., Hiral, Y., Wong, G. G., Clark, S. C., and Ogawa, M. (1988). Synergistic factors for stem cell proliferation: further studies of the target stem cells and the mechanism of stimulation by interleukin-1, interleukin-6 and granulocyte colony-stimulating factor. *Blood 72:*2007–2014.

49. Roberts, A. W., and Metcalf, D. (1995). Noncycling state of peripheral blood progenitor cells mobilized by granulocyte colony stimulating factor and other cytokines. *Blood 86:*1600–1605.

50. Ohi, S., Sakamaki, S., Matsunaga, T., et al (1995). Co-administration of IL-3 with G-CSF increases the CFU- mobilisation into peripheral blood. *Int J Hematol 62:*75–82.

51. Jacobsen, S. E., Ruscetti, F. W., Dubbios, C. M., Wone, J., and Keller, J. R. (1992). Induction of colony-stimulating factor receptor expression on hemopoietic progenitor cells: proposed mechanism for growth factor synergism. *Blood 80:*678–687.

52. McNiece, I. K., McGrath, H. E., and Quensenberry, P. J. (1988). Granulocyte colony stimulating factor augments in vitro megakaryocyte formation by interleukin-3. *Exp Haematol 16:*807–810.

53. Heyworth, C. M., Dexter, T. M., Nicholls, E., and Whetton, A. D. (1994). Combinations of colony-stimulating factors promote enhanced proliferative potential in enriched granulocyte-macrophage colony-forming cells. *Exp Hematol 22:*1089–1094.

54. Bot, F. J., van Eijk, L., Schipper, P., and Lowenberg, B. (1989). Effects of human interleukin-3 on granulocytic colony-forming cells in human bone marrow. *Blood 73:*1157–1160.

55. Paquette, R. L., Zhou, J. Y., Yang, Y. C., et al (1988). Recombinant gibbon interleukin-3 acts synergistically with recombinant human G-CSF and GM-CSF in vitro. *Blood 71*:1596–1600.

56. Takaue, Y., Kawano, Y., Reading, C. L., et al (1990). Effects of recombinant human G-CSF, GM-CSF, IL-3, and IL-1 alpha on the growth of purified human peripheral blood progenitors. *Blood 76*:330–335.

57. Krumwieh, D., Weinmann, E., Siebold, B., and Seiler, F. R. (1990). Preclinical studies on synergystic effects of IL-1, IL-3, G-CSF and GM-CSF in cynomolgus monkeys. *Int J Cell Cloning Suppl 1*:229–248.

58. Hocker, P., Geissler, K., Kurz, M., Wagner, A., and Gerharti, K. (1993). Potentiation of GM-CSF or G-CSF induced mobilization of circulating progenitor cells by pretreatment with IL-3 and harvest by apheresis. *Int J Artif Organs 16*:25–29.

59. D'Hondt, V., Guillaume, T., Humblet, Y., et al (1994). Tolerance of sequential or simultaneous administration of IL-3 and G-CSF in improving peripheral blood stem cells harvesting following multi-agent chemotherapy: a pilot study. *Bone Marrow Transplant 13*:261–264.

60. Uchiyama, H., Shimazaki, C., Fujita, N., et al (1994). Kinetics of serum cytokines in adults undergoing peripheral blood progenitor cell transplantation. *Br J Haematol 88*:639–642.

61. Lemoli, R. M., Fortuna, A., Fogli, M., et al (1995). Proliferative response of human marrow myeloid progenitor cells to in vivo treatment with granulocyte colony stimulating factor alone and in combination with interleukin-3 after autologous bone marrow transplantation. *Exp Hematol 23*:1520–1526.

62. Henon, P. R., Becker, M., Sovalat, H., et al (1995). Mobilization of peripheral blood stem cells with chemotherapy and cytokines in multiple myeloma. *Stem Cells 13*:148–155.

63. Vannucchi, A. M., Bosl, A., Laszlo, D., et al (1995). Treatment of a delayed graft failure after allogeneic bone marrow transplantation with IL-3 and GM-CSF. *Haematologica 80*:341–343.

64. Locatelli, F., Zecca, M., Ponchio, L., et al (1994). Pilot trial of combined administration of EPO and granulocyte colony stimulating factor to children undergoing allogeneic bone marrow transplantation. *Bone Marrow Transplant 14*:929–935.

65. Chao, N. J., Schriber, J. R., Long, G. D., et al (1994). A randomized study of erythropoietin and granulocyte colony-stimulating factor (G-CSF) versus placebo and G-CSF for patients with Hodgkin's and non-Hodgkin's lymphoma undergoing autologous bone marrow transplantation. *Blood 83*:2823–2828.

66. Pierelli, L., Scambia, G., Menichella, G., et al (1996). The combination of EPO and granulocyte colony stimulating factor increases the rate of haemopoietic recovery with clinical benefit after peripheral blood progenitor cell transplantation. *Br J Haematol 92*:287–294.

67. Bessho, M., Toyoda, A., Itoh, Y., et al (1992). Trilineage recovery by combination therapy with recombinant human granulocyte colony-stimulating factor (rhG-CSF) and erythropoietin (rhEPO) in severe aplastic anemia. *Br J Haematol 80*:409–411.

68. Weide, R., Lyttelton, M., Samson, D., et al (1993). Sustained trilineage response in a patient with ALG resistant severe aplastic anemia after treatment with G-CSF, EPO and cyclosporin A: association of recovery with marked elevation of serum alkaline phosphatase. *Br J Haematol 85:*608–610.

69. Ohsaka, A., Sughara, Y., Imai, Y., and Kikuchi, M. (1995). Evolution of severe aplastic anemia to myelodysplasia with monosomy 7 following granulocyte colony-stimulating factor, EPO and high-dose methylprednisolone combination therapy. *Intern Med 34:*892–895.

70. Imamura, M., Kobayashi, M., Kobayashi, S., Yoshida, K., Mikuni, C., Ishikawa, Y., et al (1995). Combination therapy with recombinant human granulocyte colony-stimulating factor and erythropoietin in aplastic anemia. *Am J Hematol 48:*29–33.

71. Hellstrom-Lindberg, E., Birgegard, G., Carlsson, M., Carnesky, J., Dahl, I. M., Dybedal, I., et al (1993). A combination of granulocyte colony-stimulating factor and erythropoietin may synergistically improve the anemia in patients with myelodysplastic syndromes. *Leuk Lymphoma 11:*221–228.

72. Negrin, R. S., Stein, R., Vardiman, J., et al (1993). Treatment of the anemia of myelodysplastic syndromes using recombinant human granulocyte colony stimulating factor in combination with EPO. *Blood 82:*737–743.

73. Maurer, A. B., Ganser, A., Seipelt, G., et al (1995). Changes in erythroid progenitor cell and accessory cell compartments in patients with myelodysplastic syndromes during treatment with all trans retinoic acid and haemopoietic growth factors. *Br J Haematol 89:*449–456.

74. Glaspy, J. A., Souza, L., Scates, S., et al (1992). Treatment of hairy cell leukemia with granulocyte colony stimulating factor and recombinant consensus interferon or recombinant interferon-alpha-2b. *J Immunother 11:*198–208.

75. Lorber, C., Willfort, A., Ohler, L., et al (1993). Granulocyte colony stimulating factor (rhG-CSF) as an adjunct interferon alpha therapy of neutropaenic patients with hairy cell leukaemia. *Ann Hematol 67:*13–16.

76. Crawford, J., Finglin, R., Chang, A., et al (1993). Phase I/II trial of recombinant human interleukin-6 (rhIL-6) and granulocyte colony-stimulating factor (rhG-CSF) following ifosphamide, carboplatin and etoposide (ICE) chemotherapy in patients with advanced non-small cell lung carcinoma (NSCLC). *Blood 82:*367a (abstr 1454).

77. Devine, S. M., Winton, E. F., Holland, H. K., et al (1994). Simultaneous administration of interleukin-6 (IL-6) and Neupogen (rhG-CSF) following autologous bone marrow transplantation (ABMT) for breast cancer (BCA). *Blood 84:*89a (abstr 343).

78. Champlin, R., Mehra, R., Kaye, J., et al (1994). Phase I study of recombinant human interleukin-11 following autologous BMT in patients with breast cancer. *Proc Am Soc Clin Oncol 13:*100 (abstr 201).

79. Haylock, D. N., To, L. B., Dowse, T. L., Juttner, C. A., and Simmons, P. J. (1992). Ex vivo expansion and maturation of peripheral blood CD34$^+$ cells into the myeloid lineage. *Blood 80:*1405–1412.

80. Shapiro, F., Yao, T. J., Raptis, G., et al (1994). Optimisation of conditions for ex vivo expansion of CD34$^+$ cells from patients with stage IV breast cancer. *Blood 84:*3567–3574.

81. Brugger, W., Mocklin, W., Heimfeld, S., et al (1993). Ex vivo expansion of enriched peripheral blood CD34$^+$ progenitor cells by stem cell factor, interleukin-1 beta (IL-1 beta), IL-6, IL-3, interferon-gamma and erythropoietin. *Blood 81*:2579–2548.

82. Henschler, R., Brugger, W., Luft, T., Frey, T., Mertelsmann, R., and Kanz, L. (1994). Maintenance of transplantation potential in ex vivo expanded CD34$^+$-selected human peripheral blood progenitor cells. *Blood 84*:2898–2903.

83. Urashima, M., Hoshi, Y., Akiyama, M., et al (1995). Ex vivo expansion of umbilical cord blood haematopoietic progenitor cells by combinations of cytokines. *Acta Paediatr Jpn 37*:160–165.

84. Chang, Q., Hanks, S., Akard, L., et al (1995). Maturation of mobilized peripheral blood progenitor cells: preclinical and phase I clinical studies. *J Hematother 4*:289–297.

19
Immunomodulatory Properties of Filgrastim (r-metHuG-CSF) in Preclinical Models

Thomas Hartung and Albrecht Wendel
University of Konstanz, Konstanz, Germany

I. INTRODUCTION

One of the key functions of the immune system is to fight infections, allowing the host to survive. To achieve this goal, a collaborative coordination of the cellular defense strategies between macrophages, granulocytes, and the various subsets of lymphocytes is needed. The initial emergency line of defense mobilizes macrophages, neutrophilic granulocytes, natural killer (NK) cells, and cytotoxic lymphocytes. When activated, these cells directly attack and eventually kill invading microorganisms. During this process, an inflammatory response is usually initiated with the characteristic symptoms of local hyperthermia or fever, swelling, redness, and pain. In pathological situations, excessive activation of the non-specific immune system overrides the control mechanisms of such inflammatory reactions, and loss of function of the host's tissue or organs can result.

To assure continuous control and balance, a variety of endogenous mediators, such as cytokines and lipid mediators, provide the fine tuning of inflammatory reactions and enable a sensible cross-talk between different leukocyte populations. The pattern of these signals regulates the strength of the contribution to the defense of major organs other than the blood system, eg, of the liver (induction of the acute phase), of mucosal epithelia (mucus production), and of the central nervous system (fever generation). In concordance with their assigned endogenous functions, these mediators, when given to laboratory animals, healthy volunteers, or patients in the form of exogenous pure compounds, act as biological response modifiers by interfering with the body's humoral signaling system.

In particular, hematopoietic growth factors are prime candidates able to modulate the mediator network because their major effect is to induce alterations in absolute, as well as relative, numbers of leukocytes. There is no doubt that increased hematopoiesis is an important prerequisite of continuous recruitment of leukocytes over the entire time span of an infectious event. For the successful eradication of infectious material, production of hematopoietic growth factors controlled from the site of infection is an efficient mechanism to guarantee an adequate supply of leukocytes. In addition to changing the average maturity of peripheral cells by preferential release of premature forms of leukocytes from the bone marrow, hematopoietic growth factors also can promote the differentiation process of peripheral white blood cells. During permanent overactivation of the immune system the dilemma arises that any signal recruiting more leukocytes and activating them represents a source of potential danger. It therefore makes sense teleologically to combine the activity of augmentation and activation of a given growth factor on a single leukocyte population with constraints of activity directed towards other leukocytes. Such a type of hierarchic regulation provides coordinated termination of activities and thereby avoids an overshoot of the inflammatory response. This chapter will provide recent evidence to show that granulocyte colony-stimulating factor (G-CSF) as an endogenous anti-infectious mediator has pharmacological properties sharing any of these four characteristics, ie, change of leukocyte numbers, recruitment of more immature leukocytes, maturation/augmentation of leukocyte functions, and counterregulatory suppression of other leukocyte populations. This profile justifies thinking of G-CSF as a biological response modifier with properties exceeding the ones of a mere hematopoietic growth factor.

II. G-CSF RECRUITS NEUTROPHILS FROM THE BONE MARROW

The average life span of a polymorphonuclear leukocyte (PMN) in peripheral blood was calculated to be about 7 hours (reviewed in Ref. 1). Due to the short half-life of these cells, healthy adult humans produce approximately 120 billion PMN per day. This exceedingly large rate can be further increased almost tenfold under stress conditions such as infection (reviewed in Ref. 2).

Granulocyte colony-stimulating factor was originally identified in vitro as a factor inducing the proliferation and maturation of granulocyte precursors: bone marrow cells plated on soft agar formed colonies of granulocytic cells when incubated in the presence of G-CSF. In an approved pharmaceutical form, recombinant human (rHu)G-CSF is available as a recombinant protein expressed in *Escherichia coli,* ie, as a non-glycosylated protein (Filgrastim), or expressed in Chinese hamster ovary cells in a glycosylated form (lenograstim). While no major differences in the pharmacodynamic activities of glycosylated and non-glycosylated rHuG-CSF were found (3), some differences in the pharmacokinetic properties of these different forms were reported.

In animal experiments, injection of rHuG-CSF resulted in neutrophilia demonstrating not only increased production but also release of PMN (4–6). In contrast to granulocyte-macrophage colony-stimulating factor (GM-CSF), G-CSF reduces the bone marrow transit time of neutrophils (6). Repetitive subcutaneous (SC) injection of Filgrastim for 7 days into rats resulted in a biphasic increase in PMN counts (4). In mice (5) or humans, no such negative feedback mechanism on PMN release from bone marrow was found.

In contrast, when endogenous G-CSF was blocked either by injecting neutralizing anti-G-CSF-antiserum (7), or by inducing the production of auto-antibodies against G-CSF in dogs (8), or by deleting the gene for G-CSF (9), neutropenia was always observed. This observation suggests that G-CSF is not only a signal for PMN formation and release in emergency reactions but also is a permanently present mediator needed to maintain physiological white cell counts. It is emphasized in this context that any overlap of functions with other hematopoietic growth factors appears to be absent for this function of G-CSF, since G-CSF knock-out mice can not compensate for the deficiency of this factor.

In healthy volunteers who receive Filgrastim, a pronounced and transient decrease in peripheral PMN counts is observed approximately 30 minutes after SC injection (10–12). This phenomenon was attributed in some studies to adherence of cytokine-activated PMN primarily to the endothelia of the lung vasculature. We made similar observations in a healthy volunteer (**Figure 1**) and in a coauthor (AW) who had received a single dose of 480 μg Filgrastim. This effect was studied in more detail by Katoh et al (11), who provided evidence that the secondary increase in PMN after the initial neutropenia is not due to a recruitment from the marginated pool of PMN attached to endothelial cells but rather to recruitment of less mature performed PMN from bone marrow as identified by low alkaline phosphatase content. The transit time of the post-mitotic pool in bone marrow was shortened after injection of 300 μg Filgrastim from 6.4 to 2.9 days (13). Sixty minutes after Filgrastim injection, the neutrophil counts returned to pretreatment values. Then, a continuous subsequent

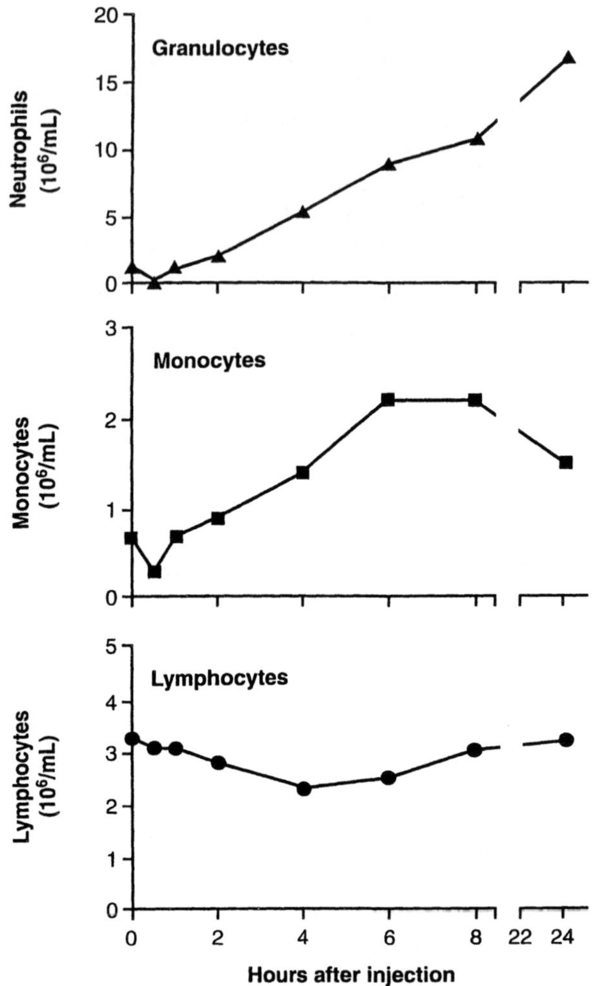

Figure 1.   Leukocyte counts after subcutaneous injection of 480 μg Filgrastim into a healthy volunteer. Blood was taken at the time points indicated. The white blood cell count was assessed by fluorescent-activated cell sorting (FACS) analysis.

increase in PMN counts occurred that peaked approximately 12 hours after injection (13). Doses of Filgrastim ranging from 1 to 60 μg/kg/day given for 5 or 6 days produced dose-dependent increases of 1.8- to 12-fold in the absolute neutrophil count (ANC) (14). In a recent volunteer study, we found a dose-dependent increase in PMN counts with a plateau lasting over the whole treatment with Filgrastim (**Figure 2,** upper panel). Blood neutrophil half-life was not significantly altered by Filgrastim treatment. Seventy-two hours after the last injection, counts equal to those before treatment were found. No significant differences in PMN recruitment were observed when comparing the variables between young and elderly subjects (13,14).

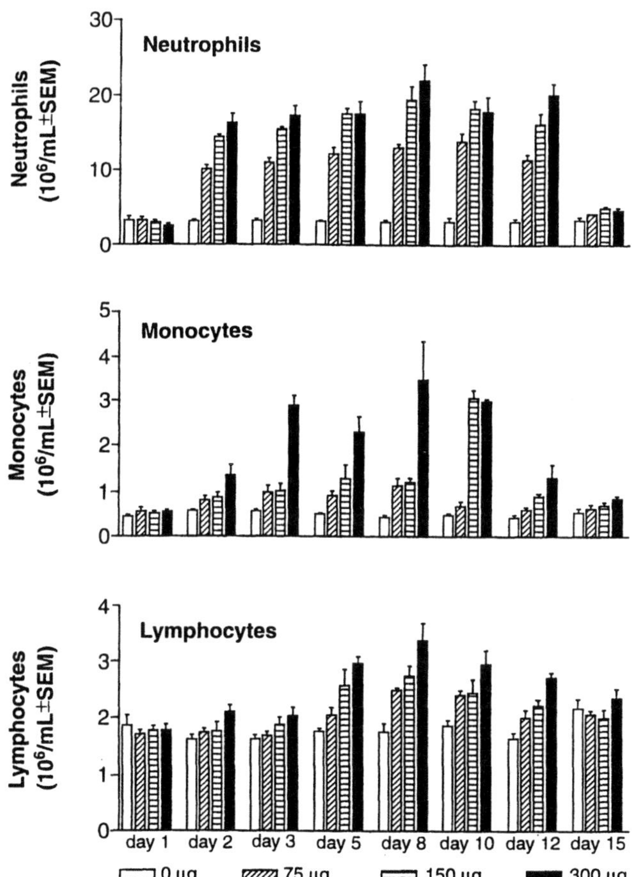

Figure 2. Time course of leukocyte counts with daily subcutaneous Filgrastim treatment of healthy volunteers for 12 days. Four groups of six healthy volunteers were treated in a double-blind manner with either placebo or 75 μg, 150 μg, or 300 μg Filgrastim. The white blood cell count was assessed by FACS analysis on the days indicated.

III. G-CSF DIFFERENTIALLY MODULATES NEUTROPHIL FUNCTIONS

Polymorphonuclear leukocytes are endowed with 300 to 1000 G-CSF receptors per cell and, already at a low level of occupancy, these receptors transmit a signal causing a biological response (reviewed in Ref. 1). This implies that G-CSF can act directly on mature neutrophils and that it should be possible to study such biological properties in vitro in cultures of purified granulocytes. In vivo, however, such actions are superimposed by activation of newly recruited precursor cells and by a rapid PMN turnover. Possible feedback regulation processes by signals to or from other leukocyte populations may easily escape detection in assays using isolated PMN.

A. Phagocytosis and Killing

Pure G-CSF has no bactericidal effect as such (15). Isolated PMN exposed in vitro to rHuG-CSF acquire an increased potency to kill *E. coli* and *Staphylococcus aureus* (16–20). Conflicting results were reported as to the in vitro efficacy of rHuG-CSF in increasing *Candida albicans* killing. Yamamoto et al (21) observed increased killing, while Roilides et al (29) and Bober et al (19) noted no effect on *C. albicans* killing but an increased *Micrococcus lysodeikticus* and yeast particle phagocytosis instead.

Using a bactericidal killing assay in whole blood, we found that the bactericidal activity of blood from Filgrastim-treated healthy volunteers was increased by several orders of magnitude (**Figure 3**). This increased killing activity that was seen ex vivo was shown to be mediated by PMN.

B. Oxidative Burst

The production of reactive oxygen species (oxidative burst) via the myeloperoxidase system is a powerful weapon of PMN to fight bacteria. Recombinant G-CSF itself failed to induce an oxidative burst of isolated PMN in suspension. In contrast, PMN adherent to laminin or plasma proteins in the presence of rHuG-CSF released reactive oxygen species to an extent that was similar to the one obtained after stimulation with tumor-necrosis factor (TNF) (22). Obviously, rHuG-CSF is capable of priming PMN in suspension for oxidative burst in vitro as well as ex vivo. In vitro, rHuG-CSF stimulates oxidative burst induced by the bacterial chemotactic peptide N-formyl-methionyl-leucyl-phenylalanine (fMLP) by 50% to 200% (3,21,23–26). Treweeke et al (27), however, found no effect on fMLP-inducible formation of reactive oxygen species. Increased oxidative burst also was found after stimulation of human PMN in the presence of rHuG-

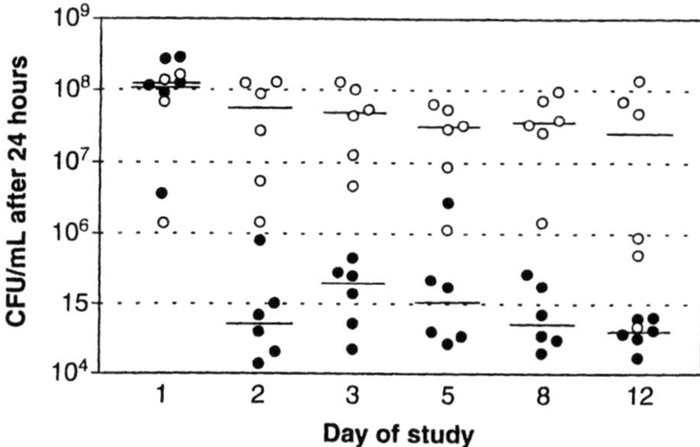

Figure 3. Effect of daily injection of 300 μg Filgrastim on ex vivo bactericidal activity of blood against *Salmonella*. Blood from the placebo- and 300-μg Filgrastim-treated volunteers (study described in the legend of Figure 2) was used for a bactericidal killing assay.

CSF in response to ionomycin (23,28) complement factor C5a (24), and *E. coli* α-hemolysin (29). The effect of rHuG-CSF was synergisticaly increased by TNF (24) and inhibited by protein kinase C inhibitors or pertussis toxin (23). In rats, rHuG-CSF only increased reactive oxygen release by PMN from neonatal but not from adult animals (30). When PMN were isolated from Filgrastim-treated individuals, a similar augmentation of ex vivo–stimulated burst was found (23,24,31–34). Interestingly, rHuG-CSF apparently fails to increase burst stimulated by phorbol ester (31; our unpublished results) or zymosan (32,33).

Recombinant HuG-CSF was also shown to restore impaired neutrophil burst in animal models: PMN from rats made diabetic by streptozocin treatment showed impaired oxidative burst in response to FMLP which was partially restored by rHuG-CSF in vitro (35). After cecal ligation and puncture, rats showed a suppressed PMA-inducible burst that was restored by rHuG-CSF treatment (36).

C. Degranulation

A variety of bactericidal proteins and small molecules are preformed and stored in the granules of PMN. These agents are ready to be immediately released via degranulation, and include proteases such as elastase and cathepsin G, the iron-binding protein lactoferrin and the bactericidal permeability increasing protein (BPI). Human PMN contain peroxidase-positive (azurophil or primary) and peroxidase-negative (specific or secondary) granules (37). In vitro, rHuG-CSF had no effect on α-hemolysin–induced release of β-glucuronidase (29) or on FMLP-inducible release of myeloperoxidase (27).

Polymorphonuclear leukocytes from patients with urothelial cancer treated with Filgrastim before chemotherapy were found to contain increased amounts of alkaline phosphatase, a marker of secondary granules (38), and PMN from Filgrastim-treated volunteers also had an increased number of primary granules compared with cells from untreated control donors (14). Increased elastase serum levels during Filgrastim therapy were reported in non-neutropenic patients with advanced malignancies (34). In ex vivo experiments done with Filgrastim-treated healthy volunteers, we found no change of degranulation of primary granules (myeloperoxidase, elastase, BPI). On the contrary, we observed a 50% reduction of secondary granule release (lactoferrin) in their blood. It appears that Filgrastim has only little impact on degranulation in vitro or ex vivo, even though it changes the granule content of PMN in vivo.

D. Chemotaxis and Adhesion

A prerequisite to efficiently fight bacteria that have invaded tissue is the recruitment of PMN specifically targeted to the site of infection, a directed movement that is known as chemotaxis when a chemical gradient provides a lead. In vitro, rHuG-CSF was shown to induce a chemotactic response of PMN at 70 ng/mL comparable to the one induced by 10^{-8} FMLP (39). Recombinant HuG-CSF also was reported to increase FMLP-but not leukatriene B_4 (LTB$_4$)–inducible chemotaxis (19). The adhesion of PMN to nylon fibers also was augmented by rHuG-CSF in vitro (3).

Ex vivo and in vivo, however, there is little evidence of increased chemotaxis and infiltration. In 12 patients receiving chemotherapy for the treatment of small-cell lung cancer, no effect of Filgrastim infusion on PMN chemotaxis was found (12). In the model of SC *E. coli* injection to rats, Filgrastim treatment resulted in less infiltration as assessed by local myeloperoxidase activity compared with control subjects (40); however, at the same time, it resulted in a better elimination of bacteria. In human volunteers, Filgrastim treatment resulted in 50% reduction of PMN in implanted skin chambers (13). Others found a minimal reduction of PMN entry into skin windows (14). Local skin abrasion in healthy volunteers did not result in supranormal skin accumulation of PMN during Filgrastim treatment (14). A possible interpretation of these observations is that it might not be the level of systemically circulating neutrophils but rather the strength of the local inflammatory stimulus that determines the actual degree of accumulation of neutrophils at the inflammatory site.

In rodents, however, rHuG-CSF treatment was shown to restore experimentally impaired chemotaxis, ie, after burn in mice (42) or after peritonitis brought about by cecal ligation and puncture in rats (36). In these small-animal models, it is possible that the observed restoration of chemotaxis might have been due rather to quick recruitment of new PMN from bone marrow rather than to rHuG-CSF effects on the rapidly turning over pool of PMN that were functionally impaired by the experimental intervention.

E. Receptors and Surface Markers

Surface markers can be easily quantified by flow cytometry and taken as a measure of the actual state of cell activation and maturation. In vitro, the absence of an effect of rHuG-CSF on FcγRI (CD64), CR-1 (CD35), CR-3 (CD11b/CD18), and ICAM-1 (CD54) was reported by Bober et al (19). There are further reports on a lack of effect of rHuG-CSF on CD11b expression (27,41). In contrast, an increased CD11b/CD18 expression (43,44); increased (41), as well as decreased (43–45), LAM-1 expression; increased CD35 (43); and altered IgA receptor expression (43) was observed.

Interestingly, rHuG-CSF increased CD11b/CD18 in vivo in neonatal but not in adult rats (30). In humans in vivo, rHuG-CSF initially increased CD11b, CD35, and CD14, followed by a subsequent downregulation of CD11b/CD18 and up-regulation of LAM-1 and CD14 (3,11,43,44,46). Others did not find changes in CD11b or LAM-1 (41). CD64 (FcγRI), which is usually restricted to mononuclear cells, was found to be expressed on PMN during Filgrastim treatment (47–49), while there was no change in CD32 (FcγRII) and CD16 (FcγRII). Simultaneously, alkaline phosphatase was initially increased after Filgrastim treatment and strongly decreased in the newly formed population. Similarly, in 10 postoperative/post-trauma patients treated with low-dose Filgrastim infusions, CD64 expression of PMN was increased by 50% (32).

In healthy volunteers treated with Filgrastim, we observed shedding of L-selectin (CD62L) with concomitantly increased serum levels of soluble L-selectin. Other PMN surface markers such as CD11b/CD16, CD14, and CD71 were little affected; however, we also observed the well-known increase in CD64 that was paralleled by decreased CD16 (unpublished results).

Taken together, the available data on PMN surface markers are not completely consistent. A factor responsible for divergent results might be that any isolation procedure of PMN is likely to influence this sensitive endpoint. On the other hand, it is difficult to compare different treatment regimens (eg, single versus repeated treatment, infusion versus SC injection) that will result in considerably different neutrophil kinetics and differentiation as well as activation. Last but not least, a number of the studies cited were performed on patients with very different underlying diseases under treatment with various drugs other than rHuG-CSF.

F. Neutrophil Survival

Granulocyte colony-stimulating factor may be the most effective cytokine in attenuating the physiological apoptosis of PMN (43,50). In vitro, rHuG-CSF prolongs PMN survival from 7 hours in control cells to more than 36 hours (51), as well as their oxidative burst capacity and CD16 expression (52). The anti-apoptotic effect of rHuG-CSF was inhibited by protein kinase C inhibitors (51). This activity of G-CSF on PMN occurs apart from its activity on the bone marrow, not only because it can be shown in isolated cells but also because G-CSF increases the survival time of PMN at concentrations considerably below those required for stimulating colony formation (1); however, rHuG-CSF treatment seemed not to prolong the survival time of PMN in the circulation. Therefore, the physiological significance of the anti-apoptotic activity of G-CSF remains to be clarified.

G. Inflammatory Mediator Release

Polymorphonuclear leukocytes release a variety of inflammatory mediators upon stimulation for instance with endotoxin, lipopolysaccharide (LPS). This, however, does not seem to be restricted to eicosanoids such as leukotrienes or to proinflammatory cytokines such as TNF-α. One one hand, LPS induces a significant G-CSF release from PMN in an amount of 170 pg/10^6 cells within 24 hours (52). On the other, these cells are also capable of contributing to the elimination of G-CSF (53). In vitro, rHuG-CSF is a weak inducer of secretion of arachidonic acid metabolites (54) and initiated in vitro a release of approximately 100 IU/10^7 cells within 12 hours of interferon (IFN)-α from PMN (15); however, rHuG-CSF had no effect in vitro on interleukin (IL)-8 or LTB$_4$ release (29).

When blood was tested ex vivo from Filgrastim-treated volunteers, LPS-inducible IL-1ra release was increased 2-fold per PMN compared with blood taken from control subjects, while the shedding of soluble TNF receptors was unaffected when calculated per PMN which, of course, were present in higher numbers in the blood of Filgrastim-treated donors (56,57). Concomitantly, the LTB$_4$ release capacity when stimulated with LPS per individual cell was decreased to approximately 20% of controls (unpublished).

These data demonstrate that even if its actions on neutrophils only are taken into consideration, G-CSF has the characteristics of a pleiotropic cytokine. Overall, G-CSF appears to induce minor PMN responses by itself, but modifies the response of these cells to subsequent stimulation. While phagocytosis and oxidative burst are augmented, G-CSF barely affected degranulation and chemotaxis. On the one hand, influences of G-CSF on surface molecule expression are attributed at least in part to the recruitment of PMN with different stages of maturation. There is a de novo expression of CD64 associated with increased antibody-dependent cellular toxicity (58). Mediator release by PMN is shifted by G-CSF in favor of an anti-inflammatory balance, ie, enhanced release of IL-1ra and IFN-α and decreased formation of the chemotactic LTB_4. The next section will discuss the fact that this differential modulation of inflammatory reactions in general is also true for leukocytes other than neutrophils.

IV. G-CSF ALSO MODULATES MONOCYTES AND LYMPHOCYTES

Originally, G-CSF was identified as a hematopoietic growth factor stimulating the growth of various leukocyte lineages and was termed "pluripoetin" for this reason (reviewed in Ref. 2). For years, research concentrated on the predominant effects of G-CSF on neutrophilic granulocytes and only more recently on its actions on monocytes/macrophages. Until recently, only very few data were available on influences of G-CSF on the numbers of circulating lymphocytes and possible modulations of their functions.

A. Effect of G-CSF on Monocytes and Macrophages

Infusion of Filgrastim (30 μg/kg) to non-neutropenic patients with advanced malignancies resulted in a transient, but nearly complete, decrease of monocytes after 15 to 30 minutes that normalized after 60 minutes (34). Similar observations were made in a human volunteer injected with 480 μg Filgrastim (**Figure 1,** second panel). At later time points and with high doses of Filgrastim (30 to 60 μg/kg), as great as a tenfold monocytosis was observed (10,38). Smaller rHuG-CSF doses resulted in a threefold increase in monocyte counts (56,59), which correlated with an increase in granulocyte macrophage colony-forming cells (CFC-GM), ie, granulocyte and monocyte precursors, in bone marrow (60). Sustained daily Filgrastim treatment of volunteers resulted in a monocytosis with a relative maximum about day 8 (**Figure 2,** second panel). It is still unclear if the effects of rHuG-CSF on monocytes/macrophages are direct or if they are caused by an intermediate mediator. Actually, a subset of human monocytes was shown to bind G-CSF at very low levels (reviewed in Ref. 2). It is also known that occupancy of only a small number of G-CSF surface receptors may be sufficient to stimulate maximal biological responses, at least in granulocytes (reviewed in Ref. 2). For these reasons, an action of G-CSF on monocytes seems theoretically possible.

Endotoxin from Gram-negative bacteria is a powerful monocyte/macrophage activator, and LPS is the most potent inducer of the expression and release of monokine TNF-α, ie, a pivotal pro-inflammatory signal of the LPS response. When various rodent macrophage populations were exposed to Filgrastim in vitro, LPS-inducible TNF was not affected. The situation was, however, different ex vivo. The LPS-stimulated TNF release of various macrophage populations prepared from donor animals pretreated in vivo with Filgrastim was markedly suppressed compared with cells from control animals (63).

Several findings in humans support also the idea that G-CSF acts on monocyte functions. When patients with refractory testicular cancer were treated with 20 mg/m^2/day Filgrastim, major histocompatibility complex (MHC)-I on monocytes was increased while MHC-II was unaffected (61). In our laboratory, we have observed a significant decrease of monocytic HLA-DR (-40%) on monocytes from Filgrastim-treated volunteers (unpublished). Wiltschke et al (61) noted that spontaneous TNF release from monocytes in culture and cell-killing activity on the tumor cell line U937 was not changed by Filgrastim treatment. In volunteers who had received Filgrastim, the ex vivo LPS-stimulated TNF-release capacity of whole blood was suppressed (56,62). This ex vivo suppression of TNF release after Filgrastim treatment in vivo also was true for a variety of stimuli other than LPS, such as preparations from gram-positive bacteria, superantigens, or phorbol ester. While Filgrastim had no effect on TNF release in cultures of isolated human monocytes, the release of this cytokine was attenuated by G-CSF when neutrophils were simultaneously present (63). This finding suggests that such an influence on a monocyte function is likely to be mediated by a yet unknown factor derived from PMN.

Excessive overactivation of macrophages by LPS in vivo results in organ damage and lethality. Mice and rats injected with high doses of LPS causing endotoxic shock were protected by rHuG-CSF pretreatment against LPS lethality (64). Similarly, liver injury in galactosamine-sensitized rodents induced by LPS was prevented by rHuG-CSF pretreatment. In rats that received Filgrastim, glucose utilization upon LPS challenge was augmented due to increased uptake by ileum, spleen, liver, and lung (65). In addition, in LPS-challenged pigs, lung injury was partially attenuated after Filgrastim pretreatment (66,67).

In some infection models, eg, *E. coli* peritonitis in dogs (68) and cecal puncture in rats (69), Filgrastim treatment also reduced serum TNF levels. In these cases, however, it cannot be decided whether this effect was due to a reduction of the systemic load with bacteria and hence less TNF production or to a direct effect of Filgrastim on TNF-producing cells.

In 18 patients with acute leukemia, rHuG-CSF was given in the recovery phase during 27 courses of consolidation chemotherapy; under these conditions, peripheral blood monocytes spontaneously produced high concentrations of IL-6 (60). In 14 patients with myelodysplastic syndrome, combined treatment with all-*trans* retinoic acid and Filgrastim resulted in increased serum concentrations of soluble TNF and IL-2 receptors, while TNF-α concentrations were already elevated before therapy and did not significantly change later (70). With rHuG-CSF therapy of cancer patients subsequent to chemotherapy, serum TNF (three of five patients) and urinary peptide-leukotriene metabolites decreased (71). In 38 neutropenic patients with gynecological cancer, rHuG-CSF treatment doubled the concentration of serum neopterin (72), which is regarded as a marker of macrophage activation.

B. Effect of G-CSF on Lymphocytes

Although monocytes/macrophages are the major producers of G-CSF, human T cells also are a source of G-CSF, eg, in response to TNF in the absence or presence of IFN-γ (73). Effects of Filgrastim treatment on lymphocyte counts in humans were noted in the very early reports (10,74). Treatment with higher doses of Filgrastim (10 or 100 μg/kg/day) was associated with a 1.5- to 2.5-fold increase in absolute lymphocyte counts (2). A relative lymphocytosis also was documented in a study done in our laboratory with volunteers repeatedly treated with Filgrastim who displayed a maximum lymphocyte count about day 8 (**Figure 2,** third panel).

Functional changes in lymphocytes by rHuG-CSF were recently described. In mice, delayed-type hypersensitivity to sheep red blood cells was enhanced by rHuG-CSF (75). T cells from rHuG-CSF–treated mice showed reduced type-1 lymphokine (IL-2, IFN-γ) release while type 2 lymphokine formation (IL-4) was augmented (76), survival was significantly improved in a murine graft-versus-host disease (GVHD) model when the grafts were from rHuG-CSF-treated mice. In vitro, rHuG-CSF had no effect on mixed lymphocyte reaction of human blood mononuclear cells, but reduced TNF secretion in this cell model (77). In line with this finding, rHuG-CSF in vitro had also no effect on murine T-cell proliferation or IL-2 production in response to staphylococcal exotoxin B (SEB) (78). In vivo, however, rHuG-CSF protected galactosamine-sensitized mice against T cell-mediated liver injury initiated by the superantigen SEB; these mice had reduced IL-2 serum levels but there was no effect on TNF release (78).

Volunteers treated repeatedly with Filgrastim had an augmented ex vivo proliferation of lymphocytes in response to anti-CD3-antibodies or phytohemagglutinin. At times later than 72 hours after initiation of treatment, we found a profound suppression of the proliferative response of lymphocytes. During the 12 days of treatment, there was no effect on natural killer NK cell function. In 19 patients who had received a high-dose chemotherapy, Filgrastim treatment caused an increase in the serum concentration of soluble IL-2 receptor (79), regarded as a marker of T-cell activation. In a case of severe congenital neutropenia, the level of endogenous G-CSF was elevated to 300 pg/mL before treatment and the lymphokines GM-CSF, IL-2, and IL-3 were slightly elevated. The patient showed clinically an only moderate response to high-dose rHuG-CSF (1600 μg/m^2/day), but a drastic increase in soluble IL-2 receptor was found in the absence of significant changes in the levels of lymphokines (80).

V. ROLE OF ENDOGENOUS G-CSF FORMATION

Being an endogenous mediator subject to a flexible regulation on demand, studying the role of endogenous G-CSF under normal and pathophysiological conditions represents a way of expanding the knowledge of the pharmacological properties of exogenous G-CSF. Possible approaches include determining the conditions under which G-CSF is formed and estimating the amounts that are released, examining the regulation of this production and the influence of physiological modulators, and studying the effect of G-CSF neutralization in experimental animals under normal and stress conditions such as infection.

A. Endogenous Formation of G-CSF

In human monocytes, a major source of G-CSF, stimulation with LPS, peptidoglycan breakdown products, phorbol ester, or cycloheximide was shown to augment the message for G-CSF by prolonging the half-life of the mRNA that codes for G-CSF (81–85). Furthermore, LPS, phytohemagglutinin, phorbol ester, human serum, GM-CSF, IL-3, and IL-4 were shown to induce G-CSF protein formation by monocytes (56,86–89). The relative potency of a number of immune stimuli to induce G-CSF release in human whole blood is shown in **Figure 4.** Notably, the Gram-positive stimuli, ie, heat-killed *S. aureus* and the lipoteichoic acid derived from this organism were the most potent inducers of G-CSF release. Alveolar macophages

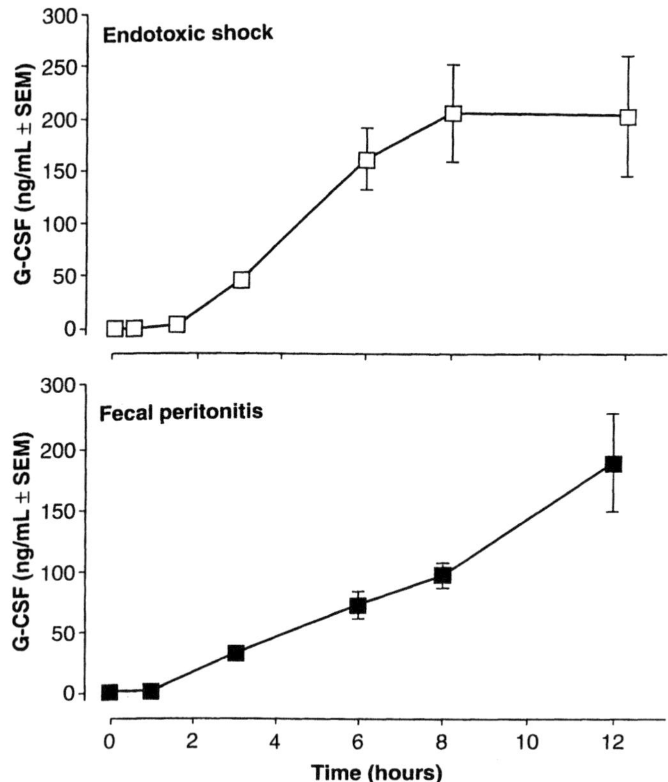

Figure 4. The course of G-CSF in serum of mice during endotoxic shock or fecal peritonitis (7,96). Male BALB/c mice were challenged with 1.5 mg/kg endotoxin from *Salmonella abortus equi* given intraperitoneally (IP) at time point 0. Three mice were killed at each time point indicated. Alternatively, female BALB/a$_2$ mice (n = 6) were IP injected with a lethal dose of a human stool suspension. Serum G-CSF was determined by an ELISA set up from antibodies raised against murine recombinant G-CSF.

were shown to produce 2- to 3-fold more G-CSF in response to LPS than an equivalent number of monocytes (1).

Beyond macrophages, a number of cell types were reported to produce G-CSF: IL-1 induced G-CSF release from human marrow stromal cells (90) and human umbilical vein endothelial cells (91). Formation of G-CSF mRNA was described in fibroblasts stimulated with either TNF or phorbol ester (92) and in human umbilical vein cells in response to IL-1 and TNF (93); TNF also was reported to release G-CSF protein from fibroblasts

(92). In human mesothelial cells, epidermal growth factor (EGF), LPS, and TNF induced G-CSF transcripts (94). A number of carcinoma cell lines were found to produce G-CSF, eg, derived from squamous cell carcinoma and hepatoma (2).

Mice challenged with *Listeria monocytogenes* showed G-CSF serum levels with a relative maximum at 48 hours that were still increased after 5 days (95). **Figure 5** shows the kinetics of serum G-CSF after LPS challenge (96) or induction of fecal peritonitis (7). Dogs injected with endotoxin showed increased G-CSF serum levels by 2 hours, peaked at 4 hours, and had not returned to normal by 24 hours after LPS (97). It appears that the G-CSF response parallels that of IL-6 and is somewhat similar in time course to IL-1 and TNF (43). Endogenous G-CSF is extractable from all major organs at levels higher than those present in the circulation and tissue production levels can be elevated rapidly (in minutes or hours) after stimulation (98).

Serum G-CSF levels also were reported in humans (reviewed in Ref. 99). Healthy humans only rarely show detectable G-CSF serum levels (7

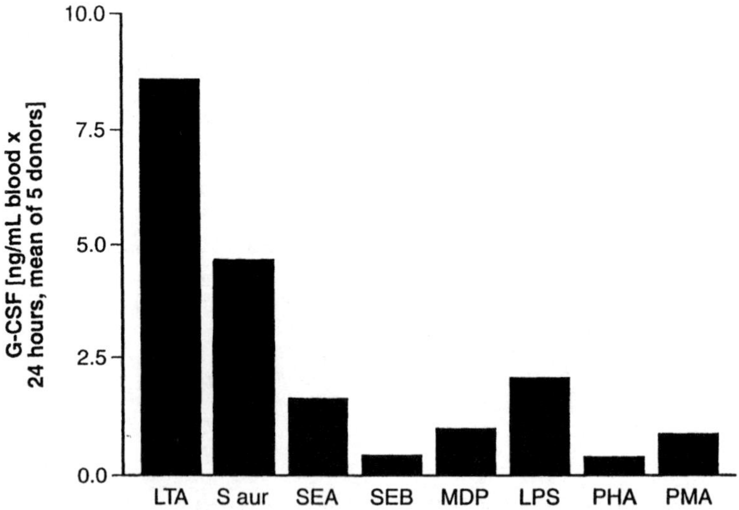

Figure 5. G-CSF formation by human whole blood in response to various stimuli. Blood from five healthy volunteers was diluted 1:4 by RPMI 1640 and incubated for 24 hours in the presence of 10 μg/mL lipoteichoic acid (LTA) from *Staphylococcus aureus,* 10 μg/mL muramyl dipeptide (MDP), 10 μg/mL endotoxin (LPS) from *Salmonella abortus equi,* 10 μg/mL phytohemagglutinin (PHA), 1 μg/mL staphylococcal exotoxin A and B (SEA, SEB), 100 nM phorbol ester (PMA), or 0.0001% cells (v/v) of heat-killed *Staphylococcus aureus.* Levels of G-CSF were determined by ELISA in the cell-free supernatant.

of 56 [100], 4 of 32 [101]; and <10≦ [43]). In a number of infectious diseases, however, elevated serum G-CSF levels were reported (102,103). In febrile patients with or without neutropenia, G-CSF serum levels are elevated (mean, 500 pg/mL; range, <10 pg to 140 ng/mL) compared with afebrile control subjects (103). In the acute phase of infection, G-CSF levels of 24 patients ranged from 30 pg to 3 ng/mL (101). In the acute phase of infection, patients with repeated infection had significantly lower G-CSF serum levels (200 pg/mL) than those with a single episode of infection (1 ng/mL) (104). In a case report of a patient who had shock and multiple-organ failure after self-administration of 1 mg LPS, 277 ng/mL G-CSF were determined in his serum 22 hours later (105).

In a variety of neutropenic disorders, G-CSF was found in serum of patients. In half of the patients with acute leukemia, G-CSF levels >50 pg/mL were found during induction chemotherapy (106). Five of six patients with aplastic anemia had serum G-CSF levels >50 pg/mL (102). For this disease, a reverse correlation between blood neutrophil count and serum G-CSF was demonstrated (100). In a patient with cyclic neutropenia, G-CSF up to 1 ng/mL was measured in parallel with a reduction in PMN numbers, which rapidly decreased to below detection limit after recovery of PMN counts (107). Fifty percent of the patients with myelodysplastic syndrome had detectable G-CSF serum levels (100). In a patient with autoimmune neutropenia, peripheral PMN counts changed almost in parallel with G-CSF levels (102). To date, it is unclear whether G-CSF is released as a direct consequence of neutropenia, which would imply a sensor for neutropenia in the host, or its formation being a consequence of infections to which these patients are prone.

Notably, the production of G-CSF by leukocytes of neonates is significantly reduced (108). For other patient groups with reduced neutrophil response to infection (eg, patients with diabetes), no studies on G-CSF production capacity are available. After injection of 10 μg/kg Filgrastim SC, serum G-CSF levels were higher than 10 ng/mL for 10 to 16 hours (109). Thus, pharmacological doses create serum concentrations of G-CSF that are in the upper range of the ones reached via endogenous production during infection. This means that the pharmacodynamic action potential of rHuG-CSF is either due to continuing an endogenous defense signal for a prolonged time or due to a substitution of an insufficient endogenous production at a very early time.

B. Regulation of G-CSF Formation

Relatively little is known about the regulation of G-CSF formation. The presence of anti-IL-1-antibodies partially blocks G-CSF release of human blood leukocytes in the presence of human serum (89) suggesting a role of endogenous IL-1 formation in augmenting G-CSF release. Lipopolysaccharide or IL-1–inducible G-CSF production, respectively, by human monocytes is increased by IFN-γ as well as prostaglandin E_2 and suppressed by IL-4 (82,83,110,111). Granulocyte-macrophage colony-stimulating factor had no effect on G-CSF formation (83).

When human whole blood was incubated in the presence of LPS, we found a significant inhibition of G-CSF formation (approximately 50%) by IL-4, IL-6, IL-13, and IFN-γ. In the same series of experiments, IL-1, IL-3, IL-8, IL-10, IL-11, and TNF had no major effect, and GM-CSF slightly increased G-CSF release (unpublished results).

On the basis of this poor present database, it appears that G-CSF formation is barely modulated by other cytokines. This fact might be taken as evidence for the crucial nature of this response.

C. Inhibition of Endogenous G-CSF

Granulocyte colony-stimulating factor "knock-out" mice show chronic neutropenia (20% to 30% of normal) and impaired ability to control infection with *L. monocytogenes* (9). Infection-induced granulopoiesis was strongly impaired in these animals. When rats were passively immunized with rabbit anti-G-CSF antibodies, intrabronchial application of *Pseudomonas aeruginosa* induced less PMN recruitment and bactericidal activity (1). Injection of antibodies did not result in changes in circulating PMN in this study. In contrast, using a goat anti-murine-G-CSF antiserum, we found severe neutropenia after injection into mice (7). These animals were much more susceptible to infection with a human stool suspension and died from otherwise sublethal fecal peritonitis. In this fulminant model of infection, injection of the anti-G-CSF-antiserum at the time point of infection had no effect on survival and early neutrophilia (3 hours) but blunted the late neutrophilia (12 hours). Dogs given Filgrastim repeatedly were reported to develop an autoimmune neutropenia (8). These findings suggest that G-CSF plays a role not only in emergency recruitment of PMN in case of infections, but also in the continuous maintenance of normal neutrophil counts.

VI. CONSEQUENCES OF IMMUNOMODULATION BY EXOGENOUS G-CSF ON COURSE OF INFECTION

Since the major function of the immune system is the anti-infectious defense, the immunomodulatory properties of G-CSF have to be seen in the context of these diseases. The bearing of G-CSF in infectious disease is subject of a number of comprehensive reviews (1,14,20,43,98). A recent compilation of literature on Filgrastim in animal infection models (Dr. Lyn Frumkin, personal communication) comprising 47 publications listed 39 studies on bacterial infections, 8 on fungal infections, and only a single study where viral and bacterial infection was combined. In 37 studies, Filgrastim treatment significantly improved the course of infection (30 bacterial and 7 fungal models), in 8 studies there was no effect (6 bacterial, 1 bacterial/viral and 1 fungal model), and only in two cases was worsening of outcome found (2 bacterial models of acute lung injury). Notably, only five studies used neutropenic animals; in these cases Filgrastim was beneficial as expected.

A. Bacterial Infections

A number of studies with neutropenic animals or impaired PMN function demonstrated the efficacy of Filgrastim treatment. Filgrastim increased survival in mice made neutropenic with cyclophosphamide and challenged with *P. aeruginosa, Serratia marcescens,* or *S. aureus* (112,113). A number of studies demonstrated improved outcome in a murine model of *Pseudomonas* burn wound infection (42,114,115). Treatment also was beneficial in mice challenged with *P. aeruginosa* pneumonia after hemorrhage (116). Newborn rats known to have impaired PMN functions challenged with group B streptococci were protected (117–119) when Filgrastim was given prophylactically or simultaneously with challenge.

From the variety of studies of Filgrastim in infections of normal hosts, only some examples will be mentioned. Filgrastim had a clear beneficial effect in a number of peritonitis models. Rats given 15 μg Filgrastim after the onset of peritonitis brought about by cecal ligation and puncture showed decreased mortality and improved histopathology in lung as well as renal and hepatic functions (120). In the same animal model, as little as 2 μg/kg Filgrastim was protective (36,121) and synergistic if combined with antibiotic treatment. Similarly, in a rat cecal-puncture model, Filgrastim decreased lethality (69). Rats were also protected by Filgrastim in combination with antibiotics when a human stool suspension was inoculated into the peritoneum (122). Similar results were obtained in an analogous mouse model without concurrent treatment with antibiotics (7). Filgrastim pretreatment was also beneficial in rat *E. coli* peritonitis (123). Sepsis after intraperitoneal (IP) implantation of an *E. coli*–infected clot in dogs was improved by Filgrastim (68). Pretreatment with Filgrastim, however, had no effect on survival of newborn rats challenged IP with group B streptococci (124).

In pneumonia models, Filgrastim effects were not unambiguously beneficial. Filgrastim pretreatment protected rats against pneumococcal pneumonia (125), but there was no effect in rats on a chronic ethanol diet. In a similar experimental infection model (*Klebsiella pneumoniae*), however, Filgrastim protected both control and ethanol-treated animals (126). Therapy with Filgrastim (24 hours after challenge) of rabbits inoculated transtracheally with *Pasteurella multocida* trended towards improved survival (77% versus 67%) but showed mild increase in inflammation in liver, spleen, and lung (127). Detrimental Filgrastim effects were described in rats after intrabroncheal application of *E. coli* in the presence or absence of hyperoxia (128).

B. Viral, Fungal, and Parasitic Infections

Surprisingly, there are only few reports on Filgrastim treatment of these types of infections. We are not aware of any study on viral or parasitic infections. Filgrastim was reported to increase in vitro the oxidative burst of human PMN in response to *Aspergillus hypae* (129). In neutropenic mice after cyclophosphamide treatment, Filgrastim protected against challenge with *Candida albicans, Cryptococcus neoformans,* or *Aspergillus fumigatus* (113,122,130,131). Polak-Wyss (132,133), however, noted no beneficial effect in neutropenic murine cryptococcosis and local candidosis.

VII. OUTLOOK: IMMUNOMODULATION BY FILGRASTIM IN FUTURE CLINICAL USE

To date, there is broad consensus on the clinical efficacy of Filgrastim in neutropenic disorders such as iatrogenic neutropenia (due to chemotherapy, radiotherapy, myelosuppressive drugs), idiopathic neutropenia, leukemic neutropenia, aplastic anemia, and agranulocytosis (14,134–138). The efficacy of Filgrastim in a variety of preclinical models encourages additional trials for future indications. A promising indication for the use of Filgrastim in non-neutropenic settings is sepsis prophylaxis and treatment in cases were the time point of a risk is known (major surgery) or can be anticipated (trauma, burn).

The results of orientating clinical investigations support the notion that an overwhelming concordance of the hitherto available set of preclinical data with the practical experiences during the world-wide usage of the drug exists. It seems appropriate to further explore the use of the pharmacological potential of Filgrastim for new indications such as non-neutropenic infection, sepsis prophylaxis, chronic inflammation, and AIDS.

REFERENCES

1. Nelson, S. (1994). Role of granulocyte colony-stimulating factor in the immune response to acute bacterial infection in the nonneutropenic host: an overview. *Clin Infect Dis 18:*S197–204.
2. Demetri, G. D., and Griffin, J. D. (1991). Granulocyte colony-stimulating factor and its receptor. *Blood 78:*2791–2808.
3. Yuo, A., Kitagawa, S., Ohsaka, A., Saito, M., and Takaku, F. (1990). Stimulation and priming of human neutrophils by granulocyte colony-stimulating factor and granulocyte-macrophage colony-stimulating factor: qualitative and quantitative differences. *Biochem Biophys Res Commun 171:*491–497.
4. Ulich, T. R., del Castillo, J., Guo, K., and Souza, L. (1989). The hematologic effects of chronic administration of the monokines tumor necrosis factor, interleukin-1, and granulocyte-colony stimulating factor on bone marrow and circulation. *Am J Pathol 134:*149–159.
5. Tamura, M., Hattori, K., Nomura, H., et al (1987). Induction of neutrophilic granulocytosis in mice by administration of purified human native granulocyte colony-stimulating factor (G-CSF). *Biochem Biophys Res Comm 142:* 454–460.
6. Tkatch, L. S., and Tweardy, D. J. (1993). Human granulocyte colony-stimulating factor (G-CSF), the premier granulopoietin: biology, clinical utility, and receptor structure and function. *Lymphokine Cytokine Res 12:*477–4887.
7. Barsig, J., Bundschuh, D. S., Hartung, T., et al (1996). Control of fecal peritoneal infection in mice by colony-stimulating factors. *J Infect Dis 174:*790–799.
8. Hammond, W. P., Csiba, E., Canin, A., et al (1991). Chronic neutropenia. A new canine model induced by human granulocyte colony-stimulating factor. *J Clin Invest 87:*704–710.
9. Lieschke, G. J., Grail, D., Hodgson, G., et al (1994). Mice lacking granulocyte colony-stimulating factor have chronic neutropenia, granulocyte and macrophage progenitor cell deficiency, and impaired neutrophil mobilization. *Blood 84:*1737–1746.
10. Morstyn, G., Campbell, L., Souza, L. M., et al. Effect of granulocyte colony-stimulating factor on neutropenia induced by cytotoxic chemotherapy. *Lancet March 26, 1988:*667–672.
11. Katoh, M., Shirai, T., Shikoshi, K., Ishii, M., Saito, M., and Kitagawa, S. (1992). Neutrophil kinetics shortly after initial administration of recombinant human granulocyte colony-stimulating factor: neutrophil alkaline phosphatase activity as an endogenous marker. *Eur J Haematol 49:*19–24.
12. Bronchud, M., H., Potter, M. R., Morgenstem, G., et al (1988). *In vitro* and *in vivo* analysis of the effects of recombinant human granulocyte colony-stimulating factor in patients. *Br J Cancer 58:*64–69.
13. Price, T. H., Chatta, G. S., and Dale, D. C. (1996). Effect of recombinant granulocyte colony-stimulating factor on neutrophil kinetics in normal young and elderly humans. *Blood 88:*335–340.

14. Dale, D. C. (1994). Potential role of colony-stimulating factors in the prevention and treatment of infectious diseases. *Clin Infect Dis 18:*S180–S188.

15. Souza, L. (1990). CSF in vivo: effects on hematopoiesis. *Int J Cell Cloning 8:*227–228.

16. Kropec, A., Lemmen, S. W., Grundmann, H. J., Engels, I., and Daschner, F. D. (1995). Synergy of simultaneous administration of ofloxacin and granulocyte colony-stimulating factor in killing of *Escherichia coli* by human neutrophils. *Infection 23:*298–300.

17. Daschner, F. D., Grundmann, H., Anding, K., and Lemmen, S. (1995). Combined effect of human neutrophils, ceftazidime and granulocyte colony-stimulating factor on killing of *Escherichia coli*. *Eur J Clin Microbiol Infect Dis 14:*536–539.

18. McKenna, P. J., Nelson, S., and Andresen, J. (1996). Filgrastim (rhuG-CSF) enhances ciprofloxacin uptake and bactericidal activity of human neutrophils *in vitro*. Abstracts of the American Thoracic Society, May 10–15. *Am J Respir Crit Care Med 153* (abstr).

19. Bober, L. A., Grace, M. J., Pugliese-Sivo, C., et al (1995). The effect of GM-CSF and G-CSF on human neutrophil function. *Immunopharmacology 29:*111–119.

20. Roilides, E., Walsh, T. J., Pizzo, P. A., and Rubin, M. (1991). Granulocyte colony-stimulating factor enhances the phagocytic and bactericidal activity of normal and defective human neutrophils. *J Infect Dis 163:*579–583.

21. Yamamoto, Y., Klein, T. W., Friedman, H., Kimura, S., and Yamaguchi, H. (1993). Granulocyte colony-stimulating factor potentiates anti-*Candida albicans* growth inhibitory activity of polymorphonuclear cells. *FEMS Immunol Med Microbiol 7:*15–22.

22. Nathan, C.F. (1989). Respiratory burst in adherent human neutrophils: triggering by colony-stimulating factors CSF-GM and CSF-G. *Blood 73:*301–306.

23. Balazovich, K. J., Almeida, H. I., and Boxer, L. A. (1991). Recombinant human G-CSF and GM-CSF prime human neutrophils for superoxide production through different signal transduction mechanisms. *J Lab Clin Med 118:*576–584.

24. Khwaja, A., Carver, J. E., and Linch, D. C. (1992). Interactions of granulocyte-macrophage colony-stimulating factor (CSF), granulocyte CSF, and tumor necrosis factor alpha in the priming of the neutrophil respiratory burst. *Blood 79:*745–753.

25. Sullivan, G. W., Carper, H. T., and Mandell, G. L. (1993). The effect of three human recombinant hematopoietic growth factors (granulocyte-macrophage colony-stimulating factor, granulocyte colony-stimulating factor, and interleukin-3) on phagocyte oxidative activity. *Blood 81:*1863–1870.

26. Kitagawa, S., Yuo, A., Souza, L. M., Saito, M., Miura, Y., and Takaku, F. (1987). Recombinant human granulocyte colony-stimulating factor enhances superoxide release in human granulocytes stimulated by the chemotactic peptide. *Biochem Biophys Res Comm 144:*1143–1146.

27. Treweeke, A. T., Aziz, K. A., and Zuzel, M. (1994). The role of G-CSF in mature neutrophil function is not related to GM-CSF-type cell priming. *J Leukocyte Biol 55:*612–616.

28. Yuo, A., Kitagawa, S., Ohsaka, A., et al (1989). Recombinant human granulocyte colony-stimulating factor as an activator of human granulocytes: potentiation of responses triggered by receptor-mediated agonists and stimulation of C3bi receptor expression and adherence. *Blood 74*:2144–2149.

29. König, B., and König, W. (1994). Effect of growth factors on *Escherichia coli* alpha-hemolysin-induced mediator release from human inflammatory cells: involvement of the signal transduction pathway. *Infect Immun 62*:2085–2093.

30. Wheeler, J. G., Huffine, M. E., Childress, S., and Sikes, J. (1994). Comparison of colony stimulation factors on *in vitro* rat and human neutrophil function. *Biol Neonate 66*:214–220.

31. Iacobini, M., Palumbo, G., Bartolozzi, S., et al (1995). Superoxide production by neutrophils in children with malignant tumors treated with recombinant human granulocyte colony-stimulating factor. *Haematologica 80*:13–17.

32. Weiss, M., Gross-Weege, W., Wernet, P., et al (1994). Filgrastim (rhG-CSF) for prophylaxis and therapy of sepsis in surgical intensive care patients: neutrophil function. *Clin Intensive Care 5*(suppl 96).

33. Weiss, M., Gross-Weege, W., Schneider, M., et al (1995). Enhancement of neutrophil function by in vivo filgrastim treatment for prophylaxis of sepsis in surgical intensive care patients. *J Crit Care Med 10*:21–26.

34. Lindemann, A., Herrmann, F., Oster, W., et al (1989). Hematologic effects of recombinant human granulocyte colony-stimulating factor in patients with malignancy. *Blood 74*:2644–2651.

35. Sato N., and Shimizu, H. (1993). Granulocyte colony-stimulating factor improves an impaired bactericidal function in neutrophils from STZ-induced diabetic rats. *Diabetes 42*:470–473.

36. Goya, T., Torisu, M., Doi, F., and Yoshida, T. (1993). Effects of granulocyte colony-stimulating factor and monobactam antibiotics (Aztreonam) on neutrophil functions in sepsis. *Clin Immunol Immunopathol 69*:278–284.

37. Borregaard, N., Lollike, K., Kjeldsen, L., et al (1993). Human neutrophil granules and secretory vesicles. *Eur J Haematol 51*:187–198.

38. Gabrilove, J. L., Jakubowski, A., Fain, K., et al (1988). Phase I study of granulocyte colony-stimulating factor in patients with transitional cell carcinoma of the urothelium. *J Clin Invest 82*:1454–1461.

39. Wang, J. M., Chen, Z. G., Colella, S., et al (1988). Chemotactic activity of recombinant human granulocyte colony-stimulating factor. *Blood 72*:1456–1460.

40. Lang, C. H., Bagby, G. J., Dobrescu, C., Nelson, S., and Spitzer, J. J. (1992). Effect of granulocyte colony-stimulating factor on sepsis-induced changes in neutrophil accumulation and organ glucose uptake. *J Infect Dis 166*:336–343.

41. Demetri, G. D., Spertini, O., Pratt, E. S., et al (1990). GM-CSF and G-CSF have different effects on expression of the neutrophil adhesion receptors LAM-1 & CD11b. *Blood 76* 178a (abstr 704).

42. Sartorelli, K. H., Silver, G. M., and Gamelli, R. L. (1991). The effect of granulocyte colony-stimulating factor (G-CSF) upon burn-induced defective neutrophil chemotaxis. *J Trauma 31*:523–530.

43. Dale, D. C., Liles, W. C., Summer, W. R., and Nelson, S. (1995). Granulocyte colony-stimulating factor—role and relationship in infectious diseases. *J Infect Dis 172*:1061–1075.

44. Yong, K. L., and Linch, D. C. (1992). Differential effects of granulocyte- and granulocyte-macrophage colony-stimulating factors (G- and GM-CSF) on neutrophil adhesion *in vitro* and *in vivo*. *Eur J Haematol 49*:251–259.

45. Ohsaka, A., Saionji, K., Sato, N., Mori, T., Ishimoto, K., and Inamatsu, T. (1993). Granulocyte colony-stimulating factor down-regulates the surface expression of the human leucocyte adhesion molecule-1 on human neutrophils in vitro and in vivo. *Br J Haematol 84*:574–580.

46. Hansen, P. B., Kjaersgaard, E., Johnsen, H. E., et al (1993). Different membrane expression of CD11b and CD14 on blood neutrophils following in vivo administration of myeloid growth factors. *Br J Haematol 85*:50–56.

47. Repp, R., Valerius, T., Sendler, A., et al (1991). Neutrophils express the high affinity receptor for IgG (Fc gamma RI, CD64) after in vivo application of recombinant human granulocyte colony-stimulating factor. *Blood 78*:885–889.

48. Kerst, J. M., de Haas, M., van der Schoot, C. E., et al (1993). Recombinant granulocyte colony-stimulating factor administration to healthy volunteers: induction of immunophenotypically and functionally altered neutrophils via an effect on myeloid progenitor cells. *Blood 82*:3265–3272.

49. Kerst, J. M., van de Winkel, J. G., Evans, A. H., et al (1993). Granulocyte colony-stimulating factor induces hFc γ RI (CD64 antigen)-positive neutrophils via an effect on myeloid precursor cells. *Blood 81*:1457–1464.

50. Liles, W. C., Waltersdorph, A. M., and Klebanoff, S. J. (1994). Regulation of apoptosis in human neutrophils: effects of proinflammatory mediators, protein kinase inhibitors, and antibodies directed aginst β2-integrins. *Clin Res 42*:148A.

51. Adachi, S., Kubota, M., Matsubara, K., et al (1993). Role of protein kinase C in neutrophil survival enhanced by granulocyte colony-stimulating factor. *Exp Hematol 21*:1709–1713.

52. Ichinose, Y., Hara, N., Ohta, M., et al (1990). Recombinant granulocyte colony-stimulating factor and lipopolysaccharide maintain the phenotype of and superoxide anion generation by neutrophils. *Infect Immun 58*:1647–1652.

53. Morstyn, G., Lieschke, G. J., Sheridan, W., Layton, J., and Cebon, J. (1989). Pharmacology of the colony-stimulating factors. *Trends Pharmacol Sci 10*:154–159.

54. Herrmann, F., Lindemann, A., and Mertelsmann, R. (1990). G-CSF and M-CSF: from molecular biology to clinical application. *Biotherapy 2*:315–324.

55. Shirafuji N., Matsuda, S., Ogura, H., et al (1990). Granulocyte colony-stimulating factor stimulates human mature neutrophilic granulocytes to produce interferon-alpha. *Blood 75*:17–19.

56. Hartung, T., Volk, H-D., and Wendel, A., (1995). G-CSF—an anti-inflammatory cytokine. *J Endotoxin Res 2*:195–201.

57. Hartung, T., Doecke, W. D., Gantner, F., et al. (1995). Effect of granulocyte colony-stimulating factor treatment on ex vivo blood cytokine response in human volunteers. *Blood 85*:2482–2489.

58. Valerius, T., Repp, R., de Wit, T. P., et al. (1993). Involvement of the high-affinity receptor for IgG (Fc γ RI; CD64) in enhanced tumor cell cytotoxicity of neutrophils during granulocyte colony-stimulating factor therapy. *Blood* 82:931–939.

59. van der Wouw, P. A., van Leeuwen, R., van Oers, R. H., Lange, J. M., and Danner, S. A. (1991). Effects of recombinant human granulocyte colony-stimulating factor on leucopenia in zidovudine-treated patients with AIDS and AIDS related complex, a phase I/II study. *Br J Haematol* 78:319–324.

60. Liu, K. Y., Akashi, K., Harada, M., Takamatsu, Y., and Niho, Y. (1993). Kinetics of circulating haematopoietic progenitors during chemotherapy-induced mobilization with or without granulocyte colony-stimulating factor. *Br J Haematol* 84:31–38.

61. Wiltschke, C., Krainer, M., Wagner, A., Linkesch, W., Zielinski, C. C. (1995). Influence of in vivo administration of GM-CSF and G-CSF on monocyte cytotoxicity. *Exp Hematol* 23:402–406.

62. Baram, D., Madara, P., Culpepper, S., et al (1996). Human neutrophils regulate TNF concentrations in monocyte cultures and whole blood. Abstracts of the American Thoracic Society, May 10–15. *Am J Respir Crit Care Med* 153 (abstr).

63. Terashima, T., Soejima, K., Waki, Y., et al (1995). Neutrophils activated by granulocyte colony-stimulating factor suppress tumor necrosis factor-alpha release from monocytes stimulated by endotoxin. *Am J Respir Cell Mol Biol* 13:69–73.

64. Gorgen, I., Hartung, T., Leist, M., et al (1992). Granulocyte colony-stimulating factor treatment protects rodents against lipopolysaccharide-induced toxicity via suppression of systemic tumor necrosis factor-alpha. *J Immunol* 149:918–924.

65. Lang, C. H., Bagby, G. J., Dobrescu, C., Nelson, S., and Spitzer, J. J. (1992). Modulation of glucose metabolic response to endotoxin by granulocyte colony-stimulating factor. *Am J Physiol* 263:R1122–R1129.

66. Kanazawa, M., Ishizaka, A., Hasegawa, N., Suzuki, Y., and Yokoyama, T (1991). Granulocyte colony-stimulating factor does not enhance endotoxin-induced acute lung injury in guinea pigs. *Am Rev Respir Dis* 145:1030–1035.

67. Fink, M. P., O'Sullivan, B. P., Menconi, M. J., et al (1993). Effect of granulocyte colony-stimulating factor on systemic and pulmonary responses to endotoxin in pigs. *J Trauma* 34:571–578.

68. Eichacker, P. Q., Waisman, Y., Natanson, C., et al (1994). Cardiopulmonary effects of granulocyte colony-stimulating factor in a canine model of bacterial sepsis. *J Appl Physiol* 77:2366–2373.

69. Lundblad, R., Nesland, J. M., and Giercksky, K. E. (1996). Granulocyte colony-stimulating factor improves survival rate and reduces concentrations of bacteria, endotoxin, tumor necrosis factor, and endothelin-1 in fulminant intra-abdominal sepsis in rats. *Crit Care Med* 24:820–826.

70. Ganser, A., Seipelt, G., Verbeek, W., et al (1994). Effect of combination therapy with all-trans-retinoic acid and recombinant human granulocyte col-

ony-stimulating factor in patients with myelodysplastic syndromes. *Leukemia 8*:369–375.

71. Denzlinger, C., Holler, E., Reisbach, G., Hiller, E., and Wilmanns, W. (1994). Granulocyte colony-stimulating factor inhibits the endogenous leukotriene production in tumour patients. *Br J Haematol 86*:881–882.

72. Marth, C., Weiss, G., Koza, A., Reibneggr, G., Daxenbichler, G., Zeimet, A. G., et al (1994). Increased production of immune activation marker neopterin by colony-stimulating factors in gynecological cancer patients. *Int J Cancer 58*:20–23.

73. Lu, L., Srour, E. F., Warren, D. J., Walker, D., Graham, C. D., Walker, E. B., et al (1988). Enhancement of release of granulocyte- and granulocyte-macrophage colony-stimulating factors from phytohemagglutinin-stimulated sorted subsets of human T lymphocytes by recombinant human tumor necrosis factor-alpha. *J Immunol 141*:201–207.

74. Kerrigan, D. P., Castillo, A. Foucar, K., Townsend, K., and Neidhart, J. (1989). Peripheral blood morphological changes after high-dose antineoplastic chemotherapy and recombinant human granulocyte colony-stimulating factor administration. *Am J Clin Pathol 92*:280–285.

75. Terashita, M., Kudo, C., Yamashita, T., Gresser, I., and Sendo, F. (1996). Enhancement of delayed-type hypersensitivity to sheep red blood cells in mice by granulocyte colony-stimulating factor administration at the elicitation phase. *J Immunol 156*:4638–4643.

76. Pan, L., Delmonte J. Jr., Jalonen, C. K., and Ferrara, J. L. (1995). Pretreatment of donor mice with granulocyte colony-stimulating factor polarizes donor T lymphocytes toward type-2 cytokine production and reduces severity of experimental graft-versus-host disease. *Blood 86*:4422–4429.

77. Kitabayashi, A., Hirokawa, M., Hatano, Y, Lee, M., Kuroki, J., Niitsu, H., et al (1995). Granulocyte colony-stimulating factor downregulates allogeneic immune responses by posttranscriptional inhibition of tumor necrosis factor-alpha production. *Blood 86*:2220–2227.

78. Aoki, Y., Hiromatsu, K., Kobayashi, N., Hotta, T., Saito, H., Igarashi, H., et al (1995). Protective effect of granulocyte colony-stimulating factor against T-cell-mediated lethal shock triggered by superantigens. *Blood 86*:1420–1427.

79. Dreger, P., Grelle, K., Eckstein, V., Suttorp, M., Muller-Rochholtz, W., Loffler, H., et al (1993). Granulocyte-colony-stimulating factor induces increased serum levels of soluble interleukin 2 receptors preceding engraftment in autologous bone marrow transplantation. *Br J Haematol 83*:7–13.

80. Shitara, T., Yugami, S., Ijima, H., Sotomatu, M., and Kuroume, T. (1994). Cytokine profile during high-dose rhG-CSF therapy in severe congenital neutropenia. *Am J Hematol 45*:58–62.

81. Ernst, T. J., Ritchie, A., Demetri, G. D., and Griffin, J. D. (1988). Regulation of granulocyte- and monocyte-colony stimulating factor mRNA levels in human blood monocytes is mediated primarily at a post-transcritopnal level. *J Biol Chem 64*:5700–5703.

82. de Wit, H., Dokter, W. H., Esselink, M. T., Halie, M. R., and Vellenga, E. (1993). Interferon-gamma enhances the LPS-induced G-CSF gene expression

in human adherent monocytes, which is regulated at transcriptional and posttranscriptional levels. *Exp Hematol 21:*785–790.

83. Hamilton J. A. (1994). Coordinate and noncoordinate colony-stimulating factor formation by human monocytes. *J Leuko Biol 55:*355–361.

84. Cluitmans, F. H., Esendam, B. H., Landegent, J. E., Willemze, R., and Falkenburg, J. H. (1993). Regulatory effects of T cell lymphokines on cytokine gene expression in monocytes. *Lymphokine Cytokine Res 12:*457–464.

85. Dokter, W. H., Dijkstra, A. J., Koopmans, S. B., Mulder, A. B., Stulp, B. K., Halie, M. R., et al (1994). G(AnH)MTetra, a naturally occurring 1,6-anhydro muramyl dipeptide, induces granulocyte colony-stimulating factor expression in human monocytes: a molecular analysis. *Infect Immun 62:*2953–2957.

86. Oster, W., Lindemann, A., Mertelsmann, R., and Herrmann, F. (1989). Production of macrophage-, granulocyte-, granulocyte-macrophage- and multi-colony-stimulating factor by peripheral blood cells. *Eur J Immunol 19:*543–547.

87. Oster, W., Lindemann, A., Mertelsmann, R., and Herrmann, F. (1989). Granulocyte-macrophage colony-stimulating factor (CSF) and multilineage CSF recruit human monocytes to express granulocyte CSF. *Blood 73:*64–67.

88. Wieser, M., Bonifer, R., Oster, W., Lindemann, A., Mertelsmann, R., and Herrmann, F. (1989). Interleukin-4 induces secretion of CSF for granulocytes and CSF for macrophages by peripheral blood monocytes. *Blood 73:*1105–1108.

89. Quesniaux, V. F., Wehrli, S., Ziegler, I., Legendre, B., Wishart, W., Fagg, B., et al (1992). Human serum stimulates the production of G-CSF, IL-1, IL-6 and IL-8 by human peripheral blood leucocytes. *Br J Haematol 82:*6–12.

90. Fibbe, W. E., van Damme, J., Billiau, A., Goselink, H. M., Voogt, P.J., van Eeden, G., et al (1988). Interleukin 1 induces human marrow stromal cells in long-term culture to produce granulocyte colony-stimulating factor and macrophage colony-stimulating factor. *Blood 71:*430–435.

91. Zsebo, K. M., Yuschenkoff, V. N., Schiffer, S., et al (1988). Vascular endothelial cells and granulopoiesis: Interleukin-1 stimulates release of G-CSF and GM-CSF. *Blood 71:*99–103.

92. Koeffler, H. P., Gasson, J., and Tobler, A. (1988). Transcriptional and post-transcriptional modulation of myeloid colony-stimulating factor expression by tumor necrosis factor and other agents. *Mol Cell Biol 8:*3432–3438.

93. Seelentag, W. K., Mermod, J. J., Montesano, R., and Vassalli, P. (1987). Additive effects of Interleukin 1 and tumour necrosis factor-alpha on the accumulation of the three granulocyte and macrophage colony-stimulating factor mRNAs in human endothelial cells. *EMBO J 6:*2261–2265.

94. Demetri, G. D., Zenzie, B. W., Rheinwald, J. G., and Griffin, J. D. (1989). Expression of colony-stimulating factor genes by normal human mesothelial cells and human malignant mesothelioma cells lines in vitro. *Blood 74:*940–946.

95. Cheers, C., Haigh, A. M., Kelso, A., Metcalf, D., Stanley, E. R., and Young, A. M. (1988). Production of colony-stimulating factors (CSFs) during infec-

tion: separate determinations of macrophage-, granulocyte-, granulocyte-macrophage-, and multi-CSFs. *Infect Immun 56*:247–251.

96. Barsig, J. (1996). The colony-stimulating factors G-CSF and GM-CSF as immunomodulators in murine sepsis models. Thesis, University of Konstanz, Hartung-Gorre, Konstanz.

97. Dale, D. C., Lau, S., Nash, R., Boone, T., and Osborne, W. (1992). Effect of endotoxin on serum granulocyte and granulocyte-macrophage colony-stimulating factor levels in dogs. *J Infect Dis 165*:689–694.

98. Metcalf, D. (1987). The role of the colony-stimulating factors in resistance to acute infections. *Immunol Cell Biol 65*:35–43.

99. Sallerfors, B. (1994). Endogenous production and peripheral blood levels of granulocyte-macrophage (GM)- and granulocyte (G-) colony-stimulating factors. *Leuk Lymphoma 13*:235–247.

100. Watari, K., Asano, S., Shirafuji, N., Kodo, H., Ozawa, K., Takaku, F., et al (1989). Serum granulocyte colony-stimulating factor levels in healthy volunteers and patients with various disorders as estimated by enzyme immunoassay. *Blood 73*:117–122.

101. Kawakami, M., Tsutsumi, H., Kumakawa, T., Abe, H., Hirai, M., Kurosawa, S., et al (1990). Levels of serum granulocyte colony-stimulating factor in patients with infections. *Blood 76*:1962–1964.

102. Omori, F., Okamura, S., Shimoda, K., Otsuka, T., Harada, M., and Niho, Y. (1992). Levels of human serum granulocyte colony-stimulating factor and granulocyte-macrophage colony-stimulating factor under pathological conditions. *Biotherapy 4*:147–153.

103. Cebon, J., Layton, J. E., Maher, D., and Morstyn, G. (1994). Endogenous haematopoietic growth factors in neutropenia and infection. *Br J Haematol 86*:265–274.

104. Kawakami, M., Tsutsumi, H., Kumakawa, T., Hirai, M., Kurosawa, S., Mori, M., et al (1992). Serum granulocyte colony-stimulating factor in patients with repeated infections. *Am J Hematol 41*:190–193.

105. Taveira da Silva, A. M., Kaulbach, H. C., Chuidian, F. S., Lambert, D. R., Suffredini, A. F., and Danner, R. L. (1993). Brief report: shock and multiple-organ dysfunction after self-administration of *Salmonella* endotoxin. *N Engl J Med 328*:1457–1460.

106. Sallerfors, B., and Olofsson, T. (1991). Granulocyte-macrophage colony-stimulating factor (GM-CSF) and granulocyte colony-stimulating factor (G-CSF) in serum during induction treatment of acute leukaemia. *Br J Haematol 78*:343–351.

107. Misago, M., Kikuchi, M., Tsukada, J., Hanamura, T., Kamachi, S., and Eto, S. (1991). Serum levels of G-CSF, M-CSF and GM-CSF in a patient with cyclic neutropenia. *Eur J Haematol 46*:312–313.

108. Cairo, M. S. (1993). Therapeutic implications of dysregulated colony-stimulating factor expression in neonates. *Blood 82*:2269–2272.

109. Lieschke, G. J., and Burgess, A. W. (1992). Granulocyte colony-stimulating factor and granulocyte-macrophage colony-stimulating factor. *N Engl J Med 327*:28–35.

110. Vellenga, E., Dokter, W., de Wolf, J. T., van de Vinne, B., Esselink, M. T., and Halie, M. R. (1991). Interleukin-4 prevents the induction of G-CSF mRNA in human adherent monocytes in response to endotoxin and IL-1 stimulation. *Br J Haematol 79*:22–26.

111. Hamilton, J. A., Whitty, G. A., Royston, A. K., Cebon, J., and Layton, J. E. (1992). Interleukin-4 suppresses granulocyte colony-stimulating factor and granulocyte-macrophage colony-stimulating factor levels in stimulated human monocytes. *Immunology 76*:566–571.

112. Ono, M., Matsumoto, M., Matsubara, S., Tomioka, S., and Asano, S. (1988). Protective effect of human granulocyte colony-stimulating factor on bacterial and fungal infections in neutropenic mice. *Behring Inst Mitt 83*:216–221.

113. Matsumoto, M., Matsubara, S., Matsuno, T., Tamura, M., Hattori, K., Nomura, H., et al (1987). Protective effect of human granulocyte colony-stimulating factor on microbial infection in neutropenic mice. *Infect Immun 55*:2715–2720.

114. Mooney, D. P., Gamelli, R. L., O'Reilly, M., and Hebert, J. C. (1988). Recombinant human granulocyte colony-stimulating factor and *Pseudomonas* burn wound sepsis. *Arch Surg 123*:1353–1357.

115. Silver, G. M., Gamelli, R. L., and O'Reilly, M. (1989). The beneficial effect of granulocyte colony-stimulating factor (G-CSF) in combination with gentamicin on survival after *Pseudomonas* burn wound infection. *Surgery 106*:452–456.

116. Abraham, E., and Stevens, P. (1992). Effects of granulocyte colony-stimulating factor in modifying mortality from *Pseudomonas aeruginosa* pneumonia after hemorrhage. *Crit Care Med 20*:1127–1133.

117. Cairo, M. S., Mauss, D., Kommareddy, S., Norris, K., van de Ven, C., and Modanloou, H. (1990). Prophylactic or simultaneous administration of recombinant human granulocyte colony-stimulating factor in the treatment of group B streptococcal sepsis in neonatal rats. *Pediatr Res 27*:612–616.

118. Cairo, M. S., Plunkett, J. M., Mauss, D., and van de Ven, C. (1990). Seven-day administration of recombinant human granulocyte colony-stimulating factor to newborn rats: modulation of neonatal neutrophilia, myelopoiesis, and group B *Streptococcus* sepsis. Blood 76:1788–1794.

119. Cairo, M. S., Plunkett, J. M., Nguyen, A., and van de Ven, C. (1992). Effect of stem cell factor with and without granulocyte colony-stimulating factor on neonatal hematopoiesis: In vivo induction of newborn myelopoiesis and reduction of mortality during experimental group B streptococcal sepsis. *Blood 80*:96–101.

120. Toda, H., Murata, A., Matsuura, N., Uda, K., Oka, Y., Tanaka, N., et al (1993). Therapeutic efficacy of granulocyte colony-stimulating factor against rat cecal ligation and puncture model. *Stem Cells 11*:228–234.

121. O'Reilly, M., Silver, G. M., Greenhalgh, D. G., Gamelli, R. L., Davis, J. H., and Hebert, J. C. (1992). Treatment of intra-abdominal infection with granulocyte colony-stimulating factor. *J Trauma 33*:679–682.

122. Lorenz, W., Reimund, K. P., Weitzel, F., Celik, I., Kurnatowski, M., Schneider, C., et al. (1994). Granulocyte colony-stimulating factor prophylaxis before

operation protects against lethal consequences of postoperative peritonitis. *Surgery 116:*925–934.

123. Dunne, J. R., Dunkin, B. J., Nelson, S., and White, J. C. (1996). Effects of granulocyte colony-stimulating factor in a nonneutropenic rodent model of *Escherichia coli* peritonitis. *J Surg Res 61:*348–354.

124. Iguchi, K., Inoue, S., and Kumar, A. (1991). Effect of recombinant human granulocyte colony-stimulating factor administration in normal and experimentally infected newborn rats. *Exp Hematol 19:*352–358.

125. Lister, P. D., Gentry, M. J., and Preheim, L. C. (1993). Granulocyte colony-stimulating factor protects control rats but not ethanol-fed rats from fatal pneumococcal pneumonia. *J Infect Dis 168:*922–926.

126. Nelson, S., Farkas, S., Fotheringham, N., et al (1996). Filgrastim in the treatment of hospitalized patients with community acquired pneumonia (CAP). Abstracts of the American Thoracic Society, May 10–15. *Am J Respir Crit Care Med 153* (abstr).

127. Smith, W. S., Sumnicht, G. E., Sharpe, R. W., Samuelson, D., and Millard, F. E. (1995). Granulocyte colony-stimulating factor versus placebo in addition to penicillin G in a randomized blinded study of gram-negative pneumonia sepsis: analysis of survival and multisystem organ failure. *Blood 86:*1301–1309.

128. Freeman, B. D., Correa, R., Karzai, W., et al. (1996). Controlled trial of rG-CSF and CD11b-directed MAb during hyperoxia and *E. coli* pneumonia in rats. *J Appl Physiol 80:*2066–2076.

129. Roilides, E., Uhlig, K., Venzon, D., Pizzo, P. A., and Walsh, T. J. (1993). Enhancement of oxidative response and damage caused by human neutrophils to Aspergillus fumigatus hyphae by granulocyte colony-stimulating factor and gamma interferon. *Infect Immun 61:*1185–1193.

130. Uchida, K., Yamamoto, Y., Klein, T. W., Friedman, H., and Yamaguchi, H. (1992). Granulocyte colony-stimulating factor facilities the restoration of resistance to opportunistic fungi in leukopenic mice. *J Med Vet Mycol 30:*293–300.

131. Hamood, M., Bluche, P. F., De Vroey, C., Corazza, F., Bujan, W., and Fondy, P. (1994). Effects of recombinant human granulocyte colony-stimulating factor on neutropenic mice infected with Candida albicans: acceleration of recovery from neutropenia and potentiation of anti-C. albicans resistance. *Mycoses 37:*93–99.

132. Polak-Wyss, A. (1991). Protective effect of human granulocyte colony-stimulating factor (hG-CSF) on Cryptococcus and Aspergillus infections in normal and immunosuppressed mice. *Mycoses 34:*205–215.

133. Polak-Wyss, A. (1991). Protective effect of human granulocyte colony-stimulating factor (hG-CSF) on *Candida* infections in normal and immunosuppressed mice. *Mycoses 34:*109–118.

134. Roilides, E., and Pizzo, P. A. (1991). Modulation of host defenses by cytokines: evolving adjuncts in prevention and treatment of serious infections in immunocompromised hosts. *Clin Infect Dis 15:*508–524.

135. Roilides, E., and Pizzo, P. A. (1993). Biologicals and hematopoietic cytokines in prevention or treatment of infections in immunocompromised hosts. *Hematol Oncol Clin North Am 7:*841–864.

136. Gabrilove, J. (1992). The development of granulocyte colony-stimulating factor in its various clinical applications. *Blood 80*:1382–1385.
137. Glaspy, J. A., and Golde D. W. (1992). Granulocyte colony-stimulating factor (G-CSF): preclinical and clinical studies. *Semin Oncol 19*:386–394.
138. Hollingshead, L. M., and Goa, K. L. (1991). Recombinant granulocyte colony-stimulating factor (rG-CSF). A review of its pharmacological properties and prospective role in neutropenic conditions. *Drug 42*:300–330.

20
Filgrastim (r-metHuG-CSF) in Pneumonia

Jeff Andresen and Hassan Movahhed
Amgen Inc., Thousand Oaks, California

Steve Nelson
*Louisiana State University Medical Center,
New Orleans, Louisiana*

I. INTRODUCTION

The polymorphonuclear leukocyte (neutrophil) is a critical component of the host defense response to bacterial infections. The mature neutrophil is highly specialized to respond to microbial invasion by means of chemotaxis, phagocytosis, intracellular killing, and extracellular release of leukocyte products including lysosomal enzymes, reactive products of oxygen metabolism, and other mediators of inflammation. As neutrophils have a limited life span of approximately 6 to 8 hours in the blood stream, they must be constantly replenished. The normal human adult produces approximately 10×10^{10} cells/day (1). The ability to increase the rate of production of phagocytic cells after a bacterial challenge is likely to be a critical factor in the host's ability to restore homeostasis and survive infection.

The generation of mature neutrophils from immature progenitor cells depends on the presence of several hematopoietic growth factors, the colony-stimulating factors (CSF) (2). Among these, granulocyte-CSF (G-CSF) primarily regulates the proliferation, differentiation, maturation, survival, and functional activity of mature neutrophils (3). The functional properties of neutrophils that are enhanced by G-CSF are primarily those related to host defense against microorganisms. These observations coupled with data demonstrating that endogenous G-CSF levels are elevated in a variety of infections (4,5) suggest that G-CSF has a major role in the immune response to an infectious challenge. The release of endogenous G-CSF would function to augment the production, mobilization, and activation of neutrophils necessary to combat invading pathogens.

The effect of G-CSF on enhancing the functional activities of mature neutrophils and to increase their numbers suggests that the administration of exogenous G-CSF may augment the host response to, and recovery from, a local or systemic infection. However, products of the neutrophil such as reactive oxygen species and proteases are potential mediators of tissue and organ damage. The potential adverse effects of G-CSF–induced proliferation and activation of neutrophils has been the subject of active investigation in non-neutropenic animal models of sepsis and the adult respiratory distress syndrome (ARDS). The concept that increasing the number and activity of neutrophils could potentially exacerbate an infectious process resulting in increased morbidity has been an important consideration in the design and monitoring of clinical trials of recombinant human G-CSF (r-metHuG-CSF, Filgrastim) in non-neutropenic patients with infectious disease.

The efficacy of Filgrastim, alone or in combination with antibiotics, has been explored in a variety of non-neutropenic infectious disease models including neonatal sepsis, pneumonia, burn wound infections, intraabdominal sepsis, and intramuscular infection. In this chapter we will focus on the effects of Filgrastim in animal models of pneumonia and the effects of Filgrastim on pulmonary function in animal models of sepsis and endotoxic shock, and will summarize the data from recent clinical studies of Filgrastim in patients with pneumonia.

II. PNEUMONIA

Pneumonia is among the leading causes of morbidity and mortality in the United States, especially in the elderly, patients with significant comorbid disease, and those admitted to the intensive care unit (ICU) requiring mechanical ventilation. Despite advances in antibiotic therapy and diagnosis, pneumonia remains the sixth leading cause of death in the United States (6). Mortality rates for patients with community- or hospital-acquired pneumonia requiring ICU admission and mechanical ventilation have been estimated to be as high as 25% to 50%, respectively (7). While the pathogenesis of bacterial pneumonia is not fully understood and is most likely multifactorial, alteration of lung host defenses and the virulence of the pathogen are recognized as important factors. Compromised respiratory defenses permit colonization of the posterior pharynx with bacteria. The subsequent aspiration of these bacteria from the oropharynx into the lower respiratory tract can result in pneumonia in a susceptible host. In addition, once a respiratory infection has been established, any preexisting or subsequent alteration in a patient's ability to respond to that infection can further increase patient morbidity and mortality. A major concern within the ICU involves the markedly increased risk of pneumonia in patients requiring ventilator support. It has been estimated that the risk of developing pneumonia is increased approximately 20-fold by mechanical ventilation (8). This is most likely a reflection of a more severely compromised host (increased severity of illness) and a more virulent spectrum of causative pathogens.

As the predominant phagocytic cells within the alveolar space of the non-infected lung, macrophages are centrally located to orchestrate the initial response to an intrapulmonary infection. In addition to their phagocytic functions, macrophages are a potent source of proinflammatory mediators, including tumor necrosis factor (TNF)-α, that play a central role in initiating host defenses against invading pathogens (9). TNF-α is an important mediator of the host defense to infection. Among the pliotropic effects of TNF-α is the induction of production and release of other cytokines, including G-CSF (10). Recent data show that human alveolar macrophages produce G-CSF when stimulated in vitro by lipopolysaccharide (LPS) and that alveolar macrophages lavaged from the lungs of patients with pneumonia spontaneously release G-CSF in culture (11). Serum levels of G-CSF also are elevated in patients with bacterial pneumonia (5) and other bacterial infections (4).

In response to factors released by pulmonary macrophages, chemotactic fragments of complement as well as products of bacteria, neutrophils rapidly infiltrate the lungs of animals exposed to a bacterial challenge. The rate of clearance of bacteria from the lung is directly related to the numbers of recruited neutrophils (12). Both the influx of neutrophils to sites of infection and the rate of clearance of bacteria from those sites are increased by treatment with Filgrastim (13,14).

III. PRECLINICAL STUDIES IN PNEUMONIA

Consumption of alcohol impairs selected functions of the normal host immune system and thereby significantly increases the susceptibility of these hosts to a variety of infections, including bacterial pneumonia (15). In studies of humans and animals, the most consistent immunologic defect induced by ethanol is an impairment in neutrophil delivery to sites of infection (16,17). In animal models of pulmonary infection, ethanol-induced suppression of pulmonary antimicrobial activity is directly related to an inhibition of neutrophil influx from the systemic vasculature into the lower respiratory tract (18).

We have investigated the effects of Filgrastim in ethanol-treated rats with experimentally induced pneumonia (14). Animals were treated with Filgrastim or inactive vehicle for 2 days before intraperitoneal (IP) administration of ethanol or saline, which was followed by an intratracheal challenge with *Klebsiella pneumoniae.* Evaluation at 4 hours after challenge showed that Filgrastim augmented the recruitment of neutrophils into the infected lungs of saline-treated rats and significantly attenuated the adverse effects of ethanol on neutrophil entry into the infected lungs of ethanol-treated rats. Filgrastim also enhanced the antibacterial defenses of the lungs of control rats that were not immunocompromised with alcohol. The percentage of the original bacterial inoculum remaining viable was 79% among recipients of the inactive vehicle, compared with only 2% among recipients of Filgrastim. Similarly, among ethanol-treated rats, Filgrastim decreased the proliferation of bacteria from 225% of the initial inoculum to 118%. Furthermore, all animals receiving vehicle before ethanol administration and bacterial challenge became bacteremic, whereas rats receiving Filgrastim had sterile blood cultures. The mortality of these bacteremic rats was 100%, while less than 10% of Filgrastim-treated rats died of infection.

Lister et al (19) examined the effect of rHuG-CSF on host defenses in a subacute model of alcohol intoxication. Animals were placed on a liquid diet for 7 days and were pretreated with rHuG-CSF on days 6 and 7. Alcohol-fed and pair-fed animals were then infected transtracheally with *Streptococcus pneumoniae* on day 8. The administration of rHuG-CSF significantly increased neutrophil recruitment into the infected lungs and increased survival in control rats; however, it did not enhance neutrophil delivery or survival in ethanol-fed rats. This model of chronic intoxication results in multifactorial changes, including hepatic abnormalities and possible impairment in neutrophil production, not seen in the acute model of alcohol intoxication.

It is widely appreciated that viral infections, including the influenza A virus (IAV), predispose the host to secondary bacterial infections often caused by *S. pneumoniae* or *Staphylococcus aureus* (20). The mortality associated with IAV infection continues to be high despite the availability of the influenza vaccine and antibiotics (21). Several in vivo and in vitro studies in animals and humans indicate that that IAV suppresses neutrophil functions (22). Among the effects of IAV on neutrophil function are suppression of adherence, chemotaxis, oxidative burst, degranulation, and antimicrobial activities (23,24). In chinchillas infected with IAV, suppression of neutrophil functions precedes the development of secondary bacterial infections (23,25). In chinchillas infected intranasally with IAV on day 1 followed by *S. pneumoniae* on day 4, administration of Filgrastim on days 3 to 9 caused a significant increase in neutrophil oxidative response compared with control animals (26). However, no difference was observed in the incidence of pneumococcal infection or mortality in animals administered Filgrastim compared with control animals in this study. Filgrastim was administered at a dose (16 μg/kg) that did not significantly increase the number of circulating neutrophils. This absence of an induced neutrophilia may explain in part these experimental results.

Even in individuals with no other detectable underlying disease, splenectomy carries an increased risk of life-threatening pneumococcal infection (27,28). In a mouse model, Filgrastim administered beginning 1 day before bacterial challenge to 3 days post-challenge was found to improve survival in splenectomized mice exposed to an aerosol of *S. pneumoniae* (29). Splenectomized treated mice had a 70% survival compared with 20% survival in saline-treated controls. Filgrastim treatment significantly reduced the number of viable bacteria recovered 24 hours after aerosol challenge from the lungs and tracheobronchial lymph nodes of splenectomized and sham-operated mice compared with saline-treated controls. In a subsequent study, Filgrastim treatment was shown to increase phagocytosis of *S. pneumoniae* by pulmonary macrophages isolated from the lungs of splenectomized or sham-operated mice compared with saline-treated control animals (30). The mechanism through which G-CSF stimulates the phagocytic ability of macrophages has not been described.

In a recent study Filgrastim plus antibiotic was evaluated in an animal model of gram-negative pneumonia and sepsis (31). Rabbits were infected with the penicillin-sensitive *Pasteurella multocida* and then randomized to receive penicillin with placebo or Filgrastim (5 to 8 μg/kg) for 5 days. The rabbits treated with Filgrastim had a significant increase in the number of peripheral white blood cells by day 4 compared with the placebo group and a trend towards improved survival compared with the placebo group (77% versus 67%). Analysis of pretreatment variables showed that the subset of animals with sepsis-induced neutropenia before treatment had significantly improved survival when treated with Filgrastim.

IV. NEUTROPHILS AND ORGAN INJURY

Although neutrophils are a critical component of host defense during infection and sepsis, these cells and their products have been implicated as mediators in the pathogenesis of ARDS. Neutrophils are present in the lavage fluid of patients with ARDS, and their numbers and the concentration of their products appear to correlate with the severity of pulmonary dysfunction (32–34). In addition to proteolytic enzymes, neutrophils are active producers of a series of reactive oxygen metabolites capable of injuring the alveolar-capillary membrane (35,36). Furthermore, neutrophil depletion has been reported to attenuate the degree of lung injury in some animal models of ARDS (37).

While neutrophils and their products have been implicated in the pathogenesis of ARDS, patients with profound neutropenias induced by cytotoxic chemotherapy and aplastic anemia have developed ARDS (38,39). The majority of these patients had peripheral neutrophil counts $<0.1 \times 10^9$/L, and the development of severe neutropenia preceded the development of ARDS. Histologic examination of the lungs at autopsy showed diffuse pulmonary damage in the absence of neutrophil sequestration. These observations suggest that neutrophils are not essential or required in the pathogenesis of the syndrome.

Interestingly, infection is the leading cause of death in patients who survive for more than 3 days from the onset of ARDS (40), and the lung is the primary source of infection in these patients. The occurrence of pulmonary infections in the presence of large numbers of intrapulmonary neutrophils suggests an impairment in the antimicrobial activity of neutrophils in patients with ARDS. Recent studies have shown several defects in the function of neutrophils isolated from humans and animals with ARDS or sepsis. Among the defects observed in human neutrophils include decreases in production of oxygen metabolites, chemotaxis, and microbicidal activity (41). Decreased phagocytosis has been reported and attributed to a loss of Fcγ receptors from neutrophils isolated from pigs with intraabdominal sepsis (42). When neutrophils from patients with septic shock were treated in vitro with Filgrastim, there was an increase in the number of FcγRI receptors and significant improvement of oxidative metabolism and degranulation (43). As G-CSF increases the functions of neutrophils, the administration of exogenous G-CSF may be able to enhance the antimicrobial functions of neutrophils and reduce infectious complications in these patients.

Filgrastim has been evaluated in several animal models of endotoxic shock, bacterial sepsis, and lung injury. In a porcine model of sublethal endotoxin challenge, recombinant bovine G-CSF (rbG-CSF) was studied to detect effects, if any, of rbG-CSF–induced neutrophilia on the pulmonary and hemodynamic responses induced by an intravenous (IV) endotoxin infusion (44). Pretreatment with rbG-CSF resulted in an approximate five-fold increase in the peripheral neutrophil count and resulted in an apparent increase in pulmonary neutrophil sequestration, as measured by myeloperoxidase activity in bronchoalveolar lavage fluid, after endotoxin challenge. However, pretreatment with rbG-CSF did not adversely effect the pulmonary or cardiovascular responses to endotoxin in this acute model of endotoxic shock.

In an ovine model of endotoxin-induced lung injury, pretreatment with rHuG-CSF, resulting in a fourfold increase in circulating neutrophils, did not exacerbate lung injury (45). The degree of pulmonary vascular permeability induced by the IV administration of endotoxin was actually reduced in sheep pretreated with rHuG-CSF. Similarly, in a model of acute endotoxin-induced lung injury in guinea pigs (46), animals pretreated with recombinant human G-CSF (lenograstim) had an approximate 2-fold increase in peripheral neutrophils before IV administration of endotoxin. The groups that received lenograstim plus endotoxin or endotoxin alone had increased infiltration of neutrophils into the lung tissue compared with saline controls. However, lenograstim pretreatment attenuated the effect of endotoxin on pulmonary edema and decreased the leakage of protein into the alveolar spaces. Interestingly, a group of animals made neutropenic by cyclophosphamide before challenge with endotoxin had decreased influx of neutrophils into the lung, as well as a decrease in pulmonary vascular permeability or edema compared with the placebo-treated animals.

Conflicting results were obtained in a model of intratracheal endotoxin-induced lung injury (47). Animals were divided into four groups: 1) saline control, 2) transtracheal endotoxin alone, 3) cyclophosphamide intraperitoneally 7 days before endotoxin challenge, and 4) cyclophosphamide at day -7 and lenograstim daily for 5 days before endotoxin. Group 3 animals were neutropenic with a mean absolute neutrophil count (ANC) of 0.75×10^9/L compared with 2.5×10^9/L in control animals before endotoxin challenge. Pretreatment with lenograstim in group 4 animals induced a substantial increase in peripheral neutrophils (mean ANC, 13×10^9/L). Compared with groups 1 and 2, group-4 animals demonstrated a marked accumulation of neutrophils in pulmonary tissues, a significant increase in pulmonary vascular permeability as measured by influx of ^{125}I-albumin, and a significant increase in pulmonary edema as measured by wet/dry lung weights. Although cyclophosphamide treatment of group 3 animals resulted in a significant decrease in intrapulmonary neutrophils in response to endotoxin, there was no attenuation of pulmonary edema or pulmonary vascular permeability in these neutropenic animals.

We have also investigated the effects of Filgrastim in a non-infectious model of pulmonary inflammation (48). In one model, acute lung injury was induced in rats by the systemic administration of α-naphthylthiourea (ANTU), an agent that causes an oxidant injury to endothelial cells. In another model, intrapulmonary hydrochloric acid was used to cause a pulmonary tissue injury. Animals treated with Filgrastim had a five- to eightfold increase in peripheral neutrophils at the time of injury. The combination treatments, ANTU plus Filgrastim or hydrochloric acid plus Filgrastim, resulted in a more severe lung injury manifested by increased wet/dry weight ratios and increased fluid permeability. Thus, in models of non-infectious lung injury, Filgrastim treatment increased lung leak.

The effects of recombinant canine G-CSF (rcG-CSF) on hematological, cardiovascular, and pulmonary parameters, as well as mortality, have been investigated in a canine model of septic shock (49). Tracheotomized and stabilized dogs were pretreated for 9 days with either 5 μg/kg/day rcG-CSF, 0.1 μg/kg/day rcG-CSF, or vehicle control. A fibrin clot containing an LD_{40-50} dose of *Escherichia coli* was implanted in the peritoneum and the animals were observed for 28 days. Neutrophilia was observed in the high-dose rcG-CSF–treated animals. There did not appear to be any deleterious effects of rcG-CSF on arterial blood gases, pulmonary function, hemodynamics, or cardiac function. Mortality was lower in animals treated with 5 μg/kg/day rcG-CSF (3/17) when compared with those treated with 0.1 μg/kg/day (7/17) or vehicle control (8/20) (p = 0.04). Blood and serum was sampled periodically during the course of the study for quantitative blood cultures, serum endotoxin levels, and TNF levels. During the first 6 hours after placement of the *E. coli*–infected clot, fewer viable bacteria were recovered from the blood of animals treated with high-dose rcG-CSF compared with the low-dose rcG-CSF and control animals (p < 0.04). The high-dose rcG-CSF group had significantly lower (p < 0.002) serum endotoxin levels compared with the low-dose and control groups during the first 8 hours of the infection. Reflecting the reduction in endotoxin levels, the high-dose group also had a reduction in peak serum TNF levels compared with control animals (p < 0.05). This effect of rcG-CSF on reducing endotoxin levels as well as reducing serum TNF levels has been observed in other models of bacteremia and septic shock (50,51).

Receptors and binding sites for endotoxin are present on the neutrophil cell membranes and within primary granules. Neutrophil-associated proteins capable of binding endotoxin include the bactericidal/permeability-increasing protein (BPI) (52,53) and CD-14 which recognizes the complex formed by the binding of LPS to the soluble lipopolysacharide-binding protein (LBP) (54). The cell surface expression of CD-14 on human neutrophils is increased by treatment of neutrophils with exogenous G-CSF. The increased rate of clearance of endotoxin observed in animals treated with Filgrastim may be mediated, in part, by the increased number of neutrophils expressing endotoxin binding sites as well as an increase in the number of sites on each cell.

In summary, in non-neutropenic animal models of bacterial infection, treatment with rG-CSF results in an increase in the number of neutrophils that migrate to the site of infection, an acceleration in clearance of bacteria, a reduction of serum endotoxin and TNF-α, and reductions in mortality. Similarly, in models of endotoxic shock, rG-CSF treatment has attenuated the effects of endotoxin and has not had an adverse effect on cardiopulmonary function. Only in models of non-infectious pulmonary injury has treatment of rG-CSF caused an exacerbation of the inflammatory process.

V. CLINICAL TRIALS OF FILGRASTIM IN PATIENTS WITH PNEUMONIA

To date, three separate studies have been performed with Filgrastim in hospitalized patients with pneumonia. Phase 1–2 studies have been conducted in patients with community-acquired pneumonia (CAP) and patients with pneumonia and sepsis. A large phase 3 randomized, double-blind, placebo-controlled study in patients hospitalized with pneumonia was recently completed. Summaries of these trials are included below.

A. Phase 1 Dose-Escalation Study in Community-Acquired Pneumonia

This was an open-label, non-controlled, sequential escalating-dose study designed to evaluate the safety of Filgrastim given at doses of 75 to 600 μg/day subcutaneously (SC) in patients with CAP (55). Patients (5 to 10 per group) were treated with Filgrastim for a maximum of 10 days in combination with antibiotic therapy for a minimum of 5 days. Filgrastim was discontinued if the ANC increased to $\geq$75 $\times$ 10^9/L. Results from this study showed that Filgrastim is safe in patients with CAP at the doses studied, as demonstrated by the lack of adverse events attributable to Filgrastim, notably ARDS, a deterioration of oxygenation (PaO$_2$/FiO$_2$), and shifts in WHO toxicity grades for laboratory values. Filgrastim produced an increase in ANC at doses $\geq$150 μg/day, and the median value of maximum ANC was 38 $\times$ 10^9/L (a 196% increase from baseline) across all dose groups. In all dose groups, pulmonary symptoms and findings improved or remained stable for the majority of patients. Clinical outcomes (duration of fever, antibiotic use, and hospitalization) appeared to be independent of dose, although the number of patients was small in each dose group. Consistant with the known safety profile. Thus, ANC more than three times the normal value were well tolerated in these infected patients.

B. Phase 3 Community-Acquired Pneumonia

This was a double-blind, placebo-controlled, multicenter study designed to evaluate the safety and efficacy of SC Filgrastim, 300 μg/day, combined with IV antibiotics on the time to resolution of morbidity (TRM), an index designed to assess normalization of temperature, respiratory rate, chest radiograph, and oxygenation (56). Patients were treated with the study drug (Filgrastim or placebo) for up to 10 days, and therapy was discontinued when the patient met TRM, when IV antibiotics were stopped, or when the white cell count reached 75×10^9/L.

In total, 756 patients were enrolled at 71 sites within the United States, Canada, and Australia. Overall, 191 Filgrastim-treated patients (50.3%) and 211 placebo-treated patients (56%) had a causative pathogen in the baseline sputum specimen, as determined by an independent infectious disease review board. The most frequently isolated pathogens were consistent with those usually seen in CAP patients, and included *Streptococcus pneumoniae, Hemophilus influenzae, Staphylococcus aureus,* and *Moraxella catarrhalis.* Bacteremia was confirmed in 52 (13.7%) Filgrastim-treated patients and 58 (15.4%) placebo-treated patients at baseline. *S. pneumoniae* was the pathogen most frequently isolated from blood cultures.

Filgrastim elevated the ANC threefold to a median of 37×10^9/L. The median length of therapy with the study drug was 4 days. **Figure 1** shows the ANC response for patients who received four doses of Filgrastim. After termination of Filgrastim therapy, ANC values returned to the normal range within 2 to 5 days.

The overall mortality was low, 6.1% for Filgrastim and 6.4% for placebo. In the 28-day treatment period, the incidence of severe adverse events, when categorized by body system, or preferred term were similar across treatment groups. A primary concern in initiating this trial was the safety of increasing blood neutrophils in patients with active lung inflammation. However, an important finding in this study was the remarkable safety profile of Filgrastim in these patients and specifically the lack of significant respiratory events. In fact, data from the study showed that patients who received Filgrastim had a lower incidence of ARDS and other sepsis-induced organ dysfunctions compared with patients who received placebo. The only treatment-related adverse event seen in this trial was the higher incidence of bone pain, which was generally mild to moderate in severity.

Results showed that TRM and length of hospital stay were not affected by the addition of Filgrastim to IV antibiotic therapy. However, intent-to-treat analyses showed a significant effect on radiologic resolution of pneumonia. Time to radiographic improvement or stability was achieved more rapidly in Filgrastim-treated patients (p = 0.03). In addition, by day 28, 30% of Filgrastim-treated patients had complete radiographic resolution of infiltrates compared with 22% for the placebo-treated patients (p = 0.005). The radiographic and clinical resolutions were associated with im-

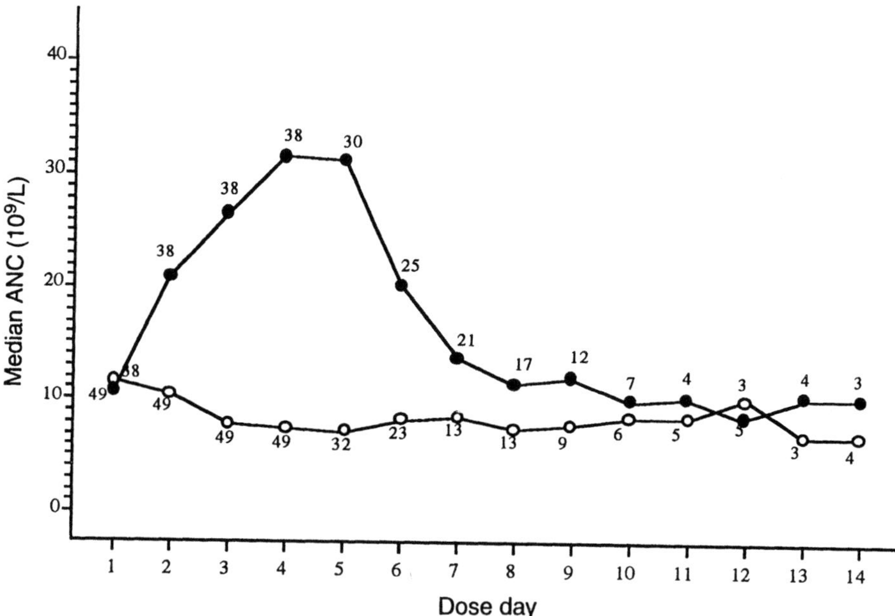

Figure 1. Absolute neutrophil count (ANC) response for patients who received four doses of Filgrastim. Open circles, control (n below symbol); closed circles, Filgrastim (n above symbol).

portant clinical benefits in the form of lower incidence of organ dysfunction, collected as adverse events in this trial. **Table 1** shows that Filgrastim-treated patients had a lower incidence of ARDS and disseminated intravascular coagulation (DIC). Overall, 32 patients receiving placebo developed either an end-organ dysfunction or empyema compared with 16 patients receiving Filgrastim. Similar trends toward lowered incidence of acute renal failure, septic shock, and empyema also were seen among patients treated with Filgrastim (**Table 1**).

Based on the benefit seen in the intent-to-treat analysis in the radiographic resolution of pneumonia in Filgrastim-treated patients, a post hoc analysis was performed in the large subgroup of patients with multilobar pneumonia. Of the 123 placebo-treated patients who had multilobar pneumonia, 21 (17.1%) developed organ dysfunction or empyema, and of the 138 Filgrastim-treated patients who had multilobar pneumonia, 8 (5.8%) developed organ dysfunction or empyema. These data suggest that the

Table 1.   Incidence of Systemic and Local Complications.

	Intent-to-treat analysis		
	Placebo (n = 376)	Filgrastim (n = 380)	p-Value
End-organ Failure (EOD)			
Adult respiratory distress syndrome	14	4	0.017
Disseminated intravascular coagulation	7	0	0.007
Acute renal failure	9	4	0.153
Septic shock	19	10	0.080
Empyema	6	1	0.068
Patients with $\geq$1 EOD or empyema	32	16	0.015

patients who had the most benefit in this trial were those who had extensive disease as determined by multilobar pneumonia at baseline.

C. Phase 2 Pneumonia with Severe Sepsis or Septic Shock

A double-blind, randomized, placebo-controlled, multicenter study was designed to evaluate the safety and efficacy of Filgrastim, 300 μg IV daily for 5 days, combined with standard therapy, in patients with pneumonia and severe sepsis or septic shock. Hypotension requiring vasopressors or two or more end-organ dysfunctions were used as the definition of sepsis for all patients. Dosing with Filgrastim or placebo was started within 24 hours of diagnosis of sepsis. Mortality (days 15 and 29) as well as incidence and onset (through day 8) and resolution (through day 29) of end-organ dysfunction were assessed in this study. Nineteen patients were enrolled; one was excluded because of death after randomization but before drug infusion. Twelve patients received Filgrastim and six received placebo. The mean age of all patients was 54 years, and the mean APACHE II scores at study entry were not significantly different between groups (Filgrastim, 26.3 $\pm$ 6.6; placebo, 28.3 $\pm$ 14.3).

All patients were admitted to the ICU. The median time to discharge from the ICU was >29 days in the placebo group and 19 days in the Filgrastim group. There were no serious attributable adverse events in patients receiving Filgrastim. In this small study, the use of Filgrastim was associated with reduced mortality and more Filgrastim-treated patients resolved their organ dysfunction by day 29, compared with placebo-treated patients (**Tables 2 and 3**).

Table 2. Incidence and (Resolution) of Organ Dysfunction by Day 29 in Patients with Severe Sepsis or Septic Shock.

Organ dysfunction	Filgrastim (n = 12)	Placebo (n = 6)
Shock	10 (9)	4 (0)
ARDS	5 (2)	4 (1)
Acute renal failure	0	3 (0)
DIC	1 (1)	1 (0)

ARDS = adult respiratory distress syndrome.
DIC = disseminated intravascular coagulation.

Table 3. Mortality in a Phase 2 Study of Patients with Severe Sepsis or Septic Shock.

	Filgrastim (n = 12)	Placebo (n = 6)	p-Value
Day 15	2/12 (17%)	4/6 (67%)	0.042
Day 29	3/12 (25%)	4/6 (67%)	0.108

V. FUTURE DIRECTIONS

Based on the results of preclinical investigations and human studies, two new clinical trials in patients with pneumonia have been undertaken. In the first study, the effect of Filgrastim will be evaluated on the reduction of mortally in hospitalized patients with bacterial pneumonia and severe sepsis. A total of 700 patients are expected to be enrolled, and the primary endpoint is 29-day all-cause mortality. Patients will receive 300 μg of Filgrastim or placebo for up to 5 days. Study-related observations consisting of vital signs, laboratory determinations, adverse events, and onset of organ dysfunction will continue through day 10, and mortality, length of ICU stay, and duration of mechanical ventilation will be monitored through day 29. This study is being conducted in centers throughout the United States, Canada, Australia, and several European countries.

The second clinical study will assess the effect of Filgrastim in hospitalized patients with multilobar pneumonia. The primary endpoint of this study is "treatment failure," defined as the development of organ dysfunction, empyema, or death from any cause. A total of 480 patients are expected to be enrolled. Patients will receive 300 μg of Filgrastim or placebo for up to 10 days. All study-related observations will be continued through day 29. The study is being conducted at centers throughout the United States, Canada, and Australia.

Results of recent trials of biologic response modifiers in patients with severe infections have not met with clinical success, and in some cases have even had results that suggested an adverse effect. The basic premise of these studies was that "rogue inflammation" was directly responsible for patient morbidity and mortality. We propose, and recent data support this hypothesis, that immune amplification with Filgrastim is a more rational treatment strategy in these immunocompromised, infected patients. The clinical trials currently underway with Filgrastim should provide more definitive evidence for this hypothesis.

REFERENCES

1. Athens, J. W., Haab, O. P., Raab, S. O., et al (1961). Leukokinetic studies. IV. The total blood, circulating, and marginal granulocyte pools and the granulocyte turnover rate in normal subjects. *J Clin Invest 40:*989–995.

2. Weisbart, R. H., Gasson, J. C., and Golde, D. W. (1989). Colony-stimulating factors and host defense. *Ann Intern Med 110:*297–303.

3. Souza, L. M., Boone, T. C., Gabrilove, J., et al (1986). Recombinant human granulocyte colony stimulating factor: effects on normal and leukemic myeloid cells. *Science 232:*61–65.

4. Cebon, J., Layton, J. E., Maher, D., and Morstyn, G. (1994). Endogenous haemopoietic growth factors in neutropenia and infection. *Br J Haematol 86:*265–274.

5. Pauksen K., Elfman, L., Ulfgren, A.-K., and Venge, P. (1994). Serum levels of granulocyte-colony stimulating factor (G-CSF) in bacterial and viral infections, and in atypical pneumonia. *Br J Haematol 88:*256–260.

6. Centers for Disease Control (1994). Pneumonia and influenza death rates— United States, 1979–1994. *MMWR 44:*535–537.

7. Niederman, M. S., Bass, J. B., Jr., Campbell, G. D., et al (1993). Guidelines for the initial management of adults with community-acquired pneumonia: diagnosis, assessment of severity, and initial antimicrobial therapy. *Am Rev Respir Dis 148:*1418–1426.

8. Campbell, G. Jr., Niederman, M., Broughton, X., et al (1996). Hospital-acquired pneumonia in adults: diagnosis, assessment of severity, initial antimicrobial therapy, and preventative strategies. *Am J Respir Crit Care Med 153:*1711–1725.

9. Sibille, Y., and Reynolds, H. Y. (1990). Macrophages and polymorphonuclear neutrophils in lung defense and injury. *Am Rev Respir Dis 141:*471–501.

10. Koeffler, H. P., Gasson, J., Ranyard, J., Souza, L., Shepard, M., and Munker, R. (1987). Recombinant human TNFα stimulates production of granulocyte colony-stimulating factor. *Blood 70*:55–59.

11. Tazi, A., Nioche, S., Chastre, J., Smiejan, J.-M., and Hance, A. J. (1991). Spontaneous release of granulocyte colony stimulating factor (G-CSF) by alveolar macrophages in the course of bacterial pneumonia and sarcoidosis: Endotoxin-dependent and entotoxin-independent G-CSF release by cells recovered by bronchoalveolar lavage. *Am J Respir Cell Mol Biol 4*:140–147.

12. Toews, G. B., Gross, G. N., and Pierce, A. K. (1979). The relationship of incoculum size to lung bacterial clearance and phagocytic cell response in mice. *Am Rev Respir Dis 120*:559–566.

13. Yasuda, H., Ajiki, Y., Shimozato, T., et al (1990). Therapeutic efficacy of granulocyte colony-stimulating factor alone and in combination with antibiotics against *Pseudomonas aeruginosa* infections in mice. *Infect Immun 58*:2502–2509.

14. Nelson, S., Summer, W., Bagby, G., et al (1991). Granulocyte colony-stimulating factor enhances pulmonary host defenses in normal and ethanol-treated rats. *J Infect Dis 164*:901–906.

15. MacGregor, R. R. (1986). Alcohol and immune defense. *JAMA 256*:1474–1479.

16. MacGregor, R. R., Safford, M., and Shalit, M. (1988). Effect of ethanol on functions required for the delivery of neutrophils to sites of inflammation. *J Infect Dis 157*:682–689.

17. Gluckman, S. J., and MacGregor, R. R. (1978). Effect of acute alcohol intoxication on granulocyte mobilization and kinetics. *Blood 52*:551–559.

18. Astry, C. L., Estry, G. A., and Jakab, G. J. (1983). Impairment of polymorphonuclear leukocyte immigration as a mechanism of alcohol-induced suppression of pulmonary antibacterial defenses. *Am Rev Respir Dis 128*:113–117.

19. Lister, P. D., Gentry, M. J., and Preheim, L. C. (1993). Granulocyte colony-stimulating factor protects control rats but not ethanol-fed rats from fatal pneumococcal pneumonia. *J Infect Dis 168*:922–226.

20. Loosli, C. G. (1973). Influenza and the interaction of viruses and bacteria in respiratory infections. *Medicine 52*:369–384.

21. Douglas, R. G., Jr. (1990). Prophylaxis and treatment of influenza. *N Engl J Med 322*:443–450.

22. Abramson, J. S. and Mills, E. L. (1988). Depression of neutrophil function induced by viruses and its role in secondary microbial infections. *Rev Infect Dis 10*:326–341.

23. Abramson, J. S., Giebink, G. S., Mills, E. L., and Quie, P. G. (1981). Polymorphonuclear leukocyte dysfunction during influenza virus infection in chinchillas. *J Infect Dis 143*:836–845.

24. Abramson, J. S., Parce, J. W., Lewis, J. C., et al (1984). Characterization of the effect of influenza virus on polymorphonuclear leukocyte membrane response. *Blood 64*:131–138.

25. Abramson, J. S., Gieink, G. S., and Quie, P. G. (1982). Influenza A virus-induced polymorphonuclear leukocyte dysfunction in the pathogenesis of experimental pneumococcal otitis media. *Infect Immun 36*:289–296.

26. Abramson, J. S., and Hudnor, H. R. (1994). Effect of priming polymorphonuclear leukocytes with cytokines (granulocyte-macrophage colony-stimulating factor [GM-CSF] and G-CSF) on the host resistance to *Streptococcus pneumoniae* in chinchillas infected with influenza A virus. *Blood 83*:1929–1934.

27. Gopal, V., and Bisno, A. L. (1977). Fulminant pneumococcal infections in "normal" asplenic hosts. *Arch Intern Med 137*:1526–1530.

28. Dickerman, J. D. (1976). Bacterial infection and the asplenic host: a review. *J Trauma 16*:662–668.

29. Hebert, J. C., O'Reilly, M., and Gamelli, R. L. (1990). Protective effect of recombinant human granulocyte colony stimulating factor against pneumococcal infections in spenectomized mice. *Arch Surg 125*:1075–1078.

30. Hebert, J. C., O'Reilly, M., Yuenger, K., Shatney, L., Yoder, D. W., and Barry, B. (1994). Augmentation of alveolar macrophage phagocytic activity by granulocyte colony stimulating factor and interleukin-1: influence of splenectomy. *J Trauma 37*:909–911.

31. Smith, W. S., Sumnicht, G. E., Sharpe, R. W., Samuelson, D., and Millard, F. E. (1995). Granulocyte colony-stimulating factor versus placebo in addition to penicillin G in a randomized blinded study of gram-negative sepsis: analysis of survival and multisystem organ failure. *Blood 86*:1301–1309.

32. McGuire, W. W., Spragg, R. C., and Cohen, A. B. (1982). Studies on the pathogenesis of the adult respiratory distress syndrome. *J Clin Invest 69*:543–553.

33. Idell, S., Kucich, U., Fein, A., et al (1985). Neutrophil elastase-releasing factors in bronchoalveolar lavage from patients with adult respiratory distress syndrome. *Am Rev Respir Dis 132*:1098–1105.

34. Weiland, J. E., Davis, W. B., Holter, J. F., Mohammed, J. R., Dorinsky, P. M., and Gadek, J. E. (1986). Lung neutrophils in the adult respiratory distress syndrome. Clinical and pathophysiologic significance. *Am Rev Respir Dis 133*:218–225.

35. Weiss, S. J., Young, J., LoBuglio, A. F., Slivka, A., and Nimeh, N. H. (1981). Role of hydrogen peroxide in neutrophil-mediated destruction of cultured endothelial cells. *J Clin Invest 68*:714–721.

36. Varani, J., Fligiel, S. E., Till, G. O., Kunkel, R. G., Ryan, U. S., and Ward, P. A. (1985). Pulmonary endothelial cell killing by human neutrophils. *Lab Invest 53*:656–663.

37. Heflin, A. C., Jr., and Brigham, K. L. (1981). Prevention by granulocyte depletion of increased vascular permeability of sheep lung following endotoxemia. *J Clin Invest 68*:1253–1260.

38. Maunder, R. J., Hackman, R. C., Riff, E., Albert, R. K., and Springmeyer, S. C. (1986). Occurrence of the adult respiratory distress syndrome in neutropenic patients. *Am Rev Respir Dis 133*:313–316.

39. Ognibene, F. P., Martin, S. E., Parker, M. M., et al (1986). Adult respiratory distress syndrome in patients with severe neutropenia. *N Engl J Med 315*:547–551.

40. Montgomery, A. B., Stager, M. A., Carrico, C. J., and Hudson, L. D. (1985). Causes of mortality in patients with the adult respiratory distress syndrome. *Am Rev Respir Dis 132*:485–489.

41. Martin, T. R., Pistorese, B. P., Hudson, L. D., and Maunder, R. J. (1991). The function of lung and blood neutrophils in patients with adult respiratory distress syndrome. *Am Rev Respir Dis 144*:254–262.

42. Simms, H. H., D'Amico, R., Monfils, P., and Burchard, K. W. (1991). Altered polymorphonuclear leukocyte FcγR expression contributes to decreased candicidal activity during intraabdominal sepsis. *J Lab Clin Med 117*:241–249.

43. Simms, H. H., and D'Amico, R. (1994). Granulocyte colony-stimulating factor reverses septic shock-induced polymorphonuclear leukocyte dysfunction. *Surgery 115*:85–93.

44. Fink, M. P., O'Sullivan, B. P., Menconi, M. J., et al (1993). Effect of granulocyte colony-stimulating factor on systemic and pulmonary responses to endotoxin in pigs. *J Trauma 34*:571–578.

45. Koizumi, T., Kubo, K., Shinozaki, S., Koyama, S., Kobayashi, T., and Sekiguchi, M. (1993). Granulocyte colony-stimulating factor does not exacerbate endotoxin-induced lung injury in sheep. *Am Rev Respir Dis 148*:132–137.

46. Kanazawa, M., Ishizaka, A., Hasegawa, N., Suzuki, Y., and Yokoyama, T. (1992). Granulocyte colony-stimulating factor does not enhance endotoxin-induced acute lung injury in guinea pigs. *Am Rev Respir Dis 145*:1030–1035.

47. Terashima, T., Kanasawa, M., Sayama, K., et al (1993). Granulocyte colony-stimulating factor exacerbates acute lung injury induced by intratracheal endotoxin in guinea pigs. *Am J Respir Crit Care Med 149*:1295–1303.

48. King, J., DeBoisblanc, B. P., Mason, C. M., et al (1995). Effect of granulocyte colony-stimulating factor on acute lung injury in the rat. *Am J Respir Crit Care Med 151*:302–309.

49. Eichacker, P. Q., Waisman, Y., Natanson, C., et al (1994). Cardiopulmonary effects of granulocyte colony-stimulating factor in a canine model of bacterial sepsis. *J Appl Physiol 77*:2366–2373.

50. Haberstroh, J., Breuer, H., Lucke, I., et al (1995). Effect of recombinant human granulocyte colony-stimulating factor on hemodynamic and cytokine response in a porcine model of *Pseudomonas* sepsis. *Shock 4*:216–224.

51. Lundblad, R., Nesland, J. M., and Giercksky, K.-E., (1996). Granulocyte colony-stimulating factor improves survival rate and reduces concentrations of bacteria, endotoxin, tumor necrosis factor, and endothelin-1 in fulminant intraabdominal sepsis in rats. *Crit Care Med 24*:820–826.

52. Marra, M. N., Wilde, C. G., Collins, M. S., Snable, J. L., Thornton, M. B., and Scott, R. W. (1992). The role of bactericidal/permeability-increasing protein as a natural inhibitor of bacterial endotoxin. *J Immunol 148*:532–537.

53. Weersink, A. J., van Kessel, K. P., van den Tol, M. E., et al (1993). Human granulocytes express a 55-kDa lipopolysaccharide-binding protein on the cell surface that is identical to the bactericidal/permeability-increasing protein. *J Immunol 150*:253–263.

54. Wright, S. D., Ramos, R. A., Hermanowski-Vosatka, X., Rockwell, P., and Detmers, P. A. (1991). Activation of the adhesive capacity of CR3 on neutrophils by endotoxin: dependence on lipopolysaccharide binding protein and CD14. *J Exp Med 173*:1281–1286.

55. deBoisblanc, B. P., Mason, C. M., Andresen, J., et al (1997). Phase 1 safety trial of Filgrastim (r-metHuG-CSF) in non-neutropenic patients with severe community-acquired pneumonia. *Respir Med 91*:387–394.
56. Nelson, S., Farkas, S., Fothringham, N., Ho, S., Marrie, T., and Movahhed, H. (1996). Filgrastim in the treatment of hospitalized patients with community acquired pneumonia (CAP). *Am J Respir Crit Care Med 153S*:A535.

21
Use of Filgrastim (r-metHuG-CSF) in Human Immunodeficiency Virus Infection

David L. Pitrak
University of Illinois College of Medicine at Chicago, Chicago, Illinois

I. INTRODUCTION

The acquired immune deficiency syndrome (AIDS) is characterized by infections with a variety of opportunistic pathogens because of impaired cell-mediated immunity that occur with progressive depletion and dysfunction of CD4$^+$ lymphocytes. It is now well recognized, however, that there are many immune abnormalities in human immunodeficiency virus (HIV) infection, including impaired humoral immunity, neutropenia, and neutrophil (polymorphonuclear leukocyte, PMN) dysfunction (1). This wide array of immune defects helps explain the predisposition to infections with common pyogenic bacterial pathogens that also occur as a consequence of HIV infection (2,3). Recurrent bacterial infections are now recognized as criteria for the diagnosis of AIDS (4). The single most common cause of febrile illness resulting in hospitalization of HIV-infected patients is bacterial infection (5). Bacterial infections also account for an increasing proportion of the morbidity and mortality caused by HIV infection (6).

Infections with certain pathogens, *Streptococcus pneumoniae,* and *Haemophilus influenzae* are usually associated with antibody deficiency or defects in the complement system, and there is evidence that these arms of the immune system may be impaired in HIV infection (7–9). There is evidence for activation of the complement system in HIV infection, although there are little data on how this may effect opsonization of bacteria (10). On the other hand, the abnormalities of the humoral immune system related to B-cell function have been well characterized (7–9,11,12). Although there is hypergammaglobulinemia, increased spontaneous B-cell proliferation, and increased numbers of circulating B cells, there is little capacity to respond to a second stimulus in many patients, especially those with advance HIV disease. The unresponsiveness can be demonstrated in vitro as decreased B-cell proliferation in response to mitogens and low antibody responses to specific antigens and in vivo by a poor antibody responses to vaccination. Clearly humoral immunity is important in the host defense against pyogenic bacteria, but this alone cannot entirely explain the high risk for development of secondary bacterial infections. The use of intravenous (IV) immunoglobulin for the prevention of bacterial infections has met with limited success. Although there is evidence from uncontrolled trials that IV immune globulin (IVIG) can decrease the number of bacterial infections in patients with a prior history of recurrent serious infections such as pneumonia and septicemia, there is little evidence to support the use of IVIG routinely in HIV-infected patients (13–15). In a large study of the efficacy of IVIG in children with HIV infection, IVIG was associated with a modest decrease in serious bacterial infections and days of hospitalization in those children with $CD4^+$ >200/μL; there was no benefit in children with $CD4^+$ <200/μL, the group at greatest risk for serious bacterial infections (14). Clearly deficient antibody production is not the only process resulting in an increased risk of bacterial infections in these patients. Antibody replacement augmenting opsonophagocytosis may not be a benefit if the function of the phagocytic cells is impaired.

Neutropenia occurs frequently in HIV infection and is associated with a significant risk for bacterial infection (16). In addition, there are infections with certain pathogens in this population that are classically associated with defects in the number of function of PMN, even in patients without absolute neutropenia. For example, there has been an increase in the frequency of serious infections with *Pseudomonas aeruginosa* in patients with advanced HIV infection, often in patents without other recognized risk factors (17). Polymorphonuclear leukocytes are important in host defense against *Pseudomonas,* and the occurrence of infections with this pathogen underscores the significance of neutropenia and impaired PMN function in this population and PMN are also important effector cells against some of the opportunistic pathogens that infect HIV patients, such as *Aspergillus* and *Candida.* Neutrophils also can effectively kill HIV-infected cells through antibody-dependent cell-mediated cytotoxicity (ADCC), as well as free HIV through the release of toxic oxygen species, although the chemical significance of the contribution of neutrophils to host defense against HIV is unknown (18–20).

II. HIV INFECTION–RELATED NEUTROPENIA

Patients infected with HIV often have hematologic problems, including anemia, thrombocytopenia, and granulocytopenia (16,21), and are often neutropenic, although severe neutropenia with absolute neutrophil counts (ANC) <0.50 × 10^9/L is uncommon as a consequence of HIV infection itself. Absolute neutropenia most frequently results from myelosuppressive therapy for secondary infectious or neoplastic HIV-related complications (16,22). Antiretroviral therapy with zidovudine (ZDV) is a major cause of neutropenia, and other antiretrovirals also have been implicated, although the risk is much lower relative to ZDV. Trimethoprium/sulfmethoxazole therapy for *Pneumocystis carinii* pneumonia (PCP) and to a lesser extent for PCP propylaxis can cause bone marrow suppression. Bone marrow infiltration by neoplasms and opportunistic pathogens, such as *Mycobacterium avium* complex (MAC), *Myobacterium tuberculosis* (MTb), cytomegalovirus (CMV), and fungi also can cause cytopenias. In one study of HIV-related bacteremias, 18.5% of neutropenias were related to HIV infection itself, 24.6% due to marrow infiltration by opportunistic pathogens or lymphoma, and 56.7% from myelosuppressive medications (23).

The pathophysiology of neutropenia in HIV infection unrelated to drug therapy or secondary complications is not completely understood. A subset of neutropenic HIV-infected patients may have autoimmune neutropenia; however, this is probably accounts for a small proportion of cases of neutropenia (24,25). A recent study of neutropenic children with HIV infection showed that while many children had circulating antineutrophil antibodies, the presence of these antibodies did not correlate with ANC (26). Myeloid precursors can be infected by HIV, and this may affect normal proliferation and development of PMN, but mature cells are not targets and express very little, if any, HIV (27). Infection of bone marrow-stromal cells may affect the microenvironment for myelopoiesis. Both HIV and CMV can infect these bone marrow-stromal cells and impair production of colony-stimulating factors such as granulocyte colony-stimulating factor (G-CSF) and granulocyte-macrophage colony-stimulating factor (GM-CSF) (28,29).

Granulocyte colony-stimulating factor is a key growth factor important in maintaining the normal number of circulating PMN. Although several growth factors can stimulate myelopoiesis, studies of the G-CSF–deficient knock-out mice indicate G-CSF is necessary to maintain normal neutrophil counts (30). There is evidence accumulating that HIV infection results in a state of endogenous G-CSF deficiency (31,32). Patients with HIV have lower serum levels of G-CSF for the degree of neutropenia compared with patients with neutropenia from other etiologies, such as immunosuppressive therapy. The source of G-CSF is activated T cells, monocytes, and macrophages, all target cells for HIV. Depletion and dysfunction of these cells as a result of HIV infection may result in decreased production of G-CSF to support normal granulopoiesis and/or neutrophilic responses to infection. We recently reported that G-CSF production by peripheral blood mononuclear cells (PBMC) isolated from patients with AIDS produce significantly less G-CSF in response to a challenge with lipopolysaccharide (LPS) (33). Local G-CSF production by bone marrow-stromal cells is likely to be very important in granulopoiesis, and there is a decreased G-CSF response by bone marrow-stromal cells from patients with HIV in culture in response to interleukin (IL)-1 and LPS (28).

The low levels of G-CSF may not only impair myelopoiesis and affect neutrophil maturation, but also reduce the circulating half-life of neutrophils. We have shown that neutrophils isolated from patients with HIV infection and advanced disease exhibit accelerated apoptosis ex vivo (34). There was also a significant decrease in viability after 18 hours in culture due an increase in the proportion of non-viable, apoptotic cells. This decrease in neutrophil survival time may, in part, account for the relative neutropenia so common in patients with AIDS. There are a number of possible mechanisms for accelerated apoptosis, including increased levels of pro-inflammatory mediators that may accelerate the process. Alternatively, growth factor deficiency could result in accelerated apoptosis. The principle way growth factors cause clonal expansion of different bone marrow precursors is by inhibiting apoptosis of these cells (35). Our data show that accelerated neutrophil apoptosis in HIV infection is reversible in vitro and ex vivo in patients receiving Filgrastim (33,34).

It must be noted, however, that Filgrastim has not been licensed for the treatment of AIDS-associated neutropenia in a majority of countries. Australia, Japan, and India have licensed Filgrastim use for this indication.

III. IMPAIRED NEUTROPHIL FUNCTION IN HIV INFECTION

A number of viral infections are associated with impaired neutrophil function (36). The function of PMN is impaired in patients with HIV infection, including defects in chemotaxis, phagocytosis, the respiratory burst, and microbicidal capacity (37–44). The results of different studies are at times conflicting, and can be explained by different methods and differences in the patient populations studied. The most important clinical variable is the stage of HIV disease as evidenced by absolute CD4[+] lymphocyte count. A number of investigators have shown that PMN are activated in vivo throughout the course of HIV infection, even in the absence of any secondary infectious complications (45,46). Early in the course of HIV infection this in vivo activation or priming may actually result in enhanced function. The in vivo activation continues throughout the course of HIV infection, but eventually the functional capacity of the PMN begins to decrease significantly. A similar mechanism has been proposed for impaired PMN function in patients with chronic hepatitis B infection (47).

We examined PMN chemiluminescence in a cohort of 78 patients with HIV infection at different stages of disease (45); those for HIV-infected patients had altered oxidative metabolism in response to opsonin receptor–dependent stimulation with zymosan opsonized with purified human complement. Patients with early HIV infection with $CD4^+$ counts >500/μL exhibited increased PMN chemiluminescence in response to opsonized zymosan compared with controls, while patients with advanced disease with low $CD4^+$ counts show significantly decreased chemiluminescence. Maximum opsonin receptor expression (MOR) is achieved by exposing PMN to quantities of proinflammatory mediators (primers) to induce a maximal number of CD11b (C3bi receptor) and CD35 (C3b receptor, complement receptor 1) on the PMN surface. Priming also may enhance PMN oxidative responses to a second stimulus by increasing the increasing affinity of opsonin receptors for their particular ligand, or enhancing intracellular signaling. This may be the result of chronic in vivo activation and metabolic exhaustion or "burnout," as PMN from HIV-infected patients at all stages of disease show increases in ratio of the chemiluminescence for unprimed PMN at circulating opsonin receptor expression (COR), to chemiluminescence for cells with maximum opsonin receptor expression (MOR), ie, an increased COR/MOR ratio. By whatever mechanism, PMN from HIV-infected patients behave as if they have been primed or activated by proinflammatory mediators in vivo. It is interesting that this activation also is seen in high-risk control subjects, ie, subjects seronegative despite risk factors for HIV infection, but these high-risk control subjects show enhancement rather than depression of MOR chemiluminescence activity.

Opsonin receptor–independent NADPH-oxidase and myeloperoxidase activities, basal and stimulated, were significantly increased for HIV-infected patients, especially those with advanced disease with $CD4^+$ <200/μL. The increase in enzyme activities of oxidase and myeloperoxidase that this represents may be the result of in vivo activation. The decrease in myeloperoxidase activity for patients with very advanced HIV infection with $CD4^+$ <100/μL, although still significantly higher than values seen in control subjects without HIV infection may be due to degranulation resulting from in vivo activation. The high-risk control patients, exhibiting increased COR/MOR ratios, did not show increases in stimulated oxidase or myleperoxidase activities.

Absolute CD4$^+$ lymphocyte count was the only patient variable that significantly correlated with opsonin-dependent PMN chemiluminescence activity according to multiple regression analysis. Despite a good correlation between ANC and CD4$^+$ count (r = 0.24, p = 0.04), ANC was not an independent predictor of impaired PMN chemiluminescence on multiple regression analysis.

In vivo activation is not the only process that contributes to impaired neutrophil function in HIV infection. As stated earlier, neutrophils from patients with advanced HIV disease undergo accelerated apoptosis or programmed cell death (34). Apoptosis of neutrophils isolated from 10 patients with advanced HIV infection and 7 control subjects was examined morphologically by fluorescent microscopy after dual staining with acridine orange and ethidium bromide, fluorescent stains that intercalate DNA. Acridine orange stains the nucleus bright green and allows visualization of the nuclear chromatin pattern, while ethidium bromide identifies non-viable cells by staining the nucleus orange. There was little apoptosis identified immediately after isolation, but over time apoptosis was observed and the proportion of apoptotic cells was significantly higher for the patients with AIDS at 3, 6, and 18 hours (**Figure 1**). Apoptotic cells eventually died, and by 18 hours there was a significant decrease in viability for the patient's neutrophils due to an increased number of non-viable, apoptotic cells. As neutrophils become apoptotic, they become functionally impaired, and the accelerated apoptosis of neutrophils ex vivo may in part explain the impairment of function seen in this patient population.

IV. RISK OF SERIOUS BACTERIAL INFECTIONS DUE TO NEUTROPENIA AND NEUTROPHIL DYSFUNCTION

Early, there was debate if HIV-related neutropenia was a significant risk factor for the development of bacterial infections and if clinicians should intervene when patients developed neutropenia (48). Several studies have now shown that neutropenia is clearly associated with a significant risk of bacterial infections (49–51). Yu and Mitsuyasu (52) reported on 48 HIV-infected patients with ANC $<0.75 \times 10^9$/L seen between 1987 and 1991. There was a significant risk for bacterial and fungal infections for the patients with ANC $<0.5 \times 10^9$/L compared with those with ANC 0.5 to 0.75×10^9/L. When patients receiving antibiotics for PCP prophylaxis and antifungal therapy were analyzed separately, there was still a significantly higher incidence of bacteremia and systemic fungal infections in the group with ANC $<0.5 \times 10^9$/L. This study demonstrates the clinical significance

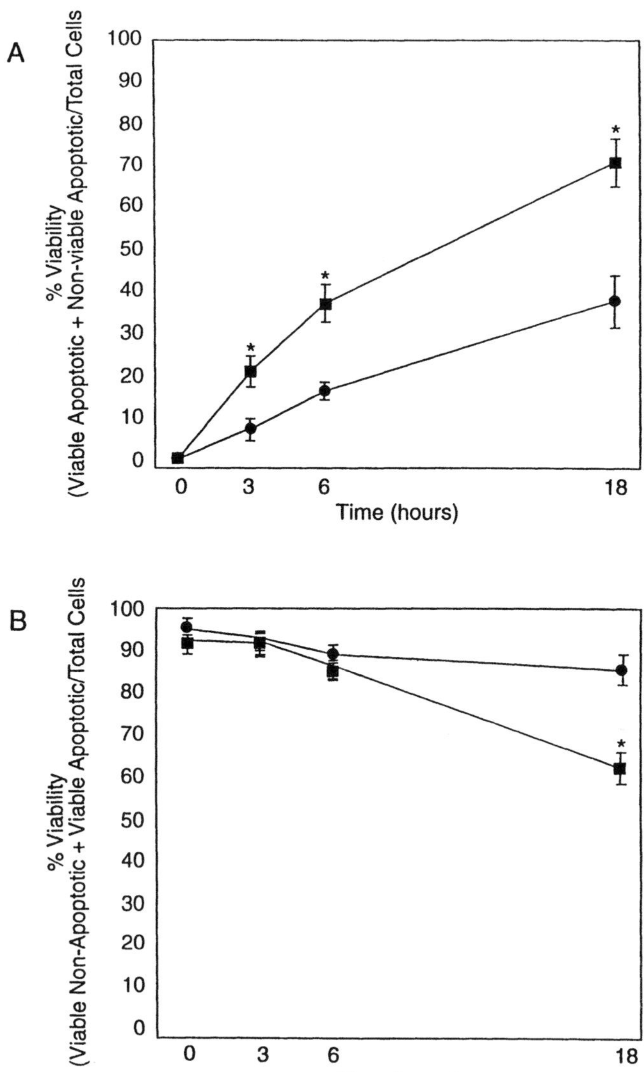

Figure 1. Apoptosis and viability in patients with AIDS compared with normal control subjects. Solid squares, control subjects (n = 7); solid circles, HIV + patients (n = 10); asterisk, p < 0.05, Student's *t*-test; bars show ± standard error of the mean. (Adapted from Ref. 34.)

of absolute neutropenia, but did not define whether or not the risk of bacterial infections occurs with less severe degrees of neutropenia.

Moore and co-workers (50) reported on the risk of bacterial infections in a cohort of neutropenic HIV-infected patients. The incidence of bacterial infections for 118 neutropenic patients with ANC $<1.0 \times 10^9/L$ was compared with that of a control group of non-neutropenic patients matched for absolute $CD4^+$ lymphocyte count, injection drug use, and duration of follow-up for HIV infection. The neutropenic and non-neutropenic groups were similar also with respect to nutritional status. Serious bacterial infections, including pneumonia, endocarditis, bacteremia, and enterocolitis, occurred more frequently in the neutropenic patients. Although the neutropenic patients were more likely to have indwelling IV catheters and be receiving therapy with the myelosuppressive agents ZDV and trimethoprim/sulfamethoxazole, adjusted relative risks for bacterial infection were significantly increased for neutropenic patients. The adjusted relative risk for bacterial infection was 2.33 (95% CI, 1.00 to 5.40) for patients with ANC $<1.0 \times 10^9/L$ and 7.92 (95% CI, 1.18 to 53.2) for ANC $<0.5 \times 10^9/L$. This demonstrates the importance of neutropenia as an independent risk factor for bacterial infections in the HIV-infected population, including patients with ANC between 0.5 and $1.0 \times 10^9/L$.

Jacobson and co-workers (51) reported on the risk of bacterial infections in a very large cohort, 2047 patients followed at San Francisco General Hospital between 1992 and 1993. All patients were stratified according to ANC into nine groups, and data on hospitalizations for serious bacterial infections were collected. The association between neutropenia and other clinical parameters and bacterial infections was determined by step-wise logistic regression analysis. The severity of neutropenia was significantly (p = 0.002) associated with serious bacterial infections, with an increased risk for patients with ANC $<1.0 \times {}^9/L$.

Since 1994 there have been seven studies published showing the association between HIV-related neutropenia and serious bacterial infections (**Figure 2**). The risk for bacterial infections for different levels of neutropenia were consistent from study to study, and comparable to the risk of infection to that seen for non-HIV–infected patients who are neutropenic from chemotherapy.

V. FILGRASTIM FOR HIV-RELATED NEUTROPENIA

Studies examining the efficacy of Filgrastim in the management of HIV-related neutropenia have reported the experience with <500 patients (53). In all studies, Filgrastim has been highly effective in reversing neutropenia

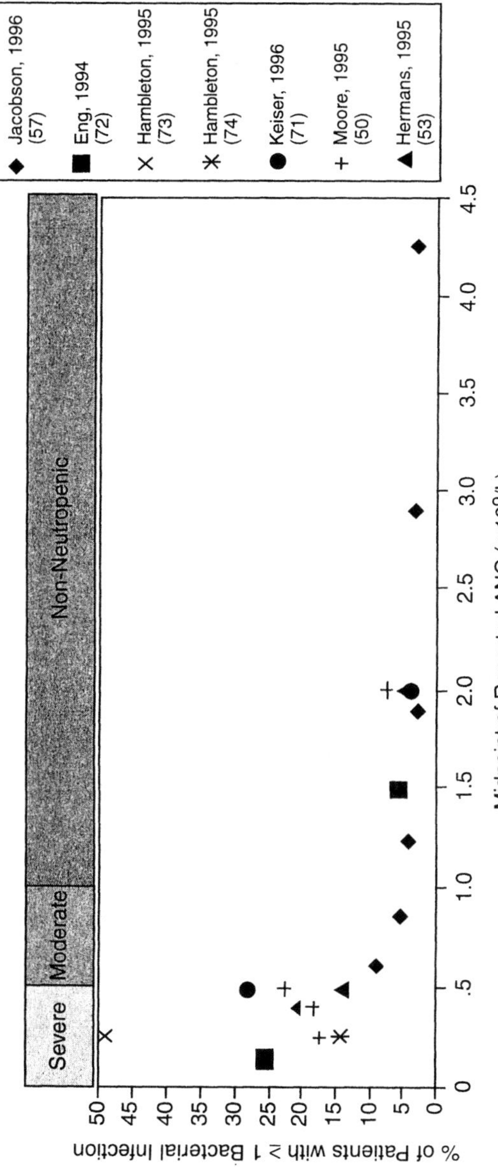

Figure 2. Consequence of neutropenia: relationship between absolute neutrophil count (ANC) and bacterial infection. Hambleton paper referenced two populations: ✕, all other infections; ✶, all bacterial infections.

and maintaining normal neutrophil counts with continuation of therapy. In the study by Hermans et al, 98% of patients with ANC $<1.0 \times 10^9$/L increased their ANC to $>2.0 \times 10^9$/L after an initial course of daily induction Filgrastim, with a median time to response of 2 days. Thereafter, ANC was maintained with dosing of 300 μg one to three times per week. Side effects in this population were minimal, and there was no evidence of bone marrow exhaustion.

Jacobson et al (54) have shown that therapy with Filgrastim one to three times a week decreases the incidence of neutropenia in patents receiving ganciclovir for CMV retinitis. This included patients receiving multiple myelosuppressive medications for other HIV-related conditions as well. The use of Filgrastim in children with HIV infection has also been studied. Mueller and co-workers (55) reported that Filgrastim supported the use of ZDV in 17/19 children who had neutropenia despite dose reduction of their ZDV to 120 mg/m^2 every 6 hours.

In all the studies of Filgrastim in management of neutropenia, there has been no evidence of bone marrow exhaustion or lineage diversion. This also is true when growth factors are used in combination. Miles and co-workers (56) reported on the simultaneous use of Filgrastim with recombinant human erythropoietin (rHuEPO) in 22 patients with anemia and granulocytopenia because of HIV and/or therapy with bone marrow-suppressive agents such as ZDV, ganciclovir, trimethoprim/sulfmethoxazole, and pyrimethamine. Bone marrow-suppressive agents were discontinued, Filgrastim administered, and rHuEPO then begun when neutropenia had resolved. Neutrophil counts normalized in all patients. In all cases there was significant increase in hematocrit with rHuEPO, but after reinstitution of ZDV, six patients developed severe anemia requiring transfusion. Rather than bone marrow exhaustion from cytokine therapy, there was in vitro evidence of extreme sensitivity of erythroid precursors from these patients to ZDV. Other studies in children demonstrate good efficacy in the treatment of HIV-related anemia and neutropenia with combination rHuEPO and Filgrastim.

A multicenter, randomized, placebo-controlled trial in the prevention of grade 4 neutropenia (ANC $<0.5 \times 10^9$/L) has been recently completed (57). Two-hundred and fifty-eight patients with ANC of 0.75 to 1.0 $\times$ 10^9/L were randomized to observation or Filgrastim therapy. Patients were treated >24 weeks, and followed for development of ANC $<0.5 \times$ 10^9/L, bacterial infections, hospitalizations, days of IV antibiotics, and survival. Data were recently published. Filgrastim was very effective in preventing grade 4 neutropenia. Additionally, although the study did not allow patients to remain absolutely neutropenic without therapy for an extended period of time, a number of secondary endpoints were met, including a decrease in the number of bacterial infections, serious bacterial infections, hospital days, and the use of IV antibiotics.

VI. FILGRASTIM AND HIV-RELATED NEUTROPHIL DYSFUNCTION

There is good evidence for improved PMN function with Filgrastim in vitro and ex vivo (18,58,59). In vitro, there is enhanced respiratory burst capacity, bactericidal and fungicidal capacity, and ADCC toward HIV-infected cells. Ottenello and co-workers (60) have shown that patients with recurrent bacterial infections evaluated for abnormal neutrophil function frequently do have impaired chemotaxis or a partial defect in the respiratory burst, while congenital neutrophil defects are relatively rare. Although the partial defects in neutrophil functions in HIV are not as severe as seen in congenital syndromes such as chronic granulomatous disease, there may still be a significant risk of infection because the defects are multiple and occur in the setting of impaired humoral immunity as well. We completed a study of the effect of Filgrastim on PMN function in 78 HIV-infected patients at different stages of disease before, during, and after administration of Filgrastim 300 μg subcutaneously (SC) daily or every other day for 8 days (33). Polymorphonuclear leukocyte chemiluminescence was measured using the whole blood already described earlier. Polymorphonuclear leukocytes also were isolated and evaluated for the ability to kill *E. coli.* Bacteria were opsonized with 10% normal human serum and mixed with PMN at a bacteria cell ratio of 1:1, and colony counts were performed after 2 hours of incubation. Filgrastim administration significantly increased basal and primed chemiluminescence in response to opsorized zymosan, even for patients with advanced HIV infection with CD4$^+$ <200/μL who had significantly impaired function pretherapy (**Figure 3**). There was a decrease in in vivo activation as evidenced by a decreased in ratio of unprimed to primed chemiluminescence for patients at all stages of disease (**Figure 4**). Finally, bacterial killing increased with Filgrastim, and the difference was significant for patients with advanced HIV infection who had impaired bactericidal capacity pretherapy (**Figure 5**). Although the clinical benefit of treating non-neutropenic HIV-infected patients with neutrophil dysfunction is unknown, there is great potential for this therapy, and clinical trials in patients at high risk for bacterial infections are warranted.

VII. DRUG TOXICITY

Filgrastim is extremely safe (61). The only frequent side effect is mild to moderate bone pain due to marrow expansion, and this occurs less fre-

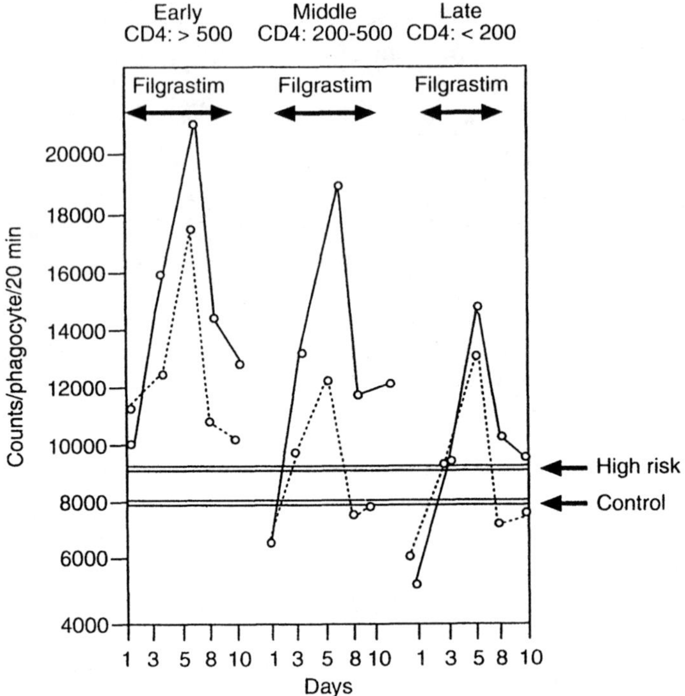

Figure 3. Filgrastim administration significantly increased basal and primed chemiluminescence in response to opsonized zymosan, even for patients with advanced HIV infection with CD4$^+$ <200/μL who had significantly impaired function pretherapy. Solid line, daily dosing; broken line, alternative daily dosing; CD4 counts given per μL.

quently in patients with advanced HIV infection who do not have as high a neutrophilic response to Filgrastim administration. This pain is usually relieved with acetaminophen and resolves quickly after discontinuation of the drug. Elevated serum uric acid due to increased cell turnover had been noted, but has been of no clinical significance.

Although certain cytokines, including GM-CSF, can upregulate HIV replication, there has been no evidence for enhanced viral replication with Filgrastim administration (56). This is consistent with in vitro experiments showing enhanced viral replication in human monocytes with IL-3, macrophage colony-stimulating factor (M-CSF), and GM-CSF, but no increase with G-CSF (62). There are in vivo data for HIV-infected patients with lymphoma receiving rHuGM-CSF or Filgrastim during CHOP (cyclophos-

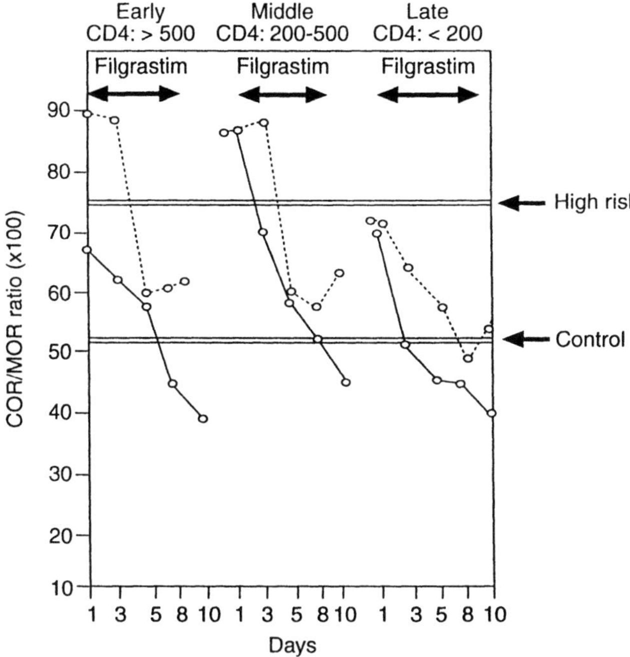

Figure 4. In vivo activation of polymorphonuclear leukocytes as measured as the ratios of unprimed (circulating opsonin receptor expression, COR) and primed (maximal opsonin receptor expression, MOR). Solid line, daily dosing; broken line, alternative daily dosing; CD4 counts given per μL.

phamide, Adriamycin®, incristine, prednisone) chemotherapy. The levels of p24 antigen were increased above baseline for the patients receiving rHuGM-CSF, while there was no increase for the patients receiving Filgrastim (56). More recently, there have been studies that do not demonstrate an increase in HIV relication due to monocyte activation by GM-CSF (63). However, the use of rHuGM-CSF has been limited by more significant side effects, including flu-like symptoms, ie, myalgias, fatigue, and asthenia, especially with long-term administration (64,65).

There are some theoretical concerns about the use of Filgrastim in non-neutropenic patients. It is possible that very high neutrophil counts can result in leukostasis, and this may result in tissue injury (62). The generation of toxic oxygen species by higher numbers of highly functional cells could cause tissue damage and/or oxidative stress that could impair the function of other cells types. There is the possibility of a myeloid malignancy, although this has mainly been a concern for children with

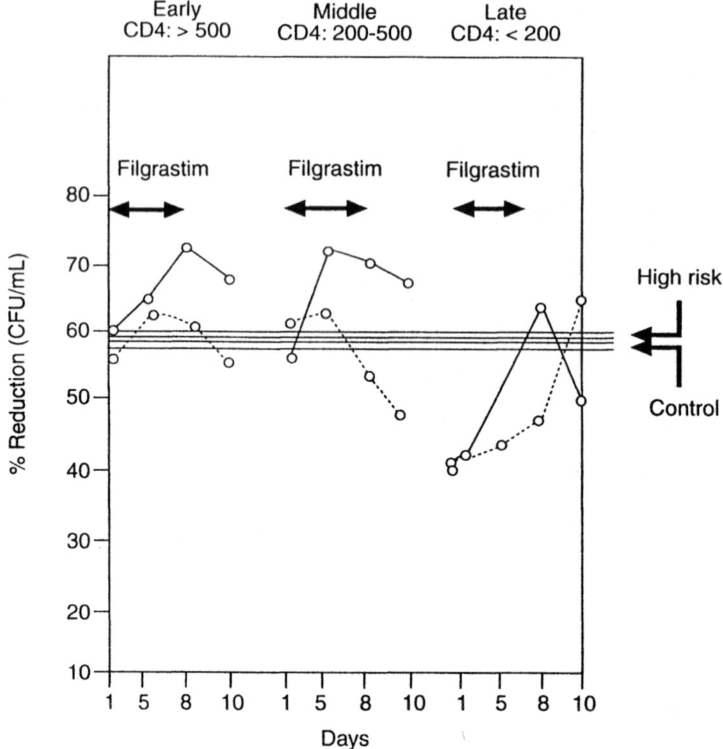

Figure 5. Bacterial killing of *E. coli* by polymorphonuclear leukocytes before, during, and after administration of Filgrastim. Solid line, daily dosing; broken line, alternative daily dosing; CD4 counts given per μL.

severe congenital neutropenia, a group of patients with a propensity for the development of myeloid leukemia as part of the underlying disease process (66,67).

VIII. USE OF FILGRASTIM IN OPPORTUNISTIC INFECTIONS

There are reasons to consider Filgrastim therapy for opportunistic infections. The first would be that Filgrastim will prevent or reverse neutropenia caused by very myelosuppressive regimens, allowing optimal dosing and preventing discontinuation of therapy. This has been studied in patients with CMV retinitis. In a retrospective study by Morgan and Strickland (68), patients receiving IV ganciclovir for CMV retinitis and Filgrastim were able to receive higher doses of ganciclovir, had less reactivation, required fewer intravitreal injection, and had a better outcome with respect to loss of vision and retinal detachment than patients not receiving Filgrastim.

Polymorphonuclear leukocytes are effector cells against a number of opportunistic pathogens that can infect HIV patients. *Aspergillus* has been reported as a complication of AIDS and can occur even in patients without absolute neutropenia (69). Most work in this area has focused on disseminated *Mycobacterium avium* complex (DMAC) infections in AIDS patients (70). Data in animals have shown the effectiveness of rG-CSF in DMAC. Clinical trials are currently underway to evaluate the safety and efficacy of Filgrastim administration as an adjunct to antimycobacterial drug therapy for DMAC infections in patients with AIDS.

IX. OTHER BENEFITS OF FILGRASTIM ADMINISTRATION

Keiser and co-workers (71) have not only shown a decrease in the number of bacterial infections, including bacteremias, in HIV patients, but also a survival advantage that is not entirely explained by the decrease in bacteremias. This survival advantage was significant, statistically and clinically, with an increase in survival time for patients receiving Filgrastim to 466 days versus 216 days for untreated patients ($p < 0.01$). This indicates that Filgrastim has beneficial effects other than a reduction in serious bacterial infections. It is possible that Filgrastim prevented opportunistic infections or allowed more aggressive therapy for HIV infection and/or HIV-related complications.

Another advantage of the use of Filgrastim is the ability to continue antiretroviral therapy. Although a number of antiretroviral agents are now available, many without any significant associated hematologic toxicity, continuation of ZDV, the agent most often associated with neutropenia, may be advantageous. Zidovudine has excellent activity in the central nervous system relative to other antiretrovirals, and recent data show that ZDV use is associated with significant decreases in immune activation and viral load in cerebrospinal fluid compared with other agents, such as dideoxyinosime (ddl). This may mean that all patients should receive ZDV despite its hematologic toxicity. Filgrastim allows continuation of optimal doses of myelosuppressive agents for therapy of opportunistic infections, such as ganciclovir therapy for CMV infections.

Cytotoxic T-lymphocyte responses are considered protection, but there are no data on the relative contribution of neutrophil ADCC to the clearance of HIV. The contribution may be significant, especially in late-stage disease when PMN compose a larger proportion of the possible effector cells in ADCC. Furthermore, neutrophils can be primed or activated without upregulating HIV replication. Viral load studies have been performed to demonstrate that there is no HIV upregulation, but there have not been any studies specifically examining whether or not Filgrastim therapy can reduce viral load by enhancing killing of HIV or HIV-infected cells.

X. FUTURE AREAS OF RESEARCH

It is clear that Filgrastim administration can increase neutrophil counts and improve neutrophil function, and this is associated with a decrease in bacterial infections and improved survival. It is still not known which patients should be treated with Filgrastim. Patients with severe neutropenia with ANC $<0.5 \times 10^9$/L are at great risk for bacterial infection, and it is reasonable to begin Filgrastim at this point, even in the absence of any active infection. Since the risk of infection begins to increase at an ANC $<1.0 \times 10^9$/L, it may also be reasonable to begin at an ANC 0.5 to 1.0×10^9/L to prevent grade 4 neutropenia, especially if the ANC is on the decline or decline is will be highly likely with the addition of any myelosuppressive therapy.

Patients with very advanced HIV infection, $CD4^+$ $<200/\mu$L, are at risk for infection due to impaired neutrophil function as well as relative or absolute neutropenia, and Filgrastim is of potential clinical benefit to them. It is important to better understand the risk factors for development of bacterial infections to identify any subset of patients who would benefit most from Filgrastim therapy. This may not only include identification of host characteristics, such as $CD4^+$ count or risk factor for HIV, but other interventions or therapies for HIV or related conditions. For example, patients receiving trimethoprim/sulmethoxazole have a decrease in the incidence of bacterial infections, and patients receiving rifabutin have a decrease in the number of staphylococcal infections. It is possible that patients receiving agents known to prevent bacterial infections will not gain additional benefit with Filgrastim, while patients receiving other agents may require Filgrastim. Work in this area will be important in formulating a rational approach the management of these patients.

XI. CONCLUSIONS

Filgrastim appears to be a safe and effective agent in the treatment and prevention of neutropenia associated with HIV infection and myelosuppressive therapy for HIV-related infections and neoplasms. There is not only an increase in the number of circulating neutrophils in patients receiving Filgrastim, but also enhanced function of these cells. There are data to show that patients receiving Filgrastim have a decreased incidence of bacterial infections and improved survival not entirely explained by the decrease in serious bacterial infections. Filgrastim has the potential to be clinically useful in the prevention and/or treatment of bacterial infections in non-neutropenic patients, adjunctive therapy for certain opportunistic infections, and perhaps as an adjunct to therapy for infection with HIV itself.

REFERENCES

1. Fauci, A. S., and Lane, H. C. Human immunodeficiency virus (HIV) disease. Aids and related disorders. In: *Harrison's Principles and Practice of Medicine.* 13th ed. pp. 1567–1618, 1994, New York: McGraw-Hill, Inc.
2. Witt, D. J., Craven, D. E., and McCabe, W. R. (1987). Bacterial infections in adult patients with the acquired immune deficiency syndrome (AIDS) and AIDS-related complex. *Am J Med 82:*900–906.
3. Nichols, L., Balogh, K., and Silverman, M. (1989). Bacterial infections in the acquired immunodeficiency syndrome. Clinicopathologic correlations in a series of autopsy cases. *Am J Clin Pathol 92:*787–790.
4. Centers for Disease Control (1987). Revision of the CDC surveillance case definition for acquired immunodeficiency syndrome. *MMWR 36* (15).
5. Barat, L. M., Gunn, J. E., Steger, K. A., Perkins, C. J., and Craven, D. E. (1996). Causes of fever in patients infected with human immunodeficiency virus who were admitted to Boston City Hospital. *Clin Infect Dis 23:*320–328.
6. Stein, M., O'Sullivan, P., Wachtel, T., et al (1992). Causes of death in persons with human immunodeficiency virus infection. *Am J Med 93:*387.
7. Fauci, A. S., Pantaleo, G., Stanley, S., and Weissman, D. (1996). Immunopathogenic mechanisms of HIV infection. *Ann Intern Med 124:*654–663.
8. Lane, H. C., Masur, H., Edgar, L. C., Whalen, G., Rook, A. H., and Fauci, A. S. (1983) Abnormalities of B-cell activation and immunoregulation in patients with the acquired immunodeficiency syndrome. *N Engl J Med 309:*453–458.
9. Lane, H. C., Depper, J. M., Greene, W. C., Whalen, G., Waldmann, T. A., and Fauci, A. S. (1985). Qualitative analysis of immune function in patients with the acquired immunodeficiency syndrome. Evidence for a selective defect in soluble antigen recognition. *N Engl J Med 313:*79–84.
10. Senaldi, G., Peakman, M., McManus, T., Davies, E. T., Tee, D. E., and Vergani, D. (1990). Activation of the complement system in human immunodeficiency virus infection: relevance of the classical pathway to pathogenesis and disease severity. *J Infect Dis 162:*1227–1232.

11. Bernstein, L. J., Ochs, H. D., Wedgwood, R. J., and Rubinstein, A. (1985). Defective humoral immunity in pediatric acquired immune deficiency syndrome. *J Pediatr 107*:352–357.
12. Nelson, K. E., Clements, M. L., Miotti, P., Cohn, S., and Polk, B. F. (1988). The influence of human immunodeficiency virus (HIV) infection on antibody responses to influenza vaccines. *Ann Intern Med 109*:383–388.
13. Mofenson, L. M., Moye, J. Jr., Bethel, J., Hirschhorn, R., Jordan, C., and Nugent, R. (1992). Prophylactic intravenous immunoglobulin in HIV-infected children with CD4$^+$ counts of 0.20 $\times$ 10^9/L or more: Effect on viral, opportunistic, and bacterial infection. The National Institute of Child Health and Human Development Intravenous Immunoglobulin Clinical Trial Study Group. *JAMA 268*:483–488.
14. The International Institute of Child Health and Human Development Intravenous Immunoglobulin Study Group (1991). Intravenous immunoglobulin for the prevention of bacterial infections in children with symptomatic human immunodeficiency virus infection. *N Engl J Med 325*:73–80.
15. Spector, S. A., Gelber, R. D., McGrath, N., et al (1994). A controlled trial of intravenous immune globulin for the prevention of serious bacterial infection in children receiving zidovudine for advanced human immunodeficiency virus infection. *N Engl J Med 331*:1181–1187.
16. Hambleton, J. (1996). Hematologic complications of HIV infection. *Oncology 10(5)*:671–680.
17. Kielhofner, M., Atmar, R. L., Hamill, R. J., and Musher, D. M. (1991). Life-threatening *Pseudomonas aeruginosa* infections in patients with human immunodeficiency virus infection. *Clin Infect Dis 14*:403–411.
18. Baldwin, G. C., Fuller, N. D., Roberts, R. L., Ho, D. D., and Golde, D. W. (1989). Granulocyte- and granulocyte-macrophage colony-stimulating factors enhance neutrophil cytotoxicity toward HIV-infected cells. *Blood 74*:1673–1677.
19. Szelc, C. M., Mitcheltree, C., Roberts, R. L., and Stiehm, E. R. (1992). Deficient polymorphonuclear cell and mononuclear cell antibody-dependent cellular cytotoxicity in pediatric and adult human immunodeficiency virus infection. *J Infect Dis 166*:486–493.
20. Klebanoff, S. J., and Coombs, R. W. (1992). Viricidal effect of polymorphonuclear leukocytes on human immunodeficiency virus-1. Role of the myeloperoxidase system. *J Clin Invest 89*:2014–2017.
21. Zon, L. I., and Groopman, J. E. (1988). Hematological manifestations of the human immunodeficiency virus (HIV). *Semin Hematol 25*:208–218.
22. Harbol, A. W., Liesveld, J. L., Simpson-Haidaris, P. J., and Abboud, C. N. (1994). Mechanisms of cytopenia in human immunodeficiency virus infection. *Blood Rev 8*:241–251.
23. Tumbarello, M., Tacconelli, E., Caponera, S., Cauda, R., and Ortona, L. (1995). The impact of bacteraemia on HIV infection. Nine years experience in a large Italian university hospital. *J Infect 31*:123–131.
24. Outwater, E., and McCutchan, J. A. (1985). Neutrophil-associated antibodies and granulocytopenia in AIDS. *Clin Res 33*:413a.

25. Murphy, M. F., Metcalfe, P., Waters, A. H., et al (1985). Immune neutropenia in homosexual men (letter). *Lancet 1*:217–218.
26. Weinberg, G. A., Gigliotti, F., Stroncek, D. F., et al (1997). Lack of relation of granulocyte antibodies (antineutrophil antibodies) to neutropenia in children with human immunodeficiency virus infection. *Pediatr Infect Dis J 16*:881–884.
27. Busch, M., Beckstead, J., Gantz, D., and Vyas, G. (1986). Detection of human immunodeficiency virus infection of myeloid precursors in bone marrow samples from AIDS patients (abst). *Blood 68*:122a.
28. Moses A. M., Heneveld, M., Nelson, J. A., Williams, S., Rarick, M., and Bagby, G. C. (1995). CD34⁺ bone marrow microvascular endothelial cells (MVEC) are consistently infected by HIV-1 in patients with AIDS: suboptimal release of IL-6 and G-CSF by infected stromal cells does not depend upon release of soluble inhibitory factors. *Blood 86*:287a (abstr 1136).
29. Lagneaux, L., Delforge, A., Snoeck, et al (1996). Imbalance in production of cytokines by bone marrow stromal cells following cytomegalovirus infection. *J Infect Dis 174*:913–919.0
30. Lieschke, G. J., Grail, D., Hodgson, G., et al (1994). Mice lacking granulocyte colony-stimulating factor have chronic neutropenia, granulocyte and macrophage progenitor cell deficiency, and impaired neutrophil mobilization. *Blood 84*:1737–1746.
31. Watari, K., Asano, S., Shirafuji, N., et al (1989). Serum granulocyte colony-stimulating factor levels in healthy volunteers and patients with various disorders as estimated by enzyme immunoassay. *Blood 73*:117–122.
32. Mauss, S., Steinmetz, H. T., Willers, R., Jablonowski, H., and Haussinger, D. (1995). Low granulocyte colony-stimulating factor (G-CSF) levels in neutropenia but not in acute febrile infection in HIV seropositive individuals: a causal role for G-CSF in HIV-associated neutropenia. Program and Abstracts of the 35th. ICAAC, San Francisco, 17–20 September. 1250a (abstr 1250)
33. Pitrak, D. L., Mullane, K. M, Bilek, M., et al (1996a). Filgrastim (r-metHuG-CSF) treatment of HIV-infected patients improves neutrophil function. Proceedings of the XI International Conference on AIDS, (abstr Th.B.4181).
34. Pitrak, D. L., Tsai, H. C., Mullane, K. M., Sutton, S. H., and Stevens, P. (1996b). Accelerated neutrophil apoptosis in the acquired immunodeficiency syndrome. *J Clin Invest 98*:2714–2719.
35. Williams, G. T., Smith, C. A., Spooncer, E., Dexter, T. M., and Taylor, D. R. (1990). Haemopoietic colony stimulating factors promote cell survival by suppressing apoptosis. *Nature 343*:76–79.
36. Abramson, J. S., and Mills, E. L. (1988). Depression of neutrophil function induced by viruses and its role in secondary microbial infections. *Rev Infect Dis 10*:326–341.
37. Valone, F. H., Payan, D. G., Abrams, D. I., and Goetzl, E. J. (1984). Defective polymorphonuclear leukocyte chemotaxis in homosexual men with persistent lymph node syndrome. *J Infect Dis 150*:267–271.
38. Ras, G. J., Anderson, R., and Lukey, P. T. (1985). Defective polymorphonuclear leukocyte migration in AIDS (letter). *S Afr Med J 68*:292–293.

39. Lazzarin, A., Ulberti Foppa, C., Galli, M., et al (1986). Impairment of polymorphonuclear leukocyte function in patients with acquired immunodeficiency syndrome and with lymphadenopathy syndrome. *Clin Exp Immunol* 65:105–111.

40. Nielsen, H., Kharazmi, A., and Faber, V. (1986). Blood monocyte and neutrophil functions in the acquired immune deficiency syndrome. *Scand J Immunol* 24:291–296.

41. Ellis, M., Gupta, S., Galant, S., et al (1988). Impaired neutrophil function in patients with AIDS or AIDS-related complex: A comprehensive evaluation. *J Infect Dis* 158:1268–1276.

42. Murphy, P. M., Lane, H. C., Fauci, A. S., and Gallin, J. J. (1988). Impairment of neutrophil bactericidal capacity in patients with AIDS. *J Infect Dis* 158:627–630.

43. Pitrak, D. L., Bak, P. M., De Marais, P., Novak, R. M., and Andersen, B. R. (1993). Depressed neutrophil superoxide production in human immunodeficiency virus infection. *J Infect Dis* 167:1406–1410.

44. Roilides, E., Mertins, S., Eddy, J., Walsh, T. J., Pizzo, P. A., and Rubin, M. (1990). Impairment of neutrophil chemotactic and bactericidal function in children infected with human immunodeficiency virus type 1 and partial reversal after in vitro exposure to granulocyte-macrophage conlony-stimulating factor. *J Pediatr* 117:531–540.

45. Mullane, K., Pitrak, D., Bilek, M., Novak, R., Allen, R., and Stevens, P. (1994). In vivo neutrophil; activation and burnout in HIV infection. *Clin Res* 42:155a.

46. Elbim, C., Prevot, M. H., Bouscarat, F., et al (1994). Polymorphonuclear neutrophils from human immunodeficiency virus-infected patients show enhanced activation, diminished fMLP-induced L-selectin shedding, and an impaired oxidative burst after cytokine priming. *Blood* 84:2759–2766.

47. Vierucci, A, De Martino, M., Graziani, E., Rossi, M. E., London, W. T., and Blumberg, B. S. (1983). A mechanism for liver injury in viral hepatitis: effects of hepatitis B virus on neutrophil function in vitro and in chidren with chronic active hepatitis. *Pediatr Res* 17:814–820.

48. Shaunak, S., and Bartlett, J. A. (1989). Zidovudine-induced neutropenia: are we too cautious? *Lancet* 2:91–92.

49. Yu, K., Mitsuyasu, R. Relative incidence of infection during neutropenia: a retrospective comparison between HIV+ patients with ANC <500 cells/mm^3 and ANC <750 cells/mm^3. (1994) *Clinical Research* 42:278A. (abstr 255).

50. Moore, R. D., Keruly, J. C., and Chaisson, R. E. (1995). Neutropenia and bacterial infection in acquired immunodeficiency syndrome. *Arch Intern Med* 155:1965–1970.

51. Jacobson, M. A., Cohen, P. T., Liu, R. C. C., et al (1996). Risk of hospitalization for serious bacterial infection (SBI) associated with neutropenia severity in patients with HIV disease. XIth Int Conf on AIDS; Vancouver, B.C. (abstr Tu.B.180).

52. Yu, K. and Mitsuyasu, R. (1994). Relative incidence of infection during neutropenia: A retrospective comparison between HIV+ patients with ANC <500 cells/mm^3 and ANC <750 cells/mm^3. *Clin Res* 42(2):278A. (abstr 255).

53. Hermans, P., Rozenbaum, W., Jou, A., et al (1995). Filgrastim (r-met-G-CSF) to treat neutropenia and support myelosuppressive medication dosing in HIV infection. Eur Conf on Clin Aspects and Treatment of HIV; Copenhagen, Denmark.

54. Jacobson, M. A., Stanley, H. D., and Heard, S. E. (1992). Ganciclovir with recombinant methionyl human granulocyte colony-stimulating factor for treatment of cytomegalovirus disease in AIDS patients (letter). *AIDS 6*:515–517.

55. Mueller, B. U., Jacobson, F., Butler, K. M., Husson, R. N., Lewis, L. L., and Pizzo, P. A. (1992). Combination treatment with azidothymidine and granulocyte colony-stimulating factor in children with human immunodeficiency virus infection. *J Pediatr 121*:797–802.

56. Miles, S. A., Mitsuyasu, R. T., Moreno, J., et al (1991). Combined therapy with recombinant granulocyte colony-stimulating factor and erythropoietin decreases hematologic toxicity from zidovudine. *Blood 77*:2109–2117.

57. Kuritzkes, D., Parenti, D., Ward, D., and Rachlis, A. (1997). Filgrastim (r-metHuG-CSF) for prevention of severe neutropenia and associated clinical sequelae in HIV-infected patients: results of a 24-week, prospective randomized, controlled trial. 4th Conference on Retroviruses and Opportunistic Infections, Jan 22–26, 1997. (abstr 365).

58. Roilides, E., Walsh, T. J., Pizzo, P. A., and Rubin, M. (1991). Granulocyte colony-stimulating factor enhances the phagocytic and bactericidal activity of normal and defective human neutrophils. *J Infect Dis 163*:579–583.

59. Vecchiarelli, A., Monari, C., Baldelli, F., et al (1995). Beneficial effect of recombinant human granulocyte colony-stimulating factor on fungicidal activity of polymorphonuclear leukocytes from patients with AIDS. *J Infect Dis 171*:1448–1454.

60. Ottenello, L., Dapino, P., Pastorino, G., Dallegri, F., and Sacchetti, C. (1995). Neutrophil dysfuntion and increased susceptibility to infection. *Eur J Clin Invest 25*:687–692.

61. Welte, K., Garbrilove, J., Bronchud, M. H., Platzer, E., Morstyn, G. (1996). Filgrastim (r-metHuG-CSF): the first ten years. *Blood 88*:1907–1929.

62. Koyanagi, Y., O'Brien, W. A., Zhao, J. Q., Golde, D. W., Gasson, J. C., Chen, I. S. (1988). Cytokines alter production of HIV-1 from primary mononuclear phagocytes. *Science 241*:1673–1675.

63. Hardy, W. D. (1991). Combined ganciclovir and recombinant human granulocyte-macrophage colony-stimulating factor in the treatment of cytomegalovirus retinitis in AIDS patient. *J AIDS 4*:S22–S28.

64. Mitsuyasu, R. T., and Golde, D. W. (1989). Clinical role of granulocyte-macrophage colony-stimulating factor. *Hematol Oncol Clin North Am 3*:411–425.

65. Scadden, D. T., Bering, H. A., Levine, J. D., et al (1991). Granulocyte-macrophage colony-stimulating factor mitigates the neutropenia of combined interferon alfa and zidovudine treatment of acquired immune deficiency syndrome-associated Kaposi's sarcoma. *J Clin Oncol 9*:802–808.

66. Dale, D. C., Bonilla, M. A., Davis, M. W., et al (1993). A randomized controlled phase III trial of recombinant human granulocyte colony-stimulating factor (filgrastim) for treatment of severe chronic neutropenia. *Blood 81*:2496–2502.

67. Rosen, R. B., and Kang, S. J. (1979). Congenital agranulocytosis terminating in acute myelomonocytic leukemia. *J Pediatr 94*:406–408.

68. Morgan, K. M., and Strickland, S. R. (1993). The use of granulocyte stimulating factor (GCSF) has improved the outcome in AIDS related CMV retinitis. *Int Conf AIDS 9*:413 (abstr PO-B16-1666).

69. Pursell, K. J., Telzak, E. E., and Armstrong, D. (1992). *Aspergillus* species colonization and invasive disease in patients with AIDS. *Clin Infect Dis 14*:141–148.

70. Bermudez, L. E., Petrofsky, M., Young, S. L., and Stevens, P. (1996). Granulocyte-colony stimulating factor treatment of mice induces anti-*Mycobacterium avium* activity ex-vivo in neutrophils and splenic macrophages. *J Clin Invest 98*:2714–2719.

71. Keiser, P., Higgs, E., and Smith, J. (1996). Neutropenia is associated with bacteremia in patients infected with the human immunodeficiency virus. *Am J Med Sci 312*:118–122.

72. Eng, R. H. K., Yen, K., Tecson-Tumang, F., et al (1994). Risk of gramnegative bacteremia during neutropenia in patients with AIDS. *Infect Dis Clin Pract 3*:373–375.

73. Hambleton, J., Aragón, T., Modin, G., et al (1995). Outcome for hospitalized patients with fever and neutropenia who are infected with human immunodeficiency virus. *Clin Infect Dis 20*:363–371.

22
Neonatal Sepsis—An Unsolved Problem

Prabhakar Kocherlakota, Edmund F. LaGamma, Magdy Ahmad, Candace Breen, Marc G. Golightly, and Howard B. Fleit
SUNY at Stony Brook, Stony Brook, New York

I. BACKGROUND ISSUES AND CONFOUNDING PROBLEMS IN STUDYING SEPSIS IN NEONATES

A. Bacterial Sepsis

Bacterial sepsis occurs in 1 to 10 of every 1000 full-term newborns (1). It is 30 to 50 times more common in extremely low birth weight infants (neonates <800 g at birth [2]) and is a major cause of morbidity and mortality in this patient population. In a recent report, 4% of deaths in the first 3 days after birth and 45% of deaths after 2 weeks of postnatal age were directly related to infections (3,4). Mortality from neonatal sepsis depends upon the virulence of the organism, the gestational age and birth weight of the neonate, the particular combination of associated neonatal risk factors, and an array of maternal antepartum complications with postnatal significance (1–4). Sepsis also increases morbidity by increasing the duration of mechanical ventilation, nutritional support, and length of hospital stay (1–5).

Whether classified as early (≤3 days postnatal age) or late (>3 days postnatal age; nosocomial) in onset, bacterial sepsis is inversely related to the birth weight (6). Predisposing factors for early-onset sepsis relate to perinatal complications such as premature rupture of amniotic membranes, prolonged rupture of membranes, maternal chorioamnionitis, and male gender (6,7). In contrast, predisposition to late-onset sepsis correlates best with postnatal factors such as the overall length of the hospital stay, the duration of mechanical ventilation, the number of vascular access devices used, and other invasive interventions common to neonatal intensive care (4,6,7). What is most disturbing to neonatal clinicians is that in the last 20 years, the mortality due to sepsis has remained relatively constant (1,2,4,5,8) despite continued improvement in management in all other forms of neonatal illnesses, including surfactant replacement, advances in neonatal ventilatory techniques, use of an extracorporeal membrane oxygenator (ECMO), and a number of other innovative treatments (eg, nitric oxide, liquid ventilation).

Among the many related patient-specific causes for an increased incidence of infection are general issues such as an immature immune system (9,10), immaturity and ineffective physical barriers, such as the skin and gut mucosa (10,11), insufficient nutrient intake with catabolism (12), frequent use of invasive procedures, prolonged exposure to the hospital environment, and chronic associated conditions arising from premature birth, such as bronchopulmonary dysplasia and short bowel syndrome (4,9,10,13). In addition, many of these problems are unavoidable, requiring time and extrauterine growth to be overcome. Thus, in view of these fixed limitations to providing care, at least one area of innovative therapy is emerging and involves augmenting the functional capacity of the immature immune system with cytokines to lessen the morbidity and mortality due to neonatal sepsis.

The neonate's immature immune system is limited in a variety of ways (9) with quantitative deficiencies of the myeloid and phagocytic systems, in general, and immaturity of neutrophil function, in particular. Deficits are thought to contribute significantly to the high morbidity and mortality from bacterial sepsis in this population (9). Quantitative deficiencies include decreased neutrophil storage and myeloid proliferative pools and disturbed release of neutrophils from the bone marrow (9,10,14,15). Neutrophil function studies have shown decreased chemotaxis, impaired phagocytosis, and ineffective intracellular killing (9,15,16). Dysfunctional receptor-opsonization capacity includes decreased surface expression, failure of up-regulation, decreased subcellular storage capacity, and increased loss of neutrophil Fcγ receptors (Fcγ RIII and Fcγ RII) and complement receptors (CR1 and CR3) (9,10,16). Each of these activities may be further impaired in the extremely preterm infant. Augmenting the number and function of neutrophils has been suggested to possibly assist in defending against bacterial infections and result in a lower neonatal morbidity and mortality because of the key role of neutrophils in infection.

With the availability of a panoply of newly discovered, cloned, and commercially available cytokines, a method of testing these assumptions on augmenting the immune system can be accomplished. Carefully designed clinical trials in infants are needed to define those clinical situations amenable to intervention with any particular cytokine therapy. In this chapter we will focus on Filgrastim and its use in neonates. We will place special emphasis on issues concerning neonatal neutropenia and neonatal sepsis. Neutropenia-related diseases such as Kostman syndrome and congenital leukemia are beyond the scope of this chapter but are addressed elsewhere in the medical literature (see, for example, Ref. 17) and in this volume (see Chapters 5 and 6).

B. Reduced In Vitro G-CSF Production in the Neonate

Granulocyte colony-stimulating factor (G-CSF) is an endogenous, lineage-specific hematopoietic growth factor that serves a critical role in the differentiation, function, and egress of neutrophils from the bone marrow (18,19). It is considered to be an essential protein as its sequence and functions are conserved across species (20). Endogenous G-CSF is produced at basal levels by monocytes, endothelial cells, fibroblasts, and stromal cells (21,22), and G-CSF can be increased in response to inflammatory mediators such as interleukin (IL)-1, tumor-necrosis factor (TNF)-α, granulocyte-macrophage colony-stimulating factor (GM-CSF), and lipopolysaccharide (LPS) (see Ref. 23 for a review). In addition, it appears that neonatal production of G-CSF is limited compared with that of adults when tested in vitro. For example, Cairo et al (24) demonstrated decreased mRNA expression and the attendant lower level of released of G-CSF protein from cultured mononuclear cells. Lee et al (25) showed that decreased levels of G-CSF mRNA were not due to transcriptional regulation but instead to reduced mRNA stability. In support of this interpretation, Cairo et al (24) and Schibler et al (26) noted lower generation of G-CSF protein by neonatal mononuclear cells compared with adult cells even after stimulation with LPS or IL-1a.

C. Higher G-CSF Levels in Neonates In Vivo

Measurement of neonatal levels of endogenously produced G-CSF are confounded by many variables including postconceptional maturational age, postnatal age, various cell types (including the placenta) that produce G-CSF, placental transfer of maternal-derived G-CSF, and the particular disease state of the neonate (26–33, 35, 85, 86). In addition, enzyme-linked immunosorbent assay (ELISA)-type assays may only detect G-CSF antigenicity while biological assays may be more clinically relevant as they measure function of the protein (**Table 1**). Interestingly, after birth, the predominant in vivo observation is that high levels of circulating G-CSF are detected

Table 1. Endogenous G-CSF Levels in Neonates.

Reference	N	Gestational age (weeks)	Postnatal age	Method	Level (pg/mL)
85	34	>36	Birth	Bioassay	41 ± 3
86	7	>36	1 to 3 days	ELISA	Nil(<50)
29	20	>37	4–7 hours	ELISA	228 ± 40
26	16	38	Not given	ELISA	114 ± 21
30	52	32	<48 hours	ELISA	>100
35	42	26 to 31	<72 hours	ELISA	180 ± 60
31	105	35	4.6 days	ELISA	280

ELISA = enzyme-linked immunosorbent assay.

(even when compared with adult normal values), contrasting sharply with the in vitro observations showing decreased production (24–26).

Although G-CSF can cross the placenta and enter the fetal circulation in the human neonate (27,28), there has been no consistent proportional relationship between maternal serum G-CSF levels and the corresponding serum values or the absolute neutrophil count (ANC) observed in their respective neonates (29–31). Moreover, high initial neonatal G-CSF values decline rapidly compared with adult levels as the infant matures ex utero over the first few weeks after birth (29,32,33). Lastly, unlike adults, levels of G-CSF in neutropenic neonates were not found to be elevated even when compared with non-neutropenic neonates (**Table 1**) (29,30).

There are several clinical conditions where G-CSF levels do show unique patterns relative to "control values" in normal very-low-birth-weight neonates. While a postconceptional increase of circulating G-CSF was noted by Gessier et al (29) and others (30), a significantly lower concentration was noted in neonates who were small-for-gestational age (SGA) compared with average-for-gestational-age (AGA) term neonates (284 ± 54 [n = 13] versus 114 ± 31 [n = 8] pg/mL; [29]). Lower concentrations of G-CSF also were noted in neonates born to women with pregnancy-induced hypertension (preeclampsia) compared with neonates of nonhypertensive women (146 ± 42 [n = 10] versus 182 ± 21 [n = 54] pg/mL [29]). On the other hand, there have been no reported effects of the use of antepartum glucocorticosteroids, perinatal asphyxia, or respiratory distress syndrome on G-CSF levels in neonates (29,30).

During sepsis, higher levels of G-CSF were measured in infected neonates compared with matched healthy patients (29,30,33). In a further refinement of these observations, Kennon et al (31) showed significantly higher levels of G-CSF in a group of neonates with proven sepsis

(2278 pg/mL), compared with a "clinical sepsis" group (no positive blood culture; 1873 pg/mL) versus noninfected neonates (280 pg/mL). In each of these circumstances, no relationship between endogenous levels of G-CSF and the ANC was shown.

In neonates, receptors for G-CSF exist on a variety of cell types and appear to be functional. Receptors are expressed at the same levels as in adult white cells (24). This suggests that neonatal white cell responses should occur unless attenuated in some way due to downstream deficits in intracellular signal transduction mechanisms or in a functional capacity such as cellular killing capacity. In support of this speculation is the observations that G-CSF can bind to its cognate receptors on neonatal white blood cells (24). Whether neonatal white cells differ in response to treatment compared with adult cells has not been definitively resolved.

D. Pharmacokinetics of Filgrastim in the Neonate

The half-life of exogenous G-CSF in adults varies inversely with the ANC and is shortened further after repeated doses (34). In neonates, steady-state levels of Filgrastim have been modeled after a first-order kinetics paradigm. Babies show peak serum concentrations within 2 to 8 hours after subcutaneous (SC) or intravenous (IV) administration, and have an effective clinical half-life of the drug of approximately 3.5 to 4.4 hours (35). Closer inspection reveals a more complex situation.

In a specific example, Gillian et al noted that the peak serum concentration was reached at 2 hours after IV administration and was not proportionately dose dependent. If 1 μg/kg/day Filgrastim was given, circulating G-CSF levels increased within 2 hours from 180 $\pm$ 60 to 2040 $\pm$ 1340 pg/mL. In contrast, when Filgrastim 10 μg/kg/day was given, a disproportionately higher ($\geq$10-fold) increase in the peak serum concentration was achieved (126,750 $\pm$ 22,570 pg/mL). The one-compartment, first-order kinetics model may not be sufficient to explain the observed phenomenon. Additional studies are necessary in this area as well as in the non-neutropenic adult to better define the pharmacological principles necessary for defining optimal clinical dosing schedules for this drug. Overall, although various doses and timing schedules have been evaluated (**Table 2**), it appears that a change in the ANC as well as a clinical benefit can be obtained at 10 μg/kg/day IV once per day as reported by Schibler et al (36), Gillian et al (35), and in our studies (37).

E. Preeclampsia-Associated Neutropenia and Sepsis

Preeclampsia, a major cause of premature delivery, is known to occur in 6% of pregnancies in the United States, but in rates as high as 24% in

Table 2. Use of Filgrastim During Neonatal Sepsis.

Reference	N	Diagnosis	Dosage (μg/kg/day)
35	42	"Presumed" sepsis	1, 5, 10, 20
51	12	Septic, neutropenic	5, 10
68	14	Septic, neutropenic	5
36	10	Septic, neutropenic	10
37	10	Septic, neutropenic, shock	10

developing countries (38). The incidence of neutropenia and its severity is directly related to the degree of intrauterine growth retardation, stage of prematurity, and severity of maternal hypertension or hypertension associated with hemolysis, elevated liver enzymes, and low platelets (ie, HELLP syndrome [39]). In neonates, pregnancy-induced hypertension is associated with ANC below the normal range in 36% to 57% of live births (39–45). This deficiency is believed to be the result of a putative placenta-derived bone marrow-inhibitory factor that suppresses stem cell division and is detected only in women with preeclampsia (46). The "inhibitor" has not yet been identified.

According to Koenig and Christensen (46), preeclampsia-associated neutropenia resolves within 72 hours in most cases. They also noted that neutropenia may last longer than 3 days in 17% of neonates, although other investigators found a much higher percentage of persistence (41,43). In our studies, we have made additional refinements to these previously described patterns and have categorized our results into three typical clinical responses (**Table 3, Figure 1**). In the first, persistent preeclampsia-associated neutropenia may have its onset at birth (immediate) or be present after the first postnatal day (delayed). Subsequently, the immediate-onset type either continues to be persistent or shows a biphasic pattern (**Table 3, Figure 1**). Why these patterns emerge is open to speculation, but based

Table 3. Proposed Classification of Preeclampsia-Associated Neutropenia.

Type of neutropenia	N	Incidence
Immediate-onset	18/27	67% of all cases
Persistent	10/18	56% of immediate-onset cases
Biphasic	8/18	44% of immediate-onset cases
Delayed-onset	9/27	33% of all cases

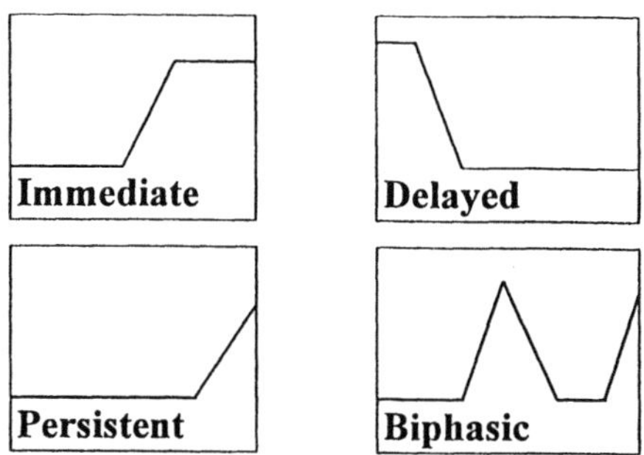

Figure 1. Patterns of the absolute neutrophil count (ANC) response in preeclampsia-associated neutropenia. In the majority of affected very-low-birth-weight (VLBW) neonates, an immediate onset of neutropenia occurs at birth, while only one-third of cases show a delayed onset (Table 2). However, about half of the early cases will persist more than 3 days after birth, while the other half shows a phasic pattern.

upon a reasonable literature, the patterns are likely to be related to the interaction between a stem cell inhibitor (46) and the "stress" of being born with the attendant catecholamine surge resulting in demargination of available white cells (45,47).

In a recent report, we demonstrated that Filgrastim could overcome the putative "pregnancy-induced hypertension inhibitory factor" by reversing the persistent neutropenia (neutropenia lasting more than three consecutive days before treatment) at a dose of 10 μg/kg/day (48). The ANC was not significantly different compared with untreated matched control patients when a 5-μg/kg/day-dose was used (49). In neonates born to women with preeclampsia, Makhlouf et al (50) showed that ANC could increase within 6 hours of administration, and others showed a sustained elevation for up to 2 weeks (51). In our series, we found that this sustained increase typically occurs after 1 week postnatal age (48,49), and presumably follows from entry of more progenitor cells into the myeloproliferative pathway.

Preeclampsia-associated neutropenia has been shown to be an additional risk factor for an increased incidence of infection in preterm neonates. In a prospective study, Koenig and Christensen (39) found a 23% incidence of nosocomial infection in preeclampsia associated neutropenia and only 3% when there was no neutropenia. Doron et al (42) also found a significantly higher incidence of early-onset sepsis in neonates with preeclampsia-associated neutropenia (6% versus 0%) as did other investigators (44).

In a non-randomized phase 1–2-type study, we found that treatment of persistent neutropenia with Filgrastim can decrease the occurrence of septicemia in the first 28 days after birth from 55% to 10% (p $<$ 0.05, power $=$ 0.8) (52). Additional prospective and blinded trials are needed to substantiate these intriguing possibilities. Moreover, whether maternally administered Filgrastim (27) can induce neonatal neutrophil responses and thus have a prophylactic role in the prevention of postnatal neutropenia or in decreasing the incidence of neutropenia-related infections through placental transfer of Filgrastim in these neonates is unknown.

II. NEONATAL SEPSIS AND PREVIOUS ATTEMPTS AT ADJUNCTIVE THERAPIES

Neonatal sepsis is commonly associated with neutropenia (53) due to a shortened neutrophil half-life from 6.3 hours to $<$4 hours (14), a limited ability to augment production of the neutrophil proliferative pool, a defective production of cytokines, and a rapid depletion of the infant's neutrophil storage pool during bacteremia (9,14,53,54). The resulting neutropenia serves as an ominous marker for an exaggerated additional risk of mortality due to sepsis which increases to as great as 90% (55).

Despite appropriate antimicrobial and optimal supportive therapy, mortality resulting from neonatal sepsis remains disturbingly high. Hand washing, isolation techniques, a better understanding of the gut mucosa and gut priming as a means to improving gut barrier functions to limit bacterial translocation (11,12), better catheter techniques, and better overall management have all failed to lower the incidence of bacterial sepsis to date (3,4). Therefore, to improve the survival, morbidity, and severity of sequelae in survivors, various forms of supportive and corrective therapy have been devised.

Granulocyte transfusions and IV immunoglobulins are two such therapies. Granulocyte transfusions are generally not done for a variety of reasons, including the risk of graft-versus-host disease (GVHD), polycythemia, fluid overload, alloimmunization, alteration in complement activity, pulmonary sequestration of leukocytes, risks of transmission of viral infections, logistical problems in timely harvesting of the white cells, and the relatively high costs of the harvesting procedure (55–57).

Intravenous immune globulin (IVIG) replacement therapy has been attempted in neonates because of their low circulating IgG levels and impaired production of antigen-specific antibodies, and IVIG has been administered both prophylactically as well as therapeutically with mixed results (58–60). Earlier studies with small numbers of patients reported beneficial effects; however, later studies, including several recent large multicenter trials, failed to establish a substantial beneficial effect in the reduction of nosocomial infections in very low birth weight (VLBW) neonates except for patients <1000 g at birth (58). In particular, *Staphylococcus epidermidis* infections were decreased when high titer antibody for this organism was used (58). Although not considered to be a continuing problem, recent reports of transmission of viral infection (hepatitis C) (61) have effectively quenched the use of IVIG in premature babies despite its importance in opsonization and alleviation of grossly deficient serum levels <300 mg/dL that persist for several months after birth (9,10,62). In all likelihood, there will be a resurgence of interest in the use of IVIG after its proper place as therapy is identified in association with augmentation of cellular immune defenses. At present, it appears that organism-specific immunoglobulin therapy will eventually return as adjunctive support with or without cytokine use. Thus, in summary, practical yet innovative therapies to aid in prevention of infection or in improvement in host defenses are desperately needed.

III. FILGRASTIM AS ADJUNCTIVE THERAPY FOR NEONATAL SEPSIS

Filgrastim has been shown to enhance stochastic entry into the granulocytic pathway, to increase rapid egress of immature neutrophils into the peripheral circulation, and most importantly, to improve the killing capacity of mature neutrophils (63). In addition, the myeloid lineage from premature infants expresses receptors for this cytokine at adult levels (33). Taken together, these observations suggest that Filgrastim replacement therapy may be beneficial in this population.

In initial animal experiments, Cairo and his colleagues (64) demonstrated that prophylactic administration of Filgrastim with antibiotics improved survival of neonatal rat pups to 91% compared with 28% observed with antibiotics alone. These authors concluded that Filgrastim had both prophylactic and synergistic effects in the treatment of neonatal sepsis in rats. In a related study, Filgrastim improved the survival from 39% to 57% in leukopenic rabbits challenged with an LD_{50} of *Pasteurella multocida* (65). Recently, several human case reports argue that administration of Filgrastim has contributed significantly to improved outcome due to neonatal septicemia even without neutropenia (35,66,67). In addition, we (52) and others (51,68) have demonstrated that the neonatal bone marrow can, in fact, respond to Filgrastim therapy by increasing the ANC even when neutropenia was persistent for more than 24 hours in neonates with septic shock (51). A similar observation is reported by Schibbler and colleagues (36).

In a randomized study of 42 non-neutropenic newborn infants treated in the first 3 days of postnatal life because of a diagnosis of "presumed sepsis," Gillian et al (35) demonstrated a nonproportional dose-responsive increase in neutrophils within 24 hours of administration of 5 or 10 μg/kg Filgrastim. They reported no evidence of toxicity immediately after the drug (Gillan et al, 1994) or at a follow-up 2 years later (69).

In a pilot study by Bedford-Russel et al (51), Filgrastim was administered to 12 ill neutropenic neonates. The ANC increased in all neonates 4 days after Filgrastim administration and may have increased survival of their patients. Drug-induced toxicity was not evident in any of their neonates except for a decrease in the platelet count observed in one of their patients who had entered this trial with thrombocytopenia already in existence. In a recent abstract, Filgrastim was administered at 5 μg/kg/day for 5 days to 14 VLBW neonates with presumed sepsis and neutropenia, and showed that 13 of them responded with an increase of the ANC from 1.19 to 12.98×10^9/L on day 5 with no adverse effects (68).

IV. FILGRASTIM THERAPY FOR NEUTROPENIA AND SEPTIC SHOCK IN NEONATES WITH EARLY- AND LATE-ONSET SEPSIS

In VLBW neonates with septic shock whose hospital course was further complicated by persistent ($\geq$24 hours) and severe ($<1 \times 10^9$/L) neutropenia, we observed that Filgrastim could increase the ANC (37,52). We entered seven neonates with early-onset sepsis plus neutropenia ($>$24 hours) identified at birth and seven neonates with late-onset sepsis plus neutropenia ($>$24 hours; patients with necrotizing enterocolitis, NEC) in the Filgrastim-treatment group. Results were compared with 11 consecutively admitted patients entered into the concurrently treated conventional therapy control group (5 early-onset, 6 late-onset [NEC]). No significant differences existed in the birth weight, gestational age, use of antibiotic therapy, magnitude of respiratory support, severity of metabolic acidosis, use of vasopressors, or other supportive therapy between the two groups.

In the Filgrastim-treated group and in the conventionally treated control group, the ANC at baseline was 0.58 ± 0.14 and $0.44 \pm 0.15 \times 10^9$/L (p = ns; mean $\pm$ SEM), respectively (**Figure 2**). The ANC increased in the Filgrastim-treated group by tenfold over baseline at 24 hours, but only twofold in the conventional-therapy group. There were no "nonresponders" in the Filgrastim group by 24 hours after the first dose of study drug. Monocyte cell counts also increased significantly in both Filgrastim and untreated groups compared with their respective baseline values by 7 days

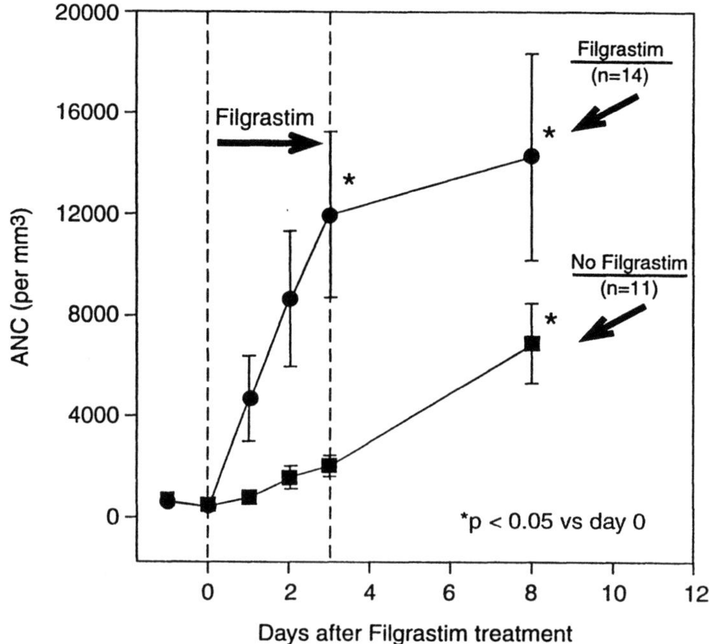

Figure 2. Neonatal ANC responses to Filgrastim during septic shock. Note the early rapid response in the Filgrastim-treated group even at 24 hours after initiating therapy. See text for additional details.

after entry into this protocol, but remained within normal range for age in all cases. No clinically significant effect on lymphocytes, erythrocytes, or platelet counts was noted in the Filgrastim group. Thirteen patients in the Filgrastim-treated group (92%; 13/14) and five in the conventionally treated group (55%; 5/11) survived ($p < 0.03$, power 0.8) to 28 days after the onset of the signs of sepsis (52).

In the literature, reports of experience in adults indicate that concurrent adjunctive Filgrastim treatment may have an advantage, even in non-neutropenic patients (47,65,70). From our studies, we concluded that Filgrastim can increase the neutrophil count in critically ill neutropenic neonates in a rapid and reproducible fashion. Thus, this finding indicates that Filgrastim may be effective in a therapeutically useful time frame to overcome bone marrow suppression or neutrophil consumption. Blinded randomized trials are needed to validate the beneficial effects of Filgrastim. These studies are currently in progress as two multicenter projects in the United States (late-onset sepsis) and Europe (early-onset sepsis).

V. DEVELOPMENTAL ISSUES REGARDING EXPRESSION OF FCγ AND COMPLEMENT RECEPTORS

Immaturity of neutrophil functions may exaggerate quantitative deficiencies of myeloid and phagocytic systems in neonates and thus may contribute significantly to a higher incidence of sepsis in neonates. Neutrophil Fcγ and complement receptors are required for efficient and effective opsonization, phagocytosis, and killing through binding of antibody and complement. Payne and Fleit (71) and others (72–75) have demonstrated decreased expression of FcγII and FcγIII receptors on human neonatal neutrophils. These investigators also have demonstrated independent effects of postconceptional age, postnatal age, and disease states on these maturational and developmental events. In addition, Falconer et al (73) and Abughali et al (76) showed decreased complement receptor density on neutrophils in neonates compared with adults.

After Filgrastim therapy, not only quantitative increases in neutrophil number in adults but also qualitative improvement in neutrophil function have been demonstrated (77). Ohsaka et al (78) showed increased CR1, CR3, FcγRI, and FcγRII on adult neutrophils in vivo. Kerst et al (79) demonstrated that FcγRI and FcγRIII increased after Filgrastim therapy in healthy human adult volunteers. The increased expression of FcγRI on neutrophils has been demonstrated by many authors (79–82). Increased C3bi receptor levels on adult cells also were demonstrated by several investigators (78,83). However, in the only study published thus far, Gillan et al (35) showed that Filgrastim increased CR3 expression on neutrophils in a subset of neonates with "presumed sepsis." Clearly additional information is needed in this area to better understand the cellular basis of the clinical effects of Filgrastim in human neonates.

VI. CELLULAR EFFECTS OF FILGRASTIM ON NEONATAL NEUTROPHIL RECEPTORS

We studied the effects of Filgrastim on the number of receptors (mean fluorescent intensity) as well as the proportion of neutrophils expressing Fcγ receptors (FcγRI, FcγRII, FcγRIII) and complement receptors CR1 and CR3 in three groups of neonatal patients: preterm (treated but not infected), preeclampsia-associated neutropenia (treated but not infected), and septic-neutropenic (treated and infected). Results of these three affected neonates were compared with "control" preterm neonates (not infected) who were not treated with Filgrastim. The demographic data of these four groups are summarized in **Table 4**. Neutrophil receptors were studied using flow cytometric analysis on blood sampled before (day 0) and after the third day of Filgrastim treatment (84).

We noted that preterm neonates had a higher expression of FcγRI on their neutrophils even before drug treatment, consistent with a previous report of Falconer et al (73). All neonates we treated (n = 21) showed an increase in FcγRI after treatment with Filgrastim, but the increase did not achieve statistical significance compared with pretreatment levels or to levels achieved in untreated control patients (n = 8). The lack of statistical significance presumably is due to the already elevated baseline expression of FcγRI in sick VLBW neonates (72–74) that may not be able to be augmented further.

Table 4. Demographic Data of Neonates Studied for Fcγ and Complement Receptor Expression.

	VLBW no Filgrastim (N = 8)	VLBW Filgrastim (N = 5)	PET Filgrastim (N = 8)	Sepsis Filgrastim (N = 8)
Birth weight (grams)				
Mean ± SD	871 ± 234	810 ± 184	732 ± 257	1146 ± 626
Range	604–1165	586–950	356–1100	534–2365
Median	735	930	670	935
Gestational age (weeks)				
Mean ± SD	28 ± 1	26 ± 2	28 ± 2	28 ± 3
Range	26–29	24–29	25–32	24–34
Median	28	27	27	29
RDS	8	5	8	7
Survanta	7	5	6	7
UAC	7	5	6	7
UVC	5	4	4	3
IVH > grade II	1	1	2	1
Sepsis	0	0	0	8

IVH = intraventricular hemorrhage.
PET = preeclamptic toxemia.
RDS = respiratory distress syndrome.
Survanta = exogenous surfactant (Abott Laboratories).
UAC = umbilical artery catheter.
UVC = umbilical vein catheter.
VLBW = very low birth weight.
Note: Values represent mean + standard deviation (+ SD) where indicated. No differences between group characteristics are statistically significant.

In contrast to the lack of significant effect on receptor density per neutrophil, Filgrastim not only increased the ANC but also significantly increased the absolute number of cells expressing each receptor subtype we studied (**Figure 3**) (84). The overall pattern of increase in the absolute numbers of neutrophils expressing these receptors was similar among the four groups. Therefore, it appears that the background disease state does not influence the ability of the bone marrow to respond to Filgrastim.

We conclude that regardless of their prior condition (VLBW neonates, preeclampsia-associated neutropenic neonates, or septic-neutropenic neonates), Filgrastim increased the ANC resulting in a phenotypically mature cell type that was indistinguishable from noninfected control patients.

VII. FUTURE DIRECTIONS

Augmenting immune function may not be considered prudent in a population of neonates whose risk of hypoxia, ischemia, and reperfusion injury may be expected to increase their risks of developing such conditions as bronchopulmonary dysplasia, retinopathy of prematurity, intraventricular hemorrhage, and necrotizing enterocolitis. On the other hand, prevention of neonatal sepsis may prove beneficial in decreasing other complications of prematurity. While it is of some theoretical value to consider possible risks, there is absolutely no evidence to date to indicate that these potential complications of augmented cellular immunity are likely to occur in practice. In our preliminary observations (unpublished) on alveolar lavage, we found no difference in the neutrophil cell count before or after treatment with Filgrastim. Nevertheless broader experience in the use of Filgrastim in the clinical setting will better define the risk-benefit ratio of using Filgrastim in this setting, and the multicenter trials currently under way investigating applications in early- and late-onset sepsis will better define the risks and benefits of neonatal therapies.

In summary, it appears that the continued improvement in long-term outcome and survival of extremely immature neonates to birth weights of ≤ 500 g and to gestational ages <24 weeks has magnified the clinical problem of the developmental immaturity of cellular and humoral host defense systems and the increased risk of nosocomial sepsis. In the future, it appears that some combination of selective neonatal colonization using commensal bacteria specified by the clinician, in association with humoral replacement therapies (organism-specific immunoglobulin, complement, fibronectin, etc), and an array of appropriate cytokines (as prophylaxis or therapy) will evolve as the clinical "cocktail" provided by neonatologists in dealing with

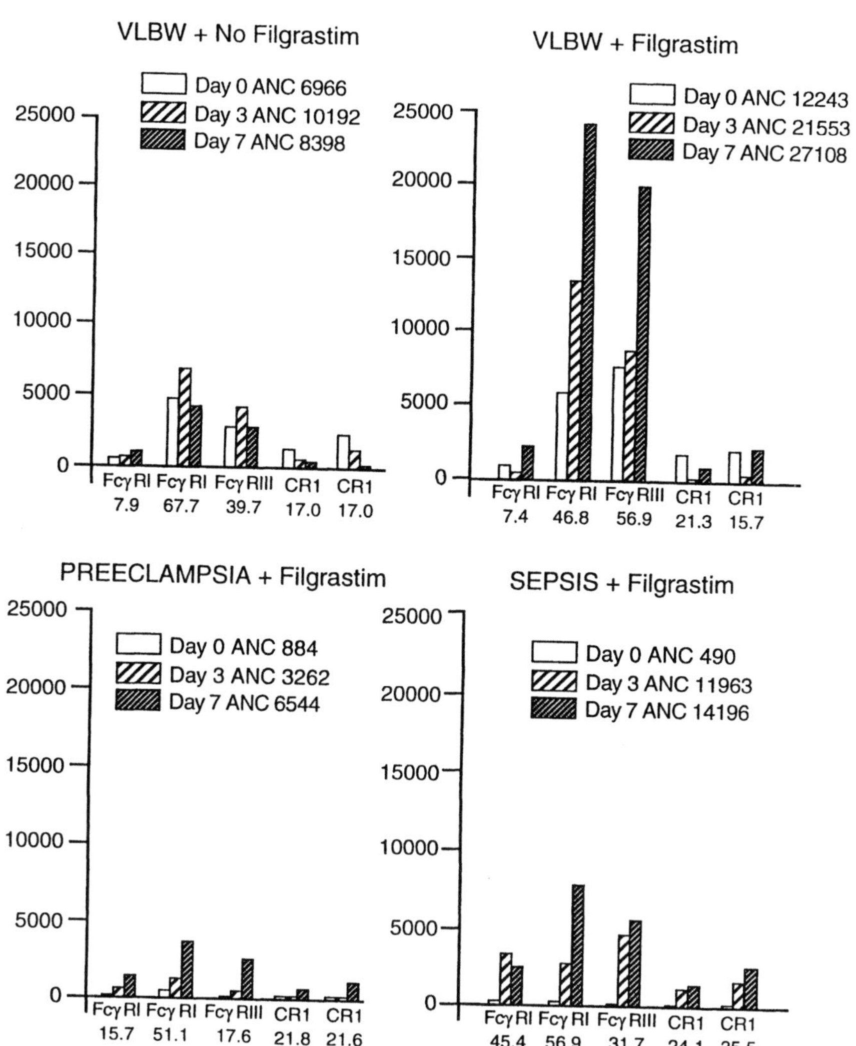

Figure 3. Effect of Figrastim on the absolute neutrophil count (ANC) of neutrophils expressing various Fcγ and complement receptors. Each bar represents the absolute number of cells expressing the phenotype indicated on the x-axis. Numbers accompanying the indicated bar code legends correspond to the ANC for that patient group. See text for additional details. VLBW = very low birth weight; ANC = absolute neutrophil count.

the desperate situation of a disturbingly high infection-related morbidity and mortality.

ACKNOWLEDGMENTS

We would like to thank Dr. Joseph D. De Cristofaro, Dr. Josie Niu, and Ms. Adrienne Combs, RN, for their helpful comments during the preparation of this manuscript.

REFERENCES

1. Siegel, J. D. and McCracken, G. H. (1981). Sepsis neonatorum. *N Engl J Med 304*:642–647.
2. La Gamma, E. F., Drusin, L. M., Mackles, A. W., Machalek, S., and Auld, P. A. (1983). Neonatal infections. An important determinant of late NICU mortality in infants less than 1,000 g at birth. *Am J Dis Child 137*:838–841.
3. Stoll, B. J., Gordon, T., Korones, S. B., et al (1996). Early-onset sepsis in very low birth weight neonates: A report from the National Institute of Child Health and Human Development Neonatal Research Network. *J Pediatr 129*:72–80.
4. Stoll, B. J., Gordon, T., Korones, S. B., et al (1996). Late-onset sepsis in very low birth weight neonates: A report from the National Institute of Child Health and Human Development Neonatal Research Network. *J Pediatr 129*:63–71.
5. Ohlsson, A., and Vearncombe, M. (1987). Congenital and nosocomial sepsis in infants born in a regional perinatal unit: cause, outcome, and white blood cell response. *Am J Obstet Gynecol 156*:407–413.
6. Klein, J. O., and Marcy, M. S. Bacterial sepsis and meningitis. In: J. S. Remington, J. O. Klein eds. *Infectious Diseases of the Fetus and Newborn Infant.* Philadelphia: W.B. Saunders Company, 1995, pp. 835–874.
7. Polin, R. A., and St. Geme III, J. W. (1992). Neonatal sepsis. *Adv Pediatr Infect Dis 7*:25–61.
8. Greenough, A. (1996). Bacterial sepsis and meningitis. *Semin Neonat 1*:147–159.
9. Lewis, D. B., and Wilson, C. B. Developmental immunology and the role of host defenses in neonatal susceptibility to infection. In: J. S. Remington, J. O. Klein, eds. *Infectious Diseases of the Fetus and Newborn Infant.* Philadelphia: W.B. Saunders Company, 1995, pp. 21–98.
10. Kemp, A. S., and Campbell, D. E. (1996). The neonatal immune system. *Semin Neonat 1*:67–75.
11. La Gamma, E. F., and Brown, L. E. (1995). Feeding practices for infants weighing less than 1500g at birth and pathogenesis of necrotizing enterocolitis. *Clin Perinatol 21*:271–306.

12. Adan, D., La Gamma, E. F., and Brown, L. E. (1995). Nutritional management and multisystem organ failure/systemic inflammatory response in critically ill preterm neonates. *Crit Care Clin 11:*751–784.

13. Beck-Sague, C. M., Azimi, P., Fonesca, S. N., et al (1994). Bloodstream infections in neonatal intensive care unit patients: results of a multicenter study. *Pediatr Infect Dis J 13:*1110–1116.

14. Christensen, R. D. (1989). Neutrophil kinetics in the fetus and neonate. *Am J Pediatr Haematol Oncol 11:*215–223.

15. Engle, W. A., McGuire, W. A., Schrelner, R. L., and Yu, P.-L. (1981). Neutrophil storage pool depletion in neonates with sepsis and neutropenia. *J Pediatr 113:*747–749.

16. Hill, H. R. (1987). Biochemical, structural, and functional abnormalities of polymorphonuclear leukocytes in the neonate. *Pediatr Res 22:*375–382.

17. Foucar, K., Duncan, M. H., and Smith, K. J. (1993). Practical approach to the investigation of neutropenia. *Clin Lab Med 13:*879–894.

18. Moore, M. A., Welte, K., Gabrilove, J., et al (1987). Biological activities of recombinant granulocyte colony stimulating factor (rhG-CSF) and tumor necrosis factor: in vivo and in vitro analysis. *Hamatol Bluttranfus 31:*210–220.

19. Welte, K., Bonilla, M. A., Gillio, A. P., et al (1987). Recombinant human granulocyte colony-stimulating factor: effects on hematopoiesis in normal and cyclophosphamide-treated primates. *J Exp Med 165:*941–948.

20. Kuga, T., Komatsu, Y., Yamaski, M., et al (1989). Mutagenesis of human granulocyte colony stimulating factor. *Biochem Biophys Res Comm 159:*103–111.

21. Vellenga, E., Rambaldi, A., Ernst, T. J., Ostapovicz, D., Griffin, J. D. (1988). Independent regulation of M-CSF and G-CSF gene expression in human monocyte. *Blood 71:*1529–1532.

22. Morstyn, G., and Burgess, A. W. (1988). Hemopoietic growth factors: a review. *Cancer Res 48:*5624–5637.

23. Welte, K., Gabrilove, J., Bronchud, M. H., Platzer, E., and Morstyn, G. (1996). Filgrastim (r-metHuG-CSF): the first 10 years. *Blood 88:*1907–1929.

24. Cairo, M. S., Suen, Y., Knoppel, E., et al (1992). Decreased G-CSF and IL-3 production and gene expression from mononuclear cells of newborn infants. *Pediatr Res 31:*574–578.

25. Lee, S. M., Knoppel, E., Van De Ven, C., and Cairo, M. S. (1993). Transcriptional rates of granulocyte-macrophage colony-stimulating factor, granulocyte colony-stimulating factor, interleukin-3, and macrophage colony-stimulating factor genes in activated cord versus adult mononuclear cells: alteration in cytokine expression may be secondary to posttranscriptional instability. *Pediatr Res 34:*560–564.

26. Schibler, K. R., Liechty, K. W., White, W. L., and Christensen, R. D. (1993). Production of granulocyte colony-stimulating factor in vitro by monocytes from preterm and term neonates. *Blood 82:*2478–2484.

27. Calhoun, D. A., Rosa, C., and Christensen, R. D. (1996). Transplacental passage of recombinant human granulocyte colony-stimulating factor in women with an imminent preterm delivery. *Am J Obstet Gynecol 174:*1306–1311.

28. Ishii, E., Masuyama, T., Yamaguchi, H., et al (1995). Production and expression of granulocyte- and macrophage-colony- stimulating factors in newborns: their roles in leukocytosis at birth. *Acta Haematol 94:*23–31.

29. Gessler, P., Kirchmann, N., Kientsch-Engel, R., Haas, N., Lasch, P., and Kachel, W. (1993). Serum concentrations of granulocyte colony-stimulating factor in healthy term and preterm neonates and in those with various diseases including bacterial infections. *Blood 82:*3177–3182.

30. Russell, A. R., Davies, E. G., McGuigan, S., Scopes, G. J., Daly, S., Gordon-Smith, E. C. (1994). Plasma granulocyte-colony stimulating factor concentrations (G-CSF) in the early neonatal period. *Br J Haematol 86:*642–644.

31. Kennon, C., Overturf, G., Bessman, S., Sierra, E., Smith, K. J., and Brann, B. (1996). Granulocyte colony-stimulating factor as a marker for bacterial infection in neonates. *J Pediatr 128:*765–769.

32. Nakayama, H., Kukita, J., Ohga, S., and Ueda, K. (1995). Granulocyte colony-stimulating factor levels in cord blood and neonatal peripheral blood. *Acta Pediatr Jpn 37:*237–239.

33. Shimada, M., Minato, M., Takada, M., Takahashi, S., and Harada, K. (1996). Plasma concentration of granulocyte-colony-stimulating factor in neonates. *Acta Paediatr 85:*351–355.

34. Lord, B. I., Bronchud, M. H., Owens, S., et al (1989). The kinetics of human granulopoiesis following treatment with granulocyte colony-stimulating factor in vivo. *Proc Natl Acad Sci USA 86:*9499–9503.

35. Gillan, E. R., Christensen, R. D., Suen, Y., Ellis, R., van de Ven, C., and Cairo, M. S. (1994). A randomized, placebo-controlled trial of recombinant human granulocyte colony-stimulating factor administration in newborn infants with presumed sepsis: significant induction of peripheral and bone marrow neutrophilia. *Blood 84:*1427–1433.

36. Schibler, K. R., Le, T. V., and Leung, L. (1996). G-CSF administration to neonates with early onset sepsis and neutropenia: a randomized, placebo-controlled trial. *Pediatr Res 39:*A291 (abstr).

37. La Gamma, E. F., and Kocherlakota, P. (1996). Reversal of sepsis associated neonatal neutropenia with recombinant human granulocyte colony stimulating factor (rhG-CSF). *Pediatr Res 39:*A1325.

38. Symonds, E. M. (1995). Hypertension in pregnancy. *Arch Dis Child 72:*F139–F144.

39. Koenig, J. M., and Christensen, R. D. (1989). Incidence, neutrophil kinetics, and natural history of neonatal neutropenia associated with maternal hypertension. *N Engl J Med 321:*557–562.

40. Brazy, J. E., Grimm, J. K., and Little, V. A. (1982). Neonatal manifestations of severe maternal hypertension occurring before the thirty-sixth week of pregnancy. *J Pediatr 100:*265–271.

41. Gessler, P., Luders, R., Konig, S., Haas, N., Lasch, P., and Kachel, W. (1995). Neonatal neutropenia in low birthweight premature infants. *Am J Perinatol 12:*34–38.

42. Doron, M. W., Makhlouf, R. A., Katz, V. L., Lawson, E. E., and Stiles, A. D. (1994). Increased incidence of sepsis at birth in neutropenic infants of mothers with preeclampsia. *J Pediatr 125:*452–458.

43. Fraser, S. H., and Tudehope, D. I. (1996). Neonatal neutropenia and thrombocytopenia following maternal hypertension. *J Paediatr Child Health 32*:31–34.
44. Cadnapaphornchai, M., and Faix, R. G. (1992). Increased nosocomial infection in neutropenic low birth weight (2000 grams or less) infants of hypertensive mothers. *J Pediatr 121*:956–961.
45. Manroe, B. L., Weinberg, A. G., Rosenfeld, C. R., and Browne, R. (1979). The neonatal blood count in health and disease. I. Reference values for neutrophilic cells. *J Pediatr 95*:89–98.
46. Koenig, J. M., and Christensen, R. D. (1991). The mechanism responsible for diminished neutrophil production in neonates delivered of women with pregnancy-induced hypertension. *Am J Obstet Gynecol 165*:467–473.
47. Dale, D. C. (1994). Potential role of colony stimulating factors in the prevention and treatment of infectious diseases. *Clin Infect Dis 18*:S180–S188.
48. La Gamma, E. F., Alpan, O., and Kocherlakota, P. (1995). Effect of granulocyte colony-stimulating factor on preeclampsia-associated neonatal neutropenia. *J Pediatr 126*:457–459.
49. Kocherlakota, P., and La Gamma, E. F. (1997). Recombinant human granulocyte colony stimulating factor (rhG-CSF) increased absolute neutrophil count (ANC) and decreased the incidence of infections in preeclampsia associated neutropenia in VLBW neonates. *Pediatr Res 41*:159A.
50. Makhlouf, R. A., Doron, M. W., Bose, C. L., Price, W. A., and Stiles, A. D. (1995). Administration of granulocyte colony-stimulating factor to neutropenic low birth weight infants of mothers with preeclampsia. *J Pediatr 126*:454–456.
51. Bedford-Russel, A. R., Davies, E. G., Ball, S. E., and Gordon-Smith, E. (1995). Granulocyte colony-stimulating factor treatment for neonatal neutropenia. *Arch Dis Child 72*:F53–F54.
52. Kocherlakota, P., and La Gamma, E. F. Recombinant human granulocyte colony stimulating factor (rhG-CSF) may improve outcome due to neonatal sepsis complicated by neutropenia. *Pediatrics 100*:1e6.
53. Al-Mulla, Z. A., and Christensen, R. D. (1995). Neutropenia in the neonate. *Clin Perinatol 22*:711–739.
54. Christensen, R. D., and Rothstein, G. (1980). Brief clinical and laboratory observations, exhaustion of mature marrow neutrophils in neonates with sepsis. *J Pediatr 96*:316–318.
55. Christensen, R. D., Rothstein, G., Anstall, H. B., and Bybee, B. (1982). Granulocyte transfusions in neonates with bacterial infection, neutropenia, and depletion of mature marrow neutrophils. *Pediatrics 70*:1–6.
56. Cairo, M. S., Worcester, C. C., Rucker, R. W., et al (1992). Randomized trial of granulocyte transfusions versus intravenous immune globulin therapy for neonatal neutropenia and sepsis. *J Pediatr 120*:281–288.
57. Wheeler, J. G., Chauvenet, A. R., Johnson, C. A., Block, S. M., Dillard, R., and Abramson, J. S. (1987). Buffy coat transfusions in neonates with sepsis and neutrophil storage pool depletion. *J Pediatr 97*:422–425.
58. Baker, C. J., Melish, M. E., Hall, X. X., Casto, D. T., Vasan, U., and Givner, L. B. (1992). Intravenous immunoglobulin for the prevention of nosocomial

infection in low-birth-weight neonates. The multicenter group for the study of immunoglobulin in neonates. *N Engl J Med 327*:213–219.

59. Fanaroff, A. A., Korones, S. B., Wright, L. L., et al (1994). A controlled study of intravenous immunoglobulin to reduce nosocomial infection in very-low-birth-weight infants. National Institute of Child Health and Human Development National Research Network. *N Engl J Med 330*:1107–1113.

60. Lacy, J. B., and Ohlsson, A. (1995). Administration of intravenous immunoglobulins for prophylaxis or treatment of infection in preterm infants: meta-analysis. *Arch Dis Child 72*:F151–F155.

61. Jonas, M. M., Baron, M. J., Bresee, J. S., and Schneider, L. C. (1996). Clinical and viriological features of hepatitis C virus infection associated with intravenous immunoglobulin. *Pediatrics 98*:211–215.

62. Bellanti, J. A., Pung, Y., and Zeligs, B. J. Immunology. In: Avery, G., Fletcher, B., MacDonald, M. A., eds. *Neonatology: Pathophysiology and Management of the Newborn.* Philadelphia: JB Lippincott Company, 1994, pp. 1000–1028.

63. Roilides, E., Walsh, T. J., Pizzo, P. A., and Rubin, M. (1991). Granulocyte colony-stimulating factor enhances the phagocytic and bactericidal activity of normal and defective human neutrophils. *J Infect Dis 163*:579–583.

64. Cairo, M. S., Mauss, D., Kommareddy, S., Norris, K., van de Ven, C., and Modanlou, H. (1990). Prophylactic or simultaneous administration of recombinant human granulocyte colony stimulating factor in the treatment of group B streptococcal sepsis in neonatal rats. *Pediatr Res 27*:612–616.

65. Smith, W. S., Sumnicht, G. E., Sharpe, R. W., Samuelson, D., and Millard, F. E. (1995). Granulocyte colony-stimulating factor versus placebo in addition to penicillin G in a randomized blinded study of gram-negative pneumonia sepsis: Analysis of survival and multisystem organ failure. *Blood 86*:1301–1309.

66. Roberts, R. L., Szelc, C. M., Scates, S. M., et al (1991). Neutropenia in an extremely premature infant treated with recombinant human granulocyte colony-stimulating factor. *Am J Dis Child 145*:808–812.

67. Murray, J. C., McClain, K. L., and Wearden, M. E. (1994). Using granulocyte colony-stimulating factor for neutropenia during neonatal sepsis. *Arch Pediatr Adolesc Med 148*:764–766.

68. Lebovitz, E. (1995). A pilot, prospective trial of recombinant human granulocyte colony stimulating factor (rhG-CSF) in the treatment of newborn infants with presumed sepsis and neutropenia. *Int Congr Chemother* (abstr 6A).

69. Rosenthal, J., Healey, T., Ellis, R., Gillan, E., and Cairo, M. S. (1996). A two year follow up of neonates with presumed sepsis treated with recombinant human granulocyte colony stimulating factor during the first week of life. *J Pediatr 128*:135–137.

70. Nelson, S. (1994). Role of G-CSF in the immune response to acute bacterial infections in the non-neutropenic host: an overview. *Clin Infect Dis 18*:S197–S204.

71. Payne, N. R., and Fleit, H. B. (1996). Extremely low birth weight infants have lower FcγRIII (CD16) plasma levels and their PMN produce less FcγRIII compared to adults. *Biol Neonate 69*:235–242.

72. Meada, M., Van Schie, R. C., Yuksel, B., et al (1996). Differential expression of Fc receptors for IgG by monocytes and granulocytes from neonates and adults. *Clin Exp Immunol 193*:343–347.

73. Carr, R., and Davies, J. M. (1990). Abnormal FcRIII expression by neutrophils from very preterm neonates. *Blood 76*:3:607–611.

73. Falconer, A. E., Carr, R., and Edwards, S. W. (1995). Neutrophils from preterm neonates and adults show similar cell surface receptor expression: analysis using a whole blood assay. *Biol Neonate 67*:26–33.

74. Payne, N. R., Frestedt, J., Hunkeler, N., and Gehrz, R. (1993). Cell-surface expression of immunoglobulin G receptors on the polymorphonuclear leukocytes and monocytes of extremely premature infants. *Pediatr Res 33*:452–457.

75. Smith, J. B., Campbell, D. E., Ludomirsky, A., et al (1990). Expression of the complement receptors CR1 and CR3 and the type III Fcγ receptor on neutrophils from newborn infants and from fetuses and Rh disease. *Pediatr Res 28*:120–126.

76. Abughali, N., Berger, M., and Tosi, M. F. (1994). Deficient total cell content of CR3 (CD11b) in neonatal neutrophils. *Blood 83*:1086–1092.

77. Dale, D. C., Liles, W. C., Summer, W. R., and Nelson, S. (1995). Granulocyte colony stimulating factor: role and relationships in infectious diseases. *J Infect Dis 172*:1061–1075.

78. Ohsaka, A., Saionji, K., Kuwaki, T., Takeshima, T., and Igari, J. (1995). Granulocyte colony-stimulating factor administration modulates the surface expression of effector cell molecules on human monocytes. *Br J Haematol 89*:465–472.

79. Krest, J. M. van de Winkel J. G., Evans, A. H., et al (1993). Recombinant granulocyte colony stimulating factor administration to healthy volunteers: induction of immunophenotypically and functionally altered neutrophils via an effect on myeloid progenitor cells. *Blood 82*:3265–3272.

80. Valerius, T., Repp, R., deWit, T. P., et al (1993). Involvement of the high-affinity receptor for IgG (FcgRI; CD64) in enhanced tumor cell cytotoxicity of neutrophils during granulocyte colony-stimulating factor therapy. *Blood 82*:931–939.

81. Repp, R., Valerius, T., Sendler, A., et al (1991). Neutrophils express high affinity receptor for IgG (FcγR1; CD64) after in vivo application of rhG-CSF. *Blood 78*:885–889.

82. Gericke, G. H., Ericson, S. G., Pan, L., Mills, L. E., Guyre, P. M., and Ely, P. (1995). Mature polymorphonuclear leukocytes express high-affinity receptors for IgG (FcγRI) after stimulation with granulocyte colony-stimulating factor (G-CSF). *J Leuk Biol 57*:456–461.

83. Yuo, A., Kitagawa, S., Ohsaka, A., et al (1989). Recombinant human granulocyte colony-stimulating factor as an activator of human granulocytes: potential of responses triggered by receptor-mediated agonists and stimulation of C3bi receptor expression and adherence. *Blood 74*:2144–2149.

84. Kocherlakota, P., La Gamma, E. F., Breen, C., Fleit, H. B., and Golightly, M. Effect of rhG-CSF on neutrophil Fcγ and complement receptors in ventilated

very low birth weight neonates. In: J. A. Bellanti, R. Bracci, G. Prindull, and M. Xanthou, eds. Neonatal Hematology and Immunology III. Amsterdam: Elsevier Science Inc., pp. 39–48.

85. Laver, J., Duncan, E., Abboud, M., et al (1990). High levels of granulocyte and granulocyte-macrophage colony-stimulating factors in cord blood of normal full-term neonates. *J Pediatr 116*:627–32.

86. Cairo, M. S,. Gillan, E. R., Buzby, J. S., van de Ven, C., and Suen, Y. (1993). Circulating steel factor (SLF) and G-CSF levels in preterm and term newborn and adult peripheral blood. *Am J Pediatr Hematol Oncol 15*:311–315.

23

Filgrastim (r-metHuG-CSF) for the Treatment of Myelodysplastic Syndromes

Robert S. Negrin
Stanford University, Stanford, California

Peter L. Greenberg
*Stanford Medical Center, Stanford, and VA Palo Alto
Health Care System, Palo Alto, California*

I. INTRODUCTION

The myelodysplastic syndromes (MDS) provide a clinical model for the study of the evolution of a relatively benign disorder to one which is frankly leukemic (ie, acute myeloid leukemia [AML]). Because this disease effects predominantly elderly individuals, its management has posed a therapeutic challenge to support the complications of the patient's dominant cytopenia without excessive toxicity. With the emergence of the colony-stimulating factors (CSF) for clinical use, these agents have been evaluated both in vitro and in vivo in an effort to improve the management of patients with MDS.

In MDS, defective proliferation of hematopoietic precursors has been suggested to be due to decreased production or responsiveness to hematopoietic growth factors. In addition, MDS precursors appear to have an increased rate of apoptosis (1,2). The identification of recurring chromosomal abnormalities, primarily involving chromosomes 5, 7, and 8 have suggested potential genetic lesions that provide unique insights into this disease. Because some leukemic cells will proliferate with the addition of CSF in vitro (3–5), there has been concern about the use of these agents in MDS patients. These in vitro findings have suggested the possible efficacy of CSF in this clinical setting and have led to therapeutic trials (6).

491

II. CLINICAL CHARACTERISTICS OF PATIENTS
WITH MDS

The MDS are characterized by chronic cytopenias and cytopathies with dysplastic morphologic features. Patients with MDS are at increased risk for infection, bleeding, and symptomatic anemia, as well as transformation to AML (7–9). This disorder has been subclassified by marrow morphologic criteria (mainly the proportion of marrow blasts) into five subtypes by the French-American-British (FAB) group which are useful for prognostic purposes (10). These subtypes include refractory anemia (RA), refractory anemia with ringed sideroblasts (RARS), refractory anemia with excess blasts (RAEB), RAEB in transformation (RAEB-T), and chronic myelogenous leukemia (CML). Some authors have suggested that CML more likely belongs in the group of myeloproliferative disorders (11). Patients with RA and RARS generally have a median survival of 2 to 4 years, a low incidence of evolution to AML, and often die of diseases unrelated to MDS (12–14). In contrast, those patients with RAEB, RAEB-T, and CML have very poor prognosis with median survival of 6 to 12 months, with approximately 50% of such patients evolving into AML.

The MDS are a disease most commonly seen in elderly patients with >70% of patients more than 60 years of age. Men and women are affected equally. The incidence of this disease is approximately 4/100,000; however, this is increasing because of the advancing age of the population in the United States and Western Europe. The cause of this disorder is unknown; however, it has been speculated that unrepaired damage to DNA may play an important role. Chromosomal abnormalities are found in 40% to 70% of patients with MDS. These abnormalities typically involve chromosomes 5, 7, or 8; with monosomy 5, 5q-, monosomy 7, and trisomy 8 being the most common findings. A variety of other chromosomal abnormalities have been found in these patients. Evidence of clonality has been demonstrated in MDS patients by X-linked genetic analysis including restriction fragment length polymorphism studies, as well as by cytogenetics and in vitro culture (15,16).

Patients with MDS are often diagnosed following a routine blood test or one performed for non-specific complaints. Patients can present with isolated cytopenias or pancytopenia that typically prompts a bone marrow examination. This usually demonstrates a hypercellular bone marrow with evidence of dysplasia in at least two of the hematopoietic cell lines on the bone marrow aspirate. The clinical complications of MDS patients are typical of the patient's dominant cytopenia and include infections, bleeding, and anemia. Red blood cell transfusions are common, as well as antibiotic and platelet support in a smaller percentage of patients. In addition to the cytopenias, MDS patients often have abnormally functioning neutrophils, platelets, and red cells (17).

III. THERAPY FOR MDS

Therapeutic options for patients with MDS are limited. The only curative approach has been allogeneic bone marrow transplantation (allo-BMT) in selected patients. The advanced age of most patients with this disorder makes this impractical for many MDS patients (18,19). Recent successes in transplanting older patients as well as the growth of the National Marrow Donor Program (NMDP), however, has increased options for patients with MDS. Cytotoxic chemotherapy has been used for patients who show signs of evolution to AML, but with generally poor results (20,21). Newer topoisomerase-1–interacting drugs, such as topotecan, have produced clinical responses in some patients (22), and various agents such as pyridoxine, androgens, danazol, and corticosteroids have been used with limited benefit in a small proportion of patients (23–27). Low-dose cytarabine and retinoids have been widely studied, but have not been shown to improve survival in clinical trials, including several randomized studies (28–33). The standard approach for the treatment of patients with MDS has been to monitor patients carefully and support them with transfusions of red cells and platelets if required, as well as aggressive treatment of infections.

IV. RATIONALE FOR THE USE OF COLONY-STIMULATING FACTORS IN MDS

The development of in vitro marrow clonogenic culture assays led to the discovery of a complex family of interacting growth factors, the CSF, which are likely to have critical physiologic roles in the control of hematopoiesis in vivo (3). Interleukin (IL)-3 and granulocyte-macrophage colony-stimulating factor (GM-CSF) have predominantly proliferative effects on immature hematopoietic precursors, whereas G-CSF and macrophage-CSF (M-CSF) have differentiative and proliferative effects on more lineage-committed precursors. Each hematopoietic cell lineage appears to be regulated by both proliferative and differentiative stimuli. In MDS, marrow cultures have generally demonstrated subnormal clonal growth and defective cellular maturation of myeloid and erythroid precursors that become more abnormal as these patients evolve toward AML (34). The presence of defective in vitro myeloid clonogenicity has been shown to have negative prognostic import in MDS (34). Erythropoietin (EPO) is a relatively late-acting factor that predominantly effects erythroid colony-forming cells (CFC-E) and a portion of the earlier erythroid burst-forming cells (BFC-E), which generate CFC-E. It acts in synergy with G-CSF and certain earlier-acting factors to enhance erythropoiesis in vitro (35). Defective in vitro erythropoiesis has been demonstrated in MDS (32).

The chronic cytopenias that characterize this disorder have led to the consideration of CSF for the treatment of MDS. The major questions include not only whether patients will respond to these agents with an improvement in blood counts but, more importantly, whether patients derive clinical benefit with a decreased need for transfusions and risk of infection. Central to the use of CSF in MDS is also whether such treatment alters survival or the pace of disease progression towards AML.

V. IN VITRO EFFECTS OF COLONY-STIMULATING FACTOR ON MARROW PRECURSORS IN MDS

The use of differentiation-inducing agents such as Filgrastim for treating myeloid clonal hemopathies including MDS is based on an evolving body of in vitro marrow culture and preclinical data indicating that such agents may diminish self-replication of abnormal cell clones concomitant with enhancing their differentiation (36–41).

We have compared the proliferative and differentiative effects of Filgrastim and recombinant human (rHu) GM-CSF on marrow hematopoietic precursors from patients with MDS using both methylcellulose and liquid suspension culture assays (42,43). Enriched immature myeloid cell populations, consisting of predominantly myeloblasts and promyelocytes and containing the hematopoietic progenitor cells, were obtained by removing the adherent and mature myeloid and erythroid bone marrow cells using anti-My8 and anti-glycophorin antibodies. These enriched cells were plated in clonogenic assays to determine myeloid colony formation (CFC-GM) and differentiation after liquid culture with Filgrastim, rHuGM-CSF, and placental-conditioned media, or without a source of CSF. Myeloid differentiation was assessed morphologically, and secondary plating was used to determine hematopoietic progenitor cell regeneration. Levels of CFC-GM stimulated with Filgrastim or rHuGM-CSF were subnormal in the majority of MDS patients (42). Recombinant HuGM-CSF exerted a stronger proliferative stimulus than Filgrastim for marrow CFC-GM growth in these MDS patients, and rHuGM-CSF stimulated increased CFC-GM growth in MDS patients with abnormal cytogenetics in comparison to those who had normal cytogenetics, whereas Filgrastim demonstrated similar and lower CFC-GM proliferative effects in patients with either normal or abnormal cytogenetics. Dose-response curves of MDS marrow cells with Filgrastim and rHuGM-CSF for CFC-GM growth demonstrated normal-to-increased proliferative sensitivity to rHuGM-CSF compared with normal-to-decreased proliferative responses to Filgrastim (42).

To assess differentiative effects of the CSF, immature marrow cells were placed in liquid culture for 7 days with or without the CSF. For marrow immature cells from both normal controls and patients with MDS, Filgrastim induced greater granulocytic differentiation than did rHuGM-CSF. Granulocytic differentiation occurred in cells obtained from most normal subjects treated with Filgrastim. The granulocytic induction was less potent for the MDS patients than for normal controls. These studies demonstrated that Filgrastim had greater granulocytic differentiative and less proliferative activity for MDS marrow cells than did rHuGM-CSF in vitro (42). This was particularly true for those patients with RAEB or RAEB-T subtypes and those with normal cytogenetics. In liquid culture, Filgrastim or rHuGM-CSF also decreased myeloid clonal regeneration in most patients, which was associated with enhanced myeloid differentiation.

Several studies have indicated that increased apoptosis of hemopoietic precursors may account for the ineffective hematopoiesis and cytopenias occurring in MDS (1,2). These studies also have shown that exposure to hematopoietic cytokines decreased the apoptosis concomitant with improving the patients' blood counts.

VI. HEMATOLOGIC EFFECTS OF FILGRASTIM TREATMENT OF PATIENTS WITH MDS

We have used *E. coli*–derived (non-glycosylated) rHuG-CSF (Filgrastim) administered subcutaneously (SC) to treat patients with primary MDS since chronic administration of the drug is likely to be required for this disease. In the initial phase 1–2 dose-escalation study Filgrastim was administered by daily SC injection at dosage levels between 0.1 to 3.0 μg/kg/day until a neutrophil response was demonstrated (44). In this study of 18 patients, 16 had an increase in white blood cell count (2- to 10-fold) and absolute neutrophil count (ANC) (5- to 40-fold). This was noted even among the 11 patients who were severely neutropenic (ANC $<0.5 \times 10^9$/L). In most patients, responses were seen at dosage levels between 0.3 to 1.0 μg/kg/day. Upon discontinuation of Filgrastim injections, blood counts returned to pretreatment baseline values in all patients after a 2- to 4-week period (44).

A longer-term phase 2 study was performed to determine whether these responses could be sustained (45). Eleven patients were treated with the dose of Filgrastim required to maintain an ANC $>1.8 \times 10^9$/L for 6-

month periods. After the initial dose-escalation trial, the blood counts returned to their pretreatment levels. Filgrastim was restarted, again by daily SC injection, resulting in a rapid increase in white cell count and ANC. After 6 months, Filgrastim was discontinued and again the blood counts returned to their pretreatment values, yet could be improved again by retreatment (45).

Ten of the 11 patients enrolled in the maintenance phase study responded to Filgrastim with a normalization of the ANC (45). These effects could be maintained for prolonged periods. In 6 patients, Filgrastim was stopped after 6 months of treatment and in all patients the blood counts returned to pretreatment levels. Red blood cell responses were also noted in 4 of the 10 anemic patients, where 2 non-transfusion–dependent patients had a greater than 20% increase in hematocrit and 2 more severely anemic red cell transfusion-dependent patients had a decrease in red blood cell requirements. Only a single patient had a sustained increase in platelet count. Two other reports have been published using Filgrastim to treat patients with MDS. In an initial report of five patients treated with intravenous (IV) Filgrastim, all patients had a short-term increase in ANC (46). A second report of 41 patients demonstrated that most patients responded with an increase in ANC after IV administration of glycosylated rHuG-CSF (lenograstim) (47). The dose required for a response was generally 2 to 5 μg/kg/day. Patients were treated for 14 days and no patients converted to AML over this short treatment period. In these studies, the response was limited to elevations of the white blood cell count (primarily neutrophils). Several of the patients had resolution of active infections during treatment.

To further evaluate the in vivo effects of Filgrastim, a multicenter, randomized phase 3 clinical trial was performed. Patients with high-risk MDS (RAEB and RAEB-T) were randomly assigned to observation or to receive Filgrastim (patients in both cohorts of the study also received standard supportive care) (48). Filgrastim administration was adjusted to maintain a normal ANC, and the drug was administered by a daily SC injection. No crossover was allowed in this study. One-hundred and two patients were randomized, 70 with RAEB and 32 with RAEB-T. The median ANC before treatment was $<0.8 \times 10^9$/L in both groups. ANC responses were noted in all patients who received Filgrastim within 2 to 3 weeks, and the median ANC was $>4 \times 10^9$/L, which was maintained for prolonged periods of treatment. There was no change in ANC for patients in the observation group (**Figure 1**). For patients with RAEB or RAEB-T in either group of the study, there were no significant differences in the incidence of transformation to AML or for survival in the RAEB-T patients. In the RAEB patient group, however, there was poorer survival in the

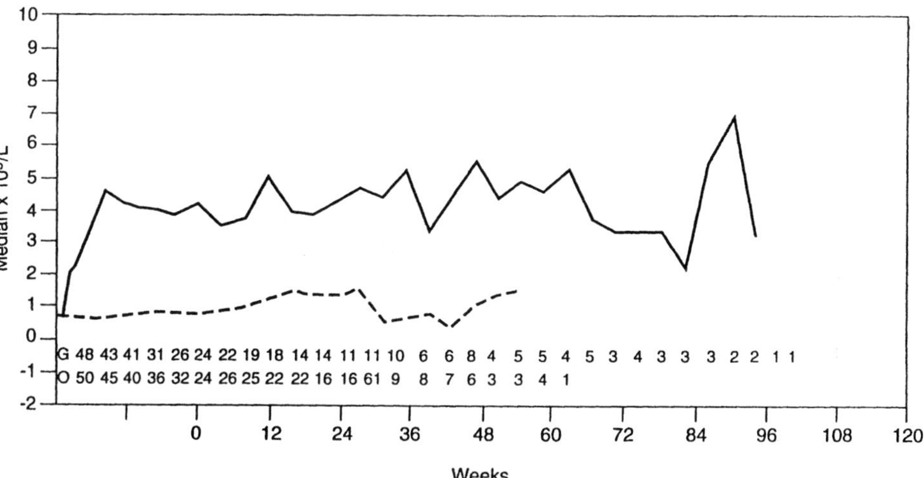

Figure 1. Neutrophil responses for the 102 patients randomized to receive Filgrastim versus observation. Solid line, Filgrastim treatment; dotted line, observation. Number of patients for treatment (G) and observation (O) given on graph.

Filgrastim-treated group, with an increase in non-leukemic disease-related deaths. In this group, patients had survival similar to those reported in prior studies that lacked growth factor intervention. In retrospect, the increased mortality in the RAEB patients treated with Filgrastim was predominantly due to a higher proportion of high-risk RAEB patients in the treatment arm. These results demonstrated no alteration in the incidence of AML evolution, indicating the relative safety of administering Filgrastim to patients with advanced forms of MDS with respect to the risk of transformation. This study underscores the need to perform randomized clinical trials to assess efficacy of treatment modalities in this disease, which has variable natural histories based on their risk group.

In this regard, an International Risk Analysis Workshop was recently convened to examine more precise features capable of predicting outcome in MDS. This workshop analyzed information from 816 patients whose data were collated and combined from a collaborative analysis. These data generated an improved International Prognostic Scoring System (IPSS) for predicting survival and AML evolution in MDS (49). This statistically weighted system incorporated a more refined marrow cytogenetic categorization, blast percentage stratification and the number of cytopenias. This approach provides an improved classification method that should enhance trial design and analysis in the setting of MDS.

VII. EFFECTS OF FILGRASTIM TREATMENT ON ALTERING INFECTION RISK

A. In Vitro Effects

A variety of studies have documented that neutrophils isolated from patients with MDS are abnormal, not only in number, but also in function (17). Morphologically, neutrophils from patients with MDS appear hypogranular. Exogenous G-CSF has been shown to enhance the function of normal neutrophils and those derived from patients with MDS with respect to respiratory burst, chemotaxis, and bacterial killing (50,51). In our studies, after Filgrastim treatment, the neutrophils frequently displayed toxic granulation and Döhle bodies. Leukocyte alkaline phosphatase score increased in four of five patients tested after Filgrastim treatment. Chemotaxis and phagocytosis was maintained or improved in six of eight patients with MDS tested as assessed by in vitro assays (44).

B. In Vivo Effects

The improvement in neutrophil counts in patients with MDS after Filgrastim therapy was encouraging. To approach the question of whether the incidence of infections could be altered with Filgrastim treatment, a retrospective analysis was performed on our MDS patients who were treated with Filgrastim in the maintenance phase 2 trial. In this analysis, the relative risk of bacterial infection was evaluated. In those patients with an ANC $<1.5 \times 10^9/L$, it was found that the infection risk was increased compared with those patients who responded to Filgrastim treatment with an ANC $>1.5 \times 10^9/L$ (45). These data suggest that Filgrastim-treated MDS patients with higher neutrophil counts may also be protected from serious infections. This observation requires confirmation in the randomized phase 3 trial, the analysis of which is ongoing.

VIII. DOES GROWTH FACTOR TREATMENT OF MDS ALTER THE RISK OF LEUKEMIC TRANSFORMATION?

A central question relating to the use of hematopoietic growth factors in patients with MDS is whether they alter the inherent risk of evolution to AML. The results from the randomized trial of Filgrastim versus observation in patients with advanced MDS (ie, RAEB and RAEB-T) indicated that there was no observed increase in transformation events in these high-risk patients who were treated chronically with Filgrastim (48). There are no similar trials reported as yet in low-risk patients (RA, RARS) at the current time, although small numbers of low-risk patients have been treated chronically with Filgrastim and transformation remains an infrequent event.

IX. TOXICITY OF FILGRASTIM THERAPY IN MDS PATIENTS

Filgrastim has been an extremely well tolerated drug in a variety of clinical settings. The major toxicity associated with treatment has been bone pain and infrequent thrombocytopenia. The SC injection of Filgrastim has been well tolerated in an outpatient setting. Several thrombocytopenic patients developed mild and tolerable ecchymoses at the injection sites. One patient with a history of psoriasis developed a flare of her skin disease while receiving Filgrastim. This was severe enough to require discontinuation of cytokine therapy, whereupon the psoriasis resolved (44). Serum samples have not indicated development of antibodies to Filgrastim.

X. EFFECTS OF FILGRASTIM ON CLONOGENIC ASSAYS BEFORE AND AFTER TREATMENT

As differentiation induction is one of the aims of Filgrastim treatment in MDS, an important question is whether the responding neutrophils are derived from the abnormal clone or from residual normal cells. Several approaches have been used to address this question. Marrow cytogenetics have been evaluated before and after treatment. In most patients, the cytogenetic pattern found before treatment persisted after treatment with Filgrastim. In one patient with trisomy 8, this was particularly instructive since this individual had all abnormal metaphases before treatment. This responsive patient had an elevated neutrophil count that persisted for 18 months. Bone marrow-cytogenetic studies performed after 2, 6, 12, and 18 months of treatment all showed trisomy 8 in virtually all metaphases (45). These findings suggested that the neutrophils present were derived from the abnormal clone.

A more direct approach to this question is to study restriction fragment length polymorphisms (RFLP) on female patients who had heterozygous expression of certain X-linked genes such as pyruvate glycerol kinase (PGK) or hypoxanthine phosphoribosyltransferase (HPRT). Four women were analyzed in this fashion, and one was found to be polymorphic at the PGK gene (46). Homozygosity was found in all bone marrow fractions before treatment, indicative of clonal hematopoiesis. After Filgrastim treatment, this patient experienced a fivefold increase in her ANC. Neutrophils isolated at the time of maximal response had the same homozygous pattern on RFLP analysis, indicating that they were derived from the abnormal clone (45). This result is in contrast to a single patient analyzed by this approach after treatment with rHuGM-CSF in which polyclonal hematopoiesis resulted (53).

XI. COMBINATION THERAPY OF MDS PATIENTS WITH FILGRASTIM AND ERYTHROPOIETIN

In vitro erythroid clonal growth in MDS demonstrated suboptimal responses of MDS erythroid precursors to EPO (34). Analysis of the relationship between EPO levels in MDS and patients' erythroid progenitors suggested that the anemia was not due to an abnormality in the capacity of EPO to induce generation of CFC-E but was more likely influenced by the size of the BFU-E population (53). These findings suggested that treatment of MDS patients with rHuEPO alone was unlikely to have clinical benefit. Such in vitro predictions have been borne out by the limited clinical erythroid responses (approximately 20%) using EPO alone (6). We have recently demonstrated that Filgrastim augments the in vitro EPO responsiveness of BFU-E in MDS (54).

Due to this synergy, phase 2 clinical trials have been performed using both Filgrastim and rHuEPO to treat patients with MDS (55,56). Fifty-three of the initial 55 patients (96%) had a neutrophil response. Forty-four patients were evaluable for an erythroid response and 21 (48%) responded with either an increase in hemoglobin level or a decrease in transfusion requirements (56). An example of one patient's response is shown in **Figure 2.** Erythroid responses were significantly more likely to occur in patients with relatively low serum EPO levels. Many of these responses continued for many months (**Figure 3**). To determine whether true in vivo synergy occurred between these two growth factors, 17 patients who had responded for at least an 8-week maintenance phase had Filgrastim discontinued. All of these patients lost their neutrophil response; however, eight patients continued to have an erythroid response to rHuEPO alone (56). In 7 of the 9 remaining patients, resumption of Filgrastim caused recurrent erythroid responses (**Table 1**). A similar study using Filgrastim plus rHuEPO was performed by Hellstrom-Lindberg and colleagues (57), who treated an additional 21 patients in whom 38% had an erythroid response. Responses to combined cytokine therapy with these two CSF were particularly evident in patients with RARS, an MDS subtype with generally poor responsiveness

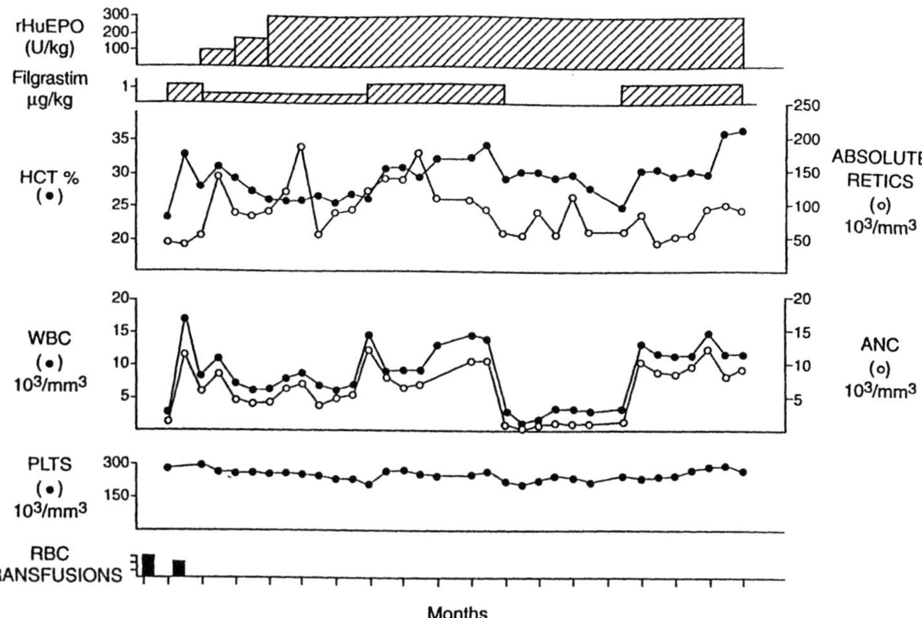

Figure 2. Clinical course of a patient treated with both Filgrastim and rHuEPO. Duration and amount of Filgrastim and rHuEPO treatment shown by cross-hatched boxes. Number and time of red cell transfusions shown by solid bars. HCT = hematocrit; RETICS = reticulocytes; WBC = white blood cells; ANC = absolute neutrophil count; PLTS = platelets; RBC = red blood cells. (From Ref. 56.)

to rHuEPO alone. The data from these two studies were analyzed together to determine which patients were most likely to respond to combined therapy with Filgrastim and rHuEPO. These results indicated that those patients with relatively low serum EPO levels (<100 U/L) and low red cell transfusion requirements (<2 units/month) had a 74% response rate to the combined therapy (58). In contrast, those patients who had relatively high serum EPO levels (>500 U/L) and higher red cell transfusion requirements (>2 units/month) had only a 10% chance of responding to combination therapy. These data indicate that combination therapy with Filgrastim plus rHuEPO may improve not only the neutropenia but also the anemia associated with this disorder in MDS patients, and that the response rate with the combined therapy is greater than that with rHuEPO alone. Therapy with rHuGM-CSF plus rHuEPO also has shown beneficial results for treating the anemia of MDS (59).

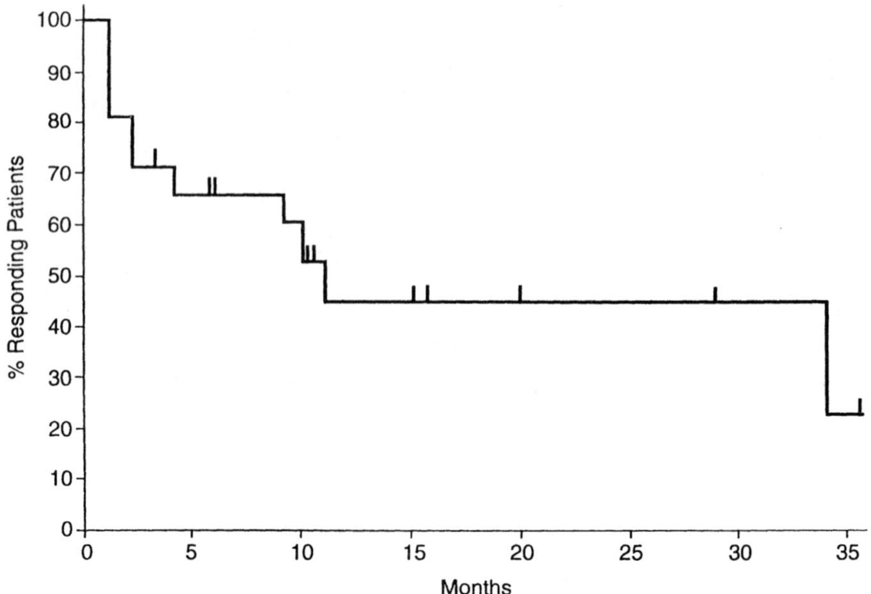

Figure 3. Duration of response for patients who responded to combined therapy with Filgrastim and rHuEPO. (From Ref. 56.)

XII.   WHAT IS THE ROLE OF COLONY-STIMULATING FACTOR THERAPY IN THE TREATMENT OF MDS?

The available phase 1–2 studies using Filgrastim to treat patients with MDS indicate that the majority of patients respond with an increase in neutrophil counts. The retrospective analysis of infection risk suggests that this treatment may protect patients from infection within the limitations of that analytic approach.

Other CSF have been used in the treatment of MDS. Recombinant HuGM-CSF has been the most widely studied and five trials have been reported (60–64). Treatment has generally been short term. Some patients, however, have been treated with daily maintenance injections. These studies are summarized in **Table 2**. Of the 45 patients reported, 38 (84%) have had improvements in ANC. In addition to an increase in neutrophils, eosinophilia has been prominent in these patients. Fourteen patients had an increase in reticulocytes and three had decreased red cell transfusion requirements. Seven patients have progressed to AML over relatively short-term treatment, generally in individuals with greater than 15% marrow

Table 1. Erythroid Response at Different Stages of Treatment with Filgrastim Plus rHuEPO.

Regimen	Erythroid responders	Erythroid non-responders
Acute phase:		
Filgrastim + rHuEPO	21 (48%)	23 (52%)
Maintenance phase:		
Filgrastim + rHuEPO	17 (81%)	4 (9%)
rHEPO only	8	9[a]
Filgrastim + rHuEPO	7	1

[a]One patient not evaluable.

blasts before treatment. A phase 3 trial performed with rHuGM-CSF in patients with less than 15% marrow blasts has been reported in abstract form (65). Most patients who received rHuGM-CSF had an increase in their ANC and decreased incidence of infections. The cross-over design of that study precludes an analysis of progression to AML or survival.

Recombinant HuIL-3 has been used to treat MDS in a limited number of patients over short treatment periods. Modest improvements in neutrophils were noted. A small number of patients also appeared to have transient red cell and platelet responses (66,67), but an increase in toxic side effects were noted in these studies. These investigations indicated that rHuIL-3 may be useful in some patients with MDS; however, the neutrophil responses were not as dramatic as those seen with Filgrastim or rHuGM-CSF, and effects on the other cell lineages were modest. There are no comparative studies between these growth factors.

These studies indicate that cytokines alone and in combination have a potentially valuable role in the management of the anemias of patients with MDS. Studies are required to provide insight into the impact of such

Table 2.   Clinical Effects of Filgrastim and rHuGM-CSF for the Treatment of MDS.

	Filgrastim	rHuGM-CSF
Short-term treatment		
	n = 18	n = 45
FAB Subtypes (N)		
RA	2	14
RAEB/RAEB-T	16	26
CMML	0	5
Duration	42–56 days	7–14 days × 1–5 courses
Daily dose	0.1–3.0 μg/kg	30–750 μg/m^2
Route	SC	SC and IV
Responses (N[%])		
Neutrophils	16 (89%)	38 (84%)
Reticulocytes	5	5
Platelets	1	1
Marrow maturation	9	16
Progression to AML	0	7
Long-term treatment		
Duration	6–28 months	2–9 weeks
Number of patients	11	5
Persistent response (%)	10 (89%)	1 (20%)
Progression to AML	5	1

SC = subcutaneous.
IV = intravenous.
Source: Modified from Ref. 68.

therapy on infectious morbidity. To date, evidence suggests that therapy of this patient population with CSF does not alter their evolution to AML. Improved selection of patients by use of stratification methods such as the recently generated international risk categorization system (IPSS) for MDS may improve our ability to determine which patients would more likely benefit from therapeutic interventions, including cytokine treatment.

REFERENCES

1.   Raza, A., Mundle, S., Iftikhar, A., et al (1995). Simultaneous assessment of cell kinetics and programmed cell death in bone marrow biopsies of myelodys-

plastics reveals extensive apoptosis as the probable basis for ineffective hematopoiesis. *Am J Hematol 48:*143–154.

2. Rajapaksa, R., Ginzton, N., Rott, L. S., and Greenberg, P. L. (1996). Altered oncoprotein expression and apoptosis in myelodysplastic syndrome marrow cells. *Blood 88:*4275–4287.

3. Metcalf, D. (1986). The molecular biology and functions of the granulocyte-macrophage colony-stimulating factors. *Blood 67:*257–267.

4. Miyauchi, J., Kelleher, C. A., Yang, Y. C., et al (1987). The effects of three recombinant growth factors, IL-3, GM-CSF, and G-CSF on the blast cells of acute myeloblastic leukemia maintained in short-term suspension culture. *Blood 70:*657–663.

5. Vellenga, E., Young, D. C., Wagner, K., Wiper, D., Ostapovicz, D., and Griffin, J. D. (1987). The effects of GM-CSF and G-CSF in promoting growth of clonogenic cells in acute myelogenous leukemia. *Blood 69:*1771–1776.

6. Greenberg, P. L. (1992). Treatment of myelodysplastic syndromes with hemopoeitic growth factors. *Semin Oncol 19:*106–114.

7. Block, M., Jacobson, L. O., and Bethard, W. F. (1953). Preleukemic acute leukemia. *JAMA 152:*1018–1021.

8. Linman, J. W., Bagby, G. C., Jr. (1978). The preleukemic syndrome (hemopoietic dysplasia). *Cancer 42:*854–864.

9. Greenberg, P. L. (1983). The smoldering myeloid leukemic states: clinical and biologic features. *Blood 61:*1035–1044.

10. Bennett, J. M., Catovsky, D., Daniel, M. T., et al (1982). Proposals for the classification of the myelodysplastic syndromes. *Br J Haematol 51:*189–199.

11. Goasquen, J. E., and Bennett, J. M. (1992). Classification and morphologic features of the myelodysplastic syndromes. *Semin Oncol 19:*4–13.

12. Foucar, K., Langdon, R. M. II, Armitage, J. O., Olson, D. B., and Carroll, T. J., Jr. (1985). Myelodysplastic syndromes. A clinical and pathological analysis of 109 cases. *Cancer 56:*553–561.

13. Kerkhofs, H., Hermans, J., Haak, H. L., and Leeksma, C. H. (1987). Utility of the FAB classification for myelodysplastic syndromes: investigation of prognostic factors in 237 cases. *Br J Haematol 65:*73–81.

14. Mufti, G. J., Stevens, J. R., Oscier, D. G., Hamblin, T. J., and Machin, D. (1985). Myelodysplastic syndromes: a scoring system with prognostic significance. *Br J Haematol 59:*425–433.

15. Janssen, J. W., Buschle, M., Layton, M., et al (1989). Clonal analysis of myelodysplastic syndromes: evidence of multipotent stem cell origin. *Blood 73:*248–254.

16. Greenberg, P. L. (1986). In vitro culture techniques defining biological abnormalities in the myelodysplastic syndromes and myeloproliferative disorders. *Clin Haematol 15:*973–993.

17. Boogaerts, M. A., Nelissen, V., Roelant, C., and Goossens, W. (1983). Blood neutrophil function in primary myelodysplastic syndromes. *Br J Haematol 55:*217–227.

18. O'Donnell, M. R., Nademanee, A. P., Snyder, D. S., et al (1987). Bone marrow transplantation for myelodysplastic and myeloproliferative syndromes. *J Clin Oncol 5:*1822–1826.

19.   Appelbaum, F. R., Barrall, J., Storb, R., et al (1990). Bone marrow transplantation for patients with myelodysplasia. Pretreatment variables and outcome. *Ann Intern Med 112*:590–597.

20.   Mertelsmann, R., Tzvi Thaler, H., To, L., et al (1980). Morphological classification, response to therapy, and survival in 263 adult patients with acute non-lymphoblastic leukemia. *Blood 56*:773–781.

21.   Armitage, J. O., Dick, F. R., Needleman, S. W., and Burns C. P. (1981). Effect of chemotherapy for the dysmyelopoietic syndrome. *Cancer Treat Rep 65*:601–605.

22.   Beran, M., Kantarjian, H., O'Brien, S., et al (1996). Topotecan, a topoisomerase I inhibitor, is active in the treatment of myelodysplastic syndrome and chronic myelomonocytic leukemia. *Blood 88*:2473–2479.

23.   Hoagland, H. C., and Linman, J. W. (1982). Pyridoxine-responsive anemia. A preleukemia manifestation? *Minn Med 55*:891–896.

24.   Cines, D. B., Cassileth, P. A., and Kiss, J. E. (1985). Danazol therapy in myelodysplasia. *Ann Intern Med 103*:58–60.

25.   Najean, Y., and Pecking, A. (1977). Refractory anaemia with excess of myeloblasts in the bone marrow: a clinical trial of androgens in 90 patients. *Br J Haematol 37*:25–33.

26.   Najean, Y., and Pecking, A. (1979). Refractory anemia with excess blast cells: prognostic factors and effect of treatment with androgens or cytosine arabinoside. Results of a prospective trial in 58 patients. Cooperative Group for the Study of Aplastic and Refractory Anemias. *Cancer 44*:1976–1982.

27.   Bagby, G. C., Jr., Gabourel, J. D., and Linman, J. W. (1980). Glucocorticoid therapy in the preleukemic syndrome (hemopoietic dysplasia): identification of responsive patients using in-vitro techniques. *Ann Intern Med 92*:55–58.

28.   Gold, E. J., Mertelsmann, R. H., Itri, L. M., et al (1983). Phase I clinical trial of 13-cis-retinoic acid in myelodysplastic syndromes. *Cancer Treat Rep 67*:981–986.

29.   Tricot, G., DeBock, R., Dekker, A. W., et al (1984). Low dose cytosine arabinoside (AraC) in myelodysplastic syndromes. *Br J Haematol 58*:231–240.

30.   Picozzi, V. J. Jr., Swanson, G. F., Morgan, R., Hecht, F., and Greenberg, P. L. (1984). 13-Cis-retinoic acid treatment for myelodysplastic syndromes. *J Clin Onc 4*:589–596.

31.   Griffin, J. D., Spriggs, D., Wisch, J. S., and Kufe, D. W. (1985). Treatment of preleukemic syndromes with continuous intravenous infusion of low-dose cytosine arabinoside. *J Clin Oncol 3*:982–991.

32.   Cheson, B. D., Jasperse, D. M., Simon, R., and Friedman, M. A. (1986). A critical appraisal of low-dose cytosine arabinoside in patients with acute non-lymphocytic leukemia and myelodysplastic syndromes. *J Clin Oncol 4*:1857–1864.

33.   Koeffler, H. P., Heitjan, D., Mertelsmann, R., et al (1988). Randomized study of 13-cis retinoic acid v placebo in the myelodysplastic disorders. *Blood 71*:703–708.

34.   Greenberg, P. L. (1992a). In vitro marrow culture studies in myelodysplastic syndromes. *Semin Oncol 19*:34–46.

35. Souza, L. M., Boone, T. C., Gabrilove, J., et al (1986). Recombinant human granulocyte colony-stimulating factor: effects on normal and leukemic myeloid cells. *Science 232*:61–65.

36. Sachs, L. (1978). The differentiation of myeloid leukaemia cells: new possibilities for therapy. *Br J Haematol 40*:509–517.

37. Lotem, J., and Sachs, L. (1981). In vitro inhibition of the development of myeloid leukemia by injection of macrophage and granulocyte-inducing protein. *Int J Cancer 28*:375–386.

38. Metcalf, D. (1982). Regulator-induced suppression of myelomonocytic leukemic cells: clonal analysis of early cellular events. *Int J Cancer 30*:203–210.

39. Hozumi, M. (1983). Fundamentals of chemotherapy of myeloid leukemia by induction of leukemia cell differentiation. *Adv Cancer Res 38*:121–169.

40. Lotem, J., and Sachs, L. (1984). Control of in vivo differentiation of myeloid leukemic cells. IV. Inhibition of leukemia development by myeloid differentiation-inducing protein. *Int J Cancer 33*:147–154.

41. Nicola, N. A. (1987). Granulocyte colony-stimulating factor and differentiation-induction in myeloid leukemic cells. *Int J Cell Cloning 5*:1–15.

42. Nagler, A., Ginzton, N., Bangs, C., et al (1990). In vitro differentiative and proliferative effects of human recombinant colony-stimulating factors on marrow hemopoiesis in myelodysplastic syndromes. *Leukemia 4*:193–202.

43. Nagler, A., Binet, C., Mackichan, M. L., et al (1990b). Impact of marrow cytogenetics and morphology on in vitro hematopoiesis in the myelodysplastic syndromes: comparison between recombinant human granulocyte colony-stimulating factor (CSF) and granulocyte-monocyte CSF. *Blood 76*:1299–1307.

44. Negrin, R. S., Haeuber, D. H., Nagler, A., et al (1989). Treatment of myelodysplastic syndromes with recombinant human granulocyte colony-stimulating factor. A Phase I-II trial. *Ann Int Med 110*:976–984.

45. Negrin, R. S., Haeuber, D. H., Nagler, A., et al (1990). Maintenance treatment of patients with myelodysplastic syndromes using recombinant human granulocyte colony-stimulating factor. *Blood 76*:36–43.

46. Kobayashi, Y., Okabe, T., Ozawa, K., et al (1989). Treatment of myelodysplastic syndromes with recombinant human granulocyte colony-stimulating factor: a preliminary report. *Am J Med 86*:178–182.

47. Yoshida, Y., Hirashima, K., Asano, S., and Takaku, F. (1991). A phase II trial of recombinant human granulocyte colony-stimulating factor in the myelodysplastic syndromes. *Br J Haematol 78*:378–384.

48. Greenberg, P., Taylor, K., Larson, R., et al (1993). Phase III randomized multicenter trial of G-CSF vs. observation for myelodysplastic syndromes (MDS). *Blood 82*:196a (abstr 768).

49. Greenberg, P., Cox, C., Le Beau, M. M., et al (1997). International scoring system for evaluating prognosis in myelodysplastic syndromes. *Blood 89*:2079–2088.

50. Fleischmann, J., Golde, D. W., Weisbart, R. H., and Gasson, J. C. (1986). Granulocyte-macrophage colony-stimulating factor enhances phagocytosis of bacteria by human neutrophils. *Blood 68*:708–711.

51. Yuo, A., Kitagawa, S., Okabe, T., et al (1987). Recombinant human granulocyte colony-stimulating factor repairs the abnormalities of neutrophils in patients with myelodysplastic syndromes and chronic myelogenous leukemia. *Blood* 70:404–411.

52. Vadhan-Raj, S., Broxmeyer, H. E., Spitzer, G., et al (1989). Stimulation of nonclonal hematopoiesis and suppression of the neoplastic clone after treatment with recombinant human granulocyte-macrophage colony-stimulating factor in a patient with therapy-related myelodysplastic syndrome. *Blood* 74:1491–1498.

53. Merchav, S., Nielson, O. J., Rosenbaum, H., et al (1990). In vitro studies of erythropoietin-dependent regulation of erythropoiesis in myelodysplastic syndromes. *Leukemia* 4:771–774.

54. Greenberg, P. L., Negrin, R. S., and Ginzton, N. L. (1991). G-CSF synergizes with erythropoietin (EPO) for enhancing erythroid colony-formation (BFU-E) in myelodysplastic syndromes (MDS). *Blood* 78:38a (abstr 141).

55. Negrin, R. S., Stein, R., Vardiman, J., et al (1993). Treatment of the anemia of myelodysplastic syndromes using recombinant human granulocyte colony-stimulating factor in combination with erythropoietin. *Blood* 82:737–743.

56. Negrin, R. S., Stein, R., Doherty, K., et al (1996). Maintenance treatment of the anemia of myelodysplastic syndromes with recombinant human granulocyte colony-stimulating factor and erythropoietin: evidence for in vivo synergy. *Blood* 87:4076–4081.

57. Hellstrom-Lindberg, E., Birgegard, G., Carlsson, M., et al (1993). A combination of G-CSF and erythropoietin may synergistically improve the anemia in patients with MDS. *Leuk Lymphoma* 11:221–228.

58. Hellstrom-Lindberg, E., Negrin, R., Stein, R., et al (1995). Efficacy of G-CSF and EPO on the anaemia in patients with myelodysplastic syndromes (MDS). *Blood* 86:338a (abstr 1340).

59. Bernell, P., Stenke, L., Wallnik, J., Hoppe, E., and Hast, R. (1996)-. A sequential erythropoietin plus GM-CSF schedule offers clinical benefit in the treatment of the anaemia of myelodysplastic syndromes. *Leuk Res* 20:693–699.

60. Vadhan-Raj, S., Keating, M., LeMaistre, A., et al (1987). Effects of recombinant human granulocyte-macrophage colony-stimulating factor in patients with myelodysplastic syndromes. *N Engl J Med* 317:1545–1552.

61. Antin, J. H., Smith, B. R., Holmes, W., and Rosenthal, D. S. (1988). Phase I/II study of recombinant human granulocyte-macrophage colony-stimulating factor in aplastic anemia and myelodysplastic syndrome. *Blood* 72:705–713.

62. Ganser, A., Volkers, B., Greher, J., et al (1989). Recombinant human granulocyte-macrophage colony-stimulating factor in patients with myelodysplastic syndromes—a phase I/II trial. *Blood* 73:31–37.

63. Herrmann, F., Lindemann, A., Klein, H., et al (1989). Effect of recombinant granulocyte-macrophage colony-stimulating factor in patients with myelodysplastic syndrome with excess blasts. *Leukemia* 3:335–338.

64. Thompson, J. A., Lee, D. J., Kidd, P., et al (1989). Subcutaneous granulocyte-macrophage colony-stimulating factor in patients with myeolodysplastic syn-

drome: toxicity, pharmacokinetics, and hematological effects. *J Clin Oncol* 7:629–637.

65. Schuster, M. W., Thompson, J. A., Larson, R., et al (1990). Randomized trial of subcutaneous granulocyte-macrophage colony-stimulating factor in patients with myelodysplastic syndrome or aplastic anemia. *Proc Am Soc Clin Oncol 9*:a793.

66. Ganser, A., Seipelt, G., Lindemann, A., et al (1990). Effects of recombinant human interleukin-3 in patients with myelodysplastic syndromes. *Blood 76*:455–462.

67. Kurzrock, R., Talpaz, M., Estrov, Z., Rosenblum, M. G., and Gutterman, J. U. (1991). Phase I study of recombinant human interleukin-3 in patients with bone marrow failure. *J Clin Oncol 9*:1241–1250.

68. Greenberg, P. L., Negrin, R., and Nagler, A. (1990). The use of hemopoietic growth factors in the treatment of myelodysplastic syndromes. *Cancer Surv 9*:199–212.

24

Use of Filgrastim (r-metHuG-CSF) in Acute Myeloid Leukemia

Gerhard Heil
Hanover Medical School, Hanover, Germany

Alan J. Barge
Amgen Ltd., Cambridge, England

I. INTRODUCTION

Acute myeloid leukemia (AML) is a rare disorder, characterized by the proliferation of immature myeloid cells (myeloblasts) in the bone marrow, leading to the suppression of normal hematopoietic activity. Anemia, neutropenia, and thrombocytopenia result, often with the presence of myeloblasts in the peripheral blood. Patients usually present with a combination of lethargy, infection, and bleeding. The incidence of AML is 1.3 cases per 100,000 in the population under 65 years of age, and 11.7 per 100,000 over 65 years (1). The etiology of de novo AML is unknown. Some cases of AML arise in patients with myelodysplastic syndromes (MDS). These usually occur in elderly patients, and are frequently associated with chromosomal abnormalities. Both MDS and AML can arise in patients previously exposed to chemotherapy or radiation, particularly with alkylating agents (2), or drugs that cross-link DNA or inhibit the nuclear enzyme topoisomerase II (3). These cases are referred to as secondary AML.

The French-American-British (FAB) group have defined eight subtypes of AML (M0-M7) on the basis of morphologic and cytochemical features (4). These are of prognostic significance. In addition, a number of chromosomal abnormalities have been identified that have been shown to be of prognostic significance (5–8).

Untreated, AML carries a very poor prognosis, with most patients dying of infection and bleeding within 3 months of diagnosis. Treatment involves very aggressive chemotherapy, with combinations of anthracyclines, such as daunorubicin, and cytosine arabinosine (Ara-C), together with additional agents such as thioguanine or etoposide (9–18). Initially, the objective of treatment is to eradicate morphological evidence of the disease from the bone marrow, to allow the remaining normal elements to regenerate, and repopulate the peripheral blood. Regeneration of normal bone marrow, and normalization of peripheral blood counts is termed complete remission, and the initial courses of chemotherapy given to achieve this are termed remission induction, or more simply induction. Usually, one or two courses of induction chemotherapy are given. Failure to achieve remission after two courses of chemotherapy usually indicates resistant disease and carries a very poor prognosis.

Once remission has been achieved, further courses of chemotherapy, designed to reduce further the disease burden, are given. These are termed consolidation. The type of consolidation chemotherapy given is dependent on prognostic factors, of which the most important is age. There is a continuous relationship between outcome and age, the greater the age, the lower the probability of a patient achieving complete remission, and the shorter the duration of the remission (19). Patients over the age of 50 or 60 years are usually unable to tolerate very high-dose chemotherapy, and are usually treated with similar chemotherapy to that used in induction (19,20). Younger patients, below the age of 50 or 60 years, are candidates for more aggressive consolidation therapy. This is usually determined by prognostic factors, and is termed risk-adjusted. Good prognostic factors include good performance status, the presence of a normal or favorable karyotype, and a white blood cell count at presentation $<25 \times 10^9$/L, as an index of disease load. Poor prognostic features include poor performance status, unfavorable chromosomal abnormalities, and a white blood cell count $>25 \times 10^9$/L (21). Of these, the most important prognostic factors are age and cytogenetics, and they are used as the stratification criteria for recent studies.

High-dose chemotherapy, with or without hematopoietic progenitor support, have been shown to prolong the duration of remission in patients with AML. Allogeneic bone marrow transplantation from an HLA-identical sibling has been shown to offer the best outcome (22–24), but is only available to a minority of patients. High-dose chemotherapy with autologous bone marrow transplantation also has been shown to improve outcome in comparison to standard-dose therapy (25). For good-risk patients, high doses of Ara-C have also been shown to be of benefit (26).

In most patients, the disease relapses after a period of remission. Induction of second and third remissions is usually undertaken. Although second remissions are sometimes achieved, they are usually of short duration, and patients die of their disease if treated with conventional chemotherapy alone. Only allogeneic bone marrow transplantation has been shown to offer the opportunity of long-term disease-free survival after relapse, and most patients eventually die of their disease.

Patients presenting with AML are usually anemic, neutropenic, and thrombocytopenic as a result of their disease, and suffer from infection and bleeding from the outset (27). Induction chemotherapy prolongs these periods of myelosuppression. As a result, patients are at risk of infection and bleeding for a long period. Infections with commensal organisms, such as skin and gastrointestinal flora, are common, resulting in bacteremia and fungemia (27–29). Local infections, such as pneumonias, cellulitis, and perianal infections are common, as are re-activations of dormant viral infections such as *Herpes simplex* or *H. zoster*. All patients are hospitalized during this period, and are often treated in reverse-isolation. Most receive prophylactic oral antibiotics and antifungals. Broad spectrum parenteral antibacterials are used empirically to treat the first sign of infection, without waiting for bacteriological information. Antifungal agents such as amphotericin-B are frequently added empirically to treatment if fevers or signs of infection persist (30).

Mortality from infection and bleeding are most frequent during the period between induction therapy and the attainment of remission (31,32). The duration of neutropenia is usually 2 to 3 weeks, and during this period virtually all patients experience fever and require treatment with anti-infectives. Patients in the elderly age group tend to tolerate this period of neutropenia less well than younger ones, and the mortality is higher in this age group during this period (31,32).

Patients who fail to attain remission do not recover their peripheral blood counts, and they are usually treated with a second course of similar chemotherapy. This prolongs further their total period of pancytopenia and increases the anticipated mortality.

Consolidation therapy, given when patients are in remission, produces similar depths of neutropenia, although for shorter periods, since patients start with relatively normal peripheral blood counts. This tends to have a lower associated morbidity and mortality. However, the application of high-dose consolidation with bone marrow transplantation, or high-dose Ara-C results in severe and prolonged neutropenia and thrombocytopenia, with appreciable associated morbidity (26).

II. THEORETICAL CONSIDERATIONS ABOUT THE USE OF FILGRASTIM IN PATIENTS WITH AML

Attempts to culture human AML cells in either semisolid media or suspension cultures have met with variable success. While recombinant multilineage (eg, interleukin [IL]-3, granulocyte macrophage-colony stimulating factor [GM-CSF]), and lineage-specific hematopoietic growth factors (eg, granulocyte-colony stimulating factor [G-CSF], erythopoietin [EPO], macrophage-colony stimulating factor [M-CSF]) stimulate an albeit limited proliferation of leukemic cells in the majority of cases, the responsiveness of AML cells to individual growth factors and their combinations is extremely heterogeneous and almost universally inferior to that observed with normal hematopoietic progenitor cells (33–37). In terms of proliferation and possibly also self-renewal, both IL-3 and GM-CSF were found to be more potent proliferative stimuli than G-CSF in most, but not all, cases of AML. Combinations of multiple cytokines were generally more effective than individual growth factors. Irrespective of the growth factor and the culture system used, growth of AML cells was self-limiting and could not be supported for extended time periods, indicating a rapid loss of self-renewal capacity under in vitro conditions, similar to normal progenitors.

The potential ability of G-CSF to induce terminal maturation of AML cells was initially suggested by experiments with the murine myelomonocytic cell line, WEHI3BD$^+$, and to a lesser degree with the human myeloid leukemia cell line, HL60 (38). However, the bulk of evidence derived from primary human leukemic samples does not support a consistent and substantial differentiating effect of either G-CSF or the other hematopoietic growth factors in AML (39–43). This inability to overcome the maturation block in AML in conjunction with the variable and generally moderate proliferative effect of G-CSF on the leukemic cells appears to be responsible for the generally poor growth of clonogenic AML progenitor cells, in contrast to normal progenitors, in which G-CSF induces orderly and very reproducible proliferative and maturational effects.

Fresh myeloblasts from patients with AML frequently express receptors for G-CSF (40–43) and GM-CSF (40–42). Binding studies using radiolabeled G-CSF demonstrated that receptor expression was within the heterogeneous range seen in normal progenitor cells. Levels of receptor expression on leukemic blasts did vary according to AML subtype. Blasts from M3 AML (promyelocytes) exhibited uniformly high receptor numbers just as normal promyelocytes do. Blasts from M1 subtype also expressed high numbers, whereas those from M2, M4, and M5 expressed low numbers. However, no correlation was observed between receptor numbers and proliferative response to G-CSF.

While the cytokine response of AML cells is clearly dependent on expression of appropriate receptors, a correlation between receptor numbers and growth factor response has not been firmly established. Thus, despite the expression of G-CSF receptors in 70 of the 72 (97%) cases of AML reported in the four series (40–42), a proliferative response to exogenously added G-CSF was only observed in 42 of 63 (67%) of the examples studied, indicating that the presence of G-CSF receptors on AML blasts is a poor index of their growth responsiveness to this factor. In an attempt to address the in vivo relevance of these observations, Baer et al (43) studied the effect of the administration of Filgrastim to patients with untreated AML. In the study, 28 patients with AML received Filgrastim (10 μg/kg/day) by continuous intravenous (IV) infusion for 72 hours before chemotherapy. Twenty-seven of the 28 patients showed an increase in circulating or bone marrow blast numbers and the number of cells in S-phase. However, despite this, there was no observed detrimental effect on the outcome of treatment. The response to chemotherapy did not appear to be significantly affected by prior G-CSF administration.

III. CLINICAL EXPERIENCE WITH FILGRASTIM IN PATIENTS WITH AML

A. Early Experience

The early clinical experience with Filgrastim in patients with AML, which was mostly limited to small studies of elderly patients, showed that Filgrastim accelerated neutrophil recovery while having no adverse impact on disease outcome (44–50). Chemotherapy-related mortality is particularly high in elderly patients with AML, and these studies showed that Filgrastim can benefit this population by ameliorating chemotherapy-associated toxicities. None of these studies provided data to suggest that Filgrastim accelerated the regrowth of leukemic cells.

Filgrastim was first used in patients with AML by Ohno et al (44), who conducted a prospective, randomized, controlled study to determine its safety and efficacy after a standard course of intensive chemotherapy. The study population of 108 patients, with either relapsed or refractory acute leukemia, included 67 patients with AML. Since this was the first trial of its kind, potential adverse effects of Filgrastim on tumor cell growth were closely monitored; eg, only AML patients with a poor prognosis were included in the trial, and bone marrow aspirations were performed at day 8, day 10, and day 12 in some patients. Furthermore, Filgrastim treatment (200 μg/m^2/day) was started 2 days after the end of the chemotherapy and was continued until the absolute neutrophil count (ANC) was $>1.5 \times 10^9$/L. Neutrophil recovery (ANC $>0.5 \times 10^9$/L) occurred approximately 1 week earlier in Filgrastim-treated patients compared with control patients (p < 0.01); platelet recovery was, however, unaffected. Although febrile episodes occurred with similar frequency in the two treatment groups, the group treated with Filgrastim experienced significantly fewer documented infections compared with control patients (p = 0.028). Fifty percent of evaluable patients from the Filgrastim-treated group entered complete remission compared with 36% in the control group, although the rate of relapse was similar in both treatment groups. Since the median age of patients in both groups was 63 years, it is probable that most had prognostically unfavorable cytogenetic abnormalities. Filgrastim treatment was shown to be safe in this small population of AML patients and it clearly did not affect the regrowth of leukemic cells.

A later prospective, double-blind, controlled study by the same group (47) examined the effect of starting Filgrastim therapy before the start of induction chemotherapy. It included 58 patients with relapsed or refractory AML. Filgrastim (200 μg/m^2/day) was administered from 2 days before the start of induction therapy (mitoxantrone [7 mg/m^2 for 5 days] and Ara-C [200 mg/m^2 for 7 days]), and was continued until either neutrophil recovery (ANC >1.5 $\times$ 10^9/L) or to a maximum of 35 days after the start of chemotherapy. Neutrophil recovery (ANC >1.0 $\times$ 10^9/L) was significantly faster in the 28 patients from the Filgrastim-treated group than the 30 patients from the placebo group (median of 25 days versus 32 days; p <0.001). Although the incidence of febrile episodes and documented infections was similar in both groups, fewer infections were documented in the Filgrastim-treated group after the third week; this difference was not, however, statistically significantly. In this study, Filgrastim neither stimulated the growth of AML cells in the bone marrow during the 2-day period before the start of chemotherapy nor accelerated the regrowth of AML cells during the 5 weeks after therapy. More patients in the Filgrastim-treated group experienced complete remission compared with patients in the placebo-treated group, but again the difference was not statistically significant (50% versus 37%; p = 0.306). Filgrastim did not influence either the rate of event-free survival of patients or disease-free survival of those patients who had achieved complete remission.

Baer et al (45) presented preliminary results from a study of Filgrastim adjunctive treatment combined with an induction protocol which delivered very high doses of chemotherapy over an exceptionally brief period of time. Previously untreated de novo (n = 20) and secondary (n = 10) patients with AML received cytarabine (12 doses of either 3 g/m^2 or 1.5 g/m^2 for patients <50 and >50 years, respectively) and idarubicin (3 doses of 12 mg/m^2). Filgrastim treatment (10 μg/kg/day) was started 24 hours after the last dose of chemotherapy and was continued until neutrophil recovery (ANC $\geq$5.0 $\times$ 10^9/L on two consecutive days). Twenty-seven patients were able to receive all the planned doses of chemotherapy. A 65% complete remission rate was obtained in de novo patients after a single course of induction therapy (90% and 40% for patients <60 and $\geq$60 years, respectively) compared with only 20% in patients with secondary AML. For the de novo AML patients who achieved a complete remission, the median times to neutrophil recovery (ANC $>0.5 \times 10^9$/L), platelet recovery (100 $\times$ 10^9/L), and day of the last platelet transfusion were 20, 28, and 23 days, respectively. The major nonhematologic toxicity was transient hyperbilirubinemia, which was observed in 64% of patients. Reversible cerebellar toxicity was seen in three patients. Full-dose idarucibin was, thus, safely administered with high-dose cytarabine (even in elderly patients). Filgrastim was associated with more rapid neutrophil recovery as well as fewer toxicities.

Estey et al (46) administered Filgrastim to elderly patients with newly diagnosed AML (n = 69) or MDS (n = 43) before, during, and after fludarabine plus cytarabine chemotherapy. Since the median age of patients in both groups was 63 years, overall this patient population most probably also had prognostically unfavorable cytogenetic abnormalities. Patients received Filgrastim (400 μg/m^2/day) starting 1 day before and/or during fludarabine (30 mg/m^2/day) and cytarabine (2 g/m^2/day) for 5 days, which was continued until the patients experienced a complete response. The control group comprised 85 newly diagnosed patients (n = 54, AML; n = 31, MDS) who had previously received fludarabine plus cytarabine chemotherapy, but not Filgrastim. Filgrastim treatment significantly accelerated recovery (ANC $\geq 1 \times 10^9$/L), the median recovery times were 21 days and 34 days for patients who did, or did not, receive Filgrastim (p = 0.0001). Complete response rates were not, however, statistically different (53% and 63%, respectively; p = 0.5) a result which may reflect the relatively high mortality rates in both groups before neutrophil recovery. Rates of infection also were similar in both groups.

Godwin et al (48) showed that Filgrastim reduced the duration of neutropenia and antibiotic use in elderly patients (median age 67 years) without altering the rate of complete remission or patient survival. This study population was heterogeneous and included patients with de novo and secondary AML which has made the data somewhat more difficult to interpret. Induction chemotherapy consisted of daunorubicin (45 mg/m^2 for 3 days) and Ara-C (200 mg/m^2 for 7 days). Blinded Filgrastim therapy was started on day 11; to be considered eligible for Filgrastim treatment, the levels of bone marrow myeloblasts in a given patient on day 10 could comprise no more than 5% of all cells in the bone marrow. Seventy-three patients received Filgrastim and 79 patients received placebo. Patients entering remission received a further course of daunorubicin (30 mg/m^2 for 2 days) and Ara-C (200 mg/m^2 for 7 days). The median time to neutrophil recovery (ANC $\leq 0.5 \times 10^9$/L) was 3 to 4 days faster in the group treated with Filgrastim compared with patients treated with placebo. This was paralleled by significant reductions in the number of days with fever (p =

0.035) and antibiotics (p = 0.024). Filgrastim treatment did not influence either the risk of fatal infection, the number of documented infections, or the duration of the first hospitalization. As expected in this elderly population, the complete remission rates were low (42% Filgrastim versus 49% placebo; p = 0.86), and the durations of the remissions were relatively short. Filgrastim did not adversely affect the outcome of the disease and produced some significantly important clinical benefits by reducing the duration of neutropenia and associated clinical consequences.

In the more recent study by Maslak et al (50), elderly patients (>60 years) received induction chemotherapy consisting of idarubicin (12 mg/m^2/day for 3 days) and Ara-C (200 mg/m^2/day for 5 days). The second course of chemotherapy consisted of mitoxantrone (12 mg/m^2/day for 3 days), etoposide (VP-16) (150 mg/m^2/day for 3 days), and Ara-C (200 mg/m^2/day for 4 days). Chemotherapy was either continued immediately for patients who failed the first induction or 1 month later for patients in complete remission. Filgrastim (10 μg/kg/day) was started 24 hours after completion of the chemotherapy and was continued until neutrophil recovery (ANC $\geq 0.5 \times 10^9$/L). Recovery was significantly faster in Filgrastim-treated patients entering complete remission (n = 26) compared with historical patients who did not receive Filgrastim (n = 28) (median 13 days versus 17 days; p = 0.008). Recovery of neutrophils to $\geq 1.0 \times 10^9$/L also was significantly faster in the Filgrastim-treated group (14 days versus 19 days, p = 0.005). Significantly fewer patients receiving Filgrastim died of chemotherapy-related toxicities compared with historical control patients (5% versus 32%: p = 0.04). Filgrastim did not affect the requirement for supportive care nor the incidence of infection-related complications. The complete remission rate was 58% in the entire study group and 71% in the de novo patients. Both disease-free survival and overall survival were comparable between the study and historical control groups. The small number of early deaths from chemotherapy-related toxicities may, at least in part, be due to Filgrastim reducing the duration of neutropenia.

Acute myeloid leukemia is an uncommon disease in the young; one of the few studies to use Filgrastim in children was conducted by the Children's Cancer Group (CCG) (49). This large study was originally designed to investigate the impact of an intensive, timed-sequential, five-drug induction regimen on disease outcome in comparison with the same five-drug combination delivered at standard intervals. After patients had been randomized to either the conventional- or timed-sequential groups (n = 296 and n = 295, respectively), an additional 136 patients were included in the study to receive Filgrastim treatment in combination with timed-sequential chemotherapy.

There was a significant improvement in event-free survival after 3 years in patients receiving sequential-treatment compared with those receiving standard treatment (43% versus 32%, respectively; p = 0.01). The 3-year post-remission disease-free survivals were 59% and 45%, respectively (p = 0.0001). Although neutrophil recovery data were not released in this preliminary communication, it was, however, reported that Filgrastim reduced the toxicities associated with the more intensive chemotherapy regimen (7% reduction in deaths) and was associated with a similar rate of remissions.

B. Multicenter Phase 3 Trial

The effects of adjunctive Filgrastim on the induction of remission in patients with de novo AML was examined in a randomized, multicenter, double-blind, placebo-controlled, phase 3 trial (51). Five hundred and twenty-one adult patients received either one or two courses of induction chemotherapy, followed by one or two courses of consolidation chemotherapy if the disease went into remission. Six days after the start of the induction chemotherapy, the patients were randomized to receive either Filgrastim (n = 259) or placebo (n = 262) after each chemotherapy course. Filgrastim therapy (5 μg/kg/day) was started 24 hours after the last chemotherapy dose and was given until neutrophil recovery (ANC $\geq 1 \times 10^9$/L for 3 days or $\geq 10 \times 10^9$/L for 1 day) to a maximum of 35 days. Throughout the study, all patients were administered prophylactic quinolone antibiotics.

The induction chemotherapy regimen consisted of daunorubicin (45 mg/m^2 for 3 days), Ara-C (200 mg/m^2 for 7 days), and etoposide (100 mg/m^2 for 5 days). Those patients who did not respond to the first course received a second induction regimen consisting of daunorubicin (45 mg/m^2 for 2 days), Ara-C (200 mg/m^{2}for 5 days), and etoposide (100 mg/m^2 for 5 days). Patients who entered complete remission after the induction therapy received the following consolidation regimen; daunorubicin (45 mg/m^2 for 3 days), Ara-C (200 mg/m^2 for 7 days), and etoposide (100 mg/m^2 for 5 days). There was no upper age limit for patients in this study, but the optional second course of consolidation therapy was different for patients >50 years and those <50 years of age. High-dose treatment with Ara-C (3 g/m^2 twice daily for 6 days) was given to the younger patients, whereas older patients received a second round of consolidation therapy that was identical to the first.

As in other clinical settings, Filgrastim treatment reduced the duration of neutropenia, fever, use of parenteral antibiotics, and hospitalization in this group of patients with AML. During each course of chemotherapy, Filgrastim treatment significantly reduced the duration of neutropenia

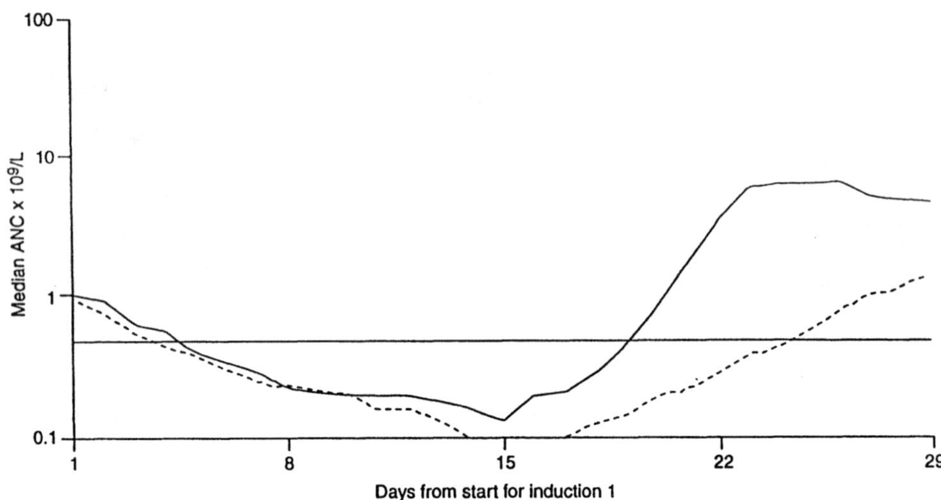

Figure 1.   Filgrastim increased neutrophil nadir. Increased neutrophil counts were first observed approximately 11 days after the start of chemotherapy. Solid line, Filgrastim; dotted line, placebo. ANC = absolute neutrophil count.

(ANC $<0.5 \times 10^9$/L). After the first round of induction chemotherapy, treatment with Filgrastim reduced the median duration of neutropenia by 5 days compared with placebo (p = 0.0001). Increased neutrophil counts were first observed approximately 11 days after the start of chemotherapy (**Figure 1**) (**Table 1**). Similar reductions in the duration of neutropenia

Table 1.   Infection-Related Endpoints During Induction 1.

	Incidence*		Duration (days)* Median (range)		
	Filgrastim	Placebo	Filgrastim	Placebo	p
Fever ($\geq$38.0°C)	91%	92%	7.0 (0–38)	8.5 (0–38)	0.009
IV antibiotics	95%	96%	15 (0–54)	18.5 (0–54)	0.0001
Hospitalization	100%	100%	20 (2–68)	25 (7–55)	0.0001
Febrile neutropenia	90%	92%	13 (0–38)	18 (0–40)	0.0001
Documented infection	39%	39%	—	—	—

* Incidence and durations calculated from day 6 (earliest time of randomization).
IV = intravenous.

were seen after the second round of induction therapy (4 days; p = 0.015); first and second rounds of standard consolidation therapy (7 days [p = 0.0001] and 5 days [p = 0.0001] respectively); and high-dose consolidation therapy (5.5 days; p = 0.001).

Neutrophil recovery also was significantly faster after the first chemotherapy course in patients treated with Filgrastim compared with those treated with placebo (20 days versus 25 days; p = 0.0001) (**Figure 2**). Improved neutrophil recovery was reflected by significant reductions in the duration of fever (7 days versus 8.5 days, respectively; p = 0.009); use of parenteral antibiotics (15 days versus 18.5 days, respectively; p = 0.0001) and hospitalization (20 days versus 25 days, respectively; p = 0.0001) although the incidence of these events did not differ significantly between the two groups (**Table 2**).

Filgrastim did not stimulate the growth of leukemia clones in vivo, nor did it affect the response to chemotherapy or the duration of remission. The complete remission rate obtained in patients treated with Filgrastim was 69%, similar to that of patients treated with placebo (68%). At a median follow-up time of 24 months, the median disease-free survival time (0.1 months and 9.4 months) and median survival time (12.5 months and 14.0 months) were not significantly different between the Filgrastim-treated and placebo-treated groups, respectively. Overall, patient survival was not significantly different from other studies using optimal chemotherapy regimens (**Figure 3**).

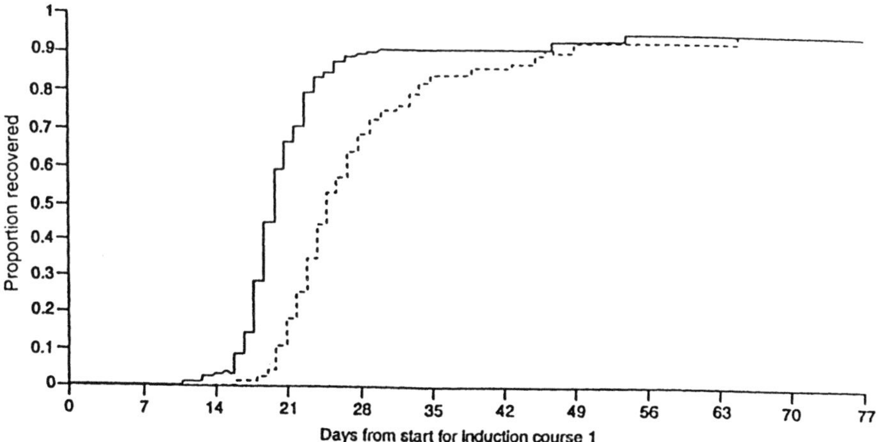

Figure 2. Neutrophil recovery also was significantly faster after the first chemotherapy course in patients treated with Filgrastim compared with those treated with placebo. Solid line, Filgrastim; dotted line, placebo.

Table 2. Infection-Related Endpoints During Consolidation 1.

	Incidence			Duration (days) Median (range)		
	Filgrastim	Placebo	p	Filgrastim	Placebo	p
Fever ($\geq$ 38.0°C)	49%	63%	0.01	0 (0–20)	2 (0–21)	0.005
IV antibiotics	44%	54%	0.08	0 (0–50)	4 (0–37)	0.04
Hospitalization	100%	100%	1.00	19 (4–58)	25 (2–42)	0.0001
Febrile neutropenia	32%	47%	0.007	0 (0–46)	0 (0–18)	0.001
Documented infection	22%	30%	0.09	—	—	—

IV = intravenous.

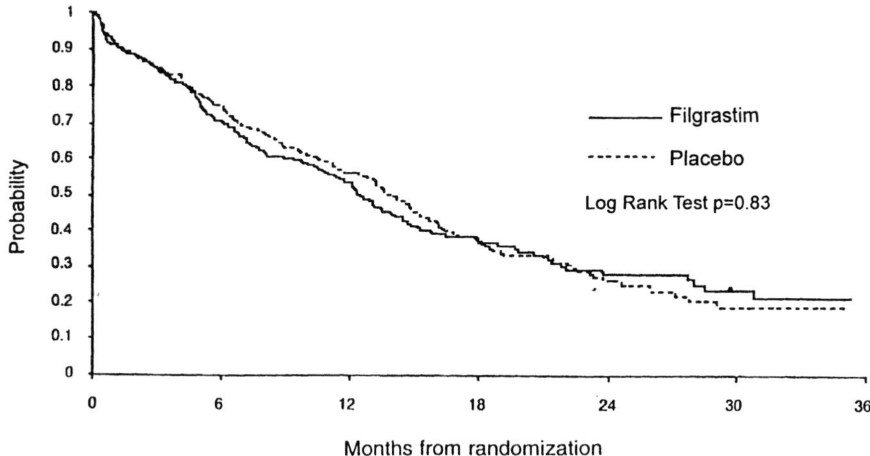

Figure 3. Patient survival was not significantly different from other studies using optimal chemotherapy regimens.

Patients reported alopecia, nausea, vomiting, and diarrhea (commonly associated with chemotherapy) as the most frequently occurring adverse events. Since the frequency and severity of these adverse events was similar in Filgrastim-treated and placebo-treated groups, Filgrastim did not exacerbate these reactions. Deaths occurred at a frequency expected for patients with AML and most of them occurred during induction therapy. Although fewer patients treated with Filgrastim died as a result of infection, the overall death rate between the two treatment groups was not significantly different; 21 patients in the Filgrastim group died compared with 25 patients in the placebo group. This result reflects, in part, the prolonged period of neutropenia experienced by all patients irrespective of treatment. Filgrastim is not expected to affect patient outcome since it has no effect on any of the known determinants of outcome in AML, ie, age and karyotypic abnormalities (19,52–55).

This trial demonstrated that Filgrastim can be used safely as an adjunctive therapy in both the induction and consolidation phases of chemotherapy for patients with de novo AML. Filgrastim did not worsen the outcome of AML, and by reducing neutropenia, accelerated neutrophil recovery, as well as reduced need for parenteral antibiotics and hospitalization, it significantly decreased the morbidity associated with the treatment.

IV. FILGRASTIM AFTER BONE
MARROW TRANSPLANTATION

Patients with AML who relapse after allogeneic transplantation have a very poor prognosis; few patients will respond to subsequent chemotherapy and almost none can survive long-term under present treatment regimens. Giralt et al (56) used Filgrastim in a group of such patients in an effort to stimulate production of normal bone marrow cells, and, thus, reinduce remission. Relapse was defined as either the presence of an abnormal clone or >5% blasts in the bone marrow. Filgrastim (5 μg/kg/day) was administered to seven patients with leukemia (one with chronic myelogenous leukemia, five with AML, and one with MDS that transformed into AML). After Filgrastim treatment, donor cells reestablished hematopoiesis in three patients. Although one of the patients who had entered complete remission experienced a relapse after 12 months, the other two were reported to still be in remission after 10 and 11 months. This preliminary study may be interpreted to show that Filgrastim may be effective in selected patients in leukemic relapse after allogeneic bone marrow transplantation and as such certainly merits further study. However, the possibility of that graft-versus-leukemia (GVL) responses may have contributed to the responses cannot be ruled out. Immunosuppressive therapy (cyclosporin and methylprednisolone) was discontinued soon after the start of Filgrastim treatment (after 7 days and 14 days, respectively).

V. CONCLUSIONS

These studies have conclusively demonstrated that Filgrastim does not worsen the outcome of the disease in patients with de novo AML. Patients receiving Filgrastim achieved the same complete remission rate as those receiving placebo and the remissions were of similar durability.

A number of studies have now been published which address the influence of myeloid growth factors in AML. **Figure 4** shows the odds ratios and 95% confidence intervals for each study. The differences in impact of growth factor on remission rate in these studies may be due to differences in sample size, patient populations, timing of randomization, timing of the administration of growth factor and timing of remission assessment. The study by Heil et al (51) using Filgrastim with 521 randomized patients and the study by Stone et al (57) using GM-CSF with 388 randomized patients are the largest undertaken to date, and show no impact of growth factor on complete remission rate. It is likely that neither r-HuGM-CSF or Filgrastim

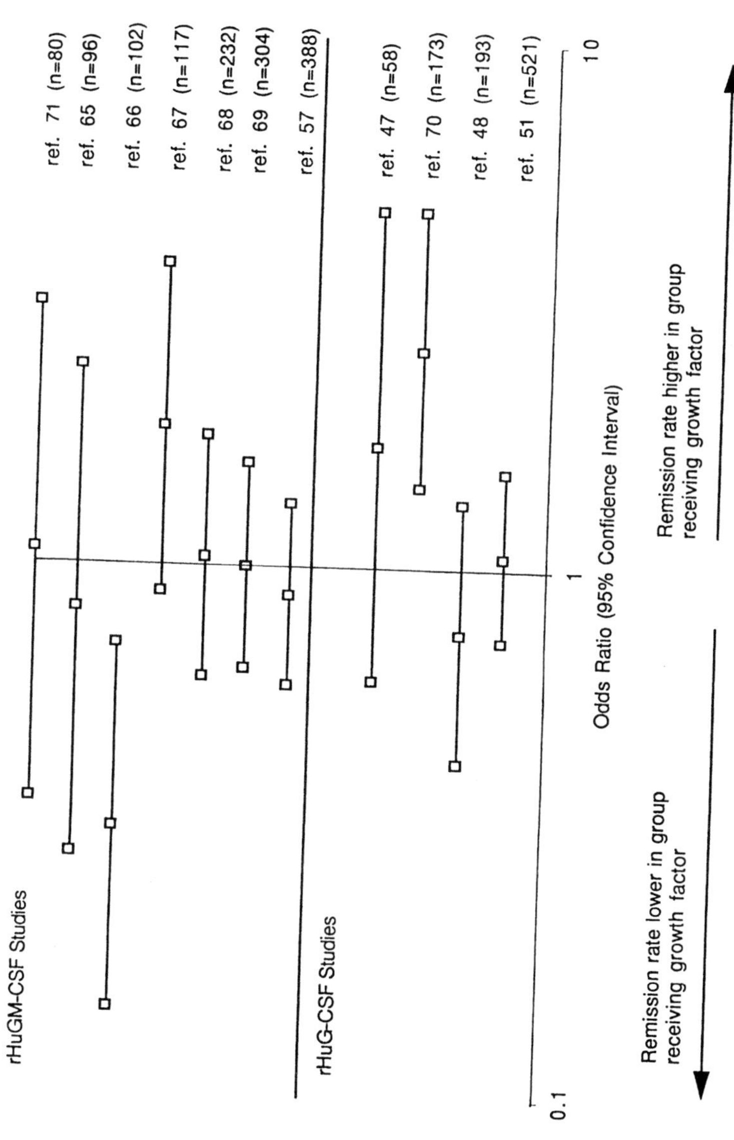

Figure 4. Odds ratio graph of complete remission rates for published studies for r-HuGM-CSF and Filgrastim in studies in AML. Solid line, Filgrastim; dotted line, placebo.

influence complete remission rate, and differences seen in some smaller studies may represent random variation.

Treatment with Filgrastim significantly reduces the duration of neutropenia and associated clinical consequences in patients with AML. During induction therapy, patients are profoundly neutropenic for approximately 18 days, before neutrophil recovery begins. During this time the majority experienced fever, and required IV anti-infectives and in-patient care. Treatment with Filgrastim does not impact the incidence of this early morbidity. However, by reducing the duration of neutropenia, it significantly reduces the duration of all of these clinical events. Patients receiving Filgrastim spend less time in hospital recovering from induction and consolidation therapy.

In this population, a major consequence of prolonged neutropenia is fungal infection with organisms such as *Aspergillus*. The incidence of fungal infection is increasing in patients with hematological malignancies (28–30). The persistence of fever for more than 48 hours after the start of broad-spectrum antibacterials usually necessitates the introduction of systemic antifungal therapy such as amphotericin-B which is associated with significant toxicity (58,59). In the study by Heil et al (51), Filgrastim significantly reduced the number of patients requiring systemic antifungal therapy. This is an important clinical benefit, not previously documented in studies of growth factors in the oncology setting, and serves to reduce the burden of toxicity in these patients.

These studies have demonstrated that Filgrastim is a safe and effective adjunct to the treatment of AML, significantly reducing neutropenia and associated morbidity, without worsening disease outcome. Filgrastim is effective when used in conjunction with all courses of chemotherapy, including standard and high-dose consolidation. Further improvements in the outlook for patients with AML may depend on the better definition of prognostic factors, and the prognostic factor adjusted application of high-dose consolidation therapy. The use of additional growth factors, such as megakaryocyte growth and development factor (MGDF) (60,61), which stimulates the production and maturation of platelets (62–64), may further reduce treatment-associated morbidity. The application of hematopoietic growth factors in this setting may enable very intensive chemotherapy to be administered with acceptable toxicity, and may allow the exploration of further intensive schedules, with or without hematopoietic progenitor cell support.

REFERENCES

1. Acute Myeloid Leukemia Incidence and Prevalence Database (IAPV) (1994). Epidemiologic. *Reviews 16*:246.

2. Shulman, L. N. (1993). The biology of alkylating-agent cellular injury. *Hematol Oncol Clin North Am 7*:325–335.

3. Ratain, M. J., and Rowley, J. D. (1992). Therapy-related acute myeloid leukemia secondary to inhibitors of topoisomerase II: from the bedside to the target genes. *Ann Oncol 3*:107–111.

4. Bennett, J. M., Catovsky, D., Daniel, M. T., et al (1985). Proposed revised criteria for the classification of acute myeloid leukemia. A report of the French-American-British Cooperative Group. *Ann Intern Med 103*:620–625.

5. Cline, M. J. (1994). The molecular basis of leukemia. *N Engl J Med 330*:328–336.

6. Rowley, J. D., Golomb, H. M., and Dougherty, C. (1977). 15;17 translocation, a consistent chromosomal change in acute promyelocytic leukaemia. *Lancet 1*:549–550.

7. Nucifora, G., and Rowley, JD. (1995). AML1 and the 8;21 and 3;21 in acute and chronic myeloid leukemia. *Blood 86*:1–14.

8. Liu, R. P., Hajra, A., Wijmenga, C., and Collins, F. S. (1995). Molecular pathogenesis of the chromosome 16 inversion in the M4Eo subtype of acute myeloid leukemia. *Blood 85*:2289–2302.

9. Gale, P. P., and Foon, K. A. (1987). Therapy of acute myelogenous leukemia. *Semin Hematol 24*:40–54.

10. Mayer, R. J. (1987) Current chemotherapeutic treatment approaches to the management of previously untreated adults with de novo acute myelogenous leukemia. *Semin Oncol 4*:384–396.

11. Mayer, R. J., Schiffer, O. A., Peterson, B. A., et al (1987). Intensive postremission therapy in adults with acute nonlymphocytic leukemia using various dose schedules of ara-C: a progress report from the CALGB. Cancer and Leukemia Group B. *Semin Oncol 14*(S1):25–31.

12. Büchner, T., Hiddemann, W., Blasius, S., et al (1990). Adult AML: the role of chemotherapy intensity and duration. Two studies of the AML Cooperative Group. In: Büchner, T., Schellong, G., Hiddemann, W., and Ritter, J., eds. *Haematology and Blood Transfusions. Acute Leukemias II 33:* pp. 261–266.

13. Rees, J. K., Gray, R. G., Swirsky, D., and Hayhoe, F. G. (1986). Principal results of the Medical Research Council's 8th acute myeloid leukaemia trial. *Lancet 2*:1236–1241.

14. Champlin, R., Ho, W., Winston, D., et al (1987). Treatment of adults with acute myelogenous leukemia: prospective evaluation of high-dose cytarabine in consolidation chemotherapy and with bone marrow transplantation. *Semin Oncol 14*(S1):1–6.

15. Jehn, U., Zittoun, R., Suciu, S., et al (1990). A randomized comparison of intensive maintenance treatment for adult acute myelogenous leukemia using either cyclic alternating drugs or repeated courses of the induction-type chemotherapy: AML-6 trial of the EORTC Leukemia Cooperative Group. In: Büchner, T., Schellong, G., Hiddemann, W., and Ritter, J. eds. *Haematology and Blood Transfusions. Acute Leukemias II 33:* pp. 277–284.

16. Petti, M. C., Broccia, G., Caronia, F., et al (1990). Therapy of acute myelogenous leukemia in adults. In: Büchner, T., Schellong, G., Hiddemann, W., and Ritter, J. eds. *Haematology and Blood Transfusions. Acute Leukemias II 33:* pp. 249–253.

17. Cassileth, P. A., Harrington, D. P., Hines, J. D., et al (1990). Comparison of postremission therapies in adult acute myeloid leukemia: preliminary analysis of an ECOG study. Eastern Cooperative Oncology Group. In: Büchner, T., Schellong, G., Hiddemann, W., and Ritter, J. eds. *Haematology and Blood Transfusions. Acute Leukemias II 33:* pp. 267–270.

18. Bishop, J. F. (1991). Etoposide in the management of leukemia: a review. *Semin Oncol 18*(S2):62–69.

19. Stone, R. M., and Mayer, R. J. (1993). The approach to the elderly patient with acute myeloid leukemia. *Hematol Oncol Clin North Am 7:*65–79.

20. Löwenberg, B., Zittoun, R., Kerkhofs, H., et al (1989). On the value of intensive remission-induction chemotherapy in elderly patients of 65+ years with acute myeloid leukemia: a randomized phase III study of the European Organization for Research and Treatment of Cancer Leukemia Group. *J Clin Oncol 7:*1268–1274.

21. Swansbury, G. J., Lawler, S. D., Alimena, G., et al (1994). Long-term survival in acute myelogenous leukemia: a second follow-up of the Fourth International Workshop on Chromosomes in Leukemia. *Cancer Genet Cytogenet 73:*1–7.

22. Thomas, E. D., Buckner, C. D., Clift, R. A., et al (1979). Marrow transplantation for acute nonlymphoblastic leukemia in first remission. *N Engl J Med 301:*597–599.

23. Appelbaum, F. R., Clift, R. A., Buckner, C. D., et al (1983). Allogeneic marrow transplantation for acute nonlymphoblastic leukemia after first relapse. *Blood 61:*949–953.

24. Clift, R. A., Buckner, O. D., Thomas, E. D., et al (1987). The treatment of acute non-lymphoblastic leukemia by allogeneic marrow transplantation. *Bone Marrow Transpl 2:*243–258.

25. Zittoun, R., Mandelli, F., Willemze, R., et al (1993). Prospective phase III study of autologous bone marrow transplantation (ABMT) v short intensive chemotherapy (IC) v allogeneic bone marrow transplantation (ALLO-BMT) during first complete remission (CR) of acute myelogenous leukemia (AML). Results of the EORTC-GIMENA AML 8A trial. *Blood 82:*10, 85a (abstr 327).

26. Mayer, R. J., Davis, R. B., Schiffer, C. A., et al (1994). Intensive postremission chemotherapy in adults with acute myeloid leukemia. *N Engl J Med 331:*896–903.

27. Bodey, G. P., Rodriguez, V., Chang, H. Y., and Narboni, G. (1978). Fever and infection in leukemic patients. A study of 494 consecutive patients. *Cancer 41:*1610–1622.

28. Horn, R., Wong, B., Kiehn, T., and Armstrong, D. (1985). Fungaemia in a cancer hospital: changing frequency, earlier onset and results of therapy. *Rev Infect Dis 7:*646–655.

29. Bodey, G. P. (1988). The emergence of fungi as major hospital pathogens. *J Hospital Infect 11:*411–426.

30. Bodey, G. P. (1993). What's new in fungal infection in leukaemic patients. *Leuk Lymphoma 11*(2):127–135.

31. Sebban, C., Archimbaud, E., Coiffier, B., et al (1988). Treatment of acute myeloid leukemia in elderly patients. *Cancer 61:*227–231.

32. Estey, E. H., Keating, M. J., McCredie, K. B., Bodey, G. P., and Freireich, E. J. (1982). Causes of initial remission induction failure in acute myelogenous leukemia. *Blood 60:*309–315.

33. Bull, J. M., Duttera, M. J., Stashick, E. D., Northup, J., Henderson, E., and Carbone, P. P. (1973). Serial in vitro marrow culture in acute myelocytic leukemia. *Blood 42:*679–686.

34. Spitzer, G., Dicke, K. A., Gehan, E. A., et al (1976). A simplified in vitro classification for prognosis in adult acute leukemia: the application of in vitro results in remission-predictive models. *Blood 48:*795–807.

35. Nara, N., and McCulloch, E. A. (1985). The proliferation in suspension of the progenitors of the blast cells in acute myeloblastic leukemia. *Blood 65:*1484–1493.

36. Coulombel, L., Eaves, C., Kalousek, D., Gupta, C., and Eaves, A. (1985). Long-term marrow culture of cells from patients with acute myelogenous leukemia. Selection in favor of normal phenotypes in some but not all cases. *J Clin Invest 75:*961–969.

37. Griffin, J. D., and Lowenberg, B. (1986). Clonogenic cells in acute myeloblastic leukemia. *Blood 68:*1185–95.

38. Souza, L. M., Boone, T. C., Gabrilove, J., et al (1986). Recombinant human granulocyte colony-stimulating factor: effects on normal and leukemic myeloid cells. *Science 232:*61–65.

39. Vellenga, E., Ostapovicz, D., O'Rourke, B., and Griffin, J. D. (1987). Effects of recombinant IL-3, GM-CSF, and G-CSF on proliferation of leukemic clonogenic cells in short-term and long-term cultures. *Leukemia 1:*584–589.

40. Vellenga, E., Young, D. C., Wagner, K., Wiper, D., Ostapovicz, D., and Griffin, J. D. (1987). The effects of GM-CSF and G-CSF in promoting growth of clonogenic cells in acute myeloblastic leukemia. *Blood 69:*1771–1776.

41. Park, L. S., Waldron, P. E., Friend, D., et al (1989). Interleukin-3, GM-CSF, and G-CSF receptor expression on cell lines and primary leukemia cells: receptor heterogeneity and relationship to growth factor responsiveness. *Blood 74:*56–65.

42. Motoji, T., Watanabe, M., Uzumaki, H., et al (1991). Granulocyte colony-stimulating factor (G-CSF) receptors on acute myeloblastic leukaemia cells and their relationship with the proliferative response to G-CSF in clonogenic assay. *Br J Hematol 77:*54–59.

43. Baer, M. R., Bernstein, S. H., Brunetto, V. L., et al (1996). Biological effects of recombinant human granulocyte colony-stimulating factor in patients with untreated acute eceptors. *Leukemia 1:*1–8.

44. Ohno, R., Tomonaga, M., Kobayashi, T., et al (1990). Effect of granulocyte colony-stimulating factor after intensive induction therapy in relapsed or re-fractory acute leukemia. *N Engl J Med 323:*871–877.

45. Baer, M. R., Christiansen, N. P., Frankel, S. R., et al (1993). High-dose cytarabine, idarubicin, and granulocyte colony-stimulating factor remission induction therapy for previously untreated de novo and secondary adult acute myeloid leukemia. *Semin Oncol 20:*6–12.

46. Estey, E., Thall, P., Andreeff, M., et al (1994). Use of granulocyte colony-stimulating factor before, during, and after fludarabine plus cytarabine induction therapy of newly diagnosed acute myelogenous leukemia or myelodysplastic syndromes: comparison with fludarabine plus cytarabine without granulocyte colony-stimulating factor. *J Clin Oncol 12*:671–678.

47. Ohno, R., Naoe, T., Kanamaru, A., et al (1994). A double-blind controlled study of granulocyte colony-stimulating factor started two days before induction chemotherapy in refractory acute myeloid leukemia. Kohseisho Leukemia Study Group. *Blood 83*:2086–2092.

48. Godwin, J. E., Kopecky, K. J., Head, D. R., Hynes, H. E., Balcerzak, S. P., and Appelbaum, F. R. (1995). A double blind placebo controlled trial of G-CSF in elderly patients with previously untreated acute myeloidleukemia. A Southwest Oncology Group study. *Blood 86*:434A (abstr 1723).

49. Woods, W. G., Kobrinsky, N., Buckley, J., et al (1996). Timed sequential induction therapy improves post remission outcome in pediatric acute myeloid leukemia (AML). *Blood 87*:4979–4989.

50. Maslak, P. G., Weiss, M. A., Berman, E., et al (1996). Granulocyte colony-stimulating factor following chemotherapy in elderly patients with newly diagnosed acute myelogenous leukemia. *Leukemia 10*:32–39.

51. Heil, G., Hoelzer, D., Sanz, M. A., et al (1995). Results of a randomised, double-blind placebo controlled phase III study of Filgrastim in remission induction and early consolidation therapy for adults with de-novo acute myeloid leukemia. (ASH). *Blood 86*:267a (abstr 1053).

52. Tashiro, S., Kyo, T., Tanaka, K., et al (1992). The prognostic value of cytogenetic analyses in patients with acute nonlymphocytic leukemia treated with the same intensive chemotherapy. *Cancer 70*:2809–2815.

53. Fonatsch, C., Gudat, H., Lengfelder, E., et al (1994). Correlation of cytogenetic findings with clinical features in 18 patients with inv(3)(q21q26) or t(3;3)(q21; q26). *Leukemia 8*:1318–1326.

54. Hoo, J. J., Gregory, S. A., Jones, B., and Szego, K. (1995). Supernumerary isochromosome 4p in ANLL-M4 myelomonocytic type is associated with favorable prognosis. *Cancer Genet Cytogenet 79*:127–129.

55. Seyger, M. M., Ritterbach, J., Creutzig, U., et al (1995). 12q13, a new recurrent breakpoint in acute non-lymphoblastic leukemia. *Cancer Genet Cytogenet 80*:23–28.

56. Giralt, S., Escudier, S., Kantarjian, H., et al (1993). Preliminary results of treatment with filgrastim for relapse of leukemia and myelodysplasia after allogeneic bone marrow transplantation. *N Engl J Med 329*:757–761.

57. Stone, R. M., Berg, D. T., George, S. L., et al (1995). Granulocyte-macrophage colony-stimulating factor after initial chemotherapy for elderly patients with primary acute myelogenous leukemia. Cancer and Leukemia Group B. *N Engl J Med 332*:1671–1677.

58. Butler, W. T., Bennet, J. E., Alling, D. W., Wertlake, P. T., Utz, J. P., and Hill, G. J. (1964). Nephrotoxicity of amphotericin B (early and late effects in 81 patients). *Ann Intern Med 61*:175.

59. Warda, J., and Barriere, S. L. (1985). Amphotericin B nephrotoxicity. *Drug Intell Clin Pharm 19*:25–26.

60. Bartley, T. D., Bogenberger, J., Hunt, P., et al (1994). Identification and cloning of a megakaryocyte growth and development factor (MGDF) that is a ligand for the cytokine receptor Mpl. *Cell 77*:117–1124.

61. Wendling, F., Marakovsky, E., Debili, N., et al (1994). c-Mpl ligand is a humoral regulator of megakaryocytopoiesis. *Nature 369*:571–574.

62. Kaushansky, K., Lok, S., Holly, R. D., et al (1994). Promotion of megakaryo-cyte progenitor expansion and differentiation by the c-Mpl ligand thrombo-poietin. *Nature 369*:568–571.

63. deSauvage, F. J. Hass, P. E., Spencer, S. D., et al (1994). Stimulation of megakaryocytopoiesis and thrombopoiesis by the c-Mpl ligand. *Nature 369*:533–538.

64. Lok, S., Kaushansky, K., Holly R. D., et al (1994). Cloning and expression of murine thrombopoietin cDNA and stimulation of platelet production in vivo. *Nature 369*:565–568.

65. Büchner, T., Hiddemann, W., Wormann, B., et al (1994). GM-CSF multiple course priming and long-term administration in newly diagnosed AML hema-tologic and therapeutic effects. *Blood 84*:27a (abstr 95).

66. Zittoun, R., Suciu, S., Mandelli, F., et al (1996). Granulocyte-macrophage colony-stimulating factor associated with induction treatment of acute myelog-enous leukemia—a randomized trial by the European Organization for Re-search and Treatment of Cancer Leukemia Cooperative Group. *J Clin On-col 14*:2150–2159.

67. Rowe, J. M., Andersen, J. W., Mazza, J. J., et al (1995). A randomized placebo-controlled phase III study of granulocyte-macrophage colony-stimulating fac-tor in adult patients (>55 to 70 years of age) with acute myelogenous leukemia: a study of the Eastern Cooperative Oncology Group (E1490). *Blood 86*:457–462.

68. Witz, F., Harousseau, J. L., Sadoun, A., et al (1995). GM-CSF during and after remission induction treatment for elderly patients with acute myeloid leukemia (AML). *Blood 86*:512a (abstr 2036).

69. Löwenberg, B., Boogaerts, M. A., Vellenga, E., et al (1995). Various modalities of use of GM-CSF in the treatment of acute myelogenous leukemia (AML). A HOVON-SAKK randomized study (HOVON-4A). *Blood 86*:512a (abstr 2035).

70. Dombret, H., Chastang, C., Fenaux, P., et al (1995). A controlled study of recombinant human granulocyte colony-stimulating factor in elderly patients after treatment for acute myelogenous leukemia. AML Cooperative Study Group. *N Engl J Med 332*:1678–1683.

71. Heil, G., Chadid, L., Hoelzer, D., et al (1995). GM-CSF in a double-blind randomized, placebo controlled trial in therapy of adult patients with de novo acute myeloid leukemia (AML). *Leukemia 9*:3–9.

25
Use of Filgrastim (r-metHuG-CSF) in Aplastic Anemia

Aruna Raghavachar and Hubert Schrezenmeier
University of Ulm, Ulm, Germany

Andrea Bacigalupo
Azienda Ospedale San Martino, Genoa, Italy

I. DISEASE DEFINITIONS

Acquired aplastic anemia is a disorder characterized by bicytopenia or pancytopenia in the presence of an aplastic or hypoplastic bone marrow. Classification of the severity of aplastic anemia rests on the peripheral blood count. Patients with aplastic anemia can be classified into three groups, nonsevere, severe, and very severe (**Table 1**), which differ in response rate to immunosuppressive treatment and survival. The major prognostic criterion is the neutrophil count at diagnosis (**Figure 1A**). Clinical features of aplastic anemia are determined by the degree of pancytopenia, and the main causes of death are infections and thrombocytopenic hemorrhage (**Figure 1B**).

II. PATHOPHYSIOLOGY

Several pathogenetic mechanisms have been proposed to account for the bone marrow failure in aplastic anemia. These include an intrinsic defect ot hematopoietic stem cells, lymphocyte-mediated suppression of hematopoiesis, dysfunction of the bone marrow microenvironment, and abnormalities in the regulation of hematopoiesis by humoral factors (1–3).

533

Table 1.   Classification of Aplastic Anemia.

Nonsevere aplastic anemia:
Cytopenia with at least *two* of the following:
  Neutrophils $<1.5 \times 10^9$/L
  Platelets $<50 \times 10^9$/L
  Reticulocytes $<60 \times 10^9$/L
Severe aplastic anemia:
Cytopenia with at least *two* of the following:
  Neutrophils $<0.5 \times 10^9$/L
  Platelets $<20 \times 10^9$/L
  Reticulocytes $<20 \times 10^9$/L
Very severe aplastic anemia:
Meets criteria of severe aplastic anemia *and*
  Neutrophils $<0.2 \times 10^9$/L (obligatory)

Source: Refs. 55, 77, 78.

Several recent reports argue against a defective microenvironment and are in favor of a stem cell defect as the cause of aplastic anemia (4–6). The number of committed hematopoietic progenitors is reduced both in patients with aplastic anemia as well as in patients in remission after immunosuppressive treatment (**Figure 2A**) (7,8). With bone marrow recovery after immunosuppressive treatment, committed progenitors increase, but not to normal values (8,9–12). In long-term cultures of bone marrow from patients with aplastic anemia, the production of nonadherent cells and granulocyte-macrophage colony-forming cell (GM-CFC) is significantly reduced (4–6,12). Marsh et al (5,6) have demonstrated a defect of hematopoietic cell function, but normal supportive stroma function by crossover experiments inoculating nonadherent bone marrow cells of patients with aplastic anemia onto preformed, irradiated normal stroma and vice versa. The hematopoietic defect could be quantitatively enumerated by limiting-dilution assays. The frequency of cobblestone area–forming cells (CAFC) or long-term culture–initiating cells (LT-CIC) (in vitro surrogates of primitive progenitors with marrow repopulating activity) is substantially reduced in bone marrow of patients with aplastic anemia compared with control subjects (**Figure 2B**) (7,13).

The frequencies of CD34$^+$, CD33$^+$, and CD34$^+$CD33$^-$ populations in bone marrow of patients with aplastic anemia are reduced compared with normal subjects (14,15), and the growth of myeloid and erythroid colonies from separated CD34$^+$ cells of patients with aplastic anemia is decreased in comparison with normal CD34$^+$ cells (14,16). Generation of colonies from aplastic anemia bone marrow can be stimulated by growth factors, in particular by *c-kit* ligand (17,18) and granulocyte colony-stimulating

A

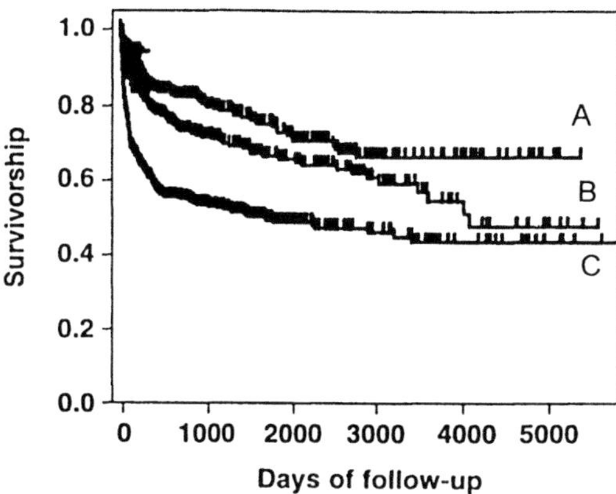

B

Causes of death in patients treated with immunosuppression

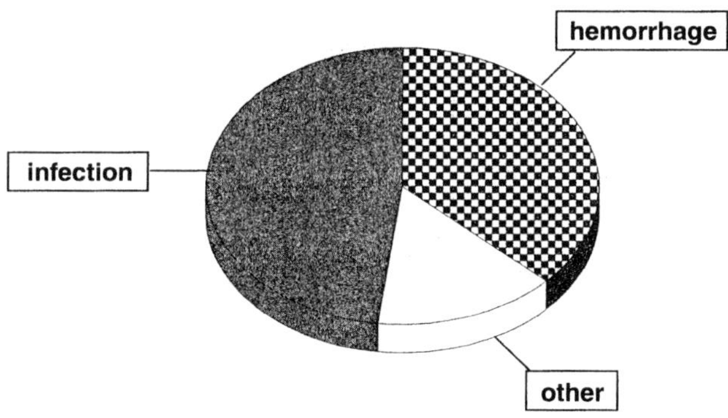

Figure 1. (A) Survival of patients with nonsevere (A; n = 319), severe (B; n = 284), or very severe aplastic anemia (C; n = 441) after immunosuppressive treatment. (B) Cause of death in patients with aplastic anemia treated with immunosuppressive treatment. ANC = absolute neutrophil count. (From the data base of the Working Party on Severe Aplastic Anemia of the European Group for Blood and Marrow Transplantation.)

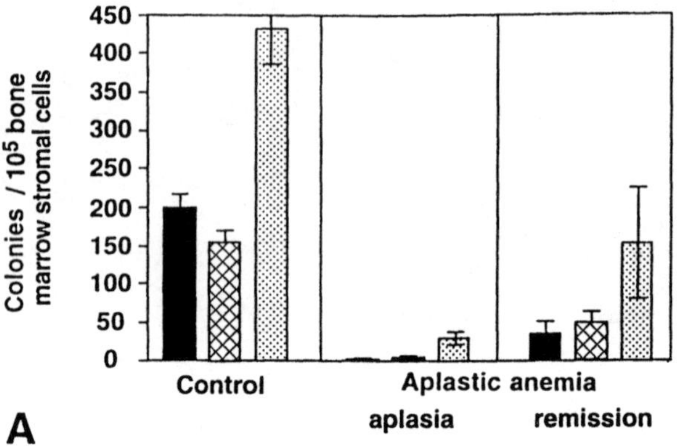

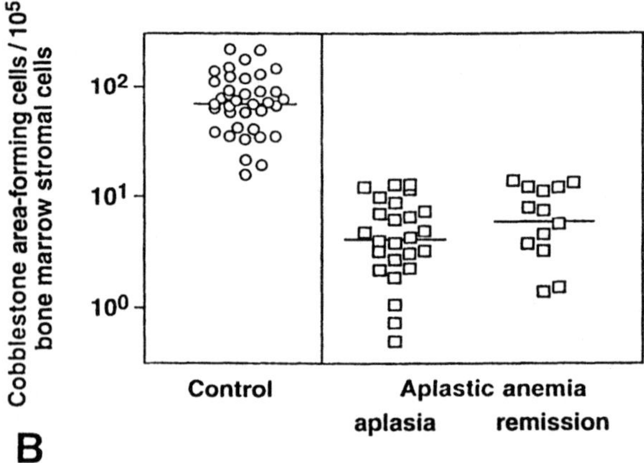

Figure 2. Hematopoietic defect in aplastic anemia. (A) Committed progenitors; (B) cobblestone area-forming cells in bone marrow of control subjects and, aplastic anemia patients in aplasia, and aplastic anemia patients in remission. AA = aplastic anemia; solid boxes, GM-CFC; hatched boxes, BFU-E; dotted boxes, CFU-E. (Panel B from Ref. 7.)

factor (G-CSF) (16). The cloning efficiency of sorted CD34[+] cells from patients with aplastic anemia could be restored by G-CSF alone or in combination with interleukin (IL-3), granulocyte-macrophage colony-stimulating factor (GM-CSF), or stem cell factor (SCF). Multiple comparisons showed that, among those cytokines, G-CSF had by far the greatest effect on colony growth from separated aplastic anemia CD34[+] cells (16).

These studies provide direct evidence that the hematopoietic defect in aplastic anemia is present at the level of primitive hematopoietic progenitor cells. The defect might be overcome, at least in vitro, by appropriate cytokine stimuli, in particular G-CSF–containing cocktails (16). Whereas there is a partial recovery at the level of committed progenitors in patients who respond to immunosuppressive treatment, the reduction of primitive stem cells is permanent in patients in remission.

A striking feature of aplastic anemia is the clinical interrelationship with clonal disorders of hematopoiesis. Aplastic anemia may evolve into paroxysmal nocturnal hemoglobinuria and vice versa. This observation prompted the search for a pathogenetic link between these two disorders. A deficiency of glycosylphosphatidyl-inositol (GPI)–anchored proteins is the pathophysiological basis of paroxysmal nocturnal hemoglobinuria and is caused by a somatic mutation of the X-linked gene PIG-A, coding for a protein that is essential for the biosynthesis of the GPI-anchor in the endoplasmatic reticulum (19). A phenotypic deficiency of GPI-anchored proteins in at least one cell lineage could be demonstrated in 30% to 60% of patients with aplastic anemia without clinical or serological evidence of paroxysmal nocturnal hemoglobinuria (20–22). Thus, the proportion of patients with aplastic anemia who show features of typical acquired aplastic anemia along with a paroxysmal nocturnal hemoglobinuria phenotype is substantially greater than recognized in the past. Response rate to immunosuppressive treatment was significantly greater in patients with normal expression of GPI-anchored proteins compared with a significant GPI-anchored protein-deficient population (21,22). The deficiency of GPI-anchored proteins might identify a subgroup of patients with aplastic anemia with different course of the disease.

In a study of the European Group for Bone Marrow Transplant (EBMT) Working Party on Severe Aplastic Anemia, the 10-year cumulative risk of developing myelodysplasia syndromes (MDS) or acute leukemia after immunosuppressive treatment for aplastic anemia was 16% (23). The increased risk of MDS/leukemia in patients with aplastic anemia prompted the question of whether aplastic anemia, in at least some cases, is a premalignant disorder with clonal hematopoiesis. One group reported that 72% of patients with aplastic anemia showed clonal hematopoiesis, but the incidence was probably overestimated due to technical reasons (24). In contrast to this report, several more recent studies reported a much lower incidence of clonal hematopoiesis as defined by polymorphic X-linked loci (<15%) (25–27). Minor clonal populations, however, may not be detected due to the limited sensitivity of X-inactivation analysis (26).

In previous reports, a number of lymphokine abnormalities in patients with aplastic anemia have been described (28,29). Zoumbos et al (28) reported significantly elevated serum levels of interferon-γ (IFN-γ) in aplastic anemia sera and excess release of IFN-γ on lymphocyte stimulation; however, the role of IFN-γ in the pathophysiology of aplastic anemia has been debated, since other investigators could find significantly elevated circulating levels of IFN-γ only in a minority of patients with the disease (30,31). Production of tumor necrosis factor-α (TNF-α) by lectin-stimulated peripheral blood mononuclear cells (PBMC) of patients with severe aplastic anemia is significantly enhanced (29). Exposure to IFN-γ and TNF-α can induce expression of Fas-antigen (CD95) on hematopoietic progenitor cells (32,33). Therefore, the cytokine overproduction might be the cause of increased expression of Fas-antigen on CD34$^+$ cells in bone marrow of patients with aplastic anemia. Apoptosis is accelerated in hematopoietic progenitors of these patients (32–34), and this may contribute to the deficiency of progenitor cells in aplastic anemia.

There is no evidence to suggest that aplastic anemia is caused by a deficiency of any known hematopoietic growth factor. Production of hematopoietic growth factors by cells from patients with aplastic anemia has usually been increased compared with healthy control subjects. Levels of endogenous G-CSF both in plasma and in bone marrow aspirates were significantly higher in patients with aplastic anemia compared with control subjects (35–37). Plasma G-CSF concentrations also were significantly higher in patients with signs of infection compared with patients without infection (**Figure 3**) (36). There was a significant negative correlation between G-CSF concentrations and absolute neutrophil counts (ANC) in patients with aplastic anemia without signs of infection (36). In bone marrow stromal cells from patients with aplastic anemia, both lipopolysaccharide (LPS)-induced G-CSF mRNA expression and spontaneous and LPS-induced G-CSF concentration in the supernatant were significantly greater than in healthy control subjects (38).

Serum levels of other hematopoietic stimulators such as GM-CSF, erythropoietin (EPO), thrombopoietin (TPO), and *flt-3* ligand also are consistently increased in the majority of patients with aplastic anemia (36,37,39–42), except for reduced serum levels of SCF (43,44). In vitro production of G-CSF, GM-CSF, and IL-6 by bone marrow stromal cells from patients with aplastic anemia is increased (45), while the mRNA expression of these factors in stromal cells and of IL-1 and SCF appears to be normal or increased (38,46,47). In contrast to the normal IL-1 mRNA expression in bone marrow stromal cultures, production of IL-1 by monocytes is reduced in patients with aplastic anemia (48,49). Production of G-CSF in long-term cultures of normal bone marrow can be stimulated by exogeneous IL-1. In contrast, approximately half of the patients with aplastic anemia showed an increase of G-CSF production in long-term marrow

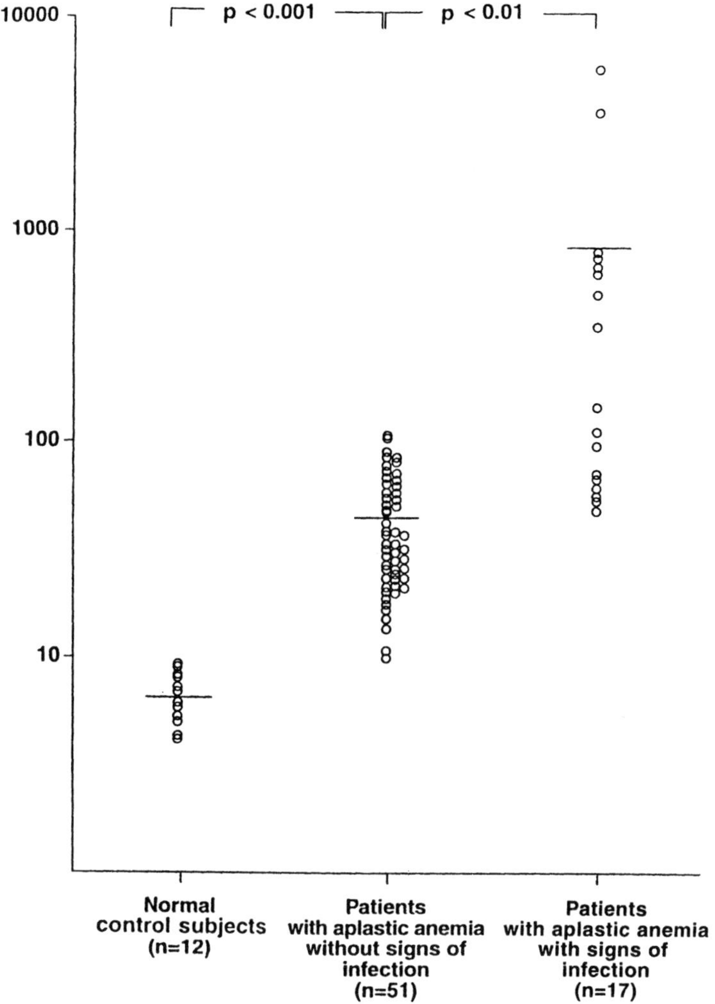

Figure 3. Endogenous plasma G-CSF concentrations in patients with aplastic anemia. Horizontal bars indicate mean values. (From Ref. 36.)

cultures, whereas G-CSF levels actually decreased in response to IL-1 (50). It is not clear whether low IL-1 production and, in some cases, the unresponsiveness to IL-1 could contribute to marrow failure in some patients with aplastic anemia. The failure of a trial with recombinant IL-1 argues against a simple IL-1 deficiency (51).

Pretreatment levels of endogenous growth factors may be useful in predicting response to immunosuppressive treatment. High levels of GM-CSF or G-CSF are associated with response to treatment (36,52), and high levels of SCF are associated with better survival (43). Using bioassays, earlier reports suggested that the ability to release high amounts of burst-stimulating activity or granulocyte colony–stimulating activity may be associated with favorable treatment outcome in patients with aplastic anemia (52–54).

III. PROBLEMS WITH STANDARD TREATMENT OF APLASTIC ANEMIA

For severe aplastic anemia, both bone marrow transplantation (BMT) and immunosuppression are effective; however, transplantation is only available to a minority of patients. One course of horse antithymocyte globulin, combined with cyclosporin A and corticosteroids, is now considered as standard regimen (55). In contrast to BMT, this intensive immunosuppression is still unsatisfactory in several points: hemopatoietic recovery is slow and often incomplete; some patients die of bleeding and/or infections because of prolonged cytopenia (**Figure 1B**); responders to immunosuppression have a significant risk of relapse and thus may not be cured (52); and evolution of aplastic anemia to another hematologic disease can occur years after successful immunosuppression (23).

IV. PHASE 1–2 TRIALS OF FILGRASTIM IN PATIENTS WITH APLASTIC ANEMIA

Patients with aplastic anemia with a very low neutrophil count at diagnosis have the poorest response to immunosuppression therapy, which translates into poor survival. Therefore, early trials focused on the effect of Filgrastim mainly in pancytopenic patients who were refractory to conventional therapy. With pharmacological doses of Filgrastim, it was possible to increase their ANC; however, this usually only occurred in patients who retained residual marrow function. In most patients, there was only a transient increase in ANC without significant increase in hemoglobin concentration or platelet count (56,57). Neutrophil response was delayed in most cases (**Figure 4**). With prolonged administration, some patients gained a trilineage response (58). A clear dose–response relationship could not be established (59).

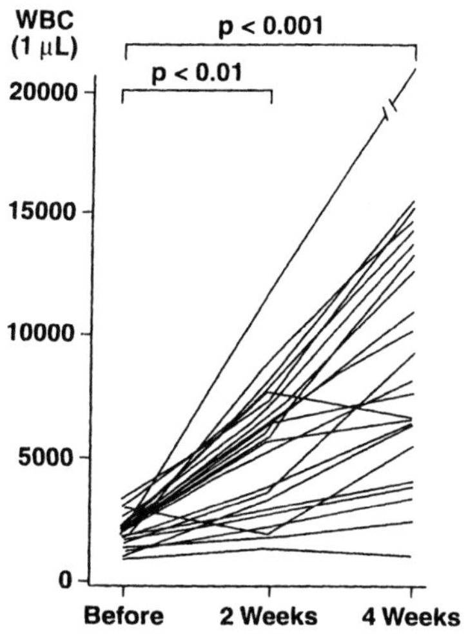

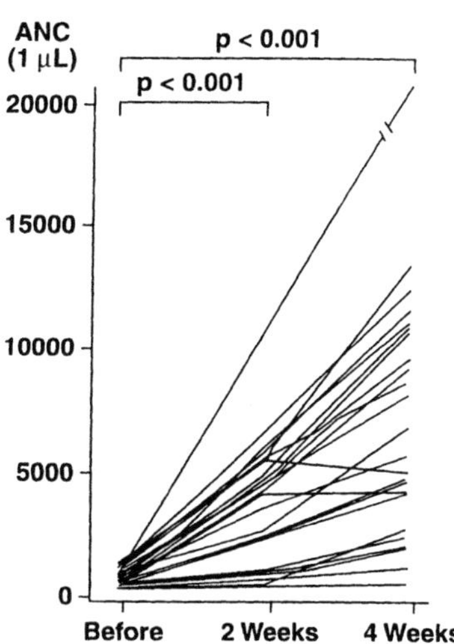

Figure 4. Responses of white blood cells (WBC) or absolute neutrophil count (ANC) to long-term administration of Filgrastim 2 to 4 weeks after the initiation of therapy. (From Ref. 58.)

V. PHASE 1–2 TRIALS OF FILGRASTIM AND ERYTHROPOIETIN

Since multilineage improvement in Filgrastim-treated patients with aplastic anemia is uncommon, some investigators began adding recombinant EPO to the Filgrastim regimen. Such a combination could also enhance the mobilizing capacity of Filgrastim (60). This could be of importance because some investigators believe that stem cell migration is a prerequisite for hematopoietic recovery in these patients. So far, the results seem disappointing with trilineage response occurring with high doses of the growth factors in only some patients after a median treatment duration of 9 months (61).

VI. FILGRASTIM IN COMBINATION WITH IMMUNOSUPPRESSIVE TREATMENT

Growth factors given with intensive immunosuppression might improve the outcome of patients with aplastic anemia (62). There are several possible effects of growth factors in combination with immunosuppressive treatment:

By reduction of early infectious complications, growth factors might increase the number of patients surviving long enough to achieve a response.

Antithymocyte globulin can induce the release of endogenous growth factors (63–68). Recombinant growth factors given along with immunosuppression might synergize with the endogenous factors released by antithymocyte globulin.

It has been hypothesized that hematopoietic response of aplastic anemia is dependent on mobilization of progenitor cells and reseeding of these cells to the bone marrow (62). It has been noted earlier that antithymocyte globulin can induce mobilization of progenitor cells in some patients (69). The combination of antithymocyte globulin and growth factors might be more potent in its mobilizing capacity. For these reasons, combinations of immunosuppressive treatment and growth factors should be investigated.

A few case reports suggested efficacy of a combination of G-CSF with cyclosporin A (70–72). In a pilot study of the EBMT Working Party on Severe Aplastic Anemia, G-CSF (5 μg/kg/day subcutaneously [SC], days 1 through 90) was combined with antithymocyte globulin, cyclosporin A, and methylprednisolone (62). Thirty-three of 40 patients (92%) had trilineage hematopoietic reconstitution and became transfusion independent at a median interval of 115 days after immunosuppressive treatment. Four patients showed no recovery and three patients (8%) died of infection. Actuarial survival at 2 years was 92% (**Figure 5**). This pilot study was continued and a total of 108 patients were treated. A recent update confirmed the high response rate (76% for previously untreated patients) with an unusually high proportion of 45% of complete responses and excellent survival (92% at 3 years) (A. Bacigalupo, unpublished data). Filgrastim in combination with immunosuppression treatment was well tolerated in this study.

Based on these encouraging data, several prospective randomized trials were designed: in an EBMT study, patients were randomized to receive antithymocyte globulin and cyclosporin A without or with G-CSF (5 μg/kg/day for 90 days). An ongoing prospective randomized trial of the German Aplastic Anemia Study Group compares antithymocyte globulin and cyclosporin A versus cyclosporin A and Filgrastim treatment. The Japanese Study Group compares different combinations of G-CSF with immunosuppressive agents. Although the survival in the EBMT pilot study is seemingly superior to survival in other published trials of immunosuppressive treatment without growth factors, careful interpretation is warranted. There is an improvement in treatment results in the last decade that might

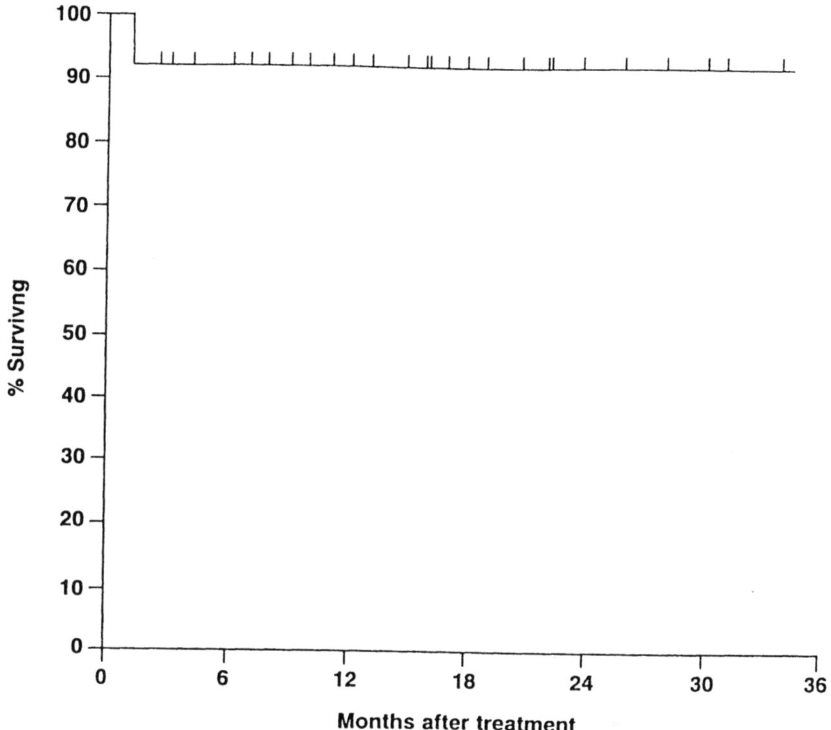

Figure 5. Overall actuarial probability of survival in 40 patients with newly diagnosed aplastic anemia and a neutrophil count $<0.5 \times 10^9$/L, who were treated with antithymocyte globulin, cyclosporin A, methylprednisolone, and Filgrastim (5 μg/kg/day for 90 days). (From Ref. 62.)

be due to a number of factors (eg, improved supportive care). The definitive role of Filgrastim in treatment of aplastic anemia can only be defined by the ongoing randomized trials.

The prolonged Filgrastim administration after antithymocyte globulin and cyclosporin A treatment resulted in mobilization of progenitors in some patients. By means of leukapheresis, a median number of 1.8 $\times$ 10^6/kg CD34$^+$ cells and a median number of 3.9 $\times$ 104/kg GM-CFC were collected (73). Growth of GM-CFC was monitored between days +33 and +77 of Filgrastim treatment (**Figure 6**). These results demonstrate that, despite reduced stem cells in the bone marrow, mobilization of progenitor cells is possible at least in some patients. Many questions concerning this phenomenon remain to be answered.

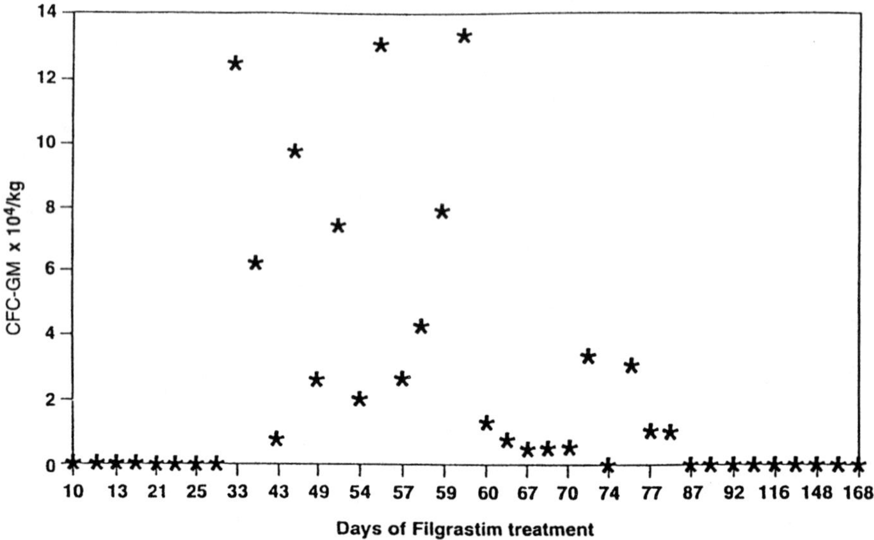

Figure 6. Numbers of CFC-GM $\times$ 10^4/kg grown for single leukaphereses plotted against time (day) of Filgrastim treatment in patients treated with antithymocyte globulin + cyclosporin A + methylprednisolone + Filgrastim. (From Ref. 73.)

It is not clear whether mobilization of progenitor cells is an epiphenomenon of treatment or whether mobilization and "re-seeding" of primitive progenitor cells is essential for response. The latter possibility is supported by the observation of Torok-Storb et al (69) that increased numbers of circulating erythroid burst-forming cells (BFU-E) correlate with response to antithymocyte globulin. Filgrastim might amplify its mobilization capacity.

Furthermore, the functional capability of mobilized cells remains to be determined. If these cells are normal, autotransplantation protocols, eg, with high-dose cyclophoshamide (74), followed by infusion of (ex vivo possibly expanded) peripheral blood progenitor cells might be feasible in aplastic anemia.

VII.   FILGRASTIM IN COMBINATION WITH STEM CELL FACTOR

Recently, an abstract reported the combined use of Filgrastim and r-metHuSCF and suggested possible benefits to the combination (75). In this study, patients who had failed antithymocyte globulin or antilymphocyte globulin therapy were treated with one of seven regimens of Filgrastim (0 to 10 μg/kg/day) and r-metHuSCF (5 to 50 μg/kg three times per week). Doses of r-metHuCSF $\geq$25 μg/kg, with or without Filgrastim, were well tolerated and produced trilineage responses in 33% of the patients who had failed previous immunosuppressive therapy. The role of Filgrastim in the responses is not well understood, although three of five trilineage responders had not responded to either Filgrastim or rHuGM-CSF earlier.

In general, Filgrastim is well tolerated in the treatment of aplastic anemia. The development of Sweet's syndrome during Filgrastim therapy in patients with aplastic anemia has to be considered (76). Of great concern is the observation of myelodysplasia and acute myeloid leukemia after Filgrastim treatment of patients with aplastic anemia (65); however, a relationship between aplastic anemia and these clonal disorders has been recognized for a long time. Approximately 12% of patients with aplastic anemia are expected to have a clonal cytogenetic abnormality at diagnosis. Therefore, a possible role of Filgrastim in the transformation of aplastic anemia to leukemic disorders can be evaluated only in controlled clinical trials with a pretreatment cytogenetic analysis. So far, the two randomized trials of G-CSF and immunosuppression do not substantiate the role of Filgrastim in the transformation of true aplastic anemia (A. Bacigalupo and A. Raghavachar, personal communication).

VIII. CONCLUSIONS

Published evidence does not support the use of Filgrastim as first treatment in aplastic anemia. It might even be harmful to those patients because prompt treatment with immunosuppression or referral for BMT will be delayed. The EBMT Working Party has discouraged the use of Filgrastim as a single agent in patients with newly diagnosed aplastic anemia (78); however, the use of Filgrastim in combination with immunosuppressive therapy seems to be promising in terms of hematopoietic response and excellent survival. Future clinical trials will focus on the mobilizing capacity of Filgrastim in patients with aplastic anemia aiming at high-dose immunosuppression with autologous stem cell support. In addition, combinations of immunosuppression, Filgrastim, and hemopoietins such as SCF or thrombopoietic growth factors are very attractive protocols for the treatment of aplastic anemia.

REFERENCES

1. Marmont, A. Aplastic anemia: good progress but still a long way to go. In: *Aplastic Anemia: Current Perspectives on the Pathogenesis and Treatment.*

Raghavachar, A. Schrezenmeier, H., and Frickhofen, N., eds., Vienna: Blackwell Scientific, 1993, pp. 2–10.

2. Young, N. S. (1992). The problem of clonality in aplastic anemia: Dr Dameshek's riddle, restated. *Blood 79*:1385–1392.

3. Nissen, C. (1991). Pathophysiology of aplastic anaemia. *Semin Hematol 28*:313–318.

4. Gibson, F. M., and Gordon-Smith, E. C. (1990). Long-term culture of aplastic anaemia bone marrow. *Br J Haematol 75*:421–427.

5. Marsh, J. C., Chang, J., Testa, N. G., Hows, J. M., and Dexter, T. M. (1990). The hematopoietic defect in aplastic anemia assessed by long-term marrow culture. *Blood 76*:1748–1757.

6. Marsh, J. C., Chang, J., Testa, N. G., Hows, J. M., and Dexter, T. M. (1991). In vitro assessment of marrow 'stem cell' and stromal cell function in aplastic anaemia. *Br J Haematol 78*:258–267.

7. Schrezenmeier, H., Jenal, M., Herrmann, F., Heimpel, H., and Raghavachar, A. (1996). Quantitative analysis of cobblestone area-forming cells in bone marrow of patients with aplastic anemia by limiting dilution assay. *Blood 88*:4474–4480.

8. Kern, P., Heimpel, H., Heit, W., and Kubanek, B. (1977). Granulocytic progenitor cells in aplastic anaemia. *Br J Haematol 35*:613–623.

9. Ruvidic, R., Jovcic, G., Biljanovic-Paunovic, L., Stojanovic, N., Mijovic, A., Pavlovic-Kentera, V. (1985). Myelopoiesis and erythropoiesis of bone marrow cells cultured in vitro in patients recovered from aplastic anaemia. *Scand J Haematol 35*:437–444.

10. Yoshida, K., Miura, I., Takahashi, T., et al (1983). Quantitative and qualitative analysis of stem cells of patients with aplastic anaemia. *Scand J Haematol 30*:317–323.

11. Nissen, C., Cornu, P., Gratwohl, A., and Speck, B. (1980). Peripheral blood cells from patients with aplastic anaemia in partial remission suppress growth of their own bone marrow precursors in culture. *Br J Haematol 45*:233–243.

12. Bacigalupo, A., Figari, O., Tong, J., et al (1992). Long-term marrow culture in patients with aplastic anemia compared with marrow transplant recipients and normal controls. *Exp Hematol 20*:425–430.

13. Maciejewski, J. P., Selleri, C., Sato, T., Anderson, S., and Young, N. S. (1996). A severe and consistent deficit in marrow and circulating primitive hematopoietic cells (long-term culture-initiating cells) in aquired aplastic anemia. *Blood 88*:1983–1991.

14. Maciejewski, J. P., Anderson, S., Katevas, P., and Young, N. S. (1994). Phenotypic and functional analysis of bone marrow progenitor cell compartment in bone marrow failure. *Br J Haematol 87*:227–234.

15. Scopes, J., Bagnara, M., Gordon Smith, E. C., Ball, S. E., and Gibson, F. M. (1994). Haemopoietic progenitor cells are reduced in aplastic anaemia. *Br J Haematol 86*:427–430.

16. Scopes, J., Daly, S., Atkinson, R., Ball, S. E., Gordon-Smith, E. C., and Gibson, F. M. (1996). Aplastic anemia: evidence for dysfunctional bone marrow progen-

itor cells and the corrective effect of granulocyte colony-stimulating factor in vitro. *Blood 87*:3179–3185.

17. Bagnara, G. P., Strippoli, P., Bonsi, L., et al (1992). Effect of stem cell factor on colony growth from acquired and constitutional (Fanconi) aplastic anemia. *Blood 80*:382–387.

18. Wodnar Filipowicz, A., Tichelli, A., Zsebo, K. M., Speck, B., and Nissen, C. (1992). Stem cell factor stimulates the in vitro growth of bone marrow cells from aplastic anemia patients. *Blood 79*:3196–3202.

19. Novitzky, N., and Jacobs, P. (1995). Immunosuppressive therapy in bone marrow aplasia: the stroma functions normally to support hematopoiesis. *Exp Hematol 23*:1472–1477.

20. Vu, T., Griscelli Bennacour, A., Gluckman, E., et al (1996). Aplastic anemia and paroxysmal nocturnal hemoglobinuria: A study of the GPI-anchored proteins on human platelets. *Br J Haematol 93*:586–589.

21. Schrezenmeier, H., Hertenstein, B., Wagner, B., Raghavachar, A., and Heimpel, H. (1995). A pathogenetic link between aplastic anemia and paroxysmal nocturnal hemoglobinuria is suggested by a high frequency of aplastic anemia patients with a deficiency of phosphatidylinositol glycan anchored proteins. *Exp Hematol 23*:81–87.

22. Schubert, J., Vogt, H. G., Zielinska Skowronek, M., et al (1994). Development of the glycosylphosphatitylinositol-anchoring defect characteristic for paroxysmal nocturnal hemoglobinuria in patients with aplastic anemia. *Blood 83*:2323–2328.

23. Socie, G., Henry-Amar, M., Bacigalupo, A., et al (1993). Malignant tumors occurring after treatment of aplastic anemia. European Bone Marrow Transplantation-Severe Aplastic Anaemia Working Party. *N Engl J Med 329*:1152–1157.

24. va Kamp, H., Landegent, J. E., Jansen, R. P., Willemze, R., and Fibbe, W. E. (1991). Clonal hematopoiesis in patients with acquired aplastic anemia. *Blood 78*:3209–3214.

25. Tsuge, I., Kojima, S., Matsuoka, H., et al (1993). Clonal haematopoiesis in children with acquired aplastic anaemia. *Br J Haematol 84*:137–143.

26. Raghavachar, A., Janssen, J. W., Schrezenmeier, H., et al (1995). Clonal hematopoiesis as defined by polymorphic X-linked loci occurs infrequently in aplastic anemia. *Blood 86*:2938–2947.

27. Josten, K. M., Tooze, J. A., Borthwick Clarke, C., Gordon-Smith, E. C., and Rutherford, T. R. (1991). Acquired aplastic anemia and paroxysmal nocturnal hemoglobinuria: studies on clonality. *Blood 78*:3162–3167.

28. Zoumbos, N. C., Gascon, P., Djeu, J. Y., and Young, N. S. (1985). Interferon is a mediator of hematopoietic suppression in aplastic anemia in vitro and possibly in vivo. *Proc Natl Acad Sci USA 82*:188–192.

29. Hinterberger, W., Adolf, G., Aichinger, G., et al (1988). Further evidence for lymphokine overproduction in severe aplastic anemia. *Blood 72*:266–272.

30. Raghavachar, A., Frickhofen, N., Digel, W., Porzsolt, F., and Heimpel, H. (1988). Lymphokines as inhibitors of hematopoietic cell proliferation. *Blood Cells 14*:471–484.

31. Torok-Storb, B., Johnson, G. G., Bowden, R., and Storb, R. (1987). Gamma-interferon in aplastic anemia: inability to detect significant levels in sera or demonstrate hematopoietic suppressing activity. *Blood 69*:629–633.
32. Maciejewski, J., Selleri, C., Anderson, S., and Young, N. S. (1995). Fas antigen expression on CD34[+] human marrow cells is induced by interferon gamma and tumor necrosis factor alpha and potentiates cytokine-mediated hematopoietic suppression in vitro. *Blood 85*:3183–3190.
33. Maciejewski, J. P., Selleri, C., Sato, T., Anderson, S., and Young, N. S. (1995). Increased expression of Fas antigen on bone marrow CD34[+] cells of patients with aplastic anaemia. *Br J Haematol 91*:245–252.
34. Philpott, N. J., Scopes, J., Marsh, J. C., Gordon-Smith, E. C., and Gibson, F. M. (1995). Increased apoptosis in aplastic anemia bone marrow progenitor cells: possible pathophysiologic significance. *Exp Hematol 23*:1642–1648.
35. Shinohara, K., Oeda, E., Nomiyama, J., et al (1995). The levels of granulocyte colony-stimulating factor in the plasma of the bone marrow aspirate in various hematological disorders. *Stem Cells 13*:421–427.
36. Kojima, S., Matsuyama, T., Kodera, Y., et al (1996). Measurement of endogenous plasma granulocyte colony-stimulating factor in patients with acquired aplastic anemia by a sensitive chemiluminescent immunoassay. *Blood 87*:1303–1308.
37. Watari, K., Asano, S., Shirafuji, N., et al (1989). Serum granulocyte colony-stimulating factor levels in healthy volunteers and patients with various disorders as estimated enzyme immunoassay. *Blood 73*:117–122.
38. Hirayama, Y., Kohgo, Y., Matsunaga, T., Ohi, S., Sakamaki, S., Niitsu, Y. (1993). Cytokine mRNA expression of bone marrow stromal cells from patients with aplastic anaemia and myelodysplastic syndrome. *Br J Haematol 85*:676–683.
39. Schrezenmeier, H., Griesshammer, M., Hornkohl, A., et al. Thrombopoietin serum levels in patients with aplastic anemia: correlation with platelet count and persistent elevation in remission. *Br J Hematol* (in press).
40. Wodnar-Filipowicz, A., Lyman, S. D., Gratwohl, A., Tichelli, A., Speck, B., and Nissen, C. (1996). Flt3 ligand level reflects hematopoietic progenitor cell function in aplastic anemia and chemotherapy-induced bone marrow aplasia. *Blood 88*:4493–4499.
41. Schrezenmeier, H., Raghavachar, A., and Heimpel, H. (1993). Granulocyte-macrophage colony-stimulating factor in the sera of patients with aplastic anemia. *Clin Invest 71*:102–108.
42. Schrezenmeier, H., Noe, G., Raghavachar, A., Rich, I. N., Heimpel, H., and Kubanek, B. (1994). Serum erythropoietin and serum transferrin receptor levels in aplastic anaemia. *Br J Haematol 88*:286–294.
43. Wodnar-Filipowicz, A., Yancik, S., Moser, Y., et al (1993). Levels of soluble stem cell factor in serum of patients with aplastic anemia. *Blood 81*:3259–3264.
44. Nimer, S. D., Leung, D. H., Wolin, M. J., and Golde, D. W. (1994). Serum stem cell factor levels in patients with aplastic anemia. *Int J Hematol 60*:185–189.
45. Kojima, S., Matsuyama, T., and Kodera, Y. (1992). Hematopoietic growth factors released by marrow stromal cells from patients with aplastic anemia. *Blood 79*:2256–2261.

46. Stark, R., Andre, C., Thierry, D., Cherel, M., Galibert, F., and Gluckman, E. (1993). The expression of cytokine and cytokine receptor genes in long-term bone marrow culture in congenital and acquired bone marrow hypoplasias. *Br J Haematol 83*:560–566.

47. Krieger, M. S., Nissen, C., and Wodnar-Filipowicz, A. (1995). Stem-cell factor in aplastic anemia: in vitro expression in bone marrow stroma and fibroblast cultures. *Eur J Haematol 54*:262–269.

48. Nakao, S., Matsushima, K., and Young, N. (1989). Decreased interleukin 1 production in aplastic anaemia. *Br J Haematol 71*:431–436.

49. Gascon, P., and Scala, G. (1988). Decreased interleukin-1 production in aplastic anemia. *Am J Med 85*:668–674.

50. Migliaccio, A. R., Migliaccio, G., Adamson, J. W., and Torok-Storb, B. (1992). Production of granulocyte colony-stimulating factor and granulocyte/macrophage-colony-stimulating factor after interleukin-1 stimulation of marrow stromal cell cultures from normal or aplastic anemia donors. *J Cell Physiol 152*:199–206.

51. Gunther, G., Bjorkholm, M., Bjorklind, A., Engervall, P., and Stiernstedt, G. (1991). Septicemia in patients with hematological disorders and neutropenia. A retrospective study of causative agents and their resistance profile. *Scand J Infect Dis 23*:589–598.

52. Schrezenmeier, H., Marin, P., Raghavachar, A., et al (1993). Relapse of aplastic anemia after immunosuppressive treatment: a report from the European Bone Marrow Transplantation Group SAA Working Party. *Br J Haematol 85*:371–377.

53. Nissen, C., Moser, Y., Speck, B., Gratwohl, A., and Weis, J. (1985). Stimulatory serum factors in aplastic anaemia. II. Prognostic significance for patients treated with high dose immunosuppression. *Br J Haematol 61*:499–512.

54. Nissen, C., Moser, Y., dalle Carbonare, V., Gratwohl, A., and Speck, B. (1989). Complete recovery of marrow function after treatment with anti-lymphocyte globulin is associated with high, whereas early failure and development of paroxysmal nocturnal haemoglobinuria are associated with low endogenous G-CSA-release. *Br J Haematol 72*:573–577.

55. Frickhofen, N., Kaltwasser, J. P., Schrezenmeier, H., et al (1991). Treatment of aplastic anemia with antilymphocyte globulin and methylprednisolone with or without cyclosporine. The German Aplastic Anemia Study Group. *N Engl J Med 324*:1297–1304.

56. Champlin, R. E., Nimer, S. D., Ireland, P., Oette, D. H., Golde, D. W. (1989). Treatment of refractory aplastic anemia with recombinant human granulocyte-macrophage colony-stimulating factor. *Blood 73*:694–699.

57. Kojima, S., Fukuda, M., Miyajima, Y., Matsuyama, T., and Horibe, K. (1991). Treatment of aplastic anemia in children with recombinant human granulocyte colony-stimulating factor. *Blood 77*:937–941.

58. Sonoda, Y., Ohno, Y., Fujii, H., et al (1993). Multilineage response in aplastic anemia patients following long-term administration of Filgrastim (recombinant human granulocyte colony stimulating factor). *Stem Cells 11*:543–554.

59. Kojima, S., and Matsuyama, T. (1994). Stimulation of granulopoiesis by high-dose recombinant human granulocyte colony-stimulating factor in children with aplastic anemia and very severe neutropenia. *Blood 83:*1474–1478.

60. Olivieri, A., Offidani, M., Cantori, I., et al (1995). Addition of erythropoietin to granulocyte colony-stimulating factor after priming chemotherapy enhances hemopotetic progenitor mobilization. *Bone Marrow Transplant 16:*765–770.

61. Bessho, M., Jinnai, I., Hirashima, K., et al (1994). Trilineage recovery by combination therapy with recombinant human granulocyte colony-stimulating factor and erythropoietin in patients with aplastic anemia and refractory anemia. *Stem Cells 12:*604–615.

62. Bacigalupo, A., Broccia, G., Corda, G., et al (1995). Antilymphocyte globulin, cyclosporin, and granulocyte colony-stimulating factor in patients with acquired severe aplastic anemia (SAA): a pilot study of the EBMT SAA Working Party. *Blood 85:*1348–1353.

63. Nimer, S. D., Golde, D. W., Kwan, K., Lee, K., Clark, S., and Champlin, R. (1991). In vitro production of granulocyte-macrophage colony-stimulating factor in aplastic anemia: possible mechanisms of action of antithymocyte globulin. *Blood 78:*163–168.

64. Barbano, G. C., Schenone, A., Roncella, S., et al (1988). Anti-lymphocyte globulin stimulates normal human T cells to proliferate and to release lymphokines in vitro. A study at the clonal level. *Blood 72:*956–963.

65. Kojima, S., Tsuchida, M., and Matsuyama, T. (1992). Myelodysplasia and leukemia after treatment of aplastic anemia with G-CSF. *N Engl J Med 326:*1294–1295.

66. Tong, J., Bacigalupo, A., Piaggio, G., Figari, O., Sogno, G., and Marmont, A. (1991). In vitro response of T cells from aplastic anemia patients to antilymphocyte globulin and phytohemagglutinin: colony-stimulating activity and lymphokine production. *Exp Hematol 19:*312–316.

67. Taniguchi, Y., Frickhofen, N., Raghavachar, A., Digel, W., and Heimpel, H. (1990). Antilymphocyte immunoglobulins stimulate peripheral blood lymphocytes to proliferate and release lymphokines. *Eur J Haematol 44:*244–251.

68. Kawano, Y., Nissen, C., Gratwohl, A., and Speck, B. (1988). Immunostimulatory effects of different antilymphocyte globulin preparations: a possible clue to their clinical effect. *Br J Haematol 68:*115–119.

69. Torok-Storb, B., Doney, K., Brown, S. L., and Prentice, R. L. (1984). Correlation of two in vitro tests with clinical response to immunosuppressive therapy in 54 patients with severe aplastic anemia. *Blood 63:*349–355.

70. Lefrere, J. J., Courouce, A. M., Girot, R., Bertrand, Y., and Soulier, J. P. (1986). Six cases of hereditary spherocytosis revealed by human parvovirus infection. *Br J Haematol 62:*653–658.

71. Kojima, S., Fukuda, M., Miyajima, Y., and Matsuyama, T. (1990). Cyclosporine and recombinant granulocyte colony-stimulating factor in severe aplastic anemia. *N Engl J Med 323:*920–921.

72. Weide, R., Lyttelton, M., Samson, D., et al (1993). Sustained trilineage response in a patient with ALG-resistant severe aplastic anaemia after treatment with

G-CSF, erythropoietin and cyclosporin A: association of recovery with marked elevation of serum alkaline phosphatase. *Br J Haematol 85*:608–610.
73. Bacigalupo, A., Piaggio, G., Podesta, M., et al (1993). Collection of peripheral blood hematopoietic progenitors (PBHP) from patients with severe aplastic anemia (SAA) after prolonged administration of granulocyte colony-stimulating factor. *Blood 82*:1410–1414.
74. Brodsky, R. A., Sensenbrenner, L. L., and Jones, R. J. (1996). Complete remission in severe aplastic anemia after high-dose cyclophosphamide without bone marrow transplantation. *Blood 87*:491–494.
75. Kurzrock, R., Paquette, A., Gratwohl, K., et al (1997). Use of stem cell factor (Stemgen, SCF) and Filgrastim (G-CSF) in aplastic anemia (AA) patients (pts) who have failed ATG/ALG therapy. *Blood 90*:173a (abstr 762).
76. Fukutoku, M., Shimizu, S., Ogawa, Y., et al (1994). Sweet's syndrome during therapy with granulocyte colony-stimulating factor in a patient with aplastic anemia. *Br J Haematol 86*:645–648.
77. Marsh, J. C., Sociè, G., Schrezenmeier, H., et al (1994). Haemopoietic growth factors in aplastic anemia: a cautionary note. European Bone Marrow Transplant Working Party for Severe Aplastic Anaemia. *Lancet 344*:172–173.
78. Bacigalupo, A., Hows, J., Gluckman, E., et al (1988). Bone marrow transplantation (BMT) versus immunosuppression for the treatment of severe aplastic anemia (SAA): a report of the EBMT SAA working party. *Br J Haematol 70*:177–182.
79. Camitta, B. M., Thomas, E. D., Nathan, D. G., et al (1979). A prospective study of androgens and bone marrow transplantation for treatment of severe aplastic anemia. *Blood 53*:504–514.

26
Use of Filgrastim (r-metHuG-CSF) in Autoimmune Disease

Roland P. Repp and Gisela Helm
University of Erlangen–Nuernberg, Erlangen, Germany

J. G. J. van de Winkel
University Hospital Utrecht and Medarex Europe, Utrecht, The Netherlands

I. INTRODUCTION

There has been much controversy about the use of recombinant human granulocyte colony-stimulating factor (rHuG-CSF) to treat patients with autoimmune diseases because of the possible induction of an exaggerated immune response. This chapter will address several aspects of neutropenia and autoimmunity.

First, we will focus on neutropenia in patients with defined autoimmune diseases. In this patient population, neutropenia can either occur as an immunologically mediated feature of the disease or because of non-immunologic, mostly treatment-related, reasons. Both aspects will be covered with regard to the use of rHuG-CSF. Second, the role of rHuG-CSF in patients with autoimmune neutropenia and no underlying autoimmune disease will be addressed.

Autoimmune neutropenia (AIN) is characterized by the immune-mediated destruction of neutrophils by antibodies directed against neutrophil antigens. Autoimmune neutropenia can be divided into primary and secondary forms. The classical examples of autoimmune diseases often known to be present with *secondary AIN* are Felty syndrome and rheumatoid arthritis associated with T-γ lymphocytosis and, to a

lesser degree, systemic lupus erythematosus. In addition, secondary autoimmune neutropenia can be found in patients with autoimmune thyroid
disease, chronic autoimmune hepatitis, hypogammaglobulinemia, hyper-
IgM-dysgammaglobulinemia, infectious mononucleosis, lymphoproliferative disease, Hodgkin's disease, or in association with autoimmune hemolytic anemia or immune thrombocytopenia. The diagnosis of *primary
autoimmune neutropenia* is made after exclusion of both these secondary
forms and also other causes of neutropenia, such as impaired myelopoiesis
in malignancy or aplastic disorders, infectious causes, metabolic or nutritional disorders, storage diseases, and drug-induced mechanisms.

The pathogenetic mechanisms of AIN include excessive neutrophil
margination, destruction of neutrophils by autoantibodies or immune complexes, and effects of suppressor T cells on myelopoiesis. The amount of
autoantibodies detected in patients with AIN does not correlate with the
degree of neutropenia. A good correlation, however, is observed with the
C3-fixing ability of the autoantibodies. Neutrophils are proposed to be
removed mainly by phagocytosis (of IgG- and C3-opsonized cells), whereas
complement lysis does not seem to play a major role in removal. Another
mechanism of neutrophil clearance is binding of immune complexes to
neutrophil Fc receptors. This is supported by the findings of Petersen and
Wiik (1), who were unable to show neutrophil binding of $F(ab')_2$ fragments
of IgG in patients with AIN, mainly with Felty syndrome. A third mechanism of neutrophil destruction represents antibody-dependent lymphocyte-
mediated granulocytotoxicity, described by Shastri and Logue (2).

The target antigen is only known in a minority of patients with AIN.
In 10% to 20% of cases of AIN in infancy, antibodies have a specificity for
NA1/NA2 (neutrophil antigen 1 and 2, located on CD16). Antibodies to
the NA allotypes also have been described in patients with primary biliary
cirrhosis, leukocyte-mediated transfusion reactions, and alloimmune neonatal neutropenia. Further antigens are CD11b/CD18, inducing functional
defects as seen in leukocyte adhesion deficiency, actin, CR1, x-hapten, or
lymphocyte-activating factor (LAF) antigens. Presence of the antigen on
myeloid precursors results in a pure white blood cell aplasia, whereas on
mature white blood cells, a maturation arrest with preserved precursor cells
is induced.

Apart from these neutropenias with autoimmune etiology, treatment-
related neutropenias must be considered. Drugs often used to treat rheumatic diseases include antimetabolites and alkylating agents, drugs that
have a direct, dose-dependent toxic effect on bone marrow precursor cells.

In addition to these predictable effects, idiosyncratic drug-induced
neutropenia and pancytopenia can be caused by virtually any drug. A
relatively high occurrence is known for drugs used in autoimmune diseases,

such as analgetics, antiinflammatory and antithyroid drugs. Pathophysiologically, immune destruction of mature circulating cells as well as bone marrow precursor cells may contribute to peripheral cytopenia. The myeloid progenitors can be virtually depleted (pure white cell aplasia) or otherwise show a relative increase of immature cells, the so-called maturation arrest.

The goal of therapy for all forms of AIN is control of infectious complications and not necessarily an increase in neutrophil counts. Although the incidence of infection increases with the severity of neutropenia, infections are less frequent in patients with immune neutropenia than in patients with a corresponding degree of neutropenia due to other causes. On the other hand, defects in neutrophil functions due to neutrophil antibodies may predispose a patient to infections despite having normal neutrophil counts.

Endogenous and recombinant forms of G-CSF increase neutrophil numbers and enhance functions such as phagocytosis, chemotaxis, and antibody-dependent cell-mediated cytotoxicity (ADCC). In autoimmune disease, the use of rHuG-CSF has been reported in rheumatoid arthritis, particularly Felty syndrome and rheumatoid arthritis–associated T-γ lymphocytosis, in antithyroid drug–induced neutropenia in autoimmune thyroid disease, and, very rarely, in systemic lupus erythematosus and some other systemic and cutaneous autoimmune disorders.

In contrast to published papers on the use of rHuG-CSF in treating patients with malignancy, only case reports are published to describe its administration in autoimmune diseases or other forms of AIN. This case will be discussed in relation to the clinical indication, dose, reversibility of neutropenia, and clinical benefit. The issue of side effects, in particular a possible exacerbation of an underlying disease, will receive special attention.

II. NEUTROPENIA ASSOCIATED WITH AUTOIMMUNE DISEASES

Neutropenia occurs in patients with autoimmune diseases and can be either disease-associated or disease-independent (3). Cytokines are important mediators of autoimmune disorders and administration of growth factors could possibly modulate the immune response. Clinically, exacerbation of rheumatic symptoms has been reported during growth factor application in some patients. On the other hand, protective effects of rHuG-CSF have been shown in preclinical models.

Both G-CSF and granulocyte-macrophage colony-stimulating factor (GM-CSF) are present in synovial fluid; GM-CSF induces HLA-DR expression in synovial cells. In addition, GM-CSF induces production of pro-inflammatory cytokines (eg, tumor-necrosis factor [TNF]-α, interleukin [IL]-1, and macrophage colony-stimulating factor [M-CSF]). The patho-physiologic role of cytokine induction is supported by the occurrence of a flare of arthritis after rHuGM-CSF administration in two patients with Felty syndrome (4,5), concomitant with high serum levels of IL-6 and TNF. In patients with autoimmune disorders treated with rHuG-CSF, no release of IL-6 and TNF-α is observed in most patients, although Canvin et al (6) described one patient with rheumatoid arthritis who had increased levels of these cytokines after rHuG-CSF administration.

T-cell activation is well established as being central in the pathogenesis of rheumatoid arthritis. After administration of Filgrastim, natural killer (NK)-cell and cytotoxic T-cell counts increased in patients with cancer (7) and acquired immunodeficiency syndrome (AIDS) (8), although lympho-cytes lack receptors for G-CSF (reviewed in Ref. 9). In resting B lympho-cytes, G-CSF does not induce a proliferative response but enhances immu-noglobulin production (10) and can trigger an increased IgA rheumatoid factor production, a potent inducer of lysosomal release by neutrophils (11).

In addition to these effects on lymphocytes, G-CSF might influence tissue damage in autoimmune diseases through neutrophil activation and increases in neutrophil numbers. Neutrophils comprise more than 90% of leukocytes in synovial fluid in rheumatoid arthritis and are considered to mediate the arthritic process through release of lysosomal enzymes and production of nitrous oxide and superoxide anion. Filgrastim activates neu-trophil chemotaxis and production of superoxide anion (reviewed in Ref. 9), and can prolong cell survival by suppression of apoptosis (12), although no association between apoptosis and synovial G-CSF or GM-CSF levels was reported by Bell et al (13).

Granulocyte colony–stimulating factor may not only provoke deleteri-ous effects, but also may attenuate the inflammatory response. Gorgen et al (14) described a reduced lethality correlated with suppressed TNF-α production in Filgrastim-treated mice challenged with endotoxin. In addi-tion, the balance of pro- and anti-inflammatory cytokines was shown to be favorably changed in ex vivo–stimulated leukocytes from healthy donors receiving Filgrastim (15). In a model of experimental immune complex-induced colitis in rabbits, Filgrastim therapy reduced tissue damage and a markedly decreased mucosal leukotriene (LT)B_4 and TXB_2 production (16).

A. Felty Syndrome

Felty syndrome is characterized by the triad of rheumatoid arthritis, spleno-megaly, and neutropenia, and occurs in 1% of all patients with rheumatoid arthritis. These patients often develop disabling or life-threatening infections that are not only attributed to low numbers of neutrophils but also to impaired neutrophil function, hypocomplementemia, and high levels of circulating immune complexes.

The cause of neutropenia in Felty syndrome is not known. In most patients, bone marrow examination reveals impaired granulopoiesis (myeloid maturation arrest). Other factors are increased destruction, excessive margination, and splenic sequestration of neutrophils. Granulocyte function may be reduced due to autoantibodies, immune complexes, or T-cell–mediated processes. Importantly, more than one mechanism may be operational.

In addition to immunosuppressive treatment, resolving the granulocytopenia (and associated infectious episodes) is an important therapeutic goal. Recombinant HuG-CSF has been used effectively in patients with Felty syndrome with various mechanisms of neutropenia, eg, maturation defect in bone marrow biopsy in three patients (17,18) or serum antineutrophil antibodies in three others (19,20). Reversal of neutropenia does not appear to completely eliminate the risk of infection, and recurrent infections despite normal absolute neutrophil count (ANC) occur in approximately one-third of patients. The capacity of peripheral blood mononuclear cells to produce G-CSF in vitro is associated with a reduction of infectious complications in patients with Felty syndrome (21), although there is no overall difference of G-CSF release between healthy donors, patients with rheumatoid arthritis, and patients with Felty syndrome.

Twenty-seven cases are reported in the literature where rHuG-CSF, often recorded as Filgrastim, was used in patients with Felty syndrome (17–20,22–43). All patients had a significant neutropenia with reported granulocyte counts from 0 to 0.86×10^9/L. All patients, except two (20,26), responded to treatment with Filgrastim within a few days when given doses between 4 and 10 μg/kg/day (n = 24) or 2 μg/kg/day (19) (n = 1). The two patients who were treatment failures received only 2 μg/kg/day, but eventually responded when either the dose was increased to 5 μg/kg/day (20) or after receiving a bolus injection of cyclophosphamide before administration of Filgrastim (26).

There are reports of Filgrastim successfully used in established neutropenic infections (17,26,30,31,33,34,35,37–39,42). Furthermore, Filgrastim seems to be effective in healing chronic skin ulcers (32), even in patients where multiple antibiotic treatments and prior skin grafts were unsuccessful (25). The clinical result remained stable after discontinuation of Filgrastim. Rarely, short-term administration of recombinant human G-CSF was used prophylactically for the prevention of infectious complications, for example, before hip replacement surgery (18). Usually, granulocyte counts rapidly decreased within the first ten days after cessation of rHuG-CSF, with the exception of one patient remaining stable for 9 months after short-term administration of rHuG-CSF (19).

Long-term therapy with rHuG-CSF for more than 4 months in patients with Felty syndrome is reported by seven authors (18,25,29,33,34,36,38) starting with Filgrastim doses between 3 and 5 μg/kg/daily and tapered to a minimum of 75 μg twice weekly. All patients maintained sufficient granulocyte counts for the entire treatment period (maximum 18 months). Yasuda et al (18) showed that the administration of 250 μg every 2 to 3 days is superior to 100 μg/day to maintain sufficient granulocyte counts. Long-term use of Filgrastim can effectively reduce the number of infectious episodes and days in hospital (33,34,36,38).

The early interest in colony-stimulating factors for Felty syndrome waned after Hazenberg's report of a severe flare of arthritis during rHuGM-CSF application (4). Reports of exacerbation of rheumatic symptoms after Filgrastim administration are less frequent and describe primarily transient arthralgias (25,31); however, several cases of arthritic flare are reported (30,35,38,40,43). These flares were mostly transient or could be sufficiently controlled by dose reduction (38), *but leading to discontinuation in one patient* (18). Arthritic flare in a patient receiving rHuGM-CSF was attributed to induction of IL-6 and TNF-α (4), but these cytokines were not found to be elevated during an exacerbation of arthritis possibly caused by administration of Filgrastim (18,31). One patient experienced an arthritic flare both after rHuG-CSF and after subsequent splenectomy, suggesting a pathophysiologic role of increasing neutrophil counts (30). On the other hand, several authors report no exacerbation of rheumatic symptoms (22,37), even during long-term administration (33,34).

A decrease in hematocrit and thrombocytes during rHuG-CSF treatment in patients with Felty syndrome was observed in three cases (17,19,29), which was pronounced in one patient receiving 10 μg/kg/day rHuG-CSF (17), but was reversible after discontinuation of the drug.

Exanthema has been described in three patients that resolved after discontinuation of Filgrastim (20,22,40). Leukocytoclastic vasculitis developed in two cases (30,31), but responded promptly to prednisone (31). Interestingly, leukocytoclastic vasculitis also is described in a patient with Felty syndrome after splenectomy (44).

In conclusion, the administration of Filgrastim to patients with acute infectious complications can increase neutrophil counts and substantially help to overcome life-threatening infections. Because of the risk of exacerbation of rheumatic symptoms, close monitoring is necessary. In selected patients with Felty syndrome and multiple serious infections, prophylactic administration of Filgrastim may be safe, effective, and more economical than frequent hospitalization. The doses should be titrated to the lowest possible to avoid adverse reactions.

B. Rheumatoid Arthritis

Neutropenia is not a feature of rheumatoid arthritis, except in context with Felty syndrome or large granular lymphocytosis. Patients with rheumatoid arthritis have been treated with Filgrastim for neutropenia related to medication and for mobilization of progenitor cells.

Drug-induced neutropenias in patients with rheumatoid arthritis can be either dose-dependent or idiosyncratic; rHuG-CSF has been used in patients with neutropenia due to gold salts (45), bucillamine (46), sulfasalazine (6,47), methotrexate (48), chlorambucil (49), or antibiotics (50). In five patients, ANC increased after rHuG-CSF administration (6,45,46,49), but three patients showed no response. One of the non-responders had sulfasalazine-induced aplastic anemia that did not respond to prednisone, anabolic steroids, or rHuG-CSF (47); the second had agranulocytosis due to flucloxacillin therapy with no response to rHuG-CSF and prednisone but a successful reversal of neutropenia after the elimination of all drugs (50). The third patient was unresponsive to rHuG-CSF treatment of gold-induced aplastic anemia. Since drug-induced agranulocytosis often remits spontaneously, it is difficult to assess the precise contribution of rHuG-CSF, especially when the thrombocyte count increases in parallel to the granulocyte count as seen in a patient with methotrexate-induced pancytopenia (48). Coincident with an increase of neutrophil counts, infections, if reported, resolved quickly (46,48).

Two patients experienced a flare of arthritis (6,46), whereas in other cases (45,48,49) no exacerbation of rheumatic symptoms were seen. An increase of arthritic symptoms has been described after administration of rHuGM-CSF to patients with rheumatoid arthritis (5).

High-dose immunosuppressive therapy with peripheral stem cell support is discussed as a potential treatment of patients with resistant rheumatoid arthritis. To assess the safety of Filgrastim for the mobilization of progenitor cells in patients with active rheumatoid arthritis, five patients premedicated with methylprednisolone received Filgrastim for 5 days without any exacerbation of rheumatic symptoms (51).

In summary, Filgrastim has been used in patients with rheumatoid arthritis for the treatment of drug-induced neutropenia and for the mobilization of progenitor cells. At present, due to the limited experience in patients with rheumatoid arthritis, no clear conclusions can be drawn. Considering that rheumatic symptoms can worsen during treatment, a close monitoring of those patients is essential.

C. Rheumatoid Arthritis Associated with Large Granular Lymphocytosis

Large granular lymphocytosis, also known as T-γ lymphocytosis, presents with severe neutropenia in 80% of cases (52) and is associated with rheumatoid arthritis in 20% of the patients (53). On the other hand, one-third of patients with rheumatoid arthritis and neutropenia (clinically resembling Felty syndrome) have large granular lymphocytosis. Antineutrophil antibodies are frequently encountered in T-γ lymphoproliferative disorders, but it is unclear whether this represents an epiphenomenon or the primary pathophysiological process.

Recurrent infections occur in approximately one-third of patients. Two patients with rheumatoid arthritis associated with T-γ lymphocytosis were treated with rHuG-CSF for fever and nonhealing leg ulcer (refractory to antibiotics) (54) or prophylactically before bilateral knee replacement surgery (55). Their ANC increased promptly and the infected leg ulcer improved and healed without difficulty after skin grafting. No side effects were seen in these two patients.

D. Systemic Lupus Erythematosus

Systemic lupus erythematosus (SLE) represents a complex autoimmune disease that may present with an involvement of nearly all organ systems. Cytopenias are well-recognized manifestations of SLE, constituting one of its 11 classification criteria. In a recent study, neutropenia was found in as many as 47% of patients at any time during the course of the disease (56). In contrast to chemotherapy-induced leukopenia or to the neutropenia of Felty syndrome, however, leukopenia in SLE is usually less pronounced. In this study, severe neutropenia (ANC $<1.0 \times 10^9$/L) was seen in only 4% of the patients.

The etiologic mechanisms of cytopenias associated with SLE have been studied extensively. Both humoral and cellular immune mechanisms have been postulated, such as increased destruction of peripheral blood cells by circulating autoantibodies, dysregulation of hematopoietic progenitor cell growth mediated by cellular mechanisms (56), and/or suppression by serum IgG (57). These different mechanisms are consistent with the variability of bone marrow findings in patients with SLE, ranging from normal or increased cellularity to hypoplastic or even aplastic bone marrow, myelofibrosis, or hemophagocytic syndrome (58).

Filgrastim has rarely been used in patients with SLE, reflecting the fact that disease-associated leukopenia in the majority of patients is moderate and usually responds to systemic corticosteroids or other immunosuppressive treatment. Indications for the use of Filgrastim have been variable.

Infections are a major source of morbidity and mortality in patients with SLE. A variety of factors in addition to neutropenia or neutrophil dysfunction contribute to a patient's susceptibility to infection, including immunoglobulin and complement deficiencies, delayed hypersensitivity, and treatment-related factors due to corticosteroids or other immunosuppressive therapy.

A German group (59–61) administered Filgrastim 5 to 8.5 μg/kg/day to nine patients who had SLE-associated neutropenia and infection despite antibiotic medication. Their ANC increased within 48 hours, and the infectious problems improved in most patients.

A favorable synergistic action of corticosteroids and Filgrastim on myeloid progenitor cells is described by Dubois et al (62). Kondo et al (63) reported on a patient with lupus-associated agranulocytosis and septic shock who was treated with methylprednisolone pulse immunosuppression and simultaneous rHuG-CSF (100 μg/day for 5 days) with a neutrophil recovery within 7 days.

Neutrophil functions were measured by Kiss et al (64) in three patients with SLE-associated neutropenia and recurrent infections, who had been treated with rHuG-CSF 1.5 μg/kg/day for 5 days. Intracellular killing of *Staphylococcus aureus,* respiratory burst, chemotaxis, and cell-surface expression of the CD64 antigen were all increased.

A different clinical setting is prophylaxis of infection. Hirashima et al (65) administered Filgrastim to a patient with SLE and severe neutropenia, but without apparent infection. With Filgrastim administered at 5 μg/kg/day, the ANC increased to $>1.0 \times 10^9$/L on day 5 and was maintained for a period of 6 weeks with doses between 2.5 and 5 μg/kg/day.

Recombinant HuG-CSF has also been used in patients with severe SLE to reduce the myelosuppressive sequelae of immunosuppressive therapy. Teske et al (66) treated three patients with high-dose pulse cyclophosphamide (up to 54 mg/kg over a period of 3 days) and rHuG-CSF. Cyclophosphamide-induced leukopenia was significantly reduced in comparison with a retrospective control group treated with 36 mg/kg cyclophosphamide, but without growth factor. No data on the long-term outcome in these patients have been published. An interesting observation has been reported by Fraser (67), who administered rHuG-CSF secondary to cyclophosphamide-induced neutropenia to patients with lupus nephritis. In addition to a reduction in infectious complications, the patients had a complete remission of their nephritis. Three patients still monitored after 3 years were in sustained remission with steroids and hydroxychloroquine only.

Adverse events possibly attributable to rHuG-CSF consisted of leukocytoclastic vasculitis in one patient and transient aggravation of cerebral symptoms in two patients with CNS lupus. The reactions resolved after cessation of rHuG-CSF and use of corticosteroids (60,61). In all other patients, no SLE exacerbation has been noted, even after administration of rHuG-CSF for six weeks (65).

In summary, Filgrastim has been administered to a small number of patients with SLE-associated neutropenia for prophylaxis of infection, as an adjunct to antibiotics in patients presenting with treatment-refractory infection and in the setting of dose-intensive immunosuppressive therapy. At present, clinical benefit with regard to infection was shown in most patients and the reported possible adverse effects were reversible. The use of Filgrastim as adjunct to high-dose immunosuppressive therapy may be of potential interest, although its clinical value has yet to be determined.

E. Autoimmune Skin Diseases and Vasculitis

Frequently seen autoimmune skin diseases are the primary bullous dermatoses, which are not associated with AIN. Only one report deals with the use of rHuG-CSF in these diseases. A peripheral blood progenitor cell donor with a history of dermatitis herpetiformis was treated with Filgrastim for mobilization of stem cells. No exacerbation of the dermatitis herpetiformis was seen. After four doses, acute iritis developed but with local treatment, it resolved within a few days (68).

Vasculitic disorders consist of a very heterogeneous group of diseases characterized by immunologically mediated inflammation and necrosis of blood vessels. The term cutaneous leukocytoclastic angiitis refers to manifestations confined to the skin (69). Pathogenetically, deposits of immune complexes in the blood vessel walls initiate an inflammatory response resulting in vascular damage. Histologically in leukocytoclastic vasculitis, a cellular infiltrate consisting of mononuclear cells and nuclear debris of polymorphonuclear leukocytes dominates in the wall of the microvasculature.

Autoimmune neutropenia is not a feature of primary vasculitic disorders. Therefore, the indications for the use of Filgrastim in these patients have been for other reasons and very rare.

In systemic vasculitis, Filgrastim 5 μg/kg/day has been given to a patient with renal impairment for the treatment of agranulocytosis attributed to captopril and/or ranitidine. Neutrophils recovered after 7 days, and no exacerbation of the vasculitis was reported (70).

Two patients with pre-existing cutaneous leukocytoclastic angiitis have been treated with Filgrastim, one at a dose of 125 μg/day, because of dapsone-induced agranulocytosis (71), and the other for the treatment of underlying hairy cell leukemia (72). Whereas the first patient did not experience an exacerbation of the vasculitis or any other toxicity attributable to Filgrastim, the other patient developed Sweet's syndrome at doses of 3 μg/kg/day, although only diagnosed on a clinical basis. After Filgrastim was discontinued, symptoms improved gradually after 2 weeks. Although the pathogenetic mechanisms underlying Sweet's syndrome are largely unknown, polymorphonuclear leukocytes are considered the primary effector cells of tissue destruction. A possible role of endogenous G-CSF in the pathogenesis of Sweet's syndrome is discussed by Reuss-Borst et al (73).

In conclusion, in the few case reports with pre-existing vasculitis, no exacerbation of vasculitis attributable to Filgrastim was seen, although one patient developed a neutrophil dermatosis. Vasculitis, probably induced by Filgrastim, was seen in some other patients with autoimmune diseases (SLE, Felty syndrome), although reversible after discontinuation of the cytokine and, in some cases, corticosteroids. Careful monitoring is indispensable, since increasing leukocyte counts may predispose to the induction of vasculitis.

F. Autoimmune Thyroid Disease

Autoimmune thyroid disease includes Graves' disease, Graves' orbitopathy, Hashimoto thyroiditis, and, possibly, postpartum thyroiditis. In Graves' disease, hematologic evaluation may indicate leukopenia, but in most patients, it is mild and does not require special consideration, although serum antibodies to neutrophils are found in approximately 50% of the patients with Graves' disease (74). Alterations in white blood cell counts are even less common in hypothyroidism than in thyrotoxicosis, but, since Hashimoto disease is associated with an increased risk of other autoimmune diseases including pernicious anemia and systemic lupus erythematosus, leukopenia can occur due to those conditions. Filgrastim was only reported to have been used for the treatment of neutropenia induced by antithyroid drug therapy.

Despite improved antibiotic treatment and supportive care, the outcome of drug-induced agranulocytosis in general is potentially fatal. Prognosis depends on a variety of clinical conditions, such as renal function and bacteremia (75). Spontaneous recovery of the granulocytopenia is highly variable and ranges from days to weeks depending on the extent of bone marrow damage (76).

Case reports about Filgrastim in antithyroid drug–induced granulocytopenia have covered 45 patients, 35 with severe neutropenia (ANC $<0.5 \times 10^9/L$) and 10 with mild neutropenia (ANC 0.7 to 0.9 $\times 10^9/L$) (77–87). Filgrastim dosing ranged from 75 μg/day (86,87) to 10 μg/kg/day (77). Neutrophil counts recovered in all patients within 1 to 14 days, with no clear relation to Filgrastim dosage. Since spontaneous recovery is common in these patients, it is, at present, not possible to assess the role of Filgrastim, especially since high levels of endogenous G-CSF have been found in patients with severe antithyroid drug–induced agranulocytosis (86–88). No randomized trial has been reported, and only Tamai et al (86) compared the outcome of 12 Filgrastim-treated patients with a historical control group. They found that Filgrastim significantly reduced time to recovery and postulated that the positive effect of Filgrastim is limited to patients with granulocyte counts of $>0.1 \times 10^9$/L. An attempt to correlate time to recovery to the extent of impaired myelopoiesis in antithyroid drug–induced agranulocytosis was made by Tamai et al (86), suggesting that patients with a decreased bone marrow myeloid-erythroid ratio had a significantly longer time to recovery. Tajiri et al (87) reported that patients with severe granulocytopenia had the longest time to recovery, in accordance to a more pronounced extent of bone marrow damage. Infectious complications were present in the majority of patients before granulocyte recovery with resolution after reconstitution, though the clinical course is not documented in all cases.

The issue of exacerbation of thyroid dysfunction in patients with pre-existing thyroid antibodies has been raised previously with various cytokines, including rHuGM-CSF (89). In the published cases of treatment with Filgrastim, no exacerbation of thyroid autoimmune disease and no induction of other autoimmune phenomena has been reported. In addition, the effect of glycosylated rHuG-CSF (lenograstim) on thyroid function has been studied in 20 patients with breast cancer treated with dose-escalating chemotherapy and rHuG-CSF 10 μg/kg/day (90). In two patients, antibodies to thyroid microsomes and, in one, additionally to thyroglobulin, were slightly elevated before rHuG-CSF administration and were found negative 1 month after treatment. The other 18 patients did not show any thyroid abnormalities before or after treatment with rHuG-CSF.

Presently, it is difficult to estimate the value of Filgrastim in antithyroid drug–induced neutropenia, since spontaneous recovery is very variable and may be dependent on the degree and duration of neutropenia and on the extent of bone marrow impairment.

G. Various Autoimmune Diseases

Few reports have been published where growth factors were administered in other autoimmune diseases. A patient with aplastic anemia attributed to gold salts administered for psoriatic arthritis was treated unsuccessfully with rHuG-CSF (1.5 μg/kg/day) for 10 days. This patient died after 2 months because of infectious complications (91).

Filgrastim 300 μg daily was administered to a patient with polymyalgia rheumatica and captopril-induced agranulocytosis. Neutrophil counts increased to normal values after 2 days; no influence on the course of the polymyalgia was mentioned (92).

One patient with ulcerative colitis has been treated with Filgrastim because of mesalazine-induced agranulocytosis. White blood cell counts normalized within 1 week after starting Filgrastim. An exacerbation of the colitis is not reported and the patient died because of a perforated duodenal ulcer (93). A second report deals with a patient with a history of HLA-B27 positive ankylosing spondylitis and ulcerative colitis developing cyclic neutropenia. Because of infectious complications, 5 μg/kg/day Filgrastim was given. In parallel to an increase in neutrophils, bone and joint pain, as well as diarrhea, worsened, and a decrease in platelet counts and hemoglobin values was observed (94).

III. AUTOIMMUNE NEUTROPENIA NOT ASSOCIATED WITH AUTOIMMUNE DISEASES

Primary AIN is defined as an immunologically mediated destruction of neutrophils in the absence of underlying disorders. Secondary AIN are mainly associated with autoimmune diseases (see previous discussion) but also occur in the presence of other diseases, for example, lymphoproliferative disorders and common variable immunodeficiency.

Two variants of primary AIN are known. Primary autoimmune neutropenia of childhood is mediated in most cases by well-characterized anti-neutrophil autoantibodies and often characterized by a benign, self-limiting course. The adult variant is much rarer and more heterogeneous than the syndrome observed in children, and most patients develop a chronic neutropenia over many years with a substantial risk for serious infections. Primary AIN is difficult to distinguish unequivocally from chronic idiopathic neutropenia, which is a mixture of conditions with different etiologies, including immune-mediated mechanisms of neutrophil destruction (95).

Rare case reports exist of rHuG-CSF administration in a variety of AIN not associated with autoimmune diseases, such as primary AIN (96–98), secondary AIN in common variable immunodeficiency (99), immunoblastic-like T-cell lymphoma (100), T-γ lymphocytosis (98), thymoma (101), liver cirrhosis (102), and after bone marrow transplantation (103). Response to rHuG-CSF treatment seems to be dependent on an effective treatment of the underlying disease. Patients with primary AIN (96–98), common variable immunodeficiency (99), and liver cirrhosis (102) responded to Filgrastim alone, even after failure with steroids (96,102); in other patients, however, a successful reversal of neutropenia could only be achieved by combining rHuG-CSF with a causal treatment of the underlying disease. A patient with AIN after bone marrow transplantation was treated successfully with steroids in addition to Filgrastim (103), and a lymphoma patient responded to chemotherapy and rHuG-CSF (100). Postiglione et al (101) reported a patient with thymoma, who was initially unresponsive to rHuG-CSF, prednisone, and high-dose IV immunoglobulin, but recovered after thymectomy, aggressive plasma exchange, and cyclophosphamide in addition to Filgrastim.

Long-term treatment was reported for six patients for up to 24 months with maintenance doses as low as 25 to 50 μg per day (97) or 150 μg thrice weekly (n = 5) (98), and 480 μg/day (n = 1) (98). Neutrophil counts remained stable during the whole period and the incidence of infections, as evidenced by the use of antibiotics, decreased in these patients (98).

The increased ANC during Filgrastim administration can lead to the adsorption of autoantibodies as shown by Kuijpers et al (99), who have treated an 8-year-old boy with AIN associated with common variable immunodeficiency. Antineutrophil autoantibodies with a specificity for FcγRIII (CD16) disappeared in parallel to the normalization of the ANC. Upon tapering Filgrastim, anti-FcγRIII antibodies reappeared together with an absolute neutropenia.

Use of Filgrastim either for the treatment of established infections or as prophylaxis to reduce the number of infectious episodes in patients with chronic severe neutropenia of autoimmune origin seems promising. Low doses of Filgrastim may be sufficient in responding patients to maintain an ANC of 1.0×10^9/L. The causal treatment of underlying diseases is important.

IV. CONCLUSIONS

In autoimmune diseases, Filgrastim has been used in patients with neutropenia of different etiology for treatment and prophylaxis of infectious complications, for dose intensification of immunosuppressive therapy, and for mobilization of progenitor stem cells.

A possible induction of an exaggerated immune response is a major concern in these patients. An exacerbation of symptoms related to the preexisting autoimmune diseases, such as arthritis or aggravation of lupus encephalopathy, has been described in various patients during administration of Filgrastim. The symptoms were reversible after discontinuation or reduction of Filgrastim or the additional administration of steroids. No reports exist on the induction of arthritis in patients without rheumatic diseases. Rarely, an induction of vasculitis and neutrophilic dermatosis was reported in patients with autoimmune diseases treated with Filgrastim, although not obviously more frequent than in patients with other diseases.

Clinical benefit of Filgrastim for treatment and prophylaxis of infections has been reported in several patients with autoimmune diseases, although, due to the limited number, final conclusions cannot be drawn. In established neutropenic infections, the correction of neutropenia with resultant improvement of clinical status probably outweighs the risk of adverse events. Prophylactic administration of Filgrastim for the prevention of infectious episodes may be of value in patients with severe neutropenia, a chronic course not influenced by causal therapy of underlying diseases, and multiple serious infections in their history. The doses should be titrated to the lowest possible to avoid adverse reactions. Filgrastim may be effective and more economical than frequent hospitalization and its accompanying costs. Because of the risk of exacerbation of autoimmune symptoms, close monitoring is necessary.

REFERENCES

1. Petersen, J., and Wiik, A. (1983). Lack of evidence for granulocyte specific membrane-directed autoantibodies in neutropenic cases of rheumatoid arthritis and in autoimmune neutropenia. *Acta Pathol Microbiol Immunol Scand C 91*:15–22.
2. Shastri, K., and Logue, G. L. (1993). Autoimmune neutropenia. *Blood 81*:1984–1995.
3. Valerius, T., Helm, G., Repp, R., Manger, B., and Kalden, J. R. (1996). Haematologic changes in rheumatic diseases. *Arthritis Rheum Munch 16*:126–131.
4. Hazenberg, B. P., van Leeuwen, M. A., van Rijswijk, M. H., Stern, A. C., and Vellenga, E. (1989). Correction of granulocytopenia in Felty's syndrome by granulocyte-macrophage colony-stimulating factor. Simultaneous induction of interleukin-6 release and flare-up of the arthritis. *Blood 74*:2769–2770.
5. De Vries, E. G., Willemse, P. H., Biemsa, B., Stern, A. C., Limburg, P. C., and Vellenga, E. (1991). Flare-up of rheumatoid arthritis during GM-CSF treatment after chemotherapy. *Lancet 338*:517–518.
6. Canvin, J. M., Crilly, A., and Field, M. (1995). G-CSF treatment associated with cytokine level elevation and exacerbation of arthritis. *Arthritis Rheum 38*:S355.
7. Kerrigan, D. P., Castillo, A., Foucar, K., Townsend, K., and Neidhart, J. (1989). Peripheral blood morphologic changes after high-dose antineoplastic chemotherapy and recombinant human granulocyte colony-stimulating factor administration. *Am J Clin Pathol 92*:280–285.
8. Stricker, R. B., and Goldberg, B. (1996). Increase in lymphocyte subsets following treatment of HIV-associated neutropenia with granulocyte colony-stimulating factor. *Clin Immunol Immunopathol 79*:194–196.
9. Frampton, J. E., Lee, C. R., and Faulds, D. (1994). Filgrastim: A review of its pharmacologic properties and therapeutic efficacy in neutropenia. *Drugs 48*:731–760.

10. Morikawa, K., Miyawaki, T., Oseko, F., Morikawa, S., and Imai, K. (1993). G-CSF enhances the immunoglobulin generation rather than the proliferation of human B lymphocytes. *Eur J Haematol 51*:144–151.

11. Blackburn, W. D., Koopman, W. J., Schrohenloher, R. E., and Heck, L. W. (1986). Induction of neutrophil enzyme release by rheumatoid factors: evidence for differences based on molecular characteristics. *Clin Immunol Immunopathol 40*:347–355.

12. Colotta, F., Re, F., Polentarutti, N., Sozzani, S., and Mantovani, A. (1992). Modulation of granulocyte survival and programmed cell death by cytokines and bacterial products. *Blood 80*:2012–2020.

13. Bell, A. L., Magill, M. K., McKane, R., and Irvine, A. E. (1995). Human blood and synovial fluid neutrophils cultured in vitro undergo programmed cell death which is promoted by the addition of synovial fluid. *Ann Rheum Dis 54*:910–915.

14. Gorgen, I., Hartung, T., Leist, M., et al (1992). Granulocyte colony-stimulating factor treatment protects rodents against lipopolysaccharide-induced toxicity via suppression of systemic tumor necrosis factor-alpha. *J Immunol 149*:918–924.

15. Hartung, T., Docke, W. D., Gantner, F., et al (1995). Effect of granulocyte colony-stimulating factor treatment on ex vivo blood cytokine response in human volunteers. *Blood 85*:2482–2489.

16. Hommes, D. W., Meenan, J., Dijkhuizen, S., Ten Kate, F. J., Tytgat, G. N., and van Deventer, S. J. (1996). Efficacy of recombinant granulocyte colony-stimulating factor (rhG-CSF) in experimental colitis. *Clin Exp Immunol 106*:529–533.

17. Wun, T. (1993). The Felty syndrome and G-CSF-associated thrombocytopenia and severe anemia. *Ann Intern Med 118*:318–319.

18. Yasuda, M., Kihara, T., Wada, T., et al (1994). Granulocyte colony-stimulating factor induction of improved leukocytopenia with inflammatory flare in a Felty's syndrome patient. *Arthritis Rheum 37*:145–146.

19. Hoshina, Y., Moriuchi, J., Nakamura, Y., Arimori, S., and Ichikawa, Y. (1994). CD4$^+$ T-cell-mediated leukopenia of Felty's syndrome successfully treated with granulocyte-colony-stimulating factor and methotrexate. *Arthritis Rheum 37*:298–299.

20. Ito, T., Miyairi, Y., Kuwabara, T., Dan, K., and Nomura, T. (1992). Granulocyte-colony stimulating factor corrects granulocytopenia in Felty's syndrome. *Am J Hematol 40*:318–319.

21. Meliconi, R., Uguccioni, M., Chieco-Bianchi, F., et al (1995). The role of interleukin-8 and other cytokines in the pathogenesis of Felty's syndrome. *Clin Exp Rheumatol 13*:285–291.

22. Bhalla, K., Ross, R., Jeter, E., Madyastha, P., and Stuart, R. (1993). G-CSF improves granulocytopenia in Felty's syndrome without flare-up of arthritis. *Am J Hematol 42*:230–231.

23. Chattopadhyay, C., Swinson, D. R., Patel, K., and Bhavnani, M. (1994). Use of G-CSF in Felty's syndrome. *Arthritis Rheum 37*:S341.

24. Choi, V., Mant, M., and Aaron, S. (1992). Treatment of refractory Felty's syndrome with G-CSF and cyclophosphamide. *Arthritis Rheum 35:*S340.
25. Choi, M. F., Mant, M. J., Turner, A. R., Akabutu, J. J., and Aaron, S. L. (1994). Successful reversal of neutropenia in Felty's syndrome with recombinant granulocyte colony stimulating factor. *Br J Haematol 86:*663–664.
26. Pixley, J. S., Yoneda, K. Y., and Manalo, P. B. (1993). Sequential administration of cyclophosphamide and granulocyte-colony stimulating factor relieves impaired myeloid maturation in Felty's syndrome. *Am J Hematol 43:*304–306.
27. Toshioka, K. (1995). Combined therapy of G-CSF and prednisolone for neutropenia in a patient with Felty's syndrome. *Am J Hematol 48:*130–131.
28. Nielsen, H., and Mejer, J. (1992). Remission of neutropenia in a case of Felty's syndrome by rhG-CSF. *Scand J Rheumatol 21:*305–307.
29. Wandt, H., Seifert, M., Falge, C., and Gallmeier, W. M. (1993). Long-term correction of neutropenia in Felty's syndrome with granulocyte colony-stimulating factor. *Ann Hermatol 66:*265–266.
30. Vidarsson, B., Geirsson, A. J., and Onundarson, P. T. (1995). Reactivation of rheumatoid arthritis and development of leukocytoclastic vasculitis in a patient receiving granulocyte colony-stimulating factor for Felty's syndrome. *Am J Med 98:*589–591.
31. Farhey, Y. D., and Herman, J. H. (1995). Vasculitis complicating granulocyte colony stimulating factor treatment of leukopenia and infection in Felty's syndrome. *J Rheumatol 22:*1179–1182.
32. Fohlman, J., Hoglund, M., and Bergmann, S. (1994). Successful treatment of chronic wound infection in neutropenia and rheumatoid arthritis with filgrastim (rhG-GSF). *Ann Hematol 69:*153–156.
33. Fraser, D. D., Sartiano, G. P., Butler, T. W., and Treadwell, E. L. (1993). Neutropenia of Felty's syndrome successfully treated with granulocyte colony stimulating factor. *J Rheumatol 20:*1447–1448.
34. Graham, K. E., and Coodley, G. O. (1995). A prolonged use of granulocyte colony stimulating factor in Felty's syndrome. *J Rheumatol 22:*174–176.
35. Hayat, S. Q., Hearth-Holmes, M., and Wolf, R. E. (1995). Flare of arthritis with successful treatment of Felty's syndrome with granulocyte colony stimulating factor (G-CSF). *Clin Rheumatol 14:*211–212.
36. Klausmann, M., Kaiser, U., Pflueger, K. H., and Havemann, K. (1994). Successful therapy and prevention of infections in Felty's syndrome with long-term low-dose G-CSF. *J Cancer Res Clin Oncol 120:*R110.
37. Kaiser, U., Klausmann, M., Richter, G., Pflueger, K. (1992). GM-CSF versus G-CSF in the treatment of infectious complication in Feltys syndrome—a case report. *Ann Hematol 64:*205–206.
38. Krishnaswamy, G., Odem, C., Chi, D. S., Kalbfleisch, J., Baker, N., and Smith, J. K. (1996). Resolution of the neutropenia of Felty's syndrome by long-term administration of recombinant granulocyte colony stimulating factor. *J Rheumatol 23:*763–765.
39. Marfeuille, M., Clerc, D., and Mariette, X. (1995). Successful treatment of agranulocytosis and sepsis with granulocyte colony stimulating factor in a case of Felty's syndrome. *J Rheumatol 22:*1803–1804.

40. McMullin, M. F., and Finch, M. B. (1995). Felty's syndrome treated with rhG-CSF associated with flare of arthritis and skin rash. *Clin Rheumatol* *14*:204–208.

41. Schots, R., Verbruggen, L. A., and Demanet, C. (1995). G-CSF in Felty's syndrome: correction of neutropenia and effects of cytokine release. *Clin Rheumatol 14*:116–118.

42. Wagner, D. R., Combe, C., and Gresser, U. (1993). GM-CSF and G-CSF in Felty's syndrome. *Clin Invest 71*:168–171.

43. Vidal, C., Wong, A. Y., Pendleton, J., Shastri, K., and Logue, G. L. (1994). Successful treatment of Felty's syndrome with granulocyte-colony stimulating factor (G-CSF) complicated with exacerbation of rheumatoid arthritis. *Blood 84*:588a (abstr 2336).

44. Comer, M., and Bucknall, R. C. (1992). Digital vasculitis after splenectomy in a patient with Felty's syndrome. *Ann Rheum Dis 51*:908–909.

45. Collins, D. A., Tobias, J. H., Hill, R. P., and Bourke, B. E. (1993). Reversal of gold-induced neutropenia with granulocyte colony-stimulating factor (G-CSF). *Br J Rheumatol 32*:518–520.

46. Nakashima, H., Kawabe, K., Ohtsuka, T., et al (1995). Rheumatoid arthritis exacerbation by G-CSF treatment for bucillamine-induced agranulocytosis. *Clin Exp Rheumatol 13*:677–679.

47. Suwa, A., Kawai, S., Hirakata, M., Nakamura, A., and Inada, S. (1995). A case of rheumatoid arthritis complicated with salazosulfapyridine-induced aplastic anemia. *Nihon Rinsho Meneki Gakkai Kaishi 18*:116–122.

48. Ellman, M. H., Telfer, M. C., and Turner, A. F. (1992). Benefit of G-CSF for methotrexate-induced neutropenia in rheumatoid arthritis. *Am J Med 92*:337–338.

49. Rosello, R., Fabregas, M. D., and Torres, C. (1995). G-CSF treatment of immunosuppressor induced granulocytopenia in rheumatoid arthritis. *Rev Esp Reumatol 22*:322–324.

50. Burton, I. E., Moussa, K. M., and Sanders, P. A. (1995). Agranulocytosis in rheumatoid arthritis associated with long-term flucloxacillin for staphylococcal osteomyelitis. *Acta Haematol 94*:196–198.

51. McGonagle, D., Morgan, G., and Emery, P. (1996). The use of G-CSF in active rheumatoid arthritis to mobilise CD34[+] haemopoietic progenitor stem cells. *Arthritis Rheum 39*:S246.

52. Loughran, T. P. (1993). Clonal diseases of large granular lymphocytes. *Blood 82*:1–14.

53. Saway, P. A., Prasthofer, E. F., and Barton, J. C. (1989). Prevalence of granular lymphocyte proliferation in patients with rheumatoid arthritis and neutropenia. *Am J Med 86*:303–307.

54. Walls, J., Dessypris, E. N., and Krantz, S. B. (1992). Case report: granulocyte colony-stimulating factor overcomes severe neutropenia of large granular lymphocytosis. *Am J Med Sci 304*:363–365.

55. Stanworth, S. J., Green, L., Pumphrey, R. S., Swinson, D. R., and Bhavnani, M. (1996). An unusual association of Felty syndrome and TCR gamma delta lymphocytosis. *J Clin Pathol 49*:351–353.

56. Quismorio, F. P. Hematologic and lymphoid abnormalities in systemic lupus erythematosus. In: Wallace, D.J., and Hahn, B. H., eds. *Dubois' Lupus Erythematosus.* 5th ed. Baltimore: Williams & Wilkins, 1997, pp. 793–816.

57. Liu, H., Ozaki, K., Matsuzaki, Y., Abe, M., Kosaka, M., and Saito, S. (1995). Suppression of haematopoiesis by IgG autoantibodies from patients with systemic lupus erythematosus (SLE). *Clin Exp Immunol 100*:480–485.

58. Lorand-Metze, I., Carvalho, A., and Costallat, L. T. (1994). Bone marrow morphology in systemic lupus erythematosus. *Pathologe 15*:292–296.

59. Euler, H. H., Schwab, U. M., and Schroeder, J. O. (1994). Filgrastim for lupus neutropenia. *Lancet 344*:1513–1514.

60. Schwab, U. M., Corzillius, M., Harten, P., and Euler, H. H. (1995). Treatment of SLE associated neutropenia with granulocyte colony stimulating factor. *Arthritis Rheum 38*:S303.

61. Schwab, U. M., Harten, P., Zeuner, R. A., Kuelper, M., and Euler, H. H. (1996). G-CSF as an adjunct to antibiotic treatment in patients with lupus-associated neutropenia and infection. *Z Rheumatol 55*:174–179.

62. Dubois, C. M., Neta, R., Keller, J. R., Jacobsen, S. E., Oppenheim, J. J., and Ruscetti, F. (1993). Hematopoietic growth factors and glucocorticoids synergize to mimic the effects of IL-1 on granulocyte differentiation and IL-1 receptor induction on bone marrow cells in vivo. *Exp Hematol 21*:303–310.

63. Kondo, H., Date, Y., Sakai, Y., and Akimoto, M. (1994). Effective simultaneous rhG-CSF and methylprednisolone "pulse" therapy in agranulocytosis associated with systemic lupus erythematosus. *Am J Hematol 46*:157–158.

64. Kiss, E., Olajos, A., Szegedi, G., and Marodi, L. (1996). Effect of rhG-CSF on neutrophil functions of granulocytes of patients with systemic lupus erythematosus. *Eur J Clin Invest 26*:A35.

65. Hirashima, K., Yoshida, Y., Asano, S., et al (1991). Clinical effect of recombinant human granulocyte colony-stimulating factor (rhG-CSF) on various types of neutropenia including cyclic neutropenia. *Biotherapy 3*:297–307.

66. Teske, E., Harten, P., Schroeder, J. O., and Euler, H. H. (1992). G-CSF following pulse cyclophosphamide in SLE. *Arthritis Rheum 35*:S152.

67. Fraser, D. D. (1995). [Reply]. *J Rheumatol 22*:1804.

68. Parkkali, T., Volin, L., Siren, M. K., and Ruutu, T. (1996). Acute iritis induced by granulocyte colony-stimulating factor used for mobilization in a volunteer unrelated peripheral blood progenitor cell donor. *Bone Marrow Transplant 17*:433–434.

69. Jennette, J. C., Falk, R. J., Andrassy, K., Bacon, P. A., and Churg, J. (1994). Nomenclature of systemic vasculitides. *Arthritis Rheum 37*:187–192.

70. Patton, W. N., Holyoake, T. L., Yates, J. M., Boughton, B. J., and Franklin, I. M. (1992). Accelerated recovery from drug-induced agranulocytosis following G-CSF therapy. *Br J Haematol 80*:564–565.

71. Miyagawa, S., Shiomi, Y., Fukumoto, T., Ishii, Y., and Shirai, T. (1993). Recombinant granulocyte colony-stimulating factor for dapsone-induced agranulocytosis in leukocytoclastic vasculitis. *J Am Acad Dermatol 28*:659–661.

72. Glaspy, J. A., Baldwin, G. C., Robertson, P. A., et al (1988). Therapy for neutropenia in hairy cell leukemia with recombinant human granulocyte colony-stimulating factor. *Ann Intern Med 109:*789–795.

73. Reuss-Borst, M. A., Muller, C. A., and Waller, H. D. (1994). The possible role of G-CSF in the pathogenesis of Sweet's syndrome. *Leuk Lymphoma 15:*261–264.

74. Weitzman, S. A., Stossel, T. P., Harmon, D. C., Daniels, G., Maloof, F., and Ridgway, E. C. (1985). Antineutrophil autoantibodies in Graves' disease. *J Clin Invest 75:*119–123.

75. Julia, A., Olona, M., Bueno, J., et al (1991). Drug-induced agranulocytosis: prognostic factors in a series of 168 episodes. *Br J Haematol 79:*366–371.

76. Patton, W. N., and Duffull, S. B. (1994). Idiosyncratic drug-induced haematological abnormalities. Incidence, pathogenesis, management and avoidance. *Drug Saf 11:*445–462.

77. Willfort, A., Lorber, C., Kapiotis, S., et al (1993). Treatment of drug-induced agranulocytosis with recombinant granulocyte colony-stimulating factor (rhG-CSF). *Ann Hematol 66:*241–244.

78. Hunault, M., Dombret, H., Gardin, C., et al (1995). Hematopoietic growth factors in drug-induced agranulocytosis. *Leukemia 9:*1286–1287.

79. Demuynck, H., Zachee, P., Verhoef, G. E., et al (1995). Risks of rhG-CSF treatment in drug-induced agranulocytosis. *Ann Hematol 70:*143–147.

80. Yokoyama, K., Sato, T., Nakajima, S., et al (1992). Successful treatment of methimazole-induced agranulocytosis by granulocyte colony-stimulating factor. *Am J Hematol 40:*76–77.

81. Teitelbaum, A. H., Bell, A. J., and Brown, S. L. (1993). Filgrastim (r-metHuG-CSF) reversal of drug-induced agranulocytosis. *Am J Med 95:*245–246.

82. Balkin, M. S., Buchholtz, M., Ortiz, J., and Green, A. J. (1993). Propylthiouracil (PTU)-induced agranulocytosis treated with recombinant human granulocyte colony-stimulating factor (G-CSF). *Thyroid 3:*305–309.

83. Magner, J. A. and Snyder, D. K. (1994). Methimazole-induced agranulocytosis treated with recombinant human granulocyte colony-stimulating factor (G-CSF). *Thyroid 4:*295–296.

84. Genet, P., Pulik, M., Lionnet, F., and Bremont, C. (1994). Use of colony-stimulating factors for the treatment of carbimazole-induced agranulocytosis. *Am J Hematol 47:*334–335.

85. Adorf, D., Grajer, K. H., Kaboth, W., and Nerl, C. (1994). Agranulocytosis induced by antithyroid therapy: effects of treatment with granulocyte colony stimulating factor. *Clin Invest 72:*390–392.

86. Tamai, H., Mukuta, T., Matsubayashi, S., et al (1993). Treatment of methimazole-induced agranulocytosis using recombinan human granulocyte colony-stimulating factor (rhG-CSF). *J Clin Endocrinol Metab 77:*1356–1360.

87. Tajiri, J., Noguchi, S., Okamura, S., et al (1993). Granulocyte colony-stimulating factor treatment of antithyroid drug-induced granulocytopenia. *Arch Intern Med 153:*509–514.

88. Fukata, S., Murakami, Y., Kuma, K., Sakane, S., Ohsawa, N., and Sugawara, M. (1993). G-CSF levels during spontaneous recovery from drug-induced agranulocytosis. *Lancet 342:*1495.

89. Hoekman, K., Von Bloomberg-van der Flier, B. M., Wagstaff, J., Drexhage, H. A., and Pinedo, H. M. (1991). Reversible thyroid dysfunction during treatment with GM-CSF. *Lancet 338*:541–542.

90. Van Hoef, M. E., and Howell, A. (1992). Risk of thyroid dysfunction during treatment with G-CSF. *Lancet 340*:1169–1170.

91. Macdonald, A. G., Capell, H. A., and Murphy, J. (1993). Gold induced aplastic anaemia unresponsive to G-CSF. *Ann Rheum Dis 52*:488.

92. Chia, H. M., Kalra, L., Lakhani, A. K., and Samaratunga, I. R. (1993). Filgrastim for low-dose, captopril-induced agranulocytosis. *Lancet 342*:304.

93. Wyatt, S., Joyner, M. V., and Daneshmend, T. K. (1993). Filgrastim for mesalazine-associated neutropenia. *Lancet 341*:1476.

94. Storek, J., Glaspy, J. A., Grody, W. W., Susi, E., and Slater, E. D. (1993). Adult-onset cyclic neutropenia responsive to cyclosporine therapy in a patient with ankylosing spondylitis. *Am J Hematol 43*:139–143.

95. Logue, G. L., Shastri, K., Laughlin, M., Shimm, D. S., Ziolkowski, L. M., and Iglehart, J. L. (1991). Idiopathic neutropenia: Antineutrophil antibodies and clinical correlations. *Am J Med 90*:211–216.

96. Takahashi, K., Taniguchi, S., Akashi, K., et al (1991). Human recombinant granulocyte colony-stimulating factor for the treatment of autoimmune neutropenia. *Acta Haematol 86*:95–98.

97. Taniguchi, S., Shibuya, T., Harada, M., and Niho, Y. (1993). Decreased levels of myeloid progenitor cells associated with long-term administration of recombinant human granulocyte colony-stimulating factor in patients with autoimmune neutropenia. *Br J Haematol 83*:384–387.

98. Krishnan, K., Bockenstedt, P. L., and Adams, P. T. (1995). Treatment of autoimmune neutropenia by recombinant granulocyte colony stimulating factor (r-met-hu-GCSF). *Blood 86*:65a (abstr 247).

99. Kuijpers, T. W., De Haas, M., De Groot, C. J., Von Dem Borne, A. E., and Weening, R. S. (1996). The use of rh-G-CSF in chronic autoimmune neutropenia: reversal of autoimmune phenomena, a case history. *Br J Haematol 94*:464–469.

100. Osawa, H., Otsuka, T., Inoue, T., et al (1993). Immunoblastic lymphadenopathy (IBL)-like T cell lymphoma associated with granulocytopenia and autoimmune hemolytic anemia. *Rinsho Ketsueki 34*:165–170.

101. Postiglione, K., Ferris, R., Jaffe, J. P., and Stroncek, D. (1995). Immune mediated agranulocytosis and anemia associated with thymoma. *Am J Hematol 49*:336–340.

102. Uoshima, N., Kimura, S., Hiramori, N., et al (1992). The successful treatment of G-CSF in autoimmune neutropenia with liver cirrhosis complicated by esophageal varices. *Rinsho Ketsueki 33*:1741–1746.

103. Klumpp, T. R., Herman, J. H., Goldberg, S. L., and Mangan, K. F. (1995). Autoimmune neutropenia following bone marrow or blood stem cell transplantation (BMT): clinical features and response to therapy. *Blood 86*:109a (abstr 422).

27

Treatment of Diabetic Foot Ulcers and Infection with Filgrastim (r-metHuG-CSF)

Andrew Gough and Michael Edmonds
Kings College Hospital, London, England

I. INTRODUCTION

Diabetes is a chronic, debilitating, and costly disease with severe complications, such as blindness, renal failure, and amputation. The impact of diabetes is massive in both human and economic terms, not only in the United States and Europe, but also in the developing world. In the United States, diabetes is present in 7% of the population and in 15% of persons over 65 years of age, and these numbers are increasing. In 1987, the economic cost of the disease was $20.4 billion.

It has been estimated that 20% of all hospital admissions among diabetic patients in the United States are for foot problems, and hospital costs alone have been estimated at $200 million in 1980 (1). Foot ulceration in combination with infection is a frequent cause of amputation, the rate of which is 15 times higher in diabetic patients compared with non-diabetic patients (2). In the United States, there are at least 60,000 limb amputations per year in diabetic patients. The problem of limb amputation is of such a serious and global nature that two targets for reduction of amputations have been set in recent years. The St. Vincent Declaration (3) in Europe aims for a 50% reduction over 5 years, and Healthy People 2000 in the United States for a 40% reduction by the year 2000. These targets are achievable provided that effective new strategies are developed to manage the diabetic foot, and that infection is recognized as the principal cause of tissue destruction leading to amputation. Filgrastim may prove to be an important new alternative in the treatment of infection.

II. THE DIABETES SYNDROME

Diabetes mellitus is a syndrome with metabolic, vascular, and neuropathic components. The metabolic syndrome is characterized by alterations in carbohydrate, protein, and fat metabolism secondary to insulin deficiency. The vascular element consists of abnormalities in both large and small vessels, and the neuropathic component is a peripheral neuropathy.

Three factors lead to tissue necrosis in the foot: neuropathy, ischemia, and infection. The diabetic foot syndrome can be divided into two categories: the neuropathic foot, with neuropathy but adequate circulation; and the neuroischemic foot, with both neuropathy and macrovascular disease in the leg leading to reduced arterial blood flow to the foot.

A. The Neuropathic Foot

The neuropathic foot is characterized by callus formation, plantar ulceration, digital necrosis, Charcots arthropathy, and, rarely, neuropathic edema. Neuropathy causes functional abnormalities of the intrinsic muscles of the foot, leading to deformity, so that the metatarsal heads are near the skin surface. This results in increased pressure and callus formation. Tissue necrosis occurs below the callus resulting in a small cavity filled with serous fluid that eventually breaks through to the surface with ulcer formation. At this stage, infection usually supervenes. If drainage is inadequate, cellulitis develops with rapid spread of sepsis to underlying tendons, bones, and joints. Superficial cellulitis is often caused by a single pathogen, frequently *Staphylococcus aureus* or streptococci. Deeper infection in the diabetic foot is often polymicrobial (4), with three to six organisms typically isolated per infection. Staphylococci, streptococci, anaerobes, and enteric facultative gram-negative bacilli are most frequently isolated. Coagulase-negative staphylococci and corynebacteria are commonly cultured, and are usually thought to be ulcer colonizers, although some evidence would suggest they may play a pathogenic role (5). Digital necrosis may occur in the neuropathic foot, when bacteria infecting the toe produce angiotoxins that damage the digital arteries, leading to septic vasculitis and in situ thrombosis, causing vascular occlusion (6).

B. The Neuroischemic Foot

The neuroischemic foot is characterized by intermittent claudication, rest pain, ulceration, and eventually gangrene. As in the neuropathic foot, trauma to the neuroischemic foot may lead to ulceration, infection, and tissue necrosis. Although the foot may be neuroischemic with reduced perfusion, the chief cause of tissue destruction is infection. To confirm our observations regarding the role of infection in tissue destruction in the diabetic foot, we conducted a histological review of the digital arteries in 48 consecutively amputated toes: 21 were from patients with diabetic neuropathy, 19 from patients with diabetic neuroischemia, and 8 from non-diabetic patients. Arteries were examined for intimal hyperplasia, septic vasculitis, and in situ thrombosis. Intimal hyperplasia was significantly more common in the toes from patients with neuropathic and neuroischemic feet than the toes from patients who did not have diabetes. Similarly, the frequency of septic vasculitis with thrombus formation was significantly higher in the diabetic specimens. Osteomyelitis was present in 90% of the toes from patients with neuropathic feet and 80% of the toes from patients with neuroischemic feet, but in only one of the toes from non-diabetic patients. Therefore, in this unselected series, infection was an important cause of amputation.

III. DIABETES AND INFECTION

There is an important association between diabetes and infection. Infection impairs blood glucose control and is more common in patients whose diabetes is poorly controlled (7). Lower extremity infections are more frequent and severe in diabetic patients (8,9), and often prolonged. The relationship between diabetes and persistent infection was demonstrated by Bessman et al (10), who studied subcutaneous (SC) abscesses in diabetic and non-diabetic mice inoculated with *Escherichia coli* and *Streptococcus faecalis, E. coli* and *Bacteroides fragilis,* and *S. faecalis* and *B. fragilis.* After 2 weeks, the bacterial count in the abscesses was significantly higher in the diabetic mice for all combinations. Furthermore, at 2 weeks, *B. fragilis,* an anaerobe commonly found in the diabetic foot, was eradicated from the non-diabetic, but not the diabetic mice. These findings may be due to alterations in neutrophil function in diabetic individuals.

A. Neutrophil Dysfunction and Diabetes

Effective neutrophil antimicrobial action depends on the generation of several oxygen-derived free radicals. These toxic species, which include the superoxide anion (O_2^-), are formed during the respiratory (or oxidative) burst activated after chemotaxis and phagocytosis. Deficiencies in neutrophil chemotaxis (11), phagocytosis (12), superoxide production (13), respiratory burst activity (14), and intra-cellular killing (15) have all been described in association with diabetes. Exact mechanisms underlying diabetes-related defects in superoxide generation are unknown. NADPH is crucial to oxidative killing, acting as an electron donor in the conversion of oxygen to superoxide, and its consumption by the polyol pathway, in which excess glucose is metabolized, depletes levels in neutrophils of diabetic patients. Indeed, polyol pathway inhibition is associated with improved killing (16). Glycosylation of proteins (17) involved in the oxidative burst, however, also may be important. Circulating neutrophils have a lifespan of 6 to 10 hours; therefore in diabetes, the greatest exposure to hyperglycemia occurs during bone marrow development (10 to 14 days). Filgrastim-induced shortened marrow maturation time (18) could therefore reduce marrow exposure to elevated glucose, with consequently less neutrophil protein glycosylation, with potentially improved superoxide production, and oxidative killing. Sato and Shimizu (19) found that rHuG-CSF primed and enhanced the production of superoxide in response to a soluble stimulus in neutrophils form streptozocin-induced diabetic rats.

IV. POTENTIAL ROLE OF FILGRASTIM IN DIABETIC FOOT INFECTION

Successful treatment of infection remains crucial if tissue destruction and amputation is to be avoided in diabetic foot sepsis. The antibiotic treatment of synergistic, polymicrobial infections, typical of the diabetic foot, however, is beset with problems. Often, the organisms are not known for 24 to 48 hours, frequently necessitating the use of several broad-spectrum antimicrobial agents, increasing the likelihood of adverse effects and development of resistance. Many strains of resistant bacteria exist already; methicillin-resistant *S. aureus* is now commonly isolated from ulcers in our unit. Diabetic nephropathy and renal impairment also may limit the choice of available antibiotics.

Filgrastim potentially represents a new approach to treating infection, especially as it may correct a bactericidal defect in diabetic neutrophils and should not aggravate the problem of antibiotic resistance.

In a randomized, double-blind study investigating Filgrastim as a possible novel strategy for treating infection, we compared Filgrastim (20 patients), 5 μg/kg/day, SC, with placebo (20 patients). Treatment period was for 7 days. Individuals recruited all required hospitalization for intravenous (IV) antibiotics for limb-threatening cellulitis and were matched for age, sex, duration of diabetes, peripheral neuropathy, and vascular disease. Both groups received similar broad-spectrum IV antibiotic regimes (ceftazidime 3 g daily, amoxil 1.5 g daily, metronidazole 1.5 g daily, and flucloxacillin 2 g daily). Blood glucose control was optimized in all participants with insulin.

Filgrastim treatment was associated with the development of significant leucocytosis at day 7 due almost entirely to a profound neutrophilia **(Table 1)**, although there was a statistically significant increase in lymphocyte and monocyte numbers. Neutrophils from both patient groups showed impaired opsonised zymosan-stimulated superoxide levels at baseline relative to non-diabetic control patients. The Filgrastim-treated group demonstrated significantly improved superoxide levels and at the end of the treatment phase (p < 0.0001), whereas in the placebo-treated group, levels were unchanged **(Table 2)**. No Filgrastim-treated patients required surgery/amputation compared with four placebo (two toe amputations, two debridements under general anesthesia) (p < 0.114). Deep tissue swabs became sterile quicker in Filgrastim-treated patients, 4 (2 to 10) days [median; (range)], versus 8 (2 to 79) (p = 0.02). Improved neutrophil function in the Filgrastim-treated group was associated also with more rapid cellulitis resolution, 7 (5 to 20) days versus 12 (5 to 93) (p = 0.03), shorter hospital stay 10.0 (7 to 31) days versus 17.5 (9 to 100) (p = 0.02), and reduced IV antibiotic requirements, 8.5 (5 to 30) days versus 14.5 (8 to 63) (p = 0.02).

Filgrastim was well tolerated by all patients. Fifteen percent developed self-limiting bone pain, and 35% a reversible rise in serum alkaline phospha-

Table 1. Total White Cell, Neutrophil, Lymphocyte, and Monocyte Changes with Treatment.

	Filgrastim	Placebo
Total white cell count ($\times 10^9$/L)*		
Baseline:	7.6 (4.8–17.1)	7.8 (3.7–11.1)
Day 7:	27.8 (10.8–41.0)[a]	6.1 (4.1–12.3)[b]
Neutrophils ($\times 10^9$/L)**		
Baseline:	5.6 (2.6–15.9)	5.5 (1.4–7.9)
Day 7:	22.4 (7.9–37.1)[a]	3.8 (1.0–6.7)[b,c]
Lymphocytes ($\times 10^9$/L)§		
Baseline:	1.8 (0.9–3.3)	1.9 (0.8–3.3)
Day 7:	2.6 (1.5–4.9)[d]	1.8 (1.0–5.1)[f]
Monocytes ($\times 10^9$/L)†		
Baseline:	0.39 (0.1–0.9)	0.47 (0.1–1.1)
Day 7:	0.69 (0.2–3.7)[e]	0.54 (0.2–1.1)[g]

Medians; Ranges in parenthesis. [a]p < 0.0001, [b]p < 0.0001, [c]p = 0.023 versus baseline value, [d]p = 0.0074, [e]p = 0.003, versus baseline value, [f]p = 0.012, [g]p = 0.044 versus Filgrastim. * normal range (NR): 4.0–10.0. **NR: 2.5–7.5. §NR: 1.3–4.0. †NR: 0.2–1.5.

Table 2. Changes in Stimulated Neutrophil Superoxide Levels.

	Non-diabetic control	Filgrastim	Placebo
Stimulated Superoxide Levels (nmol/10^6 neutrophils/30 min)			
Baseline:	18.6 (12.2–29.1)	6.6 (1.5–18.9)*	6.2 (1.7–17.2)*
Day 7:		16.1 (4.2–24.2)**	7.3 (2.1–11.5)§

Medians; Ranges in parenthesis. *$p < 0.0001$ versus non-diabetic controls. **$p < 0.0001$ versus baseline. §$p < 0.0001$ versus Filgrastim.

tase, the only adverse event significantly more common in the Filgrastim-treated group. Nausea, diarrhea, hypoglycemia, and injection-site bruising, occurred in a few patients, but with equal frequency between Filgrastim and placebo groups.

If these findings are reproduced in large-scale studies, Filgrastim could help achieve International targets for the reductions in diabetes-associated amputations.

V. SUMMARY

Foot infection remains a significant cause of morbidity and mortality in patients with diabetes, and has serious financial implications. The use of antibiotics in the treatment of this condition is complicated by polymicrobial infections and emerging bacterial resistance. Filgrastim, by improving impaired neutrophil function, associated with diabetes, may play an important role in the future management of the diabetic foot.

REFERENCES

1. Bessman, A. N. (1982). Foot problems in diabetes. *Compr Ther 8*:32–37.
2. Most, R. S., and Sinnock, P. (1983). The epidemiology of lower extremity amputations in diabetic individuals. *Diabetes Care 6*:87–91.
3. WHO/IDF (1990). Diabetes care and research in Europe. The St Vincent Declaration. *Diabetic Med 7*:360.
4. Sapico, F. L., Witte, J. L., Canawati, H. N., Montgomerie, J. Z., and Bessman, A. N. (1984). The infected foot of the diabetic patient: quantitative microbiology and analysis of clinical features. *Rev Infect Dis 6*:S171–S176.
5. Bessman, A. N., Geiger, P. J., and Canawati, H. (1992). Prevalence of *Corynebacteria* in diabetic foot infections. *Diabetes Care 15*:1531–1533.

6. Edmonds, M. E., Foster, A. V., Greenhill, M., Sinha, J., Philpott-Howard, J., and Salisbury, J. (1992). Acute septic vasculitis not diabetic microangiopathy leads to digital necrosis in the neuropathic foot. *Diabetic Med 9(suppl 1)*:34A (abstr P85).
7. Rayfield, E. J., Ault, M. J., Keusch, G. T., Brothers, M. J., Nechemias, C., and Smith, H. (1982). Infection and diabetes: The case of glucose control. *Am J Med 72*:439–450.
8. Leibovici, L., Yehezkelli, Y., Porter, A., Regev, A., Krauze, I., and Harell, D. (1996). Influence of diabetes mellitus and glycaemic control on the characteristics and outcome of common infections. *Diabetic Med 13*:457–463.
9. Lipsky, B. A., Pecoraro, R. E., and Wheat, L. J. (1990). The diabetic foot. Soft tissue and bone infection. *Infect Dis Clin North Am 4*:409–432.
10. Bessman, A. N., Sapico, F. L., Tabatabal, M., and Montgomerie, J. Z. (1986). Persistence of polymicrobial abscesses in the poorly controlled diabetic host. *Diabetes 35*:448–453.
11. Mowat, A. G., and Baum, J. (1971). Chemotaxis of polymorphonuclear leukocytes from patients with diabetes mellitus. *N Engl J Med 284*:621–627.
12. Davidson, N. J., Sowden, J. M., and Fletcher, J. (1984). Defective phagocytosis in insulin controlled diabetics: evidence for a reaction between glucose and opsonising patients. *J Clin Pathol 37*:783–786.
13. Sato, N., Shimizu, H., Suwa, K., Shimomura, Y., Mori, M., and Kobayashi, I. (1992). Myeloperoxidase activity and generation of active oxygen species in leukocytes from poorly controlled diabetic patients. *Diabetes Care 15*:1050–1052.
14. Marphoffer, W., Stein, M., Schleinkofer, L., and Frederlin, K. (1993). Evidence of ex vivo and in vitro impaired neutrophil oxidative burst and phagocytic capacity in type 1 diabetes mellitus. *Diabetes Res Clin Pract 19*:183–188.
15. Marphoffer, W., Stein, M., Maeser, E., and Frederlin, K. (1992). Impairment of polymorphonuclear leukocyte function and metabolic control of diabetes. *Diabetes Care 15*:256–260.
16. Boland, O. M., Blackwell, C. C., Clarke, B. F., and Ewing, D. J. (1993). Effects of ponalrestat, an aldose reductase inhibitor, on neutrophil killing of *Escherichia coli* and autonomic function in patients with diabetes mellitus. *Diabetes 42*:336–340.
17. Nielson, C. P., and Hindson, D. A. (1989). Inhibition of polymorphonuclear leukocyte respiratory burst by elevated glucose concentrations in vitro. *Diabetes 38*:1031–1035.
18. Chatta, G. S., Price, T. H., Allen, R. C., and Dale, D. C. (1994). Effects of in vivo recombinant methionyl human granulocyte colony-stimulating factor on the neutrophil response and peripheral blood colony-forming cells in healthy young and elderly adult volunteers. *Blood 84*:2923–2929.
19. Sato, N., and Shimizu, H. (1993). Granulocyte-colony stimulating factor improves an impaired bactericidal function in neutrophils from STZ-induced diabetic rats. *Diabetes 42*:470–473.

28
Safety Profile of Filgrastim (r-metHuG-CSF)

Karen L. Patterson, Shirley Y. Masuda, and Sherri L. Brown
Amgen Inc., Thousand Oaks, California

By 1997, approximately 1.9 million patients had been treated with Filgrastim in clinical trials and in the postmarket setting in the 6 years since its commercial launch. The safety profile from postmarket safety surveillance is consistent with that seen previously in clinical trials. Safety data from both clinical trials in newly approved indications and the postmarket experience are reviewed here.

I. CLINICAL TRIALS

Since commercial introduction of Filgrastim, clinical trials have been conducted in diverse areas exploring new indications. Phase 3 trials were conducted in three new areas of investigation leading to approval for bone-marrow transplantation (BMT), severe chronic neutropenia (SCN), and peripheral blood progenitor cell (PBPC) mobilization. Additional phase 3 trials have been conducted in human immunodeficiency virus infection (HIV) (Chapter 21), acute myeloid leukemia (AML) (Chapter 24), and infection in non-myelosuppressed patients (Chapters 20 and 21).

All adverse events are actively solicited from clinical trials including those related to underlying disease, concomitant medications, and presence of infection. Thus, the emphasis here is on those clinical trials adverse events determined by the investigators to be possibly associated with Filgrastim therapy.

It is important to note that no new safety issues were identified in these additional areas of investigation.

A. Bone Marrow Transplantation Trials

Two hundred eleven patients were enrolled in three non-randomized studies and three randomized studies. Of these, 132 were treated with Filgrastim after autologous BMT (ABMT). The designs of the studies were similar in that all patients were treated with a preparative chemotherapy regimen, 6 or 7 days later ABMT was performed, and patients were then treated with Filgrastim for a period of up to 28 days, starting either on the day of transplantation or 1 day later (1).

Of the three randomized studies, two were randomized, controlled trials. The first trial studied the effects of Filgrastim in 54 patients undergoing ABMT for lymphoma; the patients were randomized to a control group or to receive either 10 μg or 30 μg/kg/day Filgrastim. The second trial studied the effects of Filgrastim at dose levels of either 10 μg/kg/day or 20 μg/kg/day on 43 patients undergoing high-dose chemotherapy and ABMT for relapsed Hodgkin's disease (HD) or high-grade non-Hodgkin's lymphoma (NHL) in first complete remission or chemosensitive relapse. The third trial studied the effects of Filgrastim on 70 patients with myeloid and non-myeloid malignancies (lymphoma, acute lymphoblastic leukemia [ALL], AML, chronic myelogenous leukemia [CML], and aplastic anemia) randomized to receive either Filgrastim 300 μg/m^2/day or placebo after myeloablative chemotherapy and allogeneic BMT (allo-BMT).

The first non-randomized trial studied the effects of Filgrastim on 39 patients with HD after high-dose chemotherapy and ABMT; the second trial on 35 patients with NHL, ALL, and germ cell tumor; and the third on 45 patients with breast cancer or malignant melanoma. Two hundred two patients, 117 from the randomized studies and 85 from the non-randomized studies, were evaluable for safety.

1. Efficacy

Filgrastim was found to be efficacious in accelerating neutrophil recovery after ABMT or allo-BMT and in reduction of the duration of severe neutropenia after high-dose cytotoxic chemotherapy with BMT. This reduction in neutropenia reduced the duration of febrile neutropenia, hospitalization, and use of antibiotics.

2. Safety

Overall, the adverse events reported in these clinical trials were those typically seen in patients receiving intensive chemotherapy followed by

BMT. For related events in the randomized controlled trials, the body systems most frequently affected were musculoskeletal (1.7%) and skin and appendages (1.7%). The most frequently reported adverse event was bone pain. For related events in the non-randomized controlled trials, the body systems most frequently affected were skin and appendages (20%), body as a whole (20%), and musculoskeletal (13%). The most frequently reported adverse events were rash, fever, and hypervolemia.

All studies demonstrated that Filgrastim accelerates neutrophil recovery and causes a decrease in the duration of neutropenia. This resulted in a clinically important reduction in the duration of fever and the use of antibiotics. Data from the non-randomized ABMT trials also were compared with matched published historical control data with similar findings in addition to reduction in duration of total parenteral nutrition.

Data on red blood cell and platelet transfusions from two non-randomized and one randomized trial were comparable with data from historical controls. In the first non-randomized trial that studied the effects of Filgrastim administered to 39 patients with HD after high-dose chemotherapy and ABMT, the median duration of platelet support was 18 days with 30-minute Filgrastim infusions, and 20 days with continuous intravenous (IV) Filgrastim infusion; the median duration for red blood cell support was 16 and 16 days, respectively. In the second non-randomized study, the median duration of platelet support was 23 days with corresponding red blood cell support of 15 days. In the randomized study, median duration of platelet transfusion was 17 days in the Filgrastim-treated group and 18 days in the placebo group; this was not found to be statistically significant. Filgrastim did not affect red blood cell or platelet levels.

Filgrastim was very well tolerated by patients over a broad age range. Treatment with Filgrastim did not have negative effects on either the safety or efficacy of cytotoxic chemotherapy. The course of both acute and chronic graft-versus-host disease (GVHD) was not negatively affected by the addition of Filgrastim to the allogeneic transplantation procedure.

B. Peripheral Blood Progenitor Cell Mobilization

Four studies with 126 patients were conducted in the development of Filgrastim for PBPC mobilization (**Table 1**). In addition to the phase 1–2 trials listed, a randomized phase 3 trial compared post-transplant hematopoietic recovery in patients receiving Filgrastim administered at a dose of 10 μg/kg/day for 6 days to mobilize PBPC, which were harvested by leukapheresis versus harvested autologous bone marrow. Post-myeloablative chemotherapy, the 63 patients enrolled received either PBPC transplant or autologous marrow transplant. In both treatment groups, Filgrastim 5 μg/kg/day was

Table 1. Peripheral Blood Progenitor Cell Trials.

Phase	Tumor type	Filgrastim dose for mobilization	PBPC mobilization	Filgrastim dose post-PBPC	Length of Filgrastim administration after PBPC or ABMT
1–2	Breast cancer	10 μg/kg/day	7 days	10 μg/kg/day	ANC recovery or up to 28 days
2	ALL, NHL, HD	24 μg/kg/day	6 or 7 days	24 μg/kg/day	ANC recovery or up to 28 days
2	Breast cancer	12 μg/kg/day	6 days	5 μg/kg/day × 3 cycles post-chemo	ANC recovery or up to 28 days
3	NHL, HD	10 μg/kg/day	6 days	5 μg/kg/day post-PBPC or ABMT	ANC recovery or up to 28 days

ABMT = autologous bone marrow transplantation.
ALL = acute lymphoblastic leukemia.
ANC = absolute neutrophil count.
HD = Hodgkin's disease.
NHL = non-Hodgkin's lymphoma.
PBPC = peripheral blood progenitor cell.

administered post-transplant until absolute neutrophil count (ANC) recovery or for a maximum of 28 days (1).

Results from two studies were used as supportive data for efficacy only. Hematologic recovery endpoint data from a non-randomized trial conducted on 35 patients with NHL, ALL, or germ cell tumor after high-dose chemotherapy and ABMT was included. In a non-Amgen study 95 patients with NHL or HD received either unmobilized PBPC (n = 30) or PBPC and Filgrastim (n = 65) at doses of 5 μg/kg/day or 10 μg/kg/day.

1. Efficacy

Filgrastim was demonstrated to mobilize large numbers of progenitor cells, more than can be collected from leukaphereses in steady state according to published summaries. Additionally, when infused after high-dose chemotherapy, these PBPC allowed a more rapid neutrophil and platelet recovery compared with marrow or unmobilized PBPC, with decreased duration of hospitalization. The acceleration of platelet recovery in patients receiving Filgrastim-mobilized PBPC was associated with reductions in platelet transfusion requirements compared with patients treated with marrow or unmobilized PBPC.

2. Safety

No previously unknown side effects of Filgrastim were identified during the mobilization phase of the studies in 126 patients. Many of the adverse events reported were those expected as a consequence of cancer chemotherapy.

For related events in these trials, the body systems most frequently affected were musculoskeletal, metabolic and nutrition, body as a whole, and central nervous system/peripheral nervous system. The most frequently reported adverse events in decreasing order of frequency were bone pain, increased alkaline phosphatase, and headache (**Table 2**).

In the treatment phase when Filgrastim was administered to approximately 110 patients as supportive therapy after high-dose chemotherapy and PBPC transplant, adverse events reported were consistent with those seen and expected in patients after transplant-preparative chemotherapy.

For related events in the treatment phase of these trials, the body systems most frequently affected were musculoskeletal (15%), metabolic and nutrition (5%), and application site (4%). The most frequently reported adverse events in decreasing order of frequency were skeletal pain, increased alkaline phosphatase, and abnormal hepatic function.

Adverse events and laboratory data were consistent with what is expected with Filgrastim in general during harvesting procedures and during administration of Filgrastim after high-dose chemotherapy and infusion of Filgrastim-mobilized cells. Filgrastim administration for the mobilization of PBPC and then as supportive therapy after infusion of these cells following high-dose chemotherapy appears to be safe and well tolerated.

C. Severe Chronic Neutropenia Trials

Five studies involving 197 patients were conducted in the development of Filgrastim for the treatment of SCN (1). Of these, four phase 1–2 trials were open-label, non-comparative trials to evaluate the pharmacologic effect of Filgrastim on 74 patients with idiopathic, cyclic, or congenital neutropenia; the doses ranged from 1 μg/kg/day to 12 μg/kg/day.

One phase 3 trial was an open-label, randomized, controlled, multicenter trial to evaluate the efficacy and safety of Filgrastim treatment in 123

Table 2. Patient Incidence of Treatment-Related Adverse Events by Preferred Term for Peripheral Blood Progenitor Cell Studies (Mobilization Phase).

Body system and adverse event	Number of patients	Percent
Musculoskeletal	55	44
pain skeletal	48	38
pain back	7	6
arthralgia	3	2
pain limb	3	2
Metabolic/Nutrition	19	15
Alkaline phosphatase increased	19	15
hyperuricemia	2	2
Body as a whole	11	9
fever	3	2
abnormal laboratory findings	2	2
influenza-like symptoms	2	2
access erythema	1	1
access pain	1	1
fatigue	1	1
pain chest	1	1
CNS/PNS	11	9
headache	9	7
somnolence	2	2
Hematologic	7	6
thrombocytopenia	5	4
anemia	2	2
ecchymosis	1	1
leukocytosis	1	1
thrombocythemia	1	1
Application site	4	3
injection site pain	2	2
injection site ecchymosis	1	1
injection site erythema	1	1
Gastrointestinal	4	3
pain abdominal	2	2
diarrhea	1	1
nausea/vomiting	1	1
Liver and biliary	4	3
hepatic function abnormal	3	2
hepatic enzymes increased	2	2
Respiratory	1	1
dyspnea	1	1
pharyngitis	1	1
Special senses	1	1
taste perversion	1	1

N = 126.

patients with SCN. Two groups, observation and treatment, were used for comparison, and the randomization was stratified by diagnosis of idiopathic, cyclic, or congenital neutropenia. Patients randomized to group A were immediately treated with Filgrastim, and patients randomized to group B were observed for 4 months before receiving Filgrastim. In this trial, safety variables reported in the treatment stages were compared with those in the no-treatment observation period.

Safety data from the phase 1–2 trials treatment stages were compared with the treatment stages of the phase 3 trial, and the phase 1–2 trials treatment stage was compared with the observation period of the phase 3 trial.

1. Efficacy

Daily, chronic administration of Filgrastim was shown to be highly efficacious in the treatment of patients with idiopathic, cyclic, and congenital neutropenia. Filgrastim therapy in these patients resulted in highly clinically and statistically significant increases in the ANC within 1 to 2 weeks of the start of Filgrastim. The response to Filgrastim differed based upon the diagnosis of idiopathic, cyclic, or congenital neutropenia; patients with congenital neutropenia required higher doses to elevate and maintain neutrophil counts than did patients with idiopathic or cyclic neutropenia.

Clinical and statistically significant improvements associated with the increase in ANC were found in the infectious morbidity normally associated with SCN. Decreases were seen in the incidence and duration of infection-related events, antibiotic use, fever, hospitalizations, and oropharyngeal lesions. Patients in each diagnostic group showed clinical improvement. The patients with congenital neutropenia had the greatest relative improvement, most likely a result of the greater degree of neutropenia seen in these patients before treatment with Filgrastim.

2. Safety

Table 3 shows the exposure-adjusted rates for adverse events with a group difference of at least 10% for treated patients in the phase 1–2 and phase 3 trials compared with those for the untreated patients from the phase 3 trial. The exposure-adjusted event rate is defined as the total number of a specific adverse event occurring in patients in a treatment group divided by the total treatment exposure for the patients in that treatment group. Treatment exposure is defined as the total number of patient-months (28-day period) of follow-up in a specific treatment group. There were fewer adverse events reported in the treated groups when compared with the untreated group.

Table 3. Exposure-Adjusted Adverse Event Rates Group Difference $\geq 10\%$.

Adverse event	Untreated phase 3	Treated phase 1-2	Treated phase 3
Fever	77%	11%	27%
Stomatitis	60%	7%	20%
Headache	24%	11%	35%
Pain	—	—	25%
Sore throat	18%	3%	—
Diarrhea	18%	3%	—
Infection	18%	2%	—
Bone pain	—	—	17%
Fatigue	16%	6%	—
Abdominal Pain	15%	2%	—

Safety analysis of 238 patient-years showed that long-term subcutaneous (SC) Filgrastim administration caused relatively few adverse events. The most frequently affected body systems were hematologic, musculoskeletal, and skin and appendages. Most adverse events related to this therapy were mild and easily managed. Some symptoms, such as bone pain, were noted early in treatment and tended to decrease in frequency with long-term therapy. The number and duration of reported adverse events related to Filgrastim decreased with the duration of treatment.

Asymptomatic splenomegaly was reported in approximately 30% of those receiving long-term treatment and <3% of patients required splenectomy; most of those patients who had splenectomies had a pre-study history of splenomegaly. In those patients (<6%) who developed thrombocytopenia with platelet count $<50 \times 10^9$/L, most had a pre-study history of thrombocytopenia; they were managed successfully by brief interruptions of treatment or dose reductions. One percent of patients with prior history of thrombocytopenia developed clinically significant thrombocytopenia with long-term Filgrastim administration.

In 3% of patients, symptoms of cutaneous vasculitis developed simultaneously with an increase in ANC and abated when the ANC decreased. Most patients were able to continue Filgrastim at the same or at a reduced dose. Most patients with reported urinary abnormalities had developed hematuria and/or proteinuria which was either transient and minor, associated with a urinary tract infection, or explained by pre-existing systemic or renal disease.

Measurable effects on growth, sexual development, endocrine assessments, and osteoporosis were not detected. Fewer than 6% of patients were reported to have osteoporosis based on routine radiographic examinations.

Less than half of these had a pre-study history of radiographically confirmed osteoporosis or osteopenia. Filgrastim did not worsen this condition in those patients with pre-study osteoporosis.

Acute myeloid leukemia (AML) or myelodysplasia (MDS) was reported in <3% of patients. The development of AML/MDS appeared to be confined to those patients with congenital neutropenia. Some of these patients either had evidence of AML/MDS or were not fully evaluated for cytogenetic abnormalities before the initiation of Filgrastim therapy. There were no cases of MDS or AML reported in the subgroup of congenital neutropenia patients with glycogen storage disease type 1b, nor in those patients with idiopathic or cyclic neutropenia. It is unknown whether or not the development of these findings is related to the chronic daily administration of Filgrastim, or reflects the natural history of SCN. Overall, the benefits substantially outweigh the risks associated with this treatment. No new significant safety issues to change the safety profile of Filgrastim were found in these clinical trials investigating new indications.

II. POSTMARKETING SAFETY PROFILE

In the 6 years since commercial launch of Filgrastim, approximately 1.9 million patients have been treated with Filgrastim to decrease the incidence of infection, as manifested by febrile neutropenia, in patients with non-myeloid malignancies receiving myelosuppressive anti-cancer drugs associated with a significant incidence of severe febrile neutropenia, for ABMT, BMT, SCN, and mobilization of PBPC. In addition to adverse events reported in these licensed indications, postmarket reports in non-licensed clinical settings, ie, HIV infection, MDS, and non–cytotoxic drug–induced neutropenia, have been received.

A. Discussion

As shown in **Figure 1**, 55% of patients with reported adverse events were female, and the age group most frequently affected (31%) were between 41 and 60 years. The percentage of reports by indication are summarized in **Figure 2**. This displays the percentage of reports (rounded to the nearest whole percent) by indication for Filgrastim usage and shows that 64% of the adverse events reported were in the adjunct to chemotherapy indication while 9% were in the SCN indication. The reporting frequency of adverse events mirrors actual usage in the clinical setting.

The number and percentage of primary adverse event terms by body system are summarized in **Figure 3**. The body systems most frequently

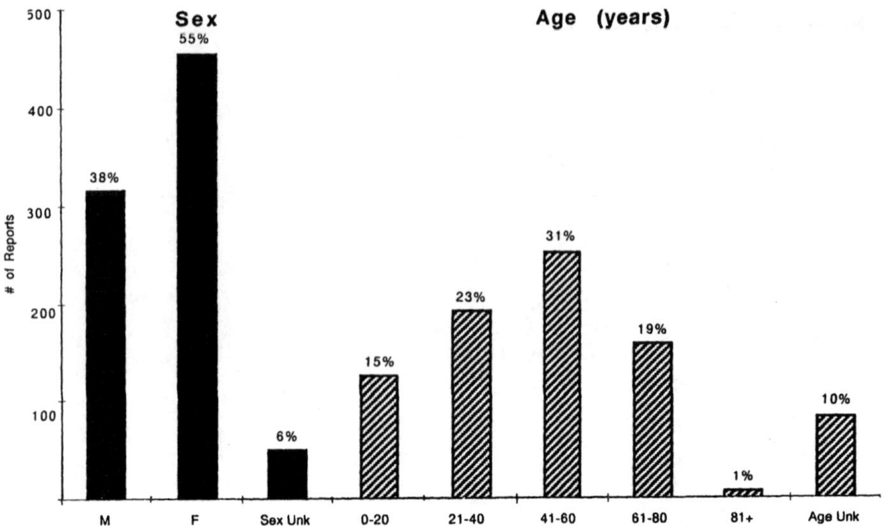

Figure 1. Summary of demographic characteristics. Black boxes, sex. F = female; M = male. Hatched boxes, age groups in years; UNK = unknown.

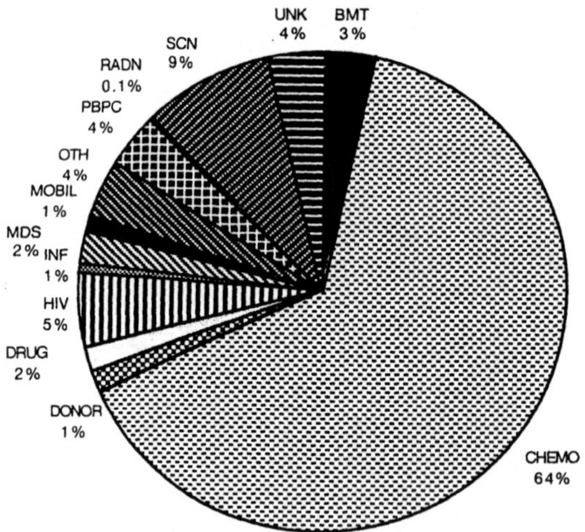

Figure 2. Percentage of reports by indication. BMT = bone marrow transplant; CHEMO = neutropenia associated with chemotherapy; DONOR = normal donor PBPC; DRUG = non-cytotoxic drug-induced neutropenia; HIV = neutropenia associated with HIV infection; INF = prevention or treatment of infection; MDS = myelodysplastic syndrome; MOBIL = cytokine only mobilized PBPC; OTH = neutropenia associated with underlying disease; PBPC = chemotherapy with PBPC support; RADN = neutropenia associated with radiation; SCN = severe chronic neutropenia; UNK = unknown.

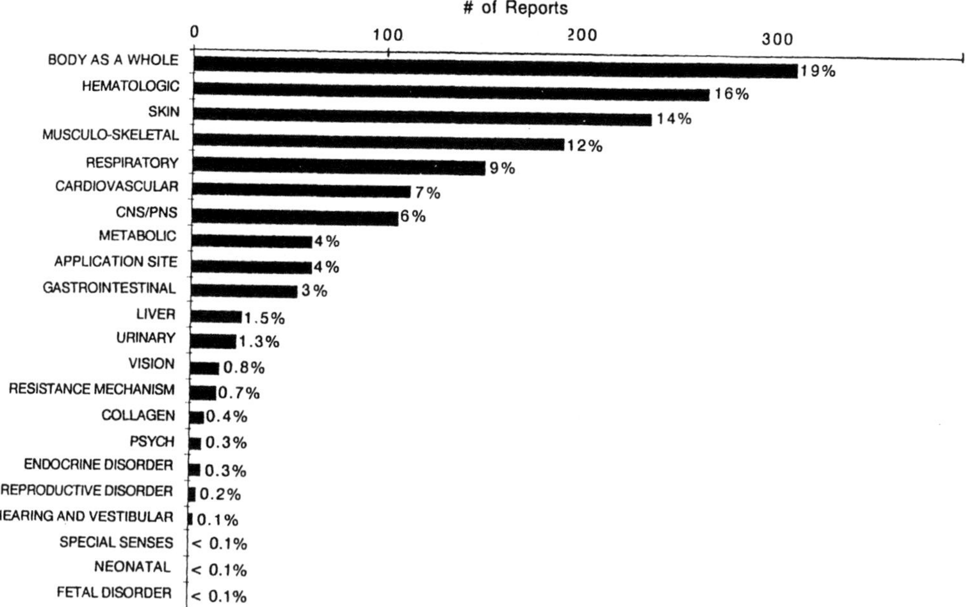

Figure 3. Number and percentage of adverse event reports by body system of primary term.

affected were body as a whole (19%), hematologic (16%), and skin and appendages (14%).

The number of adverse events by preferred term reported in at least 60 cases is listed in decreasing order of frequency on **Table 4**. The most frequently reported adverse events were fever (5%), skeletal pain (4%), dyspnea, and thrombocytopenia (3% each).

The complex nature of patients treated with Filgrastim after cytotoxic chemotherapy makes it difficult to determine the source of fever since cancer, neutrophilia, neutropenia, prophylactic antibiotics, and infection are all factors that can produce this symptom. Fever, as reported here, is not part of the symptom complex often associated with cytokine usage.

Skeletal pain was the most consistently reported adverse event across all indications. It was generally mild to moderate in severity and controlled with non-narcotic analgesics. Headache may also be skeletal in nature. Myalgia was sometimes reported concurrently with skeletal pain. Skeletal pain is noted in the current Filgrastim label.

Dyspnea was reported in patients with pre-existing pulmonary conditions, post-chemotherapy complications and/or sepsis, and those with his-

Table 4. Frequency of Adverse Events by
Preferred Term Over 6 Years in 1650 Reports
from Pool of Approximately 1.9 Million Patients.

Preferred term (n = 3860)	Frequency
Fever	5%
Pain skeletal	4%
Dyspnea	3%
Thrombocytopenia	3%
Headache	2%
Therapeutic response decreased	2%
Pruritus	2%
Nausea	2%
Myalgia	2%
Rash	2%
Pain chest	2%
Rash maculo-papular	2%

tory of allergies. Thrombocytopenia occurred during the expected nadir in many patients who received standard or high-dose chemotherapy. Other confounding variables included concomitant medications, concurrent infection, disease, and other treatment that could contribute to the frequency, duration, and severity of thrombocytopenia. Nausea is not unusual in the patient population being treated (ie, those receiving chemotherapy or concomitant medications that could induce nausea/vomiting and progression of disease). In most cases rash and pruritus were attributed to alternative causalities including history of allergies and the administration of concomitant medications.

B. Issues

The number of patients enrolled in clinical trials is small compared with those receiving marketed product. Consequently, very rare events may not be characterized during the clinical trial period. With increased market exposure, new safety issues may emerge, changing or adding to the premarket safety profile since the drug is being observed and tested under a whole range of clinical conditions. Spontaneously reported reactions are analyzed in detail by pharmaceutical manufacturers to look for potential indications of important, previously unknown safety factors. Postmarketing safety surveillance is a means for identifying these potential problems; however, in the postmarket setting, the determination of real quantification of rates of suspected reactions is not possible. It is difficult to ascertain objective causality. This is of less importance in the context of raising an early suspicion or signal. Given these limitations, selected safety issues were identified from spontaneous reports and published literature and are discussed below. These reports are from an estimated pool of approximately 1.9 million patients treated in approximately 6 years.

1. Dermatologic Events

Neutrophilic dermatosis, including Sweet's syndrome and pyoderma gangrenosum, are rare events that have been reported in patients receiving chemotherapy with Filgrastim support. With the increased use of Filgrastim, the reporting incidence of these events has not dramatically increased. Neutrophilic dermatosis reporting rate is approximately <1 in 33,000 patients. Sweet's syndrome has clearly been identified as a paraneoplastic syndrome seen in the patient population that typically receives Filgrastim. Although there appears to be a temporal association with Filgrastim in several of the cases reported, the role of Filgrastim in the pathogenesis of this syndrome is uncertain. In these reported cases, it is doubtful that Filgrastim induced the Sweet's syndrome, as these cases are similar to those reported in the literature in patients who were not receiving Filgrastim or any other cytokines, and in a few of the reported cases other drugs were identified as co-suspect. It has been suggested that the pathogenesis of Sweet's syndrome may be secondary to an inappropriate secretion of one or more endogenous cytokines (2). Whether exogenous administration of Filgrastim participates directly or indirectly in the development of Sweet's syndrome remains speculative.

2. Respiratory Events

The possibility that Filgrastim, or the rapid increase of white blood cells, might augment bleomycin pulmonary toxicity has been discussed in the literature. Bleomycin is known to cause pulmonary toxicity in approximately 10% of patients, lethal pulmonary fibrosis in about 1% to 2% of patients, and nonlethal pulmonary fibrosis in 2% to 3% of patients. The incidence of pulmonary toxicity appears to increase when bleomycin is used in combination with other drugs and administered to patients with reduced renal function (3). The possible augmentation of bleomycin pulmonary toxicity by Filgrastim has been addressed by studies in animal models and by both non-randomized and randomized clinical trials.

A rat model in which bleomycin lung toxicity could be reliably produced was studied. Exacerbation of toxicity was not found with Filgrastim or high neutrophil counts (Amgen data on file). Randomized clinical trials

(4–8) and non-randomized trials (9,10) demonstrated no increase in the known pulmonary toxicity of bleomycin when Filgrastim was added to treatment (**Table 5**).

Overall, 862 patients with malignancies including NHL (n = 482) and metastatic teratoma (n = 380) were enrolled in randomized clinical trials and received bleomycin-containing chemotherapy regimens. Of these, 745 patients were randomized to receive either Filgrastim (n = 369) or placebo/ no treatment with Filgrastim (n = 376).

Since commercial introduction, there have been rare reports (<1 in 39,000) of bleomycin-induced pulmonary symptoms being possibly potentiated, in the opinion of the reporting physician, by treatment with Filgrastim. Most of the reports involved patients with HD. The reported pulmonary toxicities have varied in severity and included the following: dyspnea or cough alone, pneumonia, pneumonitis, pulmonary infiltration, respiratory insufficiency, or a general respiratory disorder. The presence of two to four risk factors has been documented in the majority of these reported cases. Pulmonary and/or other infections had been documented in most of the cases. For more than half of these reported cases, there has been no baseline pulmonary function workup provided, in particular diffusion capacity of carbon monoxide values, the only measurement affected by bleomycin. All of these patients had received other pulmonary toxic agents, and many had received mantle irradiation. Pulmonary biopsy had been performed in one-fifth of patients. In most cases the patients' symptoms diminished or were controlled with or without treatment (oxygen, antibiotic, antifungal, and/ or steroid therapies). Treatment with bleomycin was discontinued in most cases, and the condition of some of these patients improved while remaining on Filgrastim therapy without bleomycin.

Although anecdotal case reports of pulmonary adverse events have been published, the available evidence from controlled clinical trials argues against an excess morbidity in Filgrastim-treated individuals. Determination of a causal relationship between Filgrastim and the possible augmentation of bleomycin pulmonary toxicity is complicated in these cases by confounding variables and contributory risk factors. There is no evidence to date to suggest that the safety profile of Filgrastim has changed in this respect.

3. Rheumatologic Events

Since commercial introduction of Filgrastim, the reporting rate of arthritis in patients treated with Filgrastim is <1 in 63,000 patients. Most of these reports of arthritis occurred in patients with Felty syndrome; however, other diagnoses including histories of rheumatoid arthritis, inflammatory joint disease (osteoarthritis and migratory polyarthritis), and Sjögren syn-

Table 5. Summary of Randomized and Non-Randomized Clinical Trials with Bleomycin.

		Randomized	
Ref.	No. patients	Diagnosis	Chemotherapy
8	80	NHL	VAPEC-B
6	380 (263 randomized to Filgrastim/no Filgrastim)	Metastatic teratoma	(Primary randomization) six cycles BEP/EP versus 3 cycles BOP/VIP; (secondary randomization) BOP (with or without Filgrastim on days 3–9) versus BEP or VIP (with or without Filgrastim on days 6–19)
4,17	278	NHL	Four courses unspecified chemotherapy containing bleomycin (half received Filgrastim; half received lenograstim)
5	67	NHL	P-VEBEC
7	57	NHL	Adriamycin®/cytoxan/vincristine; Adriamycin®/cytoxan/vincristine + bleomycin

		Non-Randomized	
Investigator	No. patients	Diagnosis	Chemotherapy
10	86	Germ cell tumors	Unspecified chemotherapy + bleomycin
9	20	Advanced testicular cancer	VIP/VB

NHL = non-Hodgkin's lymphoma.
See Glossary for description of chemotherapy regimens.

drome were also represented. One-fifth of patients had no previous history of arthritis.

In most cases, arthritis developed within the first 10 days of Filgrastim administration. Resolution of symptoms occurred in approximately half the patients after treatment with analgesics and/or non-steroidal anti-inflammatory agents. In rare cases, steroids were required. Filgrastim was discontinued in approximately three-fourths of patients who reported arthritis flares. Symptoms recurred in all patients who were rechallenged, but in a few patients, symptoms abated with treatment during Filgrastim therapy.

Of interest are several published case reports discussing Filgrastim use in patients with Felty syndrome that did not show an association with exacerbation of arthritis (11–14).

Whether exogenous Filgrastim plays a role in the exacerbation of arthritis or the development of arthritis symptoms remains speculative. While neutrophil activation can contribute to the anti-inflammatory response, other mechanisms may be responsible for the initiation of these events.

4. Pregnancy

When Filgrastim was first licensed, there was little information available regarding placental transfer. Since that time two publications have addressed this issue. Maternally administered Filgrastim was demonstrated to cross the placenta and specifically induce bone marrow and spleen myelopoiesis in the fetus and neonate in a preclinical rat study (15).

There are no well-controlled trials studying the use of Filgrastim in human pregnancies. However, the transplacental passage of a measurable amount of Filgrastim in women who received a single dose of Filgrastim and in whom preterm delivery was imminent was reported. Granulocyte colony-stimulating factors and neutrophil concentrations in maternal blood and umbilical venous cord blood were measured after administration of a single IV dose of Filgrastim 25 μg/kg in 11 women in whom preterm delivery was imminent. These measurements were compared with values from 34 control women and 35 cord sera. The transplacental passage of a measurable quantity of Filgrastim was noted, especially in those women who received Filgrastim at least 30 hours before delivery. According to the authors, this amount of Filgrastim can have a biologic effect on the fetus (16).

Additional information is needed from case reports to determine any possible effects which Filgrastim may have on reproductive issues.

C. Overall Safety Evaluation

Filgrastim remains well tolerated within all patient groups despite the severity of illness of the patients. Many reported adverse events received may reflect underlying illness or side effects of chemotherapy rather than adverse effects of Filgrastim.

1. Cancer Patients

Dose-dependent musculoskeletal pain remains the most consistently observed adverse event across all cancer patient populations in multiple settings including adjunct to chemotherapy, PBPC, and BMT. Medullary bone pain was reported in 14% of patients who received Filgrastim doses ranging from 0.4 to 12 μg/kg/day and in 27% of patients who received Filgrastim doses ranging from 20 to 120 μg/kg/day. Musculoskeletal pain was controlled with non-narcotic analgesia in the majority of patients. The cause of musculoskeletal pain is not known, although it appears to be related to myelopoietic stimulation. Other reported adverse events included thrombocytopenia and anemia that are consistent with complications of cytotoxic chemotherapy; the potential for receiving full doses of chemotherapy on the prescribed schedule may put the patient at a greater risk of hematologic complications. Other variables including disease, infection, and treatment can also contribute to the frequency, duration, and severity of thrombocytopenia in these patients. It is difficult to determine if Filgrastim treatment contributes to the development of thrombocytopenia in chemotherapy-treated patients. Dose-dependent, predictable, and spontaneously reversible increases in serum alkaline phosphatase, lactic dehydrogenase, and uric acid were reported. There were no serious clinical sequelae reported in association with biochemistry abnormalities.

2. Severe Chronic Neutropenia

Chronic daily administration of Filgrastim to patients with SCN causes relatively few adverse events; those that do occur are mild and easily managed. The number and duration of reported adverse events related to Filgrastim decreases with the duration of treatment. Bone pain is noted early in treatment and tends to decrease in frequency with long-term therapy. Some patients developed splenomegaly that was not progressive. A few patients underwent splenectomy; most of these patients had a pre-study history of splenomegaly. Some of the patients experienced thrombocytopenia with a platelet count $<50 \times 10^9$/L that was controlled by interrupting or reducing the Filgrastim dose. Cutaneous vasculitis occurred infrequently; generally, symptoms of vasculitis developed simultaneously with an increase in the neutrophil count and resolved when the neutrophil count decreased. Growth and development in children were not adversely affected by long-term Filgrastim treatment. Secondary AML/MDS has been reported in patients with congenital neutropenia; however, before the availability of cytokine therapy, AML had been reported to occur as a rare event in these patients. An international SCN disease registry actively monitors events in this patient population.

III. OVERALL SUMMARY

In summary, Filgrastim remains well tolerated within all patient groups treated, despite the severity of illness of the patients. Approximately 1.9 million patients have been treated worldwide since commercial introduction. No new significant safety issues have been discovered in clinical trials or in the postmarket setting. Increased patient exposure has revealed a more comprehensive safety profile for Filgrastim, which supports its continued safety in widely differing patient populations.

REFERENCES

1. Welte, K., Gabrilove, J., Bronchud, M. H., Platzer, E., and Morstyn, G. (1996). Filgrastim (r-metHuG-CSF): The first 10 years. *Blood 88*:1907–1929.
2. Reuss-Borst, M. A., Muller, C. A., and Waller, H. D. (1994). The possible role of G-CSF in the pathogenesis of Sweet's syndrome. *Leukemia Lymphoma 15*:261–264.
3. Haskell, C. M. *Cancer Treatment*, 3rd ed. Philadelphia: W.B. Saunders Company, 1990, p. 33.
4. Bastion, Y., Reyes, F., Bosly, A., et al (1994). Possible toxicity with the association of G-CSF and bleomycin. *Lancet 343*:1221–1222.
5. Bertini, M., Freilone, R., Vitolo, U., et al (1994). P-VEBEC: A new 8-weekly schedule with or without rG-CSF for elderly patients with aggressive non-Hodgkin's lymphoma. *Ann Oncol 5*:895–900.
6. Fossa, S., Kaye, S. B., Mead, B. M., et al (1995). An MRC/EORTC randomised trial in poor prognosis metastatic teratoma comparing treatment with/ without filgrastim. *Proc Am Soc Clin Oncol 14*:245 (abstr 656).
7. Ogawa, M., Masaoka, T., Mizoguchi, H., Takaku, F., and Nakashima, M. (1990). A phase II study of KRN 8601 rhG-CSF on neutropenia induced by chemotherapy for malignant lymphoma—a multi-institutional placebo controlled double-blind comparative study. *Gan To Kagaku Ryoho* (Japan) *17*(1):365–373.
8. Petengell, R., Hurney, H., Radford, J. A., et al (1992). Granulocyte colony stimulating factor to prevent dose-limiting neutropenia in non-Hodgkin's lymphoma: A randomized clinical trial. *Blood 80*:1430–1436.
9. Blanke, C., Loehrer, P., Einhorn, L., and Nichols, C. (1994). A phase II study of VP-16 plus ifosfamide plus cisplatin plus vinblastine plus bleomycin (VIP/ VB) with Filgrastim for advanced stage testicular cancer. *Proc Am Soc Clin Oncol 13*:234 (abstr 723).

10. Saxman, S. B., Nichols, C. R., Stephens, A. W., and Einhorn, L. H. (1995). Pulmonary toxicity in patients with advanced stage germ cells tumors receiving bleomycin with and without granulocyte colony stimulating factor. *Proc Am Soc Clin Oncol 14:*255 (abstr 690).

11. Bhalla, K., Ross, R., Jeter, E., Madyastha, P., and Stuart, R. (1993). G-CSF improves granulocytopenia in Felty's syndrome without flare-up of arthritis [letter]. *J Hematol 42:*230–231.

12. Fohlman, J., Hoeglund, M., and Bergmann, S. (1994). Successful treatment of chronic wound infection in neutropenia and rheumatoid arthritis with Filgrastim (rhG-CSF). *Ann Hematol 69:*153–156.

13. Graham, K. E., and Coodley, G. O. (1995). Prolonged use of granulocyte colony stimulating factor in Felty's syndrome. *J Rheumatol 22:*174–176.

14. Kaiser, U., Klausmann, M., Richter, G., and Pfluger, K. H. (1992). GM-CSF versus G-CSF in the treatment of infectious complication in Felty's syndrome—a case report. *Ann Hematol 64:*205–206.

15. Medlock, E. S., Kaplan, D. I., Cecchini, M., Ulich, T. R., del Castillo, J., and Andresen, J. (1993). Granulocyte colony-stimulating factor crosses the placenta and stimulates fetal rate granulopoiesis. *Blood 81*(4):916–922.

16. Calhoun, D. A., and Christensen, R. D. (1996). Assessment of transplacental passage of recombinant human granulocyte colony-stimulating factor in women with an imminent preterm delivery. *J Invest Med 44*(1):114.

17. Bastion, Y., and Coiffier, B. (1994). Pulmonary toxicity of bleomycin: Is G-CSF a risk factor? [letter] *Lancet 344:*474.

29
Practical Aspects of Filgrastim (r-metHuG-CSF) Administration

Jane Campbell
*Haematology Oncology Clinics of Australasia,
Brisbane, Australia*

I. INTRODUCTION

Patients with cancer have a significant, increased risk of infection. This increase may be related to either their disease (compromised host defenses and sequelae of treatment due to the absence of neutrophils, disruption to the natural barriers caused by infection, a shift in microbial flora), treatment (surgery, chemotherapy, radiotherapy), or both. Over the past years, a decrease in infections has been achieved because of the development of β-lactam and fluoroquinolone antibiotics; the increased use of prophylactic antibiotics, antifungal and antiviral drugs; and the use of colony-stimulating factors, such as Filgrastim (1).

The use of Filgrastim in the management of the patient with cancer has resulted in a reduction in the incidence, severity, and duration of neutropenia and, consequently, a reduction in infectious complications while maintaining the proper dose and schedule of chemotherapy (see other chapters in this volume).

Indeed, the role of Filgrastim has not been confined to that of a reduction in infectious episodes. The role of peripheral blood progenitor cell (PBPC) transplantation in the management of solid tumors and hematological malignancies has been enhanced by the unexpected finding that Filgrastim-mobilized PBPC shorten the time, not only of neutrophil recovery but also that of platelet recovery after myeloablative therapy (2). Further evidence has shown that in patients with lymphoma treated with

high-dose chemotherapy, infusing Filgrastim-mobilized PBPC rather than autologous bone marrow significantly reduced the number of platelet transfusions and the time to platelet and neutrophil recovery and led to earlier discharge from the hospital (2).

The successful use of Filgrastim requires knowledge about the drug, its storage, dosage, and administration technique, as well as effective patient education. It must be ensured that the health professional and the patient have the necessary knowledge to safely and effectively administer the drug and to recognize side effects that may occur.

II. NEUTROPENIA

Neutrophils comprise approximately 50% to 70% of the total white cell count. The ratio of neutrophils to the total white cell count may vary on a daily basis because of the short life span of circulating neutrophils. It is, therefore, essential when determining a patient's susceptibility to infection that a differential white cell count is taken and the absolute neutrophil count (ANC) is determined. The differential count specifically identifies the number of white cells by type (segmented, bands, lymphocytes, monocytes, eosinophils, and basophils). In addition, the differential count will identify any abnormal or premature forms of cells observed microscopically.

It is now well documented that there is an inverse relationship between the ANC and a patient's risk of infection. Despite preventive measures, if a patient sustains a significantly low neutrophil count for a sufficient period of time, an infection will invariably result, with a 100% incidence of infection if the neutrophil count remains $<1.0 \times 10^9$/L for 3 weeks (3,4). The impact of an infection related to neutropenia can vary from mild to life threatening. Therefore, a shortened "window" period for infectious opportunity stands to be of great benefit to patient outcome and hospital resources.

Neutropenia is considered to be an ANC $<2.5 \times 10^9$/L. The ANC is determined by multiplying the percentage of neutrophils and bands by the total number of white blood cells (ie, ANC = total white cell count $\times$ [% neutrophils + % bands] divided by 100) (5). An ANC of 1.0 to 1.5 $\times$ 10^9/L is not considered a significant risk to the patient. Moderate risk of infection occurs when the ANC is between 0.5 and 1.0 $\times 10^9$/L, and severe risk occurs when the ANC is $<0.5 \times 10^9$/L (6).

Neutropenia, also referred to as granulocytopenia (although the granulocyte pool includes basophils and eosinophils as well), is one of the most significant life-threatening side effects of chemotherapy. It is also one of the most reliable predictors of the potential for infection in the patient with cancer or hematological malignancy (7). Neutropenia is present during approximately 80% of febrile episodes in these patients (8). Febrile episodes may require the administration of antibiotics, antifungal, and/or antiviral drugs even in the absence of identified sources, and if they occur during neutropenia may result in the patient's hospitalization, thus potentially decreasing the patient's quality of life and increasing the costs of patient care.

III. ABOUT FILGRASTIM

Filgrastim has been shown to primarily affect neutrophil progenitor proliferation, differentiation, and selected end-cell functional activation. Filgrastim is a lineage-specific colony-stimulating factor with selectivity for the neutrophil lineage. Filgrastim is produced by *Escherichia coli* bacteria that have been transformed with plasmid DNA that includes the gene for human G-CSF. It acts by binding to specific cell-surface receptors present on the neutrophil precursor cells and has the capacity to decrease the normal time of neutrophil maturation in the bone marrow from the usual 5 to 6 days to 1 day. The mature neutrophils released into the peripheral circulation are functionally active. The bioavailability of the new Filgrastim formulation containing 5% sorbitol is unchanged from that of the old formulation containing 5% mannitol (NEUPOGEN® Package Insert).

IV. INDICATIONS FOR THE USE OF FILGRASTIM

Filgrastim is currently indicated to decrease the incidence of infection, as manifested by febrile neutropenia, in patients with non-myeloid malignancies receiving myelosuppressive anticancer drugs in doses not usually requiring bone marrow transplantation; in patients with non-myeloid malignancies receiving myeloablative chemotherapy; for reducing the duration of neutropenia and clinical sequelae after autologous or allogeneic bone marrow transplantation; for chronic administration to increase neutrophil counts and to reduce the incidence and duration of infections in patients with severe chronic neutropenia (SCN); and for the mobilization of PBPC alone, or after myelosuppressive chemotherapy, to accelerate neutrophil and platelet recovery by infusion of such cells after myeloablative or myelosuppressive therapy in patients with non-myeloid malignancies. In some countries, Filgrastim is indicated for treatment of the neutropenia of AIDS.

V. DOSING AND ADMINISTRATION

A. The Stability of Filgrastim

Filgrastim is a sterile, clear, colorless, preservative-free liquid for subcutaneous (SC) or intravenous (IV) administration. Filgrastim now contains sorbitol instead of mannitol, a modification that has enhanced the cryoprotective properties of the formulation. Filgrastim can now withstand accidental freezing and subsequent thawing at temperatures as low as −20°C. Therefore the patient can receive vials that may have been accidentally frozen first, but use of vials exposed to multiple freeze–thaw cycles or freezing for extended periods of time are not recommended. Exposure to controlled room temperature (up to 30°C) for periods up to 3 days will not affect its stability, however, product left unrefrigerated should be used first (NEUPOGEN® Package Insert). It is advisable to contact the manufacturing company if one is unsure or any of the above parameters have been extended.

In some areas of the world, Filgrastim is available in single-use prefilled syringes for self administration. The syringes contain either 300 μg in 1.0 mL or 480 μg in 1.6 mL. Storage conditions for the syringes are the same as for vials.

B. Preparation for Administration

Filgrastim is commercially available in two vial sizes: 1.0 mL (300 μg) and 1.6 mL (480 μg). Each vial is for single use only and any leftover material should be discarded. A great advantage of the preparation of Filgrastim is that it is a ready-to-use solution, simplifying the procedure for drawing up the solution. This assists not only the patient when administering the injection, but also the health professional when educating the patient and/or caregiver. If the solution is drawn up under aseptic conditions, such as a laminar flow hood, it may be stored in polypropylene syringes at 2° to 8°C (36° to 46°F) for as long as 24 hours before use.

When using Filgrastim, both the recommended dose and administration time of Filgrastim and the chemotherapy regimen should be followed. Dose reduction is not recommended. The administration of Filgrastim should be continued daily for up to 2 weeks or until an ANC >10 × 10^9/L.

Filgrastim is not compatible with saline and should be diluted with 5% dextrose with or without human serum albumin (HSA). When giving Filgrastim as an IV infusion, a flush before and after Filgrastim administration with 5% dextrose is required. It is stable in 5% dextrose solution when the concentration is ≥15 μg/mL. If less than this concentration is required, HSA should be added to reach a final albumin concentration of 2 mg/mL to prevent the protein from adhering to the container walls. Filgrastim should not be diluted to concentrations <2 μg/mL (NEUPOGEN® Package Insert).

Concurrent administration of Filgrastim within a 24-hour window before or after the administration of chemotherapy is not recommended.

It is believed that if Filgrastim is given within this time frame, the first "peak" of proliferating progenitor cells may be damaged or destroyed by the chemotherapy.

C. Dosage

The recommended starting dose after chemotherapy is 5 μg/kg/day (**Table 1**). Dosages for the mobilization of peripheral blood progenitor cells in patients receiving chemotherapy with or without bone marrow transplantation and for the treatment of SCN are given in Table 1. In patients receiving autologous bone marrow transplantation after marrow-ablative chemotherapy, the recommended starting dose is 10 μg/kg/day (**Table 2**). If individual patients require dose escalation (this may be due to either a poor response

Table 1. Dosage Recommendations After Myelosuppression Chemotherapy, for Mobilization of Peripheral Blood Progenitor Cells, or for Treatment of Severe Chronic Neutropenia.

Indication	Filgrastim dosage
Cancer patients receiving myelosuppressive chemotherapy	5 μg/kg/day as a single daily SC bolus or short IV infusion; doses may be increased in 5-μg/kg increments for each chemotherapy cycle according to duration and severity of ANC nadir
Mobilization in patients receiving chemotherapy followed by PBPC with or without BMT	Used alone: 10 μg/kg/day as a 24-hour SC infusion or single daily SC injection for six consecutive days Used after chemotherapy: 5 μg/kg/day daily by SC injection from first day after completion of chemotherapy until ANC nadir is passed and ANC is normal
Patients with SCN	Congenital: 6 μg/kg SC twice daily to start Idiopathic or cyclic: 5 μg/kg/day SC Chronic administration is required to maintain clinical benefit

ANC = absolute neutrophil count.
BMT = bone marrow transplantation.
IV – intravenous.
PBPC = peripheral blood progenitor cell.
SC = subcutaneous.
SCN = severe chronic neutropenia.

Table 2. Titration of Daily Dose Against Neutrophil Count for Patients Receiving Bone Marrow Transplantation.

Neutrophil count	Filgrastim dose adjustment
>1.0 × 10⁹/L for 3 consecutive days	Reduce to 5.0 μg/kg/day
Then, if ANC remains >1.0 × 10⁹/L for 3 consecutive days	Discontinue Filgrastim
If ANC decreased to <1.0 × 10⁹/L	Resume at 5 μg/kg/day

ANC = absolute neutrophil count.

time or an unacceptable magnitude of the neutrophil response), the dose should be increased in increments of 5 μg/kg. A maximum-tolerated dose of Filgrastim has not been identified, but doses as great as 115 μg/kg/day have been used; however, it is recommended that the patient's white cell count should not exceed 75 × 10⁹/L (NEUPOGEN® Package Insert). Doses >10 μg/kg/day have only shown benefit in children with SCN.

D. Discontinuing Therapy

Discontinuing Filgrastim before the time of the expected chemotherapy-induced neutrophil nadir is not recommended. A 50% decrease in circulating neutrophils within 1 to 2 days and a return to pretreatment levels in 1 to 7 days results from discontinuation. Due to the short half-life of the neutrophils, premature discontinuation will now allow the patient the full benefit of the action of the drug. The number of days a patient requires administration of Filgrastim will be dependent on a number of factors, including the chemotherapy regime used, concurrent conditions, neutrophil profile, and hematological recovery.

E. Autologous Bone Marrow Transplantation

The starting dose at 10 μg/kg/day should be given by either SC (bolus or infusion) or IV (infusion over a 4- to 24-hour period). After the neutrophil nadir, the daily dose should be titrated against the neutrophil count. If the ANC decreases to <1.0 × 10⁹/L at any time during the 5-μg/kg/day administration, Filgrastim should be increased to 10 μg/kg/day and the steps outlined in Table 2 followed.

F. PBPC Collection and Therapy

When used alone for PBPC mobilization, the recommended dose of Filgrastim is 10 μg/kg/day as a single daily SC injection or a continuous 24-hour infusion. Filgrastim should be given for at least 4 days before the first leukapheresis procedure, and should be continued through to the day of the last leukapheresis procedure. The actual collection of PBPC, however, should be done before the administration of Filgrastim for that day.

After myelosuppressive chemotherapy for PBPC mobilization, the recommended dose of Filgrastim is 5 μg/kg/day given daily by SC injection from 24 hours after completion of chemotherapy until the expected neutrophil nadir is passed and the ANC has recovered to the normal range. Chapters 13 and 14 and Table 1 further discuss the use of Filgrastim in PBPC mobilization and collection.

H. Routes of Administration

Recommended routes for the administration of Filgrastim include SC injection, IV short infusion, SC continuous infusion, and IV continuous infusion. There have been no clinical trials to date to determine whether Filgrastim can be given via intramuscular (IM), intravenous push (IVP), or bolus route. It is therefore not recommended at this time.

VI. PRETREATMENT ADMINISTRATION ASSESSMENT

Before the administration of any drug, a pretreatment assessment is required. Factors that should be noted are the patient's diagnosis; past history and treatment; knowledge about the current treatment plan, including medications, known allergies, laboratory results (in particular ANC); and a physical assessment of the patient. If a patient is neutropenic, it is recommended that regular monitoring of his/her temperature be done. Many institutions recommend monitoring every 4 hours, with the patient reporting any increase in temperature above 38°C to be reported to his/her physician or health-care provider.

VII. ADVERSE EFFECTS AND INTERVENTIONS

Many of the adverse effects reported in patients who are receiving Filgrastim can be attributed to the concurrent cytotoxic chemotherapy or the neutropenia associated with chemotherapy. The neutropenia nadir may occur 2 to 3 days earlier in the patient receiving Filgrastim. Hematological monitoring (complete blood count, differential, and platelet count) is recommended before commencement of chemotherapy and twice a week during Filgrastim administration.

The administration of Filgrastim may enable the patient to receive higher doses of chemotherapy; as a result of this, patients should be assessed regularly for an increased risk of the hematological and non-hematological toxicities of the chemotherapy agents used.

Filgrastim has not been associated with fevers, chills, flu-like symptoms, rashes, or dyspnea. If any of these occur, the patients should be investigated for other causes.

A. Medullary Bone Pain

Medullary bone pain is the only consistently observed side effect attributed to Filgrastim. The bone pain is thought to occur as a result of the marrow expansion that occurs from the increase in the neutrophil pool. It is believed that bone pain coincides with the period immediately before the neutrophils are released from the marrow. Once the neutrophils are released and the patient is recovering from the chemotherapy-induced nadir, the bone pain usually resolves (NEUPOGEN® Package Insert). Bone pain occurs in approximately 20% of patients with the common sites being the spine, limbs, back, pelvis, and jaw. The bone pain is generally reported to be mild to moderate and can be relieved with non-narcotic analgesia. Occasionally this pain can be severe and may require with narcotic (or codeine-based) analgesia (9). The bone pain may be felt immediately after Filgrastim administration or up to several days after administration, and generally does not last longer than 24 hours. The incidence of bone pain was greater in patients who were receiving higher doses of Filgrastim IV than in patients receiving lower doses SC (NEUPOGEN® Package Insert; 9).

B. Injection-Site Irritation

Injection-site irritation or discomfort can occur and at times may be quite uncomfortable for the patient. It appears the best method to reduce this discomfort is to give the injection slowly. Rapid injections can cause discomfort as the fluid is injected under high pressure.

Other techniques to minimize irritation and discomfort from SC injections include allowing the Filgrastim ampule to reach room temperature before injection, using a needle size consistent with SC injection (5/8", 25 gauge or smaller), rotating injection sites, and keeping the volume of Filgrastim small with the volume not exceeding 1.5 mL. In addition, ensure that no liquid is on the needle or needle tip to cause irritation. Ice may be applied to the site before injection. As with any SC injection, it is advisable to rotate the injection sites on a daily basis. Alternating sites on the arms, legs, and abdomen may be used.

VIII. PATIENT AND CAREGIVER EDUCATION

Successful patient and caregiver education is the result of a comprehensive assessment, including the patient's expectations, requirements, and intellectual capabilities. To impart accurate information and enable an effective education program to proceed, the health-care provider must determine:

1.   What information has already been given to the patient, and his/her understanding of that information
2.   What additional information should be given
3.   How much information the patient wants to know and is capable of comprehending
4.   Who the patient wishes to have this information
5.   What is the most effective method of delivering this information to the patient and/or his/her caregiver

Information on the patient's disease, treatment, and expected side effects, as well as an overview of the function of the bone marrow should be provided. In addition, education regarding preventative strategies that may reduce the risk of infection should be included.

It is important that the patient and caregiver realize that the usual signs and symptoms of infection often are decreased or muted because of a depressed immune function, although fever is usually present in most infected neutropenic patients (9).

Little is known about the factors that facilitate patients to use effective self-care behaviors, although evidence suggests that a patient's confidence in his/her capability to regulate thought processes, emotional states, and the social environment, as well as levels of behavioral attainment are important (10). It is suggested that a person's belief in his/her capability to organize and execute the action required to deal with prospective situations underpins self-care behavior as a whole and, consequently, compliance with the self-management/health-care behavior expected (11).

It is, therefore, important that health beliefs, self-efficacy, and beliefs about control should be considered when developing a patient and caregiver education plan. In addition, factors such as language barriers, the individual's response to the diagnosis, the ability to read and write, the position of the patient within the family group, and treatment outcome expectations also need to be explored before selecting an effective education program.

When the Filgrastim injections can be self-administered by the patient, or by his/her caregiver, this should be tried. Independence from the hospital or clinic setting on a daily basis can maximize the patient's independence and minimize the disruption to his/her lifestyle. A comprehensive assessment of the patient and his/her caregiver should be able to identify possible non-compliance that may be related to either low self-esteem, loss of locus of control, or poor communication. The ability and/or willingness of the patient to comply with the administration of Filgrastim will have a direct impact on the outcome of the drug's efficacy. Information will, when given to the patient at a level of his/her understanding and combined with a supportive environment, have a direct impact on the patient's compliance level. It some clinical settings, however, it will be desirable for the patient to attend a clinic to receive the injections.

It is important that any patient and caregiver education should be a consistent, coordinated team approach. Often the giving of information is started by the physician and consolidated by the primary nurse and health team. The information and education that should be fully explained includes:

Why Filgrastim has been chosen as a treatment
Why a given method of Filgrastim administration has been chosen
What is expected of the patient and his/her caregiver
What are the expectations of the patient and caregiver of the health professionals team
What supports are required and what resources are available
How to store and care for the Filgrastim vials and equipment
How to administer the injections, emphasizing the importance of maintaining a sterile technique
What to do when things go wrong during the administration (ie, a trouble-shooting session)
Where to inject, using recommended injection-site rotation
What adverse effects may occur and when to report them
Why it is crucial to always report any adverse effects such as fever >38°C, unwell feelings, or chills or rigors

Education about Filgrastim and its administration is usually not given in isolation. The administration of Filgrastim occurs generally at a time of crisis for the patient and his/her family. Patients and their caregivers will require significant support, including practical, written, and emotional resources. Dealing with the patient who is receiving treatment for cancer, and the resultant ramifications that this may have on his/her life and the lives of family members, requires a supportive, well-informed health professional who has access to appropriate resources (12).

REFERENCES

1. Wujcik, D. Infection. In: S. L. Groenwald, M. H. Frogge, M. Goodman, and D. H. Yarbo, eds. *Cancer Symptom Management,* Sudbury, MA: Jones and Bartlett Publishers, 1996, pp. 289–307.

2. Schmitz, N., Linch, D. C., Dreger, A. H., Goldstone, M. A., et al (1996). Randomised trial of filgrastim-mobilised peripheral blood progenitor cell transplantation versus autologous bone-marrow transplantation in lymphoma patients. *Lancet 347*:353–357.

3. Bodey, G. P. (1975). Infection in cancer patients. *Cancer Treat Rev 2*:89–128.

4. Carlson, A. C. (1985). Infection prophylaxis in the patient with cancer. *Oncol Nurs Forum 12*:56–64.

5. Ellerhorst-Ryan, J. M. Infection. In: S. Groenwald, M. H. Frogge, M. Goodman, G. H. Yarbo, eds. *Cancer Nursing: Principles and Practices,* Boston: Jones & Bartlett, 1993, pp. 558–574.

6. Morstyn, G., Campbell, L., Lieschke, G., et al (1991). Treatment of chemotherapy-induced neutropenia by subcutaneously administered granulocyte colony-stimulating factor with optimisation of dose and duration of therapy. *J Clin Oncol 7*:1554–1562.

7. Pizzo, P. A., Myers, J., Freifeld, A. G., et al Infections in the cancer patient. In: V. T. De Vita, S. Hellman, S. A. Rosenberg, eds. *Cancer: Principles and Practice of Oncology,* 4th ed. Philadelphia: J. B. Lippincott, 1993, pp. 2292–2337.

8. Rubin, M., Walsh, T. H., and Pizzo, P. A. Clinical approach to infections in the compromised host. In: R. H. Hoffman, E. J. Benz, S. J. Shattil, et al, eds. *Hematology: Basic Principles and Practice,* New York: Churchill Livingstone, 1991, pp. 1063–1114.

9. Wujcik, D. *Practical Questions & Answers for Oncology Nurses: Focus on G-CSF.* New York: Triclinica Communications Inc., 1992, pp. xx–xx.

10. Bandura, A. (1982). Self-efficacy mechanism in human agency. *Am Psych 37*:112–147.

11. Maibach, E., and Murphy, D. (1995). Self-efficacy in health promotion research and practice. *Health Ed Res 10*:37–50.

12. Mayer, D. K. (1990). Biotherapy: recent advances and nursing implications. *Nurs Clin North Am 25*:291–307.

30

Modeling Cost-Effective Use of Filgrastim (r-metHuG-CSF) in the Treatment of Patients with Early-Stage Breast Cancer Receiving Adjuvant Chemotherapy

Jeffrey H. Silber
University of Pennsylvania School of Medicine and Center for Outcomes Research, Children's Hospital of Philadelphia, Philadelphia, Pennsylvania

Moshe Fridman
Marshall School of Business, University of Southern California, Los Angeles, California

M. Haim Erder
Amgen Inc., Thousand Oaks, California

I. BACKGROUND AND OBJECTIVES

A. Introduction

Cost-effectiveness analysis has been increasingly used to evaluate the efficient use of health-care resources. A cost-effectiveness model has been developed to estimate the economic value of Filgrastim in the treatment of patients with early-stage breast cancer who are receiving adjuvant chemotherapy. The model incorporates clinical guidelines for chemotherapy dose reductions and Filgrastim use after episodes of severe neutropenia. The model compares the cost per year of life gained (CYLG) when Filgrastim is used according to the clinical guidelines with CYLG when Filgrastim is not used.

Until recently, most of the published data evaluating the economic value of Filgrastim were cost-minimization studies rather than cost-effectiveness studies (1–5). Cost-minimization studies assume that identical clinical outcomes can be achieved by all treatments being evaluated. For this reason, cost savings is the only criterion used in these studies to determine the relative value of alternative treatments.

These cost-minimization studies (1–5) have compared the costs of treating all patients with Filgrastim with the costs of treating no patients with Filgrastim (ie, the alternative treatment). The studies have attempted to determine if the acquisition and administration costs of Filgrastim can be offset by reduction in the use of other health-care resources such as days of hospitalization and days of antibiotic treatment. A major conclusion from these studies is that the acquisition and administration costs of Filgrastim can be offset when the risk of febrile neutropenia is >40% (6,7). This conclusion is based on the assumption that the use of Filgrastim does not affect any other health outcomes and that the impact of Filgrastim is limited to the use of health-care resources. These assumptions have been recognized by many, including authors of the Filgrastim cost-minimization studies as simplistic and insufficient. For example, Glaspy states, "there is clearly a need for further and more rigorous studies of the impact of these factors (hematopoietic growth factors) on quality of life and indirect non-health–care costs such as lost productivity of patients and their families" (8). Lyman et al (9) reported a revised economic analysis of the use of Filgrastim in which additional outcomes, such as indirect medical, non-medical, and intangible costs "incorporating quality of life considerations," were included in the analysis. This new analysis suggested that the costs associated with Filgrastim use are offset when the risk of febrile neutropenia is approximately 20%, rather than 40%.

To correctly account for the full benefit of Filgrastim use and to obtain a measure of Filgrastim's economic value, a cost-effectiveness analysis needs to be performed. Cost-effectiveness analysis is the appropriate method of economic evaluation when all treatments being compared result in the same outcome, ie, survival benefit, but are of different durations and are attained at different costs. In such cases, the value of a treatment can be determined by assessing the extra expenditures required to gain a year of life when the treatment is used. The treatment that requires fewer resources to gain a year of life is more cost effective.

Cost-effectiveness analysis has been applied recently to the economic analysis of Filgrastim. Dranitsaris et al (10) have modeled the cost effectiveness of Filgrastim in the treatment of patients with non-Hodgkin's lymphoma. Their model accounts for outcomes such as days of work gained as well as the traditional measure of days of hospitalization.

A second cost-effectiveness model developed by Silber et al (11) explores the economic implications of Filgrastim use in the treatment of patients with breast cancer where survival benefit may be affected by the dose of adjuvant chemotherapy delivered. They have shown that it is possible to predict an individual patient's risk of neutropenic complications during chemotherapy cycles 2 to 6 basing the prediction on absolute neutrophil counts (ANC) after the first cycle of chemotherapy (30). These researchers modeled the CYLG for Filgrastim use assuming that the clinical guidelines for Filgrastim use are followed and that 50% of patients with the highest risk of neutropenic complications receive Filgrastim. The CYLG in their model was approximately $34,000.

The guidelines for Filgrastim use developed by Silber et al (30) and used in the cost-effectiveness model (11) are based on patient-specific clinical data. Other guidelines have been proposed for the use of growth factors in conventional chemotherapy, but they are not based on patient-specific clinical information.

The American Society of Clinical Oncology (ASCO) promulgated guidelines for the prophylactic use of all growth factors, including Filgrastim (12). The guidelines recommend the use of Filgrastim for primary prophylaxis to reduce the risk of febrile neutropenia in those chemotherapy regimens where the risk of febrile neutropenia exceeds 40%. The guidelines coincided with the conclusions derived from the cost-minimization studies reviewed above, although they were not formally based on these economic analyses.

ASCO guidelines for the use of Filgrastim identify groups of patients classified by the expected toxicity of the chemotherapy they received. These guidelines do not offer patient-specific criteria to determine, for example, if some patients in a given cohort are at a higher risk than other members of the same cohort. Patients who are at a higher risk of febrile neutropenia than their cohorts would benefit from the administration of Filgrastim. Since ANC counts within each patient cohort are highly variable, patient-specific guidelines could be valuable.

The cost-effectiveness model we have developed explores the implication of Filgrastim use and assumes that a specific set of clinical guidelines is implemented. These clinical guidelines use patient-specific information from the patient's previous cycle of chemotherapy to manage administration of chemotherapy in the next cycle. The model compares the implications on CYLG of chemotherapy dose modifications with and without Filgrastim. For ease of presentation, and because this is a preliminary model which will require further development before implementation, we used some simple illustrative assumptions designed to demonstrate the model's potential. We hope that the preliminary results of this model will encourage further research to develop empirically based, patient-specific guidelines for the use of Filgrastim.

B. The Relationship Between Chemotherapy Treatment and Survival

The clinical literature indicates that the administration of full-dose of chemotherapy to patients with early-stage breast cancer who are receiving adjuvant chemotherapy may result in prolonged survival (13–16) (see also Chapter 8). Wood et al (13), in the largest published randomized prospective trial of this patient population, compared administration of three cumulative-dose chemotherapy levels to determine their effects on survival. The high-dose level of chemotherapy resulted in a statistically significant survival benefit at 3 years compared with the low-dose level in which only one-half the cumulative dose was delivered (for more detail see Section III.C). Some retrospective studies, although they are more susceptible to bias, also have shown benefit of full-dose chemotherapy (14). Some studies do not confirm the positive relationship between chemotherapy dose and survival (18). These studies, however, are generally small, non-randomized, retrospective studies that are difficult to interpret (19,20). A large prospective study and several large-scale retrospective studies confirm that full-dose chemotherapy is associated with prolonged survival. Our model assumes that to ensure the best survival benefit to patients with early-stage breast cancer, the goal of adjuvant chemotherapy treatment should be to administer a full dose of chemotherapy.

C. The Dilemma

Administration of full-dose chemotherapy can provide long-term survival benefits, but the cytotoxic effects of chemotherapy often are associated with short-term morbidity. Myelosuppressive chemotherapy can cause hematological toxicity that results in fever and infections requiring hospitalization. In some cases this condition, called febrile neutropenia, causes severe infections that can lead to prolonged hospitalizations or even death (21). To reduce the risk of febrile neutropenia, oncologists may choose to reduce the dose of chemotherapy or may delay the administration of chemotherapy to patients with febrile neutropenia or severe neutropenia. Wood et al (13) reported that 5% of all patients participating in their study required dose reductions of 10% or less. This percentage, however, was probably greater in the group of patients receiving high-dose chemotherapy. Data from oncologists in community-based practices indicate a much higher incidence of dose reduction and delay. Based on unpublished retrospective surveys in community-based practice, it is estimated that dose reduction occurs in 30% to 50% of all patients with breast cancer receiving chemotherapy (22).

While dose reduction or dose delay can decrease the risk of short-term morbidity, it may be associated with a decrease in long-term survival.

In some cases, dose delay or dose reductions are implemented only after patients experience their first episode of febrile neutropenia. This results not only in dose reduction but also in additional costs for treating the febrile neutropenia, as well as the subjective patient burden associated with treatment and recovery from an infection episode.

Filgrastim may be useful in the goal of delivering full-dose chemotherapy on time by reducing the risk of febrile neutropenia or severe neutropenia. Prophylactic use of Filgrastim has been shown to reduce the risk of febrile neutropenia. In a phase 3 trial, the risk of febrile neutropenia was reduced by 50% when Filgrastim was administered (23). In addition, patients administered Filgrastim had a higher ANC nadir and a shorter nadir duration than did patients not receiving Filgrastim. It appears, therefore, that the administration of Filgrastim could reduce the risk of severe neutropenia, the duration of severe neutropenia, and the incidence of febrile neutropenia. Consequently, administration of Filgrastim may facilitate the delivery of full-dose of chemotherapy by reducing the risks of these neutropenic events, which may lead to dose delays or does reductions (24).

However, not all patients require Filgrastim to receive full-dose chemotherapy. Some patients may receive chemotherapy and not experience any complications. Therefore, administering Filgrastim to all patients may cause unnecessary increases in health-care costs. To avoid extra expenditures, the simplest and most obvious solution is to adopt a "no Filgrastim" strategy. The decision to administer full-dose chemotherapy presents a clinical and economic dilemma in which short-term benefits (lower short-term morbidity and lower costs) are realized at the expense of long-term benefits (higher survival benefit). A strategy that offers no Filgrastim support to any patient has the short-term economic benefit of reducing Filgrastim cost and chemotherapy-related costs but can compromise long-term survival for patients. On the other hand, administration of Filgrastim to all patients may result in Filgrastim overutilization and higher costs (but increased survival). Health-care payers, health-care providers, and patients need to address this dilemma. Our model illustrates how economic modeling can assist in solving this dilemma.

D. A Proposed Strategy for Filgrastim Use

A reasonable solution to the dilemma of long-term versus short-term benefits would be to administer Filgrastim only to those patients who are at high risk for neutropenic events that could lead to dose delay or dose reductions. To apply this strategy, a set of conditions, ie, neutropenic complications, must be defined. When a neutropenic complication does occur, chemotherapy dose usually is reduced. Filgrastim could be used in this setting to reduce the risk of future neutropenic complications and decrease the amount of the dose reduction.

E. Study Objectives

For our study, we assumed that a set of clinical guidelines which define neutropenic complications and the rules for dose reductions and dose delays following neutropenic complications were adopted. Our objectives were to estimate the rate of neutropenic complications observed in each chemotherapy cycle and to estimate the CYLG when Filgrastim was used according to the guidelines compared with a "no-Filgrastim use" strategy.

II. ESTIMATING THE RISK OF NEUTROPENIC COMPLICATIONS BY CYCLE

A. Neutropenic Complications Defined

We defined a neutropenic complication as the occurrence of any one of the following events: a baseline ANC $<1.5 \times 10^9$/L or a baseline white cell count $<2.0 \times 10^9$/L (baseline being defined as the count on day 0 of cycle t before delivery of chemotherapy); nadir ANC $<0.25 \times 10^9$/L in cycle ($t - 1$); and/or a significant delay or reduction in cycle ($t - 1$). Delay was defined as at least 1 week, and reduction was defined as at least 15% of dose. Significant delay or reduction are those for which there was a prior low count of either an ANC $<1.5 \times 10^9$/L or white cell count $<2.0 \times 10^9$/L. Information about the first two criteria can be obtained during treatment and they define a level of neutropenia for which a change in chemotherapy is assumed. The third criterion captures the idea that some dose delays and dose reductions are performed to avoid severe neutropenia. For the purposes of this model, if a dose reduction already occurred and was preceded by a low neutrophil count, we assumed an additional dose reduction would be likely to occur in the next cycle. Therefore, we defined a dose reduction associated with low neutrophil counts as a neutropenic complication.

Our definition of neutropenic complication is illustrative, and alternative definitions at different ANC levels are possible. The important point for model construction is that a definition be provided and that the probability of a neutropenic complication can be estimated. These probabilities are needed to construct the cost-effectiveness model.

B. Retrospective Data Base

Our estimates of the frequency of neutropenic complications are based on a retrospective data set of patients with early-stage breast cancer who were receiving adjuvant chemotherapy at a time before Filgrastim was commonly used. This data set represents the complete practice of one oncologist for the years 1985 to 1993. All patients included in the sample were treated for non-metastatic, pathological stage I or II breast carcinoma. This was a non-random series of patients who met the following criteria:

Negative bone scans, normal liver function tests, and normal chest radiographs and diagnosis

Local treatment with either a modified mastectomy or conservative surgery (lumpectomy) with definitive radiation therapy

Initial treatment with standard adjuvant chemotherapy (defined below)

No chemotherapy within 2 years of the adjuvant chemotherapy for breast cancer

The study physician treated 120 patients for early-stage breast cancer. Clinical charts were available for 119 patients. Of the 119 patients, 1 received Filgrastim at the initiation of the study, data were missing for 13 patients, and 10 additional patients were excluded because they did not receive standard-dose chemotherapy as defined for the study. Thus, there were 95 patients remaining in the data set for whom neutropenic complications could be identified.

Patients included in this sample received four standard regimens of chemotherapy (25–27). Thirty-two patients received CMF (cyclophosphamide, methotrexate, 5-fluorouracil), 35 patients received CAF (cyclophosphamide, doxorubicin, 5-fluorouracil), 24 patients received radiotherapy plus CMF, and 4 patients received radiotherapy plus CAF.

C. Results: Estimated Probability of Neutropenic Complication by Cycle

Based on our definition of neutropenic complications, we identified patients who experienced a neutropenic complication in each cycle. Each patient could experience only one neutropenic complication in each cycle, but over the six cycles of chemotherapy some patients did experience more than one neutropenic complication. The frequency of patients in each cycle who experienced first, second, and third neutropenic complications is presented in **Table 1**. To simplify the model we did not calculate the probability of neutropenic complications beyond the third episode.

Of the 95 evaluable patients, 46 patients had a neutropenic complication between cycles 1 and 5, a rate of 48%. Of these 46 patients, 24 had a second neutropenic complication, a rate of approximately 52%. Of these 24 patients, 11 had a third neutropenic complication, a rate of 45%. If we

Table 1.　The Ratio of Patients with Neutropenic Complications[a] by Cycle.

Cycle number	First neutropenic complication	Second neutropenic complication	Third neutropenic complication
1	5/95	—	—
2	15/90	3/5	—
3	12/75	8/17	2/3
4	8/63	8/21	1/9
5	6/55	5/21	8/16
Total number of events	46	24	11

[a] A neutropenic complication is defined as the occurrence of any one of the following events: an absolute neutrophil count (ANC) $<1.5 \times 10^9/L$ or a white blood cell count $<2.0 \times 10^9/L$; a nadir ANC $<0.25 \times 10^9/L$ in the previous cycle; or a delay of at least 1 week or a dose reduction of at least 15% of planned dose in the previous cycle.
Source: Adapted from Ref. 30.

assume that neutropenic complications are associated with dose delays and dose reductions, then 46 of the 95 patients would have had their chemotherapy doses reduced for a total of 81 dose reductions. In practice the rate of dose delays and reductions may be higher or lower than the rate in our study and depend on local practice patterns, but our assumed clinical guidelines result in a rate of dose reductions or delays approximately the same as the rate observed in community practice (22). It is, therefore, interesting to explore the implications of our definition of neutropenic complications on the cost-effectiveness of Filgrastim use.

III.　COST-EFFECTIVENESS MODEL FOR FILGRASTIM USE

A.　Cost-Effectiveness: An Overview

It is assumed that two treatment strategies can be developed around the definition of neutropenic complications: no Filgrastim use or Filgrastim use after the occurrence of a neutropenic complication. This assumption raises several questions including what would be the expected effect of Filgrastim use on survival: What is the CYLG using Filgrastim compared with no Filgrastim use, and how does the CYLG for Filgrastim compare with CYLG of other treatments commonly used?

To answer these questions, a simplified cost-effectiveness model was constructed. Cost-effectiveness analysis is a variation of cost-benefit analysis (28). Originally, cost-benefit analyses were developed by economists to assess the economic value of public goods such as national defense and public parks. Since access to public goods is free, no market price is available to express the economic value (price) of these public goods. In a cost-benefit analysis, the challenge is to provide estimates of the economic value of public goods expressed in monetary terms. Once the monetary value of public goods is estimated, the costs and benefits of any public good can be compared with the costs and benefits derived from any other public goods. Assuming there are two investments in public good projects, the total costs associated with each project are denoted Ca and Cb and the total monetary return from the investment is denoted Ba and Bb. The marginal cost benefit of project "a" over "b" will be defined as:

$$\text{Marginal cost benefit} = \frac{\text{Ca} - \text{Cb}}{\text{Ba} - \text{Bb}}$$

Cost-benefit analysis has been applied to health care not because health care is a public good but because, like a public good, it has no real market price. Patients do not pay out of pocket for health services they receive. Most patients are insured and pay for health-care services through their monthly insurance premium and through copayments. Therefore price or out-of-pocket payments do not serve as a valid criterion of economic value in this setting. When the cost-benefit theory is applied to health care, there are no readily available health benefits measured in monetary terms. Thus, instead of measuring health benefits in monetary terms, economic evaluations of health-care treatments use measures of medical effectiveness as a measure of benefit. In this cost-effectiveness model, the years of life gained (YLG) are the measure of benefit.

The CYLG in one treatment can be compared with that of other treatments and will indicate the additional dollars that must be spent to obtain 1 year of survival. If the CYLG is low compared with other treatments, it might be concluded this treatment strategy provides good value per dollar of health-care resources used.

In the United States, patients with end-stage renal disease (ESRD) receive publicly funded care at a cost of approximately $50,000 per YLG (31). Since this investment in ESRD patients was approved by the nation's political institutions, it has been widely used as a bench mark against which the value of other treatments are evaluated. Thus some will argue that treatments that cost less than $50,000 per year of life saved provide good economic value because they provide a YLG at lower cost than society has revealed itself to be willing to pay.

Cost-effectiveness models should be constructed using the best data available, but sometimes assumptions must be used to construct a baseline case when data necessary to the model are not available. The results of any cost-effectiveness model must be carefully analyzed with respect to their dependency on the assumptions and the quality of the data used.

The results of the baseline case are subjected to sensitivity analysis in which assumptions used in the baseline case are varied, and the impact of the new assumptions on the estimates of the CYLG are analyzed. When the outcomes of the analysis are not affected to a great degree by the change in assumptions, it may be concluded that the model is robust with regard to the base case assumptions. When results are highly sensitive to changes in model assumptions, it may be concluded that the outcome of the study is highly dependent on the assumptions used. Model assumptions should also be evaluated with respect to other dimensions such as the extent to which they reflect reality and the likelihood that current conditions are likely to change.

B. Constructing a Cost-Effectiveness Model of Filgrastim Use

1. Treatment Strategies

Two strategies for the treatment of patients with early-stage breast cancer receiving adjuvant chemotherapy have been modeled. The first strategy assumed that Filgrastim would not be used at all during the administration of adjuvant chemotherapy, the no-Filgrastim strategy (NFS), but that chemotherapy dose would be reduced after neutropenic complications. The second strategy allowed the use of Filgrastim only to those patients who experienced a neutropenic complication in the previous cycle. This is the Filgrastim-use strategy (FUS). In the FUS strategy, no Filgrastim is administered in the first cycle of chemotherapy, and only patients who experience neutropenic complications after the administration of chemotherapy will receive Filgrastim. The chemotherapy dose reduction in the FUS will be smaller than the dose reduction in the NFS. It should be noted that this model addressed the use of Filgrastim after the first cycle chemotherapy and did not predict which patients would have a neutropenic complication in cycle 1. The use of Filgrastim as primary prophylaxis in cycle 1 was left to the judgment of the treating physician and was not addressed in this model.

2. Filgrastim's Marginal Cost-Effectiveness

To calculate the cost-effectiveness of Filgrastim, it was necessary to know what additional investment in the FUS would result in one additional YLG compared with the NFS. To answer this question, the model calculated the marginal health-care expenditures required to obtain one additional year of life gained using the FUS compared with the use of the NFS. The cost-effectiveness measure in this model depended on two variables that were calculated for each strategy: expected costs and expected survival.

3. Modeling Expected Survival

Expected survival was calculated as the probability of survival 3 years after chemotherapy multiplied by the duration of survival after 3 years. This definition was used because reliable clinical data on the probability of survival at 3 years of follow-up were available (13). To capture the full benefit of the treatment, it was important to account for survival after 3 years of follow-up. This was captured by the variable duration of survival. For simplicity, this variable took a constant value and was applied to all patients who survived, irrespective of the treatment strategy. A more complete analysis of survival duration can be found in Silber et al (11). The probability of survival, on the other hand, was assumed to depend on the expected cumulative dose of chemotherapy delivered to each patient. In our model, the cumulative dose of chemotherapy delivered to a patient was considered the sole predictor of the probability of survival at 3 years of follow-up.

4. Expected Cumulative Dose of Chemotherapy

We assumed that full-dose chemotherapy is associated with increased survival. For modeling purposes, we used the cumulative dose of chemotherapy as a measure of full dose. The expected cumulative dose of chemotherapy is a key variable in the model, it is a decision variable controlled by the oncologists, and it is affected by the use of Filgrastim. We assumed that the higher the cumulative dose of chemotherapy delivered to the patient, the higher the patient's 3-year probability of survival. We assumed that the relationship between cumulative dose and survival was linear (see Section III.C).

In our model, four variables affected the cumulative dose of chemotherapy delivered to each patient: the number of neutropenic complications experienced by each patient, the risk of neutropenic complication in each cycle, the adjustment in chemotherapy dose after the occurrence of each neutropenic complication, and the use of Filgrastim after the first neutropenic complication. In the model, the number of neutropenic complications experienced by the patient directly affected the cumulative dose of chemotherapy delivered. Some patients did not experience a neutropenic complication and received full-dose chemotherapy. Other patients experienced a neutropenic complication and had their dose reduced in the subsequent cycles. If they experienced an additional neutropenic complication (up to a maximum of three), additional chemotherapy dose reductions of larger magnitude were assumed.

Each patient is at risk for neutropenic complication in each cycle of chemotherapy. In this model, the group risk was expressed as a probability associated with each cycle of chemotherapy; this probability could change in each cycle. Different probabilities applied to the first, second, and third neutropenic complications. The higher the risk of a neutropenic complication, the lower the expected cumulative dose of chemotherapy that could be delivered. Because of the risk of neutropenic complications cumulative doses of chemotherapy, costs and survival were expressed as expectations in this model.

Reductions in chemotherapy dosing were assumed after the occurrence of a neutropenic complication. The magnitude of the change in chemotherapy dosing depended on whether this was the first, second, or third neutropenic complication. In our model, neutropenic complications caused the reduction of doses, and patients experiencing greater numbers of neutropenic complications had larger reductions in their doses of chemotherapy.

The magnitude of the change in chemotherapy dosing after a neutropenic complication depended on our assumptions regarding Filgrastim use. The model assumed that the use of Filgrastim was associated with a smaller reduction in the dose of chemotherapy and with a lower probability for the next neutropenic complication. The use of Filgrastim (FUS) resulted in a lower reduction in doses of chemotherapy compared with no use of Filgrastim (NFS).

5.   Modeling Expected Costs

Our cost model considers the costs of chemotherapy, Filgrastim, and hospitalization for febrile neutropenia. The cost of Filgrastim depended on the costs per dose and the number of doses used. Assumptions regarding the use and costs of Filgrastim applied only to FUS. It was assumed that Filgrastim would be used in cycles subsequent to the first neutropenic complication for as long as chemotherapy was given. For example, a patient who experienced a neutropenic complication in cycle 2 only received chemotherapy and Filgrastim in all subsequent four cycles.

The model calculated the expected dose delivered, and the cost of chemotherapy was calculated for the expected dose delivered. In this analysis, it was assumed that the maximum chemotherapy dose consisted of six cycles of chemotherapy.

Patients who experience febrile neutropenia are usually hospitalized and treated with parenteral antibiotics. Patients with early-stage breast cancer usually have a low incidence of febrile neutropenia, so the impact of hospitalization in this analysis was not large. Silber et al (11) have observed that their models were not sensitive to assumptions about hospitalization costs. To simplify the model and to ensure that our results were conservative, we did not include the costs of hospitalization in the base case.

C. Model Assumptions

It is necessary to describe the data and assumptions used in each of the components of the cost-effectiveness model and to document their sources. We will summarize assumptions affecting expected survival and expected costs.

Assumptions affecting expected survival include the dosing adjustment after each neutropenic complication in each strategy, the probability of neutropenic complications in each strategy, and the relationship between the cumulative dose of chemotherapy and the expected survival (ie, probability of survival at 3 years and the duration of survival).

It was assumed that chemotherapy dose would be reduced with each neutropenic complication. For each of the treatment strategies (NFS and FUS), several stages in chemotherapy dosing were considered (**Table 2**).

Table 2. Chemotherapy Dosing Rules After Neutropenic Complications by Complication Stage.

Stage	No Filgrastim	Filgrastim use
Cycle 1 to first neutropenic complication	Full dose of chemotherapy	Full dose of chemotherapy, but no Filgrastim
First neutropenic complication to second neutropenic complication	75% of full dose of chemotherapy	90% of full dose of chemotherapy plus Filgrastim
Second neutropenic complication to third neutropenic complication	50% of full dose of chemotherapy	50% of full dose of chemotherapy plus Filgrastim
After third neutropenic complication	Stop chemotherapy	Stop chemotherapy and stop Filgrastim

In the no-Filgrastim strategy, Filgrastim is never given and dose reduction is the only response to a neutropenic complication. In the Filgrastim-use strategy, Filgrastim is used in conjunction with chemotherapy dose reduction after a neutropenic complication.

Table 3. Probability[a] of Neutropenic Complications by Cycle for the No-Filgrastim Strategy.

Cycle number	First neutropenic complication	Second neutropenic complication	Third neutropenic complication
1	0.0263	0	0
2	0.0833	0.3	0
3	0.0800	0.2353	0.3333
4	0.0635	0.1905	0.0556
5	0.0545	0.1190	0.2500

[a] The probabilities reflect the likelihood of a neutropenic complication assessed at the end of each chemotherapy cycle.

Our assumptions regarding the probability of neutropenic complications in each cycle (**Table 3** and **Table 4**) were based on frequency of neutropenic events in our retrospective database (Table 1). We used our retrospective data base as our best estimate of neutropenic complications for the NFS (Table 3). The rate of neutropenic complications in the FUS strategy is reduced by 50% based on Crawford et al (23). We assumed that patients could experience a maximum of three neutropenic complications during the six cycles of chemotherapy. This assumption was applied to both treatment strategies.

The benefit from chemotherapy was defined as the likelihood for disease-free survival at 3 years of follow-up. Based on Wood et al (13), it was assumed that patients receiving a total cumulative dose equivalent to six cycles of full-dose chemotherapy would have a 74% chance of disease-free survival. (The disease-free survival was reported in the full-dose group

Table 4. Probability of Second and Third Neutropenic Complication by Cycle for the Filgrastim-Use Strategy.

Cycle number	First neutropenic complication	Second neutropenic complication	Third neutropenic complication
1	0.0263	—	—
2	0.0833	0.15	—
3	0.0800	0.117	0.1666
4	0.0635	0.0952	0.0278
5	0.0545	0.0595	0.125

in Wood et al [13].) A further assumption is that patients receiving a total cumulative dose equivalent to three full cycles of chemotherapy will have a 63% chance of disease-free survival. (The disease-free survival reported in the low-dose group in Wood [13].) For doses between three cycles of chemotherapy and six cycles of chemotherapy, it was assumed that the probability of 3 years of survival is linearly related to cumulative dose. We did not analyze other assumptions for the relationship between cumulative dose and survival benefit. (For a complete analysis of this relationship, see Silber et al [11].) It was assumed that any patient disease-free at 3 years of follow-up would survive an additional 10 years, irrespective of the treatment. The 10-year survival duration was assumed to apply to a 55-year-old woman.

In the baseline model, it was assumed that each cycle of Filgrastim administration would be 10 injections per cycle as specified in the package insert. Patients could receive up to five cycles of Filgrastim (no Filgrastim in cycle 1 of a six-cycle regimen). The cost of Filgrastim would be $150 per injection based on 1995 Red Book (29) prices.

Expenditures on chemotherapy were modeled for cyclophosphamide, doxorubicin, and 5-fluorouracil (CAF). Total expenditures on CAF for one cycle was estimated to be $640/cycle (29). The costs of CMF are much lower. Thus, our cost estimates of chemotherapy are conservative with respect to the comparison between the NFS and the FUS.

IV. RESULTS

A. Baseline Case Analysis

The baseline case analysis indicated that the expected survival benefit in the NFS was 7.1817 years compared with 7.3392 years in the FUS. The incremental benefit was 0.1575 years, a difference of 57.5 days. To obtain this benefit, the expected cost in the FUS was $4,922 compared with the expected cost in the NFS $3,449, resulting in an incremental cost of $1,473. Therefore, the incremental cost per life year saved was $9,355. This cost-effectiveness can be compared with other medical interventions (**Table 5**). The FUS is at the lower range of the therapies and well within the $50,000 mark.

B. Sensitivity Analysis

In the baseline case, it was assumed that Filgrastim would be administered for 10 days. In clinical practice, however, the common dosing schedule is

Table 5.   Costs[a] Per Year of Life Gained in Selected Medical Interventions.

Item	Cost	Reference
Use of Filgrastim in patients with breast cancer receiving adjuvant chemotherapy	$9,355	This study
Screen blood donors for HIV	$13,544	31
Annual mammography and breast exam (vs just exam) for woman age 40–64	$17,089	31
Lavastatin for men age 55–64 with heart disease and ≥250 mg/dL	$19,427	31
Two-vessel coronary artery bypass graft surgery (vs. medical management)	$27,959	31
AZT for people with AIDS	$26,261	31
Antihypertension drugs for patients age 40 and 95–104 mmHg	$32,030	31
Estrogen for menopausal woman age 50	$42,262	31
Dialysis for end-stage renal disease	$50,542	31

[a] In U.S. dollars for 1993.

between 7 and 9 days. Therefore, a sensitivity analysis was conducted allowing for 8 days of Filgrastim use during each cycle.

This assumption changes only the expected costs in the FUS. Total expected costs with 8 days of Filgrastim use per cycle are $4,684 instead of $4,922, resulting in $7,845 CYLG. A 20% reduction in Filgrastim use resulted in a 16% reduction in CYLG.

The costs of chemotherapy varied from a low of $100/cycle to a high of $1,000/cycle. In the NFS, the total expected costs were $539 and $5,388, respectively, rather than the base case expected cost of $3,449. In the FUS, costs were $1,773 and $7,022, respectively, compared with $4,992 for the base case expected cost. When chemotherapy costs are $100 per cycle the incremental CYLG is $7,833 compared with $10,370 when the costs are $1,000 per cycle. Reduction in chemotherapy costs of 85% results in a decrease of CYLG of 16%, while an increase in chemotherapy costs of 56% resulted in an increase in CYLG of 11%.

In the base case, duration of survival was assumed to be 10 years, and the sensitivity analysis investigated the impact of 5 years versus 15 years of survival. In the sensitivity analysis, only the expected survival benefit changed. In the FUS, expected survival at 5 years and 15 years was 3.6696 years and 11.0088 years, respectively, compared with 7.3392 in the base case. In the NFS, expected survival at 5 years and 15 years was 3.5908 years and 10.7725 years, respectively. These changes in expected survival had an impact on CYLG. Calculating for 5 years of survival, the CYLG were $18,710 compared with $9,355 in the base case. When 15 years of survival are assumed, the CYLG is $6,237. When survival duration increases by 50%, CYLG decreases by 33%; when the survival benefit decreases by 50%, however, CYLG increases by 100%.

The probability of survival associated with the appropriate treatment strategy was varied in the sensitivity analysis from 72% to 76%. In the NFS, reducing the probability of 3 years of survival reduced the expected survival benefit to 7.0214 compared with 7.1502 in the FUS, resulting in a marginal CYLG of $11,434. If the survival benefit had increased by 2%, the CYLG would been reduced to $7,916. When the probability of survival was increased by 2.7%, the CYLG was reduced by 15%; when the probability of survival decreased by 2.7%, the CYLG increased by 22.2%.

Our analysis assumed that the use of Filgrastim would reduce the risk of the next neutropenic complications by 50%. In the sensitivity analysis, this probability was varied to 25% and 75%. When the effectiveness of Filgrastim was assumed to prevent a second neutropenic complication in 25% of the patients, the expected survival benefit associated with this level of effectiveness was 7.2896 at a CYLG of $17,112. When Filgrastim was assumed to prevent a second neutropenic complication in 75% of the patients, the expected survival benefit was 7.3765 years at the CYLG of $5,049.

In this sensitivity analysis, a 50% decline in the effectiveness of Filgrastim resulted in a 82% increase in CYLG; a 50% increase in the effectiveness of Filgrastim resulted in 46% decrease in CYLG.

V. DISCUSSION

Increased financial pressures on health-care providers and payers has increased the interest in development of rational criteria to assist in choosing among competing health-care treatment strategies. Cost-effectiveness analysis can be a useful tool to aid clinical decision making when the effectiveness of alternative treatment strategies can be clearly defined. In addition, disease-stage management techniques and clinical guidelines have been developed to help practitioners implement treatment strategies that ensure proper care and conserve resources. Rarely have clinical guidelines and cost-effectiveness methods been integrated to improve decision making in the provision of health care.

Our model incorporated a set of clinical guidelines into a cost-effectiveness model, demonstrating the feasibility of using clinical information to improve health-care and resource allocation decisions. The clinical guidelines used in the analysis help identify those patients treated for early-stage breast cancer who are at higher risk for neutropenic complications. If Filgrastim is administered only to patients at risk, the analysis suggests that the CYLG by use of Filgrastim in accordance to the proposed guidelines is $9,355. This cost compares well with other acceptable and commonly used treatments and is well below the $50,000 mark for socially acceptable modalities for survival-prolonging treatments.

Our sensitivity analysis indicates that our results are most sensitive to two factors: the probability of survival at 3 years and the survival duration. While our model shows proportional sensitivity particularly to our survival assumption, this sensitivity is quite limited in its upper range. In particular, our model is quite robust to the $50,000 CYLG criteria. For our model to increase the CYLG to $50,000 or more will require making assumptions which will be considered unlikely. For example, survival duration would need to be reduced to 1.8 years from 10. If we assume the probability of survival at 3 years to be 68% rather than 74%, a reduction of more than 50% in benefit compared with half dose (63% at 3 years), the CYLG will not reach $50,000—it will be $20,581. If the efficacy of Filgrastim was 0.2 instead of 0.5 as assumed in the base case, CYLG will not exceed $50,000, but will be $43,382. The implication of this analysis is that Filgrastim use in this setting is most likely to provide good economic value. The risk that CYLG will be extremely high, ie, above $100,000, is very low.

Interestingly, the economic value of Filgrastim use in this model does not depend on savings in hospital costs as in the early cost-minimization studies. The economic value is due to patient benefit and to the guidelines governing the use of Filgrastim. Our model did not account for the costs of recurrence. However, it is unlikely that these costs would have affected the results of the base case. Silber et al (11) accounted for recurrence costs and found the model to be insensitive to these costs.

This model has some limitations that require consideration. First, the clinical guidelines used were assumed. While they provide useful information because they approximate the rate of dose delay and dose reduction observed in the community, the model would benefit from careful development of clinical guidelines based on clinical evidence. Silber et al (30) have developed a predictive model that could be used for the development of such clinical guidelines.

The second limitation relates to the estimates of neutropenic complications. Our estimates of the risks of neutropenic complications are based on a relatively small retrospective sample reflecting one physician's experience. This data set should be expanded and validated; validation studies are planned.

Our survival probabilities reflect 3-year survival. Data on 5-year survival are expected shortly, and the model should be re-evaluated when 5-

year survival data are available. Most importantly, we have assumed a linear relationship between cumulative dose of chemotherapy and survival probabilities at 3 years after treatment. The relationship between cumulative dose and survival is not well understood. This relationship should be further explored. The better the survival data, the more convincing will the model be.

VI. CONCLUSIONS

In this preliminary analysis we demonstrated that it is feasible to develop cost-effective guidelines for the use of Filgrastim. We have shown that development of such guidelines is not only feasible but is also likely to result in cost-effective use of Filgrastim and improved treatment outcomes. The analysis also indicates that Filgrastim may have a significant role in the treatment of patients with early-stage breast cancer. Further work to develop and validate the model appears warranted to implement a better treatment strategy for patients with early-stage breast cancer. Work on guideline development should be based on patient-specific clinical information that can be attained routinely during treatment.

REFERENCES

1. Faulds, D., Lewis, N. J., and Milne, R. J. (1992). Recombinant granulocyte colony-stimulating factor (rG-CSF). Pharmacoeconomic considerations in chemotherapy-induced neutropenia. *PharmacoEconomics 1*:231–249.
2. Leese, B., Collin, R., and Clark, D. J. (1994). The costs of treating febrile neutropenia in patients with malignant blood disorders. *PharmacoEconomics 6*:233–239.
3. Glaspy, J. (1994). Economic effect of myeloid growth factors on cancer treatment. *Clin Immunother 2*:192–205.
4. Uyl-de Groot, C. A., Richel, D. J., and Rutten, F. F. (1994). Peripheral blood progenitor cell transplantation mobilized by r-metHuG-CSF (Filgrastim); a less costly alternative to autologous bone marrow transplantation. *Eur J Cancer 30A*:1631–1635.
5. Smith, T. J., Hillner, B. E., Schmitz, N., et al (1997). Economic analysis of a randomized clinical trial to compare Filgrastim-mobilized peripheral-blood progenitor-cell transplantation and autologous bone marrow transplantation in patients with Hodgkin's and non-Hodgkin's lymphoma. *J Clin Oncol 15*:5–10.
6. Glaspy, J. A., Bleecker, G., Crawford, J., Stoller, R., and Strauss, M. (1993). The impact of therapy with Filgrastim (recombinant granulocyte colony-stimulating factor) on the health care costs associated with cancer chemotherapy. *Eur J Cancer 29A*:S23–S30.

7. Lyman, G. H., Lyman, C. G., Sanderson, R. A., and Balducci, L. (1993). Decision analysis of hematopoietic growth factor use in patients receiving cancer chemotherapy. *J Natl Cancer Inst 85*:488–493.

8. Glaspy, J. A. (1995). Economic outcomes associated with the use of hematopoietic growth factors. *Oncology 9*:93–105.

9. Lyman, G. H., Kuderer, N. M., and Balducci, L. (1996). Thresholds for the use of recombinant human colony stimulating factors (CSFs) based on revised cost estimates incorporating indirect medical, nonmedical and intangible cost considerations. *Blood 88*:346a (abstr 1373).

10. Dranitsaris, G., Altmayer, C., and Quirt, I. (1997). Cost-benefit analysis of prophylactic granulocyte colony-stimulating factor during CHOP antineoplastic therapy for non-Hodgkin's lymphoma. *PharmacoEconomics 11*:566–577.

11. Silber, J. H., Fridman, M., Shpilsky, A., et al (1997). The cost-effectiveness of G-CSF in early stage breast cancer. *Proc Am Soc Clin Oncol 16*:412a (abstr 1470).

12. American Society of Clinical Oncology. (1994). Recommendations for the use of hematopoietic colony-stimulating factors: evidence-based, clinical practice guidelines. *J Clin Oncol 12*:2471–2508.

13. Wood, W. C., Budman, D. R., Korzun, A. H., et al (1994). Dose and dose intensity of adjuvant chemotherapy for stage II, node-positive breast carcinoma. *N Engl J Med 331*:1253–1259.

14. Bonadonna, G., Valagussa, P., Moliterni, A., Zambetti, M., and Brambilla, C. (1995). Adjuvant cyclophosphamide, methotrexate, and fluorouracil in node-positive breast cancer: the results of 20 years of follow-up. *N Engl J Med 332*:901–906.

15. Bonadonna, G. and Valagussa, P. (1981). Dose-response effect of adjuvant chemotherapy in breast cancer. *N Engl J Med 304*:10–15.

16. Geller, N. L., Hakes, T. B., Petroni, G. R., Currie, V., and Kaufman, R. (1990). Association of disease-free survival and percent of ideal dose in adjuvant breast chemotherapy. *Cancer 66*:1678–1684.

17. Bonadonna, G. and Valagussa, P. (1996). Primary chemotherapy in operable breast cancer. *Semin Oncol 23*:464–474.

18. Ludwig Breast Cancer Study Group. (1984). Randomized trial of chemoendocrine therapy, endocrine therapy, and mastectomy alone in postmenopausal patients with operable breast cancer and axillary node metastasis. *Lancet 1*:1256–1260.

19. Churchill, D. N., Lemon, B. C., and Torrance, G. W. (1984). A cost-effectiveness analysis of continuous ambulatory peritoneal dialysis and hospital hemodialysis. *Med Decision Making 4*:489–500.

20. Redmond, C., Fisher, B., and Wiegand, H. S. (1983). The methodologic dilemma in retrospectively correlating the amount of chemotherapy received in adjuvant therapy protocols with disease-free survival. *Cancer Treat Rep 67*:519–526.

21. Pizzo, P. A., Hathorn, J. W., Hiemenz, J., et al (1986). A randomized trial comparing ceftazidime alone with combination antibiotic therapy in cancer patients with fever and neutropenia. *N Engl J Med 315*:552–558.

22. Fine, S. and Erder, M. H. (1997). An analysis of the incidence of neutropenic complications in chemotherapy treated patients with HD, NHL, or breast cancer. [Unpublished data on file.]

23. Crawford, J., Ozer, H., Stoller, R., et al (1991). Reduction by granulocyte colony-stimulating factor of fever and neutropenia induced by chemotherapy in patients with small-cell lung cancer. *N Engl J Med 325:*164–170.

24. de Graaf, H., Willemse, P. H., Bong, S. B., et al (1996). Dose intensity of standard adjuvant CMF with granulocyte colony-stimulating factor for premenopausal patients with node-positive breast cancer. *Oncology 53:*289–294.

25. Fisher, B., Brown, A. M., Dimitrov, N. V., et al (1990). Two months of doxorubicin-cyclophosphamide with and without interval reinduction therapy compared with 6 months of cyclophosphamide, methotrexate, and fluorouracil in positive-node breast cancer patients with tamoxifen-nonresponsive tumors: results from the National Surgical Adjuvant Breast and Bowel Project B-15. *J Clin Oncol 8:*1483–1496.

26. Tancini, G., Bonadonna, G., Valagussa, P., Marchini, S., and Veronesi, U. (1983). Adjuvant CMF in breast cancer: comparative 5-year results of 12 versus 6 cycles. *J Clin Oncol 1:*2–10.

27. Aisner, J., Weinberg, V., Perloff, M., et al (1987). Chemotherapy versus chemoimmunotherapy (CAF v CAFVP v CMF each ± MER) for metastatic carcinoma of the breast: a CALGB study. *J Clin Oncol 5:*1523–1533.

28. Mishan, E. J. *Cost Benefit Analysis,* 4th ed. London: Unwin Hyman, 1988, pp. 110–114.

29. Sifton, D. W. *1995 Red Book: Pharmacy's Fundamental Reference.* New Jersey: Medical Economic Data, 1995, p. 371.

30. Silber, J. H., Fridman, M., DiPaola, R. S., Erder, M. H., Pauly, M. V., and Fox, K. R. (1995). Prediction of neutropenia, dose reduction or delay in breast cancer therapy. *Proc Am Soc Clin Oncol 14:*534 (abstr 1764).

31. Tengs, T. O., Adams, M. E., Pliskin, J. S., et al (1994). Five-hundred lifesaving interventions and their cost-effectiveness. Center for Health Policy Research and Education, Duke University [Unpublished manuscript].

Glossary

5-FU	5-fluorouracil
6-MP	6-mercaptopurine
ABMT	autologous bone marrow transplantation
ABV	chemotherapy with doxorubicin, bleomycin, and vincristine
ABVD	chemotherapy with doxorubicin, bleomycin, vinblastine, and dacarbazine (DTIC)
ACT-D	actinomycin-D
ADCC	antibody-dependent cellular cytotoxicity
AFM	chemotherapy with doxorubicin, 5-FU, and methotrexate
AGA	appropriate for gestational age
AIDS	acquired immune deficiency syndrome
AIN	autoimmune neutropenia
ALL	acute lymphoblastic leukemia
allo-BMT	allogeneic bone marrow transplantation
AML	acute myeloid leukemia
ANC	absolute neutrophil count
ANTU	α-naphthylthiourea
Ara-C	cytarabine

ARDS	adult respiratory distress syndrome
AST	aspartate transaminase (SGOT)
ATP	adenosine triphosphate
ATRA	all-*trans* retinoic acid
AUC	area under the curve
b	bovine
BACOP	chemotherapy with bleomycin, doxorubicin, cyclophosphamide, vincristine, and prednisone
BEAC	chemotherapy with carmustine (BCNU), etoposide, cytarabine, and cyclophosphamide
BEAM	chemotherapy with carmustine (BCNU), etoposide, cytarabine, and melphalan
BEP	chemotherapy with bleomycin, etoposide, and cisplatin
BFC-E	erythroid burst-forming cell
BFC-MK	megakaryocyte burst-forming cell
BID	twice daily
BMT	bone marrow transplantation
BOP/BEP	chemotherapy with bleomycin, vincristine, and cisplatin followed by bleomycin, etoposide, and cisplatin
BOP/VIP	chemotherapy with bleomycin, vincristine, and cisplatin followed by etoposide, ifosfamide, and cisplatin
BPI	bactericidal permeability increasing (protein)
BRM	biological response modifier
c	canine
CAE	chemotherapy with cyclophosphamide, doxorubicin, and etoposide
CAF	chemotherapy with cyclophosphamide, doxorubicin, and 5-FU

CAFC	cobblestone area-forming cell
CALGB	Cancer and Leukemia Group B
CAP	community-acquired pneumonia
CAP-BOP	chemotherapy with cyclophosphamide, doxorubicin, procarbazine, bleomycin, vincristine, and prednisone
CAV	chemotherapy with cyclophosphamide, doxorubicin, and vincristine
CDC	Centers for Disease Control and Prevention
CDE	chemotherapy with cyclophosphamide, etoposide, and doxorubicin
CEOP	chemotherapy with cyclophosphamide, epirubicin, vincristine, and prednisone
CFP	chemotherapy with cyclophosphamide, 5-FU, and prednisone
CHAP	chemotherapy with cyclophosphamide, hexamethylmelamine, doxorubicin, and cisplatin
CHOP	chemotherapy with cyclophosphamide, doxorubicin, vincristine, and prednisone
CLL	chronic lymphocytic leukemia
CMC	chemotherapy with cyclophosphamide, methotrexate, and CNU
CMF	chemotherapy with cyclophosphamide, methotrexate, and 5-FU
CMFVP	chemotherapy with cyclophosphamide, methotrexate, 5-FU, vincristine, and prednisone
CML	chronic myelogenous leukemia
CMV	cytomegalovirus
CNU	chloroethylnitrosourea
CNV	chemotherapy with cyclophosphamide, mitoxantrone, and vincristine
CNVp	chemotherapy with cyclophosphamide, mitoxantrone, and etoposide

CODE	chemotherapy with cisplatin, vincristine, doxorubicin, and etoposide
COMLA	chemotherapy with cyclophosphamide, vincristine, methotrexate, calcium leucovorin, and cytarabine
COP BLAM	chemotherapy with cyclophosphamide, vincristine, prednisone, bleomycin, doxorubicin, and procarbazine
COR	circulating opsonin receptor expression
CPB	chemotherapy with cyclophosphamide, cisplatin, and carmustine
CSF	colony-stimulating factor
CT	computed tomography
CTX	cyclophosphamide (Cytoxan®)
CVB	chemotherapy with cyclophosphamide, etoposide, and carmustine (BCNU)
CYLG	cost per year of life gained
ddl	didanosine
Dexa-BEAM	chemotherapy with dexamethasone, carmustine, etoposide, cytarabine, and melphalan
DHAP	chemotherapy with dexamethasone, cytarabine, and cisplatin
DIC	disseminated intravascular coagulation
DMAC	disseminated *Mycobacterium avium* complex
DTNB	5,5′-dithiobis (2-nitrobenzoic acid)
E-BFC	erythroid burst-forming cell
EBMT	European Group for Blood and Marrow Transplantation
EBV	Epstein-Barr virus
E-CFC	erythroid colony-forming cell
ECMO	extracorporeal membrane oxygenation

ECOG	Eastern Cooperative Oncology Group
EGF	epidermal growth factor
ELISA	enzyme-linked immunosorbent assay
EORTC	European Organization for Research of Treatment of Cancer
EOS-CFC	eosinophil colony-forming cell
EP	chemotherapy with etoposide and cisplatin
EPO	erythropoietin
ES	embryonal stem (cell)
ESRD	end-stage renal disease
EVB	chemotherapy with epirubicin, vinblastine, and bleomycin
FAB	French-American-British Cooperative Group
FACS	fluorescent-activated cell sorting
FDA	Food and Drug Administration
FMLP	N-formyl-methionyl-leucyl-phenylalanine
G-CSF	granulocyte colony-stimulating factor
G-MP	G-mercaptopurine
GIMEMA	Gruppo Italiano Malattie Ematologiche dell' Adulto
GM-CFC	granulocyte-macrophage colony-forming cell
GM-CSF	granulocyte-macrophage colony-stimulating factor
GPI	glycosyl-phosphatidyl-inositol
GRR	IFN-γ response region
GVHD	graft-versus-host disease
GVL	graft-versus-leukemia
HD	Hodgkin's disease
HDS	high-dose sequential chemotherapy

HELLP	hemolysis, elevated liver enzymes, and low platelet count (syndrome)
HEPA	high-efficiency particulate air (filter)
hGH	human growth hormone
HIV	human immunodeficiency virus
HPP-CFC	high proliferative-potential colony-forming cell
HPRT	hypoxanthine phosphoribosyltransferase
HSA	human serum albumin
Hu	human
IAV	influenza A virus
IBMTR	International Bone Marrow Transplant Registry
ICU	intensive care unit
IFN	interferon
IFX	ifosfamide
IL	interleukin
IP	intraperitoneal, intraperitoneally
IPSS	International Prognostic Scoring System
ITAM	immunoreceptor tyrosine-based activation motif
itim	immunoreceptor tyrosine-based inhibition motif
IU	international unit
IV	intravenous, intravenously
IVIG	intravenous immunoglobulin
IVP	intravenous push
kbp	kilobase pairs
kD	kilodalton
kir	killer-cell inhibitory receptor
KSHV	Kaposi's sarcoma-associated herpes virus
LAF	lymphocyte-activating factor

LAK	lymphokine-activated killer
LBP	lipopolysaccharide-binding protein
L-CFC	leukemia colony-forming cell
LDH	lactic dehydrogenase
LIF	leukemia-inhibiting factor
L-PAM	L-phenylalanine mustard (melaphalan)
LPS	lipopolysaccharide, an endotoxin
LTA	lipoterchoic acid
LTB_4	leukotriene B_4
LT-CIC	long-term culture-initiating cell
MAC	*Mycobacterium avium* complex
m-BACOD	chemotherapy with methotrexate, calcium leucovorin, bleomycin, doxorubicin, cyclophosphamide, vincristine, and dexamethasone
M-CFC	macrophage colony-forming cell
M-CSF	macrophage colony-stimulating factor (aka: CSF-1)
MACOP-B	chemotherapy with methotrexate, doxorubicin, cyclophosphamide, vincristine, prednisone, and bleomycin
MDP	muramyl dipeptide
MDR	multidrug resistance
MDS	myelodysplastic syndromes
MHc	major histocompatibility class
MIP	macrophage inflammatory protein
MIRR	multichain immune recognition receptors
MIU	million international unit
Mix-CFC	mixed colony-forming cell
MK-BFC	megakaryocyte burst-forming cell
MK-CFC	megakaryocyte colony-forming cell

MM	multiple myeloma
MNC	mononuclear leukocyte
MOPP	chemotherapy with mechlorethamine, vincristine, procarbazine, and prednisone
MOR	maximum opsonin receptor expression
MRI	magnetic resonance imaging
MRSA	methicillin-resistant *S. aureus*
MTb	*Mycobacterium tuberculosis*
MTD	maximal tolerated dose
NA	neutrophil antigen
NCI	National Cancer Institute
NEC	necrotizing enterocolitis
NED	no evidence of disease
NHL	non-Hodgkin's lymphoma
NK	natural killer
NMDP	National Marrow Donor Program
NSABP	National Surgical Adjuvant Breast and Bowel Project
NSCLC	non-small–cell lung cancer
PΔ	plastic-adherent delta assay
PBMC	peripheral blood mononuclear cell
PBPC	peripheral blood progenitor cell
PBPCT	peripheral blood progenitor cell transplantation
PCP	*Pneumocystis carinii* pneumonia
PCR	polymerase chain reaction
PCV	chemotherapy with procarbazine, lomustine, and vincristine
PEC	chemotherapy with cisplatin, epidoxorubicin, and cyclophosphamide

PEG	poly[ethylene glycol]
PGK	pyruvate glycerol kinase
Pgp	P-glycoprotein
PHA	phytohemagglutinin antigen
PMA	phorbol ester
PMN	polymorphonuclear leukocyte (ie, neutrophil)
POMB-ACE	chemotherapy with cisplatin, vincristine, methotrexate, bleomycin, actinomycin-D, cyclophosphamide, and etoposide
ProMACE-CytaBOM	chemotherapy with cyclophosphamide, doxorubicin, etoposide, prednisone, bleomycin, cytarabine, methotrexate, and leucovorin
PVB	chemotherapy with cisplatin, vinblastine, and bleomycin
P-VEBEC	chemotherapy with prednisone, vinblastine, etoposide, bleomycin, and cyclophosphamide
PVeBV	chemotherapy of high-dose cisplatin, vinblastine, etoposide, and bleomycin sulfate
r	recombinant
RA	refractory anemia
RAEB	refractory anemia with excess blasts
RAEB-T	refractory anemia with excess blasts in transformation
RARS	refractory anemia with ringed sideroblasts
rb	recombinant bovine
rc	recombinant canine
RDI	relative dose intensity
RFLP	restriction fragment-length polymorphism
rHu	recombinant human
rr	recombinant rodent
RT-PCR	reverse transcription polymerase chain reaction

SC	subcutaneous, subcutaneously
S-CFC	spleen colony-forming cells
SCLC	small-cell lung cancer
SCF	stem cell factor
SCN	severe chronic neutropenia
SCNIR	Severe Chronic Neutropenia International Registry
SEB	staphylococcal exotoxin B
SECSG	Southeastern Cancer Study Group
SGA	small for gestational age
SLE	systemic lupus erythematosus
SWOG	Southwest Oncology Group
TBI	total body irradiation
TCR	T-cell receptor
TGF	transforming-growth factor; tumor growth factor
TMN	tumor, node, metastasis classification
TMP-SMX	trimethoprim-sulfamethoxazole
TNF	tumor necrosis factor
TPO	thrombopoietin
TRM	time to resolution of morbidity
VAB-6/EP	chemotherapy with cisplatin, vinblastine, bleomycin, cyclophosphamide, and dactinomycin followed by etoposide and cisplatin
VAPEC-B	chemotherapy with vincristine, doxorubicin, prednisone, etoposide, cytotoxan, and bleomycin
VATH	chemotherapy with vinblastine, doxorubicin, thiotepa, and fluoxymesterone
VBAP	chemotherapy with vincristine, carmustine, doxorubicin, and prednisone

VeIP	chemotherapy with vinblastine, ifosfamide, and cisplatin
VLB	vinblastine
VLBW	very low birth weight
VMCP	chemotherapy with vincristine, melphalen, cyclophosphamide, and prednisone
ZDV	zidovudine

Index

ABMT (*see* Bone marrow
    transplantation)
Abnormalities
    biochemistry, 599
    chromosomal, 139–141,
        491–492, 512
    cytogenetic, 139
    electrolyte, 108
    lymphokine, 537
ABO compatible, 156–157
Abortion, elective, 140
Absolute neutrophil count
        (ANC), 149–157, 244–246,
        254–256, 472–479,
        515–520, 557–566, 586–590
ABV (doxorubicin, bleomycin,
        and vincristine therapy),
        328
ABVD (doxorubicin, bleomycin,
        vinblastine, and
        dacarbazine therapy), 176,
        183, 316, 332
Aclarubicin, 373
Acquired immune deficiency
        syndrome (AIDS), 11, 69,
        329, 416, 450–453, 461,
        556, 630
Actinomycin, 309–310, 328, 554
Action potential, 413

Activation
    immune, 461
    macrophage, 409
    mast cell, 379
    T-cell, 410, 556
Activity
    antimicrobial, 103
    antitumor, 86, 341–345,
        348–349
    bactericidal, 403
    fungicidal, 82
    myeloperoxidase, 434, 451
    phagocytic, 78
    synergistic, 104
Acute lymphoblastic leukemia
        (ALL), 14, 178, 226, 230,
        253–254, 584
Acute myeloid leukemia (AML),
        138–140, 491–497,
        502–504, 511, 523–525,
        583–584
ADCC (*see* Antibody-dependent
        cellular cytotoxicity)
Adrenal cancer, 238
Adriamycin®, 200, 245, 459, 599
Adult respiratory distress
        syndrome (ARDS),
        430–440
Aerodigestive tract cancer, 345

**

About the Editors

GEORGE MORSTYN is Vice President of Medical and Clinical Affairs in Amgen Inc., Thousand Oaks, California, and Clinical Associate Professor at the UCLA School of Medicine, Los Angeles, California. The author or coauthor of more than 120 professional papers that reflect his research interests in colony-stimulating factors, he is a Fellow of the Royal Australasian College of Physicians and a member of the American Association for Cancer Research, the American Society for Clinical Oncology, the American Society of Hematology, and the Clinical Oncology Society of Australia, among other organizations. Dr. Morstyn received the B.Med.Sci. (1972) and the M.B., B.S. (1975) degrees from Monash University, Clayton, Victoria, Australia, and the Ph.D. degree (1982) in experimental hematology from the University of Melbourne, Parkville, Victoria, Australia.

T. MICHAEL DEXTER is Director of the Paterson Institute for Cancer Research, Christie Hospital, Manchester, England. The author, coauthor, or coeditor of over 250 professional publications, including *Colony-Stimulating Factors: Molecular and Cellular Biology* and *Hematopoietic Lineages in Health and Disease* (both titles, Marcel Dekker, Inc.), he is an editorial board member of many journals in his field of expertise. He is also a Fellow of the Royal Society (F.R.S.) and the Royal College of Pathologists (F.R.C. Path.), as well as an honorary member of the Royal College of Physicians (Hon. M.R.C.P.). Dr. Dexter received the B.Sc. degree (1970) in biology from the University of Salford, England, the Ph.D. degree (1973) in biology from the University of Manchester, England, and the D.Sc. degree (1982) in biology from the University of Salford, England.

MARYANN FOOTE is Associate Director of Medical Writing, Amgen Inc., Thousand Oaks, California. The author or coauthor of over 50 journal publications and book chapters, she is a Fellow of the American Medical Writers Association as well as a member of the American Society of Hematology, the American Society of Clinical Oncology, and The European Haematology Association, among others. Dr. Foote received the B.S. (1974) and M.S. (1976) degrees from Fairleigh Dickinson University, Teaneck, New Jersey, and the Ph.D. degree (1983) from Rutgers University, New Brunswick, New Jersey.

ISBN 0-8247-0057-0